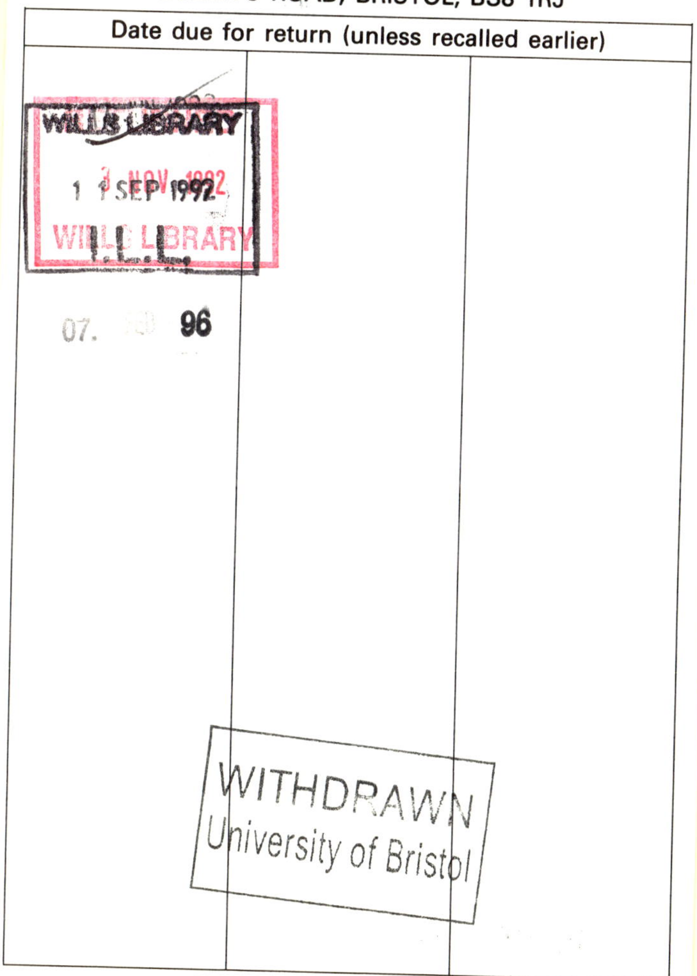

Rock engineering and excavation in an urban environment

Rock engineering and excavation in an urban environment

Proceedings of the conference held in Hong Kong from 24 to 27 February, 1986, and organized by the Institution of Mining and Metallurgy and the Hong Kong Local Section of the Institution in association with the Hong Kong Institution of Engineers

The Institution of Mining and Metallurgy

Published at the office of the
Institution of Mining and Metallurgy
44 Portland Place
London W1
England

ISBN 0 900488 87 5

Printed by The Chameleon Press Limited, London SW18 4SG

Foreword

The impetus for the conference 'Rock engineering and excavation is an urban environment', of which this volume constitutes the proceedings, arose out of the formation of the Hong Kong Local Section of the Institution of Mining and Metallurgy in late 1984. The hosting of such a major international conference is an excellent demonstration of the way in which local sections can work with the IMM to the benefit of both bodies.

The conference was also remarkable for an IMM meeting in the diversity of topics that are covered by the theme. Institution members in Hong Kong are primarily engaged in civil engineering activities of tunnelling, large cavern excavations, major rock slopes and basement excavations. The papers cover a correspondingly wide range of disciplines from engineering geology, geophysics, hydrogeology, rock mechanics, tunnel engineering, quarrying and excavation technology to monitoring of blast vibrations and cavern displacements. The fact that actual mining operations are discussed in only four of nearly forty papers is a clear demonstration of the way in which the IMM's members are now integrated into a greater range of activities than would have been the case a decade ago.

The conference proceedings include papers from thirteen countries (Saudi Arabia, United Kingdom, Hong Kong, South Africa, Austria, the U.S.A., Italy, India, Brazil, Australia, Czechoslovakia, the People's Republic of China and Finland). In addition, the many formal and informal contributions to the discussions were a feature of the conference and are summarized in this volume. The level of interest in this theme augurs well for the establishment of a further series of IMM conferences on this topic.

The conference organizers were fortunate in having Dr. Evert Hoek give a keynote address entitled 'Practical rock mechanics – developments over the past 25 years'. This was particularly appropriate in view of his own significant contributions to the science over this period.

A further feature of the conference was the extensive series of technical tours, organized by the local section: these ranged from thermal and nuclear power sites to major mass transit excavations, sewage tunnels and extensive rock excavations. These tours vividly demonstrated to delegates the formidable problems that are being routinely dealt with in this intensely developed area and highlighted the reasons for the selection of venue for this first international conference.

Particular acknowledgement should be made of the initiative and enthusiasm of the local organizing committee, who worked closely with the IMM and the United Kingdom organizing committee to ensure the success of the conference. The IMM would also like to thank all authors and speakers for their technical contributions to the meeting. A number of firms generously provided financial sponsorship: they are listed elsewhere in these proceedings and the IMM gratefully acknowledges their assistance.

Laurie Richards
Chairman, Organizing Committee
April, 1986

Organizing Committee

Dr L.R. Richards (Chairman) (United Kingdom)
K.C. Bennett (Hong Kong)
A.M. Carbray (Hong Kong)
A.W. Clover (Hong Kong)
A. Cook (United Kingdom)
M. Douglas (United Kingdom)
B.G. Fish (United Kingdom)
I. Gray (Hong Kong)
K.W. Lee (Hong Kong)
G.E. Pearse (United Kingdom)
P.A. Roberts (Hong Kong)
P.G.D. Whiteside (Hong Kong)

Acknowledgement

The Institution of Mining and Metallurgy wishes to acknowledge generous financial support for the 'Rock engineering and excavation in an urban environment' conference by the following organizations:

Electronic and Geophysical Services, Ltd.
Freeman Fox & Partners (Far East), Ltd.
Leighton Contractors (Asia), Ltd.
Mitchell, McFarlane, Brentnall & Partners
MAA Engineering Consultants (HK), Ltd.
Scott Wilson Kirkpatrick & Partners
Vianini Lavori SpA

Contents

Practical rock mechanics — developments over the past 25 years

Evert Hoek D.SC., F.I.M.M., F.ENG.
Golder Associates, Vancouver, Canada

Keynote address delivered on 24 February, 1986

INTRODUCTION

We tend to think of rock mechanics as a modern engineering discipline and yet, as early as 1773, Coulomb included results of tests on rocks from Bordeaux in a paper read before the French Academy in Paris[1,2].

In 1884 French engineers started construction of the Panama Canal and this task was taken over by the US Army Corps of Engineers in 1908. In the half century between 1910 and 1964, 60 slides were recorded in cuts along the canal and, although these slides were not analysed in rock mechanics terms, recent work by the US Corps of Engineers[3] shows that these slides were predominantly controlled by structural discontinuities and that modern rock mechanics concepts are fully applicable to the analysis of these failures. In discussing the Panama Canal slides in his Presidential Address to the first international conference on Soil Mechanics and Foundation Engineering in 1936, Karl Terzaghi[4,5] said "The catastrophic descent of the slopes of the deepest cut of the Panama Canal issued a warning that we were overstepping the limits of our ability to predict the consequences of our actions ... ".

In 1920 Josef Stini started teaching what was called "Technical Geology" at the Vienna Technical University and before he died in 1958 he had published 333 papers and books[6]. He founded the journal *Geologie und Bauwesen*, the forerunner of today's journal *Rock Mechanics*, and was probably the first to emphasize the importance of structural discontinuities on the engineering behaviour of rock masses.

Other notable scientists and engineers from a variety of disciplines did some interesting work on rock behaviour during the early part of this century. von Karman[7] (1911), King[8] (1912), Griggs[9] (1936), Ide[10] (1936), Terzaghi[11] (1945) all worked on the failure of rock materials. In 1921 Griffith[12] proposed his theory of brittle material failure and, in 1931 Bucky[13] started using a centrifuge to study the failure of mine models under simulated gravity loading.

None of these persons would have classified themselves as rock mechanics engineers - the title had not been invented at that time - but all of them made significant contributions to the fundamental basis of the subject as we know it today. I have made no attempt to provide an exhaustive list of papers related to rock mechanics which were published before 1960 but the references given above will show that important developments in the subject were taking place well before that date.

The early 1960's were very important in the general development of rock mechanics world-wide because a number of catastrophic failures occurred which clearly demonstrated that, in rock as well as in soil, "we were overstepping the limits of our ability to predict the consequences of our actions .."[4].

In December 1959 the foundation of a concrete arch dam at Malpasset in France failed and the resulting flood killed about 450 people. In October 1963 about 2500 people in the Italian town of Longarone were killed as a result of a landslide generated wave which overtopped the Vajont dam. These two disasters had a major impact on rock mechanics in civil engineering and a large number of papers were written on the possible causes of the failures[14].

In 1960 a coal mine at Coalbrook in South Africa collapsed with the loss of 432 lives. This event was responsible for the initiation of an intensive research programme which resulted in major advances in the methods used for designing coal pillars[15].

The formal development of rock mechanics as an engineering discipline in its own right dates from this period in the early 1960s and I will attempt to review these developments in the remainder of this paper. I consider myself extremely fortunate to have been intimately involved in the subject since 1958. I have also been fortunate to have been in positions which required extensive travel and which have brought me into personal contact with most of the persons with whom the development of modern rock mechanics is associated.

ROCKBURSTS AND ELASTIC THEORY

Rockbursts are explosive failures of rock which occur when very high stress concentrations are induced around underground openings. The problem is particularly acute in deep level mining in hard brittle rock. The deep level gold mines in the Witwatersrand area in South Africa, the Kolar gold mines in India, the nickel mines centred on Sudbury in Canada, the mines in the Coeur d'Alene area in Idaho in the USA and the gold mines in the Kalgoorlie area in Australia, are amongst the mines which have suffered from rockburst problems.

As early as 1935 the deep level nickel mines near Sudbury were experiencing rockburst problems and a report on these problems was prepared by Morrison[16] in 1942 . Morrison also worked on rockburst problems in the Kolar gold fields in India and describes some of these problems in his book, *A Philosophy of Ground Control*[17].

Early work on rockbursts in South African gold mines was reported by Gane et al[18] and a summary of rockburst research up to 1966 was presented by Cook et al[19]. Work on the seismic location of rockbursts by Cook[20] resulted in a significant improvement of our understanding of the mechanics of rockbursting and laid the foundations for the microseismic monitoring systems which are now common on mines with rockburst problems.

A characteristic of almost all rockbursts is that they occur in highly stressed, hard brittle rock. Consequently, the analysis of stresses induced around underground mining excavations, a key in the generation of rockbursts, can be dealt with by means of the theory of elasticity. Much of the early work in rock mechanics applied to mining was focused on the problem of rockbursts and this work is dominated by theoretical solutions which assume isotropic elastic rock and which make no provision for the role of structural discontinuities. In the first edition of Jaeger and Cook's book *Fundamentals of Rock Mechanics*[21], mention of structural discontinuities occurs on about a dozen of the 500 pages of the book. This comment does not imply criticism of this outstanding book but it illustrates the dominance of elastic theory in the approach to rock mechanics of those associated with deep-level mining problems. Books by Coates[22] and by Obert and Duvall[23] reflect the same emphasis on elastic theory.

This emphasis on the use of elastic theory for the study of rock mechanics problems was particularly strong in the English speaking world and it had both advantages and disadvantages. The disadvantage was that it ignored the critical role of structural discontinuities on the behaviour of rock masses. The advantage was that the tremendous concentration of effort on this approach resulted in advances which may not have occurred if the approach had been more general.

Many mines and large civil engineering projects have benefited from this early work in the application of elastic theory and most of the modern underground excavation design methods have their origins in this work.

DISCONTINUOUS ROCK MASSES

Stini was one of the pioneers of rock mechanics in Europe and he emphasized the importance of structural discontinuities in controlling the behaviour of rock masses[6]. Stini was involved in a wide range of near-surface civil engineering works and it is not surprising that his emphasis was on the role of discontinuities since this was obviously the dominant problem in all his work. Similarly, the text book by Talobre[24], reflecting the French approach to rock mechanics, recognized the role of structure to a much greater extent than did the texts of Jaeger and Cook, Coates and Obert and Duvall.

A major impetus was given to this work by the Malpasset dam failure and the Vajont disaster mentioned earlier. The outstanding work by Londe and his co-workers in France[25,26,27] and by Wittke[28] and John[29] in Germany laid the foundation for the three-dimensional structural analyses which we have available today.

ENGINEERING ROCK MECHANICS

Civil and mining engineers have been building structures on or in rock for centuries and the principles of engineering in rock have been understood for a long time. Rock mechanics is merely a formal expression of some of these principles and it is only during the past few decades that the theory and practice in this subject have come together in the discipline which we know today as rock mechanics. A particularly important event in the development of the subject was the merging of elastic theory, which dominated the English language literature on the subject, with the discontinuum approach of the Europeans. The gradual recognition that rock could act both as an elastic material and a discontinuous mass resulted in a much more mature approach to the subject than had previously been the case. At the same time, the subject borrowed techniques for dealing with soft rocks and clays from soil mechanics and recognized the importance of viscoelastic and rheological behaviour in materials such as salt and potash.

I should point out that significant work on rock mechanics was being carried out in countries such as Russia, Japan and China during the 25 years covered by this review but, due to language differences, this work was almost unknown in the English language and European rock mechanics centres and almost none of it was incorporated into the literature produced by these centres.

GEOLOGICAL DATA COLLECTION.

The corner-stone of any practical rock mechanics analysis is the geological data base upon which the definition of rock types, structural discontinuities and material properties is based. Even the most sophisticated analysis can become a meaningless exercise if the geological information upon which it is based is inadequate or inaccurate.

Methods for the collection of geological data have not changed a great deal over the past 25 years and there is still no acceptable substitute for the field mapping and core logging carried out by an experienced geologist. The emergence of geological engineering or engineering geology as recognized university degree courses has been an important step in the development of rock mechanics. These courses train geologists to be specialists in the recognition and interpretation of geological information which is significant in engineering design. These geological engineers, following in the tradition started by Stini in the 1920s, play an increasingly important role in modern rock engineering.

Once the geological data have been collected, computer processing of this data can be of great assistance in plotting the information and in the interpretation of statistically significant trends. Unfortunately, the development of computer software has lagged behind developments in hardware and it will be several years before a selection of high quality software for the interpretation of geological information becomes widely available for use on the ubiquitous micro-computer. These programs will eventually permit the plotting and contouring of structural data on spherical projections, the production of diamond drill core logs, the construction of plans, sections, developed and isometric views of excavations in rock and the analysis of potential failures in the rock. Similar programs will permit interpretation of collected data on in situ stresses, subsurface groundwater pressure and flow and on laboratory tests on rock specimens.

Surface and down-hole geophysical tools and devices such as borehole cameras have been available for several years and the reliability and usefulness of these tools has gradually improved as electronic components and manufacturing techniques have been improved. Current capital and operating costs of these tools are high and these factors, together with uncertainties associated with the interpretation of the information obtained from these tools, have tended to restrict their use in rock engineering. It is probable that the use of these tools will become more widespread in years to come as further developments occur.

LABORATORY TESTING OF ROCK

There has always been a tendency to equate rock mechanics with laboratory testing of rock specimens and hence laboratory testing has played a disproportionately large role in the subject. This does not imply that laboratory testing is not important but I would suggest that only about 10 to 20 percent of a well balanced rock mechanics program should be allocated to laboratory testing.

Laboratory testing techniques have been borrowed from civil and mechanical engineering and have remained largely unaltered for the past 25 years. An exception has been the development of servo-controlled stiff testing machines which permit the determination of the complete stress-strain curve for rocks. This information is important in the design of underground excavations since the properties of the failed rock surrounding the excavations have a significant influence upon the stability of the excavations.

ROCK MASS CLASSIFICATION

A major deficiency of laboratory testing of rock specimens is that the specimens are limited in size and therefore represent a very small and highly selective sample of the rock mass from which they were removed. In a typical engineering project, the samples tested in the laboratory represent only a very small fraction of one percent of the volume of the rock mass. In addition, since only those specimens which survive the collection and preparation process are tested, the results of these tests represent a highly biased sample. How then can these results be used to estimate the properties of the in situ rock mass ?

In an attempt to provide guidance on the properties of rock masses upon which the selection of tunnel support systems can be based, a number of rock mass classification systems have been developed. Typical of these classifications are those published by Bieniawski[30,31] and by Barton, Lien and Lunde[32]. These classifications include information on the strength of the intact rock material, the spacing, number and surface properties of the structural discontinuities as well as allowances for the influence of subsurface groundwater, in situ stresses and the orientation and inclination of dominant discontinuities.

These rock mass classification systems have proved to be very useful practical engineering tools, not only because they provide a starting point for the design of tunnel support but also because they force users to examine the properties of the rock mass in a very systematic manner. The engineering judgements which can be made as a result of the familiarity and understanding gained from this systematic study are probably as useful as any of the calculations associated with the classification systems.

ROCK MASS STRENGTH

One of the major problems confronting designers of engineering structures in rock is that of estimating the strength of the rock mass. This rock mass is usually made up of an interlocking matrix of discrete blocks. These blocks may have been weathered to varying degrees and the contact surfaces between the blocks may vary from clean and fresh to clay covered and slickensided.

Determination of the strength of an in situ rock mass by laboratory type testing is generally not practical. Hence this strength must be estimated from geological observations and from test results on individual rock pieces or rock surfaces which have been removed from the rock mass. This question has been discussed extensively by Hoek and Brown[33] who used the results of theoretical[34] and model studies[35,36] and the limited amount of available strength data, to develop an empirical failure criterion for jointed rock masses. Hoek[37] also proposed that the rock mass classification systems, described in the previous section of this paper, can be used for estimating the rock mass constants required for this empirical failure criterion. Practical application of this failure criterion in a number of engineering projects has shown that these estimates are reasonably good for disturbed rock masses but that, in tightly interlocked undisturbed rock masses such as those which may be encountered in tunnelling, the estimated strength values are too low.

Further work is required to improve the Hoek-Brown and other failure criteria for jointed rock masses. It may be necessary to develop a geological classification system specifically for this purpose since use of classification systems developed for the design of tunnel support has been found to have some disadvantages when used for estimating rock mass strength parameters.

IN SITU STRESS MEASUREMENT

The stability of deep underground excavations depends upon the strength of the rock mass surrounding the excavations and upon the stresses induced in this rock. These induced stresses are a function of the shape of the excavations and the in situ stresses which existed before the creation of the excavations. The magnitudes of pre-existing in situ stresses have been found to vary widely, depending upon the geological history of the rock mass in which they are measured[33]. Theoretical predictions of these stresses are considered to be unreliable and, hence, measurement of the actual in situ stresses is necessary for major underground excavation design.

During early site investigations, when no underground access is available, the only practical method for measuring in situ stresses is hydrofracturing[38] in which the hydraulic pressure required to open existing cracks is used to estimate in situ stress levels. Once underground access is availaible, overcoring techniques for in situ stress measurement[39,40] can be used and, provided that sufficient care is taken in executing the measurements, the results are usually adequate for design purposes.

GROUNDWATER PROBLEMS

The presence of large volumes of groundwater is an operational problem in tunnelling but water pressures are generally not too serious a problem in underground excavation engineering[41]. Exceptions are pressure tunnels associated with hydroelectric projects. In these cases, inadequate confining stresses due to insufficient depth of burial of the tunnel can cause serious problems in the tunnel and in the adjacent slopes.

Groundwater pressures are a major factor in all slope stability problems and an understanding of the role of subsurface groundwater is an essential requirement for any meaningful slope design[42,43]. While the actual distributions of water pressures in rock slopes are probably much more complex than the simple distributions normally assumed in slope stability analyses[44], sensitivity studies based upon these simple assumptions are generally adequate for the design of drainage systems[45]. Monitoring of groundwater pressures by means of piezometers[43] is the most reliable means of establishing the input parameters for these groundwater models and for checking upon the effectiveness of drainage measures.

In the case of dams, forces generated by the water acting on the upstream face of the dam and water pressures generated in the foundations are critical in the assessment of the stability of the dam. Estimates of the water pressure distribution in the foundations and of the influence of grout and drainage curtains upon this distribution have to be made with care since they have a significant impact upon the overall dam and foundation design[46].

The major advances which have been made in the groundwater field during the past decade have been in the understanding of the transport of pollutants by groundwater. Because of the urgency associated with nuclear and toxic waste disposal in industrialized countries, there has been a concentration of research effort in this field and advances have been impressive. The results of this research do not have a direct impact on conventional geotechnical engineering but there have been many indirect benefits from the development of instrumentation and computer software which can be applied to both waste disposal and geotechnical problems.

ROCK REINFORCEMENT

Safety during construction and long term stability are factors which have to be considered by the designers of excavations in rock. It is not unusual for these requirements to lead to a need for the installation of some form of rock support. Fortunately, practical developments in this field have been significant during the past 25 years and today's rock engineer has a wide choice of support systems[33,41,47].

In tunnelling, there is still an important role for steel sets and concrete lining in dealing with very poor ground but, in slightly better ground, the use of combinations of rockbolts and shotcrete has become very common. The use of long untensioned grouted cables in underground mining[48,49,50,51] has been a particularly important innovation which has resulted in significant improvements in safety and mining costs in massive ore bodies. The lessons learned from these mining systems have been applied with considerable success in civil engineering and the use of untensioned dowels, installed as close as possible to the advancing face, has many advantages in high speed tunnel construction. The use of untensioned grouted cables or reinforcing bars has also proved to be a very effective and economical technique in rock slope stabilization. This reinforcement is installed progressively as the slope is benched downward and it is very effective in knitting the rock mass together and preventing the initiation of ravelling.

The design of rock support systems tends to be based upon empirical rules generated from experience[52,53,54] and currently available analytical models are not very reliable. Some interesting theoretical models, which provide a very clear understanding of the mechanics of rock support in tunnels, have been developed in recent years[55,56,57,58]. These models have to be used with caution when designing actual tunnel support since they are based upon very simple assumptions and rock conditions underground may vary from these assumptions. With the development of powerful numerical models such as that described by Lorig and Brady[59], more realistic and reliable support designs will eventually become possible but it will be several years before these models can be used as design tools.

EXCAVATION METHODS IN ROCK

As pointed out earlier in this paper, the strength of jointed rock masses is very dependent upon the interlocking between individual rock pieces. This interlocking is easily destroyed and careless blasting during excavation is one of the most common causes of underground excavation instability. The following quotation is taken from a paper by Holmberg and Persson[60] :

'The innocent rock mass is often blamed for insufficient stability that is actually the result of rough and careless blasting. Where no precautions have been taken to avoid blasting damage, no knowledge of the real stability of the undisturbed rock can be gained from looking at the remaining rock wall. What one sees are the sad remains of what could have been a perfectly safe and stable rock face'.

Techniques for controlling blast damage in rock are well known[61,62,63] but it is sometimes difficult to persuade owners and contractors that the application of these techniques is worth while. Experience in projects in which carefully controlled blasting has been used generally shows that the amount of support can be reduced significantly and that the overall cost of excavation and support is lower than in the case of poorly blasted excavations[41].

Machine excavation is a technique which causes very little disturbance to the rock surrounding an underground excavation. A wide range of tunnelling machines have been developed over the past 25 years and these machines are now capable of working in almost all rock types[64,65]. Further development of these machines can be expected and it is probable that machine excavation will play a much more important role in future tunnelling than it does today.

ANALYTICAL TOOLS

Analytical models have always played an important role in rock mechanics and the earliest models date back to closed form soltions such as that for calculating the stresses surrounding a circular hole in a stressed plate, published by Kirsch in 1898[66].

The development of the computer in the early 1960s made possible the use of iterative numerical techniques such as finite element[67], boundary element[68], discrete element[69] and combinations of these methods[70,59] which have become almost universal tools in rock mechanics. The computer has also made it much more convenient to use powerful limit equilibrium methods[71,72,73,74] and probabilistic approaches[75,76,77,78] for rock mechanics studies.

The recent advent of the micro-computer and the rapid developments which have taken place in inexpensive hardware have brought us close to the era of a computer on every professional's desk. The power of these machines is transforming our approach to rock mechanics analysis since it is now possible to perform a large number of sensitivity or probabilistic studies in a fraction of the time which was required for a single analysis a few years ago. Given the inherently inhomogeneous nature of rock masses, such sensitivity studies enable us to explore the influence of variations in the value of each input parameter and to base our engineering judgements upon the rate of change in the calculated value rather than on a single answer.

In using these powerful numerical tools, it is necessary constantly to remind oneself that the answers produced are only as good as the input information. A sign above my own computer reads *'It is the duty of an engineer to judge soundly rather than to compute accurately'.*

CONCLUSIONS

Over the past 25 years, rock mechanics has developed into a mature subject which is built on a solid foundation of geology and engineering mechanics. Individuals drawn from many different disciplines have contributed to this subject and have developed a wide range of practical tools and techniques. There is still a great deal of room for development, innovation and improvement in almost every aspect of the subject and it is a field which will continue to provide exciting challenges for many years to come.

References

1. Coulomb, C.A. Essai sur une application des règles de maximis & minimis à quelques problèmes de statique, relatifs à l'architecture. *Mémoires de Mathématique & de Physique*, 7, 1776, 343-82.
2. Heyman, J. *Coulomb's Memoir on Statics.* (Cambridge: at the University Press, 1972).
3. Lutton, R.J., Banks, D.C. and Strohm, W.E. Slides in the Gaillard Cut, Panama Canal Zone. In *Rockslides and Avalanches,* Voight, B. ed., (New York: Elsevier, 1979), vol. 2, 151-224.
4. Terzaghi, K. Presidential Address. *Proc. 1st international conference for Soil Mechanics and Foundations Engineering, Cambridge, Mass.,* 1, 1936, 22-3.
5. Terzaghi, R. and Voight, B. Karl Terzaghi on rockslides: the perspective of a half-century. In *Rockslides and Avalanches,* 1979 Part 2 Voight, B. ed. (New York: Elsevier, 1979), 111-31.
6. Müller, J. Josef Stini. Contributions to engineering geology and slope movement investigations. In *Rockslides and Avalanches.* Part 2. Voight, B. ed. (New York: Elsevier, 1979), 95-109.
7. Karman, Th. von. Festigkeitsversuche unter allseitigem Druck. *Zeit d Ver Deutscher Ing.* 55, 1911, 1749-57.
8. King, L.V. On the limiting strength of rocks under conditions of stress existing in the earth's interior. *J. Geol.,* **20**, 1912, 119-38.
9. Griggs, D.T. Deformation of rocks under high confining pressures. *J. Geol.,* **44**, 1936, 541-77.
10. Ide, J.M. Comparison of statically and dynamically determined Young's modulus of rock. *Proc. Nat. Acad. Sci., USA,* **22**, 1936, 81-92.
11. Terzaghi, K. Stress conditions for the failure of saturated concrete and rock. *Proc. Am. Soc. Test. Mater.,* **45**, 1945, 777-801.
12. Griffith, A.A. The phenomenon of rupture and flow in solids. *Phil. Trans. Roy. Soc., London,* **A221**, 1921, 163-98.
13. Bucky, P.B. Use of models for the study of mining problems. *Am. Inst. Min. Metall. Engrs.,* Technical Publication 425, 1931, 28 p.
14. Jaeger, C. *Rock Mechanics and Engineering.* (Cambridge: at the University Press, 1972). 417 p.
15. Salamon, M.D.G. and Munro, A.H. A study of the strength of coal pillars. *J. S. Afr. Inst. Min. Metall.,* **65**, 1967, 55-67.
16. Morrison, R.G.K. Report on the rockburst situation in Ontario mines. *Trans. Can. Inst. Min. Metall.,* **45**, 1942.
17. Morrison, R.G.K. *A philosophy of ground control: a bridge between theory and practice.* rev. edn (Montreal: Dept. Min. and Metall Engng, McGill University, 1976), 182 p.
18. Gane, P.G., Hales, A.L. and Oliver, H.A. A seismic investigation of Witwatersrand earth tremors. *Bull. Seism. Soc. Am.,* **36**, 1946, 49-80.
19. Cook, N.G.W., Hoek, E., Pretorius, J.P.G., Ortlepp, W.D. and Salamon, M.D.G. Rock mechanics applied to the study of rockbursts. *J. S. Afr. Inst. Min. Metall.* **66**, 1966, 435-528.
20. Cook, N.G.W. The seismic location of rockbursts. In *Rock Mechanics, proc. 5th rock mechanics symposium, Univ. of Minnesota, 1962* Fairhurst C. ed. (Oxford: Pergamon Press, 1963), 493-516.
21. Jaeger, J.C. and Cook, N.G.W. *Fundamentals of Rock Mechanics.* (London: Methuen, 1969). 513 p.
22. Coates, D. *Rock Mechanics Principles,* Dept. Mines and Technical Surveys, Ottawa, 1966.
23. Obert, Leonard and Duvall, Wilbur I. *Rock Mechanics and the Design of Structures in Rock* (New York: Wiley, 1967), 65 p.
24. Talobre, J. *La mecanique des roches.* (Paris: Dunod, 1957).
25. Londe, P. Une méthode d'analyse à trois dimensions de la stabilité d'une rive rocheuse. *Annales des Ponts et Chaussées,* **135**, no. 1, 1965, 37-60.
26. Londe, P., Vigier, G. and Vormeringer, R. The stability of rock slopes, a three-dimensional study. *J. Soil Mech. Foundns Div., ASCE.,* **95**, no. SM 1, 1969, 235-262.
27. Londe, P., Vigier, G. and Vormeringer, R. Stability of slopes - graphical methods. *J. Soil Mech. Foundns Div., ASCE.,* **96**, no. SM 4, 1970, 1411-1434.
28. Wittke, W.W. Method to analyse the stability of rock slopes with and without additional loading. (in German), *Felsmechanik und Ingerieurgeologie,* Supp. 11, Vol. 30, 1965, 52-79. English translation in *Imperial College Rock Mechanics Research Report,* no. 6, July 1971.
29. John, K.W. Graphical stability analyses of slopes in jointed rock. *Proc. Soil Mech. Found. Div., ASCE,* **SM2**, paper no. 5865, 1968.
30. Bieniawski, Z.T. Engineering classification of jointed rock masses. *Trans. S. Afr. Inst. Civ. Engrs,* **15**, 1973, 335-44.
31. Bieniawski, Z.T. Geomechanics classification of rock masses and its application in tunnelling. In *Advances in rock mechanics,* vol. II, part A: *proceedings 3rd congress International Society for Rock Mechanics, Denver, 1974* (Washington, D.C.: National Academy of Sciences, 1974), 27-32.

32. Barton, N,R., Lien, R. and Lunde, J. Engineering classification of rock masses for the design of tunnel support. *Rock Mechanics,* **6**, 1974, 198-239.
33. Hoek, E. and Brown, E.T. *Underground Excavations in Rock.* (London: IMM, 1980), 527 p.
34. Hoek, E. Brittle failure of rock. In *Rock Mechanics in Civil Engineering Practice,* Stagg, K.G and Zienkiewicz, O.C. eds. (London: John Wiley and Sons, 1968), 99-124.
35. Brown, E.T. Strength of models of rock with intermittent joints. *J. Soil Mech. Foundns Div., ASCE.,* **96**, 1970, 1935-49.
36. Ladanyi, B. and Archambault, G. Simulation of shear behaviour of a jointed rock mass. *Rock mechanics - theory and practice, proc. 11th symposium on rock mechanics, Berkeley, 1969* (New York: SME-AIME, 1970), 105-25.
37. Hoek, E. Strength of jointed rock masses. *Geotechnique,* **33**, no. 3, 1983, 187-223.
38. Haimson, B.C. The hydrofracturing stress measuring method and recent field results. *Int. J. Rock Mech. Min. Sci.,* **15**, 1978, 167-78.
39. Leeman, E.R and Hayes, D.J. A technique for determining the complete state of stress in rock using a single borehole. *Proceedings 1st congress International Society for Rock Mechanics, Lisbon, 1966,* **2**, 17-24.
40. Worotnicki, G. and Walton, R.J. Triaxial 'hollow inclusion' gauges for determination of rock stresses *in situ. Proceedings International Society for Rock Mechanics symposium on investigation of stress in rock,Sydney.* Suppl. 1976 (Sydney: Instn. Engrs. Aust., 1976), 1-8.
41. Hoek, E. Geotechnical considerations in tunnel design and contract preparation. *Trans. Instn Min. Metall.,* (Sect. A: Min. industry), **91**, 1982, A101-9.
42. Hoek, E. and Bray, J.W. *Rock Slope Engineering.* 3rd edn (London: IMM, 1981), 402 p.
43. Brown, A. The influence and control of groundwater in large slopes. In *Stability in surface mining,* Brawner, C.O. ed. (New York: Society of Mining Engineers, AIME., 1982), 19-41.
44. Freeze, A.R. and Cherry, J.A. *Groundwater.* (Englewood Cliffs, NJ: Prentice-Hall Inc., 1979), 604 p.
45. Masur, C.I. and Kaufman, R.I. Dewatering. In *Foundation Engineering,* Leonards, G.A. ed. (New York: McGraw-Hill, 1962),241-350.
46. Soos, I.G.K. Uplift pressures in hydraulic structures. *Water Power and Dam Construction,* **31**, no. 5, 1979, 21-24.
47. Farmer, I.W and Shelton, P.D. Review of underground rock reinforcement systems. *Trans. Instn Min. Metall.,* (Sect. A: Min. industry), **89**, 1980, A68-83.
48. Clifford, R.L. Long rockbolt support at New Broken Hill Consolidated Limited. *Proc. Australian Inst. Min. Metall.,* no. 251, 1974, 21-6.
49. Fuller, P.G. The potential for support of long hole open stopes with grouted cables. *Proc. 5th. international congress for rock mechanics, Melbourne, 1983* (Rotterdam: A.A. Balkema, 1983), vol. **2**, D39-44.
50. Hunt, R.E.B. and Askew, J.E. Installation and design guidelines for cable dowel ground support at ZC/NBHC. *Proc. underground operators conference, Broken Hill, 1977,* 113-22.
51. Brady, B.H.G. and Brown, E.T. *Rock Mechanics for Underground Mining.* (London: George Allen and Unwin, 1985), 527 p.
52. Lang, T.A. Theory and practice of rockbolting. *Trans. Soc. Min. Engrs, Am. Inst. Min. Metall. and Petrolm Engrs.,* **220**, 1961, 333-48.
53. Endersbee, L.A. and Hofto, E.O. Civil engineering design and studies in rock mechanics for Poatina underground power station. *J. Inst. Engrs (Australia),* **35**, 1963, 187-209.
54. Cording, E.J., Hendron, A.J and Deere, D.U. Rock engineering for underground caverns. *Proc. symp. underground rock chambers, 1971* (New York: ASCE, 1971), 567-600.
55. Rabcewicz, L. Stability of tunnels under rock load. *Water Power.* **21**, No. 6-8, 1969, 225-229, 266-273, 297-304.
56. Daeman, J.J.K. Problems in tunnel support mechanics. *Underground Space,* **1**, 1977, 163-72.
57. Brown, E.T., Bray, J.W., Ladanyi, B. and Hoek, E. Characteristic line calculations for rock tunnels. *J. Geotech. Engng, Am. Soc. Civ. Engrs* **109**, 1983, 15-39.
58. Brown, E.T. and Bray, J.W. Rock-support interaction calculations for pressure shafts and tunnels. In *Rock mechanics: caverns and pressure shafts, proceedings of the International Society for Rock Mechanics symposium, Aachen, 1982* (Rotterdam: A.A. Balkema, 1982), vol. **2**, 555-65.
59. Lorig, L.J. and Brady, B.G.H. A hybrid computational scheme for excavation and support design in jointed rock media. *Proc. International Society for Rock Mechanics symposium on design and performance of underground excavations, Cambridge, 1984* Brown, E.T. and Anderson, J.A. eds (London: Brit. Geol. Soc., 1984), 105-12.
60. Holmberg, R. and Persson, P.-A. Design of a tunnel perimeter blasthole pattern to prevent rock damage. *Trans. Inst. Min. Metall.,(Sect. A: Min. industry),* **89**, 1980, A37-40.
61. Svanholm, B.O., Persson, P.-A. and Larsson. B. Smooth blasting for reliable underground openings. In *Rockstore 77: Storage in excavated rock caverns, proceedings of the first international symposium, Stockholm, 1977,* **3**, 37-43.
Bergman, M. ed. (Oxford: Pergamon Press, 1977), **3**, 573-9.
62. Langefors, U. and Kihlstrom, B. *The modern technique of rock blasting,* (New York: Wiley, 1963), 405 p.
63. Hagan, T.N. Understanding the burn cut - a key to greater advance rates. *Trans. Inst. Min. Metall. (Sect. A: Min. industry),* **89**, 1980, A30-6.
64. Robbins, R.J. Mechanized tunnelling - progress and expectations: Twelfth Sir Julius

Werhner memorial lecture. In *Tunnelling '76*. Jones, M.J. ed. (London: IMM, 1976), xi-xx.
65. McFeat-Smith, I. Survey of rock tunnelling machines available for mining projects. *Trans. Inst. Min. Metall. (Sect. A: Min. industry)*, **91**, 1982, A23-31.
66. Kirsch, G. Die theorie der elastizitat und die bedurfnisse der festigkeitslehre. *Zeit d Ver Deutscher Ing.*, **42**, 1898, 797-807.
67. Clough, R.W. The finite element method in plane stress analyses. *Proc. 2nd. ASCE conference on electronic computation, Pittsburgh, 1960*, 345-78.
68. Crouch, S.L. and Starfield, A.M. *Boundary element methods in solid mechanics.* (London: George Allen and Unwin, 1983), 322 p.
69. Cundall, P.A. A computer model for simulating progressive large scale movements in blocky rock systems. In *Rock fracture, proceedings of the international symposium on rock fracture, Nancy, 1971*, Paper 2-8.
70. von Kimmelmann, M.R., Hyde, B. and Madgwick, R.J. The use of computer applications at BCL Limited in planning pillar extraction and the design of mining layouts. *Proc. International Society for Rock Mechanics symposium on design and performance of underground excavations, Cambridge, 1984* Brown, E.T. and Anderson, J.A. eds (London: British Geotech. Soc., 1984), 53-64.
71. Sarma, S.K. Stability analysis of embankments and slopes. *J. Geotech. Engng Div., ASCE.*, **105**, no. GT12, 1979, 1511-24.
72. Brown, E.T. and Ferguson, G.A. Progressive hanging wall caving at Gath's mine, Rhodesia. *Trans. Instn Min. Metall.* (Section A: Min. industry), **88**, 1979, A92-105.
73. Gen Hua Shi and Goodman, R.E. A new concept for support of underground and surface excavations in discontinuous rocks based on a keystone principle. *Rock mechanics from research to applications, proceedings 22nd. U.S. symposium for rock mechanics, Cambridge, 1981* (Cambridge, Mass.: Massachusetts Institute for Technology, 1981), 290-96.
74. Warburton, P.M. Vector stability analysis of an arbitrary polyhedral rock block with any number of free faces. *Int. J. Rock Mech. Min. Sci.*, **18**, 1981, 425-28.
75. McMahon, B.K. A statistical method for the design of rock slopes. *Proceedings 1st Australia-New Zealand conference on geomechanics, Melbourne*, 1, 1971, 314-321.
76. Morriss, P. and Stoter, H.J. Open-cut slope design using probabilistic methods. *Proceedings 5th. congress International Society for Rock Mechanics, Melbourne, 1983* (Rotterdam: A.A. Balkema, 1983), vol. 1, C107-C113.
77. Priest, S.D. and Brown, E.T. Probabilistic stability analysis of variable rock slopes. *Trans. Instn Min. Metall.*, (Sect. A: Min. industry) **92**, 1982, A1-12.
78. Read, J.R.L. and Lye, G.N. Pit slope design methods, Bougainville Copper Limited open cut. *Proceedings 5th congress International Society for Rock Mechanics, Melbourne, 1983* (Rotterdam: A.A. Balkema, 1983), C93-C98.

Formation of a high rock slope at Ap Lei Chau, Hong Kong

C.J. Beggs M.SC., M.I.GEOL., M.H.K.I.E., F.G.S.
Binnie & Partners (Hong Kong), Hong Kong
D.P. McNicholl M.SC., C.ENG., F.I.C.E., M.H.K.I.E., F.G.S.
Hong Kong Housing Authority, Hong Kong

SYNOPSIS

The rapid increase in population in Hong Kong has led to the development of public housing estates on platforms cut into rock hillsides. This paper records the investigation, design and construction of a 70 m high rock slope at Ap Lei Chau and then critically reviews the methods used. The importance of rock joint surveys for design and the need to review the design as construction proceeds, are emphasised.

Ap Lei Chau Rock Slope

INTRODUCTION

Since World War II, the population of Hong Kong has grown from about half a million to over 5 million. The public housing programme was started in the mid 1950's and the Hong Kong Housing Authority now houses about 2.5 million people, in 152 completed estates. The housing programme continues with a target of 40,000 flats being completed each year.

Because of the very steep terrain in Hong Kong, there have been very few good building sites available and the intense building development programme has relied on sites formed on platforms cut into hillsides. The resulting excavation material has been used to reclaim more land from the sea for complementary development.

Many of these site formation projects have involved the construction of high slopes in rock. They have required a 'high risk' design and construction philosophy which recognises the risks involved in allowing high density public housing development close to the slopes. This philosophy is markedly different from that adopted in highway construction or by the mining industry when developing open-cast mines, where lower factors of safety can be used because of the lower risk to life. The design philosophy must also recognise tight construction programmes which, in a matter of 5 or 6 years, develop fully self contained estates of 40,000 to 50,000 people from virgin hillside as well as the high cost of producing the building platforms.

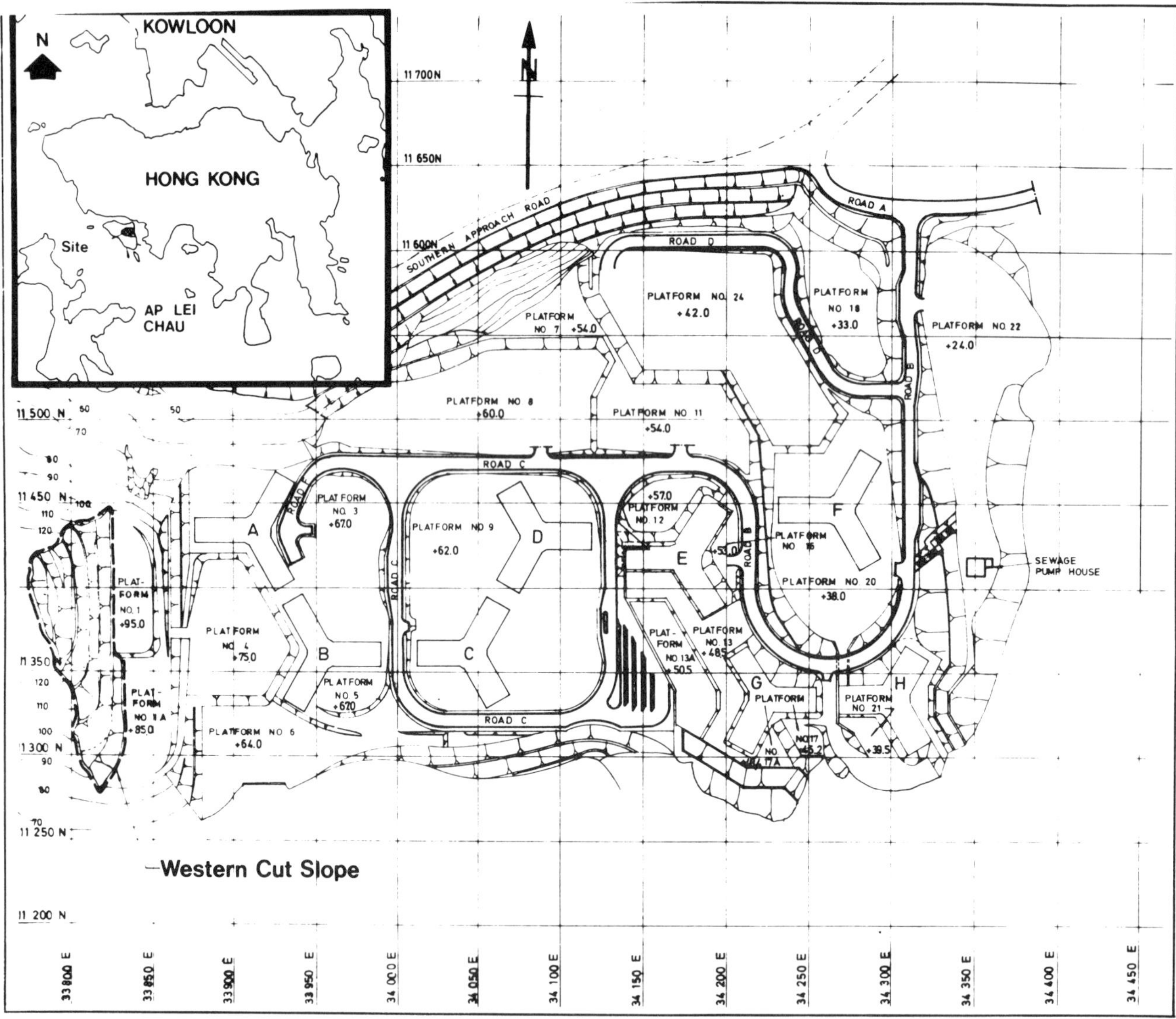

Fig. 1 Ap Lei Chau Site B - Location Plan

Description of the Site

In 1978 investigation and design was started for the site formation for a major public housing development on Ap Lei Chau, an island off the town of Aberdeen on the southern side of Hong Kong Island (Fig. 1). The scheme involved the excavation of over 2 million cubic metres of soil and rock from a steeply sloping site and the formation of a number of platforms and rock slopes. The highest slope is 70 m and forms the western limit of the site. Construction commenced early in 1982 and the site formation was finished in October 1984.

GEOLOGY

Rock Types

The Ap Lei Chau site is composed of hard fine grained pyroclastic rocks. The predominant rock type is a fine grained very strong, dark grey tuff. Also occurring are coarse grained lapilli tuffs, aphanitic to glassy basalt and welded ash flow deposits (ignimbrites).

A soil mantle of in-situ decomposed rock overlies the bed-rock with the interface between the two being quite sharp. The soil is 3-4 metres thick and comprises a yellow to reddish brown silty sand containing gravel size rock fragments.

Some banding is seen occasionally within the volcanic rocks with some tight folds probably indicating local slumping soon after deposition. Allan and Stephens (1971) suggested that some of the tuffs on Ap Lei Chau and in the Aberdeen area are water-laid and it is possible that these banded and folded rocks represent such an horizon.

Geological Structure

(i) Faults

Two almost conjugate fault systems cross the site. One system (F3,F4) strikes NW-SE and dips steeply towards the south; the other system (F5) strikes ENE-WSW with steep dips about the vertical.

Faults belonging to both systems cross the main slope with the F3,F4 faults being more continuous (Fig. 2). The F3,F4 faults traverse the full width of the slope. They are up to 8 m wide in the upper part of the slope where they are close together. Lower down they diverge to form narrow, separate zones which are 0.5 m to 1.5 m wide. The fault material is dark brown and yellow and consists of clayey sand with harder mylonite, broken rock zones and clay partings.

The F5 faults are present in the upper part of the slope but die out at about the 100 m PD level. They are 0.5 m to 2.0 m wide and consist of heavily iron stained moderately decomposed tuff with some thin clay partings. At their upper and lower ends they degenerate into joints or zones of close jointing.

(ii) Joints

There are three major joint sets and two minor sets present in the main rock slope. Fig.3a. The sets measured in drillholes and in outcrop during the site investigation stage were seen to persist with depth during the formation of the slope but with the concentrations varying with depth. The three main sets are:

Set A - well developed and persistent sub-horizontal set. The dips are generally 10°-15°, mostly towards the south across the slope face with a continuity up to 20 m. The joints in this set are frequently very closely spaced.

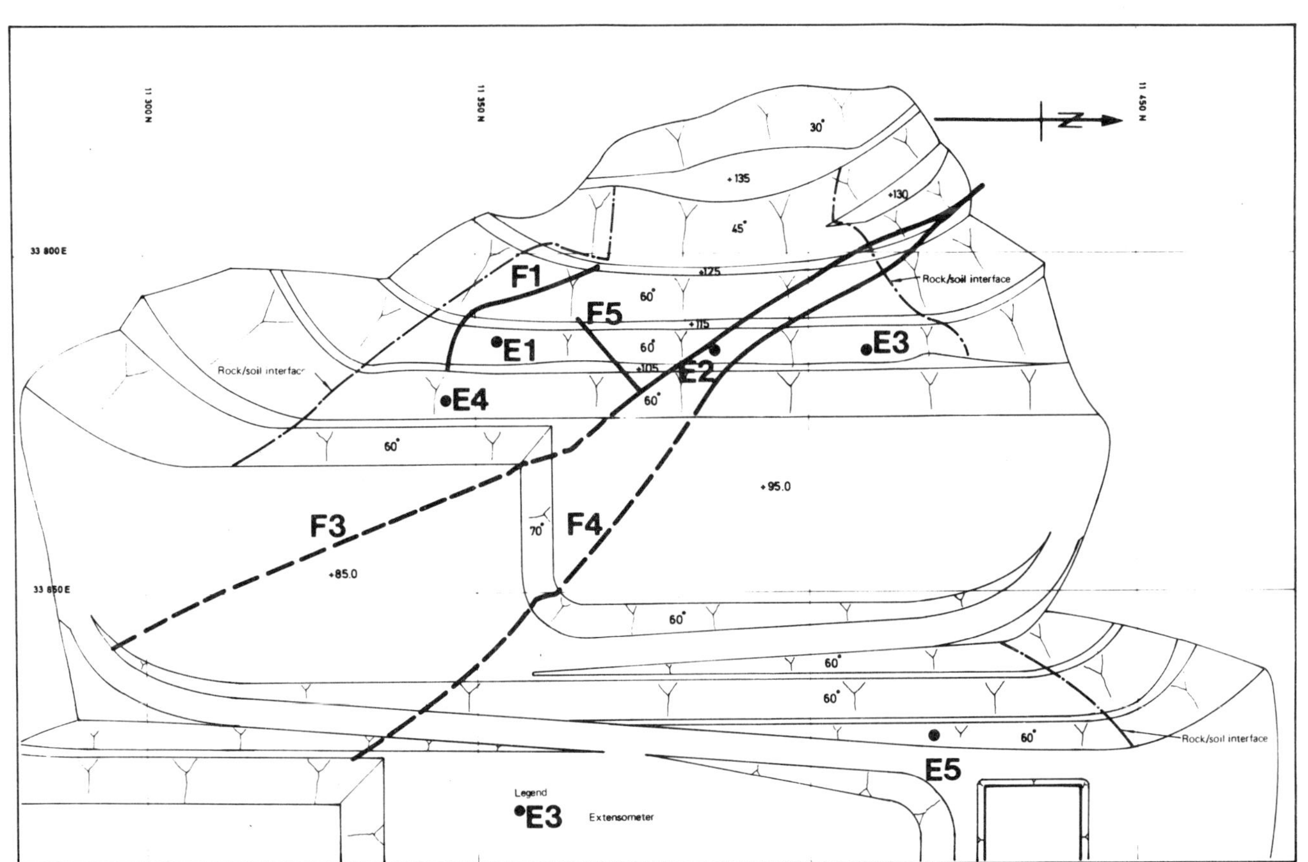

Fig. 2 Geological Structure in Main Rock Face

Set B - well developed set of steeply dipping (75°-85°) joints. The dip direction is either southeast (obliquely out of the slope) or northeast. These joints occur throughout the slope and have a continuity of 2-10 m. They are smooth and planar.

Set C - this is conjugate to set B but is less well developed at lower levels in the slope. The joints are smooth to slightly rough and planar. Continuity is 1-5 m.

Groundwater

The groundwater table is well below the slope and platform levels throughout the site. Following heavy rain there is a temporary build-up of water in the rock mass causing small flows from weepholes in sprayed concrete surfaces. During the construction period, the presence of large unpaved platform "catchment" areas caused the build-up to be higher than expected and some high flows were experienced from rock joints in the lower slopes. In one or two cases this resulted in an opening up of the joints.

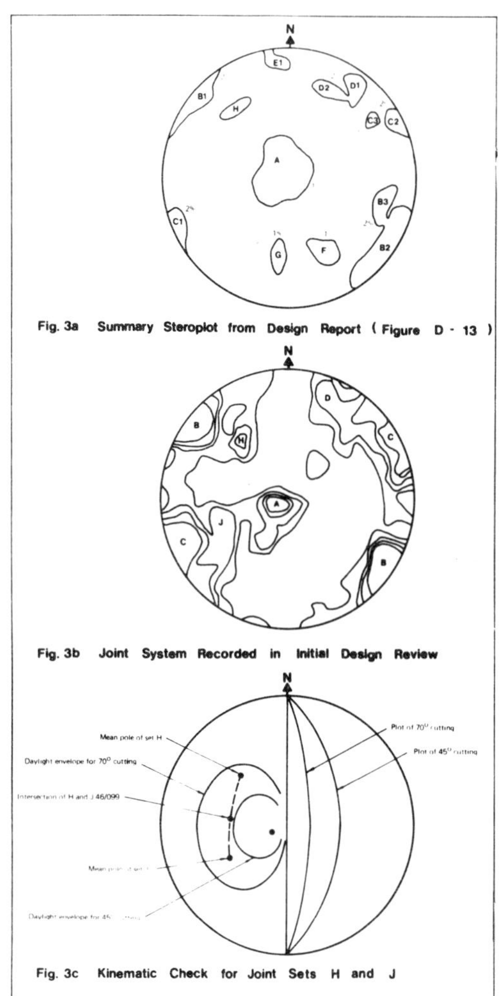

Fig. 3a Summary Steroplot from Design Report (Figure D - 13)

Fig. 3b Joint System Recorded in Initial Design Review

Fig. 3c Kinematic Check for Joint Sets H and J

SITE INVESTIGATION

Site investigation was carried out in two stages; the first between November 1978 and March 1979 and the second between September 1979 and November 1979.

During the first stage 11 holes were drilled, five of which were to investigate rock conditions for the 70 m rock slope. In the second stage a further two holes were drilled in the vicinity of the high rock slope.

All the holes were drilled using TNW core barrels. The depth of the drillholes was between 30 m and 65 m. Core recovery was generally greater than 90% despite low RQD values.

Thorough joint surveys were performed on rock outcrop within the site and around the site perimeter. These provided the basic information for the rock slope design and they were supplemented by borehole joint measurements taken using a borehole impression device to provide data on which the slope was designed.

Direct shear tests were performed on rock core specimens to provide friction angle values for the slope design. Friction angles of 40° for peak strength (rough surfaces) and 30° for residual strength (polished surfaces) were obtained.

Design

During the design process, slopes cut at 60°, 70° and 80° were analysed using stereoplot and wedge and plane failure analyses. The results of the joint survey from surface outcrop and drillholes and were plotted and the three main joint sets defined. In addition to the main sets the stereoplot also showed a wide scatter of randomly orientated joints dipping at intermediate to steep angles. While it was recognised that combinations of random joints and major joints could form wedges it was concluded that the number of such wedges would be small and each could be stabilized individually. The analysis showed that toppling was the only possible mode of failure on the major joint sets and dowels were proposed to stabilize these.

The design assessment that only toppling failures could occur considered only the main joint sets. A design review, carried out early in the construction included the minor joint sets (Fig. 3b).

Joint spacing varied very considerably. The design assessment was that major joint set spacing was very wide even though RQD values measured in drillholes was generally less than 60% and was frequently zero.

The design adopted a 70° face with only nominal berms, formed using a pre-split blasting technique. An 8 m wide rock trap area was proposed at the toe of the slope.

3.2 Blasting Trial

A blasting trial was carried out in an area east of the main rock slope. The rock in the trial area was slightly weathered to fresh moderately to closely jointed fine grained tuff which was fairly typical of the rock in the main slope.

For the trial eighteen 64 mm diameter holes were drilled in a single line, in three groups of six, and loaded with Tovex T90 in 400 mm x 25 mm plugs with 0.18 kg per plug. Detonation was by Primacord fired with No. 0 delay detonator with 1.5 m stemming.

Pattern A Six holes at 500 mm spacing loaded with full plugs at 900 mm centres. Two plugs at the bottom of the hole.

Pattern B Six holes at 500 mm spacing loaded with ½ plug at 600 mm centres. One plug at bottom of hole.

Pattern C Six holes at 1000 mm spacing loaded with full plugs at 600 mm centres. Two plugs at bottom of hole.

The best result was obtained using drilling and loading pattern B, and this was used for the whole of the main slope face.

The blasting trial served the important secondary purpose of giving more information on the rock mass fabric and the size of rock which would be produced.

Construction and Ongoing Design Review

The slope was formed in 10 m high lifts using a pre-split method, to produce a smooth slope face which would require little face stabilisation. The slope design was for a face cut at 70°. Preliminary site formation work and access road construction produced a number of rock exposures which emphasised the closely jointed nature of the rock mass and also provided the first opportunity to examine the continuity of the joints.

The steeply dipping joint sets B and C were continuous up to 5 m, but the minor joint sets appeared to have step offsets with no joint continuous for more than about 2 m.

At the top of the slope a 5 m high face was formed at 30° in the generally weaker highly weathered rock containing open joints. Because the rock at the toe of this upper face was still open and weak, the face between 125 m PD and 135 m PD was cut at 45° rather than the design slope of 70°. This face contained a number of joints belonging to the minor sets H and J which were planar, smooth and commonly coated with up to 5 mm of clay. The joint continuity was up to 4 m and the spacing was 1 m to 2 m.

The minor joint sets H and J occur irregularly throughout the slope and, in combination, gave rise to some small wedge failures in the upper slope, between 115 and 125 m PD. On their own they are also capable of giving rise to plane failures but, in general, the slope direction is such that these failures could not occur (Fig. 3c). Elsewhere on the site, plane failures have occurred where these joints are present in north facing slopes.

The geology and slope design was reviewed, based on detailed mapping of rock surfaces created during the early construction work. This review showed that the rock conditions in the upper part of the slope were less favourable than had been assumed in the design in that (a) the typical joint spacing was closer and (b) the minor joint sets were better developed and could combine to form wedge failures. If the slope was cut at the design slope of 70°, wedge failures involving 2.5 cu.m (7 tonnes) of rock could occur if the persistence of the set H or J joints was the maximum observed length of 4 m. If the persistence was much greater and a wedge failure involved a full interberm face then the volume of rock involved could be up to 100 cu.m (270 tonnes).

A modified slope design was adopted and subsequent faces were formed at 60° to reduce the volume of potential wedges. This also reduced the spalling caused by the close joint spacing. In addition, each face was carefully logged and stabilisation measures carried out before the formation of subsequent faces.

The original design had required only nominal 200 mm wide interface berms. This proved impractical for construction since it allowed insufficient working space for the drilling equipment to be set up. A 1 m berm width was therefore adopted.

Four small wedge failures occurred in the face between 115m PD and 125m PD and occasionally on berm edges. These were stabilised using rock dowels and concrete buttresses. A few instances of poor drilling alignment led to areas of "overbreak" and "underbreak" which required small concrete buttresses or trimming.

The flattening of the slope angle together with the increased berm width, reduced the space at the toe of the slope to less than that required. The 5 m wide "rock trap" at the toe of the slope was therefore omitted and greater emphasis given to preventing minor rock falls by face treatment.

Engineering Geological Logging

Immediately after each face was formed it was photographed and scaffolding was erected. Loose rock was scaled from the face before logging by the site engineering geologist. A classification system based on rock quality (weathering and joint orientation, spacing and condition) was used to divide the rock mass into regions. These classes were used in defining face stabilization measures. The face logs show all the major structural features present in the rock face and also record discontinuity orientations. An example of a face log is given in Figure 4. Following the engineering geology logging the areas and types of face stabilization measures were marked on the rock surface using spray paint. A final log of the stabilization work was produced as a record drawing.

Face stabilization

The face stabilization measures consisted primarily of sprayed concrete with weepholes installed at 1.2 m centres or closer where weak seams were covered. Where wider seams of weak highly decomposed rock, such as fault zones, were to be protected from erosion, a 100 mm square weld steel mesh was pinned to the rock face using 500 mm long dowels, before being covered with sprayed concrete. 25 mm diameter mild steel dowel bars, with a length between 2 m and 4 m, were installed to retain blocks of rock where scaling the rock would have given rise to additional unstable blocks. Small concrete buttresses were constructed where overhanging rock occurred due sometimes to adverse combinations of joints and at other times to misalignment of the presplit drillholes.

Difficulty was experienced in forming sharp crest and berm edges. It was found that the rock in the upper 1-1.5 m of the face was invariably very loose and broken, particularly with an opening up of the sub-horizontal joint set A. The problem appeared to relate to the dissipation of the blast gases into the sub-horizontal joints in the stemmed section of the pre-split holes. A possible solution would be to underdrill each row of holes so that the disturbed portion at the top of the subsequent row would be in a portion of rock to be removed. At Ap Lei Chau the most disturbed rock along the crest or berm edge was removed by hand and the remainder was stabilised with mesh and sprayed concrete extending approximately one metre back and one metre down from the crest.

On the main slope a total area of 4 800 sq.m of sprayed concrete was applied which represents 46% of the total area of the face. This is a greater area than was

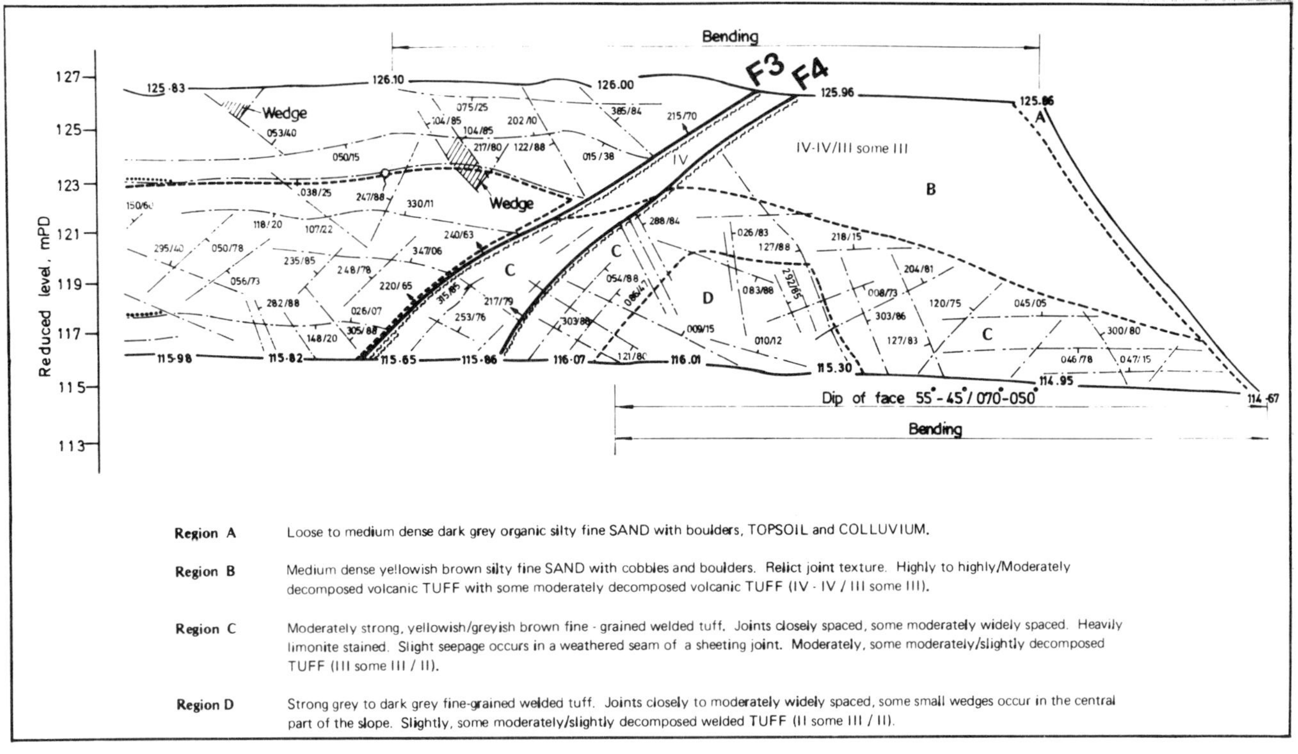

Fig. 4 Example of Engineering Geology Face Log

originally estimated and was due to the close jointing of the rock mass and only a slight improvement in rock quality with depth. This meant that stabilization measures were continued to lower levels than was expected. The exclusion of the 8 m wide rock trap at the toe of the slope and the need to reduce future maintenance to a minimum also led to a greater attention to minor areas of close jointing than would otherwise have been the case.

Instrumentation

Five 3-rod extensometer units were installed in the slope face as the excavation proceeded, to monitor the relaxation in the rock mass due to the removal of support. The position of the extensometers is shown in Figure 2.

The extensometer units were supplied by Soil Instruments Ltd and each consisted of three rods (2 m, 10 m and 15 m long) grouted into a hole drilled into the rock normal to the slope face. Movement of the rock face relative to the end of each rod was measured from a reference pad on the rock face.

Three extensometer units were installed at the 106.5 m PD level in the north, central and southern partsof the face in May 1983. Other units were installed in the southern part of the slope, at 97.0 m PD, in September 1983 and in the northern part of the slope, at 77 m PD, in February 1984.

The upper extensometers showed an outward movement in all cases, with the maximum movement registered at 4.5 mm. This was on a 15 m long rod and represents a strain of 0.03%. Most of the movement is related to a dilation in the outer 2-3 m of the face, where the maximum strain recorded was 0.1%. Readings from the three upper extensometers showed a virtual cessation of movement in the outer 2 m of the face after 10 months but continuing small movement in the 2-10 m "zone".

The lower extensometers showed less movement, especially the unit at 77 m PD which is near the base of the slope, where measurements taken were less than the measuring accuracy of the micrometer measuring device.

The slope is behaving very predictably and the movements that were recorded are well within the limits that might be expected in a slope of this size.

Results of extensometer monitoring are shown on Fig. 5.

REVIEW OF INVESTIGATION ANALYSIS AND DESIGN METHODS FOR ROCK SLOPES

As a result of the rock slope formation at Ap Lei Chau and at other sites in Hong Kong, the investigation, design and construction methods for rock slopes have been critically examined and certain key indices and procedures identified.

The procedure strategy for investigation, analysis and design of rock slopes in Hong Kong is similar to practice elsewhere and is summarised as:

(a) examine all available data;

(b) establish an hypothesis for the site geology;

(c) plan and execute site investigations;

(d) analyse results and transform data into a visualistion or model of ground conditions;

(e) plan and design the excavation;

(f) monitor progress and confirm assumptions and findings during excavation.

In Hong Kong, background data is generally available in the form of Geological Area Study Reports (Brand et al. 1982) and aerial photographs. Following interpretation and a walk over survey, an engineering geological map of the site is prepared with inferred structure and weathering patterns. The site investigation would normally include core drilling, sometimes with impression packer tests as described by Barr and Hocking (1976), or using a core orientation technique. Geophysical surveys may be used to delineate weathering changes and to locate major discontinuity concentrations. Discontinuity

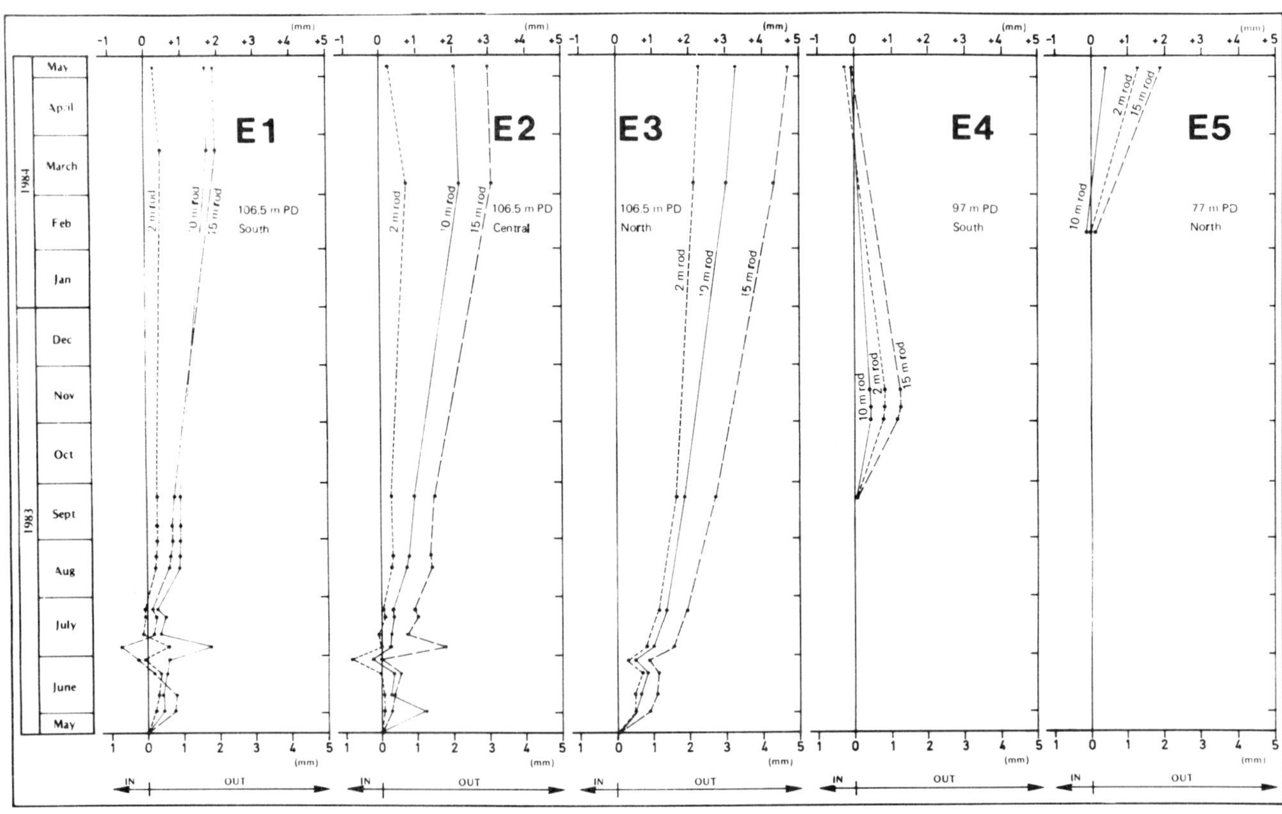

Fig. 5 Extensometer Monitoring

orientation and spacing are measured by traditional geological mapping methods. (Geological Society Working Party Report 1970.

Some kind of model must be used to catalogue and group the measured rock mass characteristics for analysis. The model has to be easily understood and communicable to the other disciplines involved in construction. Graphic models in the form of engineering geological plans and sections are commonly used. Computer modelling techniques allow the viewing orientation to be varied rapidly (Brimicombe (1983)). Physical models are often used to show specific situations.

The analysis of jointing using stereographic projections identifies potential failure modes prior to mechanical analysis of potential wedge and planar failures. While stereographic projection analysis is an essential step in the design of rock slopes, the limitations of the method have recently be highlighted by Hencher (1985). Whatever the modelling technique, the rock mass should be divided into domains of reasonably uniform properties which imply a particular engineering behaviour. Rock Mass Classification (RMC) may be used to describe the mass with a stronger implication of mechanical properties, mass fabric, structure and dimensions giving a real indication of engineering behaviour for preliminary design.

PRACTICAL ASPECTS OF INVESTIGATION AND ANALYSIS METHODS

Orientation of Discontinuities

Directional bias is a major source of errors in joint surveys because some joint sets may be so aligned that unidirectional boreholes or scanlines miss or under represent them. This is generally the case where there are subvertical joint sets which may be missed by vertical boreholes. (Terzhaghi (1965), Steffen (1976), Hudson and Priest (1983)).

Experience in Hong Kong indicates that very few geologists apply the so called "Terzhaghi" corrections and very rarely are discontinuity surveys carried out in three directions as recommended. The best way of removing bias is to check regional joint patterns first and then to take joint measurements all around the site on all available exposures before commencing investigations within the site. This approach was recommended by Prokapovich (1972) and McNicholl (1984). If the perimeter survey includes differently orientated exposures, then subjective discontinuity surveys (as defined in the Geological Society Working Report) are satisfactory. If the perimeter survey does not include data from differently orientated scanlines then the internal site investigation should include borehole orientation devices and high class sampling.

Sheeting joints are a common feature, particularly in the granites of Hong Kong. Their presence can lead to errors because they are usually aligned at flat angles or are inherently domed. (Sekula (1984)). The doming can be quite significant across the site and may result in the orientation of the exposures in the location surveyed being different from that in the area of the engineering works where adverse dip may cause localised failures.

Rock Mass Classification

Rock classification systems have been developed in an attempt to consolidate the data collected during site investigation and to enumerate a rock mass in terms of stability (Bieniawsi (1974, 1976, 1979), Barton et al (1974)). Each of the proposed systems selects a number of measurable features of the rock mass and provides indices, ranging from very good to very poor, for each feature. Generally, the sum of these indices gives an overall classification of the rock mass stability.

Experience in designing and constructing rock slopes in Hong Kong shows that a rock mass classification can be useful at the investigation/design stage, in providing a yardstick for measuring cut slope angles, excavation quantities and cost. However, because of the variable nature of rock masses generally and the tiny sample upon which measurements can be made, when compared to the volume of rock involved, a broad-brush approach to classifying rock is suggested at the investigation and design stage. The classification may be refined as experience at a particular site is gained during the slope formation.

The easily measureable features in a rock mass which have been found significant are:

- Weathering
- Discontinuity Orientation
- Discontinuity Spacing
- Discontinuity Condition

Weathering

Weathering is measured on the 4 zone scale proposed in the Geotechnical Manual for Slopes, Government of Hong Kong (1984). Some care is needed in interpreting weathering zones B and C from drill cores, particularly where soil material around corestones is washed away in the drilling process.

The weathering classification will indicate whether the failure mode will be rock or soil and will give an idea of safe slope angles at feasibility study stage. Extra investigation or a geophysical survey may be justified to firm up the limits of the intermediate weathering zones as changes to expected conditions here are always likely and will usually have a financial implication.

Table I WEATHERING ZONES, AFTER GOVERNMENT OF HONG KONG (1984)

Zone A	Structureless sand, silt and clay
Zone B	Residual material with core stones. Rock forms less than 50%. Core stones are rounded and not interlocked.
Zone C	Core stones with residual material. Rock forms 50-90%. Core stones are rectangular and interlocked.
Zone D	More than 90% rock. Minor residual material along discontinuities which may be stained.

Discontinuity Orientation

The graphical analysis of joint data using stereonets as described by John (1968), Hoek and Bray (1977) has become a standard method in rock slope design.

The use of computer or manual methods for the contouring of joint data to select representative poles is subject to a great deal of error. The data for frequently observed joints may be so dense that it suppresses important variations of less sampled joints. Moreover, the counting intervals may be so broad as to mask the existence of a different joint set. Some of the suppressed information although statistically insignificant may be critically orientated in relation to the engineering works, Hencher (1985). At Ap Lei Chau the statistically insignificant joint sets H and J were found to have a major influence on stability, because of their continuity.

Hencher & Martin (1982) characterised joint patterns in seven classes. McNicholl (1984) reviewed this classification taking into account experience of cost overruns in rock slope excavations and suggested a revised classification system for joint orientation as an index of rock mass. This is presented in Table II.

The joint orientation index will indicate whether adverse conditions are likely and will show the need for extra joint strength investigation as well as the need for instability analysis of potential sliding blocks or toppling instability. A high classification will indicate the need for care and special clauses in the blasting specification. It will also point to extra provisional quantities for dowels and rock bolts in domains with especially high ratings. A class 1 rating would indicate little or no stabilization needed whereas sprayed concrete and mesh would almost certainly be needed in class 5 and 6 conditions. For high ratings or even intermediate ratings it may be economical to seek a revision of the slope alignment.

TABLE II JOINT ORIENTATION AS AN INDEX OF ROCK MASS

POTENTIAL FAILURE MODES	CLASS
Multiple Toppling + wedge + plane	6
Toppling + wedge Toppling + plane several wedges + plane	5
Toppling several wedges wedge plus plane	4
Wedge	3
Plane	2
None or minor plane	1
Optional: If random joints persist in the mass uprate by one class	

Spacing and continuity of discontinuities

There are several ways of expressing the typical spacing of discontinuities. Many engineers and geologists use RQD as an index of rock mass and rock quality but it is sensitive to drilling quality and particularly drilling malpractices. Considerable bias is experienced in RQD obtained from vertical boreholes, particularly where the main regional joints are near vertical as in many areas of Hong Kong.

A more useful index is Discon-tinuity Frequency which is a measure of the number of discontinuities encountered per unit length of line in particular direction through a rock mass. (Priest and Hudson (1976) and Hudson & Priest (1983)).

Priest and Hudson discuss theoretical distributions of discontinuity frequency in terms of RQD*, a theoretical RQD calculated from the spacing data and capable of being used for different minimum spacings other than 0.1 m. The discontinuity spacing can be characterised in terms of RQD*, $\overline{x}$ the mean spacing in mm and λ the number of discontinuities per metre.

The relationship is

$$RQD^* = -3.68\lambda + 110.4$$

for $6 < \lambda < 16$.

The important advantages of using this relationship, suitably amended for local use are:

(i) RQD from vertical and inclined holes can be combined by converting into λ values.

(ii) Raw spacing data from boreholes can be pooled with scanline information to obtain a composite $\overline{x}$ value and hence a less biased value for RQD*.

(iii) A subjective figure can be obtained by comparing RQD from boreholes with RQD* from scanlines.

Providing differently orientated exposures are combined, the theoretical relationship has been shown to hold up for two case histories in Hong Kong (McNicholl, (1984)) and was tested also by Wallis and King in Canada (1982).

McNicholl (op sit) explored the use of Discontinuity Frequency using pooled data, as an index of rock mass and by reference to works by Bieniawski (op sit) & Deere (1968) produced Table III.

The discontinuity frequency rating will show the need for extra wide berms for high ratings and possibly rock fall zones. A high rating may indicate that machine excavation is possible. It would also indicate that special face alignment (say curved or varying slopes) would tend to instability.

TABLE III DISCONTINUITY FREQUENCY AS AN INDEX OF ROCK MASS

RQD or RQD* (pooled or calculated)	DISCONTINUITY FREQUENCY	MEAN SPACING $\overline{x}$ mm	CLASS No.
90 - 100	5.5 - 1.5	180-660	1
75 - 90	9.6 - 5.5	100-180	2
50 - 75	16.8 - 9.6	60-100	3
25 - 50	26.9 - 16.8	40- 60	4
5 - 25	47 - 26.9	20- 40	5

Sprayed concrete with mesh reinforcement would be needed for conventional slopes with a high rating and berm drainage would need to include extra quantities for making up the edges of blast damaged berms.

The continuity of joints is a major factor controlling the volume of rock likely to be involved in a failure. If joints in one set are only continuous for 1-2 metres, the volume of rock likely to fail may be quite small and sprayed concrete would be sufficient to stabilize the face.

Joint Condition and Joint Roughness

Joint roughness is a major parameter within the NGI Rock Mass Classification system Barton et al (op sit) and joint condition is a constituent of other RMC systems. Both are influenced by weathering.

If joint condition and roughness could be measured conveniently in the field they would serve as an index of joint friction angle and joint strength. It is unfortunate that classical methods of measuring these indices are time consuming and difficult. The best approach is to design the rock slope based on precedent and to carry out sensitivity checks in analysis. During excavation the joint condition and roughness can be measured regularly as exposures are developed and the rock slope design modified if needs be. Bieniawski (1979) sets down a rational index which combines both indices and this is shown slightly modified in Table IV below.

TABLE IV COMBINED INDEX FOR JOINT CONDITION AND JOINT ROUGHNESS

DESCRIPTION	CLASS
Very rough surfaces of limited extent; hard rock wall.	1
Slightly rough surfaces; aperture 1 mm; hard wall rock.	2
Slightly rough surfaces; aperture 1 mm; soft wall rock.	3
Smooth surfaces, OR gouge filling, 1-5 mm thick, OR aperture of 1-5 mm thick; joints extend several metres.	4
Open joints filled with more than 5mm of gouge, OR open more than 5 mm joints extend more than several metres	5
For very persistent joints in classes 1, 2 & 3	+1

The joint condition and roughness index would be important if combined with unfavourable orientation ratings and would dicate the need to investigate flatter slopes. At intermediate ratings there may be a tendency to loose definition at the edges of berms.

CONCLUSIONS

The Ap Lei Chau rock slope formation showed the usefulness of a thorough geological survey of discontinuity orientations around the site in designing the rock slope. It also emphasises the care that must be taken in analysing the joint data on a statistical basis. Significant faulting was missed by the site investigation which, with a slightly different orientation could have had a major effect on slope stability.

The trial blast gave very useful information on rock fabric, ahead of the main slope construction. A careful design review early on in the slope formation was instrumental in showing that severe stability problems could arise if the original design slope angle of 70° was maintained. A judgement was required to choose between an absolutely safe slope angle and an accetably safe angle, which might require stabilization measures locally. This judgement was backed up by extensometers which showed there was very little relaxation of the rock mass taking place and this was decreasing with time.

Experience from this construction has identified the key rock mass indices which need to be measured and reviewed during construction leading to a greater degree of confidence in designing and forming rock slopes in high risk locations.

Acknowledgements

The authors wish to thank the Director of Housing, Hong Kong Government for permission to publish this paper. Thanks are also due to Binnie & Partners (Hong Kong) for their support. Special mention is also due to Colin Dutton and K.W. Lai for their primary role during the design review.

References

Allan, P.M. & Stephens, E.A. (1971). Report on the Geological Survey of Hong Kong. Hong Kong Government Press.

Barr, M.V. & Hocking, G. (1976). Borehole structural logging employing a pneumatically inflatable impression packer. Proc. Symposium on Exploration for Rock Engineering, Johannesburg, November 1976.

Barton, N, Lien, R. & Lunde, J. Engineering classification of rock masses for the design of tunnel support. Rock Mechanics, Vol. 6 No. 4 pp 189-236.

Bieniawski, Z.T. (1974). Geomechanics classification of rock masses and its application in tunnelling Proc. 3rd Int. Cong. on Rock Mechanics, ISRM, Denver Vol. 11A pp 27-32.

Bieniawski, Z.T. (1976). Rock mass classification in rock engineering. Proc. Symposium on Exploration for Rock Engineering, Johannesburg. Vol. 1 pp 97-106.

Bieniawski, Z.T. (1979). The Geomechanics classification in rock engineering applications Proc. 4th Int. Cong. on Rock Mechanics.

Brand, E.W., Burnett, A.D., & Styles, K.A. (1982). The Geotechnical Area Studies Programme in Hong Kong. Proc. of 7th S.E. Asian Geotechnical Conference, Hong Kong Vol. 1 pp 107-123.

Geological Society Engineering Group Working Party Report, 1970. The logging of rock cores for engineering purposes. QJEG Vol. 3, No. 1 pp 1-25.

Government of Hong Kong (1984). Geotechnical Manual for Slopes, Geotechnical Control Office, Engineering Development Department.

Henscher, (1985). Limitations of stereographic projections for rock slope stability analysis, Hong Kong Engineer Vol. 13, No. 7 July 1985.

Henscher, S.R. & Martin, R.P. (1982). The description and classification of weathered rocks in Hong Kong for engineering purposes. Proc. 7th S.E. Asian Geotechnical Conf. Hong Kong. Vol. 1 pp 125-142.

Hoek, E. and Bray, J.W. (1977). Rock Slope Engineering. Institution of Mining and Metallurgy, London, 2nd Edition.

Hudson, J.A. & Priest, S.D. (1983). Discontinuity frequency in rock masses Int. Jour. Rock Mech. Min. Sci. & Geomech. Abstracts Vol. 120 No. 2 - pp. 73-89.

John, K.W. (1968). Graphical stability analysis of slopes in jointed rock. Journal of Soil Mechanics and Foundation Division ASCE Vol. 94 No. SM2 pp 497-526. Discussion in Vol. 95 No. SM6, 1969 pp 1541-1545.

McNicholl, D.P. (1984). The assessment of rock mass characteristies prior to the excavation of large cut slopes in Hong Kong. M.Sc dissertation Univ. of Leeds.

Priest, S.D. & Hudson, J.A. (1976). Discontinuity spacings in rock, Int. Jour. Rock Mech. Min. Sci. and Geomech. Abstract 13.

Prokapovich, N.P. (1972). A comparison of surface and subsurface structural patterns at Spring Creek, Conduit Dam, Trinity River Diversion Central Valley Project California.

Sekula, J. (1964). The deterioration of rock slopes in Hong Kong. Proc. Conf. on Geological Aspects of Site Investigation, Hong Kong Geological Society, Hong Kong.

Steffen, O.K.H. (1976). Research and development needs in data collection for rock engineering Proc. Symposium on Exploration for Rock Engineering. Johannesburg November 1976.

Terzhaghi, R.P. (1965). Sources of Error in Joint Surveys, Geotechnique Vol. 15 pp 287-304.

Wallis, P.F. & King, M.S. (1982). Discontinuity spacings in a crystalline rock. Int. Jour. Rock Mech. Min. Sci. and Geomech. Abstract. 17.

Role of blasting control in excavation works near a pre-existing tunnel

P. Berry
E.M. Dantini
Università di Roma 'La Sapienza', Facoltà di Ingegneria, Rome, Italy

SYNOPSIS

This paper looks into the blasting control problems and the relevant solutions adopted during the driving of a tunnel which crosses a penstock serving a nearby hydro-electric plant. Mechanical excavation should have been used owing to the proximity of the two tunnels, but the feasibility studies had shown this solution to be uneconomical and so the drill and blast method had to be used in the end.

Measurements were carried out to make a preliminary assessment of the vibrations caused by the rounds in the rock mass and on the tunnel walls. During construction the effects of blasting were constantly monitored. Amongst other things this research has shown that the criterion of maximum particle velocity is inadequate - for close-in blasts - in guaranteeing the safety of the penstock. It furthermore pointed to the importance of the vibrations induced by the blast air waves and to the different laws governing seismicity in the rock mass and around the tunnel. As a consequence, different blasting procedures were adopted depending on the distance of the working face of the penstock, in moving towards and away from it.

INTRODUCTION

Tunnelling in urban area is confronted with a number of problems related essentially to the excavation technique used, to the geostructural nature of the rock mass and its geomechanical characteristics, to the geometric and mechanical characteristics of surrounding man-made works and buildings to be safeguarded.

In particular, the drill and blast method demands that such problems be solved as those linked to underground waters, and to the geomechanical stability of the rock mass around the tunnel; on the other hand, unlike other excavation techniques, account is to be kept of induced fracturing and seismicity caused by the action of the explosive.

In considering the dynamic stresses, safety problems take on different aspects, depending on whether the structures and the man-made works to be protected are located at ground surface or underground. With regard to the interference of surface blasting with the buildings, reference can be made to an extensive body of rules which set the physical parameter of seismicity which needs to be controlled, as well as its maximum values, which are not to be exceeded. Even though these regulations are well consolidated, it does appear to be necessary to further studies and researches on the effects induced by blasts occurring in the vicinity of man-made facilities. In such conditions, the maximum particle velocity criterion - as can be inferred from the existing body of rules - appears to the indefinite and possible unrealistic. For example, energetically weak seismic phenomena whose source is close to the structure to be protected, may produce locally extremely high velocities of the particles, but owing to attenuation, the structure as a whole will not be involved and, consequently, its stability is not jeopardized.

Much less attention is given to the seismicity induced in rock masses by the use of explosive charges located at depth and to their consequences on surrounding engineering works - both underground and at the surface. And so, generally, in these cases too, the criterion of maximum particle velocity is used on the basis of analogy and also because once the characteristic impedance of the medium involved by the seismic waves is known, in some specific cases, particle velocity can be directly correlated to the dynamic state of stress. There being no specific regulations, the degree of risk can be assessed only on the basis of an analysis of the behaviour of the works when subjected to vibrations. But such an analysis presents aspects which require among other things the knowledge of parameters which are not always detectable.

Such problems have been dealt with in the case of the driving of a tunnel which is part of a broad scheme of hydraulic works which are under way, and which are expected to expand the water supply capacity to the town of Naples.

DESCRIPTION OF THE TUNNELS

Tunnel A crosses a built-up area close to the town of Cassino (Latium).

Most of the tunnel is excavated inside a formation which, for the area under study, is mainly dolomitic and homogeneous from the geological and structural points of view. The orientation and spacing of the layers and of the joints require a modestly thick spritz-beton lining and non - intensive bolting of the roof.

The tunnel is some 5.5 km long and its section is 47 m^2. Fig. 1 presents a schematic section of the tunnel and a scheme of the concrete works envisaged by the project. The overburden in the built-up areas is such that there are no particular safety-related problems caused by induced seismicity.

The penstock (tunnel B) built in 1952 and now operated by the National Electricity Board (ENEL), which is encountered by tunnel A (Fig. 1) has a circular section of about 23 m^2 and according to the project design it should have a 60 cm thick concrete lining, the outer 30 cm being reinforced, whereas the inner surface is lined with a 1 cm thick cement plastering.

Where they meet, the axes of the tunnels form an angle of 82° roughly and, according to the design, the floor of tunnel B will be uncovered for a maximum thickness of 30 cm. In such conditions, the stability problems deriving from the excavation technique adopted become very important and they are further aggravated by the fact that we do not know to what extent the design specific-

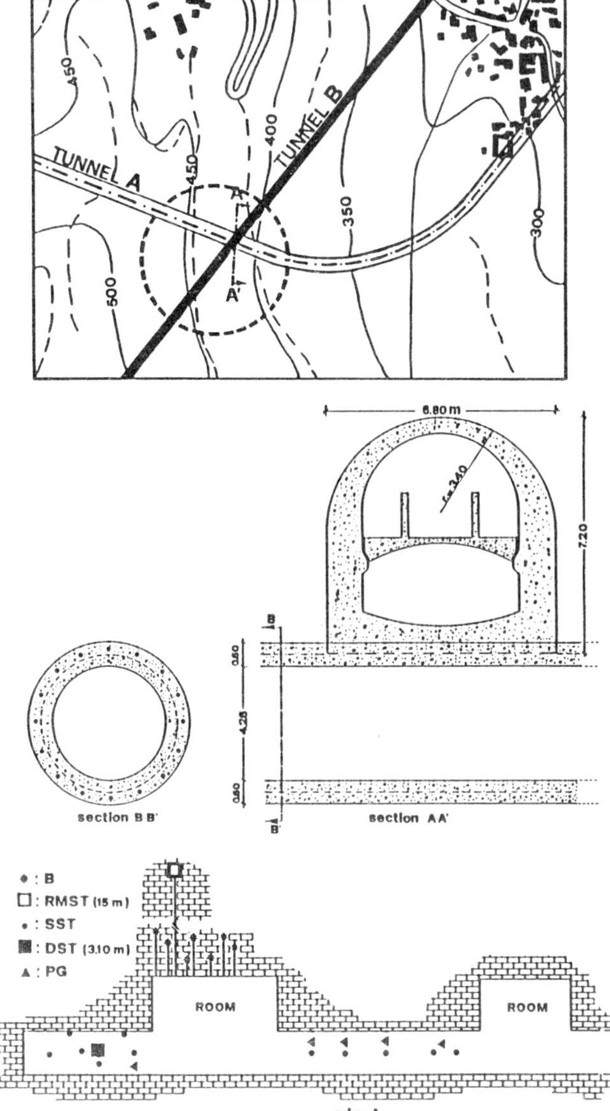

Fig. 1 - Scheme of the intersection of the two tunnels and scheme of the measurement stations in Tunnel A (plant): B - explosion holes; RMST - transducer in a 15 m long horizontal hole; SST - velocity transducers under the floor and on the side walls of the tunnel; DST - velocity transducers in a 3.1 metre long hole under the tunnel floor; PG - pressure transducers

ations were complied with, especially the thickness of the crown lining. In fact the thickness may be believed to be less than 60 cm because of the difficulties encountered during the fifties owing to the techniques used at the time to pump the concrete. Moreover we have no knowledge of the characteristics of the 33 year old concrete and of the state of repair of the plaster. In addition, it is very likely that the mechanical characteristics of the rock mass around tunnel B are very poor owing to the blasting techniques used during the fifties, at the time of its construction.

On the basis of these indications, the Construction Management Division in agreement with ENEL has set a maximum particle velocity value of 3 cm/s - a very cautious limit - for each of the three components (vertical, transverse and longitudinal) so as to secure safety conditions for tunnel B.

SEISMIC INVESTIGATION

For the design of a controlled blasting procedure, in an early stage, the seismicity induced by the rounds being used in tunnel A was evaluated through statistical elaborations of the values of the maximum particle velocities obtained[1] within a granite rock mass by camouflet explosions using different types of explosives (Fig. 2).

The correlation between particle velocity and scaled distance provided by the analysis was used to set the minimal distance (107 m) between the two tunnels; below this value the volley scheme would have had to be abandoned.

In order to assess the influence of the type of explosive being used in the tunnel, of the mechanical characteristics of the rock mass and of the different explosion configurations, tests were carried out along a side-wall of the tunnel (Fig. 1) where charges placed inside 1-to-2 m long holes at variable distances from a geophone (RMST) located at a depth of 15 m, were exploded.

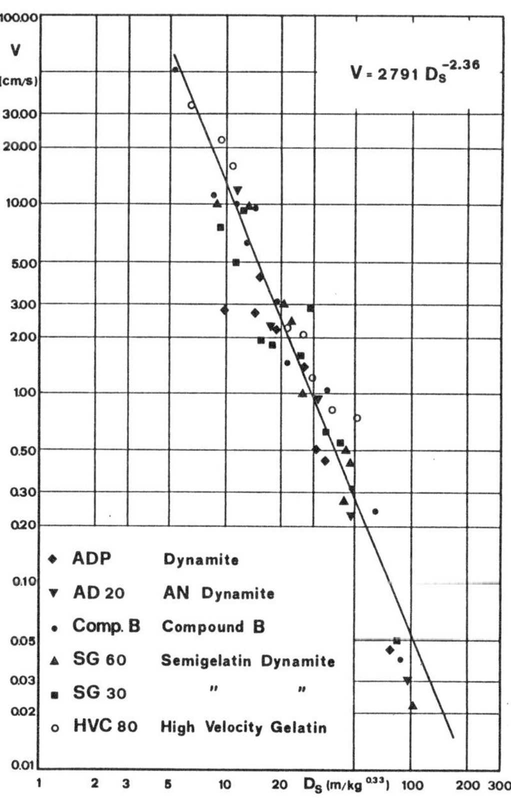

Fig.2 - Maximum particle velocity as a function of scaled distance (camouflet explosions)

An equation was thus obtained (Fig.3) between the pseudovector of the maximum particle velocity and the scaled distance, which substantially confirms the validity of the preliminary evaluation and provided sound information which has made the design of the rounds to be used at decreasing distances bewteen the tunnels more reliable.

The problems induced by seismicity in tunnel B when it is being approached by the excavation face of tunnel A, and when the latter proceeds away from tunnel B are substantially different in nature. In order to study this aspect and acquire the necessary knowledge to decide the design of the volleys when the face of tunnel A is moving away from tunnel B, geophones (SST in Fig.1) were placed at 50 cm below the floor along the axis of tunnel A; other geophones (SST) were bolted to the rock of the walls with steel brackets (scheme in Fig.4) and another geophone (DST in Fig.1) was placed at 3.1 m underneath the floor.

A typical example of the seismograms obtained by the surface and the wall geophones is given

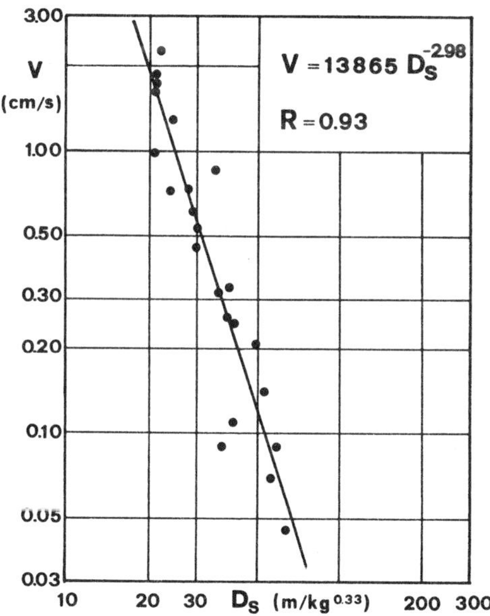

Fig.3 - Pseudovector of maximum particle velocity measured (23 tests) inside the rock mass in a 15 m deep horizontal hole as a function of scaled distance. R is the correlation coefficient

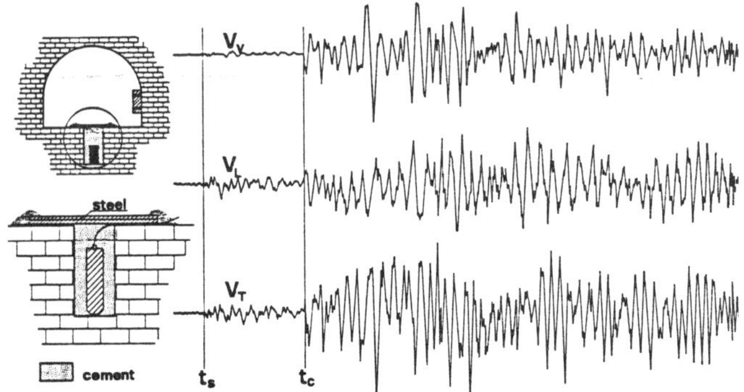

Fig.4 - Seismograms of vertical, longitudinal and transverse velocities obtained with geophones (SST) placed under the tunnel floor or bolted to the side walls. t_s is arrival time for seismic waves; t_c is arrival time for compression waves

in Fig.4. It clearly shows the presence of seismic waves followed by the compressional waves having greater amplitude induced by air blast loading.

The former come first and propagate throughout the rock mass at a speed of 2.2 - 2.7 km/s. A statistical analysis of the arrival times of the compressional waves and of the pressure waves(the latter are measured by means of transducers placed in the tunnel as shown (PG) in the scheme in Fig. 1) was used to calculate the average velocities of the air waves which were found to be equal to 0.342 km/s (Fig.5). Moreover, since the latter arrive simultaneously with the compressional waves, the presence of noticeable elastic precursors must be ruled out, consistently with the findings obtained for surface explosions[2].

The correlations between the scaled distances and the values of the maximum particle velocities induced by the seismic waves and by the compressional waves were analysed statistically, and the results are shown in Fig.6. Each point in the diagram refers to a different round and the values of the particle velocity were obtained by calculating the pseudovector using the vertical (V_v), the longitudinal (V_l) and the transverse (V_t) components on the basis of the following relationship:

$$V = \sqrt{V_v^2 + V_l^2 + V_t^2}$$

As a confirmation to the close tie between compressional waves and pressure pulse, it must be pointed out that the exponent at which the seismicity linked to the compressional waves attenuates

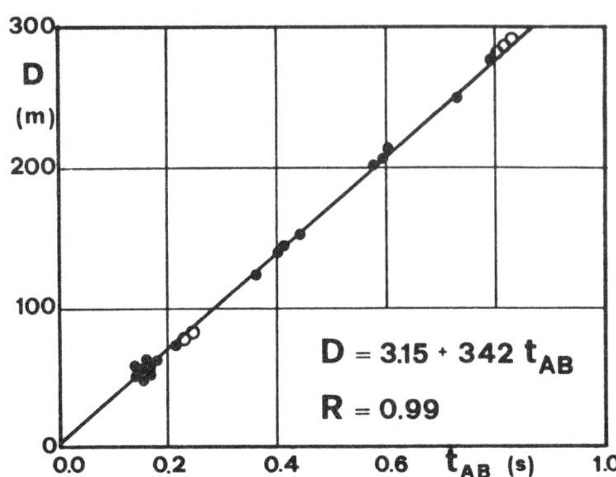

Fig.5 - Arrival time of compression waves (solid dots) and of air blast pressure (open dots)

as the scaled distance increases is of the same order of magnitude as the exponent of the law which governs the attenuation of the overpressures in air along the tunnel[3,4].

The greater scatter of the velocity values relating to the compressional waves may very likely be attributed to:

a) the position of the measurement point with respect to the geometry of tunnel A along which rooms have been excavated in order to allow for the manoeuvres of the drilling, loading and haulage equipment (Fig.1);

b) the variability of the phenomena determining the expansion of hot gases released by the explosion. Indeed, under identical round designs, they depend on the accuracy in implementing the design and in particular on the degree of stemming of the holes; on the direction of the holes with respect to the face; on the success-

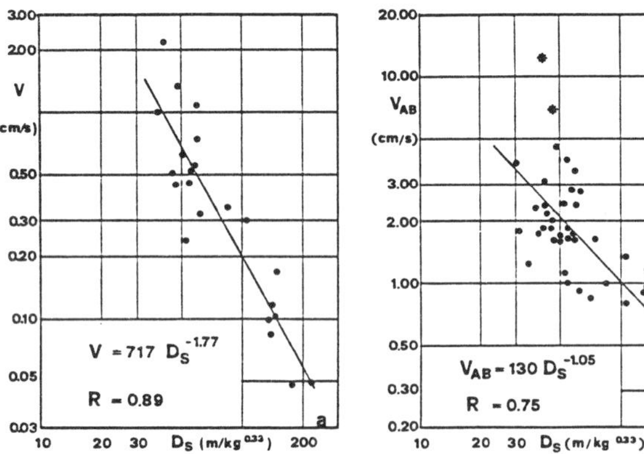

Fig.6 - Pseudovector of maximum particle velocity and scaled distance. The values of the seismic waves (a) and of the pressure waves (b) have been measured in 48 rounds with geophones placed according to the scheme in Fig.4

ion of breakage processes in the round; on the variability of the geostructural characteristics of the rock mass (joints, layers, lithologic dishomogeneities);

c) the criterion of scaling the distance with the cubic root of the amount of explosive; indeed, this criterion for a tunnel holds only for distances more or less equal to the diameter[3].

For the values obtained during this research (Fig.6b) this criterion was adopted just the same because account was kept about the fact that the amount of explosive for each delay differed by a ratio of 1 : 3.3. Even though this variation may be small, it is however, source of a further scatter of the values. In order to acquire further information about this phenomenon,it was deemed useful to start a systematic research - which is still under way - to establish a correlation between types of explosives, geometry of the openings, trend of the overpressures along the tunnel, and induced seismicity.

The influence of the geometry, point a), of the rooms present in the tunnel (Fig.1) is emphasized in the seismograms by the variation in the relationship between the values of the different velocity components; for example, in the case of geophones placed immediately underneath the tunnel floor, the vertical component prove to be greater when the geophones are placed between the face and the first of the rooms present behind the face it-

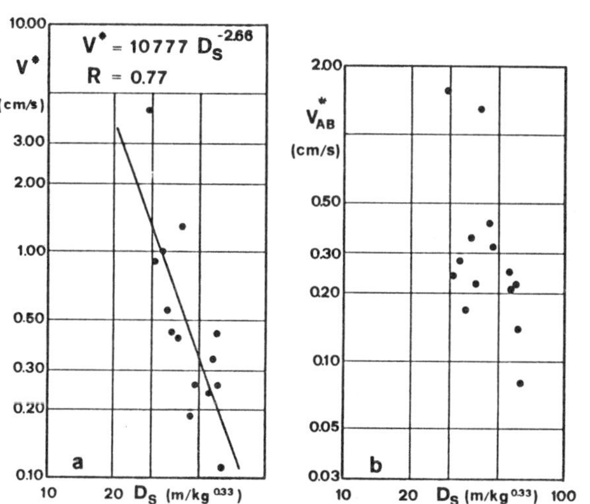

Fig. 7 - Pseudovectors of maximum particle velocity and scaled distance. The values of the seismic waves (a) and of the compressional waves (b) have been measured in 14 rounds with a geophone placed at 3.1 m underneath the tunnel floor (DST)

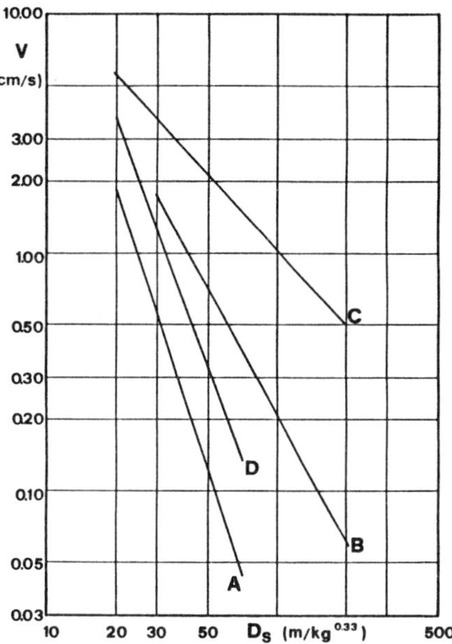

Fig.8 - Pseudovector of maximum particle velocity and scaled distance. For comparison, the equations of the pseudovector of the maximum particle velocity measured at 15 m (A), and of the seismic waves at the surface (B) and of the pressure waves at the surface (C) and of the seismic waves in depth (3.1 m) (D) are presented

self; when the geometry of the empty space determines a lateral expansion of the gases (owing to the presence of one of more rooms) the transverse component value of the seismicity is higher.

The influence of the expansion modality of the gases, point b), is well highlighted in Fig. 6b by the higher values of the velocity pseudovector; indeed, tests have shown that such values may be ascribed to fly rock phenomena (asterisks in the figure) and to the progressive reduction of the tunnel section required to get passed tunnel B.

The attenuation of seismicity, as the radial distance from the tunnel walls (Fig.7) increases, was studied by measuring the velocity at a depth of 3.1 m below the floor.

Within the range of scaled distances, it proves to be lower than the values measured along the tunnel surfaces (curves B and C in Fig.8), and it is greater than the seismicity induced in the rock mass (curve A). It must be pointed out that the values of seismicity related to the compressional waves prove to be highly scattered as would be expected (Fig.7b). On the contrary, if the values are correlated to the physical source which has determined them, namely the compressional waves measured at the surface along the vertical line passing through the geophone placed at 3.1 m, the scatter declines dramatically (Fig.9).

Fig. 10 shows the ratios between the values of the particle velocities (compressional waves/ seismic waves) related to the scaled distance, both in the case of measurements along the surface of the tunnel and for the measurements made at 3.1 m depth. The laws governing the ratios to the scaled distances underline that in order to secure safety conditions in the penstock, account is to be kept of the greater weight gained by the compressional waves vis-à-vis the seismic waves as the face moves away from tunnel B.

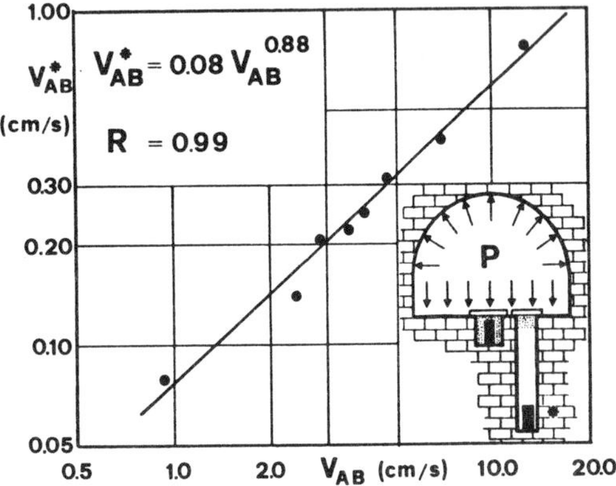

Fig. 9 - Pseudovectors of maximum particle velocity determined from compressional waves. Values measured with a surface geophone on the X axis, and values measured at 3.1 m depth under the tunnel floor on the Y axis

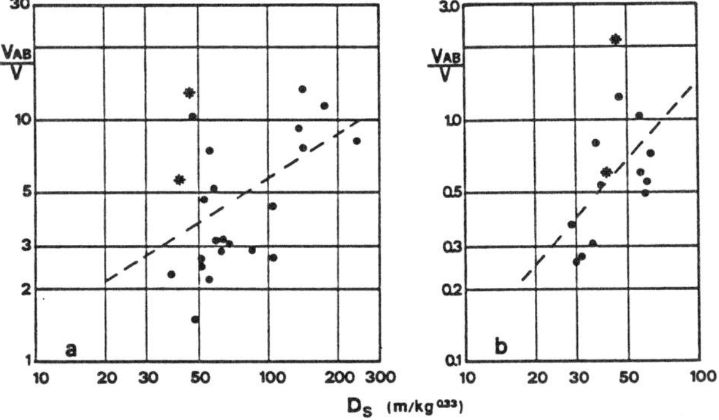

Fig. 10 - Ratios between the pseudovectors of the velocities (compressional waves/seismic waves). Values measured at the surface (a) and at a depth of 3.1 m (b)

EXCAVATION TECHNIQUES

The results of the experimental tests provided the elements for designing the most appropriate excavation phases for crossing tunnel B without impairing its mechanical and structural integrity.

Far away from the penstock, a V-cut round was adopted for the excavation and millisecond and normal delay detonators were used, producing advances of 4 m. Fig. 11a shows a schematized section of the tunnel and of the position and orientation of the holes of the cut. At 107 m from tunnel B, a fan-cut was adopted (Fig. 11b) which secures a better distribution of the charge and induces less seismicity in the rock mass. This very precautionary measure, based on the equation in Fig.2, was observed until the results of the experimental investigations were made available. For this type of round a maximum amount of explosive equal to 9.8 kg for each delay was used, and at the beginning advance of 3.5 m were achieved, which dropped progressively to 1.50 m at a distance of about 30 m from the axis of tunnel B (Fig. 11 c and d) using a maximum charge per delay equal to 0.9 kg.

Starting from this point, and making a decision on the basis of the equation in Fig.3, the height of tunnel A was reduced with three consecutive rounds, from 7.2 m to 4.5 m (Fig. 11 e) so as to leave, during this early stage of the crossing over tunnel B, a 2.4 m protection over the top of the penstock, to be cleared later. The round was later modified per force by adopting a Canadian cut with a 200 mm central hole and advances of only 1 m.

Finally, starting from a distance of approximately 8 m, a three-phase pattern was adopted (zone F in Fig.11). In stage I a 3 m high heading was excavated, some 2.8 - 3.0 m wide at its base, and 2.5 - 3.0 m long; advances of 0.8 m per round were achieved by using a maximum charge of 0.07 kg per hole; successively, (stage II), the heading was widened at the crown and along the walls by using the same method as in stage I; and finally, (stage III), still the same method was used, but the charge per hole was equal to 0.05 kg, and the foot of the heading was removed and therefore the tunnel height was increased to 4.5 m. Having reached the face of the heading, the three stages would be started over again; the advance rate was about 2.5 m per day with three shifts.

When the tunnel face (h = 4.5 m) went passed the penstock, at 10 m from its axis, the excavation

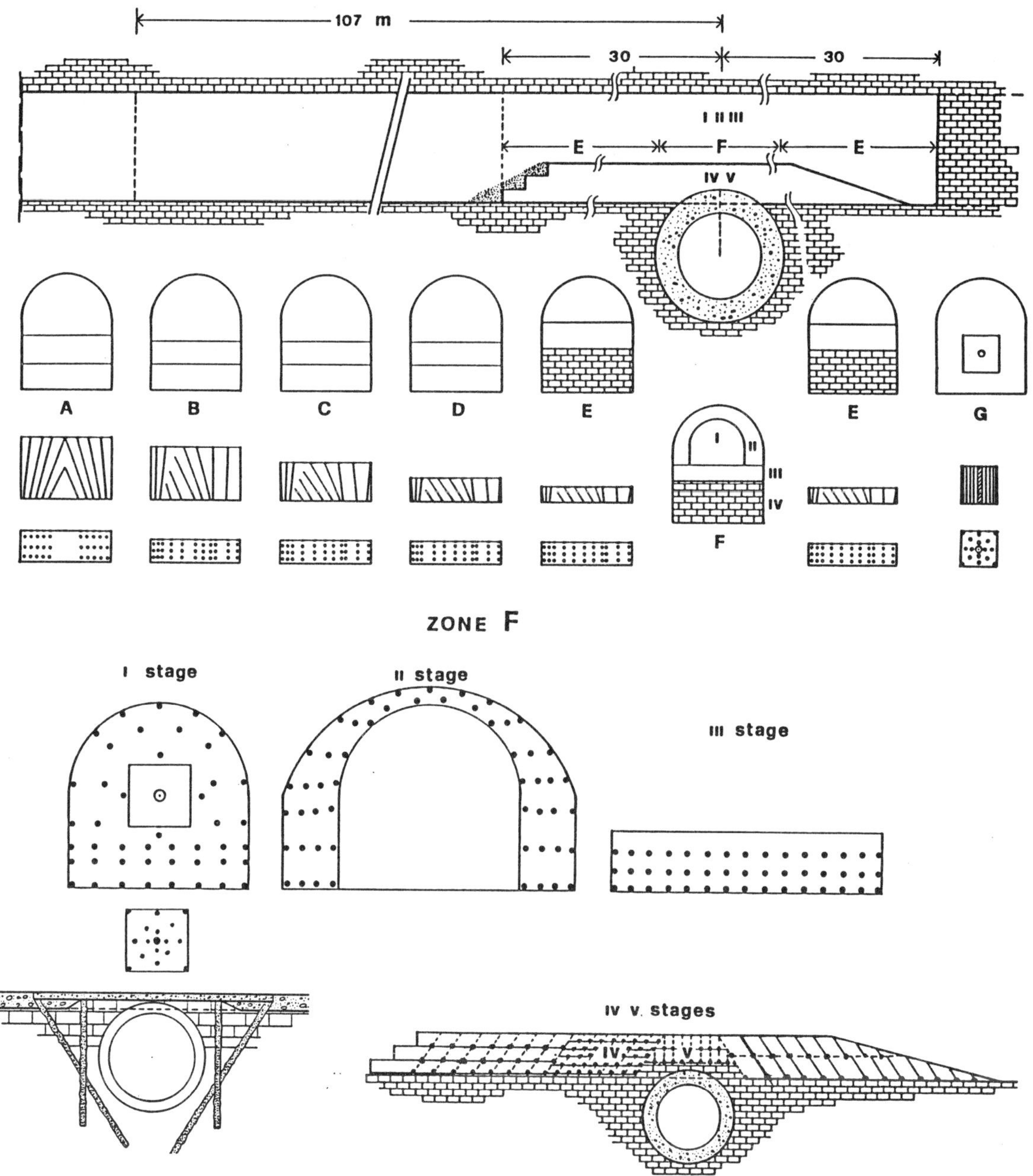

Fig. 11 - Scheme of successive tunnelling phases and of reinforcement works for tunnel B (groutings and concrete slab)

pattern E was resumed, with the charges increasing gradually while the number of blast-holes were reduced. At a distance of 16 m, the height of the tunnel was gradually increased, reaching its starting height at a distance of about 30 m.

After this, the protection slab was removed in successive stages. For the first 15 m, moving towards the penstock, horizontal charges were used following a scheme similar to that adopted for stage III. Four rows, each consisting of thirteen 2-metre-long blastholes and each loaded with 0.28 kg of explosive, were used per round

From 15 m to about 4 m from the penstock, five rows of fifteen blastholes each, their length gradually dropping from 1.5 m down to 0.8 m were used with the explosive being cut down gradually and being no more than 0.05 kg per hole, near the penstock.

Here, a 1.40 m high and 10 m long residual slab to protect the penstock had been left. The residual protecting rock mass was removed with the following scheme: on the top of the penstock tests were initially carried out with the 1140 kg hydraulic hammer of the Montabert BRH type which delivers 10 blows per second; measures of the seismicity induced by the blows suggested that pieces of detonating cord in vertical 50 cm-long holes with a maximum charge of 0.02 kg should be used.

This method enabled to remove the limestone down to 0.5 m from the upper surface of the penstock lining, the yields being good and the vibrations being lower than those induced by the hydraulic hammer. Then the lining of the penstock was uncovered by removing the remaining 0.5 m using a light hammer. The residual slab between the penstock and the working face of the tunnel (h = 7.2 m) was removed by means of 1 metre-long vertical charges, the amount of explosive increasing with the distance from the penstock.

After this last stage, the rest of the tunnel will be excavated by using a Canadian cut (Fig. 11 G), the advance rate per round being initially 1.5 - 2 m, to be increased up to 4 m.

Finally, the rock mass around tunnel B will be consolidated by means of cement groutings (following the scheme in Fig. 11) throughout the width of tunnel A and for such a depth that the lower portion of the penstock will be surrounded; a reinforced concrete slab will be placed over the crown of the penstock.

STABILITY OF TUNNEL B

In conjunction with the starting of the excavation of zone F of the tunnel, the penstock was put out of operation for maintenance works previously scheduled by ENEL. A preliminary inspection was thus made to verify the repair conditions of the penstock; a velocity transducer was bolted to the side wall closest to the excavation face (Fig. 12 a) at the intersection between tunnel B and the plane of vertical symmetry of tunnel A.

The recording instruments were placed at 2.6 km from the velocity transducer in the vicinity of the closest maintenance point of access.

When face A had gone a few metres beyond the axis of the penstock, the transducer was shifted to the opposite side (Fig. 12 b). These measurements were taken regularly until the penstock was put back into operation when the maintenance works were completed.

The values of the pseudovector of the velocities obtained through such measurements during the ex-

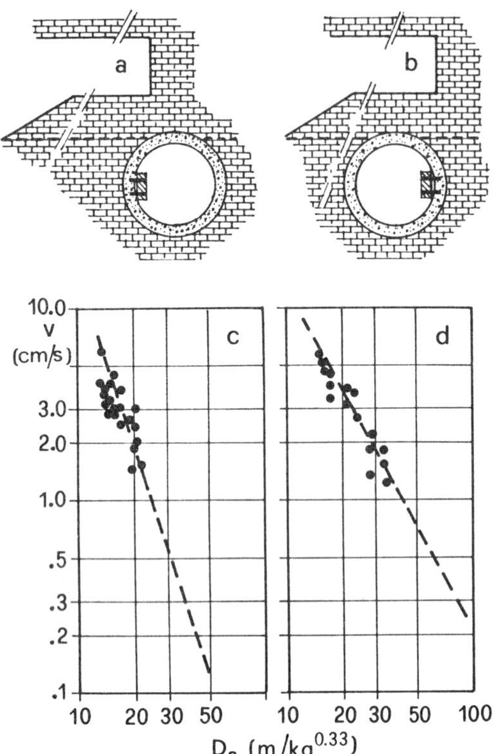

Fig.12 - Pseudovector of maximum particle velocity (c) measured in tunnel B by utilizing a geophone anchored to the side wall of the penstock, as shown in schemes (a) (when tunnel A is reaching the penstock) and (b) (when tunnel A is over the penstock). The dashed line represents the equation obtained from the preliminary experiments. The values of Fig.(d) were obtained for situation (b) when the working face of the heading was moving away from the penstock.

cavation of the heading before and immediately after the penstock are shown in Fig. 12 c, where, for comparison, the equation from Fig. 3 has been included (dashed line).

The little influence of the void (penstock)-in this type of geometric situation and for this type of volley - is stressed by the high consistency of the values obtained for the rock mass (dashed line) with the measured values.

This could be very likely attributed to the fact that the incident waves stress a circular structure having a rather marked curvature.

According to this rationale, it was to be expected that, as the working face of the heading moved away from the penstock, the induced seismicity would follow the law obtained from the preliminary tests with the geophone located at 3.1 m under the

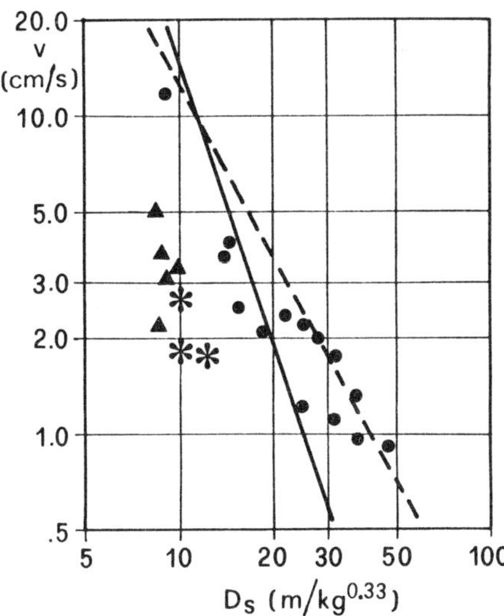

Fig. 13 - Pseudovector of maximum particle velocity measured in tunnel B during removal of protecting slab

tunnel floor (Fig. 7a); but, it is very likely that owing to the excavation geometry the influence of the surface waves prevailed over that of the direct waves (Fig. 12d); the vibrations induced in the limestone layer over the void may have followed such a pattern that, at the measurement point the velocities obtained were analogous to the surface velocities (dashed line from Fig. 6b).

The seismicity induced in the penstock is equal to or lower than that measured in tunnel A, but this is in contrast with the observations made by other authors[5], who measured lower values in a tunnel being excavated and higher ones in an existing tunnel located at 32 m above the other.

During the excavation of the slab, velocity pseudovectors were obtained (Fig. 13) which substantially follow the rationale described above.

Indeed, the velocities (solid dots in the fig.) follow the trend of the surface seismicity (dashed line), when the working face is at 10 metres or more (horizontal distance) from the geophone; the velocities obtained when the rounds are closer to the measurement point (from 10 to 5 m from the geophone and from 2 m onwards, moving away from it) tend to cluster following the law of seismicity in the rock mass (solid line).

With reference to the rounds between +5 m and -2 m, the complex phenomena of reflection and refraction (related to the geometry of the void which was protected by a thin limestone layer over the penstock lining) are a restriction to the measurements on the penstock walls (triangles in the figure) which appear to cluster along a line parallel to the seismicity law line referring to the rock mass, but with markedly lower pseudovectors. These phenomena were not highlighted during the stage III excavation of zone F, when the limestone above the penstock was only two metres higher than the layer protecting the penstock during the last explosions.

It is obvious that the seismicity in this case is closely correlated to the position of the measurement point, and that very likely the values of the pseudovector (triangles) should have clustered according to the two laws indicated in the fig. if the geophone had been placed on the crown of the penstock.

It was estimated that on the crown of the lining of the penstock, the pseudovector of the velocities may have reached a maximum value of 15 cm/s.

Also some of the values of the pseudovector, obtained during the previous excavation stage (Fig.12c, d) indicated that one or more components have gone beyond the imposed limit of 3 cm/s; however, for very close distances between the blast holes and the penstock, the criterion of the maximum particle velocity was felt to be too restrictive, and so, instead of controlling the structure safety, the strength of the concrete lining was kept under control, and consequently it was decided not to exceed the limit values of the concrete tensile and shear strength. The state of stress acting locally on the penstock was estimated by means of the relationships:

$$\sigma_{t,c} = \rho c_1 v$$

$$\tau = \rho c_2 v$$

where ρ is the density of the concrete and c_1 and c_2 are the velocities of the P and S waves in the lining of the penstock and v is the maximum particle velocity.

The frequent inspections made inside tunnel B did not show the presence of lesions, not even in the plaster; this proves that the criterion adopted was absolutely well-grounded.

During the excavation of the heading and above all during the removal of the slab, no appreciable effects of the seismicity induced by the pressure waves were detected. This may be ascribed to the

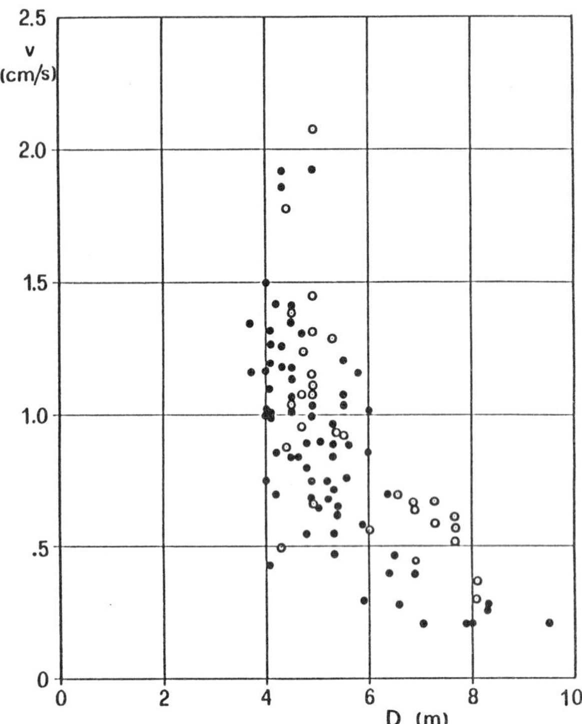

Fig. 14 - Pseudovector of maximum particle velocity (as a function of distance from geophone) induced by hydraulic hammer . Solid circles: values when coming close to the geophone; empty circles: values obtained over the void when geophone has be passed

fact that the amounts of the explosive used were low compared with the size of the tunnel cavity.

The seismicity values obtained by using the Montabert hydraulic hammer (Fig. 14) were found to be higher than those induced by 0.02 kg charges of detonating cord (asterisks in Fig.13). Moreover, even though the decay law of the pseudovector emerges very clearly, the seismicity is highly scattered owing to the modality followed in using the hammer, and to the mechanical characteristics of the rock. It is more aleatory to predict the seismic velocity values induced by mechanical excavation, and so the rock on the top of the penstock (except for a small layer of limestone cleared later with a light hand hammer) was removed by using detonating cord in vertical holes.

At this stage, after a final accurate inspection the penstock was put back into operation, the water pressure being equal to 0.2 MPa. The absence of water leakage along the portion of the penstock exposed by the removal of the limestone layer, shows that the penstock has not been damaged.

CONCLUSIONS

Tunnelling in built-up areas must face and solve safety problems related to the utilization of explosives. Account of such problems must be kept right from the design stage,and care must be taken during implementation to adopt changes according to the results provided by the safety measurements during the progress of the works.

In particular, this research has shown that it is necessary to identify the laws governing the propagation of the seismicity throughout the rock mass beyond the working face and around the tunnel excavated so far, because if this is disregarded[6] the seismic values will be even more scattered and consequently prediction on the basis of experimental data becomes aleatory.

In fact seismic control problems are numerous and highly complex and it has highlighted the fundamental importance and validity of carrying out preliminary measurements. Indeed, in this way the dynamic behaviour of the rock mass can be predicted both before and during excavation. Moreover, the importance of the seismicity induced by the air blast wave was also pointed out, also in relation to the safety of surface works when the overburden on top of the excavation is not great.

It was also found that the maximum particle velocity criterion may take on a meaning which departs considerably from its generally accepted formulation where the blasts are very close to the works to be protected.

ACKNOWLEDGEMENTS

The authors wish to thank the COGEFAR Company and in particular Eng.S.Sbardellati - Site Supervisor - for his kind help during the research, and Eng. L. Panicali for his helpful assistance in designing the rounds and the excavation methods and in collecting the experimental data.
This research was partially supported by C.N.R. (Contract CTB 83.00558.05) and by the Ministry of Education (RS 40, 1984)

REFERENCES

1. Nicholls H.R. and Hooker V.E. Comparative Study in Granite. U.S.B.M.R.I., 6093, 1965.
2. Zhu-Rui-Geng and Li Zhen. In situ Measurements of Blast Seismic Waves and their Safety

Distance. Int. Conf. on Recent Advances in Geotech. Earthquake Engineering and Soil Dynamics, Rolla, Missouri, Vol. I, 1981, p.43-46.

3. Hanna N.E. and Zobetaki M.G. Pressure Pulses produced by Underground Blasts. U.S.B.M. Q147, 1968.

4. Henrych J. The Dynamics of Explosion and its Use. Elsevier S.P.C., New York, 1972, p.167-170.

5. Sakurai S. and Kitamura Y. Vibration of Tunnel due to Adjacent Blasting. Int. Symp.Field Measurements in Rock Mechanics. Zurich, 1967.

6. Studer J. and Suesstrunk A. Swiss Standard for Vibrational Damage to Buildings. Int. Conf. on Soil Mech. and Found. Eng., Stockholm, Vol.III 1981, p. 307-312.

Subsurface investigations of abandoned limestone workings in the West Midlands of England by use of remote sensors

P.A. Braithwaite B.SC., M.SC., D.I.C., C.ENG., M.I.C.E., M.H.K.I.E., F.G.S.
Ove Arup & Partners, Birmingham, England
K.W. Cole B.SC., M.SC., C.ENG., F.I.C.E.
Ove Arup & Partners, London, England

SYNOPSIS

The Industrial Revolution has left the West Midlands of England with a legacy of dereliction. In order to aid regeneration of the region the Government is funding, through the local authorities, a programme of investigations and treatment of abandoned limestone mines. Investigation techniques using remote sensors have been adapted to determine the extent and condition of these mines. A case history illustrates the use of these techniques in determining the condition of a partly collapsed mine, and in enabling the collapse mechanism to be deduced. Recommendations for further work and options for remedial works are outlined.

INTRODUCTION

The West Midlands of England, during the 19th Century, were at the heart of the Industrial Revolution. All of the essential materials - coal, ironstone, limestone and water - were readily available to feed the new burgeoning iron working industries. Factories and furnaces were so common to the north of Birmingham, and their chimneys belched out so much smoke, that the area became known as 'The Black Country'.

Today, the Black Country has declined to a region of empty factories, overgrown canals, abandoned railways and high unemployment. The legacy of its industrial past is evident in the hundreds of hectares of derelict land. About 250 hectares of land lie derelict as a consequence of the long abandoned limestone mining industry. Unlike coal mines, which are the inherited responsibility of the National Coal Board, there is no single authority responsible for limestone mines. Ownership has passed through many hands and land sold off in parcels so that present day responsibility is nearly impossible to define.

The Government, through the Department of the Environment, is financing through the Black Country Boroughs, the investigation and treatment of old limestone mines to aid the regeneration of the region. The investigations carried out over the first four years of this project have led to the development of the investigation techniques discussed in this paper.

HISTORY OF LIMESTONE MINING

Historical records indicate that the limestone extraction industry was of Roman origin. These records indicate that stone dug from shallow quarries was used directly for building. Lime for building mortar and agricultural use was obtained by burning the stone in kilns.

Between 1750 and 1900 an iron making industry developed in the Black Country. Limestone was an essential material, being used as a flux in the smelting of iron ore. The iron making and processing industries necessitated factories, structures for the storage and transport of materials and houses for the work force. Hence, the demand for limestone as a building material also increased enormously during this period. It is estimated that, in all, some 19million tonnes of limestone was extracted by quarrying and mining within the area shown in Figure 1, the majority of extraction taking place between 1750 and 1900.

When abandoned, the general practice was to leave the mine as it stood with no attempt to backfill the workings. As limestone is strong, and the rocks above and below are also quite strong, the cavities remain "open" for many years, unlike coal and iron ore mines where the workings generally collapse within a few years. Limestone mine shafts may have been filled with debris or, as was quite usual at the time, a staging was constructed near to the ground surface and only the top section of the shaft filled. Mines below the water table gradually filled up with water.

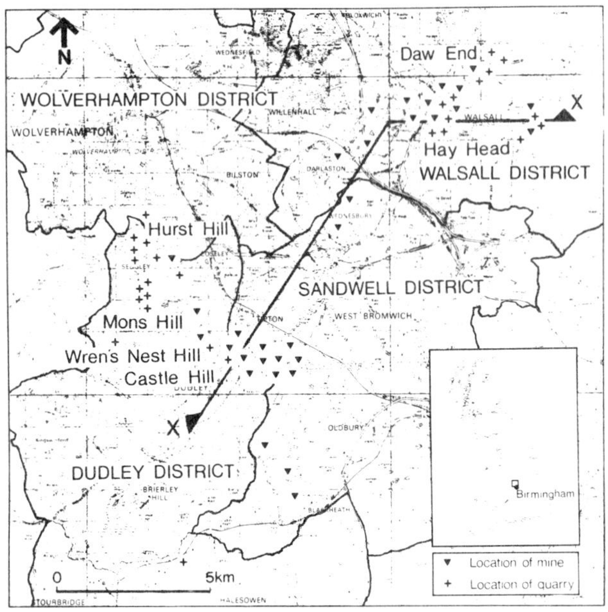

Fig.1 - Plan of the Black Country showing locations of mines and quarries.

Only after 1873 did it become a legal requirement to make proper surveys of a limestone mine at intervals and upon abandonment. Many of the early mines are not recorded by plans and it was not mandatory, even after 1873, to deposit plans with a central authority. Consequently, any plans made have become widely scattered, and it is probable that some have been lost. In some cases, the only record of existence of a mine is a description in a book or newspaper, or an account of the financial transactions concerning a mine.

THE LIMESTONE PROJECT

The existence of abandoned workings has always been known, and development over shallow mines (with up to 30m of overburden cover) has generally been avoided because of potential instability problems. Prior to 1978 it had been supposed that collapse of mine workings which were deeper than about 60m would not affect surface structures, and so no development controls existed above these deep mines. However, in the autumn of 1978, subsidence due to a partial collapse of the 145m deep Cow Pasture Mine at Wednesbury resulted in serious structural damage to warehouses and industrial plant, and some minor damage to nearby housing.

As a result of this, the Department of the Environment and the Boroughs of Dudley, Sandwell and Walsall, together with the West Midlands County Council, commissioned Ove Arup & Partners to undertake a two year study of limestone workings in the Black Country[1].

The Study

The objectives of the study, which commenced in June 1981, were to assess the risk of collapse and the need for remedial works to the mines.

An extensive data search was carried out to discover what information concerning limestone mines had survived and to collate all of these records into one data base. This exercise identified about 30 mines which needed further investigation.

The mining methods and mechanisms of collapse associated with each method were studied. This included ground investigations at four sites, where the techniques for investigating flooded mines described in the following case study were first developed.

Consideration Zones

For each mine a 'Consideration Zone', within which structures and services may be damaged by the collapse of a mine was identified. Within these zones, any new development has been virtually frozen until the condition of the mine has been determined, and any necessary remedial measures effected. The buying and selling of private properties has also been severely disrupted.

The boundary to a Consideration Zone is defined as the distance from the mine beyond which horizontal ground strains resulting from a collapse of the mine will be smaller than 0.2%. Outside the Consideration Zone boundary normal structures are unlikely to suffer more than negligible damage, as defined by Burland et al[2]. Long structures and very sensitive equipment may suffer damage at distances of up to twice that of the Consideration Zone boundary from the mine edge.

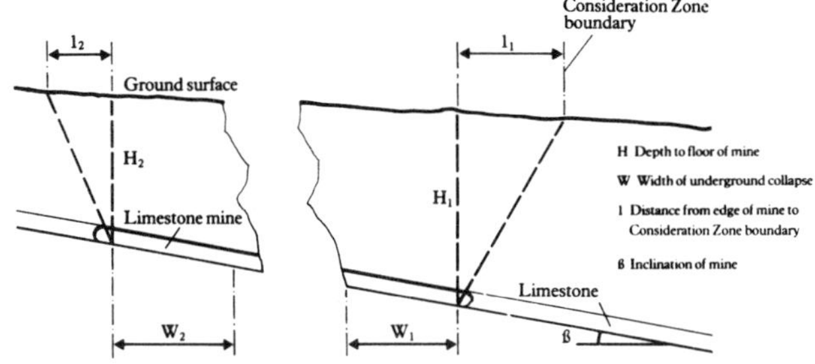

Fig.2 - General principle of determining a Consideration Zone boundary.

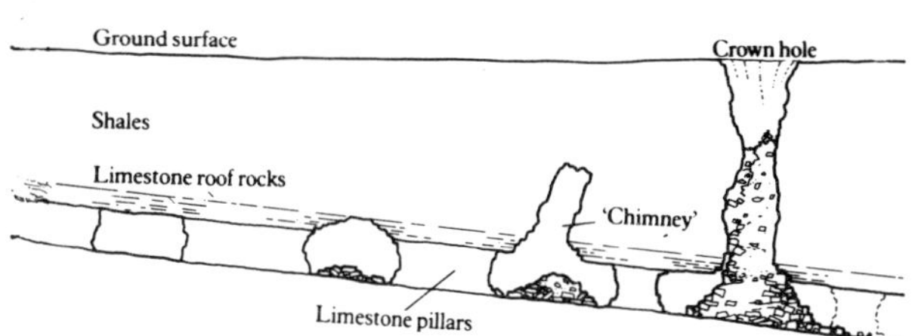

Fig.3 - Chimney and crownhole formation.

The principle for determining the Consideration Zone boundary is based on the work of Orchard[3]. This method has been adapted to meet the particular problems of the Silurian limestone strata rather than the Coal Measures for which it was devised, see Figure 2. It has shown good predictive agreement with the subsidence that occurred at Wednesbury in 1978.

Relative Risk

As part of the Study, consideration was given to evaluating the risk of collapse of a mine affecting the surface, either as a crownhole, see Figure 3, or as a general disturbance, see Figure 4. Because of the small amount of data available, it was not possible to indicate the absolute level of risk. Instead, each mine was ranked as to its relative potential for collapse, and the importance in land use terms of the Consideration Zone was evaluated. By combining potential for collapse with importance, a ranking of relative risk to the Consideration Zone was obtained. This ranking enabled each Borough to evolve its own strategy for dealing with the limestone dereliction problem.

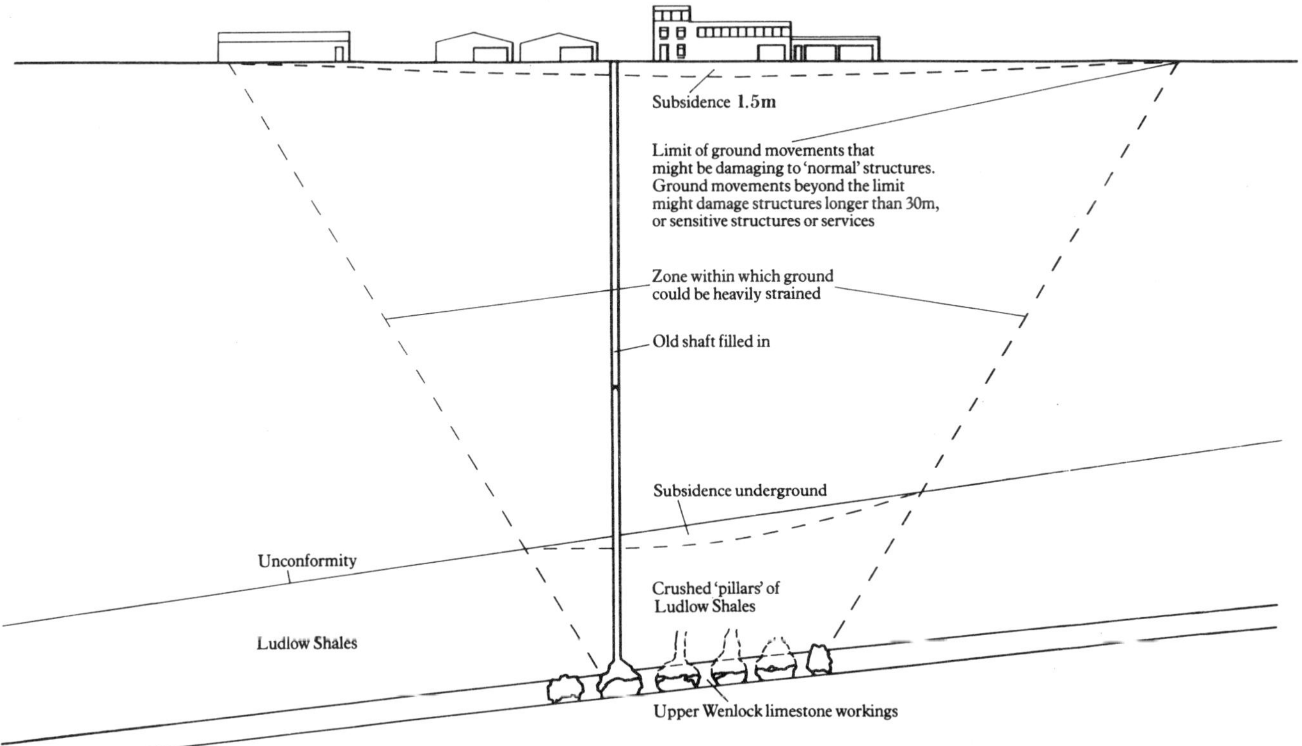

Fig.4 - General subsidence over a mine deeper than 100mm.

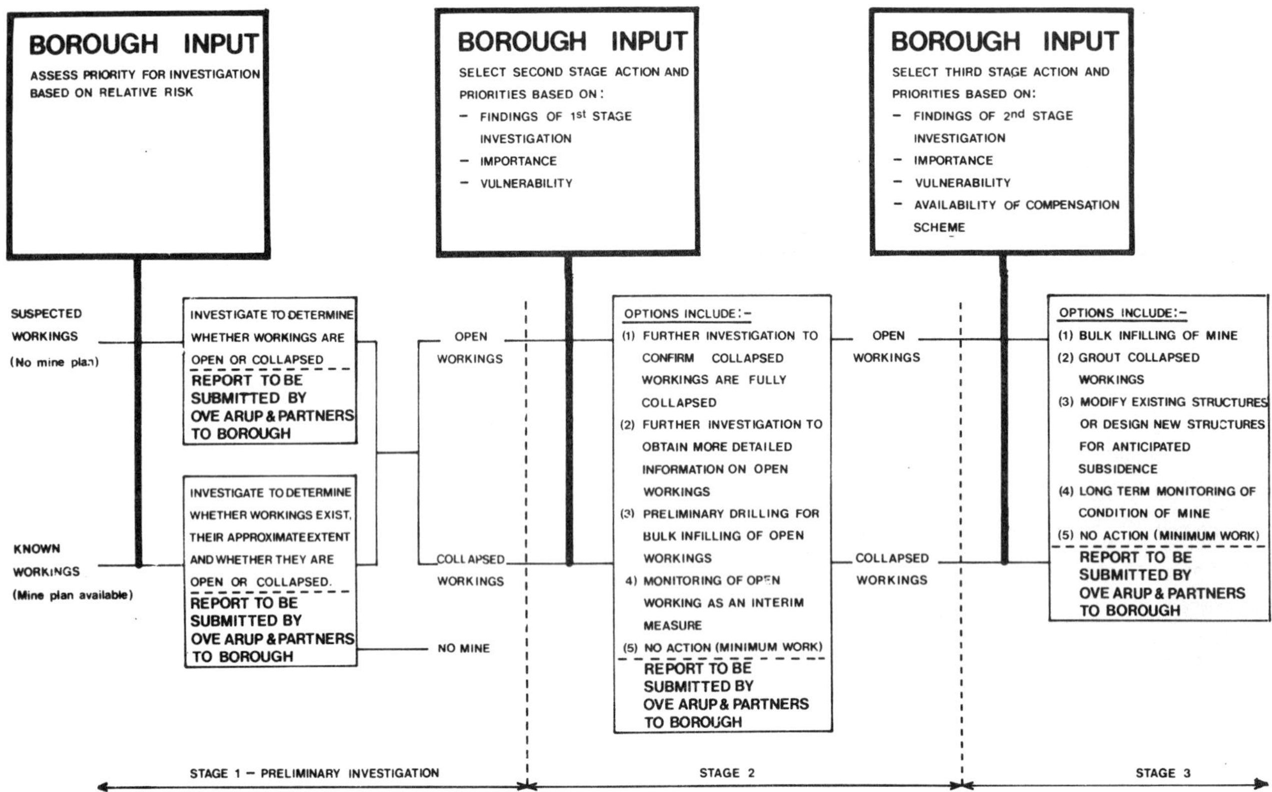

Fig.5 - Strategy for action adopted by the Boroughs.

Follow-on Work

Following publication of the Study Report[1], the three main Boroughs concerned - Dudley, Sandwell and Walsall - developed their own strategies for investigating and treating the mines. As Figure 5 shows, there are three stages of the work:

i) Stage 1 is the preliminary site investigation to prove the existence of the mine and its general condition. Sufficient work is done to enable the Stage 2 work to be designed;

ii) Stage 2 works are more detailed investigations designed to provide enough data to decide on the Stage 3 option;

iii) Stage 3 is the treatment works, which can range from minimal work in rural areas through to infilling the mine.

Two years after publication of the Study Report, Stage 1 works have been completed at 15 of the 31 mines identified at the time of the Study. The design of Stage 3 infilling has started for one mine, and one mine has been shown to be totally collapsed and the Consideration Zone removed.

The case study discusses Cow Pasture Mine, Wednesbury, which partially collapsed in 1978, causing extensive general subsidence.

INVESTIGATION TECHNIQUES

Unlike site investigations carried out prior to the construction of buildings, tunnels and other civil engineering works, the investigations for abandoned limestone mines is not primarily concerned with determining design parameters of soils and rocks. The in-situ condition of the rock is important, but the primary objective of the investigations is to determine the extent and condition of the mine cavities. In order to gain the specific information required, particular site investigation techniques were developed during the Study. Since then, these techniques have been reviewed and modified to ensure the economy and suitability to the purpose of each technique.

Drilling Methods

Large rotary drilling rigs of the type shown in Figure 6a were used. Most drillholes are fitted with a quick acting blow-out- preventor, see Figure 6b, to ensure that any excess water pressure encountered on penetrating the mine

Fig.6a- Large mobile drilling rig with drill rod carousel is commonly used.

Fig.6b- A quick acting blow-out- preventor is installed to ensure that water under pressure cannot escape from a mine through the drillholes.

cannot escape through the drillhole. It was feared that the sudden release of water pressure may initiate a collapse in the mine if it was near to failure.

High quality 92mm diameter rock core samples were recovered by using a double barrel sampler with an inner plastic "Mylar" lining and clean water as the flushing medium. The Mylar lining was extracted from the sampler, with the core inside, similar to a split inner tube of a triple barrel sampler. Cores were photographed and then logged at the rig side, using descriptions based on BS5930[4], and Rock Quality Designation and Fracture Index were also recorded. Table 1 shows the terminology used to describe the ranges of fracture state.

Careful logging of the rock core, together with details of drilling penetration rates, water flush returns and in-situ permeability tests enabled the condition of the rocks and degree of fracturing to be assessed. Examples of typical summary records are given in Figure 7

Flushing mediums other than water have been considered. However, the use of foam or drilling muds inhibit the performance of some geophysical sondes. Drilling with air causes air entrainment of the water in the mine at the vicinity of the drillhole, which interferes with the ultrasonic wave travel during ultrasonic surveys.

Table 1 - Fracture Index and Fracture State.

Degree of Fracturing	Fracture Spacing mm	Fracture Index mm	Fracture State
Very widely spaced	> 2000	< 0.5	Scarcely
Widely spaced	600-2000	0.5-1.7	Slightly
Medium spaced	200-600	1.7-5.0	Moderately
Closely spaced	60-200	5.0-17	Highly
Very closely spaced	20-60	17-50	Very Highly
Extremely closely spaced	> 20	< 50	Intensely

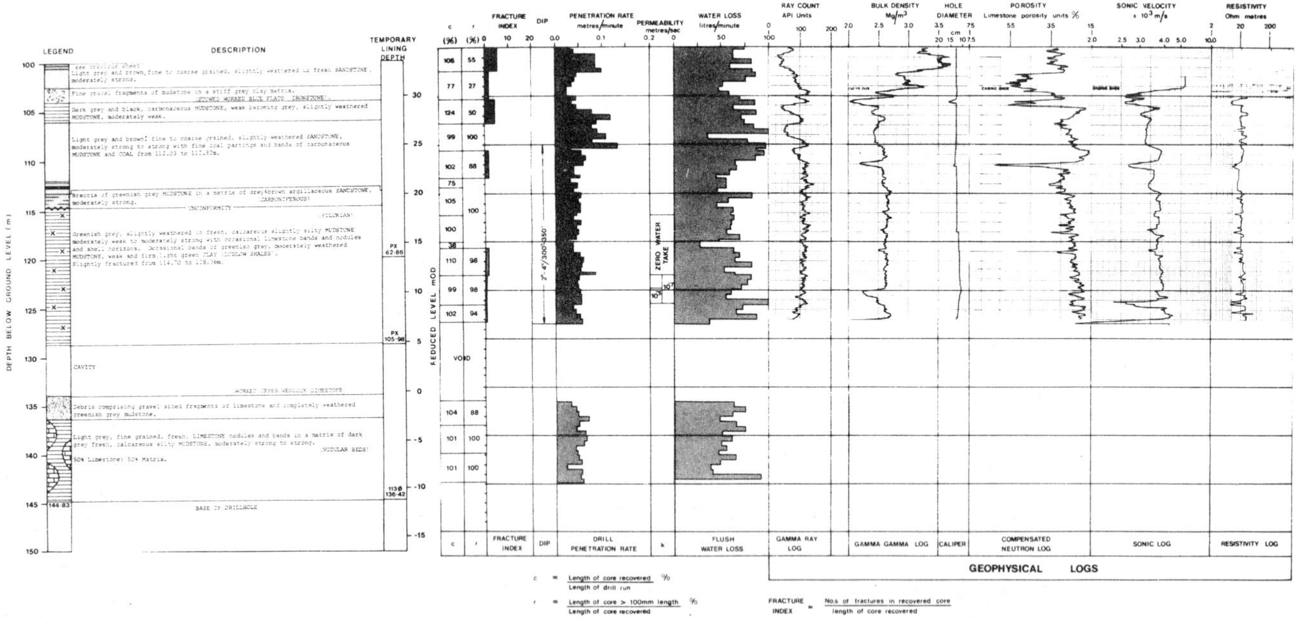

Fig.7 - Typical drillhole summary record.

On completion of drilling, a series of additional techniques were used to provide further information relating to the condition of the rocks and mine cavities.

Geophysical Logging

Geophysical logging was carried out in all drillholes by lowering the various sondes into the hole on a wireline. The full range of tests comprised gamma-ray, gamma-gamma, neutron, caliper, multi-channel sonic, focussed electronic (referred to as "resistivity") and temperature. Each log provides an indication of a rock property which can be correlated with the core and drilling data. In the absence of core the geophysical logs can be used to identify changes in lithology and stratigraphy. Interpretation of a combination of logs can provide a relationship with density of fractures and uniaxial compressive strength in Coal Measures rocks, as described by Halker et al. As more information becomes available for the Silurian strata, it should be possible to develop similar correlations for these rocks. The main use of the geophysical logs for this project has been in distinguishing between fractured rock, rock with greater than a moderate degree of fracturing (disturbed rock) found usually, but not always, near mine workings, and undisturbed rock. Figure 8 shows related resistivity and drillhole logs, from which the areas of disturbed Ludlow Shales can be clearly identified.

Core Orientation

Various methods of core orientation have been tried with varying degrees of success. The method finally adopted is to use a dipmeter log which measures the angle of dip of the strata in an open drillhole. The dipmeter is run in combination with a verticality log, which records the tilt and azimuth of the hole so that the dip direction can also be determined. The dipmeter sonde has three equi-spaced spring loaded arms which project outwards against the sides of the drillhole. Mounted on each arm is a plate fitted with a small resistivity electrode. As the sonde is raised through the drillhole, inclined beds will be recorded at least at two different levels by the three electrodes. This depth information is then integrated with the azimuth data to give the dip direction, as shown in Figure 9, where the dominant dip over the depth range 115m to 125m is about 3°, in a dip direction of about 300°.

The correlation between the dipmeter dip and dip of the in-situ rock has been checked at one site, where rock is near the surface, by digging trial pits and measuring dips on the exposed faces. Results indicate that both the angle of dip and dip direction from the dipmeter show very good correlation with the physical

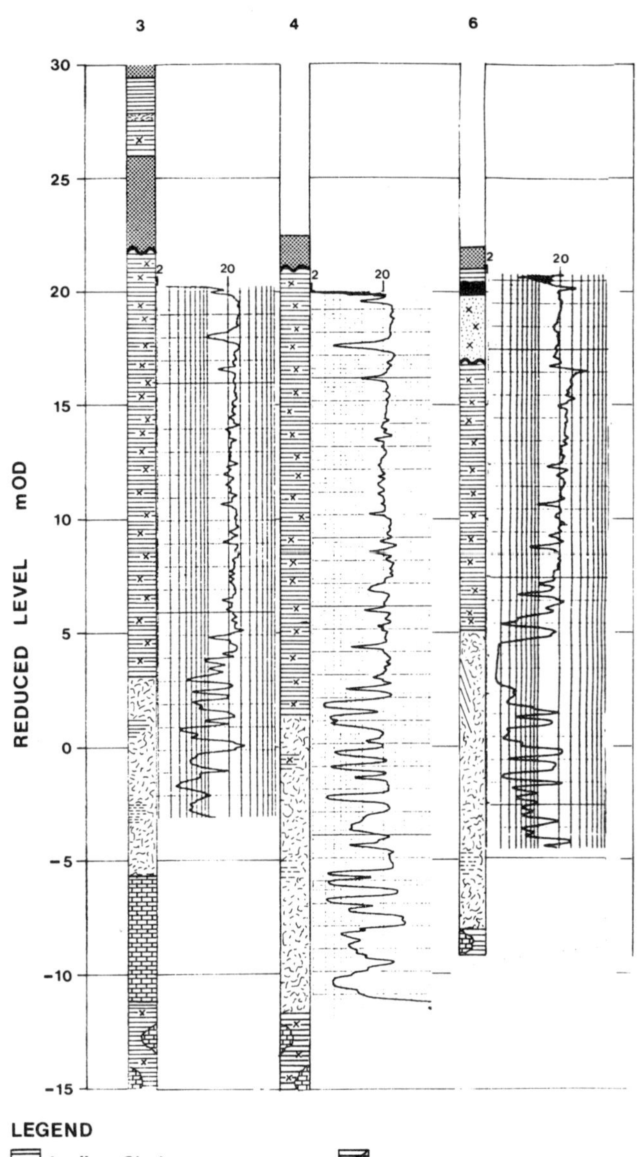

Fig.8 - Comparison of resistivity log and degree of rock fracture.

measurements. In other examples the dip and dip direction correlates well with the dip and dip direction of the mine roof obtained from the results of ultrasonic surveys, see below.

Directional Survey

A gyroscopically stabilised multishot survey instrument is used to relate the position of the drillhole at the level of the mine relative to the surface. The instrument consists of a pendulum tiltmeter, a gyroscope with a stabilising compass and two cameras. The first camera takes a picture of the tiltmeter and the second a picture of the compass. An electronic timing mechanism controls the operation of the cameras.

Photographs are taken every 20 seconds as the tool is lowered down by wireline. Two stabilisers secure the location of the tool along the central axis of the drillhole during the survey. The position of a fixed reference point at the drillhole is determined on the surface by a surveying instrument, and all positional information is related to this point. The provision of a stabilising compass and gyroscope allows the position survey to be carried out inside temporary steel casing or close to materials which would influence a magnetic compass. However, periodic checks and corrections have to be made to allow for drift of the gyroscope.

Directional surveys are more usually carried out in the oil industry, where drillholes can have angular deviations in excess of 45°. For investigation of limestone mines, drillholes are between about 30m to a maximum of 250m deep with a verticality tolerance of 1% of depth. Hence, the directional survey instrument needs to be more sensitive than is required for many oil drilling applications. The tool used in the limestone investigations is accurate to 1' of inclination over a range of 0° to 2°.

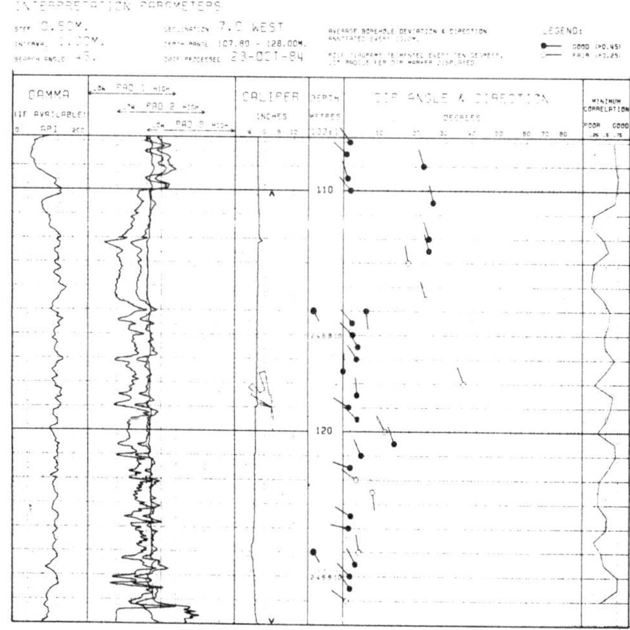

Fig.9 - Example of presentation of dipmeter results.

Television Surveys

Closed circuit television cameras (CCTV) have been lowered into workings and video recordings made of the views seen. However, the currently available cameras and lighting methods have restricted clear viewing to about 2m to 3m. This is of little benefit in showing the extent and condition of a mine in which the nearest surface may be 20m away. The video recordings have had limited use in showing the in-situ condition of joints and fractures in drillhole walls and areas of the roof and floor of the mine in close proximity to the camera. A development programme is underway to increase the performance of television surveys to view distances of about 20m, by the use of stronger lighting, image intensifiers and computer enhancement.

Ultrasonic Surveys

Every drillhole which has entered a void greater than 1m deep at the level of the mine has had a plastic lining tube grouted into it to enable ultrasonic surveys to be carried out during the investigation and at any time thereafter.

This technique has proved to be a very powerful method of remotely mapping water filled workings.

The tool used for surveying limestone workings is 65mm in diameter which enables it to go down the 92mm diameter permanent casing. At the base of the tool there are two ultrasonic emitter/receiver transducers. Each transducer emits a narrow beamed sound pulse which is reflected from a solid surface back to the receiver. The time taken for the pulse to return is a measure of distance. For Cow Pasture Mine, where surfaces were up to 50m from the survey tool, frequencies of 600kHz and 1000kHz were used.

The survey head, shown in Figure 10, rotates about its vertical axis at step intervals of 3°. At each step a sound pulse is emitted and received. The head can also tilt from the vertical to the horizontal at 5° intervals. As the head is rotated with a constant angle of tilt, the succession of pulses describes a conical surface. The tilt facility is particularly useful for providing information relating to the roof and floor of a mine, and for surveying steeply dipping mines.

At the heart of the survey system is a desk top computer with visual display screen. The computer package includes a printer and magnetic disc store.

The complete survey sequence and movement of the survey tool are controlled by computer and all of the measured values are stored in the computer memory.

A survey is carried out in two runs. The calibration run determines the depth of the roof and floor of the mine working and also measures the temperature and velocity of sound through

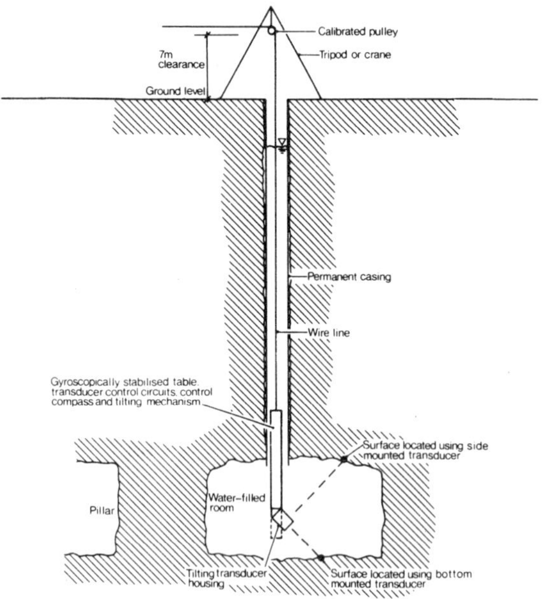

Fig.10 - Manner of conducting an ultrasonic survey.

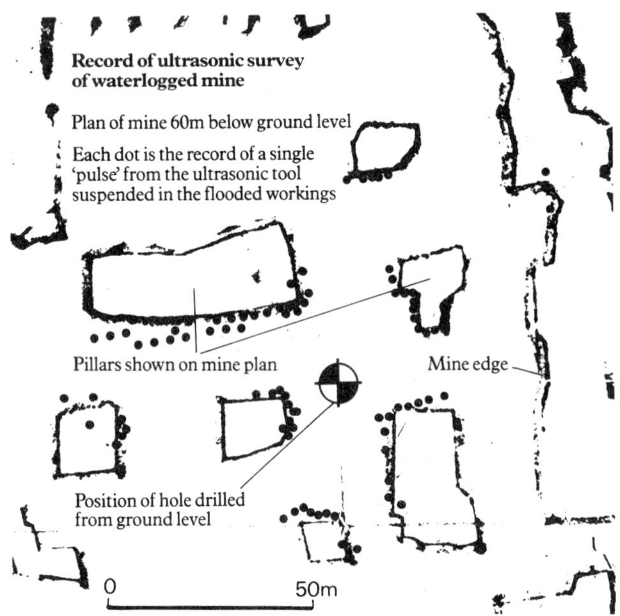

Fig.11 - Record of ultrasonic survey of waterfilled mine.

the water in the mine. Once calibration is complete, the second run commences. Horizontal scans are made at 1m intervals over the height of the workings and then tilting scans at 15° increments from vertical to horizontal are made.

The received "echo" of each pulse, in the form of an amplitude/time echogram, is seen on the computer screen. The operator can select and comit to store up to four amplitude peaks, usually the ones of greatest amplitude. At the completion of each horizontal scan, a plan view can be plotted in a polar co-ordinate system starting at north. This plan is used by the operator to determine if additional horizontal or tilting scans are required.

The computer can also produce vertical sections on completion of a survey.

On completion of sitework, the data cassettes are processed to identify the pulses which has the greatest amplitude. From an interpretation of the processed data and the echograms, horizontal plans and vertical sections at 15° intervals are produced. Using this information, workings within about 60m of the drillhole can be mapped, as shown in the example in Figure 11.

COW PASTURE MINE

Cow Pasture Mine is a pillar and room working, about 145m below the surface, in the Upper Wenlock Limestone. The Upper Wenlock Limestone is a deposit laid down during the Silurian period. The mine, shown in outline in Figure 12 was opened in 1855 and was abandoned in 1887.

A detailed mine plan is shown in Figure 14. Within the mine area approximately 75% of the available limestone was extracted, leaving the remaining 25% as pillars to support the roof. Since abandonment, the mine has flooded with water and the shafts are presumed to be collapsed.

In 1978 an area of the mine, about 90m by 140m in size collapsed to produce a saucer-shaped depression at the surface of about 200m by 300m in extent, see Figure 12. At the centre of the depression the total settlement was about 1.2m, and several single storey factory units were badly damaged. The rate of subsidence was slow at first, increasing to reach a maximum rate of about 50mm a day and then decreasing gradually over a period of several months.

Two investigations of the mine have been carried out. The first was part of the Study and was confined to the area of collapse. The

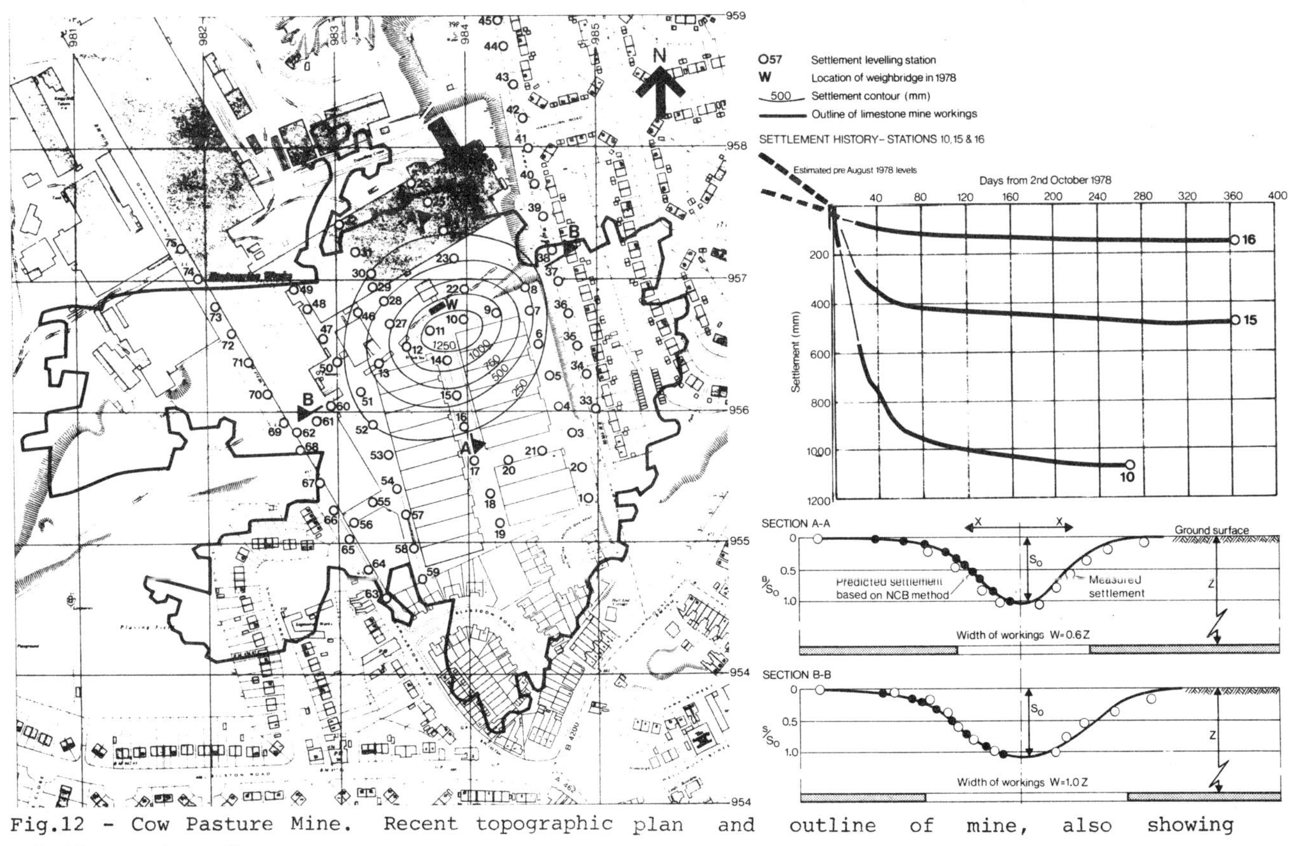

Fig.12 - Cow Pasture Mine. Recent topographic plan and outline of mine, also showing subsidence at surface.

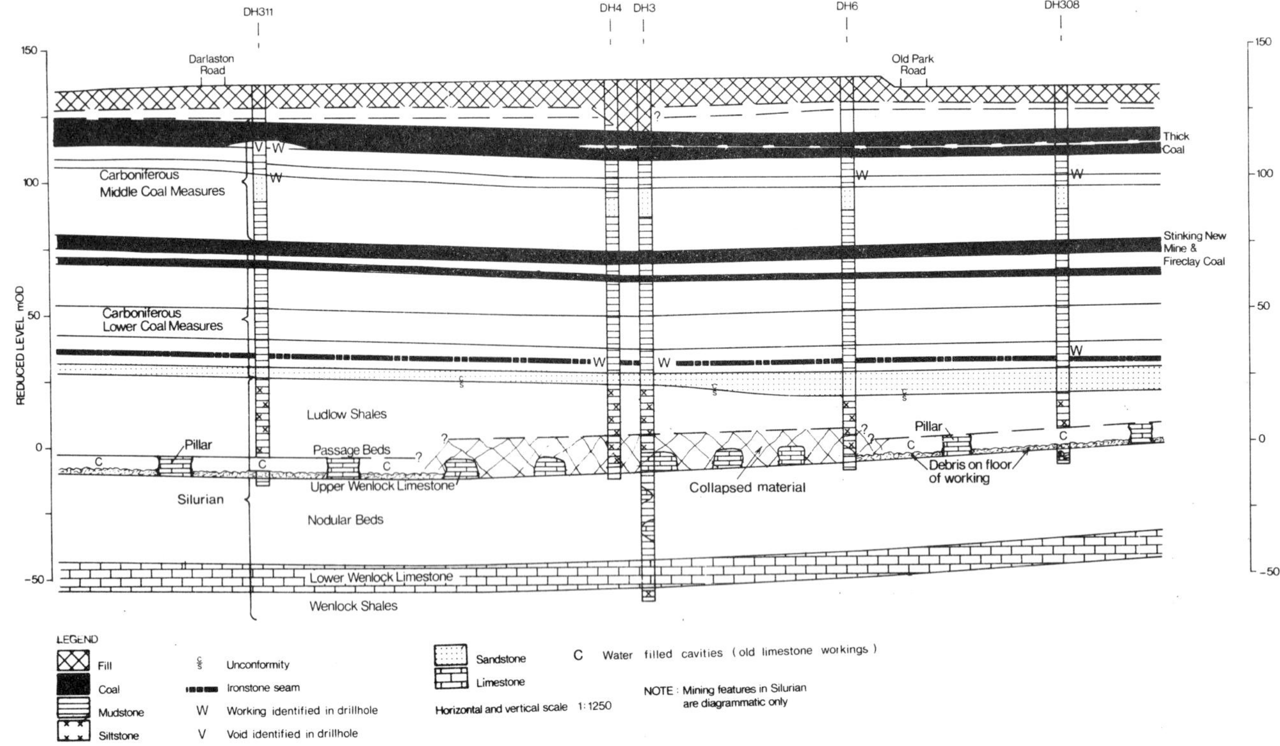

Fig.13 - Vertical section showing ground conditions on Section X-X.

second investigation was carried out in 1984 and covered the remaining areas of the mine. Both investigations used the techniques developed during the Study to investigate flooded mines up to about 300m below the surface.

Ground Conditions

The section in Figure 13 shows there to be about 100m of Carboniferous rocks which lie unconformably on the Silurian strata. Several coal seams, notably the Thick Coal, Stinking, New Mine and Fire Clay Coal, and one ironstone seam have been worked below the site. Below the Carboniferous rocks, the Silurian strata comprise, in downward succession, about 20m of Lower Ludlow Shales, 6m of Upper Wenlock Limestone, about 35m of Nodular Beds and 10m of Lower Wenlock Limestone, below which lie the Wenlock Shales.

The Ludlow Shales are a greenish grey, slightly calcareous silty mudstone, with occasional limestone bands, shelly layers, and bands of light green clay. Undisturbed Ludlow Shales at this site are moderately weak to moderately strong, with slight fracturing. They are generally fresh to slightly weathered, with bands of moderately to highly weathered mudstone. Strength tests[1,6] gave values of uniaxial compressive strength of 39MPa and above, with an average of 50MPa. Interpretation of test results was made using Hoek & Brown[7].

The Upper Wenlock Limestone is coloured buff brown and grey, and contains laminae and bands of mudstone. Fresh limestone is moderately strong to strong, with an average uniaxial compressive strength of around 80MPa.

The groundwater table is about 15m below surface.

Findings of the Investigation

Drillholes in the collapsed area showed that the Ludlow Shales were highly to intensely fractured (see Table 1) for several metres above the levels of the tops of the original limestone pillars. They also showed that the limestone pillars were themselves not crushed.

Drillholes outside the collapsed area gave core samples and geophysical records which showed that the Ludlow Shales were not as fractured as within the collapsed area. Most of the drillholes passed through cavities, as intended, and the position surveys and ultrasonic surveys confirmed that the location and plan dimensions of the mine were accurately recorded on the mine plans.

As shown in the plan at mine level in Figure 14, the ultrasonic tool was able to measure the shapes of the cavities of the mine. From the survey in Drillholes 308, it detected the face of a pillar at the eastern edge of the collapsed area of the mine. This confirmed that the surface subsidence in 1978 initiated from the collapse of a limited area of the mine, estimated to be 90m by 140m in extent. It was also evident that the 1978 event may have been the first time that collapse of the Cow Pasture limestone mine had caused surface subsidence. All overlying coal and ironstone seams were found to be collapsed, so the area has a past history of surface subsidences.

By comparing the results from the drillholes and the ultrasonic surveys, it could be shown that the majority of the cavities are no longer at the level of the limestone which had been mined, but exist within the Passage Beds and the lowest levels of the Ludlow Shales.

The drillhole and ultrasonic survey information from Drillhole 309 showed clearly that collapse of the roof rocks had progressed both upwards and outwards. The drillhole fortunately passed through a cavity in the Ludlow Shales and then into the intact limestone

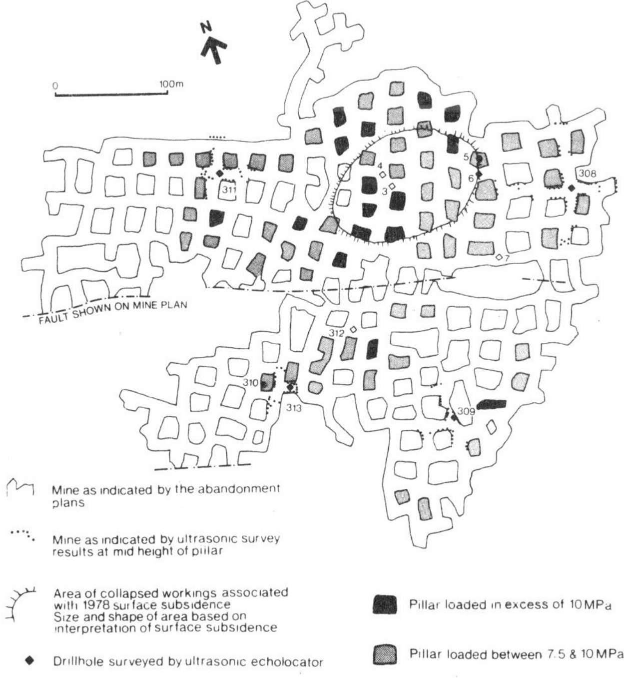

Fig.14 - Plan of Cow Pasture Mine showing extent of voids and collapsed workings.

of the pillar, as shown in Figure 15. The calculated pillar stresses in the part of the mine near Drillhole 309 are lower than in the area which collapsed in 1978, and there was no evidence of failure in the Ludlow Shales.

It thus appears that, over most of the area of the mine, the original limestone pillars lie submerged within the fallen roof rocks, and the upwards extensions of those pillars into the Passage Beds and Ludlow Shales are supporting the overburden.

Mechanism of Collapse

It can be shown that, for a given mine layout and rock condition (the degree of jointing of the Ludlow Shale in particular) there is a critical height of a Ludlow Shales pillar at which the overburden pressure will cause it to crush. When such a pillar crushes, it is likely that crushing of nearby Ludlow Shales pillars will also be induced, providing they are close to the critical height. Crushing may thus propagate and include more and more pillars, and only be arrested when pillars well below the critical height are reached.

Current Situation

So far, only preliminary investigations have been made at the Cow Pasture Mine, and these have clearly identified the collapse mechanism. By means of the permanent plastic lined casings placed in the drillholes, provision has been made to make repeated observations by ultrasonic surveys, to enable the erosion of the margin of safety in the mine to be examined. A margin of safety is defined as the difference in height of a pillar between the current height and that at which crushing failure occurs. The rate at which that margin is eroded as the Ludlow Shales weather and fall away, is of considerable interest, as much of the eastern and southern areas of the mine are overlain by housing estates which are at risk if underground collapse progresses into surface subsidence.

As a precaution, level surveys using a precise level and invar staff are being made at six week intervals. This interval has been selected, after examining the rate at which subsidence occurred during the event in 1978, as being likely to detect the onset of subsidence before the damaging effects are apparent and in

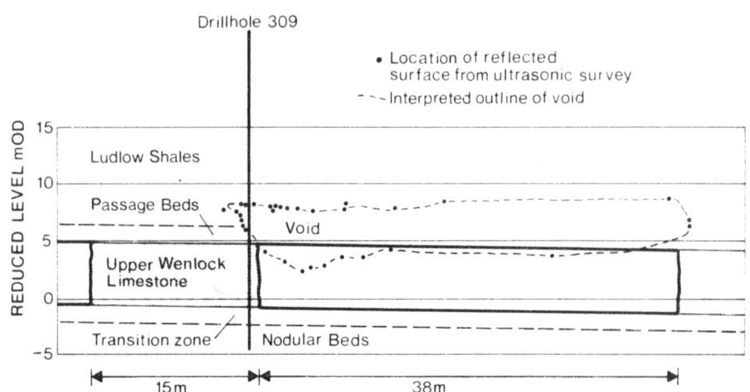

Fig.15 - Section through mine at drillhole 309.

sufficient time to take precautions by evacuating residents and closing down services, particularly gas mains.

REMEDIAL MEASURES OPTIONS

It is known from the investigation to date that about 60% of the plan area of the mine is still open, although in three out of four areas examined the "open" (water filled) cavities are above the level of the original mine roof. It is presumed that, in due course, these open areas will collapse and give rise to surface subsidences. When and where such collapses will occur, and what form they will take, cannot be accurately predicted from the information available to date.

Options for treating the mine are severely hampered by its depth and by the extent of existing surface development, which include factories, housing estates, major highways and all services, including gas mains. The options being considered include doing nothing except to maintain periodic checks on surface levels and underground events, and only take precautionary action if a subsidence event is forecast. At the other extreme is the option of filling the mine to prevent future collapses, but this is made more difficult as the cavities are no longer at mine level and may be discontinuous. Other remedies being considered include collapsing the mine by explosive charges and making good afterwards the damage done.

Two options for the next stage of investigation have been offered for consideration. In one a few drillholes would be made with the same very detailed logging, geophysical and ultrasonic surveying as for the preliminary investigations to give wider coverage of the mine. If the newly developed enhanced C.C.T.V. is available, that would also be used to increase the amount of information obtained. The information from the new and existing investigations would then be combined, and a programme of surveys of the mine using the ultrasonic tool and C.C.T.V. put in hand.

The other option offered is that the cheapest drilling method available could be used (generally not taking core samples) to enable 20 or more mine cavities to be entered and surveyed by ultrasonic tool, (and by enhanced C.C.T.V. if available). This will be done with a view to establishing the mine geometry prior to designing infilling or collapsing by means of explosives.

CONCLUSIONS

The high quality drilling methods and geophysical logging used during the project have been invaluable in aiding the determination of the present state of stress in the rocks overlying limestone workings.

Remote sensors, particularly the ultrasonic survey tool, have also proved to be very successful devices for "seeing" into mines, which would otherwise not have been possible. The ultrasonic survey tool has not previously been used for the purpose of determining the layout of mines, for comparison with mine plans. The quality of such data has proved to be good although there are clearly areas where improvements are desirable.

Closed circuit television is valuable in reinforcing the ultrasonic survey "picture" with visual pictures. At present a system is being developed that will enable pictures to be obtained underwater at distances of up to 20m.

Geophysical logging has proved useful in identifying the extent of the disturbed rock around mine workings. The dipmeter log has been particularly useful in determining the dip and dip direction of in-situ rock.

Data from the site investigations has enabled the mechanism of collapse, as well as the margin of safety, of the water filled abandoned limestone mine to be determined, and recommendations for remedial works to be made.

References

1. Ove Arup & Partners, A Study of Limestone Workings in the West Midlands. A report for the Department of the Environment, the Metropolitan Boroughs of Dudley, Sandwell and Walsall, together with the West Midlands County Council - April 1983.

2. Burland J.B., Broms B.B. and de Mello V.F.B., Behaviour of foundations and structures. Building Research Establishment Current Paper, No.51/78, Building Research Establishment, Watford, England - June 1978.

3. Orchard R.J., Prediction of the magnitude of surface movements, Colliery Engineering, Volume 34 - November 1975 - p.455-462.

4. BS5930, Code of practice for site investigations, British Standards Institution, London, England - 1981.

5. Halker A., Kuzmir N.J., Mellor D.W. and Whitworth K.R, The synthesis of fracture/ strength logs using borehole geophysics: a new geotechnical services, Quarterly Journal of Engineering Geology, Volume 15 - 1982 - p.15-28.

6. Ove Arup & Partners, Investigation of Abandoned Limestone Mines, Cow Pasture Mine, Wednesbury Old Park. A report for the Metropolitan Borough of Sandwell - February 1985.

7. Hoek E. and Brown E.T., Underground excavations in rock, The Institution of Mining & Metallurgy, London - 1980.

Acknowledgements

The authors thank the Department of the Environment, who funded both the initial study and the various follow-up works through Derelict Land Grant aid, and the Metropolitan Boroughs of Dudley, Sandwell and Walsall, for permission to publish this paper.

The Main Contractor for the Cow Pasture site investigations was Geotechnical Engineering (Northern) Limited. Geophysical logging was carried out by BPB Instruments Limited and Sperry-Sun (UK) Limited did the directional surveys. Ultrasonic surveys were undertaken by Prakla-Seismos GmbH of Hannover, West Germany. We would like to thank all of these organisations for their co-operation in developing their techniques to cope with the special problems of investigating limestone workings.

We would also like to thank all of our colleagues who have been involved with this project, both on site and in the office, for their help and encouragement.

Remedial works to a major rock slope in Hong Kong

Stephen Buttling PH.D., B.SC. (ENG.), A.C.G.I. C.ENG., M.I.C.E.
Scott Wilson Kirkpatrick & Partners, Hong Kong (now, Mass Rapid Transit Corporation, Singapore)

SYNOPSIS

A mixed rock and soil slope of granitic origin on the south side of Tsing Yi Island, Hong Kong, see figure 1, was formed in the 1960's in order to create a level platform for a power station site and a proposed major road. Several failures of parts of the slope occurred between 1973 and 1982, and remedial works were designed and instituted on an ad hoc basis.

A major scheme of repair was proposed in 1983 which involved trimming back large parts of the slope and the construction of three rows of interconnecting drains to achieve groundwater lowering of about 10 to 15m.

In the summer of 1984, while the upper parts of the slope had been trimmed to final profile but not finished, two failures occurred affecting an area of about 60m by 60m. A detailed investigation was carried out involving geological and geotechnical mapping, borehole drilling both vertical and inclined using the air foam technique and impression packer testing.

Failure was determined to have occurred where a particular set of joints co-existed comprising a 'sheeting' joint and two sub-vertical release joints. Within the joints a very soft clay infill with very low shear strength was found which had not been detected in the earlier site investigations.

Remedial works proposed comprised over 100 permanent anchors stressed against the slope at more than 1000kN each through a grillage of ground beams.

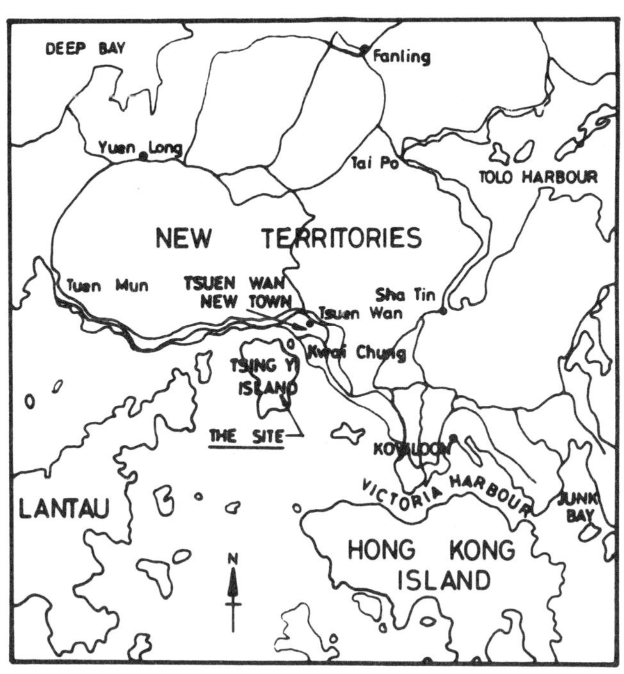

FIGURE 1. KEY PLAN

INTRODUCTION

The novel aspects of the case are considered to be these:-

i) The use of an 'aerial' photograph as an aid to mapping and geotechnical interpretation.
ii) The inability of the conventional stereographic technique to predict the failure mode.
iii) The way in which factors of safety have been adjusted to allow economic design by considering overall risk.

The slope was formed initially at about 5(V) on 3(H), as was the practice at the time for cut slopes, and failures occurred in July 1973, May 1975, August 1976, Autumn 1978 and August 1982. Various remedial works schemes were carried out including cutting back of the slope, installing horizontal drains and local buttressing. Scott Wilson Kirkpatrick & Partners (HK) were asked to investigate the slope in November 1978 and prepared a report for the Hong Kong Government in April 1982. While this was being considered, another failure occurred initially in May and progressively through to August of 1982. A further study was therefore undertaken leading to another report in August 1983. The recommendations of the August 1983 report were that the slope should be cut back as much as was practical, and that groundwater lowering should be carried out by forming rows of interconnecting drainage caissons. These recommendations were included in a contract which was let in October 1983. Because of the nature of the topography, cutting the slope back had the effect of markedly increasing its height, and the crest of the slope formed was at about 130mPD compared with the road at the toe at about 15mPD.

By June 1984 the new slopes had been formed and partially protected from 130mPD down to 90mPD, and the first caisson drains were under construction in an arc from about 90mPD in the west to about 120mPD in the east. Between sections 1 & 2 (see figure 2) confirmatory ground investigation had shown a relatively high rock head level and 51 caissons had been deleted from the contract out of a total of 158. The chunam surface protection layer (a locally used soil/cement/lime plaster) had been applied to the soft parts of the cut slope, but local remedial works to the rock slope and surface cover with sprayed concrete had not been carried out. On June 16 1984, following 133mm of rain in 24 hours recorded at Tsing Yi, about 150m^3 of rock slipped down towards the 90mPD berm, taking away a short section of berm at 100mPD. The slip was about 3m deep, roughly triangular in plan and about 20m wide at the base on the 90mPD berm. Temporary remedial measures comprising dowels through potentially unstable blocks were carried out, and the slip debris was then removed. On August 5 1984, by which time the face had been cut down to 80mPD and following 95mm of rain in 24 hours recorded at Tsing Yi, a second similar slip occurred from 90 down to 80mPD, slightly to the east of the June slip. In continuing rain up to August 11 1984, the lower slip extended a little to the west and considerably to the east. In addition fresh cracking of chunam plaster, in a series of lines parallel with the berms, was noted between the 100 and 110mPD levels.

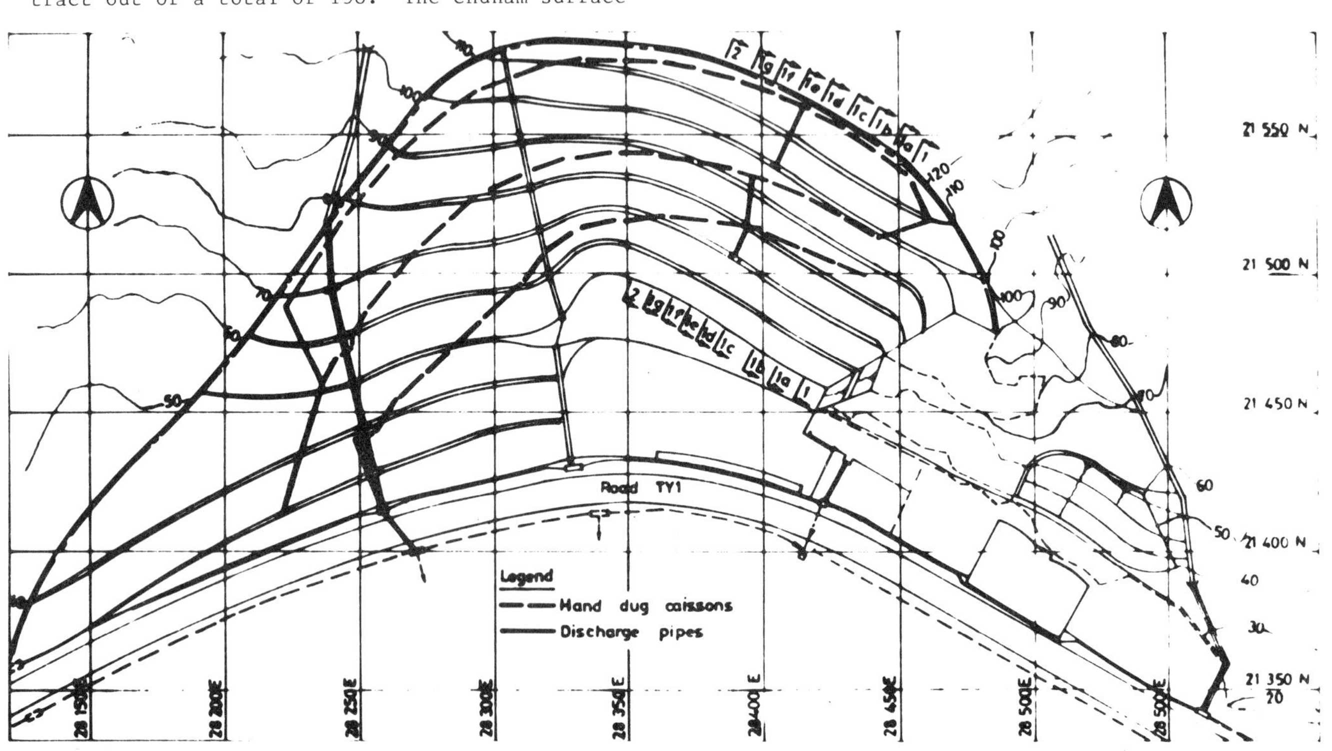

FIGURE 2. DETAIL OF SLOPE SHOWING CAISSON DRAINS AND SECTION LINES

INVESTIGATION

Surface inspection

From a detailed inspection of the failure mass it was apparent that, in both cases, the sliding material was highly to moderately decomposed rock failing along pre-existing joint planes. Many of these were infilled with several millimetres of kaolin together with a black deposit of iron or manganese oxide. Striations in these black and white layers indicated earlier movements which were not always in the same orientation as the latest slips. It was also apparent in the failure areas that there was a deposit of a soft brown clay in some joints. It appeared as a very thin layer in some of the sheeting joints, while there was evidence of a progressive build up of laminae in the sub-vertical joints. This suggested that the material had been deposited to fill up the vertical joints following repeated small downhill movements. In order to determine the real extent of the clay infilled joints, and to investigate the structure of the failure, the slope surface was extensively mapped. A photograph of the subject area was taken from a helicopter and this was enlarged to 500mm square and used on site as a basis for the surface mapping. Three overlays to the photograph were produced. The first showed the grades of weathering identified from a detailed examination of the exposed face. It was apparent that the profile was very varied, with both gradual and sudden changes of weathering grade in all directions. On the second the discontinuities were mapped and eleven major features were identified. From the drillholes and other information it was also discovered that in places the weathering could be in bands which extended to depths of 10 metres or more below the surface. Joints were mapped on all available exposures. Where they were infilled with kaolin or soft brown clay this was noted together with the extent of the opening. Based on this information an assessment was made of the stability of the various parts of the study area as a result of which three categories were identified:-

(i) stable

(ii) unstable - already moved

(iii) potentially unstable,

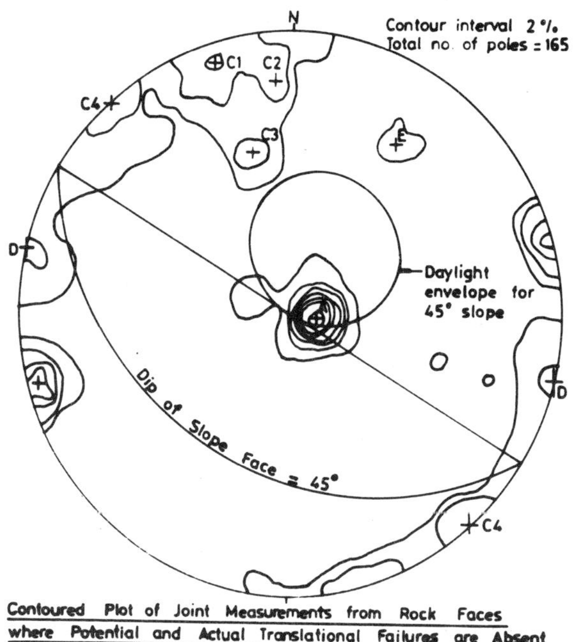

FIGURE 3. EQUAL AREA STEREO PROJECTION FOR STABLE AREAS

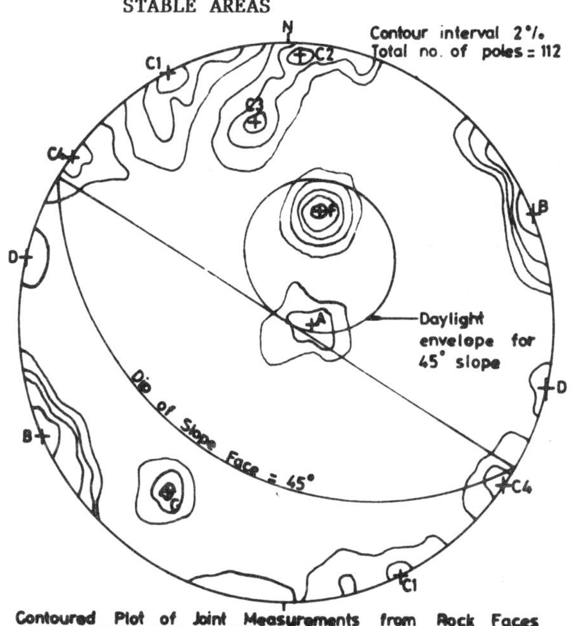

FIGURE 4. EQUAL AREA STEREO PROJECTION FOR UNSTABLE AND POTENTIALLY UNSTABLE AREAS

and these were indicated on the third overlay.

The slope surface was divided into rectangles and the dip and dip direction, spacing, persistance, roughness and infilling of each joint were mapped separately, and the results have been considered in conjunction with the above assessment. In figures 3 and 4 are shown the equal area stereo nets with the poles to the 277 discontinuities mapped for the stable areas in figure 3 and for the unstable and potentially unstable areas in figure 4. It can be seen that the subvertical joint sets B, C1, C2, C4 & D and set C3 are common to all areas. In addition there is a concentration of poles comprising set A in figure 3 which represents joints dipping at a low angle to the horizontal (about 9^0) in a direction roughly parallel with the berms of the new slope. Since these are well within any reasonable friction cone they do not pose any stability problem. In figure 4 there is an additional concentration of poles designated as joint set F centred at 34^0 and having a range from 20^0 to 46^0, the dip direction being within 10^0 of the dip direction of the slope. Thus the potential for failure exists where low shear strength on these joints is combined with a set of subvertical release surfaces such as B, C & D. The nature of the joints in set F is such that they are quite widely spaced in many areas, and they do not form the common type of wedge failure. However, combined with the subvertical release joints a planar-type failure becomes kinematically possible, and only one sheeting joint is necessary to form the sliding surface. In this case the statistical stereographic analysis of a large number of joints over the area may not draw attention to the potential failure mechanism. (Hencher 1985)

Ground Investigations

Various ground investigations have been undertaken at different times associated with the series of failures that have occurred and some of the remedial schemes. Following the second rock slope failure it was considered necessary to determine the depth below the ground surface of the clay infilled joints whose low shear strength was thought to be a prime cause. Since it was believed that the flushing water of conventional drilling would be likely to erode any surface deposit from open joints, air foam drilling was used in conjunction with triple-tube core barrels. Fourteen holes were drilled at angles of up to 45^0 to the vertical, and impression packer traces were taken at certain locations in the vertical holes to determine joint orientation. The rock cores from the existing drillholes, and from the new air foam drilling holes together with the mazier samples in decomposed rock, were all examined in detail for joints and spacing, infilling, roughness and dip were all logged. Where joints were not infilled, they were frequently iron-stained and moderately rough to rough in the grade II/III rock. Infill, where it existed, was generally no deeper than 5m below rockhead level and was of kaolin and black minerals of iron or manganese oxide, or of soft brown clay.

Laboratory testing

Nine samples of brown clay infill material were tested in direct shear under soaked conditions at different normal loads. The results are plotted in figure 5. Two groupings, one with a higher cohesion and lower friction angle than the other, are apparent. It is noticeable that the higher friction angle parameters are close to those frequently obtained for completely decomposed granite by triaxial testing, and therefore probably represent infill comprising less completely weathered material. Some samples of the soft brown clay infill were also tested for particle size distribution and Atterberg limits. One test showed that more than 96% of the sample was within the clay fraction, while the liquid limit was 145% and plastic limit 33%. In another sample the liquid limit was 100% with a plastic limit of 38%. X-ray diffraction of the latter sample was not conclusive in determining its origin or mode of resorting/deposition.

Groundwater Monitoring

There was no useful piezometer information available which was relevant to this particular problem, but it was noticed on site inspections in June and August 1984, during or immediately after the rain, that surface channels which were incomplete were full. It is therefore believed that surface

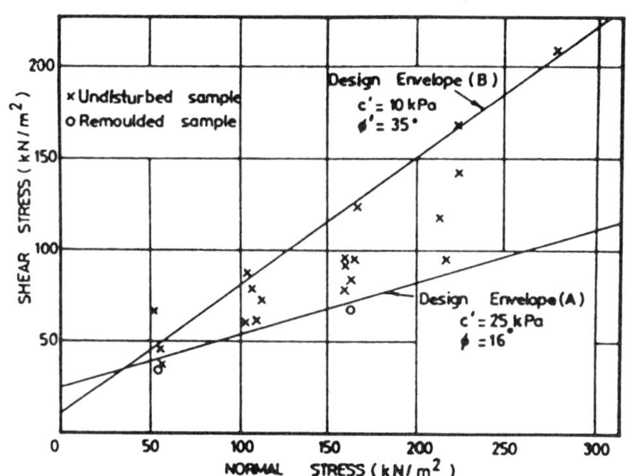

FIGURE 5. DIRECT SHEAR TEST RESULTS ON SAMPLES OF BROWN CLAY INFILL TO JOINTS

water would have penetrated any existing cracks to full depth, creating significant hydrostatic pressures which were not related to the normal groundwater level.

GEOTECHNICAL ASSESSMENT

Geology

The geology of the site is generally Cheung Chau granite overlain by colluvium. In the subject area there are also noticeable intrusions of feldspar porphyry. In places the contact is clean, but in others the gradation is so smooth as to be barely discernible. Combining the information from the surface mapping and all available drillholes, 9 geological sections were drawn up along section lines 1-1 to 2-2 (see Figure 2). Using old maps, the previously existing ground profiles were plotted approximately on the sections, though changes from Imperial to metric units and revisions to the HK Grid made exact correlation difficult. Three major features were apparent.

Firstly there are 11 major discontinuities or decomposed zones crossing the study area. These were numbered on the sections and on the first overlay. It appeared that the depth of weathering was considerable within these zones, giving a predominantly subvertical trend to the weathering profile.

Secondly, the clay filled joints existed near to rockhead level. It seems likely that they have resulted from very fine soil particles from the decomposed granite being transported by groundwater into open joints in the rock and deposited there. This is consistent with the mineralogy, and the high liquid limits and plasticity indices. It is suggested that resorting has taken place, and only the finest particles have been deposited at these particular depths. It was also noted that the dry densities of the clay were about $1Mg/m^3$ or less, and that a sample for direct shear testing from about 15m below ground level consolidated so much that it could only be tested at the first load increment of 150kPa, indicating that the infill is extremely loose.

Thirdly, if anchors are used to increase overall stability, the free anchor lengths necessary to ensure that the fixed length is in sound rock will vary significantly from location to location because of the varying weathering profile.

STABILITY

Required factors of safety

The basic requirement for factor of safety was at least 1.2. However the rigid application of this requirement, together with other conditions which are possible but unlikely such as water filled joints or tension cracks and the additional factors of safety inherent in such measures as permanent rock anchors, would produce very high compounded factors of safety which are uneconomic and unnecessary. The required factor of safety was therefore adjusted to be greater than unity in the ultimate condition, and greater than 1.2 in conditions considered likely to occur. Other states in between were designed for intermediate values.

Preliminary Stability Analyses

In order to understand the failure mechanism better, two dimensional analyses considering a simple planar failure were carried out using a method similar to that in Hoek & Bray (1977). Failure planes at 15^0 to 35^0 to the horizontal were considered, with water in a tension crack from 3 to 5m deep. The curves in figure 6 show the relationship between c' and ϕ' for a factor of safety of unity. Also shown on figure 6 are the shear strength parameters determined from the direct shear tests, apart from two results for which the apparent cohesion was greater than 40 kPa but including a remoulded test on one of those samples for which the cohesion dropped to 25kPa. It was evident that the measured shear strengths lay close to the line representing limit equilibrium on a 30^0 failure plane. It was therefore concluded that the parameters represented by that line were a reasonable estimate of the in-situ soil strength, and values for stability analyses were selected from both ends of the range as follows:-

Parameters	c' (kPa)	ϕ' (Deg)
A	25	16
B	10	35

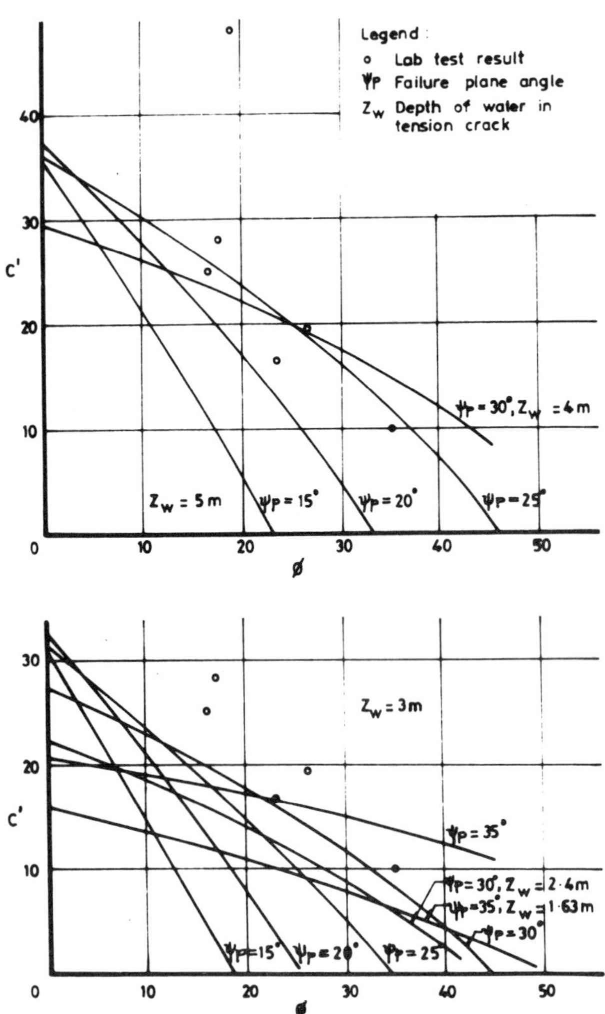

FIGURE 6. RELATION BETWEEN C' & Ø' FOR FACTOR OF SAFETY=1 AND PARAMETERS FROM DIRECT SHEAR TESTS

REMEDIAL WORKS

General

On the basis of the geotechnical assessment it was deduced that the instability was confined to a surface layer approximately 3-5m deep. This would tend, under certain circumstances, to be subject to planar type failure on sheeting joints dipping at about 35^0 out of the slope face. For ease of analysis this was considered as a sliding failure, with a failure surface roughly parallel with the ground profile and about 5m below it. It was decided that cutting the slope back further was not possible as the crest line would approach the top of the hill at about 330mPD very rapidly. Steepening the slope locally so as to remove the unstable surface material was also suggested, but it was held to be impracticable to recommence earthworks in that portion of the site at that stage. Ground anchors were therefore selected as the most likely technical solution, since they could be distributed over the area and could be adjusted easily for local variations. Retaining walls were suggested as an alternative, but the nature of the failure surfaces was such that if a wall was placed near the toe to intersect a failure then an alternative surface would be created emerging at the top of the wall. Also the nature of the rock did not lend itself to forming a good socket for a cantilevered wall system. A scheme comprising permanent prestressed ground anchors in conjunction with a grillage of primary and secondary beams was therefore proposed in order to provide the additional support necessary to give adequate factors of safety. The Janbu's Generalised Procedure of Slices was used in a program suite which allowed up to 10 rows of ground anchors to be specified.

Detailed Analyses

Analyses were carried out for one to five berm high sections of the slope, using A & B soil parameters with various sliding surfaces, water tables and ground anchor forces. These are summarised below.

Single berm

Planar failures at 25^0 to 35^0 to the horizontal were first examined using full depth of water in the tension crack and a water table following the ground surface, i.e. piezometric pressure on the slip surface reducing to zero at the slope toe (see Figure 7). The critical conditions were on the 25^0 plane with B soil parameters, but the factor of safety was improved to better than 1.6 by the introduction of two rows of anchors providing a working load of 100kN/m each. Other calculations showed lower factors of safety for circular slip surfaces, but it was considered to be unreasonably conservative to assume that the low strength joints could combine to provide a circular slip, and these were therefore discarded.

Two berms

A failure surface comprising a 4m deep tension crack based on field observations, and a curvilinear slip about 5m deep below the ground profile was postulated (see Figure 8). With a full tension crack and water table near the surface the factor of safety was about 0.7 with A parameters and about 0.65 with B. 4 rows of anchors at 150kN/m were sufficient to provide a factor of safety of 1.2 with the B parameters, but 2 at 150 and 2 at 200kN/m were necessary to achieve the same factor of safety with A parameters. Since two single berm failures had already occurred, and during the last one movements over 3 berms (between 80 & 110mPD and sections 1b - 1c) were noted, the assumptions regarding slip surface and water table for these analyses were considered to be reasonable.

Three berms

A failure surface similar to that for two berms but extended was postulated (see Figure 9). Under these conditions a water table near the ground surface over the whole length of the slip was considered to be unlikely to occur. It was therefore considered over conservative to take such a condition together with a factor of safety of 1.2 on the slope when the permanent anchors would also have a factor of safety of 2 on ultimate pullout. Considering water table WT1 as shown on Figure 9, which has about 4m of head down the failure plane, a factor of safety of 1.2 on the slope was obtained for the A parameters with an anchor load of 250kN/m which was equivalent to a factor of safety of 1.6 on the ultimate of 400kN/m. Alternatively for an anchor load of 200kN/m (F of S = 2) the factor of safety on the slope reduced to 1.04. With the B parameters and 250kN/m (F of S = 1.6) the factor of safety for the slope was 1.5. For the water table WT2 as shown on Figure 9 and an anchor load of 200kN/m the factor of safety on the slope was 1.2.

Five berms

The section and failure surfaces are shown in Figure 10. Initially the non circular surface was analysed and, considering this as an extreme case using the A soil parameters and water table WT3 a factor of safety of 1 was obtained with an anchor load of 350kN/m which was equivalent to a factor of safety of 1.2 on ultimate. Alternatively with water table WT4, anchor loads of 270kN/m (i.e. F of S = 1.5) and the A parameters, the slope factor of safety improved to 1.06, and with the B parameters this was about 1.8. Since the failure surface for such a long slip is unlikely to be

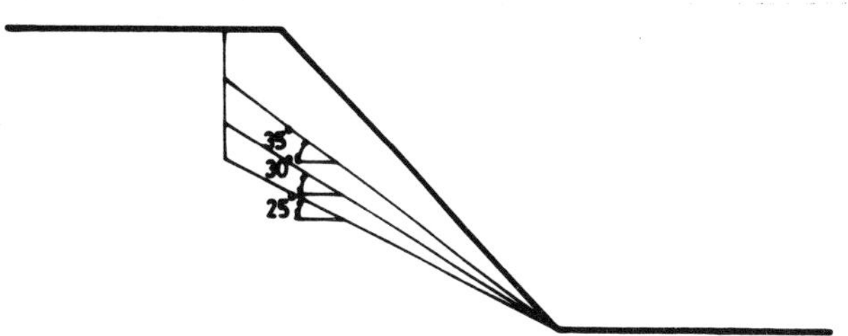

FIGURE 7. STABILITY ANALYSIS: SINGLE BERM

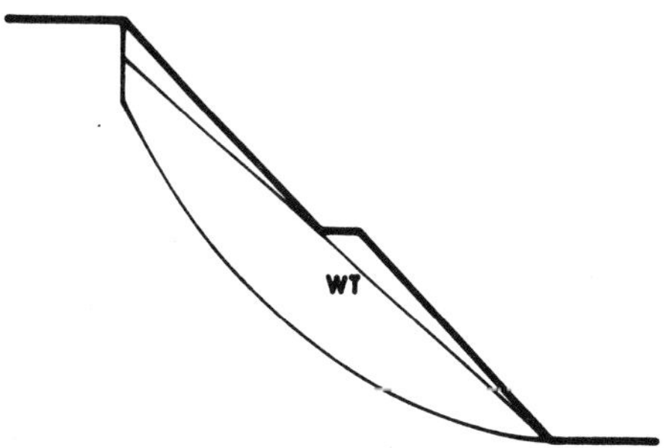

FIGURE 8. STABILITY ANALYSIS: TWO BERMS

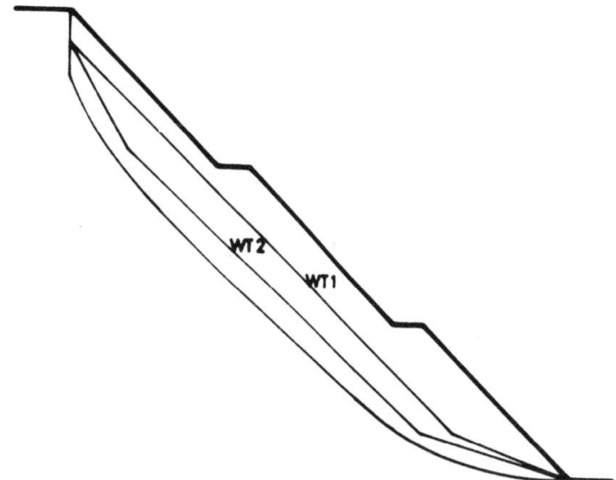

FIGURE 9. STABILITY ANALYSIS: THREE BERMS

continuous, analyses were also carried out assuming an interconnecting series of vertical and sheeting joints. For simplicity this was analysed as a single vertical joint and sheeting joints at 25^0, 30^0 and 35^0, with 4m of water both in the tension crack and down the sloping joint, reducing to zero at the slope face. For the 25^0 failure plane and A parameters the factor of safety on the slope was 1.2. For a 30^0 failure surface and B parameters the factor of safety was 1.7.

Conclusions

The failures occurred because of the combination of a set of sheeting joints of very variable vertical spacing together with sub-vertical release joints. This was very much exacerbated by the presence of a clay infill within the joints of very low shear strength considered to have been fine particles from the weathered rock washed down into joints and fissures.

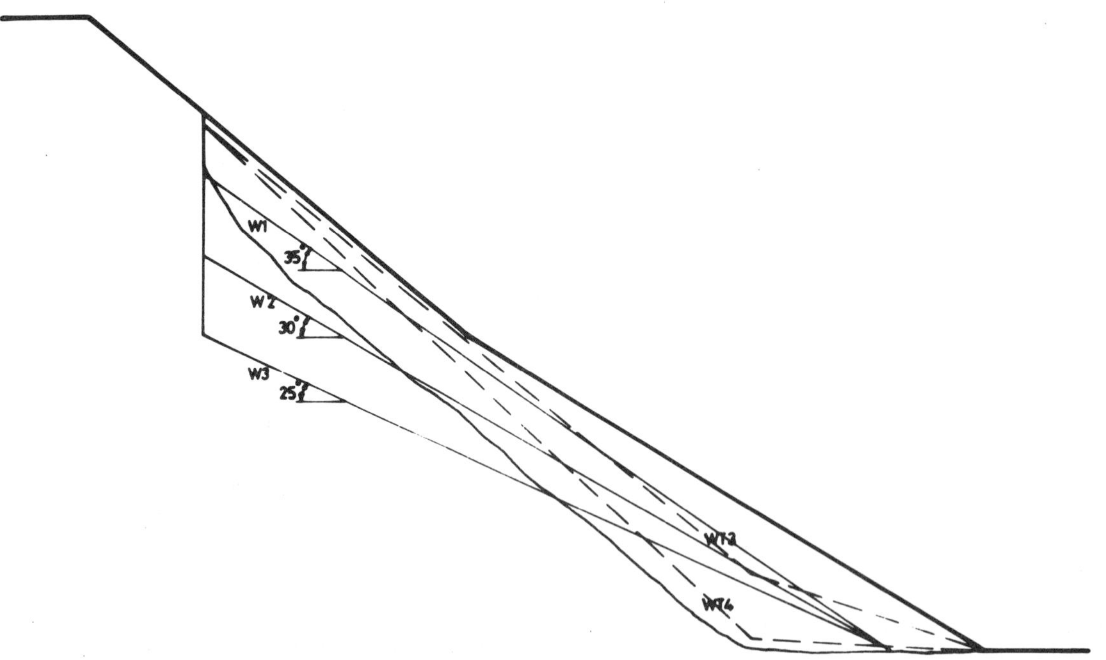

FIGURE 10. STABILITY ANALYSIS: FIVE BERMS

The temporary extreme water table conditions, which occurred when the rock faces were exposed but not treated combined with heavy rainfall, created excessive water pressures on shallow sliding surfaces.

The 'aerial' photograph was most useful in mapping the geological features in the field and discussing the engineering interpretation.

A factor of safety of 1.2 on a failure surface postulated for the whole slope together with exceptionally adverse water tables and ground anchors with a factor of safety of 2 would produce a wholly uneconomic design. A more realistic approach has been adopted by providing a factor of safety greater than unity under the worst conditions envisaged, and 1.2 on the slope with 2 on the anchors under conditions which are reasonably expected to occur. For other possibilities in between intermediate factors of safety have been used.

On the basis of the above considerations, a scheme of remedial works comprising a grillage of primary and secondary reinforced concrete beams held in place by permanent prestressed anchors has been proposed. Two rows of anchors on each of six faces spaced at 6m centres horizontally have been designed with working loads of 600, 900 and 1200kN. These works are in hand at the time of writing this paper and are expected to be completed by mid-1986.

References

Hoek E & Bray J (1977). Rock Slope Engineering, London, Institution of Mining and Metallurgy. p184.

Hencher S.R. (1985). Limitations of stereographic projections for rock slope stability analysis. Journal of the Hong Kong Institution of Engineers. 13,7.

Acknowledgements

The author gratefully acknowledges the permission of the Government of Hong Kong, and of Scott Wilson Kirkpatrick & Partners, to publish this paper.

Construction and design of 'Island Line' rock tunnels for the Hong Kong Mass Transit Railway

David Caiden B.SC., C.ENG., M.I.C.E., M.H.K.I.E.
Charles Haswell & Partners (Far East), Hong Kong
Paul Haines C.ENG., M.I.C.E.
Mass Transit Railway Corporation, Hong Kong
Victor D. Turner B.SC., C.ENG., M.I.C.E., M.H.K.I.E.
Charles Haswell & Partners (Far East), Hong Kong

SYNOPSIS

The paper gives a general description of the Island Line and explains how the design philosophy has changed from previous HKMTR Lines in order to achieve minimal environmental disruption.

An account of geological conditions is given and the difficulties of changing ground and mixed faces are explained. A description of the excavation techniques used in the rock tunnels is given together with an account of primary support methods.

An outline of the design philosophy for the permanent rock tunnel linings is included.

A specific construction contract is described to illustrate the typical problems encountered and solutions used. The difficulties of working from a cramped urban site are well highlighted by this contract which made use of decking to provide a double level site having access from roads at the different levels.

From the access shaft at this site four running tunnel drives were simultaneously in progress. The contract also called for the construction of a large diameter crossover tunnel with different sized approach tunnels, two tapered sections being included. The difficulties of hauling excavated rock away in large volumes are explained. The sequencing of blasting the large diameter crossover tunnel and tapered sections is covered. Also mentioned are the several techniques used in excavating the different running tunnels.

THE ISLAND LINE

General Description

Government approval for construction of the Island Line was given at the end of 1980. At this stage only preliminary route planning had been completed. The route was essentially that envisaged in the original consultant's report which recommended a four line system for the territory. At the time of approval the second line, the Tsuen Wan Extension, was still being constructed.

It was a condition of approval that surface disruption during construction would be kept to a minimum. This was particularly important along the busy north shore of Hong Kong Island which had the added complication of an existing tramway along much of the route. For this reason station construction in large cut-and-cover boxes, as had been used on the previous lines, was ruled out.

The solution, used for most stations, was to have driven tunnels for station platforms below the road with concourse boxes excavated off-line and connected to platforms by driven passenger adits. The disruptive element of station construction is thus made no worse than ordinary

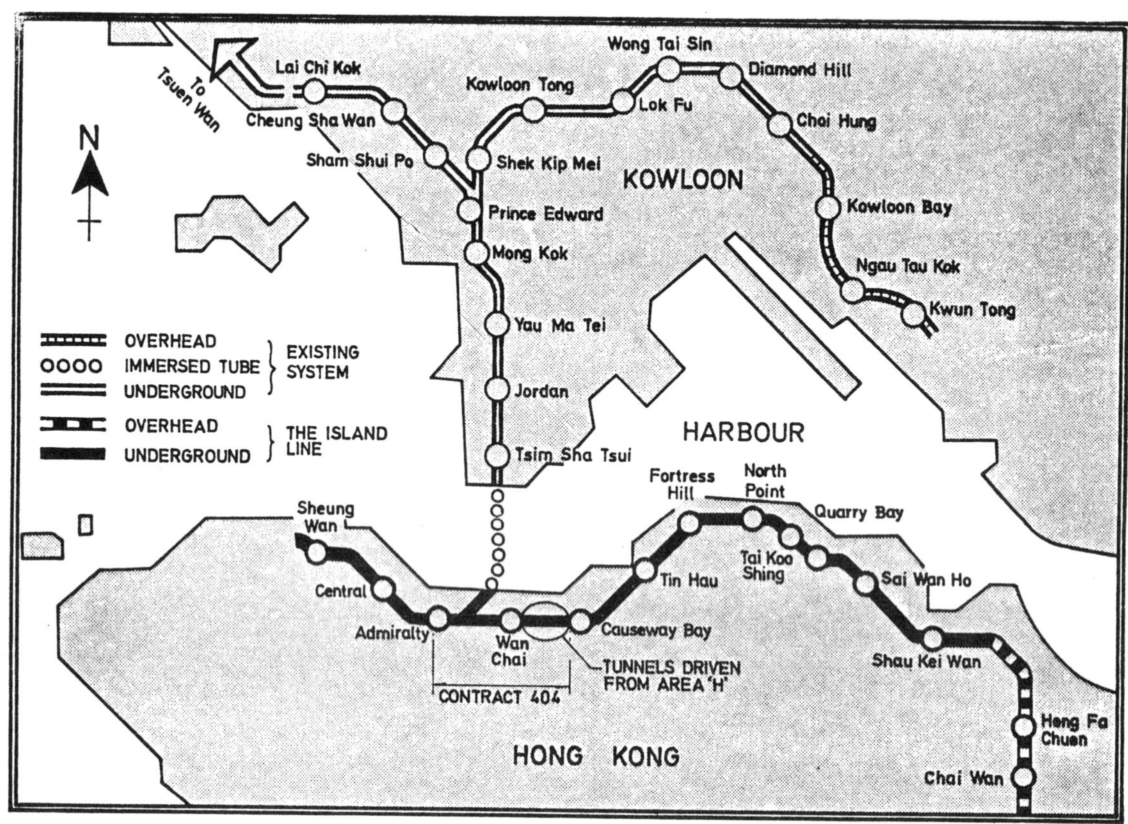

FIG. 1 LOCATION MAP

off road development projects. In the event, traffic along the main roads was kept running smoothly; a notable difference from the construction work on the previous MTR lines in the territory.

By local standards, the site areas available for tunnelling works have been very small. This often created logistics problems in organizing sites to accommodate the throughput of spoil and materials necessary for progress. This was particularly evident on sites where tunnelling required compressed air plant. A specific case is described in the section on Contract 404 below.

The tunnelling work was divided into nine main contracts with four advance contracts being let for the construction of access shafts to enable the tight construction programme to be achieved. The first advance shaft contract was out to tender by April 1981 and awarded in June the same year. The first tunnelling contract was out to tender in June 1981 and awarded in early December 1981. Contract 404, the second contract out to tender, was also awarded at that time.

The first stage of the line between Admiralty and Chai Wan Stations was opened to the public on 31 May 1985. With the completion of the next phase (Admiralty to Sheung Wan) the line route length, including overhead sections, will be about 12.5 kilometres. This brings the system route length to over 39 kilometres.

Route Geology

The route was selected by using the traditional approach of identifying a congested transport corridor and attempting to relieve congestion by providing an immediately adjacent alternative. Thus an attempt was generally made to position stations at centres of existing population concentration. It was also necessary, of course, to consider existing buildings and foundations and the availability of areas for both site access and concourse developments.

The net result has been a route which is aligned in plan along the main corridor which occupies the old northern shore area of Hong Kong Island. Much of the route at the west end of the line is beneath reclaimed land while at the east end it is close to the former sea cliff.

The country rock is known as Hong Kong Granite and is from the Upper Jurassic period. In its fresh form it is a hard pinkish granite. However, fully weathered (and known as CWG or completely weathered granite) it is a granular soil with low cohesion. The water table is very close to ground level and the CWG cannot be tunnelled without compressed air or ground treatment. There are gradations of weathering between the fresh granite and CWG. Sometimes boulder zones are encountered, boulders being completely surrounded by CWG. At other times a sudden transition arises producing mixed faced conditions with CWG at the crown and fresh granite in the invert.

Deposits grading from colluvium to alluvium are found above the CWG. These often contain gravel and cobbles but closer to the sea tend to become silty fine sands. Above this zone marine deposits of clays, silts, fine and course sands are found. The fill material in the reclaimed land can also vary but is usually a combination of CWG and granite boulders.

The situation was further exacerbated at the western end by the need to tunnel below old jetties and sea walls. The route passed near to many old buildings which were on timber piles and sometimes passed between piles. On one contract it was actually necessary to remove the lower ends of piles.

There was a limit to the depth which could be chosen for vertical alignment to avoid the unfavourable upper strata. Legal and safety aspects, and also costs, restrict compressed-air pressures. Furthermore, there is a need to minimize passenger route distances between entrances and platforms. It was therefore generally preferred to keep the maximum depth to about 30m below ground level.

The overall outcome of the alignment was that about 8.3 km of tunnels were in soft ground/mixed face conditions and about 13.2 km of tunnels were in rock. From the tunnelling point of view drives of one type (rock or soft ground) have been too short to enable fast rates of progress or economies of scale to be made.

A tunnel engineer could perhaps claim that moving the whole line inland to provide easier tunnelling would have cut the civil engineering costs by over half. A bridge or viaduct designer could make a similar argument for moving the alignment seaward. This approach of reducing infrastructure costs and letting development and population patterns grow to suit is sometimes adopted in new town planning. It is, however, seldom accepted by the transportation/planning engineers or governing bodies of existing cities.

Rock Tunnel Excavation and Primary Support

Rock tunnel diameters varied and a schedule of the main types of lining sections is given in Figure 2 of the design section below. The greatest lengths of tunnel driven were the 5.1 m nominal internal diameter running tunnels.

Because of limited drive lengths in consistant ground all contractors ruled out the use of tunnel boring machines. Also, the granite was too hard to utilize rotatary head boom cutters (road-headers).

Traditional drill and blast techniques were therefore used on all contracts. Some further details on drilling patterns and explosive quantities are given in the section below describing part of Contract 404. Island Line rock excavation combined with the Kornhill development has meant that an unprecedented quantity of explosives has been travelling the roads of Hong Kong. No explosives accidents have been recorded in the construction of the line.

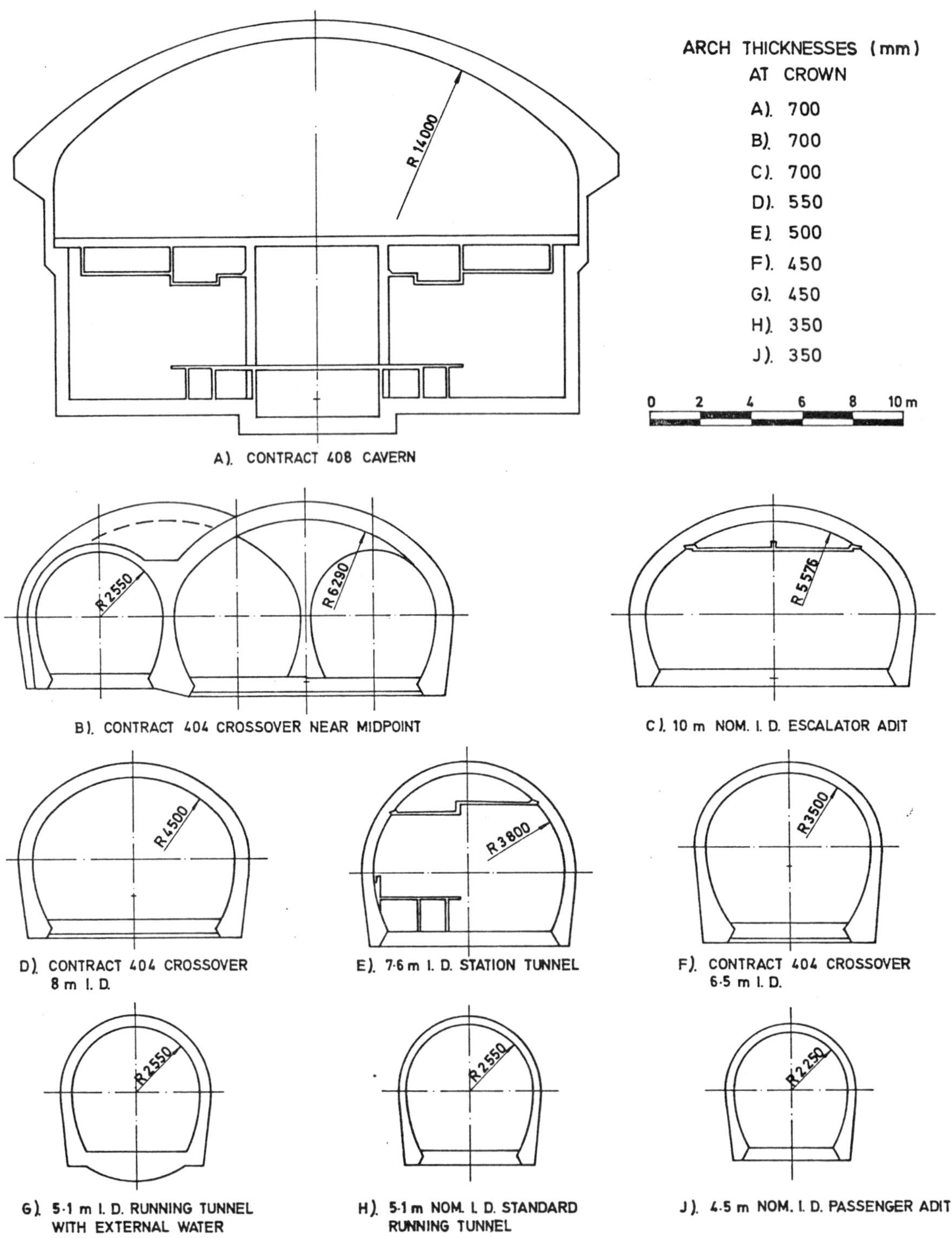

FIG. 2 EXAMPLE TUNNEL SECTIONS

MAIN TYPES SHOWN IN ORDER OF DECREASING DIAMETERS. EXCLUDING STUB ADITS AND SHORT CONNECTIONS OVER TWENTY DIFFERENT TYPES OF SECTION WERE USED.

For running tunnels in sound granite, full face blasting was used. Pulls varied on the different contracts from one to three metres. This was dependent upon the proximity of buildings and the extent of the good rock cover.

In less competent rock bench excavation was carried out, sometimes using a ring cut where arches were erected. Running tunnel progress rates typically peaked at about 30 metres per week.

On contracts having only rock excavation, diesel loaders have generally been used to shovel spoil into diesel powered lorries. On contracts which also incorporated soft ground tunnelling, rail mounted air powered rocker shovels were usually chosen. These loaded spoil to rail wagons pulled by battery powered locomotives.

On one contract lorries and spoil were taken up the shaft in a lorry lift. However, the usual method was to raise spoil to shaft tops in more compact muck hoists. From the shaft tops the excavated material has generally been hauled by road vehicles to barges at special loading jetties. A large proportion of the rock has been used for reclamation works.

Primary support required has of course been as varied as the competancy of the rock. Good fresh granite has required no support. The degree of fissuring, faults and decomposition has demanded use of one or more of the following at various locations:

1) Rockbolting
   a) random localized
   b) patterned without mesh
   c) patterned with mesh

2) Sprayed Concrete
   a) single pass
   b) multiple pass
   c) reinforced with random steel 'fibres'

3) Steel Colliery Type Arches
   a) unlagged
   b) random lagging
   c) close lagged

Where appropriate forepoling was also used.

Lagging to arches has been in steel. This has been due to the stringent specification that no timber except hardwood footblocks and hardwood wedges may be concreted in behind the permanent lining.

Probing ahead to prove the ground and locate water sources and strata interfaces was a continual requirement which had also been listed in the specification.

Lining Design

Varying ground has required varying design approaches for the insitu concrete linings which have usually been unreinforced. The main types are set out below.

1. Standard tunnels in fresh granite with at least four metres fresh granite cover.

   In these cases rock loading has been based on Terzaghi's Rock Pressure Theory (see p226 of ref. No. 7). Soft ground perched above the rock has been considered as surcharge loading. Hydrostic head can be assumed to reduce linearly from full head at the crown to 25% head at invert.

   A frame analysis computer program was then used to analyze a one metre strip of tunnel. It was considered that invert to arch construction joints were incapable of transmitting moments (that is, these joints are hinges). The passive resistance of the rock mass is simulated by a series of rock springs at each node point. The analysis is a reiterative process whereby negative deformation springs are ultimately deleted to achieve the final model.

   The finally selected lining thickness is that which will work as a masonry arch. If it was not possible to reinforce the flat invert for the total reduced water pressure uplift (steel areas become impracticable in large diameter tunnels) weep holes were provided in the invert.

2. Standard tunnels with less than four metres fresh granite cover or in fissured water bearing rock.

   The problem in these cases is that an increased water pressure will be acting on the lining. The flat inverts are weak against uplift pressures in excess of eight metres water head. Weep holes, which are popularly misunderstood, have their limits. See ref. No. 1 for an interesting explanation of the effect of leakage in tunnels.

   Where necessary flat inverts had to be thickened, more heavily reinforced and ultimately arched to resist high water pressure uplifts. The difficulty is always how to establish the actual water pressure which will be experienced by the lining; or, conversely, how to assess the head loss occurring through the continually changing rock.

   If compressed air is required to allow excavation, or because the drive adjoins a soft ground section, a direct measure of water pressure is available as the face proceeds. Measurement of water inflow quantities can also be used as a guide.

   However it is usually necessary to have an agreed design drawn before work is undertaken. In these circumstances the only safe assumption to make is that full head uplift pressure exists on the invert.

3. Tunnels with abnormal loads and point loads due to existing or proposed buildings.

   In these special cases the point loads were added in the computer analysis; sometimes hand calculations were carried out.

   Linings were either thickened or reinforced for the point loads and in extreme cases thickened and reinforced.

   On one contract excessively high foundation loads were predicted for a future development directly over the tunnel. This necessitated the standard station tunnel lining being increased to 1.2 metres thickness all round with an arched invert.

4. Caverns and tunnels with very large span arches.

   For the very large diameter tunnels (such as crossover tunnels) a reinforced arch was required in order that linings did not become unreasonably thick.

CONTRACT 404

General Description

Compared to other Island Line Contracts, Contract 404 is characterized by having the longest running tunnels combined with the highest potential air pressures required in extremely varied ground conditions.

The contract called for twin tunnels to be constructed from Admiralty in the old Naval Dockyard site just to the east of Central District, eastwards under the urban artery of Hennessy Road through Wanchai District and finally connecting to an adjacent construction contract some 50 metres west of Causeway Bay Station at Pervical Street. Included in Contract 404 was the construction of twin 7.6 metre diameter station tunnels at Wanchai with adit connections to an offstreet concourse.

Two running tunnel drives were constructed eastwards from an extension box at Admiralty Station which was already operational as part of the Modified Initial System. These drives junctioned into twin shafts constructed for the Wanchai Station drives which were also driven eastwards.

The longest single stretch of running tunnel

lay eastwards of Wanchai Station to Causeway Bay and it is this section of tunnel which will be the subject of detailed description.

All the running tunnel drives on Contract 404 involved a mixture of soft ground and hard rock tunnel to be driven in free and compressed air with associated rock/soft interfaces to be negotiated.

The distance between the east end of Wanchai Station and the west end of Causeway Bay Station was around 900 metres and due to the programming restraints together with the geology of the section, it was necessary to tunnel from approximately midway with twin drives in each direction.

The midpoint was beneath the Tonnochy Road junction with Hennessy Road, where the fresh granite level was very high. This also made the location ideal for construction of a trailing crossover which is desirable for operational purposes at regular intervals along the line.

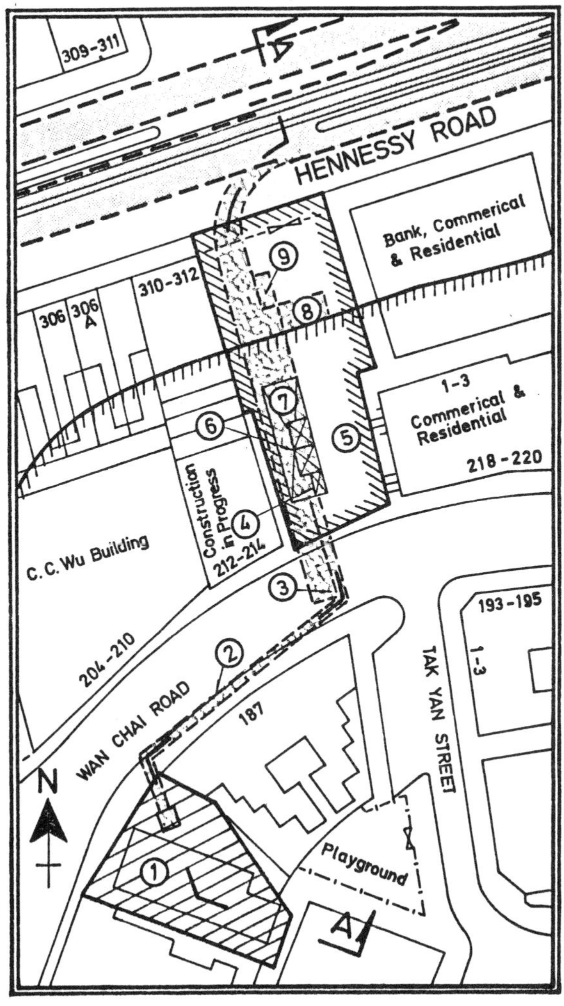

KEY

- ACCESS ADIT SHAFT & TEMPORARY WORKS TUNNELS
- PERMANENT WORKS TUNNELS
- AREA 'J'
- AREA 'H'
- 1841 SHORELINE

① COMPRESSOR HOUSE
② SERVICE TUNNEL
③ BACKSHUNT
④ MAN HOIST
⑤ COVERED MUCK BIN
⑥ MUCK HOISTS Nos 1 & 2
⑦ ACCESS SHAFT
⑧ BATTERY CHARGING ADIT
⑨ SUMP

FIG. 3 PLAN OF SITE AREAS H & J - CONTRACT 404

Works Area

A suitable 700 m^2 site for tunnelling access was available in the form of a sitting out area and playground.

The site had only one entrance from Hennessy Road and this was located at a busy controlled junction. The southern site boundary was Wanchai Road where the rapidly rising rockhead outcropped. Owing to the original topography this was some 4.8 metres above the Hennessy Road level.

Original Government guidelines for site usage were that vehicles would not be allowed to reverse out of the site into Hennessy Road and that no access would be allowed from Wanchai Road. This would have made working arrangements within the site extremely difficult.

To overcome these difficulties the Contractor proposed installing a secondary working deck over 80% of the site at the elevated level of Wanchai

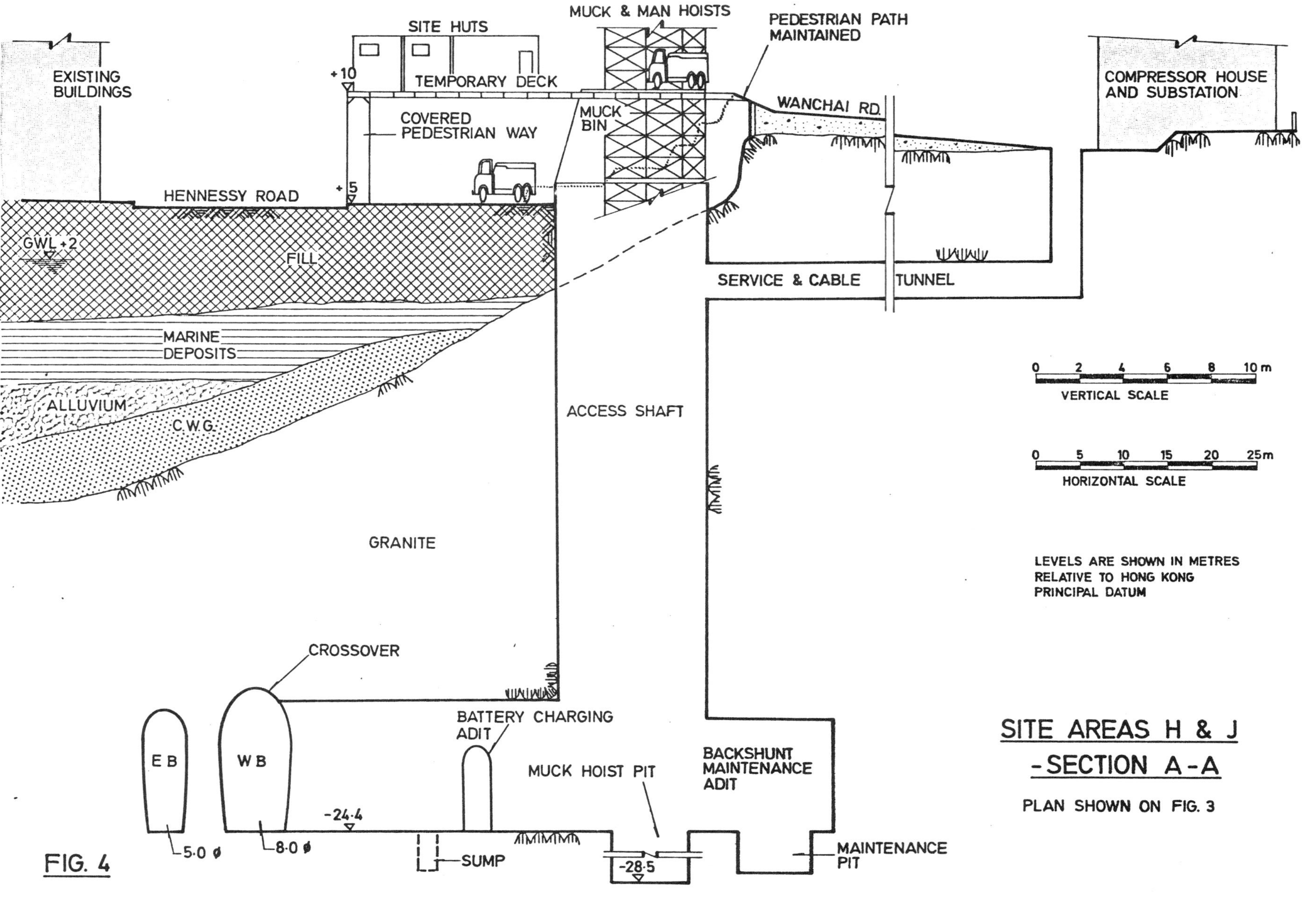

SITE AREAS H & J -SECTION A-A

PLAN SHOWN ON FIG. 3

FIG. 4

Road. The proposal was fortunately accepted and access was allowed from Wanchai Road to this decking for materials delivery.

Permission was also given for reversing of vehicles into the area from Hennessy Road using flagmen and manoeuvring during the red light phase.

These arrangements eased the still severe restrictions caused by the limited site area.

Construction Noise

The works area was surrounded by residential apartments. The Engineer and Contractor were therefore apt to receive complaints, although these were at relatively low noise levels.

Relatively is stressed because ambient noise levels due to traffic along Hennessy Road were of the order of 80 dB(A). Construction noise was not substantially higher than this. It is generally considered (as with methods for assessing industrial noise) that the problem was due to the erratic nature of the noise rather than its amplitude.

The main sources of noise nuisance, together with some methods used to alleviate the problems, are listed below using numbered headings.

1. Discharge from tubs into muck hoist

   As an open shaft was necessitated for access and ventilation reasons this was a continuing problem. Some success was achieved by muting the steel muck hoist buckets with timber and rubber. However, continual maintenance problems due to wear and tear were inevitable.

2. Discharge from muck hoist into muck bin

   This was successfully overcome by entirely enclosing the bin and hoist with timber sheeting above ground level.

3. Loading from muck bin to lorries

   Day time loading only was allowed in order to limit this problem.

4. Compressor operations

   Compressors were muffled and enclosed in a cavity blockwork clad compressor house. Apart from emergency units compressors were electrically powered and therefore inherently quieter than diesel units.

5. Cooling tower operations

   Air cooling towers were situated on the roof of the two storey compressor house building. No specific complaints were received in respect of these units, presumably because similar continuous air conditioning noise is common in Hong Kong and being of a consistent nature is less offensive to the ear than erratic noise. A baffle system could have been utilized to reduce the noise had this been considered necessary.

6. Blasting noise

   This was the cause of much complaint during the early stages of construction and where possible blasting was limited to the period 8.00 a.m. to 11.00 p.m. A timber bulkhead and door was built in the access adit soon after adit completion. Closing this during blasting enabled use of explosives to continue at night.

Excavation Statistics

The total volume of excavation for permanent works on this section of tunnels and crossover was 54,700 m^3. The total volume of rock excavated for the permanent works was 21,800 m^3 and of this 9,600 m^3 was in the crossover.

In addition to the permanent works excavation, temporary work excavation was carried out for the shaft, access adits and service tunnel giving an additional volume of rock excavation of 4,750 m^3.

In the running tunnels the rock excavation nominal face area was 30.4 m^2 and the best week's rate of advance was 30 lin.m i.e. 912 m^3/week (measure). Average overbreak on excavation was around 10% which gave an increase in required concrete volume of 45% above measure.

Including temporary works and overbreak the total rock excavation in this section is estimated at 28,250 m^3. This was excavated over a period of 100 weeks but some excavation was concurrent with soft ground excavation.

The maximum measured volume of excavation during any week was 1,460 m^3 and using 10% overbreak and a bulking factor of 1.6 this gives a handled volume of 2,570 m^3/week. This maximum handled rate was achieved when drive lengths were relatively short but mucking out facilities at night were being reduced by noise limitations.

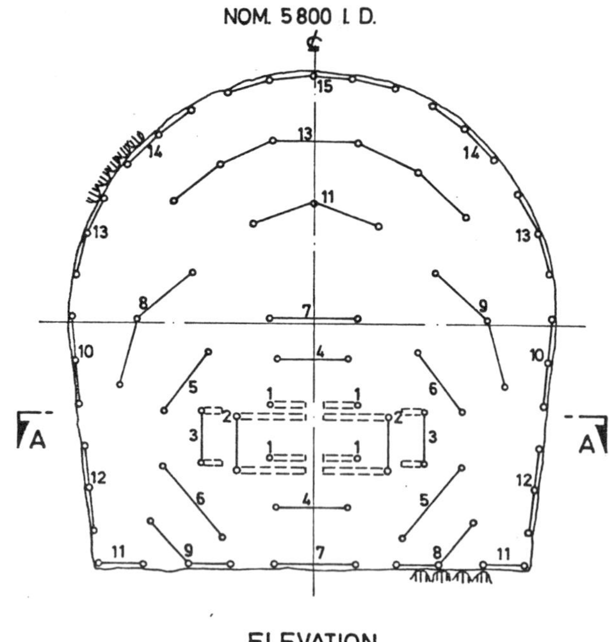

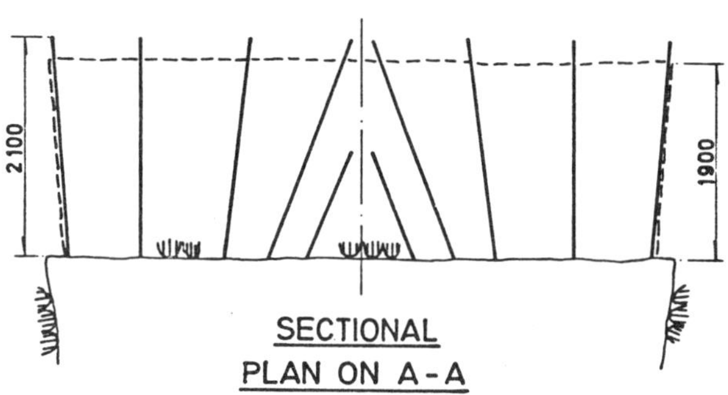

FIG. 5 BLASTING PATTERN FOR RUNNING TUNNELS

Explosives

The main explosive used was Kiri 3 which is an ammonia gelatine dynamite with a propogation velocity of 6200 m/s and a density of around 1.4 g/cm^3. The cartridge size was 25mm diameter and 150mm long giving a stick weight of 100 grammes.

To control overbreak, peripheral holes were usually charged with SB Iremite, a water gel explosive with a propogation velocity of 2700 to 3000 m/s. The diameter of these cartridges was 22mm and the length was 620mm, giving a weight per cartridge of 200 grammes. The density of this explosive was around 0.83 g/cm^3.

25 millisecond delays were generally used in the cut with peripheral blasting using 1/4 second delays. No. 8 strength detonators were used in all cases. Hole drilling size was 38mm and a proprietary glued sand stemming was employed.

Typical charging for a nominal 5.8 m diameter horseshoe face of 30 m^2 with a 1.9 m pull was 55 kg of Kiri 3 (80%) and 7 kg of SB Iremite using 79 detonators spread over 15 delays. See Fig. 5 for the drilling pattern. This means the average charge was around 4.1 kg per delay but heavier densities were inevitable for the cut.

The Specification imposed a vibration limit of 25 mm/s ppv at the closest structure. This level was never approached, typical values being around 6 mm/s. The closest distances to buildings were of the order of 25 m vertically and 12 m horizontally.

Mucking plant

Generally, running tunnel spoil was loaded away by means of pneumatic rocker shovels which had integral belt conveyors. The face pull averaged 1.9 m or about 90 m^3 of bulked rock spoil per blast. This was mucked away in around one and a half hours on an unhindered shift into 4.5 m^3 side tipping rail tubs requiring a cycle time of one tub every 3 minutes.

Rail track gauge was 750 mm and the maximum haulage distance to the pit bottom was approximately 360 m for a fully productive all rock drive. 10 tonne twin battery locomotives were used to haul a single skip up a maximum grade of 3%.

Pit bottom rail level was 23.2 m below PD and surface level of the muck bin was 4.5 m above PD. To facilitate tipping the vertical winching distance was 41 m. A twin 4.5 m^3 bucket, cable winched hoist was installed with a maximum lifting capacity of 11.4 tonnes. The average lifting load with a single 4.5 m^3 rail tub contents was around 8.0 tonnes and the lifting speed was in the region of 30 metres/minute. The return cycle time was about 50 seconds.

With use of the backshunt, the two hoists working together had a total handling capacity in the order of 200 m^3/hour.

However, if one hoist was out of action, the mucking out cycle was delayed by around one hour giving an output cycle of approximately 40 m^3/hour/hoist. The main system restriction was manoeuvring of skips in the bottleneck caused by the single track access.

Rock Excavation at Soft Ground Interfaces

The least economical rock excavation was necessarily in these sensitive zones. The contract tender had indicated use of shields in all four soft ground compressed air drives. Alternative hand drive methods, using combinations of ground treatment, arches, ground bolts and sprayed concrete were however accepted for both the deeper westbound drives. These drives were shorter and also in better material than the two upper shield drives where shield chambers were constructed in rock close to the soft ground interface. During construction of these chambers probing ahead was carried out to ensure adequate unweathered rock cover and to test the water ingress.

The contractor endeavoured to remove as much of the rock ahead of each shield chamber as possible before installing the shield. This was done by using a series of ribbed headings which decreased in size as the rock level lowered. Inevitably some rock remained ahead of the shield as it advanced along the headings.

The headings also assisted with shield guidance which can become difficult in hard rock. Prior to advancing the shield, an accurate concrete invert was placed for its support. This preformed invert made shield steering through the rock more predictable. Sometimes steel rails set to exact levels were used and this further improved results.

Problems did still occur due to the reluctance of miners to clear out the entire invert ahead of the shield after blasting. This meant that the shield was sometimes shoved when underbreak existed. Even with minor underbreak the departure from true line or level was rapid, making it impossible to build the segmental lining ring within tolerance. Subsequent hand rebuilding was therefore required in some locations.

Affects on Ground Water

Many of the buildings along the route had private wells; some very old and some of recent construction. Those in use were mainly for provision of the coolent in air conditioning heat exchangers.

A survey of wells close to the tunnel route was made prior to construction work, mainly to identify potential air loss problems during compressed air phases of soft ground drives. Wells adjacent to rock tunnels were also

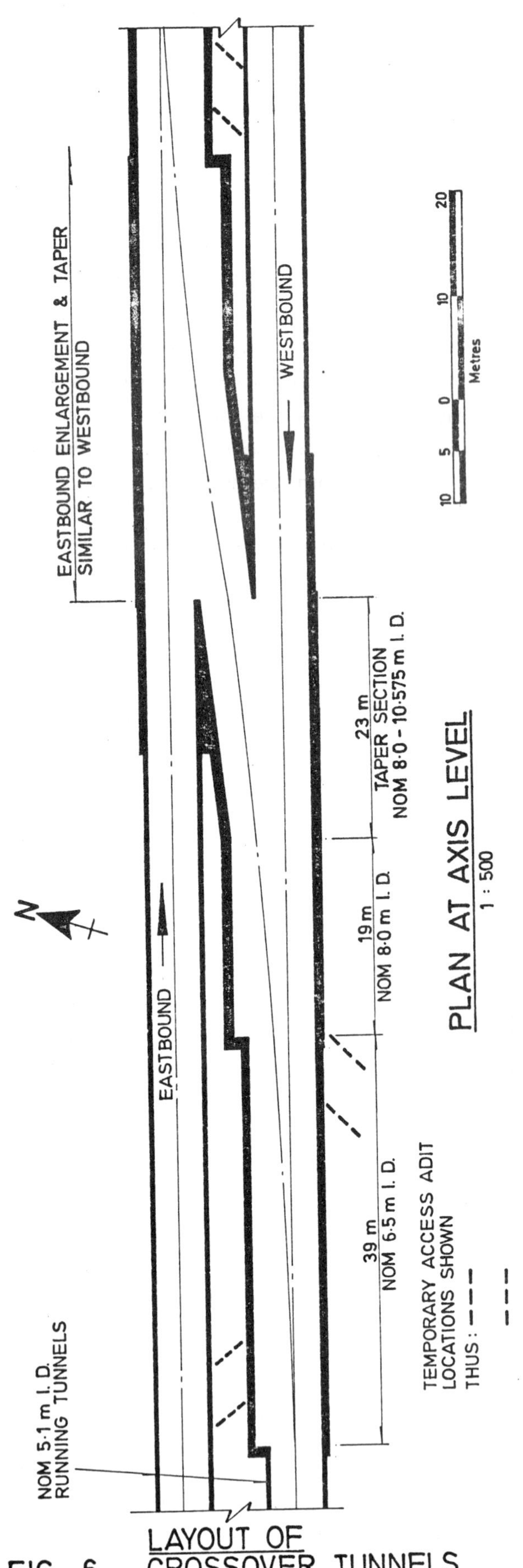

FIG. 6 LAYOUT OF CROSSOVER TUNNELS

considered because compressed air bulkheads were to be installed within rock excavations in order to allow adequate working space for commencement of soft ground tunnelling.

Air loss through wells did not prove to be a major problem; when it occurred the wells were capped. If subsequent reuse of the well was required a temporary, rather than permanent, cap was installed.

The granite through which the tunnels passed had moderately spaced tight joints, some of which were clay filled. Dolerite dykes varying from a few millimetres in thickness up to 1.2 metres thick crossed the tunnel line, dipping a few degrees from the vertical and striking ENE.

When rock cover exceeded five metres water ingress was minimal but closer to the soft ground interface, and in the unlined shaft, this was not the case. However, since buildings were generally founded on the rock, settlement over rock tunnels due to drawdown was not a problem. The drawdown itself did give rise to complaints in connexion with reduction in flows from some of the private wells. In some instances cooling water had to be supplemented by using the fresh water mains; special licences were required in such cases.

Crossover Excavation Sequence

Excavation of the four running tunnel drives and the crossover tunnels was carried out from a single access adit of 6 metres nominal diameter. The curving adit (see Fig. 3 for location) could only accommodate a single temporary rail track. The programme demanded that driving and lining of running tunnels be continuously carried out whilst enlargement and lining of the crossover was underway.

Fig. 6 shows the permanent works arrangement of the crossover tunnels. The excavation stages explained below are shown on the opposite page in Fig. 7.

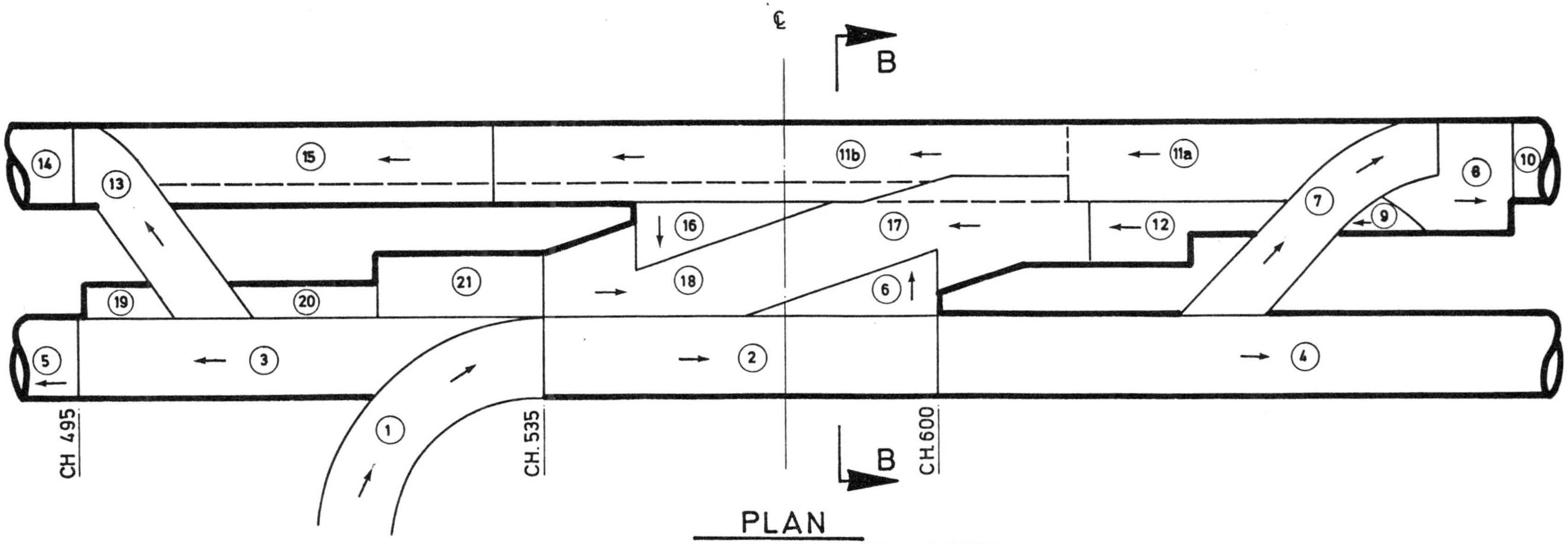

PLAN

→ DIRECTION OF ADVANCE

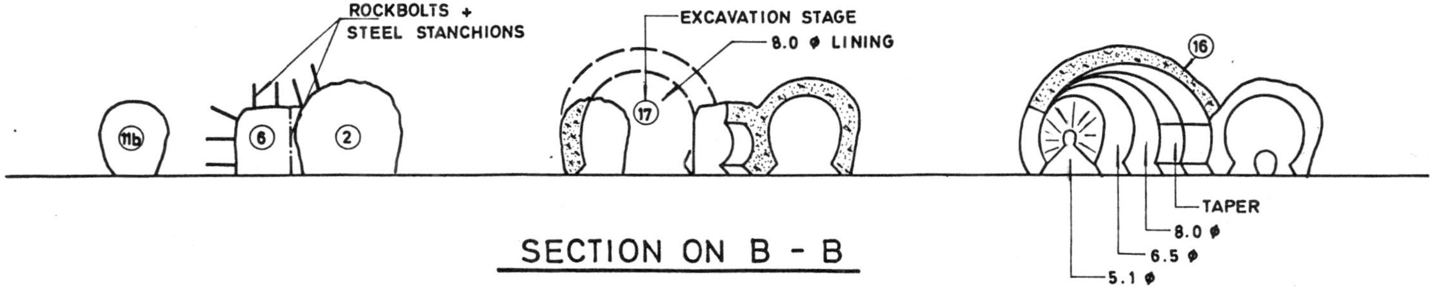

SECTION ON B - B

CROSSOVER - DIAGRAM OF EXCAVATION STAGES

N.T.S.

FIG. 7

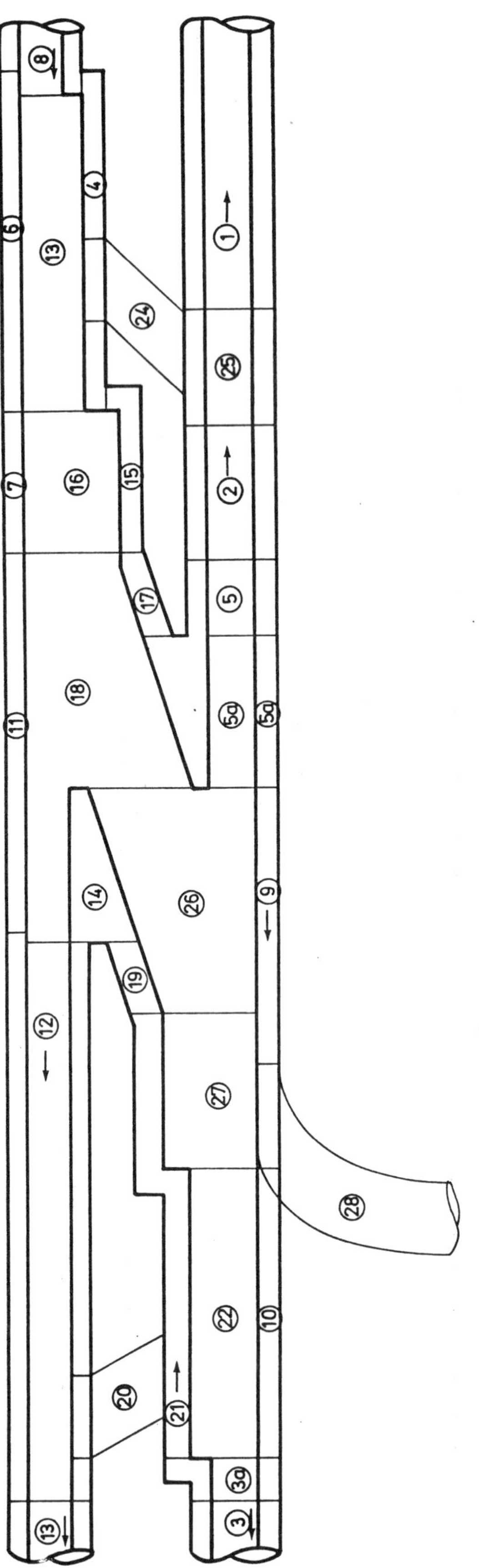

CROSSOVER - DIAGRAM OF LINING STAGES

N. T. S.

FIG. 8

Pilot tunnels of similar excavation section to the running tunnels were initially driven (Stages 2 to 4) to gain access to the westbound running tunnel drives. Temporary adits at each end of the crossover (Stages 7 & 13) provided access to the eastbound drives and enabled a bypass system to operate; this allowed concurrent construction of crossover sections and running tunnels.

The eastbound 6.5 mid tunnel (Stage 8) was driven full width with a single bench since a three and a half metre wide rock pillar existed between this and the previously lined adjacent running tunnel. A reduced pull length with limited charges ensured that acceptable vibration levels were achieved on the mass concrete lining.

Stages 6 and 16 were enlarged sideways from the pilot tunnels in order to construct the taper section central apex walls prior to the full taper enlargement stages 17 & 18.

The excavation of Stage 6 necessitated the local reduction in area of eastbound crossover pilot stage 11b; this ensured a minimum four metre thick competent rock wall separation. Pull was limited to one linear metre with 1.2m drilling length and using a double wedge cut the maximum charge per delay was reduced to about 2.5 kg.

The adjacent westbound tunnel excavation was unlined during blasting of the eastbound pilot and minor falls of rock occurred.

Stage 11a excavation was curtailed due to the requirement for the westbound apex wall concrete to be in place prior to passing of the eastbound pilot face. Stage 11b and 15 eastbound tunnel were excavated by use of a nominal 5.0 mid pilot with subsequent south wall enlargement, thus limiting vibration and rockfalls in the adjacent unlined westbound section.

Lining of the 8.0 mid and 8.5 to 10.6 mid taper sections followed behind the enlargement excavation as closely as practicable. The concreting sequence is indicated in Fig. 8 on the opposite page.

Acknowledgements

The authors wish to express their thanks to the following organizations for their permission to publish this paper and for their assistance in its production:

Maeda Construction Company Limited,
Contractor, Contract 404

Mass Transit Railway Corporation, Hong Kong

References

1. Atkinson, J.H. and Mair, R.J. (1983)
Loads on leaking and watertight tunnel linings, sewers and buried pipes due to groundwater.
Geotechnique Vol. XXXIII No. 3 and subsequent discussion in Vol. XXXV No. 1 (1985).

2. McFeat-Smith, I. and Haswell C.K. (1985)
Preliminary studies for tunnel projects in Hong Kong.
Tunnelling '85 Symposium Brighton, IMM.

3. Cater, R.W. et al (1984)
Tunnels constructed for the Hong Kong Mass Transit Railway.
Hong Kong Engineer (October), HKIE

4. Edgley, R.K. (1981)
The Hong Kong Mass Transit Railway
Hong Kong Engineer (September), HKIE

5. McIntosh, D.F. et al (1980)
Hong Kong Mass Transit Railay, Modified Initial System
Design and Construction of underground stations and cut-and-cover tunnels.
Proc. ICE, Part 1, 68, Nov., 599-626

6. Haswell, C.K. et al (1980)
Hong Kong Mass Transit Railway, Modified Initial System
Design and construction of the driven tunnels and immersed tube.
Proc. ICE, Part 1, 68, Nov., 627-655

7. Szechy, K. (1973)
The Art of Tunnelling
Akademiai Kiado, Budapest

Blasting and support problems in the Monte d'Oro tunnel, Trieste, Italy

M. Carastro MIN. ENG.
Rome, Italy
R. Ribacchi
Università di Roma, Rome, Italy
G. Soccol
Di Penta S.p.A., Rome, Italy

SYNOPSIS

The Monte d'Oro tunnel is a railway tunnel 884 m long, excavated at a maximum depth of 55 m below the Aquilinia village on the outskirts of Trieste. The tunnel crosses a formation of sandstones-mudstones alternations; for wide stretches the ordered setting of the formation was destroyed by intense shear deformations and the rock mass shows a scaly structure.

Because of the vibration danger to buildings, the tender required that the excavation should be carried out by means of mechanical methods only, and a roadheader was initially utilized. However, the presence of some very hard sandstone layers within the formation hindered the excavation and markedly reduced the rate of advance. The use of expansive materials (Bristar) proved to be unsatisfactory.

The permission was therefore obtained to carry out trial blasts, carefully monitored, to investigate the possibility to excavate the tunnel by means of drilling and blasting. The results proved satisfactory and finally this method of excavation was accepted, with limitations on the maximum charge per delay.

During this phase of the excavation, the vibration level was measured at various critical locations on the surface. The results of these measurements are discussed.

Finally the paper describes the prereinforcement techniques, the excavation method and the support system adopted in a zone where the tunnel underpassed a preexisting road tunnel with only 3 m of interposed rock.

INTRODUCTION

The railway tunnel underneath the Monte d'Oro hill (fig. 1) is a single track tunnel 5 m wide and 884 m long (fig. 2), located on the outskirts of Trieste between the Aquilinia and Noghere stations. It is to be utilized to link the port of Trieste and the stocking and handling area for containers in Valle delle Noghere. The client was Ente Zona Industriale Trieste (EZIT).

The maximum height of the tunnel overburden

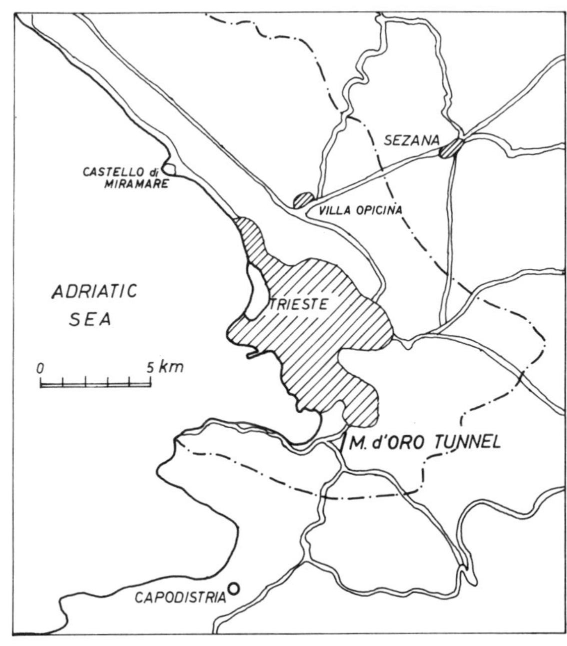

Fig. 1 - Location of the Monte d'Oro Tunnel

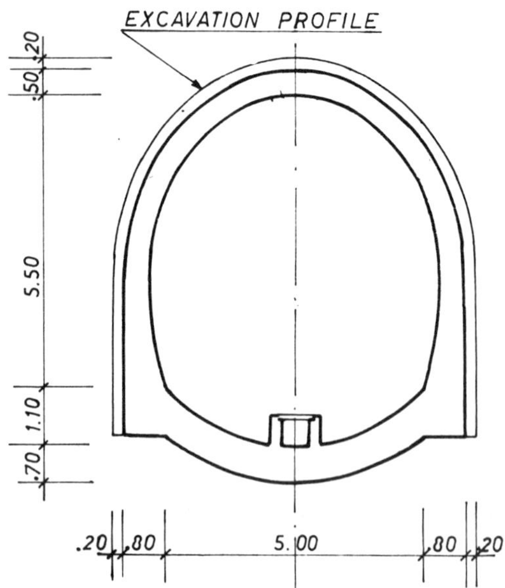

Fig. 2 - Section of the tunnel

is 55 m (fig. 3). The hill is densely covered by the houses of the Aquilinia village, which spreads out especially over its southern slopes, where the overburden is lower (5-30 m).

At a distance of about 300m from the southern entrance, the tunnel crosses, at angle of about 37°, a pre-existing road tunnel 10.9 m wide. The vertical distance between the floor of the road tunnel and the crown of the railway tunnel is only 3.6 m.

Both entrances and the corresponding initial parts of the tunnel for a total length of 90 m had been excavated some 15 years ago. The contractor for the completion of the tunnel was Di Penta S.p.A.; the works began in summer 1981 and finished in December 1983.

CHARACTERISTICS OF THE ROCK MASS

The tunnel crosses alternations of sandstones and mudstones belonging to a thick flysch formation of upper Eocene age. The geological situation observed during the excavation of the tunnel is shown in fig. 3; the rock mass is formed by domains of regularly bedded layers separated by bands of intensely sheared material 10 to 80 m thick (represented by undulating lines in Fig.3).

In the ordered domains, the bedding is usually almost horizontal, the dip being lower than 10° and having a SE direction. The attitude of the layers is somewhat distorted only near the boundary with the sheared bands.

The distribution of the lithotypes crossed by the tunnel in the ordered domains is shown in fig. 4; some variation in the sandstone/mudstone ratio along the tunnel length can be clearly seen. In the first stretch, from the Noghere entrance, the mudstone lithotypes are more abundant (about 50%), whereas, at a greater distance, the rock mass is mainly formed by sandstone layers, 0.1-0.3 m thick, separated by thin mudstone intercalations (2-5 cm thick).

In the ordered domains, practically no faults were met. However, the sandstone layers are cut by systematic joints; most of them are nearly vertical (dip > 75°) and belong to two main families (fig. 5), the first one having a strike parallel to that of the bedding; the second having a strike transversal to the bedding. The

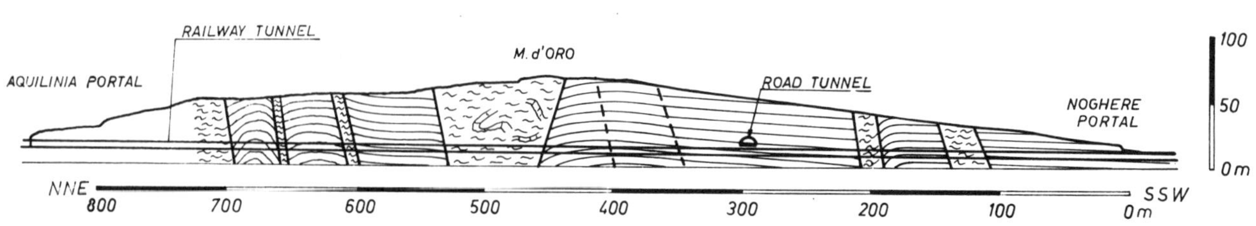

Fig. 3 - Geological situation. The dip of the beds in the ordered sequences is indicated.

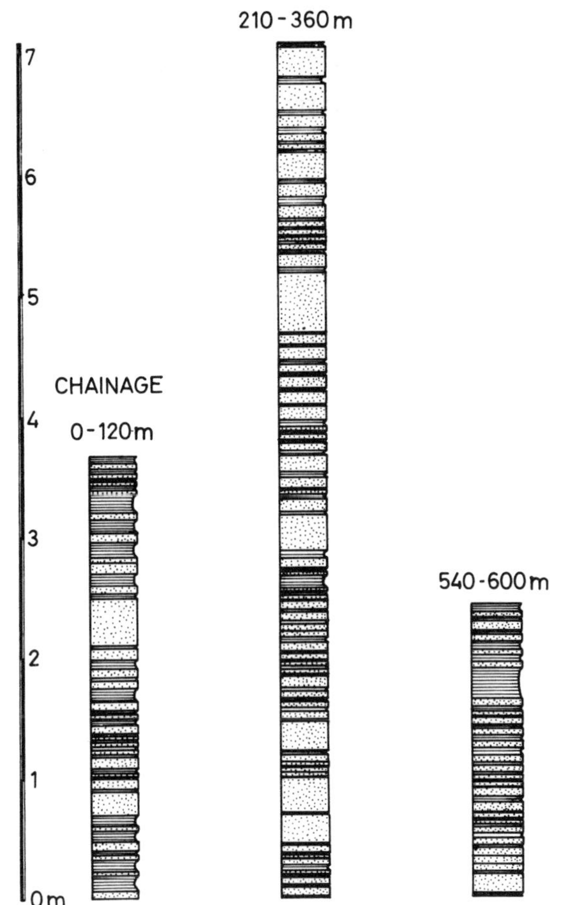

Fig. 4 - Lithotypes in the ordered sequences of the flysch formation

spacing of the joints of each family is about 0.3-0.5 m; often the spacing increases with the bed thickness. In most cases the joints affect only the sandstone layers stopping at the mudstone intercalations. The joints are usually tightly closed and sometimes partially healed or only potential.

As to the rock mass characteristics in the ordered zone, an RQD value of 85-95% was obtained in two exploratory boreholes.

In the sheared zones the rock mass is composed mainly of scales having a lenticular shape and often slickensided and polished surfaces; sheared calcite veins are present; in the less disturbed zones fractured and/or contorted fragments of layers are sometimes found.

The orientation of the sheared belts is transversal with respect to the tunnel axis.

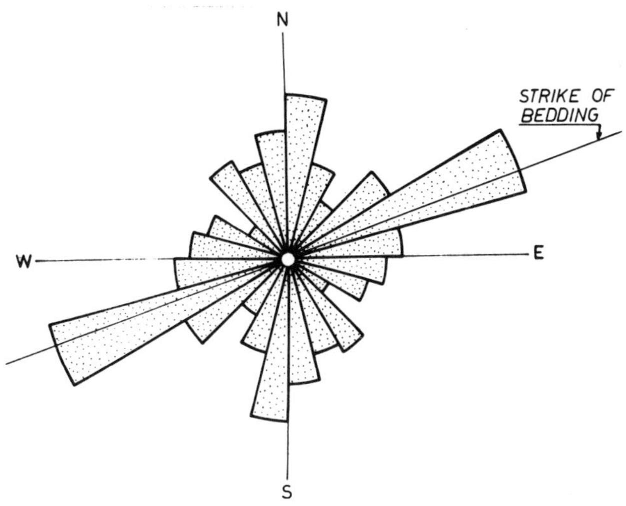

Fig. 5 - Joint directions observed in the tunnel walls

Fracturing is less systematic and always intense; however, the rock mass is often well packed and interlocked.

As to the physical and mechanical characteristics of the lithotypes, the mudstones are composed of calcite (about 24%), quartz (38%) and clay minerals. Diffractometric analyses show that among the clay minerals illite is predominant (65%), the remaining being chlorite (20%) and kaolinite (15%). Illite and chlorite include a significant percentage of smectite interlayers.

Strength characteristics were evaluated by means of the NCB cone indenter, which gives an index which is particularly suitable to indicate rock cuttability by machine.[1-2] For most of the samples the standard cone indenter number I_s is comprised between 0.80 and 1.2, a value corresponding to an estimated uniaxial compressive strength of about 25 MPa. The rock is markedly susceptible to slaking when subjected to humidity variations.

Sandstone characteristics are somewhat variable from one layer to another. The rock is usually fine-grained or medium grained; the grains are mostly calcite, quartz (or chert) and subordinately feldspars; the fabric is closely packed. In some of the layers the calcite grains are prevailing and $CaCO_3$ content rises to 65-75%.

The marked strength variations of this rock

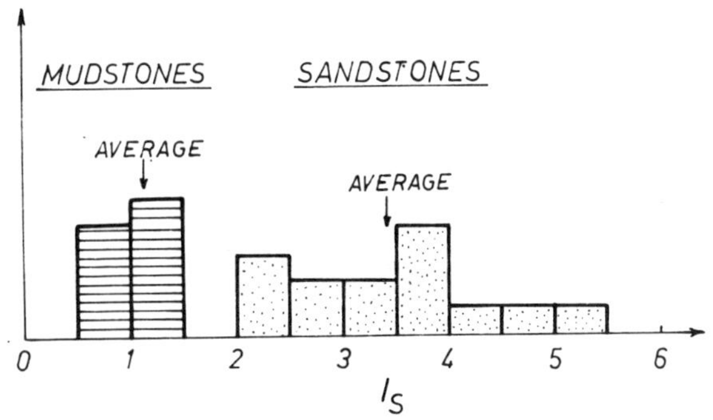

Fig. 6 - Histogram of the NCB cone indentor number I_s in the mudstone and in the sandstone layers

are evidenced in the histogram of fig. 6. I_s varies within the range 2.2-5.2, the highest values corresponding to "very strong" rock types in the NCB classification.

Uniaxial compression tests in some of the stronger layers gave values (for samples having a height/diameter ratio equal to 2) ranging between 140 and 240 MPa. It is to be noted that such high strength values are rather unusual in the typical Italian arenaceous-pelitic formations, where the typical strengths of the sandstone lithotypes falls in the range of 90-120 MPa.[3] It was observed, however, that the toughest layers were those containing the highest $CaCO_3$ content and which could be really classified as strongly diagenized calcarenites.

Finally, it should be observed that, at the time of the tender, only the logs of two exploratory boreholes near the Noghere entrance were available; later another borehole near the intersection with the road tunnel was drilled by the contractor. Therefore, the amount of mudstones in the formation was initially considered higher than actually found to be later during the excavation. Also the presence of the sheared zones within the rock mass was not indicated by the drillings nor discovered in the surface surveys; in fact the sheared material is less resistant to weathering and its outcrops are therefore frequently masked by detrital and eluvial material.

EXCAVATION

The tender required that the excavation be carried out without blasting. On the basis of the available data the contractor decided to excavate at first the top heading up to a height of 3.7 m (section being about 30 m^2) and then the 2 m high bench and the invert.

For the heading an Alpine AM50 boom type roadheader,[4] having a 100 kW power at the boom and equipped with heavy duty round picks was utilized; the hard sandstone layers were to be broken by a hydraulic jackhammer BRH JOIL,mounted on an RH-6 excavator. The excavation of the bench was to be carried out by means of the jackhammer and a ripper. A three-shift schedule was necessary to complete the work in the required time.

The operation of the roadheader proved more difficult than expected, because of the hardness of some sandstone layers. Intense vibrations of the boom and failures in the picks frequently occurred; therefore, most of the rock breaking was carried out by means of the hammer and the roadheader was mainly utilized for the final profiling. The rate of excavation was only 3 m^3/h for the roadheader and 3.3 m^3/h for the hammer.

It was feared that even worse conditions would be found at greater distances from the portal.

Moreover, as soon as nightshifts started, violent complaints came from the residents of the area. The vibration levels deriving from the hydraulic hammer were very low: for instance a p.p.v. (peak particle velocity) of 0.2 mm/s was measured at the surface at a distance of 32 m from the face. However, while the effects could be scarcely noticed during the day, they were high enough to be a serious nuisance in some of the dwellings during the night and therefore the complaints were really justified.

Finnaly a Council Ordinance, based on old noise regulations, forbade any noise producing activities from 8 p.m. to 7 a.m. and from 1 p.m. to 3 p.m.

Owing to these limitations, the application of expansive cements[5] in boreholes was tried; the aim was to obtain a preliminary fracturing of the toughest sandstone layers, thus improving the rate of advance of the hydraulic hammer in the successive day shifts.

The tests were carried out in a layer of sandstone, 0.4 m thick, at the base of the tunnel.The expansive cement, commercially known as Bristar, was loaded in vertical boreholes having a diameter of 40 mm, with a spacing of 0.35-0.40 m. Some effects were obtained by using the Bristar type 200 (the Bristar 150 had proven to be unsuccessful). Fracturing and dislocation occurred along pre-existing weakness planes (not along the alignment of the holes) and this markedly eased the action of the hydraulic hammer. However, the expansive effect required at least 36 h and therefore the Bristar could not be used during the night shift only.

Subsequently, Cardox cartridges in Ø 70 mm boreholes were tried out; the cartridges contain liquid CO_2 which, when electrically heated, act as a weak explosive. They were used, in some cases, to dislodge soft materials such as coal;however, they proved unsuccessful in the hard sandstone layers of the Monte d'Oro tunnel.

Therefore, the contractor, in full agreement with the client, tried to obtain permission to complete the excavation of the tunnel by means of drilling and blasting. The tunnel had, at that time, reached chainage 215 and was entering a rock mass mainly consisting of hard sandstone layers.

It must be pointed out that in Italy the use of explosives in urban sites is virtually discouraged, in that the preliminary technical investigations required are lengthy and restrictions on use are particularly burdensome.

The problems to be solved at the Monte d'Oro site were:

- to prove that the excavation could be safely carried out without danger to the structural safety of the buildings at the surface;
- to reduce anyway to a minimum the complaints from the residents, because otherwise it was feared that the permission of blasting might be withdrawn at any moment by the government.

To obtain the permission, a series of carefully monitored blasting tests had to be performed. A first series consisted of individual charges varying from 0.3 kg to 1.2 kg which were detonated in boreholes drilled in the excavation face; subsequently, various rounds were testes with a maximum charge per delay of 1.2 kg, trying to obtain an advance of at least 1.5 m. The effects of the blasts were recorded at the ground surface by means of 3-component velocity transducer ("geophones"); the observations were carried out by experts of the "Osservatorio Geofisico di Trieste", which acted also as adviser for the government Agency which was to grant the permission.

There is no official Italian Code regarding the maximum vibration levels allowed and therefore the decision had to be taken on the basis of engineering judgement, by having recourse also to the experience and regulations in force in other Countries.

An examination of esisting regulations shows that they are not always comparable, owing to the different local conditions, to the subjective opinions of the compilers, and to the different emphasis on environment protection.

There is now a general agreement that peak particle velocity, v, is the parameter of the blast induced vibrations which is most closely related to any possible damage to structures founded on the ground surface. Most regulations consider various classes of buildings or structures, for which different limit velocity values are allowed; there is in general no difficulty to apply this classification on the site.

Some of the best known European regulations are shown in the following Tables.

Table 1 - DDR Standards - KDT Richtlinie[6]

Building class	Frequency bandwidth (Hz)	ppv, vertical component (mm/s)
I Building of historic interest	2 - 30 30 - 100	2 2 - 14
II Trellis Buildings	2 - 30 30 - 100	5 5 - 36
III Buildings with masonry or concrete walls	2 - 30 30 - 100	10 10 - 71
IV Steel or reinforced concrete buildings	2 - 30 30 - 100	30 30 - 215

Table 2 - BRD Standards - DIN 4150[6]

Building class	Frequency bandwidth (Hz)	ppv, resultant vector (mm/s)
I Industrial or commercial buildings	< 10 10 - 50 50 - 100	20 20 - 40 40 - 50
II Residential buildings	< 10 10 - 50 50 - 100	5 5 - 15 15 - 20
III Sensitive structures	< 10 10 - 50 50 - 100	3 3 - 8 8 - 10

Table 3 - Swiss Standards for buildings[7]

Building class	Frequency bandwidth (Hz)	ppv (mm/s)
I Steel or reinforced concrete buildings	10 - 60 60 - 90	30 30 - 40
II Buildings with foundation walls and floors in concrete, walls in masonry or concrete	10 - 60 60 - 90	18 18 - 25
III Buildings with masonry walls and wooden ceilings	10 - 60 60 - 90	12 12 - 18
IV Buildings of historic interest or sensitive structures	10 - 60 60 - 90	8 8 - 12

It should be noticed that the meaning of the velocity values indicated in the tables is not homogeneous. For instance the values given by the German Standard DIN 4150 are by far more restrictive that those of the East German KDT Richtlinie.

However, the values of the Standard DIN 4150 should be considered as norm values ("Anhaltswerte"), and not as threshold limits for damage. A strict application of such values would virtually prevent any blasting work in urban sites, but it was suggested,[8] on the basis of vast experimental evidence, that design criteria 50% higher can be safely utilized without inducing damage.

The KDT Richtlinie values are threshold values which must not be normally exceeded; however, the authorization for exceeding them can be obtained when no other excavation method is technically feasible, provided that any damage produced is reimbursed.

Many recent criteria now take into account the influence of the predominant frequency of blasting waves; higher particle velocities are allowed for higher dominant frequencies. This is based both on (scarse) experimental data and on theoretical evaluations of the intensity of induced stresses as a function of the frequency

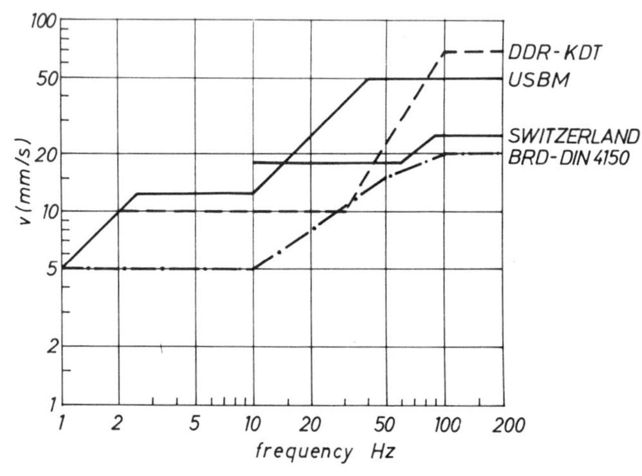

Fig. 7 - Limit values for peak particle velocity as a function of the predominant frequency of waves. The curves are valid for ordinary residential buildings (class 3 in the KDT Richtlinie and class 2 for DIN 4150)

of the forcing waves and of the natural frequency of buildings.

This factor is particularly important in urban sites, where higher frequencies are usually observed because of the short distances from the source, in comparison with typical conditions of quarry blasting. Fig. 7 shows how various regulations take into account the influence of frequency in the case of typical residential buildings. The values suggested by recent U.S. Bureau of Mines evaluations[9] are also included in the figure.

In the Monte d'Oro hill the wave frequencies were typically around 130-150 Hz. This high value is probably related also to the good quality of the rock mass, in which the propagation velocity of the longitudinal waves is 4.2 - 4.6 km/s.

During the technical discussion for the evaluation of the blasting trials, it was agreed that the KDT Richtlinie indications should be adopted as guide-lines. The buildings rising on the hill clearly belonged to group 3 of the KDT Richtlinie. On the basis of the results of the preliminary tests, the drilling and blasting excavation was finally authorized, provided that a maximum charge per delay of 1.2 kg be used and a maximum charge per round of 28 kg.

As from the discussion in the next paragraph, this would ensure a maximum value for peak particle velocity equal to 35 mm/s. However, in order to reduce the complaints of the residents, the maximum charge was maintained usually below the allowed limit, except where the overburden was thickest; as a consequence, the maximum velocity measured in the subsequent monitoring of the works was only 20 mm/s.

As it was observed that the maximum vibration level derived from the more strongly confined holes of the cut, very often, when the rock conditions allowed it, a discharge cavity for the surrounding charged holes was excavated at the centre of the face by means of the roadheader. Often, completely satisfactory rounds could not be achieved; the final profiling and some advance were sometimes performed by means of the roadheader. On average, 3 rounds per day could be blasted, with an average rate od advance of 5 m per day. Consumption of explosives was about 0.7 kg/m^3.

The effect of the rounds was always controlled near the epicentre by means of a portable seismometer; at some critical sites the vibrations were registered with 3D geophones.

The excavation of the bench, 7 m wide and 2 m high, was carried out by means of vertical holes. The spacing of the hole in each row was 1.3 m and the burden was 1.0 m. In general, the vibration intensity was markedly lower than in the excavation of the heading.

Where the overburden was lower, that is from chainage 215 to 50, the maximum charge per delay was reduced to 0.5-0.6 kg. In the last 50 m long stretch near the Noghere entrance, where the overburden varied from 20 to 2 m only, the excavation of the bench was divided into two slices, each being 1 m high, with a maximum charge per delay of 0.26 kg.

Vibration intensity was systematically measured at the surface, during the progress from chainage 215 to the Noghere portal.

ANALYSIS OF THE BLAST VIBRATIONS MEASUREMENTS

With the three-dimensional geophones the peak particle velocity components were determined in the vertical direction (v_V), in the longitudinal direction, that is along the horizontal line connecting the shot to the geophone location (v_L), and in the transversal direction (v_T).

The empirical relationship which is usually utilized for the analysis of the influence of the distance D (m) and of the charge weight W (kg) on particle velocity v (mm/s) is

$$v = K\,(D/W^c)^{-b} \qquad (1)$$

where D/W^c is known as the scaled distance.

When a large amount of experimental data with

different charges is available,it is often possible to evaluate both b and c by means of a standard logarithmic regression. However, in most cases the scaling as a function of the charge is done by assigning an a priori value to the coefficient c (usually 1/2 or 1/3).

For quarry or surface blasts, extensive experimental evidence indicates that the square root scaling (c = 1/2) usually gives the best results. When one is concerned with the effects of underground blasting on the ground surface, the situation is less clear, possibly also because different wave types are effective at different distance from the epicentre. Therefore, we should be careful when applying to urban tunnel

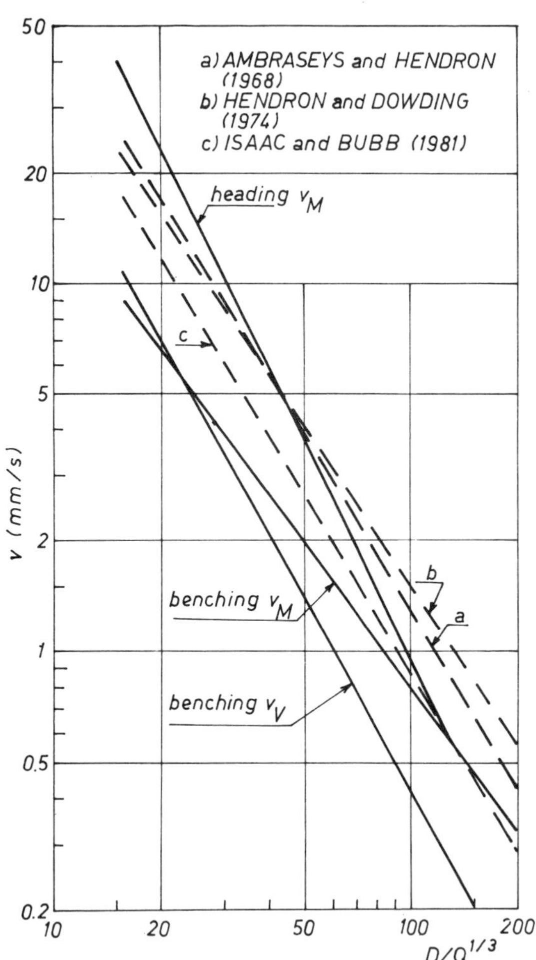

Fig. 8 - Relationships between particle velocity and scaled distance. Full lines indicate the values obtained at the Monte d'Oro tunnel,dashed lines those suggested by various Authors

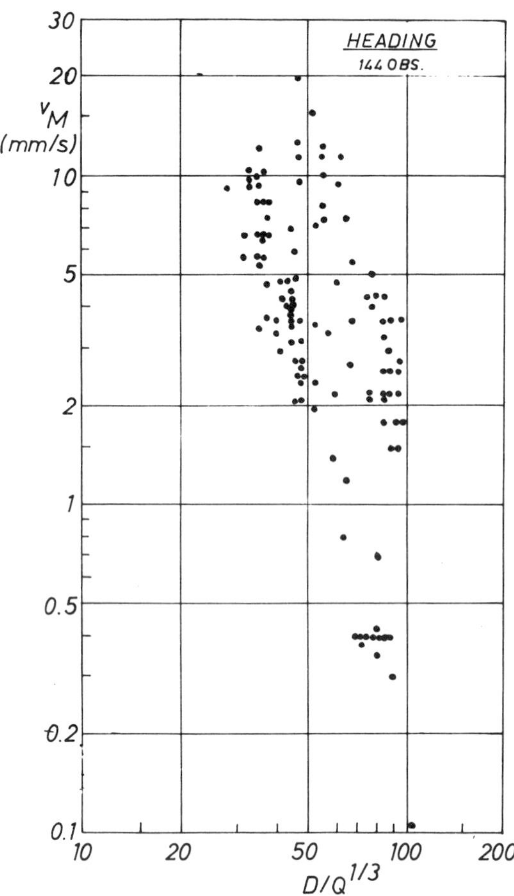

Fig. 9 - Peak particle velocity v_M as a function of the scaled distance determined in the trial blasts and in the rounds for the excavation of the heading

conditions the velocity-scaled distance relationships suggested by various authors, mainly on the basis of surface blasts observations.

Theoretical considerations[10-11] and also some experimental data[12-13] suggest that for underground blasting the cube roof scaling could be preferred.

The value of coefficient b in rel. (1) is usually comprised between 1.4 and 2.0. Fig. 8 shows some of the relationships between particle velocity and scaled distance obtained experimentally be various Authors (dashed lines).

Even though in many practical cases only one velocity component (usually the vertical one) is measured, the danger to structures could be related to the resultant vector of particle velocity v_R. Its direct determination would be difficult in field testing, because of the phase difference of various components; instead we can utilize either the maximum value v_M of the peak values of the three components or the "pseudo vector" v_{RE}, obtained assuming that peak values occur at the same phase for the three components. Measurements carried out on surface blasts[13] have shown that on average v_M underestimates v_R by 8% and v_{RE} overestimates v_R by about 24%.

For the case of the Monte d'Oro tunnel, where the buildings extend in practice all over the overlying ground surface, the critical site will be always near the epicentre and the critical component in this zone is probably the vertical one.

However, in general, the utilization of the vertical component only in the analysis of the data could induce errors, because the peak vertical velocity would show a steeper decrease (that is a higher b value) with the scaled distance, than the vectorial resultant.

The experimental data, collected during the trial tests in the Monte d'Oro hill, could not allow to reach a statistically significant decision as to the choice of the best charge scaling coefficient, and therefore a cube root scaling was assumed. Fig. 9 shows the peak velocity values v_M observed by means of 3-component geophones, both in the trial tests and in the subsequent monitoring during the excavation of the heading.

A linear regression analysis carried out on the logarithms of the particle velocity and of the scaled distance, gives the following best fit attenuation law:

$$v_M = 9100\ (D/Q^{1/3})^{-1.99} \qquad (2)$$

Comparisons with other suggested attenuation

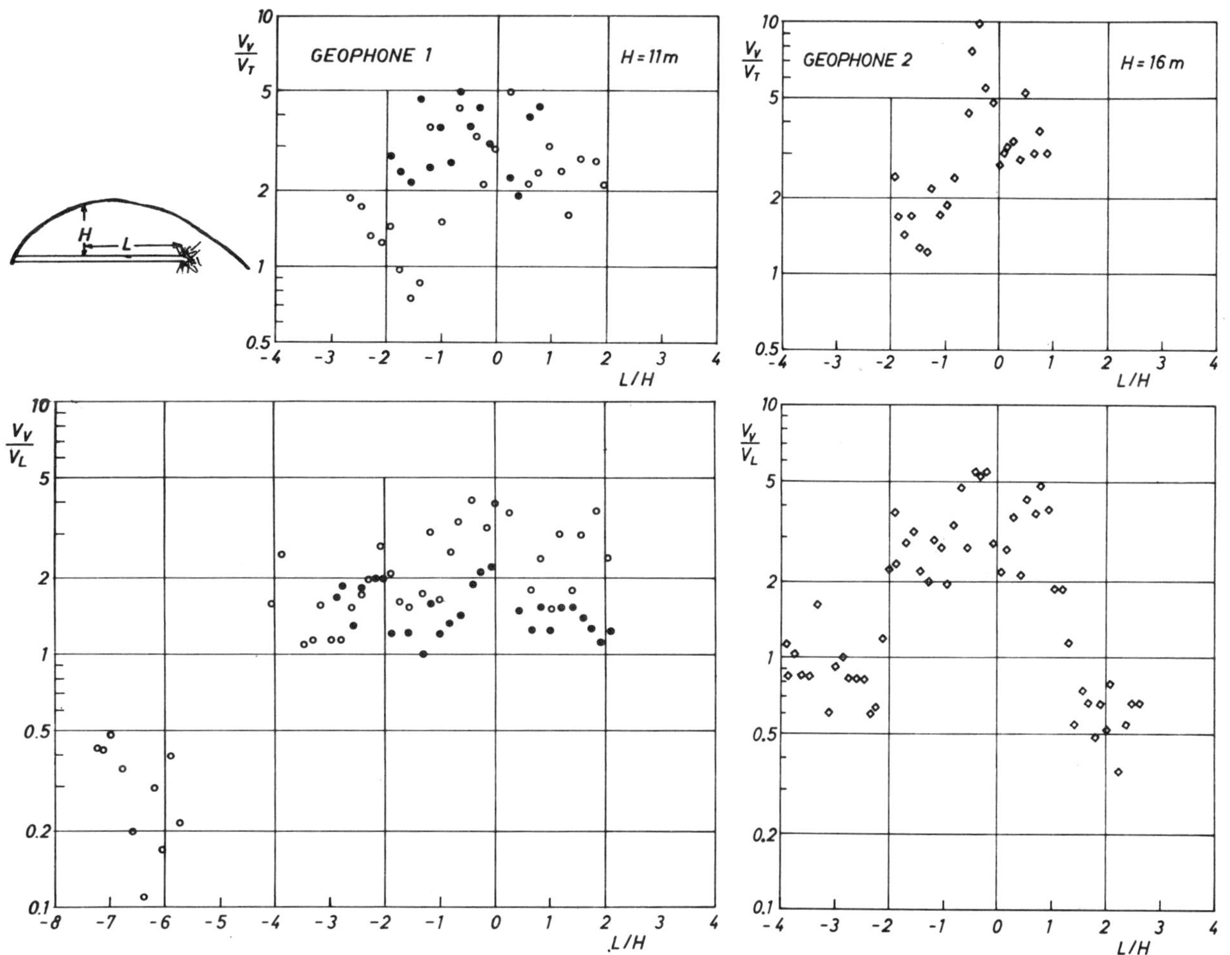

Fig. 10 - Relative importance of the various components, v_V, v_L, v_T, of the particle velocity at the ground surface as a function of the relative position of the blasting site and the observation point

relationships (Fig. 8) show that slightly higher velocity values were induced at the Monte d'Oro tunnel for the relevant scaled distance range.

The scatter of the measured values is high, as it shown by the values of the correlation coefficient r, of the standard deviation of the slope s(b) and of the standard deviation of the estimate s $(\hat{y})$ (where $y = \log v_M$)

$r = 0.66$ $s(b) = 0.19$ $s(\hat{y}) = 0.36$

For 142 degrees of freedom the value of the Student's distribution at 5% level is equal to 1.64 and $t.s\ (\hat{y}) = 0.62$. This means that the value of v_M which has a 5% probability to be exceeded by any individual blast is 4.1 times greater than the average value estimated by rel. (2). At

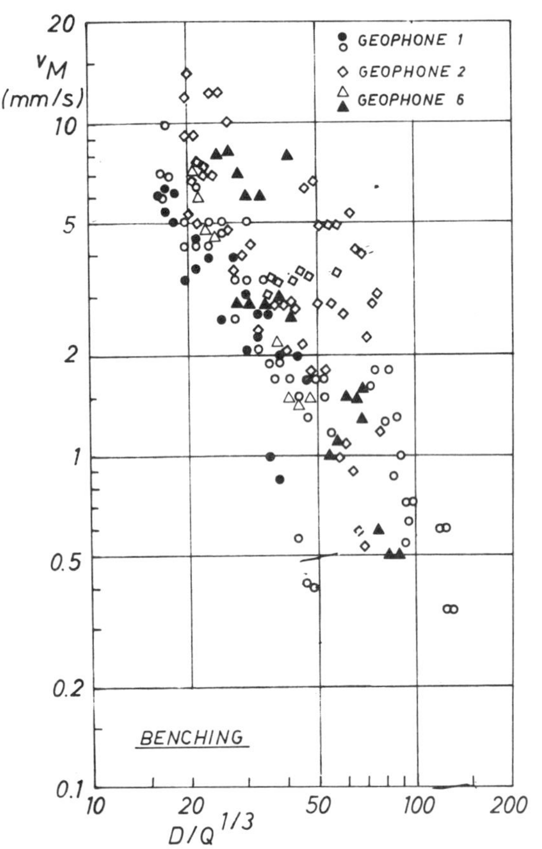

Fig. 11 - Particle velocity (highest component v_M) as a function of the scaled distance during the blasts for the excavation of the bench. Open symbol refer to the benching on the whole height or to the top slice of the bench; full black symbols to the blasting of the lower slice of the bench

a scaled distance equal to 33 $m/kg^{1/3}$, this upper limit value is therefore equal to 35 mm/s.

The corresponding diagram and the best fit relationship for the vertical component v_V are very similar to that found for v_M, because most of the measurements were taken near the epicentral area, that is at horizontal distances less than twice the overburden; therefore, the vertical component was usually the greatest one.

The observations made during the benching down in the last stretch of the tunnel near the Noghere entrance (from chainage 215) are summarized in Fig. 10 ÷ 12.

The ratios between the components of the particle velocity measured at two sites are shown in Fig. 10 as a function of the horizontal distance L between the blast and the geophone and of the overburden depth H.

The general trend of the data, although somewhat different for the two sites, shows that the v_V/v_L ration is maximum just above the blasting site and that the vertical component is the largest one up to horizontal distances equal to 4 times the overburden depth. It seems that the v_V/v_L ratio is not really symmetric with respect to the vertical axis, but that is falls at a greater rate when the blasting site has passed beyond the measurement site. A similar trend was observed also by Sakurai and Kitamura[14].

Peak particle velocity as a function of scaled distance is shown in Fig. 11. It is apparent that vibration intensity is much lower than that deriving from the excavation of the heading; this is undoubtedly due to the less confined conditions of the charge. The influence of the local site conditions is also apparent: notice the markedly higher values obtained for site 2.

The best fit coefficients of the relationships between particle velocity v_M and scaled distance for the three measuments sites - both separately and globally considered - are shown in the following Table 4.

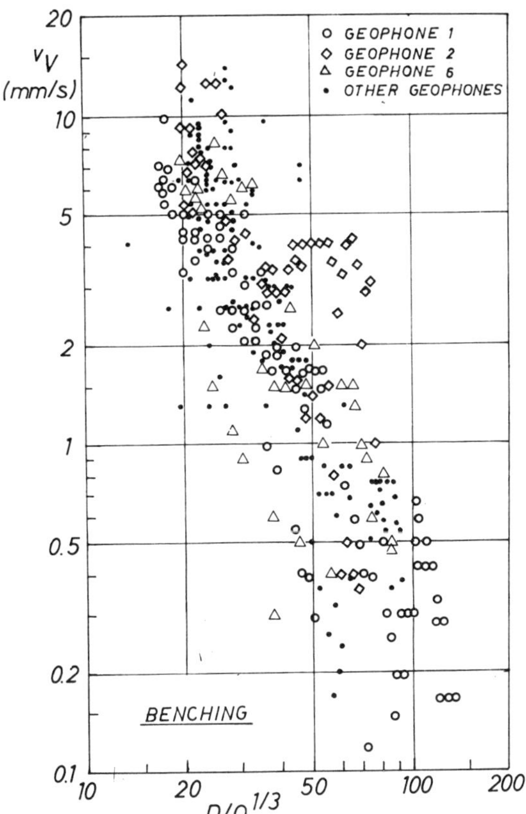

Fig. 12 - Vertical component of particle velocity as a function of the scaled distance during the excavation of the bench

Table 4 - Attenuation laws for the maximum component v_M

Geophones	K	b ± s (b)	r	s($\hat{y}$)
1	200	1.24 ± 0.08	.88	0.17
2	310	1.20 ± 0.16	.72	0.23
6	4100	1.97 ± 0.16	.91	0.18
1+2+6	340	1.32 ± 0.09	.78	0.24

Table 5 - Attenuation laws for the vertical component v_V

Geophones	K	b ± s (b)	r	s($\hat{y}$)
1	930	1.73 ± 0.07	.93	0.20
2	640	1.43 ± 0.19	.71	0.27
6	390	1.46 ± 0.23	.73	0.29
1+2+6	970	1.67 ± 0.09	.82	0.28
All sites	1250	1.74 ± 0.06	.82	0.27

The global best fit regression line is also indicated in Fig. 8.

The variation of the vertical component, v_V, with the distance both for the three-directional geophones, 1, 2, 6 and for all the geophones, is shown in Fig. 12; the correspondent best fit parameters are given in Table 5.

In this case, as was expected, the variation with the scaled distance is steeper for the vertical component than for the estimated value, v_M, of the resultant vector.

SUPPORT AND PREREINFORCEMENT

In the initial stretch of the tunnel, up to chainage 200, the tender required a provisional support by means of steel arches; the same technique was to be adopted for a length of 40 m at the crossing with the road tunnel. In the remaining part a preliminary support was to be effected by means of resin grouted bolts, wire mesh and shotcrete. The final lining (Fig. 2) was to be concreted after the stabilization of the convergence.

The characteristics of the rock mass relevant for the support are shown in the following table.

Table 6 - RMR Classification of the rock mass[15]

Partial ratings	Ordered sequences	Sheared zones
1. Uniaxial compressive strength	6	2
2. RQD	18	10
3. Joint spacing	12	8
4. Joint conditions	25	10
5. Groundwater	10	10
- Discontinuity orientation	-5	-5
	66	35

On the whole the behaviour of the rock mass was quite satisfactory and no stability problem arose, even in the sheared zones. Systematic bolts Ø 26 mm spaced 1.2 m and 3 m long were adopted. Complete stabilization of the convergence occurred as soon as the face influence was no longer felt.

Special problem arose for the intersection with the preexisting road tunnel (fig. 13). This

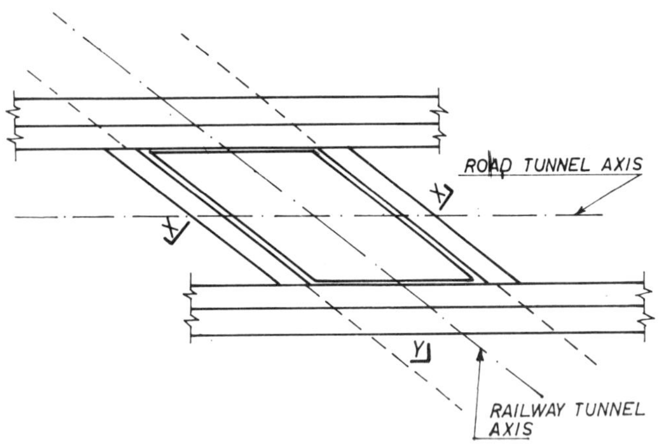

Fig. 13 - Plan of the intersection between the road and the railway tunnel

tunnel has an internal width of 10.9 m; the lining is 0.6 m thick at the crown and 1.65 m at the base; there is no invert.

Before the underpassing, a reinforcement system was built working from the road tunnel; it consisted of small diameter (100 mm) bored piles reinforced with steel pipes ("micropiles"). Four sets of micropiles (A, B, C, D) were built, as shown in Fig. 14; their characteristics are indicated in Table 7.

Table 7 - Characteristics of the micropiles

Set	Length (m)	Pipe diameter (mm)	Pipe thickness (mm)
A	6	60.3	5.9
B	15	70.0	8.8
C	12	60.3	5.9
D	8	70.0	8.8

The reinforcement was carried out in the following sequence.

I Two trenches were opened from the road tunnel floor along the internal side of its lining, down to the position of the crown of the railway tunnel. In these trenches a beam and a diaphragm wall were built (Fig. 15), the beams being supported by the micropiles of set A.

II Horizontal micropiles (set B), parallel to the crown surface of the railway tunnel,were installed on the two sides of the intersection. During the subsequent excavation of the railway tunnel these micropiles would support the load applied by the foundation of the road

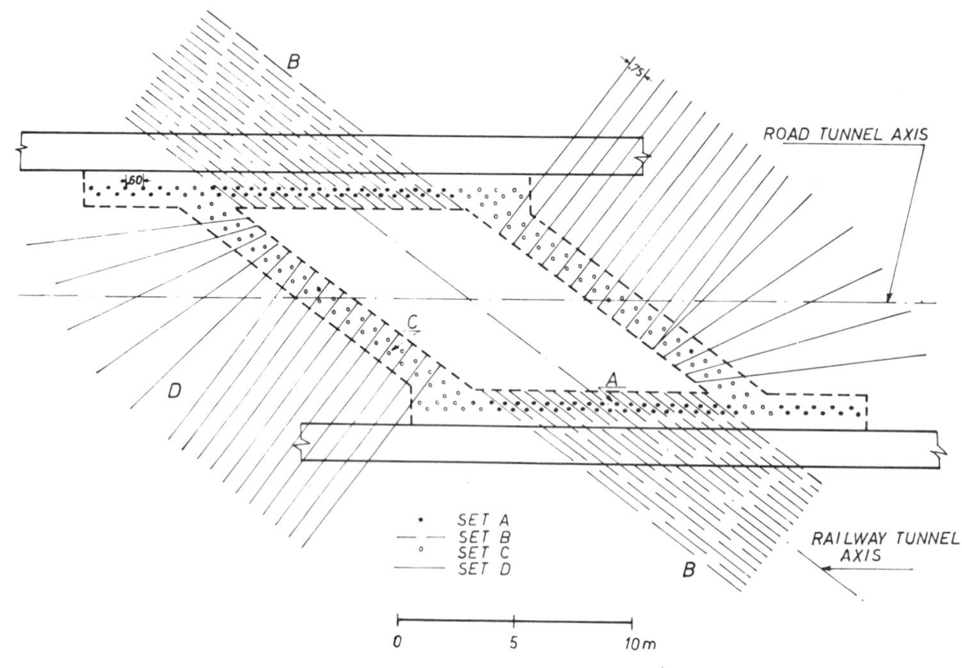

Fig. 14 - Scheme of the prereinforcement by means of micropiles at the intersection

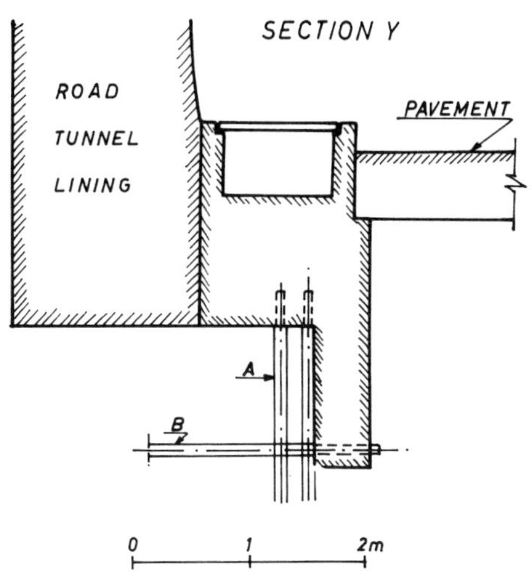

Fig. 15 - Beam and diaphragm along the road tunnel lining (section Y in Fig. 13)

tunnel lining. The design was based on the assumptions that the open space between the face and the provisional support of the rail way tunnel would be 0.6 m and that the load applied by the road tunnel lining foundation would derive from the self weight plus a 3 m height of rock (that is 25% of the tunnel width).

III Two reinforced concrete beams parallel to the

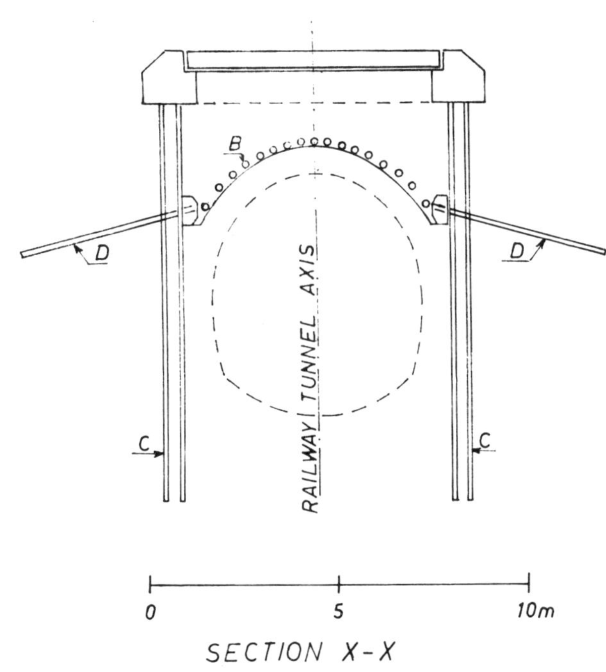

Fig. 16 - Section transversal to the railway tunnel at the intersection (XX in Fig. 13)

railway axis were built. They were founded on the vertical micropiles of set C. The beams support a skew reinforced concrete slab which forms the floor of the road tunnel at the intersection (Fig. 16).

IV Trenches along these beams were excavated and horizontal micropiles (set D) were installed; they are designed to support the horizontal thrust of the ground in the transient phase of the excavation of the railway tunnel, before the construction of its concrete lining (Fig. 16).

V The railway tunnel was finally excavated under the protection of the micropiles of set B on the crown wnd of sets C and D at the sides. The piles of set A were cut during the crossing.

No difficulties arose in the prereinforcement and in the excavation at the intersection and no damage was caused to the road tunnel lining. In the writers' opinion the adopted prereinforcement system would have been fully required if the intersection has occurred in one of the sheared zones, but it is probably overconservative within the ordered sequences of the flysch formation.

CONCLUSION

The technical problems deriving from blasting operations in an urban site could be satisfactorily solved in the case of the Monte d'Oro tunnel. The adoption of a drilling and blasting scheme for the excavation led to great savings, both in the time required for the completion of the tunnel and in the global cost.

Beside the technical problem, public relation problems are of the utmost importance in these situations. During the excavation of the tunnel good relations were maintained with the residents, who were informed of the blasting times and were shown the intensity of the measured vibrations.

It was deemed advisable to make a complete

survey of the buildings to ascertain the presence of pre-existing cracks or fissures; in fact, very often in perfectly good faith, the preoccupation for the effects of the blasts leads the residents to discover in their dwellings old cracks, which had hitherto been totally unnoticed.

This policy yielded good results; no legal sues were lodged during the excavation, and only a few were presented after the completion of the tunnel.

REFERENCES

1. Gaye F. Efficient excavation. Tunnels & Tunnelling, March 1972, p. 135-143.
2. Mc Feat-Smith I. Rock property testing for the assessment of tunnelling machine performance. Tunnels & Tunnelling, March 1977, p. 29-33.
3. Associazione Geotecnica Italiana. Some Italian experiences on the mechanical characteristics of structurally complex formations. 4th Congress International Society Rock Mechanics, Montreux, Volume I, 1979, p. 827-846.
4. Gehring K.H. Solution to problems with road-header installation in tunnelling projects. Rapid Excavation & Tunnelling Conference, Volume 2,1981, p. 1045-1061.
5. Dowding C.H., Labuz J.F. Fracturing of rock with expansive cement. Journal Geotechnical Division, Proc. ASCE, Volume 108, GT10, 1982, p.1288-1299.
6. Wustenhagen K. Die im Rahmen des Immissionsschutzes in der DDR gültige Richtlinie der Kammer der Technik. Wirkung von Sprengungen auf Gebaude. Nobel Hefte, Volume 44, 1, 1978, p. 29-35.
7. Wiss J.F. Construction vibrations: state of the art. Journal Geotechnical Engineering Division Proc. ASCE, Volume 107, GT2, 1981,p.167-181.
8. Arnold K. Neue Regelungen des Sprengerschütterungsimmissionsschutzes. Nobel Hefte, Volume 50, 2, 1984, p. 106-109.
9. Siskind D.E., Stagg M.S., Kopp J.W., Dowding C.H. Structure response and damage produced by ground vibration from surface mine blasting. U.S. Bureau of Mines, Report of Investigations 8507, 1980.
10. Ambraseys N.N., Hendron A.J. Dynamic behaviour of rock masses. In: Rock Mechanics in Engineering Practice edited by K.G. Stagg and O.C. Zienkiewicz, 1968, p. 203-236.
11. Hendron A.J., Dowding C.H. Ground and structural response due to blasting. 3rd Congress International Society Rock Mechanics, Denver, Volume IIB, 1974, p. 1359-1364.
12. Isaac I.D., Bubb C. A study of blast vibrations. Tunnels & Tunnelling, July 1981, p. 35-41 and September 1981, p. 61-64.
13. Chapot P. Etude des vibrations provoquées par les explosifs dans les massifs rocheux. Laboratoire Ponts et Chaussées, Research Report n. 105, 1981.
14. Sakurai S., Kitamura Y. Vibration of tunnel due to adjacent blasting operation. International Symposium Field Measurement in Rock Mechanics, Zurich, Volume I, 1977, p. 61-74.
15. Bieniawskii Z.T. The geomechanical classification in rock engineering applications. 4th Congress International Society Rock Mechanics, Montreux, Volume 2, 1979, p. 41-48.

ACKNOWLEDGEMENTS

Thanks are due to ing. R. Folchi and to ing. P. Tommasi for their cooperation.

One of the authors (R. Ribacchi) was supported by a CNR contract (05/85).

Seismic effects near blasting operations

F. Čermák
Building Research Institute,
Technical University of Prague,
Prague, Czechoslovakia

The Building Institute of the Technical University, Prague, is not the only organization which has, in one form or another, to solve the problems of seismic effects due to technical blasting in quarries or in the course of other, e.g. construction, operations. The role of fundamental importance is played, naturally, by the provisions of the Czechoslovak Standard ČSN 73 0036, from which it follows that either a dynamic assessment should be carried out or an experimental blast used. From an experimental blast of minor extent (minor intensity) it is possible to estimate, by analogy, the effect of the planned, major blast. On the basis of the coefficient β we can estimate the amplitude even without an experimental blast. The coefficient β includes the effect of environment between the blasting site and the place under consideration: $A = \beta \sqrt{m_n}\, L^{-1}$ (coefficient, weight of the charge, distance). The standard covers also the possibility of incorporating the effect of milisecond priming, should the charge be distributed into a major number of boreholes, in which case the reduction of the weight of the charge by means of coefficient ρ is carried out. When delayed priming is used, the time of delay is important. If the interval is shorter than 0,25 s, double the charge is considered. If delayed-action detonators are used (such as DeM detonators), an unfortunate orientation of production tolerances might result in simultaneous blast of two subsequent charges. The relation between destruction and vibration intensity (vibration velocity) is determined by a table, the damages being divided into four categories, in accordance with the vibration speed. On the basis of assumed damage we can also estimate the safe distance

$$L = K \cdot (m_n)^{2/3}$$

(K - coefficient characterizing the damage). Some of the present day blasts, which often have an adverse effect on their environment, have a major power, so that the interval recommended by the standard (5 kg - 1000 kg) is not sufficient. Also the upper limit of distance from the charge, to which the formula

$$L_{max} = 20\ (m_n)^{1/2}$$

can be considered as valid, need not be sufficient. Further limitations follow from the characteristics of geological medium (the layer of surface formations, the ground water table, ect.) Faulty zones, marked geological boundaries, as well as every major ground depression may cause complications, which result in inaccurate assessment of effects.

However, the standard ČSN 73 0036 was heavily commented upon. For this reason the Building Institute of the Technical University, Prague, The Building Geology Prague, and the Research Institute of Civil Engineering, Bratislava, oriented some of their research work to the revision of the afore mentioned standard ČSN 73 0036. The seismic effects of blasting operations were elaborated in some detail particularly by Dr.A. Dvořák, who specified the extent of damage in greater detail as follows:

0 - no damage,

1 - first signs of damage, haircracks up 1 mm in width at the contact of unbound building elements and in the joints between floors and walls,

2 - minor damage, cracks up to 5 mm in width in plaster, partitions and chimney masonry, falling-off of mortar, loosening of roofing materials,

3 - serious damage, cracks over 5 mm in width in partitions and load-bearing walls, not endangering stability, falling off of rooling and parts of chimneys.

4 - dangerous damage, cracks in load--bearing walls and lintels, endangering their load-bearing function, collapse of partitions, in-filling mesonry and chimneys, impairing of the stability of the building or structure, cracks in plain concrete.

5 - destruction, collapse of brick masonry buildings or structures or other parts with principal load-bearing members, cracks even in reinforced concrete.

Also the types of buildings and structures were classified as follows:

A - chalets, houses not satisfying building codes, as a rule, ruins, etc.,

B - standard brickwork buildings, isolated or terrace houses of a floor area inferior to 200 sq.m, of 3 storeys at the most,

C - big building of bricks or blocks, well reinforced and braced, panel buildings buildings erected of precast R.C. components, designed to withstand wind load, masonry with cement mortar

D - Buildings of in-situ reinforced concrete, steel, wood and framed buildings with good bracing, plain concrete,

E - R.C. and steel structures, underground structures of plain and reinforced concrete, forming an enclosed unit or rings.

Foundation soils are classified into three classes in accordance with the magnitude of standard loads and depth of the ground water table below the foundation base. For these data, in dependence on the vibration velocities from 5 $mm.s^{-1}$ to 400 $mm.s^{-1}$ (10 zones) the frequency of damage to buildings is determined (up to 5%, from 5% to 50%, over 50% from the overall number).

Particular attention must be afforded to millisecond priming, which can produce even the primary modes of building vibrations (excitation by a series of waves following one another). On the basis of this draft it is possible to determinate safe distanc of buildings (in accordance with the afore mentioned classes.

The draft of the revision of the standard gives a formula for the calculation of the safe distance as

$$L = k \cdot N^{\varkappa}$$

where k depends on the type of building and the category of damage and $\varkappa$ is defined for the distance range. In the case of milisecond priming 2N are considerd, once again. The applicability of the formula is determined by the intervals of charge weight defined as follows: up to 1 kg; 1 kg to 1000 kg; (10^3 to 5.10^3) kg ; (5.10^3 to 2.10^4) kg.

The relatively objective criterion is represented by the performed work, which covers the situation in both low nad high frequences. It follows from frequent measurements carried out in the Building Geology, national corporation, that it is sufficient to consider the effect of frequency on the vibration velocity. For the frequencies of $f < 10$ Hz the velocity zones will be shifted with regard to damage zones by one zone lower (the extent of damage pertains to lower velocities) for $f > 50$ Hz, on the other hand, the velocity zones will be shifted with regard to the zones and categories of damage by one zone higher. So much for the generalization of measurement results carried aut by Dr. A. Dvořák in the framework of the research problem coordinated by the Building Institute of Technical University,Prague, and further partical problems solved by the Building Geology. National Corporation, Prague.

In 1977 the Czechoslovak Standard ČSN 73 0032 - "The Calculation of Building Structures Loaded by Dynamic Effects of Machines" was issued. In this standard the structures are classified in accordance with their resistance to vibrations. Class A - the building with visible damage, cracks in plaster, structures of unprocessed stone, hollow blocks with binding material of poor load-bearing stability; Class - B - standard brick masonry structures, structures of prefabricated large size blocks, components,framed walls, worked stone, buildings without damage in good technical conditions; Class C - well reinforced buildings.

For these classes the table of ultimate values under momentary shock load is given in the standard as follows:

Table 1. Ultimate Velocities and Accelerations (ČSN 73 0032)

Class	f ultimate quantily dimension	1-10 Hz acceleration $mm.s^{-2}$	10-100 Hz velocity $mm.s^{-1}$
A		125	2
B		250	4
C		630	10

The ultimate values of velocity and acceleration are structures and their members are due to temporary impact load which does not last longer than 1 hour per one working shift.

There is an obvious difference between the two afore tioned standards.The last mentioned standard is stricter and is more generally valid, since it does not concern only the excitation by seismic waves produced by a technical blast.

In foreign literature these problems are dealt with by in the works by Morris, Crandell, and others. They are of an earlier date and their results have been generally integratted with the afore mentioned materials. Hiwever, interesting references can be found also in Soviet literature, particularly in the Handbook (Spravochnik) by Afonin,which gives not only formulas, but also tables of safe distances.

In the blasting operations in guarries, according to Mironov, Shchuplatsev and Pyatnnin, the seismic effects are divided into three zones, viz. the near zone (5-50 m), the intermediate zone (50-100 m), and the far zone. The vibration velocities in these subsequent zones are determined by the formulas

$$v = K_1 \cdot \sqrt{Q L^{-1}} \, e^{-\alpha x} x^{-1},$$
$$v = K_2 \cdot \sqrt{Q L^{-1}} \, L \, x^{-1},$$
$$v = K_3 \sqrt{Q x^{-3/2}}$$

where v - overall velocity of vibrations of the mass, K_1, K_2, K_3 - coefficients characterizing geological conditions, Q - weight of the charge, L - length of disintegrated block, α - damping coefficient (= 0,03), x - distance of observation point. In the case of milisecond priming a function, dependent on the number of charges, is introduced, viz

$$f(n) = 0.33 n^{-\frac{1}{3}} + 1.58 n^{-\frac{1}{2}} - 0.91 n^{-1}$$

or approximately,

$$f(n) \approx n^{-\frac{1}{3}}$$

The function f(n) is actually a reduction coefficient, with which the right hand side of the afore mentioned formulas is multiplied. The authors further recommend the observance of the following rules:

a) avoid charges of insufficient power,

b) possibility of use of oversize charges and changes of vibration frequency,

c) division of the charge,

d) reduction of desnity of the charge

e) use of milisecond priming.

For the sake of clarity the afore mentioned expressions are given in tables.

Sensitive approach is required by special geological conditions or major works. In these cases considerable experience is reguired and sufficiently accurate theory in every particular case.

The actual process of rock disintegration and the origin of impact waves can be solved in a epherically symmetrical system. The solution concerns the state immediately following the blast, when the process does not have any major volume. The equation of motion in the spherically symmetrical system may have the form of

$$\rho \dot{u}_R = \frac{\partial \sigma_s}{\delta R} + \frac{4}{3}\frac{\partial \sigma_t}{\delta R} + 4\frac{\sigma_t}{R}.$$

The solution is carried out in time steps. The medium is divided into spherical layers (R - radius, ρ - density). The equation defines the relation between the stress and the velocity of motion. A computer programme is available in the Building Institute, which works in accordance with the following algorithm. To determine the velocity the changes of geometric quantities, volume etc. are calculated in dependence on the time step. On the basis of calculated deformations and the results of the triaxial test the stress and velocity are determined. The triaxial test affords the data on the behaviour of both undisturbed material andyon the material with disturbed structure. We express mathematically the relation

$$\sigma_{s_{j+\frac{1}{2}}}^{n+1} = F\left[(mu)_{j+\frac{1}{2}}^{n+1}, e_{j+\frac{1}{2}}^{n+1}\right],$$

where $(mu)_{j+\frac{1}{2}}^{n+1}$ is the relative change of volume in time (n+1) and the layer (j + $\frac{1}{2}$). The other quantity is energy, its indexes being based on the same principle as explained before. The transition between the undisturbed and disturbed structure is determined by the achievement of a certain energy level. The effect of quasistatic pressure, in dependence on the magnitude of the explosion void, is considered as an adiabatic change. For the pressure in excess of the critical value of (p_k) is $\varkappa = 3$, for the pressures below this limit is $\varkappa = 4/3$. In the solution the boundary effect (wave rebound, etc.) must be taken into account.

In this way input data are obtained for major analyses. In the Building Institute the NONSAP programme system has been modified for this purpose. The system carries out two variants of calculations, viz the TL variant and the UL variant, with which also the basic equations correspond:

$$({}_0^tK_L + {}_0^tK_{NL})\,u = {}^{t+\Delta t}R - {}_0^tF - M\,{}^{t+\Delta t}\ddot{u}\,(-C\,{}^{t+\Delta t}\dot{u})$$

and

$$({}_t^tK_L + {}_t^tK_{NL})\,u = {}^{t+\Delta t}R - {}_t^tF - M\,{}^{t+\Delta t}\ddot{u}\,(-C\,{}^{t+\Delta t}\dot{u}).$$

The symbols are those in current use in the respective literature. The indices denominate the time step, the reference state time and linear (non-linear) relations. The basic problem of all calculation systems are the material charakteristics. The computers have brought about particularly the development of approximate methods. Unfortunately, the field of materials obviously is not such an attractive region. Moreover, their investigations require more or less specialized apparatuses. Aparat from that, the properties of the individual types of rocks and soils are so different that it is impossible to use one model only, as a rule. In some tipes simple models, which originated in the far distant past, are sufficient, although they have been long surpassed for classical materials (such as steel) The construction of a universal model is not successful, as a rule. However, it is possible to find certain generalizations, if the exceptinality of certain types of materials is incorporated into constants. In (3) some relations are given used for this purpose. The relation between the static tensile and compressive strengths on the one hand and the respective dynamic quantities on the other hand is expressed by the constant k , which equals

$$k = \sigma_p(\dot{\sigma}_n)/\sigma_p(0.1)$$

(the reference value corresponds with the velocity of 0.1 $MPas^{-1}$). Then for compression

$$k = 0.002(lg\dot{\sigma}_n)^3 + 0.004(lg\dot{\sigma}_n)^2 + 0.06(lg\dot{\sigma}_n) + 1.06$$

and, for tension,

$$k = 0.11\, lg\,\dot{\sigma}_n + 1.12.$$

The interval of approximate validity is given as $(10^{-5} - 10^{-6})MPas^{-1}$. For analogous relations hold, viz for compression

$$k = 0.002(lg\dot{\varepsilon})^3 + 0.044(lg\dot{\varepsilon})^2 + 0.323\,lg\dot{\varepsilon} + 1.634$$

and the tension

$$k = 0.006(lg\dot{\varepsilon})^3 + 0.094(lg\dot{\varepsilon})^2 + 0.546\,lg\dot{\varepsilon} + 1.942$$

If we introduce analogously the coefficient k for the moduli E

$$k = E_i(\dot{\sigma}_n)/E_i(0.1), \qquad (i = t, tl),$$

then for compression,

$$k = 0.002(lg\dot{\sigma}_n)^3 + 0.004(lg\dot{\sigma}_n)^2 + 0.05\,lg\dot{\sigma}_n + 1.06$$

and for tension

$$k = -0.02(lg\dot{\sigma}_n)^2 + 0.15\,lg\dot{\sigma}_n + 1.17$$

For the prevailing majority of assessed blasts the Czechoslovak Standard ČSN 73 0036 should be sufficient. Its validity, however, has its boundaries. The increase of the weight of partial charges results in the necessity of assessment of the seismic effect even beyond the boundaries given in the standard. In this respect the valid standard should be revised so as to extend its validity. In the case of special blasting operations, special geological conditions or special structures the individual assessment will be unavoidable, whether it be made theoretivally or - in the majority of cases - by combined experimental and theoretical method. The first steps in this direction have already been taken and have proved the optimum cooperation of the Building Geology, Praguem, the Research Institute of Civil Engineering, Bratislava (Branch office in Brno), and the Building Institute of Technical University, Prague.

Bibliography

/1/ Čermák F.: Means of blasting technology in tunnel driving (in Czech) Report No. 254/74, The Building Institute, Prague, 1974

/2/ Čermák F.: Wave propagation in geological medium I, Report No.547/83

The Building Institute, Prague, 1983 (in Czech)

(3) Čermák F.: Wave propagation in geological medium II (in Czech), Report No. 538/83, The Building Institute, Prague, 1983

(4) Čermák F.: Explosion waves in geological medium (in Czech), Stavebnicky časopis No. 11/1981

(5) Dvořák A.: Final report on the propagation of seismic waves and their transmission into building structures (in Czech), Research report., The Building Geology, Prague, 1980

(6) Czechoslovak Standard ČSN 73 0036

Design of a boulder fence in Hong Kong

Y.C. Chan B.SC., M.SC., D.I.C., M.H.K.I.E., M.I.C.E., M.I.STRUCT.E.
C.F. Chan B.SC., M.SC., M.H.K.I.E., M.I.C.E., M.I.STRUCT.E.
S.W.C. Au B.SC., M.PH., M.H.K.I.E., M.I.C.E.
Geotechnical Control Office, Hong Kong

SYNOPSIS

In 1982, the Geotechnical Control Office of the Hong Kong Government as part of its Landslip Preventive Measures Programme started on a pilot scheme of boulder stabilization on a large boulder field of an area of 45,000 m². The stabilization basically comprises a boulder fence along the lower boundary to catch the small boulders, plus insitu treatment of larger boulders. The design of the fence was preceded by mapping of the boulders and evaluation of the probable boulder velocities. The fence adopted is of collapsible type, composes of mild steel posts of square hollow section on a concrete foundation, connected together by steel wire ropes. This paper gives a brief description of the field mapping, the prediction of boulder velocity, the fence design and the associated testings.

INTRODUCTION

Although Hong Kong is immensely urbanised, a large proportion of its territory still remains in its natural state. Among these areas, many are steep natural slopes strewn with boulders. The boulders are either remains of in-situ decomposition of the basal rock, remaining in place or being transported from their original positions, or fallen debris from natural degradation of cliff faces higher up. Although less frequent than other forms of landslips, boulder fall is not uncommon in Hong Kong. In the years 1981-1983, the number of failure incidents involving boulders dealt with by the Geotechnical Control Office exceeded 20 each year.

THE SITE

The boulder field treated under this scheme is the natural hillside above Conduit Road in the Mid-Levels Area of Hong Kong Island about 300 metres wide (in E-W direction) and extends from the rear of the buildings (contour 150 mPD) southwards up to the toe of Seymour Cliffs (contour 280 mPD) (See Fig. 1 and Plate 1). The area is of an average gradient of 35° and is covered by a layer of boulderly colluvium of 0-20 m thick. The colluvium is largely volcanic in origin and its boulder content is as high as 75% to 100%. (See Plate 2). The boulders are believed to have been dislodged from the Seymour Cliffs during the last Ice Age.

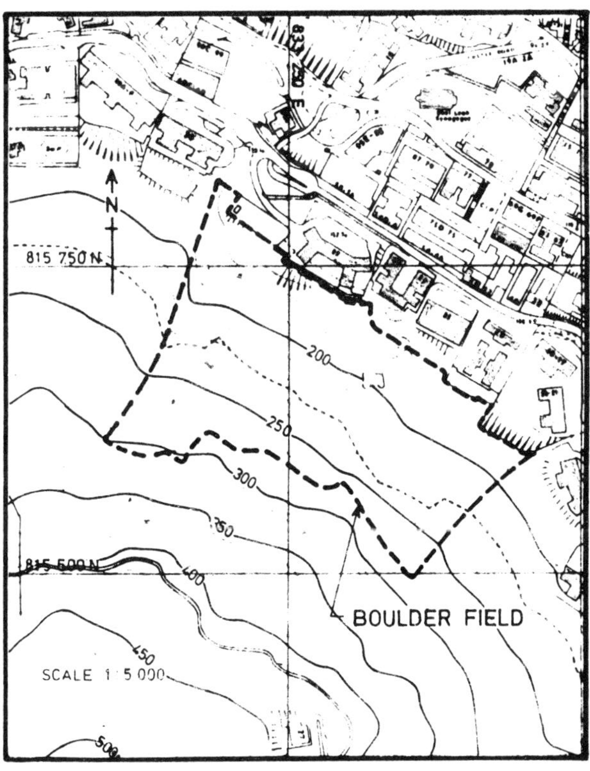

Fig. 1 Site Plan of the Boulder Field

Plate 1 Aerial View of the Boulder Field

Plate 2 A Close-up View of the Boulder Field

PRE-DESIGN BOULDER MAPPING

For the purpose of obtaining necessary information for design and cost estimation, a pre-design boulder mapping exercise was undertaken. A method of mapping in strips was employed in order to keep the job within a manageable scale and to limit the undergrowth clearance.

A 30 x 25 metres grid was set out on site and a metre wide undergrowth cleared along the grid lines. The ground profile as well as the characteristics of any boulders intercepted by the grid lines were then recorded. These boulder characteristics recorded included their location, shape, angularity, the three principal dimensions, support conditions etc. The information collected was used to estimate the size distribution of the boulders and to obtain an idea of their characteristics in general. The results indicated that in the boulder field there are about 4 500 boulders having a size greater than 2 tonnes and most of the boulders are either angular or sub-angular.

SCHEME ADOPTED

In view of the large number of boulders in the boulder field, to opt for merely insitu boulder stabilisation is unacceptable in terms of both construction cost and staff resources required in design and construction supervision. On the other hand, big boulders as heavy as 50 tonnes or more are not uncommon. To use barriers to prevent the down-slope developments from being hit by these large boulders in case of failure would be impractical. These consideration led to adoption of a combined treatment of works involving insitu stabilisation of big boulders and construction of a barrier at a lower level to arrest the smaller boulders.

The choice of barrier type is very much dictated by the difficult site access and the lack of reasonably flat areas for massive structures at the toe of the slope. The adopted fence is of collapsible type which by virtue of its deformability absorbs the energy of an impacting boulder. With such a design, members of the fence are of light weight. Also, heavy and complicated foundations are not required.

PROBLEMS IN DESIGN OF BOULDER BARRIERS

Two aspects of knowledge are required for the rational design of boulder barriers. The first relates to the energy or velocity of the boulders. The second is the boulder-barrier interaction.

Unfortunately, information relating to these two aspects in the existing literature is scarce and mostly relates to the first aspect. Also, the limited literature that is available is not sufficiently comprehensive and direct application of the described experience to our design is impossible. Ritchie[1] & Beggs et al[2] measured the trajectory of rock fragments falling down slopes but their velocities were not measured. For the purpose of designing rock shelters, Piteau[3] used a computer to simulate movement of rock fragments down a slope to establish the rock fragment distribution at the toe of slope. The programme however treated the fragments as point masses which is obviously not applicable to the case of rolling fragments. Benitez et al[4] used classical mechanics to evaluate the variation of kinetic energy of a boulder sliding down a complex slope. For the case of a rolling boulder, a spherical body was assumed. This gives the upper bound kinetic energy solution and may be very conservative in our case.

There is very limited mention in the existing literature on barrier design philosophy and procedure. Those barriers having been reported seem to be based on an empirical approach (e.g. Bhanderi & Sharma[5], Mercer R[6]).

In an attempt to overcome these problems, mathematical models were developed based on Newtonian mechanics to estimate boulder velocity and to predict boulder-barrier interaction.

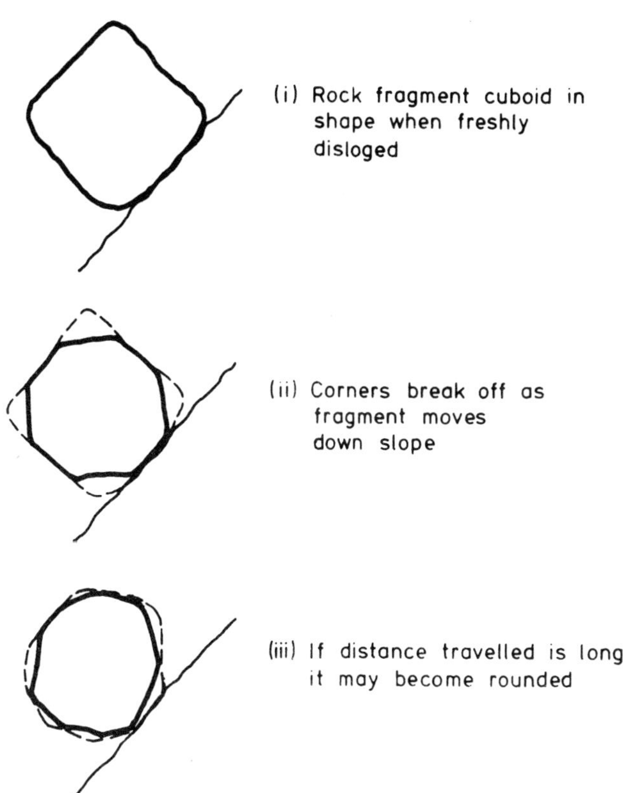

Fig. 2 Development of Shape of Boulder down a Slope

BOULDER VELOCITY

As mentioned earlier, the boulders in the area under study were derived from the Seymour Cliffs. The fragments when freshly dislodged are usually cubical in shape due to the orthogonal jointing pattern of the rock at the Cliff faces. On rolling down the slope, the corners and edges of the boulders break on impact, thus reducing the angularity of the boulders. (See Figure 2). The degree of reduction in angularity depends on the distance travelled. As the site is not far from the foot of the cliff, most of the boulders are angular to subangular.

On consideration of this process, it is reasonable to approximate the shape of the boulders to an octagonal prism in the mathematical model. This may still be a very crude starting point for evaluating probable velocities of boulders rolling down an actual slope but it nevertheless is a better simulation than the assumption of spherical boulders.

Regarding the rolling motion of a prism down a plane, it will not normally continue to rotate in its simplest form once it is set into rotation. By virtue of the impact between the plane and its sides during the rotation, other modes of motion such as sliding and bouncing will usually come into play. Their occurrences of course depend on the magnitude of the momentary velocity at the time of impact, the coefficient of restitution, the coefficient of friction, to name a few. To facilitate modelling, three types of motions were isolated for consideration. These were (i) simple rotation about its corner edge, (ii) rotation with sliding about its corner edge and (iii) bouncing. (See Figure 3). Solution to the mathematical model was obtained with the aid of computers.

Figure 4 shows a typical time-displacement relationship of an octagonal prism and its comparison with that for a cylinderical prism.

BOULDER-FENCE INTERACTION

There are in general two approaches to predict boulder-fence interaction, i.e., the force approach and the energy approach. In the force approach, the force on a fence by a boulder is found by Newton's 2nd law which says that force is the rate of change of momentum. Simple as it may sound, it is however difficult to apply because the time taken to decelerate the boulder

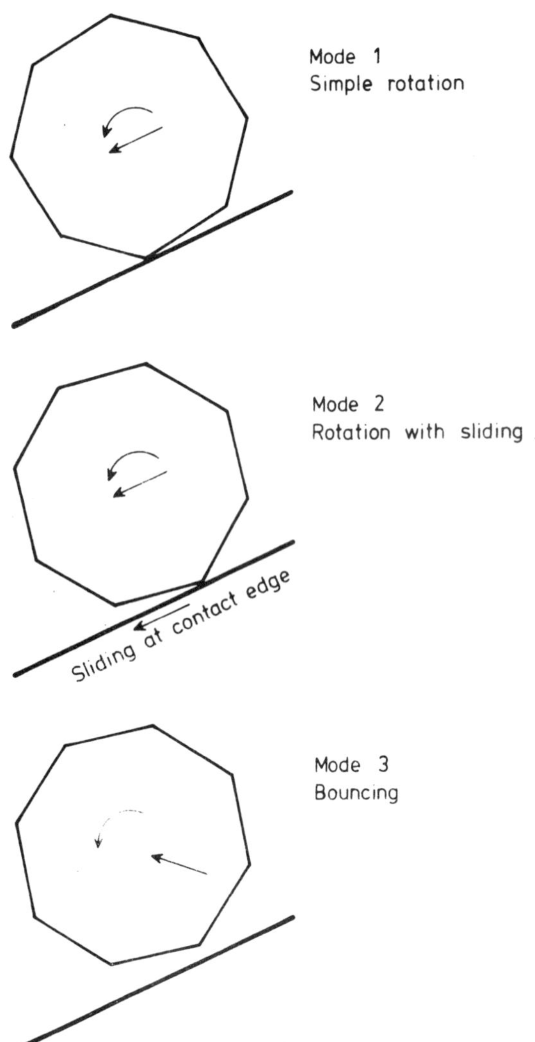

Fig. 3 Types of Boulder Motions

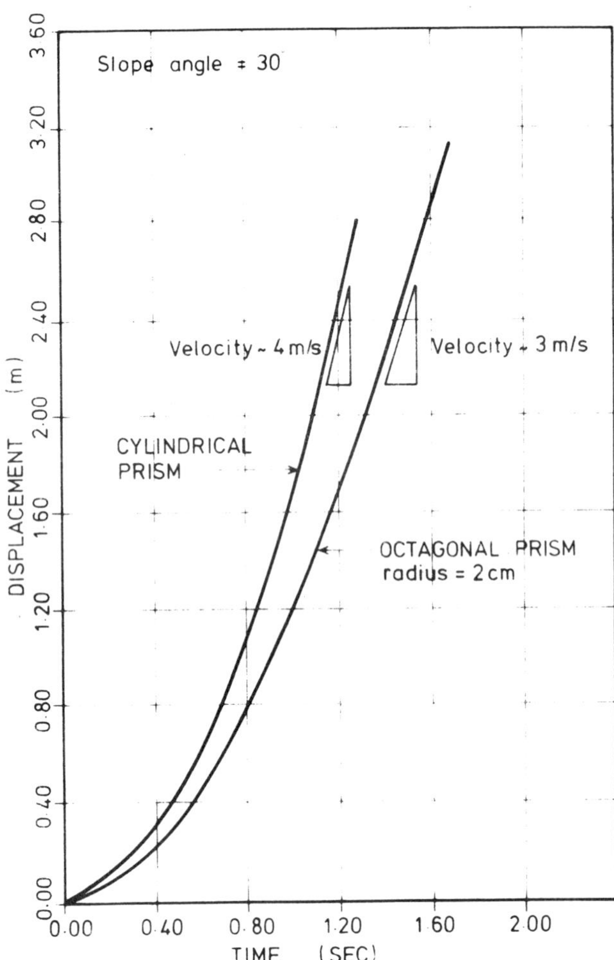

Fig. 4 Typical Time-Displacement Relationship for Octagonal and Cylindrical Prisms

depends on a lot of factors in a complicated manner.

In the energy method, work done to deform the fence can be directly related to the kinetic energy of the impacting boulder. The fence is regarded as satisfactory if the design boulder causes an acceptable degree of deformation. The designer will usually choose to use permanent deformation (plastic method) to absorb energy from the boulder, though use of temporary deformation (elastic method) is a possible alternative. A fence designed by the elastic method calls for heavier members and foundations but requires less repairs. For fences designed by the plastic method the reverse is true.

The fence used in this case was designed by the plastic method. To illustrate the design principle, it is perhaps worthwhile to consider a hypothetical fence consisting of an array of posts connected by a single row of wire rope at the top. (See Figure 5). Upon impact, the wire rope moves forward inducing forces on the post. This, in its extremity, leads to the formation of plastic hinges at the base of the posts thus allowing them to rotate and absorb energy (Figure 5 c). As the first pair of posts rotate, the adjacent pair of posts are pulled till plastic hinges form also at their bases (Figures 5 d & e). The process may repeat for the further pair of posts until the kinetic energy of the boulder is completely absorbed.

The actual fence was extended from this hypothetical case by the provision of multi-row wire ropes between the posts, and another similar but smaller, less substantial fence at the up-hill side to arrest smaller boulders. It is believed that this arrangement is most suitable for this site as the lightweight

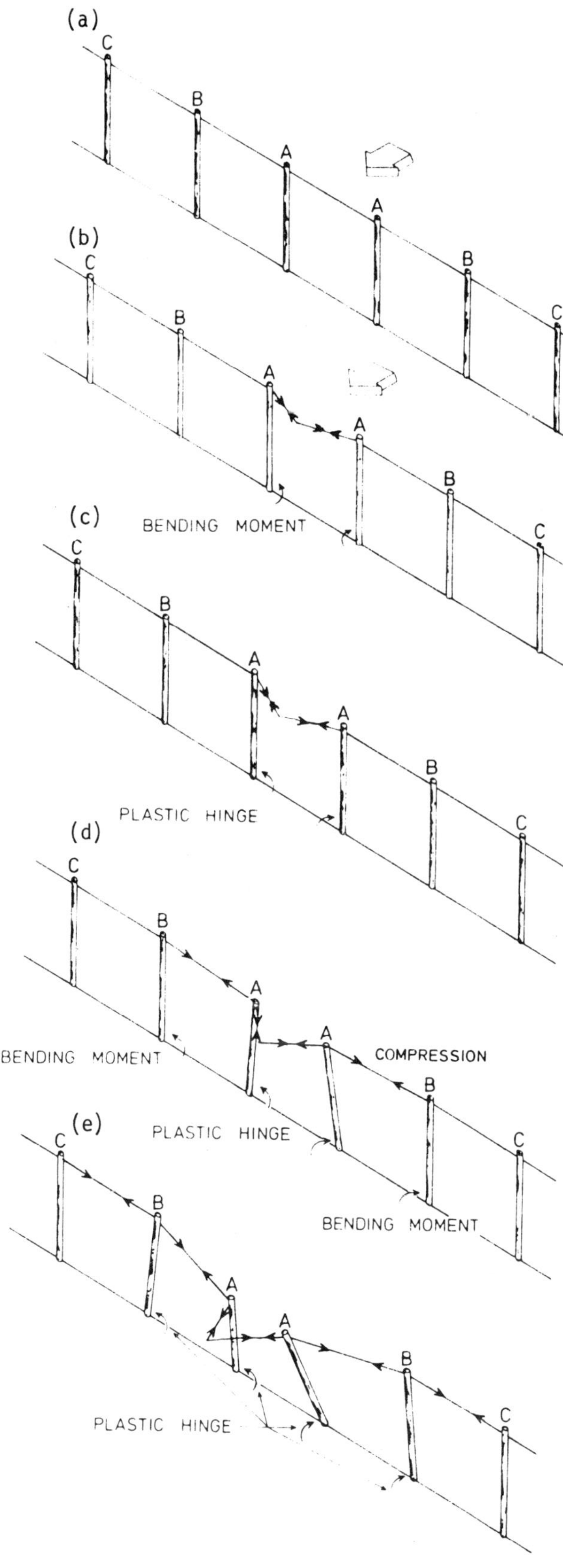

Fig. 5 Boulder Impact on a Hypothetical Fence

members permit transportation through the difficult access (foot paths in most case). Due to the limited loads transmitted, the foundation will be simple spread footing sitting either directly on the slope surface or supported by micropiles where the ground conditions are unfavourable.

A computer programme was again developed to evaluate the member deformation and force relationship. The most effective combinations of steel post and wire rope dimensions and configurations were examined. The final downhill fence comprises 3 m high SHS 150 x 150 x 10 steel posts at 3 m c/c connected together by Ø14 steel wire ropes at 200 c/c and 350 c/c at the bottom and top half of the fence respectively plus wire netting infill. The upper fence comprises 1.5 m high SHS 80 x 80 x 6.3 steel posts at 3 m c/c, linked by Ø9 steel wire ropes at 350 c/c plus wire netting infill.

Figure 6 is an isometric view of the fence. The fence can absorb 100 kN-m energy when a total of 6-7 posts are permanently deformed.

CONFIRMATORY EXPERIMENTS

Following the preliminary design, experiments and tests to check on the validity of the design were carried out. A total of three such tests/ experiments have been carried out of which two were executed in cooperation with the University of Hong Kong. They include the experiments on boulder velocity, fence/bounder interaction and loading tests on steel hollow sections. The details are described below.

EXPERIMENT ON BOULDER VELOCITY

The purpose of this experiment was to verify the mathematical derivations. It consisted of rolling polygonal prisms down a 5 metre long inclined plane. The motions of the prisms were monitored by a video camera with frame spacing of $\frac{1}{25}$ sec. The time-distance relationship was then measured from the television screen when the tapes were played back. Five variables have been considered in the experiment i.e., gradient of the plane, number of sides and diameter of prisms, coefficient of restitution and dynamic coefficient of friction. A total of four prisms

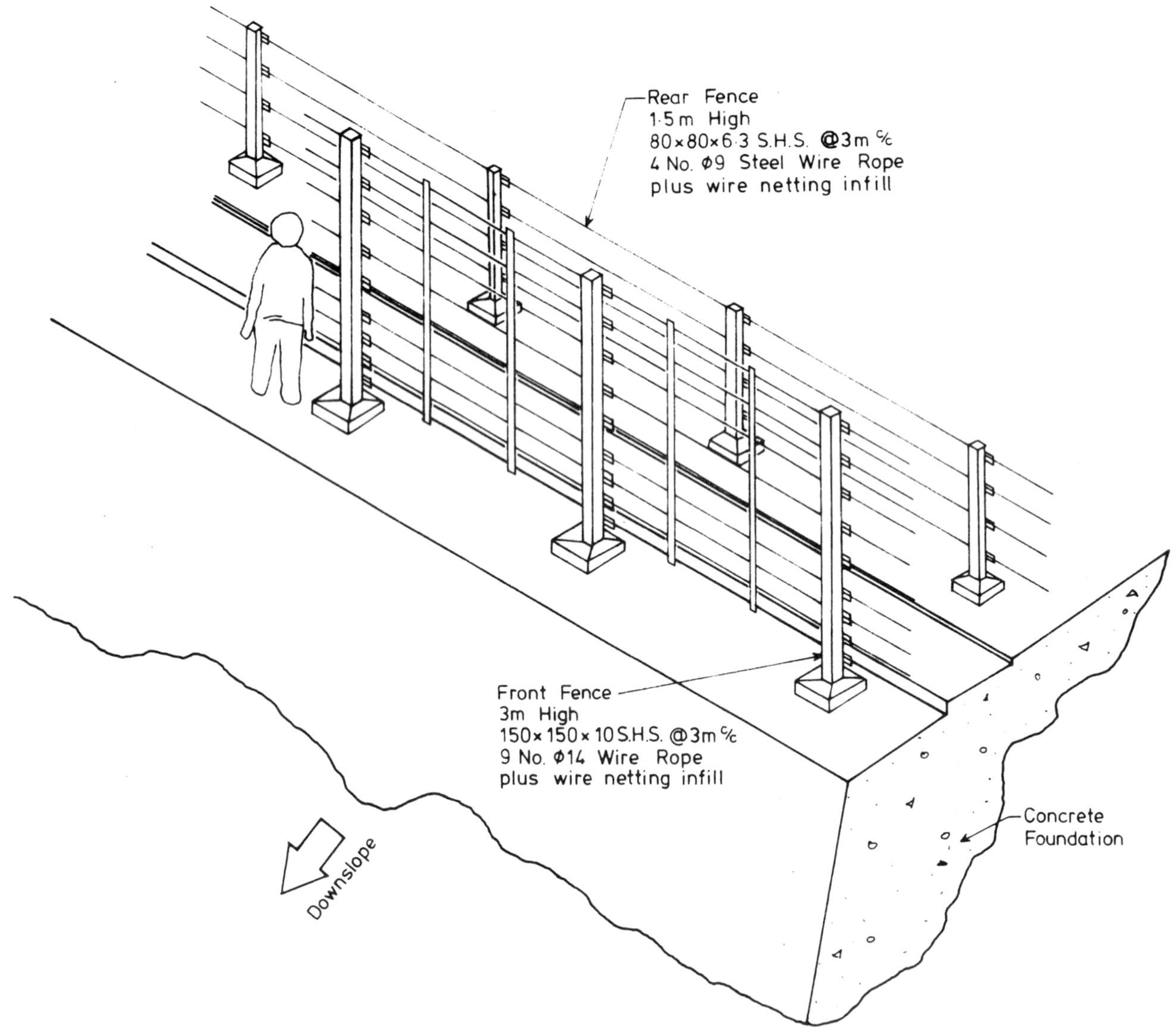

Fig. 6 Isometric View of the Boulder Fence

made of hard wood were used. Three of them were 90 mm in circum-diameter and were of regular hexagon, octagon and 16-sided polygon in cross-section. The other one was an octagonal prism with a 50 mm circum-diameter. The relatively small boulder size was used to ensure an adequate size/travel-distance ratio. Different coefficients of friction were introduced by covering the prism surfaces with papers of different texture. The timber plane was also covered with different materials to give different coefficient of restitution. The gradient of the inclined plane could conveniently be varied by adjusting the support height. Over 1 000 runs involving 29 combinations of variables were made. The results generally tie in well with the mathematical prediction. Figure 7 gives some typical results.

EXPERIMENT ON FENCE-BOULDER INTERACTION

The objective of this experiment was again to verify the mathematical model. It was carried out by Mr. W.H. Lo under the supervision of Dr. D. Ho of the Civil Engineering Department of the University of Hong Kong. Figure 8 is a schematic diagram showing the experimental set-up. Figure 9 & 10 show the close agreement between the results from the model testing and that predicted by mathematical simulation.

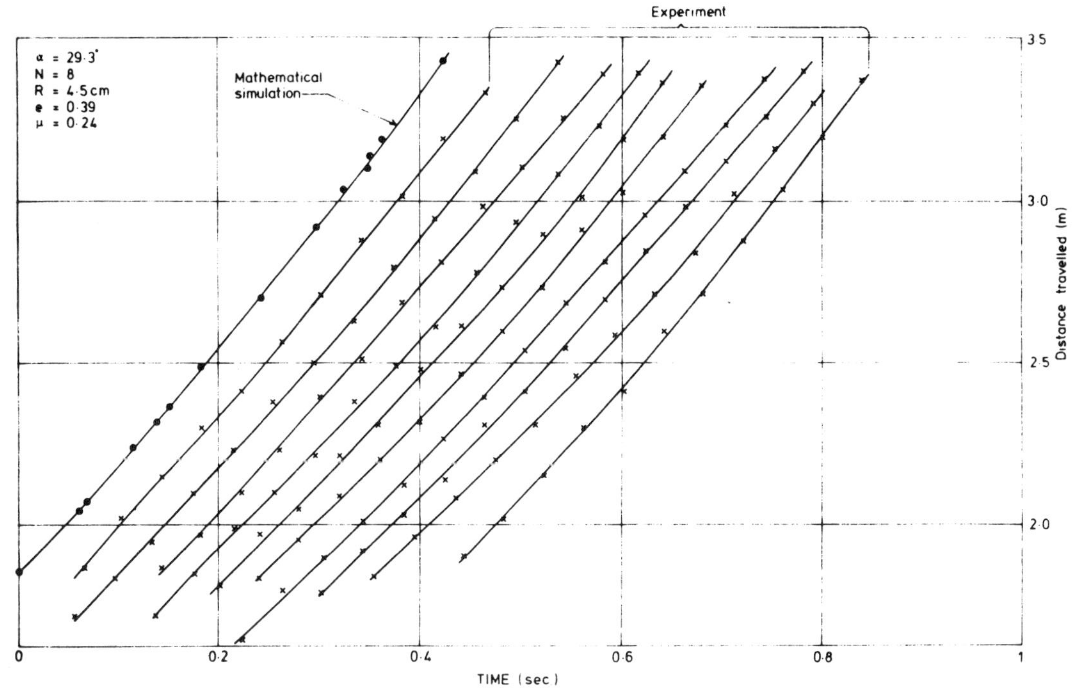

Fig. 7 Typical Experimental Results compared with that predicted by Mathematical Simulation

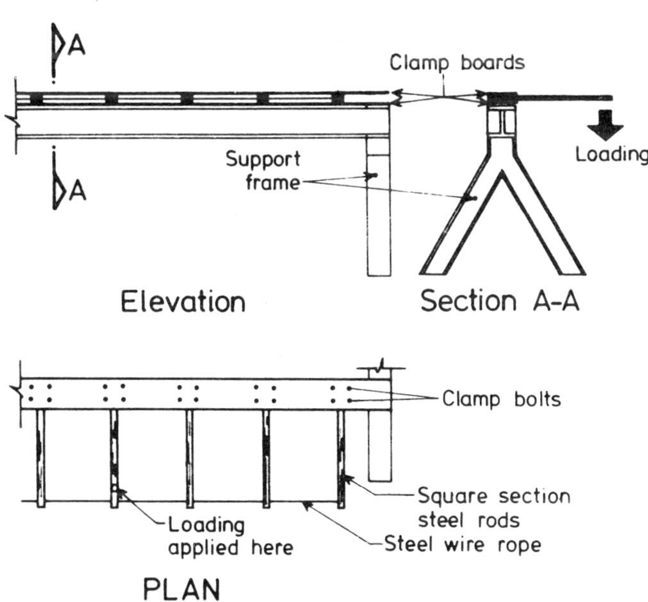

Fig. 8 The Set-up for Experiment on Fence-Boulder Interaction

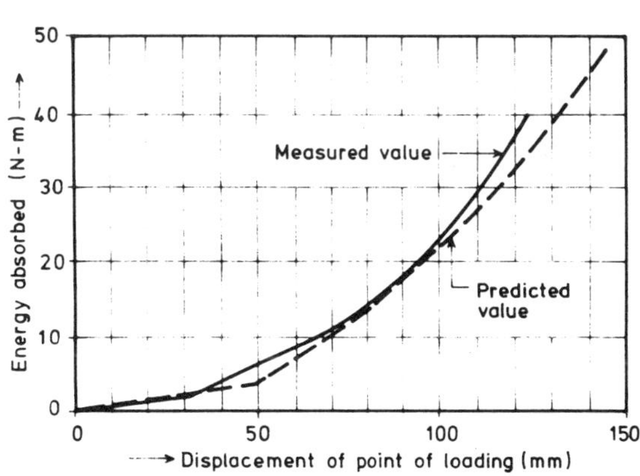

Fig. 9 Energy Absorption by Model Fence and Its Prediction by Mathematical Simulation

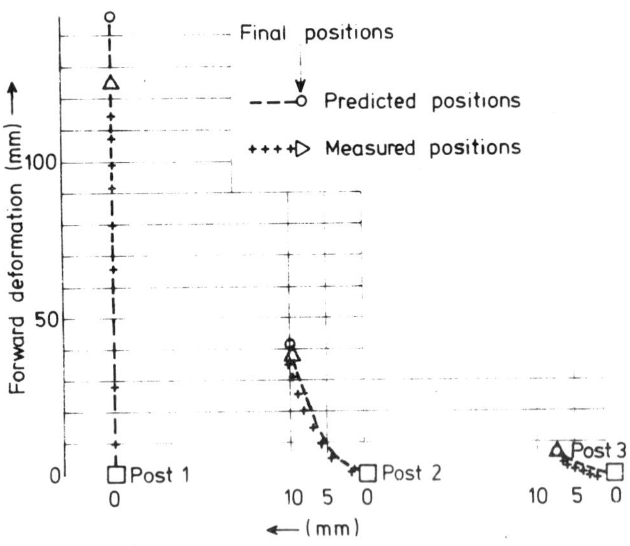

Fig. 10 Typical Experimental Results and the Prediction by Mathematical Simulation

Plate 3 Steel Hollow Section under Loading Test

TEST ON LOCAL BUCKLING

The experiment is intended to establish the local buckling behaviour of steel hollow sections at large deformation and the methods of strengthening the members.

The ability of the design hollow section to deform without buckling to the required 30° plus was doubtful. A literature review for information on this aspect proved to be of little use. The only practical solution was to actually load the design members.

Plate 3 shows the set-up of the loading apparatus. The testing indicated that local buckling occurred at less than 10° rotation of the plastic hinge which led to a substantial drop in the plastic moment. Plate 4 shows the local buckling at large deformations (approx. 50° rotation). Drop in strength due to local buckling was more pronounced in rectangular hollow sections (RHS) than in square sections (SHS). To prevent the occurrence of local buckling at an early stage of deformation different measures such as infilling the hollow sections with weak concrete and the provision of internal or external steel stiffeners were looked into. Members with such modifications were tested. The results showed that square hollow sections with concrete infill gave the highest resistance to local buckling (See Figure 11 for some of the test results).

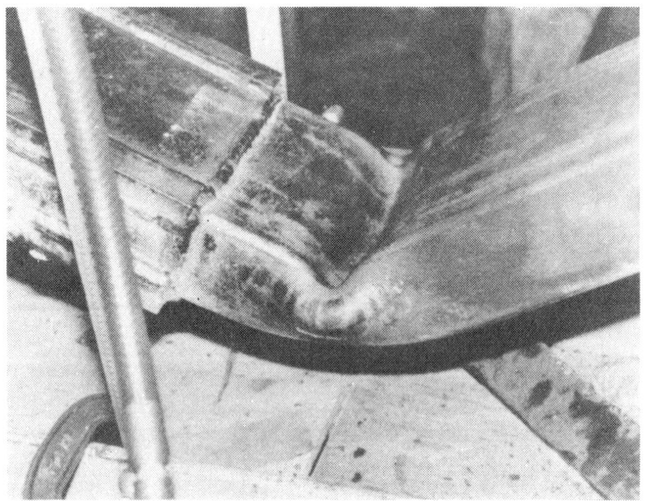

Plate 4 Local Buckling of a Rectangular Hollow Section

ENGINEERING JUDGEMENTS

The mathematical models and testing described in the preceding sections form the basis for the design of the boulder fence. In the same way as with other geotechnical engineering design, engineering judgement is indispensible in achieving a successful result. This is especially true in the design of the boulder fence.

The theoretical models for boulder velocities

and boulder/fence interaction prediction, no matter how refined they are, can only be regarded as a guide to aid the engineer in the formulation of his design. This is because of the inherent difficulties in simulating in the models, with high degree of accuracies, the various factors that will affect the motion of a boulder down a slope. These factors may include the actual shape, angularity and mass of boulders, their probable movement paths before reaching the fence, the slope angle, deformability and irregularity of the ground along the movement paths, the likelihood of the boulders being broken up in the course of motion, etc.

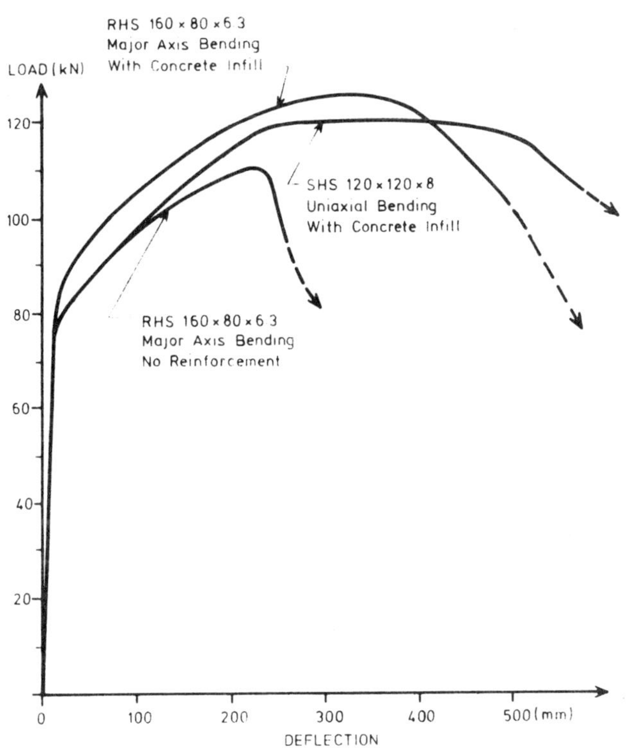

Fig. 11 Test Results on Hollow Sections

The fence is able to absorb an impact energy of 100 kN-m. Any boulder which will attain a velocity sufficient to give rise to an energy greater than 100 kN-m will be stabilised insitu. The velocity model only gives a crude idea of the relationship between the velocity and distance travelled. A close on-site examination has to be made to assess the factors which may affect the boulder motions as those described above before a decision can be made on which boulder can be retained by the fence and which should require insitu stabilization. At this particular point, the application of sound engineering judgement is most crucial.

CONCLUSION

This is one of the first attempts in Hong Kong to carry out boulder stabilization on such a scale and under such a restrictive site condition. Past experiences in both design and construction of this kind of work are scarce and considerable effort has been made to overcome this deficiency by developing the mathematical models and carrying out tests. Needless to say, considerable experience has been gained through the design already completed and the construction now underway. It is anticipated that through monitoring the performance of the fence at a later date, more valuable information can be obtained. We hope, with the construction of this prototype structure, apart from achieving its primary purpose of safe-guarding the conditions of the boulder field under treatment, we shall be able to advance another step forward towards the better understanding and appreciation of dynamic boulder behaviour and problems associated with the stabilization works.

ACKNOWLEDGEMENTS

This paper is published with the permission of the Director of Engineering Development, Hong Kong Government.

Acknowledgements are due to Mr. H.B. Phillipson of the Geotechnical Control Office for his valuable advice on the project, and to Dr. D. Ho of the University of Hong Kong for her contributions in the formulation of the mathematical models and the execution of the experiments.

REFERENCES

1. Ritchie A.M. (1963). The Evaluation of rockfall and its control. Highways Record, vol. 17, 1963, pp 13-28.

2. Beggs, C.J., Threadgold, L., Blomfield, D. & Chan, Y.C. (1985). Rockfall and its control. (Notes on a Geotechnical Group Meeting). Hong Kong Engineer, vol. 12, no. 4, pp 41-43.

(Discussion, vol. 12, no. 6, pp 12).

3. Piteau D.R. (1978). Slope stability analysis for rockfall problems : The computer rockfall model for simulating rockfall distribution Workshop on Rock Slope Engineering part D. pp D62-D84, Dept. of Civil & Structural Engineering Hong Kong Polytechnic.

4. Benitez M.A.H., Bollo, Ml.F., Rodriguez, M.P.H. (1977). Bodies falling down on different slopes - Dynamic study. Proc. of the IX I.C.S.M. F.E. 1977, Tokyo, vol. 2, pp 91-94.

5. Bhandari R.K.; Sharma S.K. (1976). Mechanics and Control of rockfalls. Journal of Institution of Engineering (India) vol. 57, pp 13-24.

6. Mercer R. (1982). The stabilisation of Craig-y-Dref, Treamadog, Ground Engineering pp. 28-35.

Blasting studies in the excavation of an underground water reservoir in granitic rock

R. Ciccu DR. ENG.
P.P. Manca DR. ENG.
G. Massacci DR. ENG.
Department of Mining and Mineral Engineering and C.N.R. Centre, University of Cagliari, Cagliari, Italy
C. Pedemonte DR. ENG.
E.S.A.F. Sardinian Body for Water Supply and Sewerage, Cagliari, Italy

SYNOPSIS

In order to not alter the landscape of an area located in proximity of an important residential center of South-Western Sardinia, the option of the underground installation of some reservoirs belonging to the hydraulic network of the Island was made.

The excavation with explosives of these works and the vibrations consequently transmitted to nearby buildings raised some problems.

A series of measurements was planned and carried out with the aim of assessing the influence of the presence of an important fault and understanding the source of damages observed in some constructions located in the area. The results obtained point out the strong vibration damping effect produced by the presence of geological and structural discontinuities.

INTRODUCTION

An underground water reservoir hosted in granite rock and consisting of three high section tunnels has been recently excavated on the outskirts of the town of Villacidro (Cagliari). Excavations were carried out using high-strength explosives in controlled rounds in order to prevent damage to surrounding buildings. This paper is concerned with the problems encountered and their solution in relation to the literature data and experimental measurements carried out.

Damage to buildings through blast-produced vibrations depends on the weight of explosive, the distance from the blast point, the structural features of the ground and the type of structure built thereon. Finding a general expression that relates these variables is further complicated by the type of rock blasted and its geological complexity.

Blast damage to buildings can be caused both by ground vibrations and by the effect of air blast. The latter phenomenon is controlled by atmospheric and environmental conditions which are often difficult to assess.

Lastly, it should be recalled that the reduction of vibration level affects the economy of blasting operations: for instance a reduction in vibration level of 20% results in a twofold increase in costs.[1]

The method adopted here is based on the use of formulae which relate some of the variables involved and allow to estimate, in conformity with safety standards, the charge weight required for excavations.[2,3] However, as it was not possible to take into account the actual elastic properties of the rock and its tectonic disturbances - the region is traversed by a large fault - it was also necessary to control the effect on the structures of the buildings concerned.[4]

GENERAL

Water system and underground reservoirs

Three tunnels have been excavated in the granite relief overlying the town of Villacidro (Cagliari) which serve as reservoirs for supplying the entire water system indicated as Scheme 35 in the master water distribution plan of Sardinia (PRGA) as well as the head and regulation reservoir for the town of Villacidro itself.

The water scheme, shown in Fig.1, ensure water supplies to eight municipalities for an overall present-day population of 84,000 rising to an estimated 90,000 by the year 2015. The system utilizes water from a catchment area upstream from the river Oridda with a flow rate of 240 l/s and that from an irrigation canal downstream for a flow rate of 160 l/s. Both flows supply a potabilization plant upstream from the underground reservoirs, where water from the canal Mulargia is also collected. The scheme also com-

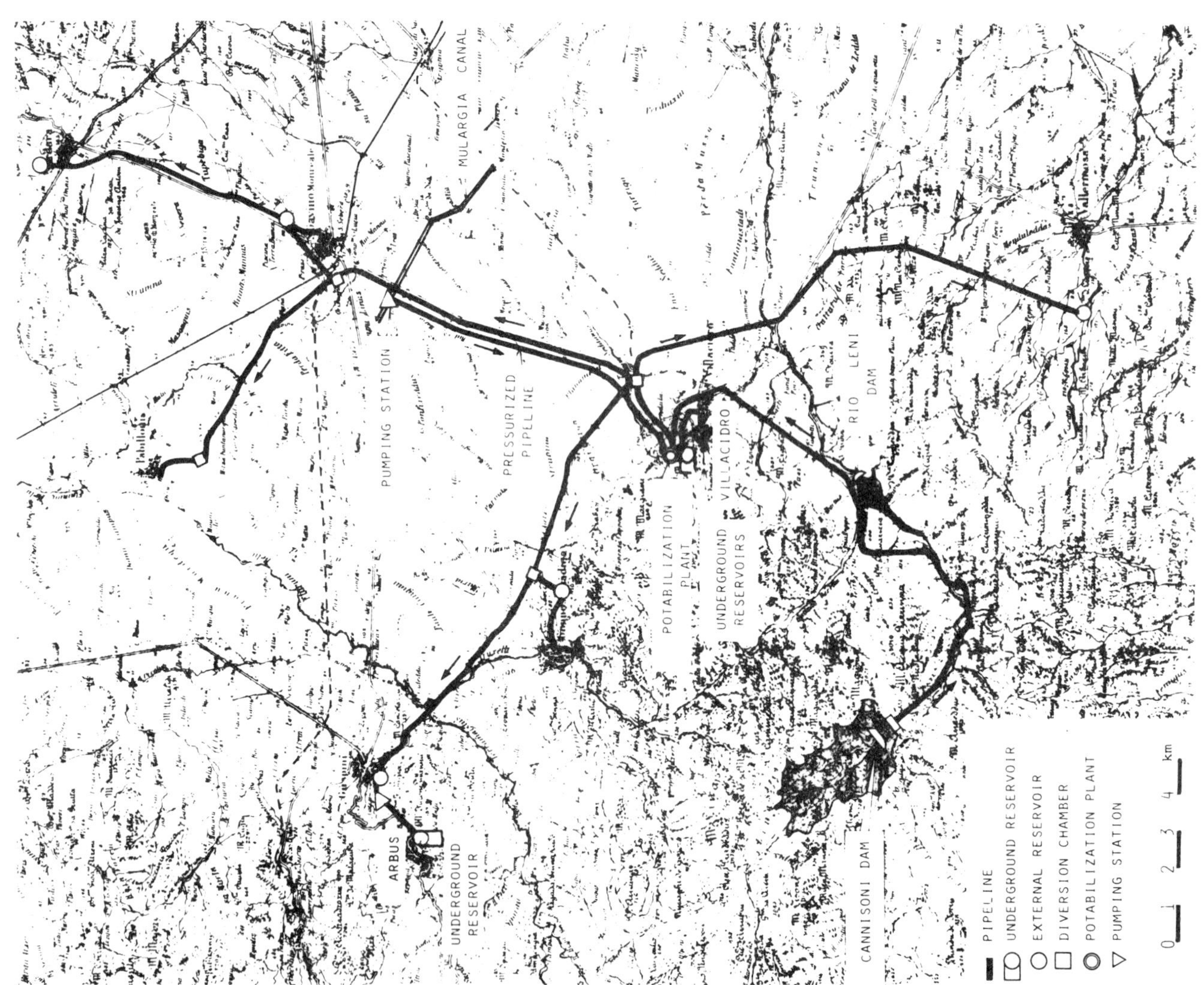

Fig. 1 Scheme of the water supplying system in the Villacidro district

Fig. 2 Panoramic view of the site where the underground reservoirs are located

prises the intake of the Mulargia canal, the pumping station and the penstock, the underground reservoirs storing drinking water as well as the pipes and reservoirs supplying built up areas.

Originally the three reservoirs in the vicinity of the potabilization plant were to be open cut. However this solution would have entailed extensive overburden removal, spoiling the countryside in an area of considerable tourist attraction (Fig.2). Consequently it was decided to place them underground. Despite being more burdensome, this solution allowed to minimize the environmental impact of the construction.

The total cost of the works, commenced in 1973 and completed in 1980 by a contractor company, namely Gecopre SpA, was 12 GLit, including pumping station and potabilization plant. The pipes extend over 72 km and there are 16,600 m^3 of underground reservoirs and 13,000 m^3 of earth-covered reservoirs.

Two of the three tunnels, excavated from a working level at 332 m a.s.l. bear eastwards whilst the third strikes N. They have identical cross-section with an excavation area of about 121 m^2. The first two are 87.20 m long and the third 33.10 m with a capacity of 6,150 m^3 and 2,300 m^3 respectively. All three tunnels have been excavated in two stages, as described later in the paper.

The layout of Fig.3 shows the underground reservoirs, the working level as well as the site of the potabilization plant, at a slightly higher level.

Design criteria

The master plan drawn up by ESAF for the water

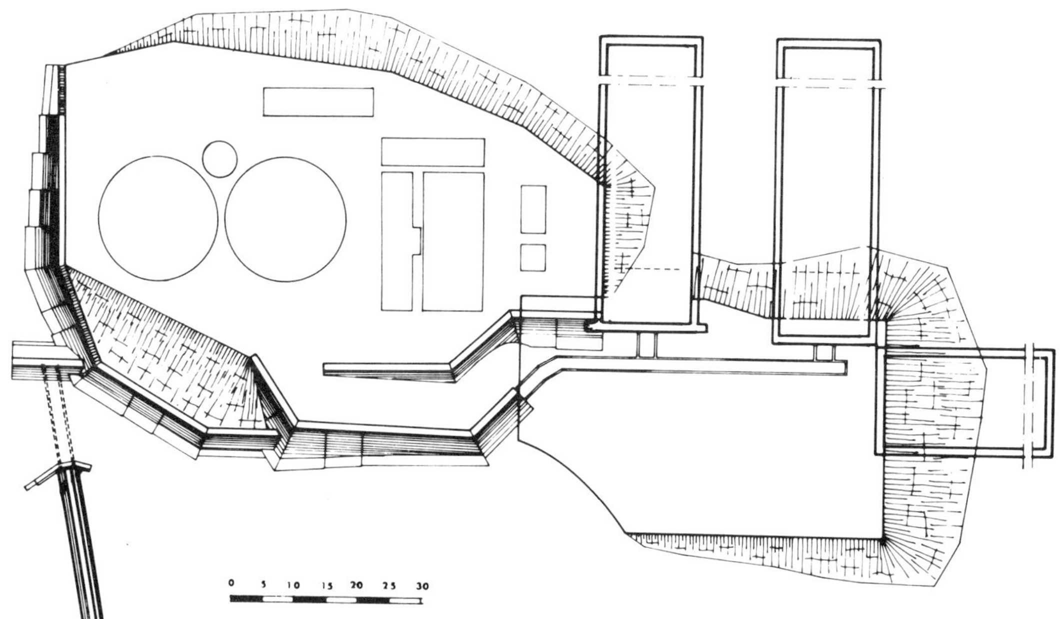

Fig. 3 Layout of the underground reservoirs and the potabilization plant

supply to certain regions of Sardinia foresees, as already mentioned, the construction in the immediate vicinity of built-up areas of underground reservoirs connected to the main catchment areas and to the distribution network downhill.

In the ambit of the above project the water system of Villacidro was a prototype to be built, where feasible, in similar situations. In fact subsequently another similar size plant was constructed in the urban area of Arbus, much nearer to the residential structures than the first. This was made possible by a better understanding of the problems involved and taking advantage of the experience gained in overcoming them.

Some caution was required in the planning and execution of the works especially in view of the adverse effects on the environment. In the light of the above the site chosen was situated about 300 m from the outskirts of the town, as Fig.4 shows. In any case, at about 150 m from the site stands a hotel and at about 100 m a private house, both occupied throughout the duration of the works. In addition there is a low voltage power line at about 25-30 m. Apart from exercising a certain care in the choice of site, additional precautions were taken, illustrated in the sequel, in selecting size of blast rounds according to internationally recognized codes for keeping annoyances within acceptable limits.

At the commencement of works the population was fairly well-disposed to put up with the inconvenience caused by excavations, if only to see

Fig. 4 Plan of the urban area involved in the excavation of the water reservoirs showing also the main geological features

the major problem of ensuring water supply resolved and no further convincing was deemed necessary.

Geology

From a geological point of view the region is dominated by a contact between granite and metamorphic cover.

The granite, the outlying ramifications of the Linas pluton, exposed by erosion, appears to be affected by fracture systems due to cooling. Some of these fractures are occupied by small crystalline quartz veins, with sporadic iron and copper ores.

The metamorphites (cornubianites) are covered by talus and alluvial terraces. They also seem to be affected by jointing systems one of which is quite pronounced striking roughly NS, dipping W, as well as by quartzose veins sometimes rich in iron oxides. The contact between these two formations is tectonic, consisting of a fault detectable on a large scale (1:100,000) with throw of the order of magnitude of a few hundred metres.

The true strike of this fault is about N30° W with an easterly dip of about 70°.

As figure 4 shows, the fault traverses the E-W tunnels excavated for the reservoirs. In fact it has been observed at about 40 m from the mouth, accompanied by a layer of clay and extends to a group of houses situated on the outskirts of the town.

In addition here and there are areas covered by talus which attain a thickness of a few metres in the zone surveyed. This talus becomes increasingly thicker down the road between the trench and the built-up area, and eventually completely covers the crystalline rock.

The two houses east of the fault also seem to be built at least in part on detritus, as can be observed from the narrow gap between them.

Clearly the geology strongly influenced the technical choices made in the design and execution of excavation works: in the first place the choice of explosive and blast round size as well as the design of drilling and blasting pattern.

In fact, the major role of structural features

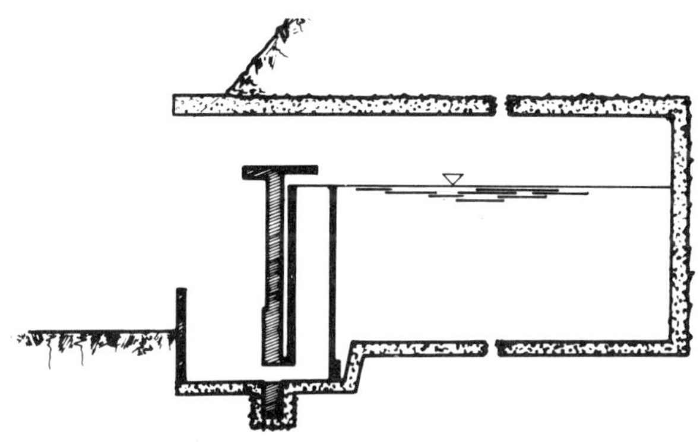

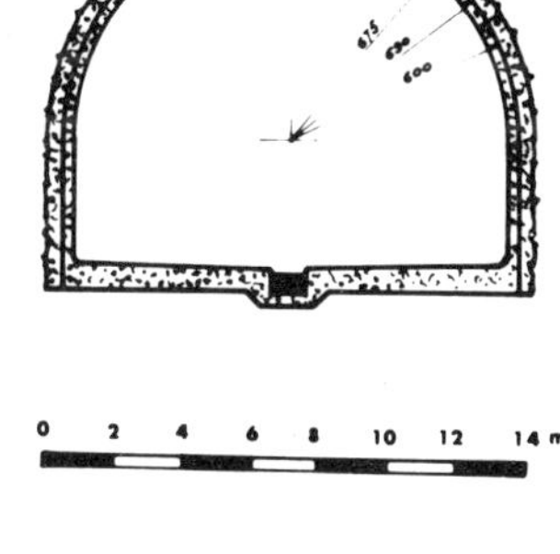

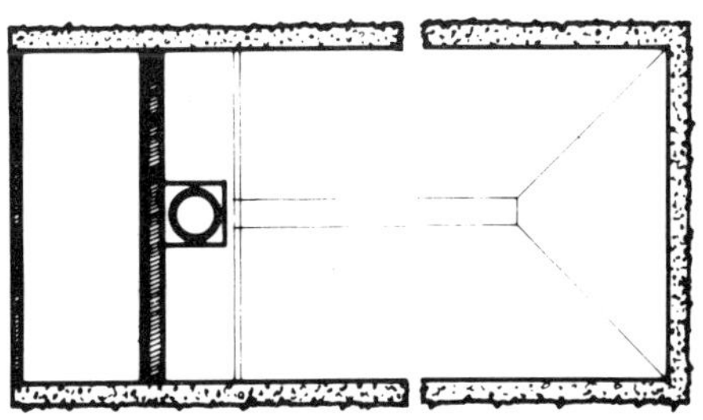

Fig. 5 Plan and cross-section of the three tunnels

of the rock in the transmission of elastic vibrations in the ground was taken into account. Moreover, the type of rock also affects the instantaneous vibration velocity produced by the explosion of a given charge Q at a distance R from the firing point.

Theoretically, in consolidated resistant rock such as granite, wave motion is characterized by higher vibration velocity due to the poor damping effect, whilst in unconsolidated rocks damping occurs at closer distances.

This has a direct influence on the instantaneous blast round size and the blasting pattern.

OVERBURDEN REMOVAL

Organization of site

In figure 5 the plant and cross sections of the tunnels are depicted. The reservoirs were excavated half way up the rock in an area with slope of 30°-45°.

This necessitated major preliminary excavation works to remove about 10,000 m^3 of overburden between the levels 332 and 355 m a.s.l. This operation was carried out in two stages. The first consisted in removing about 5,000 m^3 down to an intermediate level, about 5 m above the final tunnel foot. Subsequently, once the top heading had been excavated, overburden removal was completed.

The overburden was bench blasted with parallel boreholes, 63 mm in diameter, drilled with a 60 kg pneumatic drill.

These preliminary works took a total of about 50 working days to complete, equally divided between the two stages. Charge distribution was calculated in accordance with guide-lines laid down for cautious blasting. The charges were distributed along the borehole rather than loaded at the bottom in order to minimize vibrations of the rock mass and fly rock throw.

In addition the firing pattern was designed such that the rock was blasted on one side of the bench first, creating a new free surface and reducing charge confinement.

Air blast problems

The major problem in surface blasting was controlling air blast effects. In fact, as blasting was carried out over relatively large free surfaces, a consistent portion of the explosive energy was scattered in the atmosphere whilst the remainder, transmitted in the form of elastic waves to the ground, was more contained.

The following precautions were taken with a view to minimizing blast damage and nuisance caused by peak overpressures:

- the use of unconfined surface charges was avoided. This was possible because the size of fragmented rock was small enough not to require further blasting;
- detonating fuse was not used for firing rounds or connecting the charges. Instead electric blasting caps were employed throughout;
- the holes were stemmed with lightly compacted drill cuttings;
- bench blasting was used with sufficiently large burden;
- the total charge was limited and split up into a series of small charges;
- blasting was done in favourable atmospheric conditions so as to prevent focusing of air blast waves.

The latter precaution was of major importance owing to the strong prevailing NW winds which could create abnormal concentrations of energy on the town downwind.

Accordingly, blasting was done in the early afternoon so as to avoid incurring the danger of temperature inversion.[4,5]

As a result of the precautions taken air blast effects were rationally minimized despite the difficulties in controlling highly variable situations.

Actually, apart from a few isolated complaints, no blast damage was recorded and nuisance was confined to glass rattle.

Precautions against rock fly danger

Open-cut overburden excavation also had to be carefully controlled due to the danger of rock fly. In fact, although the rock was generally consolidated, in some places the granite was highly fractured through weathering and tectonization.

In addition to the usual precautionary measures taken to control air blast, the orientation of the bench face was also studied in order to ensure that nearby buildings were not in the trajectory of rock fly. Moreover, as already mentioned, the firing pattern consisted in series firing from top to bottom so as to provide a free outlet surface of the blast and as a result direct the rock-fly uphill.[6]

As a matter of fact during surface excavations only sporadic rock fly occurred, and no damage was caused.

Hauling and dumping of blasted rock

The fragmented rock was loaded with crawler-mounted shovels or smaller loaders into 15 t dump trucks which hauled the rock to the dump.

Some problems did arise in dumping, connected with landscaping (the area is a beauty spot) and nuisance created by heavy vehicles (noise, dust, vibrations).

The rubble was dumped in a remote spot and arranged in such a way as to facilitate revegetation.

In order to avoid heavy vehicles passing through the town, a haul road was built. However, in the final stages of excavation works another dump site had to be found and access was only possible via public roads and using vehicles with a higher payload (about 30 t) which caused some inconveniences.

With cycle times of 20-30 minutes haulage was completed in 3-4 hours with 15-20 trips per day. Inevitably a number of complaints were lodged, but as there was no alternative these were moderated.

As regards vibrations, geophysical velocity and frequency measurements were carried out in houses in the vicinity of roads where heavy vehicles passed. Values of the order of a few μm per second were recorded - practically insignificant.

EXCAVATION OF THE TOP HEADING

Control of blasting vibrations

The excavation of the tunnels was carried out in two stages. The first consisted in excavating the top heading with semicircular section and radius of 6.75 m. In the second stage, once the outside working level had been suitably lowered, the tunnel bench, 4 m high, was excavated.

The major problems posed by blast design was minimizing the intensity of elastic vibrations in the rock. In fact, in tunnel blasting a consistent portion of explosive energy is transmitted to the rock mass whilst only a minor portion is scattered in the atmosphere. Moreover, provisions had to be made for the fault where, as mentioned, wave trains appear to be channeled, resulting in a local magnification of vibrations.[4]

As little information has been published to date on this effect (Hattezburg effect) more stringent safety criteria were applied. As a result, the following precautions were taken in the blast design:

- the diameter of the boreholes was kept small in order to ensure a well-distributed column charge over a larger number of holes. In this way burden was decreased and consequently blast confinement limited, giving the rock greater possibility of expanding in the short time of effective work performed by the explosive;
- the advance per round was limited;
- delaying devices were used for the purpose of minimizing the extent of simultaneously detonating charge. This factor is no doubt decisive in determining vibration intensity and in particular displacement velocity. In fact, as far as work

performed by the explosive is concerned it is the total weight of explosive that counts, whereas in regard to vibrations only the part that detonates simultaneously is important;

In the design of the firing pattern, the minimum delay $\tau = 3/f$ beyond which wave trains do not overlap was calculated where f is the specific vibration frequency of the rock.[7]

Prudentially f = 50 Hz, instead of 100 Hz suggested by the literature for granite, was assumed and consequently a value roughly corresponding to three numbers of the short-delay series was found. However, half-second delays were finally opted for to make the most of the favourable effect produced by the interaction of the charges in relation to the statistical dispersion of effective priming times of the same numbers.

It is a well-known fact that in these types of detonators standard deviation from the average 0.5 s is about 100 ms: for these a coordination factor within delay of 1/6 is assumed, which means that the total charge distributed over holes of the same number is equal, as far as vibrations are concerned, to 1/6 of the same charge loaded into a single hole. This represented another safety factor.

In standard blasting, the explosive charge was loaded as indicated in the following table:

No. delay	No. holes	Charge per hole (kg)	Charge per delay (kg)
0	8	0.51	4.10
2	8	0.51	4.10
4	8	0.51	4.10
6	7	0.57	4.00
8	16	0.59	9.50
9	13	0.60	7.80
10	19	0.60	11.40

Other technical data are reported below:

- section of the blast: 68.32 m^2
- advance per round: 1.60 m
- blasthole diameter: 34 mm
- diameter of cartridge: 25 mm
- decoupling ratio: 0.73
- number of holes: 79
- total drilled length: 134 m
- specific drilling: 1.22 m/m^3
- overall charge: 45 kg
- specific charge: 0.411 kg/m^3

Application of damage criteria

In designing the charges the well-known damage criteria recommended in the literature and now widely accepted were taken into account. The formula[8]

$$V = K \sqrt{Q/R^{1.5}}$$

was used for predicting V (mm/s) peak particle velocity where:

Q = simultaneously detonating charge (kg);

R = distance from firing point (m);

K = coefficient depending on type of rock (for consolidated granite K = 400).

At a distance of 100 m assuming radial propagation of seismic waves, an explosive charge of 11.4 kg, corresponding to the maximum charge per delay exploded in standard blasting, would produce, taking into account a coordination factor of 1/6, a peak particle velocity of the order of 18 mm/s, well below the threshold of 50 mm/s recommended by several authors, amongst whom Langefors.[8]

For a distance of 300 m the same formula yields a peak particle velocity of about 7.6 mm/s.

The presence of the fault was the only aspect difficult to assess as it could be justifiably assumed to channel the seismic waves generated by blasting. However, compared to the prediction for a homogeneous medium no indication of magnification was detected by the geophysical measurements, as will be illustrated later on.

The explosive used, GD1 (dynamite class) had

an impedance of the same order as that of the rock to be fragmented (granite).

Despite the close spacing of the holes especially at the top heading face, the firing sequence was respected. In fact, with an interval of 1 s between the series, successive explosions could be clearly distinguished. This was also supported by the geophysical recordings.

The specific vibration frequency f was also taken into account as wave length $\lambda = c/f$ is frequency dependent given the seismic wave propagation velocity c.

In fact the higher the wave length of the vibratory motion, the lesser is the risk of damage to buildings. The risk increases for wave lengths λ smaller than the size of the building in the propagation direction whilst it should gradually diminish with greater wave length.

The latter observation is true for each of the three components of motion: vertical, radial, transverse. In this respect, harder and more consolidated rocks should as a general rule be at less of a risk. For consolidated but locally fractured granite like that hosting the reservoirs, typical frequency is of the order of 100 Hz while propagation velocity attains 5,000 m/s. The resulting wave length is thus of the order of 50 m.

This value appears somewhat reassuring as far as the differential vibratory motion between the various points of the structure is concerned, at least for residential structures occupying a limited area. This is also supported by experience gained in Sweden, pioneer in research on the effects of rock blasting.[1,8]

For instance, to produce the same extent of damage in compacted rock such as granite, rock vibration velocities four times higher than for unconsolidated rock are required.

The frequency assumed here (100 Hz) was overestimated. In fact, successive geophysical recordings indicated an initial vibration frequency of 55 Hz.

EXCAVATION OF THE TUNNEL BENCH

Bench-blasting design and technical data

The problems encountered in excavating the lower part of the tunnel were more easily resolved, especially as far as blasting pattern was concerned.

In fact the bench-blasting technique used in this stage permitted lower strength explosives (quarry type) to be employed. In any case, to avoid damaging the concrete lining of the tunnel walls, carried out as the top heading excavation advanced, smaller charges had to be used.

Parallel slightly inclined (15°) boreholes were drilled with 60 Kg crawler-mounted drills, as in overburden removal. Usually two rows of holes were blasted at a time, sometimes three. About 200 m^3 of rock was blasted daily.

The following technical data are of interest:

- blast face area: 64 m^2;
- advance per round (2 rows of blastholes): 2.8m;
- burden: 1.4 m;
- blasthole spacing: 2.0 m;
- blasthole diameter: 34 mm;
- total drilled length per blast: 75 m;
- specific drilling: 0.42 m/m^3;
- specific charge: 0.240 Kg/m^3.

CONFIRMATION OF VALIDITY OF SAFETY CRITERIA

Blast damage complaints

Despite the precautions taken in blasting design and execution, based on recommendations suggested in the literature, towards the end of excavation works blast damage complaints were lodged by some houseowners, especially during the excavation of the third N-bound tunnel.

The buildings concerned were 3 or 4 storeys high and situated on the outskirts of the town about 300 m from the site, on the same side as the tunnel with respect to the fault plane. Some were constructed with supporting walls and in a poor state of repair, others in concrete of more recent construction.

A survey of damage revealed that cracks had appeared in the floors and ceilings and outside walls at the joints with girders. This type of damage is typical of settlement in structures with badly designed foundations.[1]

However due to the effect of seismic waves these cracks can spread to an extent such as to jeopardize the safety and stability of the building. Moreover, it could not be excluded that the presence of the fault might have created a preferential route for the transmission of explosive energy released by the blast and hence caused or aggravated cracks.

Geophysical measurements of blast vibrations

In order to dispel any doubts and suppress at the origin cause for discontent among the residents, it was decided to carry out a campaign of geophysical measurements in order to obtain an objective picture of the magnitude of the characteristic parameters of elastic waves and verify them in the context of safety factors.

Horizontal and vertical measurements of vibration amplitude and frequency in the ground and buildings concerned were recorded using electrodynamic transducers connected to the measuring and recording instruments, namely Philips PR 9260 and PR 9250/01 respectively. The sensor range was:

- frequency 10 - 100 Hz;
- amplitude 0.5 - 1,500 μm (displacements);

  50 - 10,000 cm/s^2 (accelerations).

The system is sensitive to displacement velocity but as it is fitted with integrating and derivating devices, alternatively amplitude and acceleration can be measured.

Three different transducers were used for measuring simultaneously the three variables of motion in one point, or the same variable (displacement, velocity or acceleration) in three different points.

Analysis of the traces allowed to obtain the shape of the wave trains as well as the respective frequency response, indicating the presence of harmonics.

Overall measurement accuracy was approximately 10% taking into account all portions of the signal transmission chain. Several series of measurements were repeated, some recording amplitude, others velocity under different conditions of amplification and system sensitivity.

The above measurements were carried out during firing in the tunnels for excavating the top heading. The firing pattern consisted of 7 series of blastholes with half-second delays for a total charge weight of 45 kg.

As was to be expected, the motion observed was complex and cannot be expressed by means of a simple damped harmonics model. In fact staggered waves of different amplitude and frequency overlapped and they proved difficult to separate and identify.

Moreover 7 separate wave trains were distinguished, one for each group of blastholes with the same delay, the first five at 1 s intervals, the last two at 0.5 s intervals. The first three wave trains exhibited the highest peak velocities, in particular the third, for which frequency of 55 Hz and initial velocities of 90 μm/s were recorded.

In all cases frequency, from an initial value of around 44-55 Hz, settled at around 25 Hz. In the most unfavourable case, referred to the top floor of the building, the following data were obtained:

- initial frequency: 55 Hz;
- final frequency: 23 Hz;
- maximum displacement: 0.36 μm;
- maximum velocity: 90 μm/s;
- maximum acceleration: 3.1 cm/s^2.

Conclusive remarks

In conclusion, the physical quantities measured yielded values much lower than expected and in any case well below the permissible danger limits. These findings supported the validity of the design criteria followed.

It is interesting to note that measured velocity was about 80 times lower than computed velocity for a distance of 300 m. This discrepancy could be explained by the fact that the buildings concerned stood on filling material and the granite mass is highly fractured locally. Both these elements contribute to greatly attenuating the transmission of elastic waves.

Hence, the value of 400 assumed as a first approximation for the coefficient K was overcautious in the light of the experimental results.

The measurements of the effects of attenuation in transmission of elastic waves through rock masses of complex lithology and structure, therefore proved fundamental in predicting the environmental impact of blasts and in the optimum design of the latter.

This is particularly true in those cases where the splitting up of the charge is not dictated by the need for regularly contoured excavations, contrary to the case at hand.

References

1. Swedish blasting technique. Rune Gustafsson. Gothenburg: SPI, 1973. 328pp.

2. Safe use of explosives in the construction industry. In: British Standards Institution, BS 5607:1978.

3. Fornaro M. Vibrazioni dannose per le costruzioni in conseguenza dello sparo di mine: esame di quanto ammesso dalle norme e suggerito dall'esperienza. Notiziario della Associazione Mineraria Subalpina, Sept.-Dec. 1980, p. 18-28.

4. Ball M. Vibrations from blasting. Problems and solutions. Paper presented at a special study course. Nottingham: 14-15 Dec. 1976. Repr. from Quarry Management and Products.

5. Ground and air vibrations from blasting.David E. Siskind. In: SME Mining Engineering Handbook. New York: Society of Mining Engineers of AIME, 1973. Vol. 1, p. 11/99-11/111.

6. Caratteristiche ed effetti delle vibrazioni prodotte da esplosioni. In: Lavori con esplosivo. Milano: Italesplosivi, 1977. p. 111-136.

7. Handbook of surface drilling and blasting. Tamrock, 1982. p. 221-239.

8. The modern technique of rock blasting. U. Langefors and B. Kihlstrom. New York: John Wiley & Sons, 1963. 405pp.

Appraisal of potential effects of surface blasting on a rock tunnel, with particular reference to the Tai Lam Chung water tunnel, Hong Kong

A.W. Clover B.SC., M.SC., C.ENG., M.I.C.E., M.I.M.M., F.G.S.
Freeman Fox & Partners (Far East), Hong Kong

SYNOPSIS

An investigation into the possible effects of surface blasting on an existing water pressure tunnel situated deep in fair to good quality granite bedrock is described. A crude theoretical assessment of critical frequencies and peak particle velocities is presented which suggests that with respect to this particular situation the imposed Hong Kong blast vibration standards would be conservative. This view is substantiated by the results of a literature search for documented effects. As no trial blasting was included in the investigation, a data search for blast records of other similar major excavation projects was also undertaken. Analysis of the resultant data is presented, from which proposed square-root scaled distance relationships for the preliminary design of blasts in the granites of Hong Kong are developed.

INTRODUCTION

In 1984, a study was commissioned to investigate the possible effects of large-scale borrow operations above an existing 30 years old water pressure tunnel. The borrow operations were to provide 9 million m^3 of fill for a major new reclamation scheme in Hong Kong Harbour. These borrow operations would in part, involve the removal of approximately 3 million tonnes of overburden within a zone defined by a 45° arc either side of the tunnel centreline. A maximum thickness of 44m of overburden would be removed, leaving a minimum cover of 93m to the tunnel crown. The tunnel is situated within granite, which has up to a 30m thick decomposed soil profile beneath the ground surface.

Whilst some of the rock above the tunnel may be rippable, it was obvious that a significant proportion would require blasting. This blasting would need to be undertaken on a substantial and regular basis if the fill production rates, necessitated by the proposed reclamation programme, were to be met.

As with any blast, no matter how well designed the blasting may be, the blast will generate some unwanted energy that will radiate away from the blast in the form of vibration through the ground and the air. It was of concern that the ground vibrations emanating from the necessary blasting at Sham Tseng should not adversely affect the tunnel. This paper summaries the results of the investigation undertaken to determine the possible effects of blasting on the tunnel.

LOCATION AND ENGINEERING GEOLOGY

The area for which the study was undertaken is located in the North West New Territories of Hong Kong to the east of Sham Tseng village, as indicated in Figure 1. The main rock type in the vicinity of the tunnel comprises Cheung Chau Granite. The Cheung Chau Granite is the most extensively exposed granite in the New Territories. It is considered to represent the

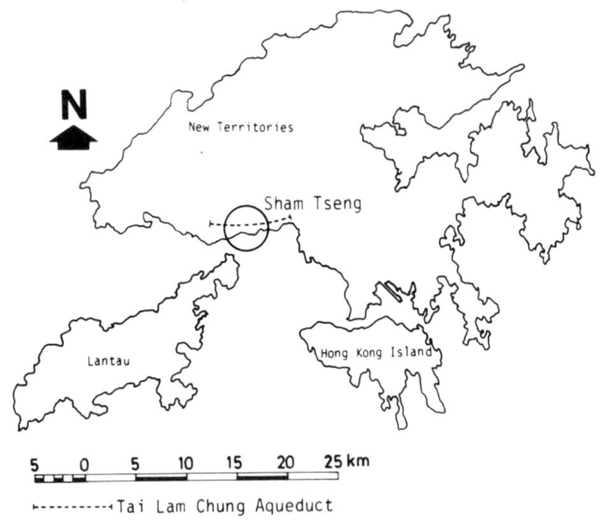

Fig. 1 Site Location

main phase of intrusive activity in Hong Kong, which occurred in the early Upper Jurassic Period approximately 155 million years ago[1]. The fresh rock is typically light grey or pale pink, is medium grained (1 to 5mm) and sparingly porphyritic, with phenocrysts of feldspar[13].

As part of the study a total of five deep, cored drillholes (up to 170 m deep) were drilled to the side of the tunnel centreline, which strikes bearing 088°. A cross-section along the line of the tunnel showing the conditions encountered in the drillholes is shown in Figure 2. A discontinuity orientation survey indicated the following three regional concentrations :

o	Set A	dip 83°	dip direction 145°
o	Set B	dip 57°	dip direction 038°
o	Set C	dip 80°	dip direction 077°

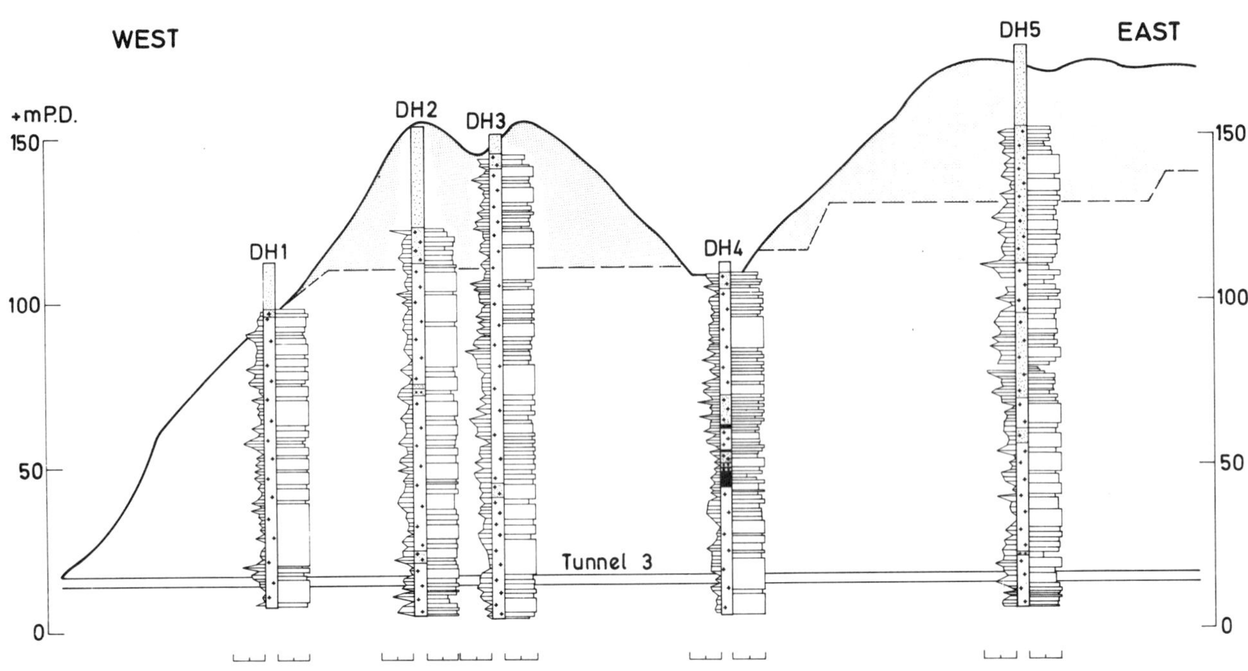

LEGEND:

SOIL – residual to highly weathered granite

CHEUNG CHAU GRANITE – moderately to slightly weathered

CHEUNG CHAU GRANITE – slightly weathered

DOLERITE – moderately to slightly weathered

Material to be excavated during proposed borrow operations

F.I. Fracture Index – number of natural discontinuities per linear metre

RQD Rock Quality Designation – percent

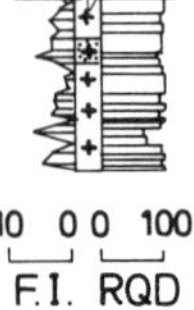

Horizontal Scale:

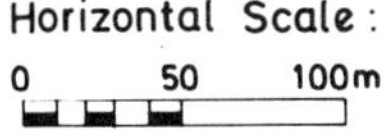

Fig. 2 Rock mass conditions about Tunnel 3

The R.Q.D. (Rock Quality Designation) and F.I. (Fracture Index) of the rockmass intersected in the drillholes, are also indicated in Figure 2. R.Q.D. values are generally in excess of 90%, with the lowest values found in DH4 in association with a dyke zone and in DH 5 in association with a phase change in the Granite. The average F.I. is in the range 2.1 to 3.6.

Detailed analysis of discontinuity spacing was also undertaken in line with the methods proposed by Priest and Hudson[18]. This analysis confirmed that the discontinuity spacing within the rock mass follows a negative exponential distribution i.e. it is a random event. The mean spacing of the discontinuties is 0.391m, with approximately 75% of the spacing values being defined by an upper limit in the range 0.35m to 0.7m. Consequently the rock mass can be generally described as containing closely to medium spaced discontinuities.

THE TAI LAM CHUNG TUNNEL

The Tai Lam Chung water supply scheme comprises a 10.4 km long west to east series of tunnels and pipelines. The scheme is currently the major source of raw water supply to Tsuen Wan, a major industrial area of Hong Kong. It is, therefore, regarded as a life-line supply. The current average flow rate through the scheme is about 225 Ml per day under a maximum static head of 48m (470kPa). Construction of the scheme was completed in March 1957 and the scheme was commissioned the same year. The limited officially recorded information concerning the scheme includes no specific geotechnical information.

The section of tunnel beneath the proposed borrow site is designated Tunnel 3. Except for a flat mass concrete invert, Tunnel 3 was designed to be primarily an unlined approximately 2m diameter rock tunnel. A concrete lining has only been provided in those sections of the tunnel where there is less than about 9m cover of 'undisturbed rock', or where 'bad ground' was encountered. Directly beneath the proposed borrow site 10% of Tunnel 3 has a concrete lining, the remainder is unlined except for some gunite. Typical cross-sections through Tunnel 3 are shown in Figure 3.

Tunnel 3 was excavated by drilling and blasting. Nearly all sources of information on Tunnel 3 make reference to the occurrence of substantial overbreak during blasting. Limited measurements undertaken at that time, indicate an average 50% increase over the design area i.e. an increase in the tunnel diameter to approximately 2.5m. There were, however, no major falls during construction and no falls occurred after the tunnelling operation had been completed.

EXISTING STABILITY OF TUNNEL 3

In the study of the rock mass conditions about Tunnel 3, three classifications techniques were utilised : the CSIR Rock Mass Rating, the NGI Tunnelling Quality Index and the Hoek-Brown Failure Criterion[8]. The results of these separate techniques were similar and indicated that the majority of the rockmass intersected by the drillholes could be regarded as fair to good quality.

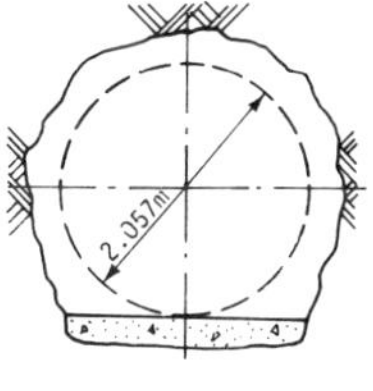

Detail 1

Cross-section of the tunnel in those sections WITHOUT a concrete lining

(cross-section shows an overbreak of 50% over the design area)

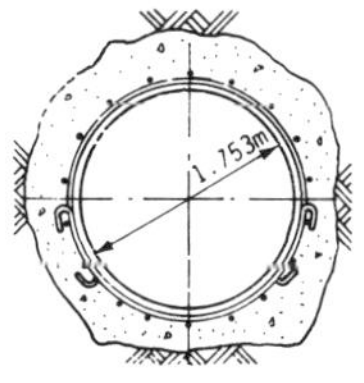

Detail 2

Detail of concrete lining : usually unreinforced, but reinforced in areas of low overburden or 'bad' ground (cross-section shows single reinforcement)

Fig. 3 Detail of Tunnel 3 cross-section

The existing stability of Tunnel 3 was also examined both in terms of these classification techniques and structurally controlled stability. Utilising published correlations of support requirements with respect to the rock mass classification techniques, an analytical approach based on the rock mass failure criteria and a kinematic feasibility appraisal of potential rock blocks, it was concluded that :

- o the majority of Tunnel 3 would have required little if any permanent support;
- o where support was required, this would have been primarily to prevent localised structurally controlled failure i.e. movement of blocks, wedges, weak zones etc.;
- o if Tunnel 3 were to be constructed today, it would be built to a generally similar form to that which already exists.

Following publication of the study report, concern was expressed that the stability appraisal of Tunnel 3 was primarily based on a theoretical approach supported by limited factual data. The logistically difficult and critical operation of a complete shut-down of the Tai Lam Chung system was, therefore, arranged to enable an inspection of Tunnel 3. The last time such a shut-down occured was in 1971. The inspection took place on a Sunday and was limited to 2.5 hours for the 1.5km of tunnel of interest. A video record of the inspection was made and the inspection confirmed that Tunnel 3 was in good condition and showed no signs of distress.

BLASTING

Existing Restrictions on Blasting near Tunnel 3

The existing restrictions on blasting near Tunnel 3 are contained in the Hong Kong Water Supplies Department (WSD) Provisional Standing Order no. 1038 (Revised January 1985) and can be summarised as follows :

- o the maximum particle velocity and amplitude of ground movement at the tunnel are to be restricted to 13mm/sec and 102 microns respectively;
- o no blasting shall be carried out within a distance of 55m horizontally or vertically from any tunnel;
- o the design of any blast is to be decided by the Hong Kong Commissioner of Mines on the basis of trial blasts at the site.

It was established that these limitations had not been defined on the basis of an engineering study and appraisal of the potential effects of blasting on a tunnel. Rather, the limits had been set in the early 1970's by the application of a reduction factor of approximately 4 to the United States Bureau of Mines (USBM) then recommended safe maximum particle velocity of 2 inches/sec for residential structures[14]. The distance restriction of 55m is applied to ensure that during blasting, tunnels will not be affected by :

- o the extension of existing discontinuities in the ground;
- o the creation of new fractures in the ground;
- o the general loosening of the ground;
- o the venting of gases along existing and new fractures.

The Potential Effects of Blast Vibrations on Tunnel 3

The potential effects of blast vibrations on Tunnel 3 can be considered under two broad categories, namely :

- o dynamic loading and strain, and potential fracture of the intact rock around the tunnel and the tunnel lining, due to the stresses and displacements induced in the intact material during the passage of the blast vibrations;

- o shaking and differential movement between the blocks of rock around the tunnel, giving rise to gravity induced rock fall.

After a review of various dynamic mechanisms, it was concluded that due to the distance of the tunnel from the proposed blasting, dynamic mechanisms would not have any adverse effects on the tunnel. i.e. the likely maximum instantaneous explosive charge weights required for quarry blasting would give rise to vibration levels at the tunnel of an magnitude insufficient to cause dynamic damage. [10, 11, 20, 21].

Since a rock mass is a discontinuous medium, there is the potential for the straining of discontinuities due to blast vibrations causing differential acceleration of adjacent rock blocks. This mechanism will occur when the motions of adjacent blocks are out of phase. Rock fall and/or sliding of blocks into an opening will then occur in unlined sections of the tunnel if the straining is of a magnitude sufficient to either reduce or destroy the mechanism resisting gravity eg. discontinuity shear strength, block interlock.

The minimum and maximum block sizes which can fall are governed by the minimum and maximum discontinuity spacing and the tunnel diameter. Motions in the rockmass will be out of phase at distances equal to one quarter of the shear wave wavelength[3]. Therefore, from the relationship $f = v/\lambda$ (where f = frequency, v = shear wave propogation velocity and λ = wavelength), the critical frequency for a particular block can be crudely estimated by substituting a value for wavelength equivalent to four times the block size. On the basis of a mean and maximum discontinuity spacing of 0.25m and 3m respectively, a tunnel diameter of 2.5m and a shear wave velocity of 1800 m/s, the critical frequencies of vibration at Sham Tseng would be in the range of 1.8kHz to 150Hz - the higher frequencies being with respect to the smaller blocks. Simultaneous application of the P.S.O. no. 1038 maximum particle velocity and amplitude limits into the equations for simple harmonic motion results in a critical frequency of about 20Hz. This frequency corresponds to a critical block dimension of 22m i.e. well outside the upper limits defined by the tunnel diameter and maximum discontinuity spacing. Maintaining the maximum amplitude criterion, the critical peak particle velocities associated with the mean and maximum discontinuity spacings are approximately 1154mm/s and 95mm/s, respectively.

Documented Effects of Blast Vibrations and Vibration Limits

The theoretical appraisal suggested that with respect to Tunnel 3, the blast vibration limits required by P.S.O no. 1038 were conservative. With this in mind, a literature search for documented effects of blasting and specified limits was undertaken to determine whether there was a factual basis for this suggestion.

The results of this search are summarised in Table 1. From a total sample of 88 case histories presented in 5 technical papers, the lowest measured peak particle particle velocity at which damage has occurred is 125 mm/s. This was in a unlined cavern beneath a quarry, where one small fall occurred when this level was initially appproached. Subsequent blast vibrations of a similar level caused no further problems[16].

By far the largest group of case histories (71 no.) is from a study into the effects of earthquake shaking on the stability of tunnels[3, 4]. In this work it is acknowledged that the frequencies at which peak earthquake motions occur, are considerably lower than those associated with blasting - 0.4Hz to 10Hz compared with 20Hz to 200Hz respectively. However, this is considered to be compensated by the fact that the total duration, number and amplitude of earthquake vibrations are generally much greater than those induced by a blast.

Of the three limits presented in Table 1, two are statutory limits and one is a site limit. The lowest statutory limit is the Swiss limit of 30mm/s followed by the Swedish limit of 70mm/s. The observed Swedish threshold for the on-set of

LOCATION	PPV (mm/s)	TYPE	ROCK	COMMENTS	SOURCE
Underground Explosion Tests (UET), U.S.A.	460	INDUCED	Sandstone	Some fall of rock in 1 out of 14 tests of 2m to 10m dia. unlined tunnels subjected to high explosive shock.	(11)
Koi Tunnel, JAPAN	338	INDUCED	-	Recorded minor cracking in concrete lining.	(19)
SWEDEN	300	THRESHOLD	Granite	On-set full of stones in unlined tunnels.	(12)
California and Alaska, U.S.A. and JAPAN	200	INDUCED	Various	No falling stones in unlined tunnels or cracking in lined tunnels below this level, in tunnels subjected to earthquake shock.	(3, 4)
Dworshak Dam U.S.A.	125	INDUCED	Granitic gneiss	Apart from one small fall as this level was initially approached, no subsequent falls at this level in a cavern beneath a quarry.	(16)
SWEDEN	70	LIMIT	Hard Rock	Enforced limit for Underground Caverns subjected to short duration construction blasting.	(17)
Norad, U.S.A.	56	INDUCED	Granite	No detrimental effects noted in cavern due to adjacent tunnel blasting.	(15)
SWITZERLAND	30	LIMIT	Various	Enforced limit for underground chambers and tunnels with and without concrete lining.	(21)
Dinorwic, U.K.	45	LIMIT	Slate	Initially enforced limit primarily to prevent damage to reinforcement concrete and installed equipment.	(10)

Table 1 Summary of Documented Effects and Vibration Limits

'falling stones' in unlined tunnels implies a factor of safety of 4.29 in the Swedish statutory limit.

During the construction of the underground chambers for the Dinorwic Power Station in the U.K., a limit of 45mm/s was initially specified in the contract documents[10]. However this was primarily intended to prevent damage to reinforced concrete structures and equipment. Measurement and analysis undertaken during the work indicated that this level could be increased, and levels as high as 187mm/s were later recorded without ill effect.

Probable Ground Response to Blasting above Tunnel 3

The generally accepted relationships for estimating peak particle velocities at a particular site, are heavily reliant on the determination of the appropriate site specific coefficients. There are, however, a number of published relationships that have been established from a wide range of blast data from various sources, e.g. Dupont[5], ICI[9], USBM[14], Oriard[16]. Such relationships are often used without question in the early stages of a project, until the specific site coefficients can be confirmed or established from the results of trial/production blasts.

Due to the circumstances of the investigation and the existing site and geological conditions at Sham Tseng, it was considered neither practical nor appropriate to include trial blasting. Furthermore, due to the particularly critical nature of this aspect, it was also considered that it would be unacceptable to proceed with the proposals on the basis of an existing published relationship, without at least some justification that that relationship was appropriate to Sham Tseng.

Consequently, the investigation also included a data search for blasting records of other major excavation projects in Hong Kong, which had been undertaken in similar geological and topographical environments to those at Sham Tseng. In addition, the technical literature was searched for blasting information particularly relevant to the proposals, i.e. blasting above an existing tunnel. The obtained sources of local information are shown in Table 2.

The Hong Kong Mines Division currently utilises the square-root scaled distance relationship (as developed by the USBM[15]) to predict peak particle velocities. This approach was therefore adopted for the study. Regression analyses of data from the various sources, on the basis of square-root scaled distance, is

Site	Geology	Source
Kornhill	Hong Kong Granite	Mass Transit Railway Corporation : Hang Lung Development/Freeman Fox & Partners (Far East)
Lam Tin	Hong Kong Granite	Hong Kong Housing Department/Halcrow Asia Partnership
Shek Lei	Needle Hill Granite	Hong Kong Housing Department/Scott Wilson Kirkpatrick & Partners
Castle Peak Power Station B	Cheung Chau Granite	China Light & Power/L.G. Mouchel & Partners (Asia)
Lamma Power Station	Sung Kong Granite	Hong Kong Electric/Fugro (Hong Kong) Ltd.

Table 2 : Hong Kong Sources of Blasting Data

presented in Figure 4. Analysis on the basis of cube-root scaled distance was also been undertaken, but this gave no improvement in correlation. It can be seen that at the 50% confidence level the results are similar to, or more optimistic than, the Dupont Normal line. There is, however, significant deviation about the regression lines of the Hong Kong data.

To obtain an 'average' Hong Kong response for use in the preliminary prediction of vibration levels at Sham Tseng, the various Hong Kong regression lines were combined. In order to overcome any tendency of bias due to the differing number of data points, this was undertaken by simply 'averaging' the coefficients and confidence limits. The resultant regression line is shown in Figure 5. At the 50% confidence level this line is generally slightly more optimistic than the Dupont Normal Line

The regression lines shown in Figures 4 and 5 are all based on vibration measurements undertaken at the surface for surface blasts. It is, therefore, possible that these relationships may not correctly predict the vibration levels at some point at depth within the rockmass. The satisfactory resolution of this aspect with respect to Sham Tseng and

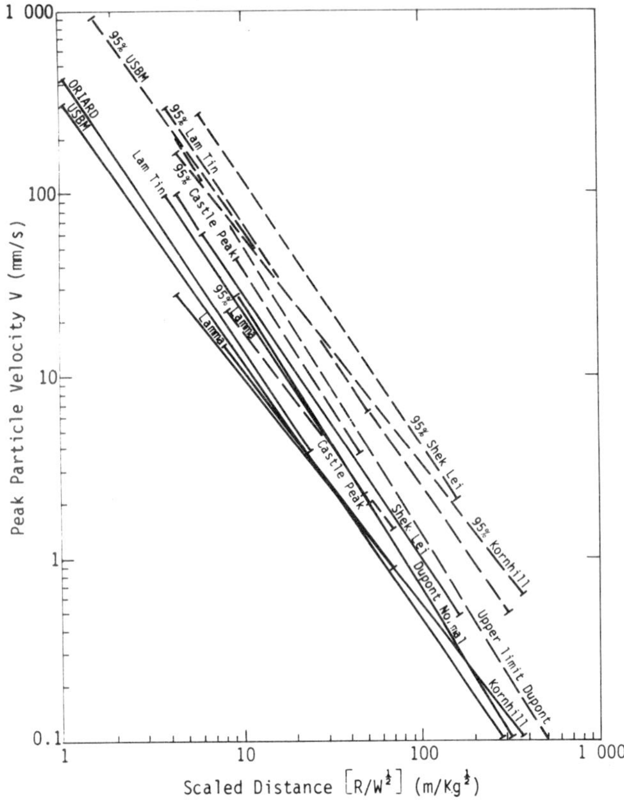

Site	R Range (m)	A	B	95% Conf. (Log_{10})	No. of Data Points
Kornhill	60-200	199.50	-1.255	0.7663	609
Lam Tin	50-100	791.74	-1.495	0.4631	53
Castle Peak	60-200	963.54	-1.561	0.2439	32
Shek Lei	60-200	913.25	-1.485	0.6518	71
Lamma	-	231.31	-1.308	0.2170	32
Dupont Normal	-	1 143	-1.600	0.3163	-
Oriard	-	464	-1.500	-	-
USBM	30-300	1 244	-1.450	0.5854	-

Fig. 4 Regression analysis of vibration data for surface blasting in the granites of Hong Kong

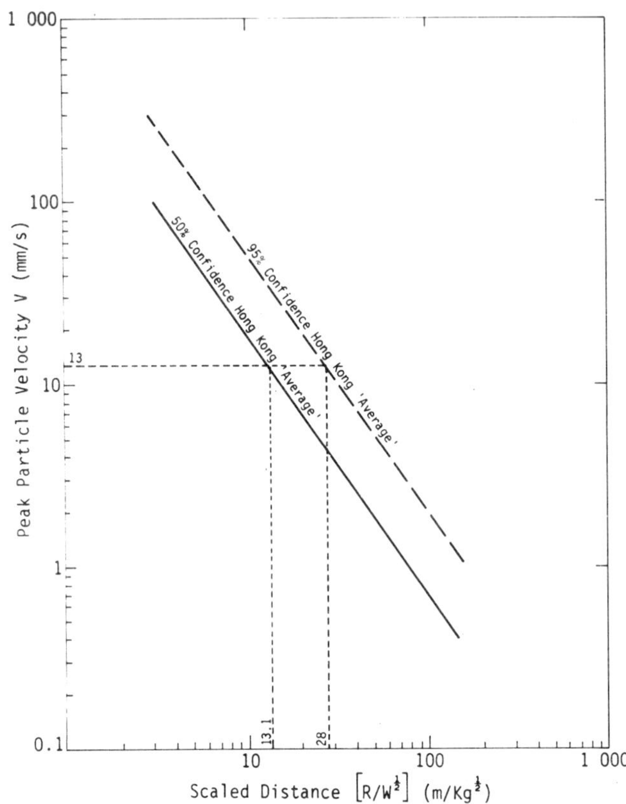

Site Parameter	Confidence Level 50%	95%
A	503	1 479
B	-1.421	-1.421

R = distance from blast (metres)

W = maximum charge weight of explosive in a detonator delay interval (kilogrammes)

Fig. 5 Suggested scaled distance relationship for the preliminary design of surface blasts in the granites of Hong Kong

Tunnel 3 was of particular importance, since :

o it would not be possible to directly measure vibration levels within the functioning tunnel;

o if blasting was to be controlled by vibration monitoring undertaken at or near the ground surface, there had to be some prior indication that such measurements could be used to assess conditions at the tunnel.

Three examples of vibration monitoring in tunnels were obtained during the data search - two examples of underground monitoring of vibrations originating from surface blasts, and one example of underground blasting/vibration monitoring[15, 19]. The regression lines for these examples together with their 'average' and the Hong Kong surface 'average' are shown in Figure 6. It can be seen that the use of the underground mean regression would result in the prediction of vibration levels lower than those predicted by the surface mean. Consequently, it was considered that the Hong Kong surface 'average' could be satisfactorily used for blast preliminary design at Sham Tseng - there is the indication that this approach would result in a pessimistic estimation of vibration levels at the tunnel. This has the implication that surface vibration monitoring at a distance from a blast similar to the shortest distance between the blast and the tunnel, would indicate levels slightly higher than those occurring at the tunnel. Consequently, surface monitoring on sound rock would represent a satisfactory means of assessing vibration levels at the tunnel.

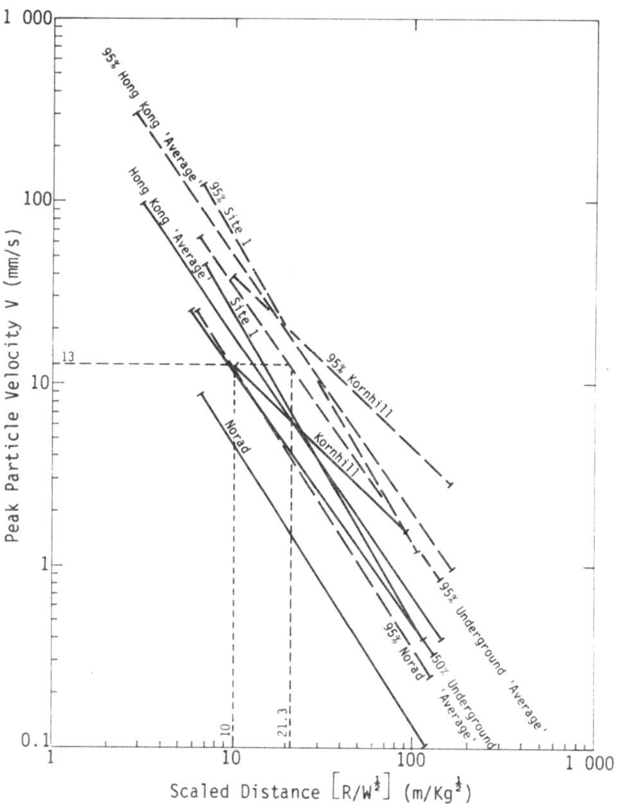

Site	Relationship	R Range (m)	A	B	95% Conf. (Log_{10})	No. of Data Points
Kornhill	Surface to Cavern	60-147	124.32	-0.95	0.504	41
Norad	Tunnel to Tunnel	60-150	179.69	-1.54	0.434	71
Site 1	Surface to Tunnel	15-60	1 433.3	-1.69	0.438	63
Mean values		-	317.54	-1.39	0.459	-

Fig. 6 Underground blasting scaled distance relationships

It must, however, be stressed that this conclusion was reached only with respect to surface blasts affecting the tunnel. From experience, it is known that vibration levels originating from confined underground blasts can be considerably greater than would be predicted from surface derived relationships. This is particularly true for points in the rock mass ahead of the blasted face[19].

The above appraisal of underground vibration presented the opportunity to establish the approximate vibration levels, to which completed sections of Tunnel 3 were subjected during construction. Based on published underground to underground correlations[15, 19], the level of construction blast vibration was estimated. The results are presented in Figure 7.

From Figure 7 it can be seen that at distances of less than 5m from the construction face, the completed section of tunnel would have been subjected to peak particle velocities of greater than 60mm/s. The P.S.O no. 1038 limit of 13mm/s would have been exceeded at all points less than 17m from the blasted face. On the assumption of an advance of 1.4m per blast, this means each point along the tunnel had already experienced vibration in excess of 13mm/s on at least twelve separate occasions. As discussed earlier, there were no major falls during the construction of Tunnel 3.

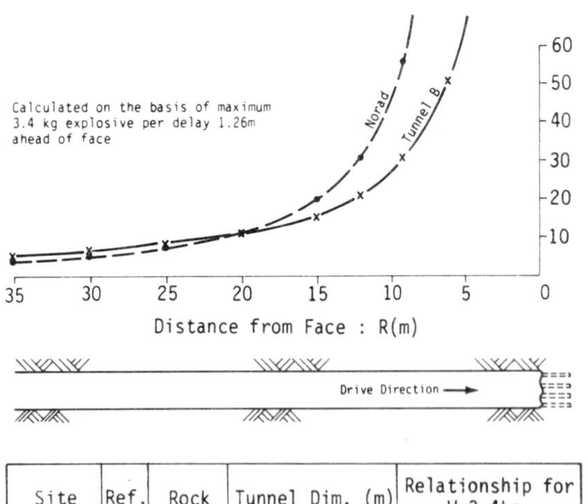

Site	Ref.	Rock	Tunnel Dim. (m)	Relationship for W=3.4kg
Norad	(15)	Granite	2.5 x 3.0	$V=4\ 937\ R^{-2.04}$
Tunnel B	(19)	Granite	4 x 3.5	$V=522\ R^{-1.3}$

Fig. 7 Indicated vibration levels during the construction of Tunnel 3

Implications with respect to Blast Preliminary Design

The investigation indicated that the P.S.O. no 1038 maximum particle velocity of 13mm/s was probably an extremely conservative restriction with respect to a hard rock tunnel, such as Tunnel 3. However, since it would not be possible to actually monitor the vibration levels and its effects from within Tunnel 3 without disrupting the supply, the vibration limit of 13mm/s was endorsed as the preliminary design vibration level.

On the basis of the 'average' Hong Kong particle velocity/scaled distance relationship presented in Figure 5. and a peak particle velocity of 13mm/s, the maximum permissible charge weights per delay for various distances are presented graphically in Figure 8. At the 50% and 95% confidence level, the lowest maximum charge weights per delay (i.e. those not to be exceeded at points within the borrow area closest to Tunnel 3) are 31kg and 6.8kg respectively.

Whilst the low charge weight of 6.8kg is within the capabilities of current Hong Kong blasting practice, its use would give to rise to inefficient excavation i.e. additional drilling and lower daily excavation rates. Consequently, it was recommended that careful monitoring of trial and early production blasts at Sham Tseng should be undertaken to establish whether the application of Hong Kong 'average' 95% confidence limit site parameters is actually warranted[2].

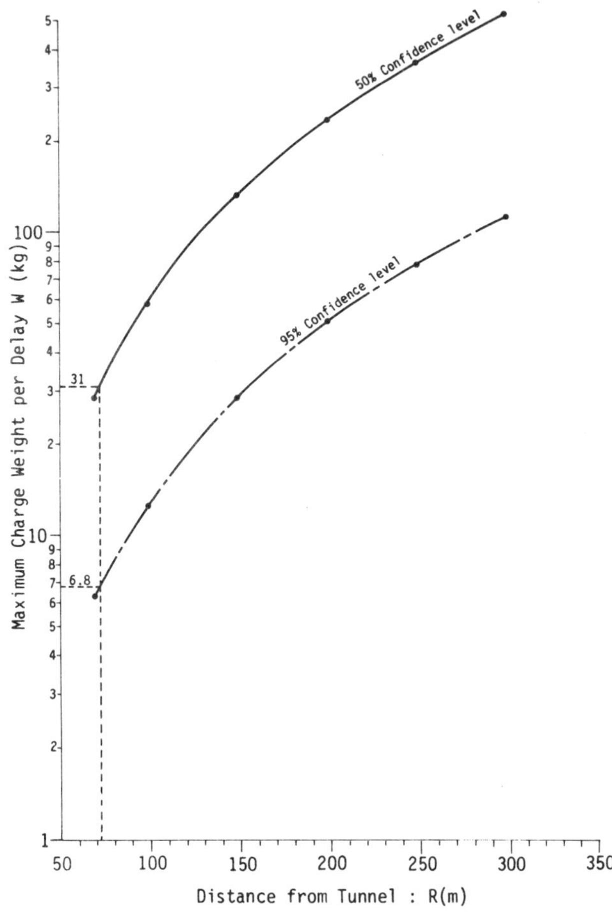

Basis of curves

Confidence Level	Square Root Scaled Distance $\left(\frac{R}{\sqrt{W}}\right)$
50%	13.1
95%	28
Refer also to Figure 5	

Fig. 8 Preliminary charge-weight per delay criteria for blasting above Tunnel 3

Potential Mitigation Measures

Two of the factors which affect the generation of blast vibrations cannot be mitigated at Sham Tseng, since they are inherrent to the proposals. These factors are the distance from the blasting to Tunnel 3 and the intervening geological conditions. The other factors which can be manipulated have been widely discussed by others and readers are referred to several papers by Hagan which cover these aspects in detail[6, 7].

The orientation of the blast face with respect to the alignment of Tunnel 3 will be of importance. Whilst not directly limiting the generation of the vibrations, the effect of these vibrations on the tunnel can be diminished if blasting faces are generally oriented perpendicular to the tunnel alignment. If the blasting face is parallel to the tunnel, a greater length of tunnel will be simultaneously subjected to vibrations of the similar magnitude. Since the tunnel is more 'rigid' in cross-section than longitudinal section, it is considered there will be greater potential for adverse affects with the parallel arrangement.

CONCLUSION

The study concluded that, provided proper controls and monitoring were implemented, blasting could be undertaken without adversely affecting the stability and function of Tunnel 3. However, blasting was only one of several aspects that were considered with respect to their potential effect on Tunnel 3. After completion of the study, it was decided that because Tunnel 3 carried a life-line supply, it could not be put at risk however slight that risk. As part of the borrow operations it has, therefore, been proposed that a duplicate tunnel would be constructed parallel to, and some 150m south of, Tunnel 3. It is anticipated that should the project proceed, the existence of the duplicate will present an ideal opportunity to confirm or otherwise, the conclusions presented in this paper.

ACKNOWLEDGEMENTS

This paper summarises part of the work undertaken for the Hong Kong Government under an extension of Agreement no. C.E. 37/82 and has been published with the kind permission of Government. The provision and permission to publish blasting vibration data from the organisations listed in Table 2 is also gratefully acknowledged. The interpretation of that data was however the responsibility of the author and no acceptance of the interpretation by those organisations should be implied. Furthermore, the opinions and conclusions expressed in this paper are not necessarily those of the Hong Kong Government or those organisations. Thanks are due to Anita Tam for typing the manuscript and to Andrew Cheng for preparing the figures.

REFERENCES

1. ALLEN, P.M. and STEPHENS, E.A. Report on the Geological Survey of Hong Kong 1967 - 1969. Government of Hong Kong 1971.

2. BROADHURST, K., WILTON, T. and HIGGINS, D. Review of standards for blasting noise and vibration. Hong Kong Contractor, June 1984, pages 15-18. (based on a paper entitled Review of current standards and recommendations for vibration and noise. Transactions Institution of Mining and Metallurgy (Section A : Mining Industry), Volume 93, Oct. 1984 A210-A213).

3. DOWDING, C.H. Lifeline viability of Rock Tunnels : Emperical Correlations and Future Research Needs. Proceedings Technical Council on Lifeline Earthquake Engineering. Speciality Conference on the Current State of Knowledge of Lifeline Earthquake Engineering, Los Angeles, August 1977, pages 320-337.

4. DOWDING, C.H. and ROZEN, A. Damage to Rock Tunnels from Earthquake Shaking. Proceedings American Society of Civil Engineers. February 1978, pages 175-191.

5. DUPONT Company. Blasters' Handbook - 175yh Anniversity (16th) Edition. E.I. du Pont de Nemours & Co (Inc) Wilmington, Delaware, 1980.

6. HAGAN, T.N. The Effects of Some Structural Properties of Rock on the Design and Results of Blasting Proceedings 3rd Australia - New Zealand Conference on Gcomechanics,. Wellington, New Zealand, May 1980, pages 2:205.

7. HAGAN, T.N. and KENNEDY B.J. A practical approach to the reduction of blasting nuisances from surface operations. Australian Mining, Vol. 69 no. 11 1977, pages 36-43.

8. HOEK, E. and BROWN, E.T. Underground Excavations in Rock. Institution of Mining and Metallurgy, London, 1980.

9. I.C.I. Blasting Practice. Fourth Edition Nobel's Explosives Company Ltd. 1972.

10. ISAAC, I.D. and BUBB, C. Engineering Aspects of Underground Cavern Excavation at Dinorwic Part 4 : A study of blast vibrations Part 2, Tunnels & Tunnelling September 1981, pages 61-65.

11. LABRECHE, D.A. Damage Mechanisms in Tunnels Subjected to Explosive Loads. Seismic Design of Embankments and Caverns Proceedings of Symposium at American Society of Civil Engineers National Convention, Philadephia, May 1983, pages 128-141.

12. LANGEFORS, U. and KIHLSTROM, B. The Modern Technique of Rock Blasting, Third Edition. Halstead Press, John Wiley & Sons, Inc., New York, 1978.

13. LUMB, P. Engineering Properties of Fresh and Decomposed Igneous Rocks from Hong Kong. Engineering Geology, Vol. 19, 1982/83, pages 81-94.

14. NICHOLLS, H.R., JOHNSON, C.F., AND DUVALL, W.I. Blasting Vibrations and their Effects on Structures. United States Bureau of Mines Bulletin 656, 1971.

15. OLSON, J.J., FOGELSON, D.E., DICK, R.A. and HENDRICKSON, A.D. Ground Vibration from Tunnel Blasting in Granite. RI 7653, Bureau of Mines Report of Investigations. U.S. Department of the Interior, 1972.

16. ORIARD, L.L. Blasting Effects and their control in Open Pit Mining. Procedings 2nd International Conference on Stability in Open Pit Mining. Vancouver 1971. AIME, New York, 1972, pages 197-222.

17. PERSOON, P.A., HOLMBERG, R. LANDE, G. and LARSSON, B. Underground Blasting in a City. Proceedings International Symposium on Subsurface Space, Rockstone 1980, Stockholm, Vol.1, pages 199-206.

18. PRIEST, S.D. AND HUDSON, J.A. Estimation of discontinuity spacing and trace length using scanline surveys. International Journal of Rock Mechanics, Mining Science and Geomechanics Abstracts 18, 1981, pages 183-197.

19. SAKURAI, S. and KITAMURA, Y. Vibration of Tunnel due to Adjacent Blasting Operation. International Symposium on Field Measurements in Rock Mechanics, Zurich, 1977, pages 61-74.

20. SPEARS, R.E. Effect of blasting, jarring, and other transitory vibrations on fresh concrete. Concrete Construction, May 1983, pages 406-407.

21. WISS, J.F. Construction Vibrations : State-of-the-Art. Journal of the Geotechnical Engineering Division, Proceedings American Society of Civil Engineers, Vol.107, Feb.1981, pages 167-181.

Slope stability on a site in the volcanic rocks of Hong Kong

A.W. Clover B.SC., M.SC., C.ENG., M.I.C.E., M.I.M.M., F.G.S.
Freeman Fox & Partners (Far East), Hong Kong

SYNOPSIS

A site formation into a volcanic tuff hillside in Hong Kong is described. Part way through the site formation, a landslide occurred in association with an intense rainstorm. The site investigation and analysis undertaken to determine the cause of the failure are described and discussed. It is concluded that the most likely cause of the failure was block sliding on an unexpected weak zone below rockhead.

The site formation was subsequently re-designed to incorporate permanent ground anchors. The design method and the influence of Hong Kong Government engineering requirements on the design of the anchors are discussed and an approach is presented, whereby the apparent anomaly of calculated factors of safety less than unity for a standing slope was overcome. The results of some 18 months regular load monitoring on installed anchors are appraised, from which it is apparent that the loads had stabilised some 400 days after initial lock-off.

INTRODUCTION

The site is situated on the south side of Hong Kong Island. Prior to construction, it comprised a well-vegetated natural hillside of volcanic soil and rock sloping at about 32° to the horizontal. As there was a height restriction on the proposed development, the necessary building area of some 2700m^2 was to be created by cutting a level platform into the slope, resulting in the formation of a new back slope at least 30m high. A geotechnical investigation was carried out, site formation plans prepared and approved, and work commenced on the site in July 1981.

As excavation proceeded, drainage and rock reinforcement measures were being installed. The work was partially complete when excavation revealed the presence of an unexpected layer of weaker material. This layer was examined, mapped and sampled and remedial measures were being designed when on 29th May 1982, after an extremely heavy rainstorm, part of the upper section of the partially formed slope slipped into the site.

This paper presents the results of the investigation into the landslide and describes the remedial and preventive measures undertaken on the slope. The site formation was completed in September 1983 and the site is now occupied by 3 residential blocks.

SYNOPSIS OF EVENTS

The year of 1982 was the wettest so far recorded in Hong Kong. The total rainfall at the Royal Observatory amounted to some 3247.5mm, which is 46% above average[12]. The rainfall at the site for the storm of 28 May to 2 June 1982 was 794mm. This corresponds to an event of approximately 1 in 60 years return period. However, the landslide occurred on the morning of the 29 May, 1982 and it was estimated that some 525mm of rain had fallen up to that time. This corresponds to an event of approximately 1

Plate 1 View of the 29 May 1982 landslide scar

in 23 years return period[11].

The landslide occurred at 7.05am. Fortunately, no persons were involved with or injured by the landslide. A view of the site after the 29 May landslide is shown in Plate 1. The slide involved an estimated 3,800m^3 of material, mainly comprising closely jointed weathered rock. Approximately 70m length of the partially completed site formation were affected. The maximum extent of transgression outside the site was 12m, and the maximum breadth and height of the scar were both 21m. Debris extended up to a maximum of 32m horizontally and 17m vertically from the toe of the scar. Small areas at the extreme northern and southern parts of the landslide came to rest as relatively intact blocks, while the remainder of the landslide mass completely disintegrated. Extensive cracking appeared on the slope surface immediately south of the landslide and significant water flow continued for several hours from a number of discrete seepage points.

Subsequent to the landslide, all slide debris was removed from the site and formation works were recommenced in a restricted area in the northern part of the site. An investigation of the landslide was also instigated. Following a further rainstorm of some 500mm on 17 August 1982, a second landslide occurred. This slide involved approximately 700m^3, in the southern area which had been disturbed by the first landslide. The scope of the then on-going investigations were subsequently expanded to include this second landslide. However, since those investigations showed the prime future potential failure mechanism at the site to be similar to the 29 May landslide, only that landslide is discussed in detail in this paper.

INVESTIGATION OF THE LANDSLIDE

Investigation of the landslide comprised the following elements:

- o topographic surveying of the landslides areas at a scale of 1 in 100;
- o engineering geological mapping of the landslides;
- o 4 additional boreholes, with associated sampling, insitu testing, and installation;
- o installation of maximum ground water level indicators (Halcrow Buckets British Patent no. 1538487), in selected existing and new standpipes;
- o three trial pits and associated probing;
- o the collection of large block samples and driven-tube samples from the first landslide failure surface;
- o in-situ shear box testing using portable equipment.

General Geology

The site is located within the type area of the Repulse Bay Formation. Rock types comprise welded tuffs of Middle Jurassic Age. A major photogeological lineament trends north-east wards along the valley in which the site is situated[1]. Joints are closely to widely spaced, and there is evidence of faulting. The weathering profile is irregular and generally increases in thickness towards the southern end of the site.

For convenience, the materials at the site were sub-divided into three main groupings, namely :

- Soil - this group included colluvium, completely weathered and some highly weathered tuff. The lower boundary was generally taken as the level at which rotary rock-core drilling commenced, and is here defined as 'Rockhead'. Rockhead contours were sub-parallel to the natural ground surface. The characteristics of the materials forming this group were :

  Colluvium - loose, brown, gravelly, sandy SILT containing rock fragments and boulders in various states of decomposition. This material was either totally absent or only up to 1 metre thick.

  Completely, and Highly weathered tuff - medium dense, yellowish grey, sandy SILT containing moderately weathered rock fragments. The thickness of this material varied from 0 to 5.5 metres.

- Fractured Rock - this was defined as material which although generally requiring rotary rock-core drilling to obtain samples, still contained a significant proportion of soil. As a consequence of this, much of the recovered rock core was not of a full disc cross-section. In addition the rock was heavily ochre stained, both on joint surfaces and internally. The material mainly comprised closely jointed, yellowish-grey, highly to moderately weathered tuff, containing open (very narrow to narrow) and clay in-filled joints. There was a significant thickening of this material along both the eastern and southern boundaries of the site. In particular on the eastern boundary, the Fractured Rock occurred in a dome-like feature.

- Sound Rock - this material mainly comprised grey to dark grey, slightly weathered tuff, with medium to moderately spaced joints, some ochre-stained. It was considered that this material represented Zone D as per the definition of the then edition of the Geotechnical Manual for Slopes, as published by the Geotechnical Control Office (GCO), Engineering Development Department, Hong Kong[7,8]. The Fractured Rock/Sound Rock interface was not entirely sub-parallel to the natural ground contours. The interface was closest to the natural ground surface at the north-east corner of the site and dipped across the site becoming deeper towards the south-west.

In total, approximately 500 discontinuity measurements were made at various times at the site. The analysis of the discontinuity data suggested five main orientations, but the majority exhibited a north-westerly dip direction. This was particularly significant with respect to the site formation, since these discontinuities dipped in the same direction as the excavated slopes. The kinematic assessment is shown in Figure 1.

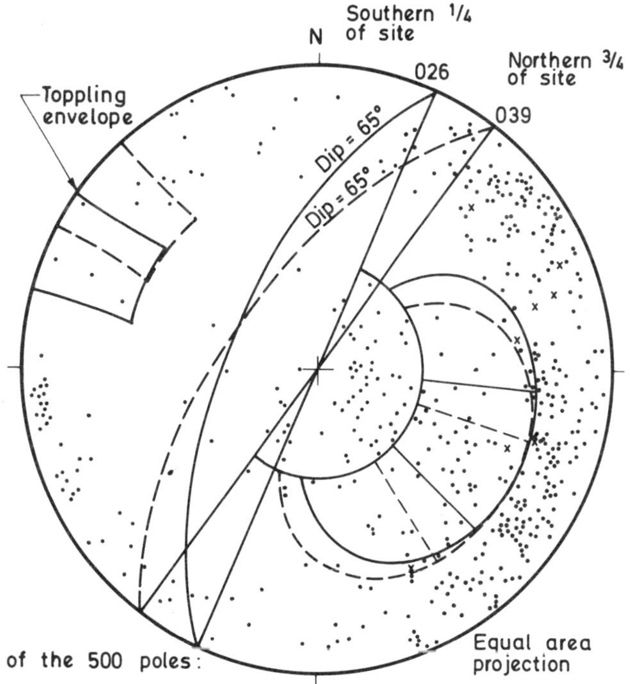

Fig. 1 Kinematic assessment of rock discontinuity orientations.

Geology of 29 May, Landslide

Soil cover varied from 0 to approximately 2m thick. The landslide back-scar was primarily defined by sub-vertical joint planes in the Fractured Rock and the lower-most section of the failure surface was defined by the interface between the Fractured Rock and the Sound Rock. In the northern part of the site this interface was formed by a 0.7m thick weak zone comprising yellow-brown, clayey, sandy, SILT, with numerous angular rock fragments and pockets/lenses of firm yellow-brown clay. Towards the south it appeared that the thickness of this zone gradually decreased.

Over most of the breadth of the landslide, the back scar and weak zone were separated by an irregular transition which passed through Fractured Rock. This was partly formed by joint planes, partly by the clay infill between open-joints and partly by shear planes through the weaker fractured rock blocks. The surface of the underlying Sound Rock was extremely rough. Roughness surveys using baseplates of 110mm, 210mm and 420mm diameter, indicated amplitudes in the range 200mm to 250mm. This was, however, considerably less than the observed thickness of the majority of the weak zone. The mean dip of the exposed surface was approximately 30°/290°. Excavation during the revised site formation later revealed that the weak zone was only a continuous planar feature over the northern half of the slope, varying in dip from 14°/323° to 40°/287°.

Following the uncovering of the weak zone by the landslide, the original site investigation boreholes and core photographs were reviewed to establish whether they contained any features which could be attributed to this zone. By apply hindsight it was possible to determine a pattern between the boreholes. However, core recovery was generally so poor in the Fractured Rock layer that it was not surprising that this feature had escaped earlier detection.

Groundwater

Since Halcrow Buckets had not been installed in the standpipes prior to the 29 May landslide, there were no records of the maximum groundwater levels associated with that failure[4]. A first indication of lower-bound levels was, however, available from dipped readings taken prior to the landslide and in the fews days following the landslide. Later investigation of the 17 August landslide showed that the intensity and total rainfall immediately prior to both landslides were almost identical, and the site formation and landslides had apparently had little influence on base groundwater levels. Consequently, it was considered that the levels associated with the 29 May landslide were probably similar to those recorded by the Halcrow Buckets at the time of the 17th August landslide. These measured levels implied either 4.4m to 20m rise above base level, or an up to 12.4m deep perched water table. The maximum value in both cases was considerably greater than would usually be anticipated in Hong Kong.

Following further measurements undertaken in the new boreholes, it was established that the permanent groundwater table existed deep within the Sound Rock. In association with storm events of greater than 50mm in 24 hours, this level could increase by up to 4.5m. For a similar event, a perched water table up to 2.5m deep formed above the weak zone or, where this was absent, above the interface between the Fractured Rock/Sound Rock.

Using all the rainfall and groundwater level data for the site, the methods outlined by Endicott were used to appraise the groundwater response with respect to rainfall[6]. This analysis showed that groundwater levels exhibited an approximately bi-modal response. Rapid and significant rise in level water was apparently associated with the first 40mm to 50mm rainfall for storms of 1 or 2 days duration, but the effect of additional rain in this period or over an extended period generally did not increase the level in the same proportion.

Material	Ø'	c'(kPa)	Comment
Soil : Completely and highly weathered tuff	37°	8	Grouped together since no significant difference in results.
Fractured Rock : joint infill and weathered zones	43°-49°	9-21	High values a reflection of the gravelly nature of the material.
Weak Zone :	32°-39°	0-14	Lower results from direct shear tests.
Rock joints :	30°	-	mean roughness 10°±7°.

Table 1 : Summary of test derived shear strength parameters.

This phenomenon had particular significance with respect to the later selection of design groundwater levels. In accordance with the Geotechnical Manual for Slopes, the slope was to be designed to a factor of safety of 1.4 for a 1 in 10 year rainstorm event and have at least a factor of safety of 1.12 for a 1 in 1000 event. Based on Jenkinson's Maximum Likelihood method, the 1 in 10 years and 1 in 1000 years return period 2 days rainfall are 429mm and 950mm, respectively[11]. Both these points fell on the flatter portions of the response curves. It was, therefore, estimated that there would generally only be a slight increase from the 1 in 10 years levels to the 1 in 1000 years levels, and that both would be of a similar magnitude to the levels measured at the time of 17 August landslide.

Shear Strength

All interpretation of laboratory shear strength test data from the original site investigation was re-appraised. To this were added data from triaxial and direct shear tests undertaken as part of the landslide investigations. In total, the results from 28 single-stage and 14 multi-stage triaxial tests on Soil, 4 single and 4 multi-stage direct shear tests on Soil and 3 multi-stage direct shear tests on rock joints were available for analysis. Except for one consolidated undrained test, all tests were under consolidated drained conditions. A summary of the interpreted results of these tests is presented in Table 1 .

Four in situ shear tests using the portable shear box described by Brand et.al were also undertaken on the material forming the weak zone[3]. This equipment ideally suited the situation at the site since it was readily transported upto and erected on the landslide scar. Unfortunately, the gravelly nature of the material gave rise to unsatisfactory box alignment and ill-defined shear surfaces, and resulted in what were considered unrealistically high shear strength values. These data were, therefore, not utilised in the appraisal, although the possible interlocking effect of the gravel was noted.

BACK ANALYSIS OF THE LANDSLIDE

To undertake a comprehensive back-analysis of any landslide, information is required on the following :

- o geometry of failure mass and failure surface;
- o geological and engineering conditions at the landslide;
- o groundwater conditions at the time of the failure;

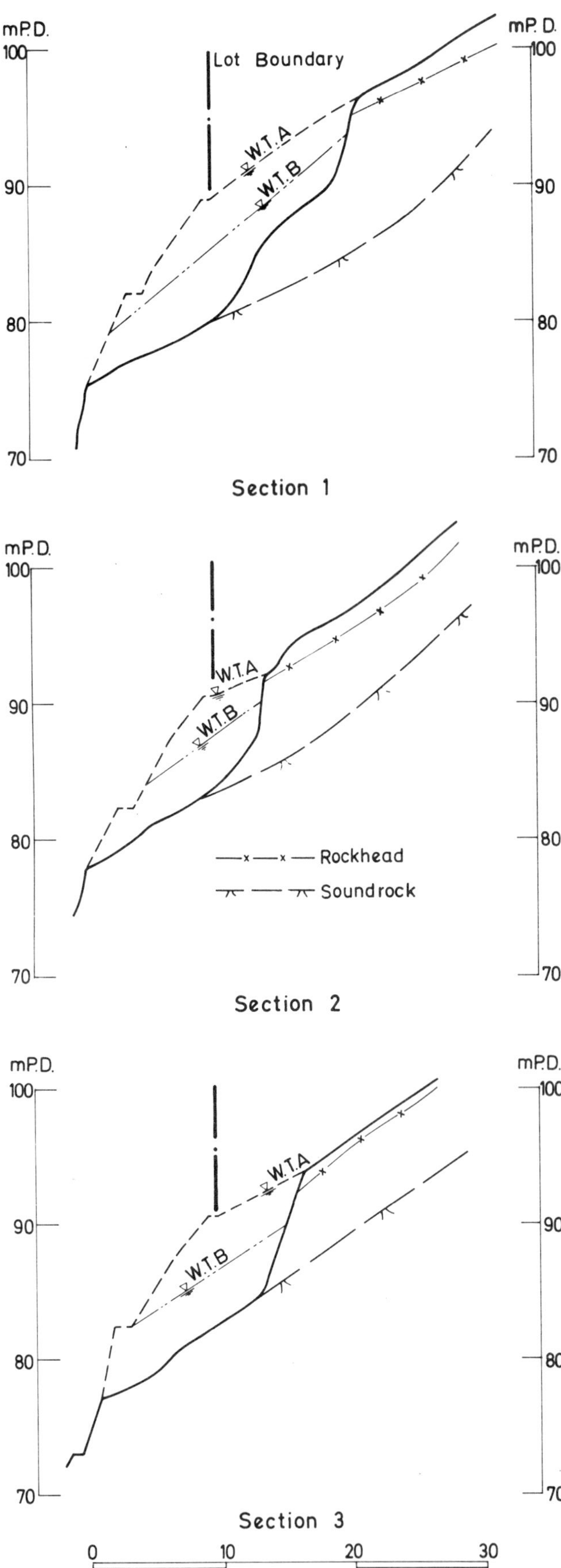

Fig. 2 Cross-sections utilised in back-analysis of the landslide.

o failure mechanism.

Where precise data on any of these factors is unknown, assumptions have to be made based on judgement. The effect of variation of these assumptions can then be tested by sensitivity analysis, in order to determine whether they are critical. Even where data are available, additional simplifying assumptions are sometimes necessary in order to bring the mathematical manipulation of the data to meaningful and manageable proportions. The prime factor influencing these assumptions will be the ultimate aim of the analysis. In this particular case, this was the determination and implementation of a conservative but sensible and practical modified site formation design, that would ensure the future long-term stability of the slopes surrounding the site.

Three cross-sections were used in the back-analysis. These are shown in Figure 2. It can be seen that the failure surface can be divided into a number of planar elements. For this reason, both non-circular rotational and translational block failure mechanisms were considered. The non-circular analysis was undertaken using Janbu's Simplified method[10]. Block analysis was undertaken using appropriate limit equibrium equations derived in a manner similar to an approach presented by Barton[2]. In the Janbu analysis only homogeneous ground conditions were considered i.e the variation in geology along the failure surface was ignored. This major simplication significantly aided the process of analysis, but resulted in the determination of 'averaged' shear strength values. In the block analysis, sliding was assumed to take place primarily in the weak zone with separation occurring on the steeper planes within the transitional material and at the sub-vertical back scar.

Various groundwater configurations within the failure mass were considered, to cater for different draw-down conditions at the slope face. Since an over-estimation of the groundwater levels associated with the failure would result in the determination of optimistic

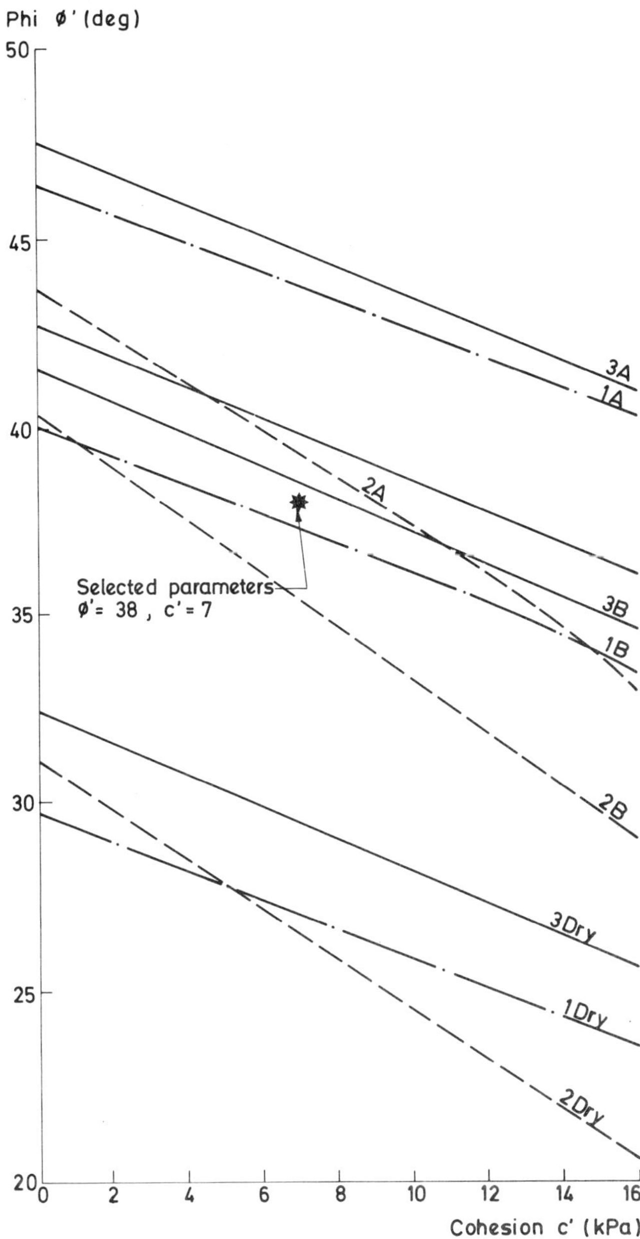

Fig. 3 Limit equilibrium shear strength parameter envelopes.

design values for the shear strength parameters, back-analysis for dry conditions was also be undertaken for comparison.

Based on the results of the analysis, block sliding on the weak zone was identified as critical mechanism but there was no common point of intersection on the critical strength envelopes for each cross-section. Typical results are presented in Figure 3. On the basis of a comparison between the results of the back analysis and shear strength tests, average values of $\phi'=38°$ and c'=7kPa were eventually chosen as the design parameters for the weak zone.

DESIGN OF THE REMEDIAL/PREVENTIVE WORKS

Application of the re-evaluated design parameters to existing conditions showed that the slope as standing and as then proposed did not have the required factor of safety. It was therefore apparent that a major revision of the site formation design was required.

An overall reduction in slope angle was not possible since a common boundary with an adjacent developed site was formed by almost vertical rock slope. The construction of a restraining structure was therefore the only practical mechanism for positively ensuring future stability. Several possible types of structure were examined.

The concept of a massive retaining wall along the lines presented by Vail and Holmes was considered but was abandoned for reasons of timing and extent of required temporary works, and the relationship with the proposed development[13]. The flexibility of a ground anchor system made it well suited to the type of variable conditions at the site. i.e. required degree of support and the orientation, thickness and location of the potential failure surface. The main disadvantage of the use of anchors was that, in view of the consequences of failure, careful monitoring and maintenance would probably be required for the life of the structure. However, detailed examination of the alternatives lead to the conclusion that the most practical solution was a permanent anchoring system.

It was proposed that the Soil and Fractured Rock slopes would be restrained by anchor loads transmitted through discrete reinforced concrete beams extending vertically both above and below potential failure surfaces. The beams would be regularly spaced across the slope, following the major potential failure plane, and would be linked by a non-structural system of concrete planks, which would retain soil for planting.

The Sound Rock slope would be cut back to an overall slope angle of 65°, thus eliminating a large proportion of the potential planar/wedge failures. In addition to the anchors, the drainage on the slope would be improved by means inclined drains. The likely drawdown effects of these drains would not, however, be taken into account in the determination of required anchor forces.

A typical section through the stablised slope is shown in Figure 4. For a given section, the required stablising force(s) were calculated on the basis of a block that had its up-slope limit defined by the point at which infinite slope analysis gave the required factor of safety. Due to the extreme variability of the geological conditions, design sections were taken at 10m intervals along the slope. Blocks with bi-planar failure surfaces were divided into sub-blocks to establish whether one or two levels of anchors were appropriate. The prime consideration of this exercise was to decrease the potential for instability in the temporary condition. Boundaries between sub-blocks were assumed to be vertical with the inter-block force acting parallel to the base of the upper block. If two levels of anchors were considered necessary, the force on the lower level was made at least equal to the upper force, regardless of whether stability calculations indicated that a lesser force would suffice.

The embedded length of an anchor was determined as shown in Figure 5, with the intention that the loaded Sound Rock block should not pull into the Fractured Rock. The assumptions made in this calculation were as follows :

- o the upper end of the fixed length to be at least 2m below the potential failure plane;
- o resistance to pull-out provided by frictional resistance on the lower surface of a pyramidal block;
- o resisting surface defined by a horizontal or upwards inclined plane from the fixed length end to the base of the anchor beam;
- o back surface of pyramidal block defined by a vertical plane from the fixed length end;
- o frictional resistance derived from weight of pyramidal block and the weight of the Soil and Fractured Rock in the zone vertically above the projection of the pyramid base on the Sound Rock surface;
- o submerged weights of Soil etc. utilised in calculations;
- o free length to be such that frictional resistance equal to or greater than total anchor working load on section.

Not withstanding the above, the anchor fixed length was to be founded in area comprising a minimum 90% grade I and II rock material, as indentified from chips during the percussion drilling of the free length. Regularly spaced inclined cored boreholes were also to be drilled to confirm the chip interpretation.

Whilst the design was being undertaken, the GCO were concurrently preparing their model specification for ground anchors. A draft copy of that document was obtained (now published) and that, together with the B.S.I. Draft for Development - Ground Anchorages, was used to prepare a tight method specification for the anchors[5, 9].

Approval of the revised site formation design was obtained in June 1983. At that time there was an unstated Government moratorium on the use of permanent anchors in Hong Kong, whilst the GCO formulated its model specification and methods of control. This situation had been brought about by past poor workmanship and a concern over corrosion. It was, however, accepted that an anchor system represented the most suitable solution for the particular circumstances, but approval was conditional on the following :

- o a commitment to a long-term regular monitoring programme of anchor loads and the anchor head corrosion protection system;

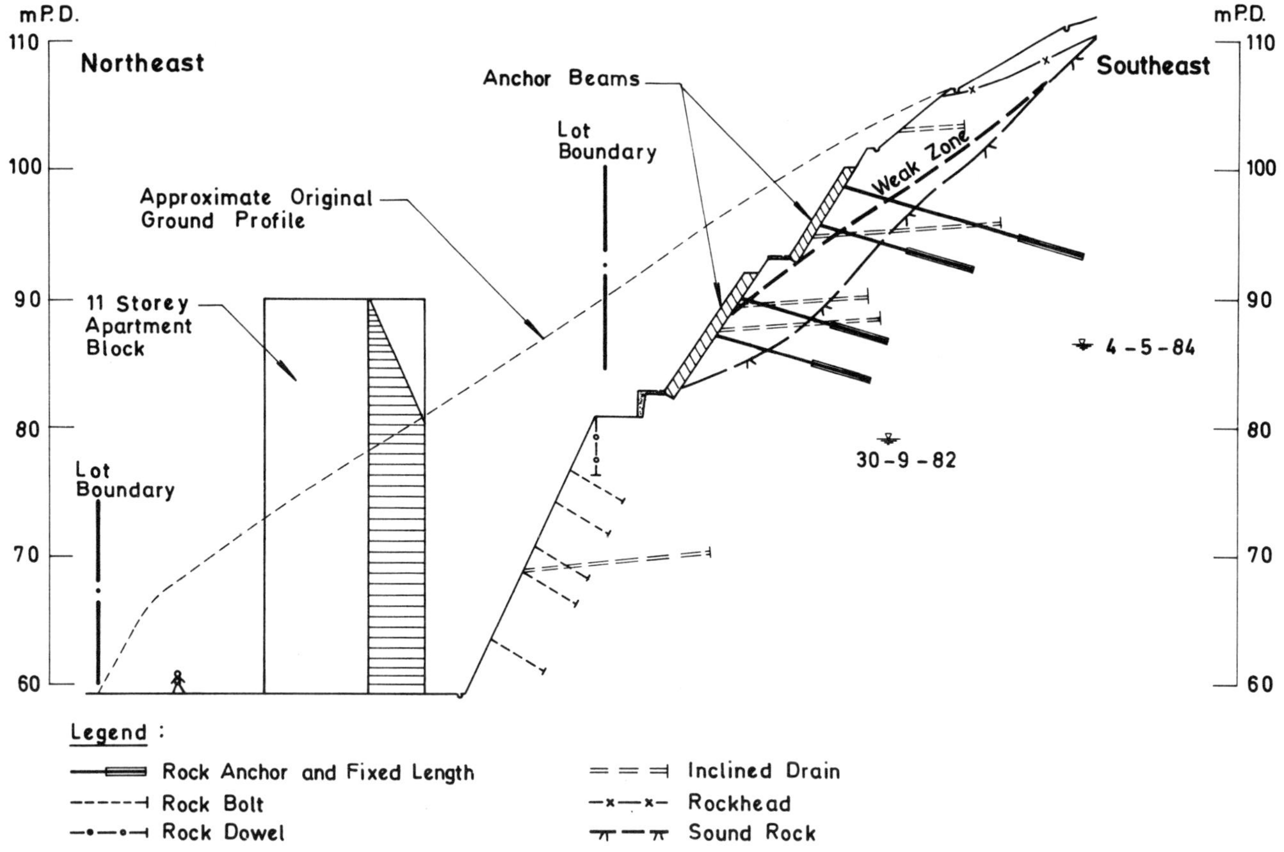

Fig. 4 Typical section through the stabilised slope.

- o regular inspection, checking and reporting of the works during the construction by a senior member of the design team;

- o on completion of the works, the submission of a 'Performance Review' in the form of a report which stated and justified that the works had been inspected and monitored during construction and that the geotechnical design assumptions upon which the completed site formation had been based were valid;

- o on completion of the anchor defects liability period, the submission of an 'Anchor Report' containing information on the monitoring undertaken to that date, any maintenance works undertaken, the criteria for assessment of monitoring results, the action to be taken should the stated criteria be exceeded and the form and frequency of future monitoring requirements.

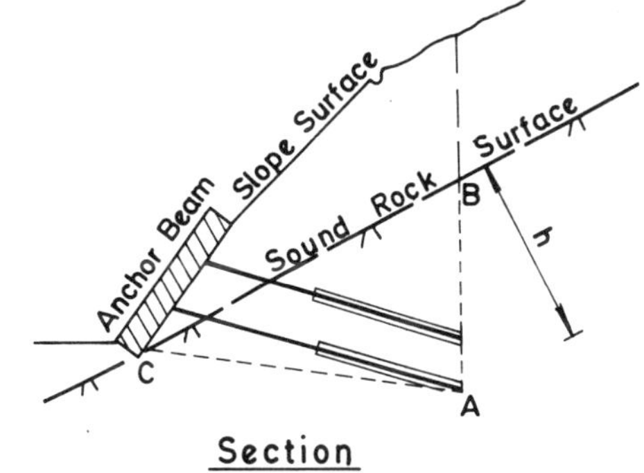

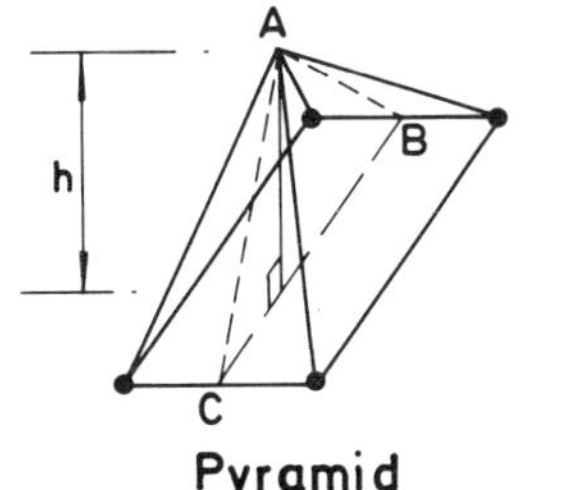

Fig. 5 Geometry used in the determination of anchor pull-out resistance.

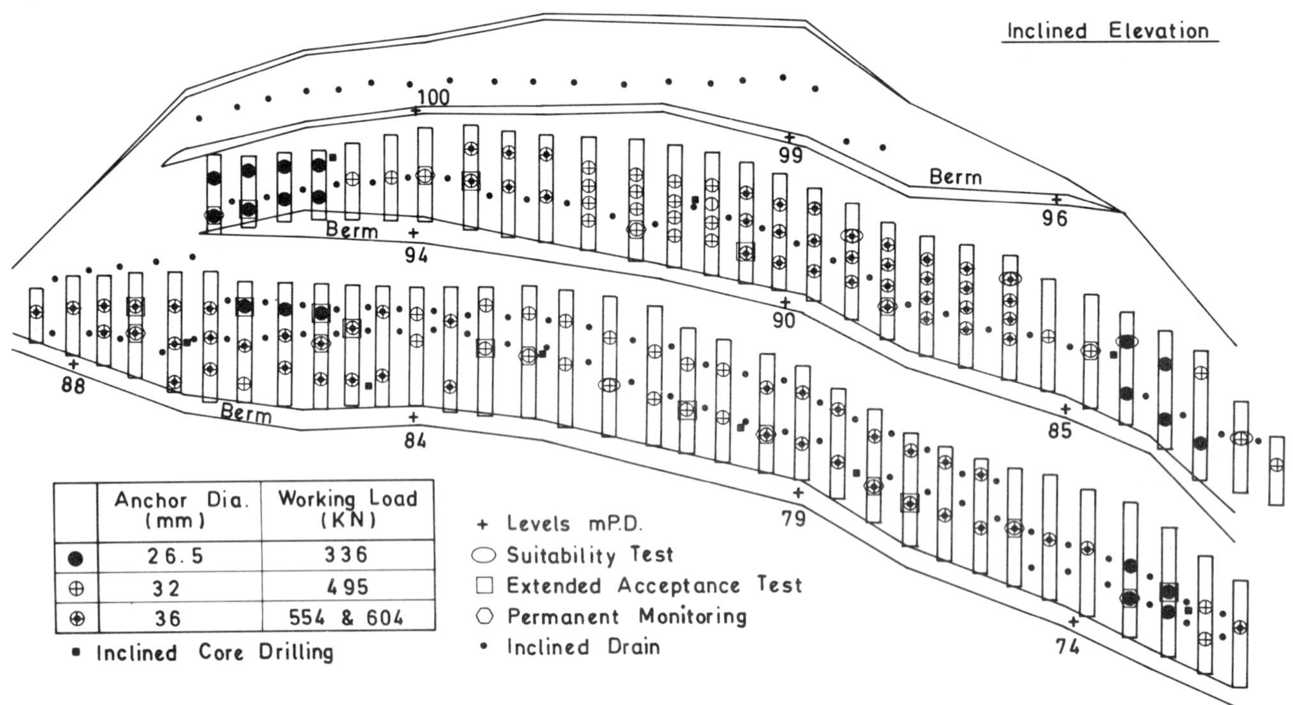

Fig. 6 Distribution of the installed threadbar anchors.

PARTICULAR ASPECTS OF THE WORKS

A view of the completed site formation is shown in Plate 2 and the distribution of the anchors in Figure 6. In total, 139 single threadbar anchors with double corrosion protection were installed. Working loads range from 336kN (26.5mm dia.) to 604kN (36mm dia.), with free lengths from between 5m to 18m. All anchors have been designed to a minimum factor of safety of 2. Allowable bond strengths of 0.99N/mm^2 and 0.31N/mm^2 for bar/grout and grout/rock respectively were implied in the design, for a cement grout with a water cement ratio of 0.39 plus 0.1% plasticizing/expansion admixture by weight cement. Prior to installation of the working anchors, two trial anchors were installed vertically downwards in massive slightly weathered tuff and stressed to confirm these strengths. Based on uniform stress distribution, these anchors indicated that ultimate bond strengths of at least 5N/mm^2 (bar/grout) and 2.8N/mm^2 (grout/rock) could be mobilised, since neither anchorage failed during the test. During the works, 7 Suitability Test anchors were installed with purposely shortened fixed lengths in order to further confirm the design bond values. These tests were successful and the anchors were later accepted as working anchors. All working anchors were subjected to an Acceptance Test of 1.5 times the working load, acceptance being based on both load and extension characteristics.

Plate 2 View of the completed site formation

Bar anchors were chosen over multi-strand anchors for the following two main reasons :

o since components of the bar system are

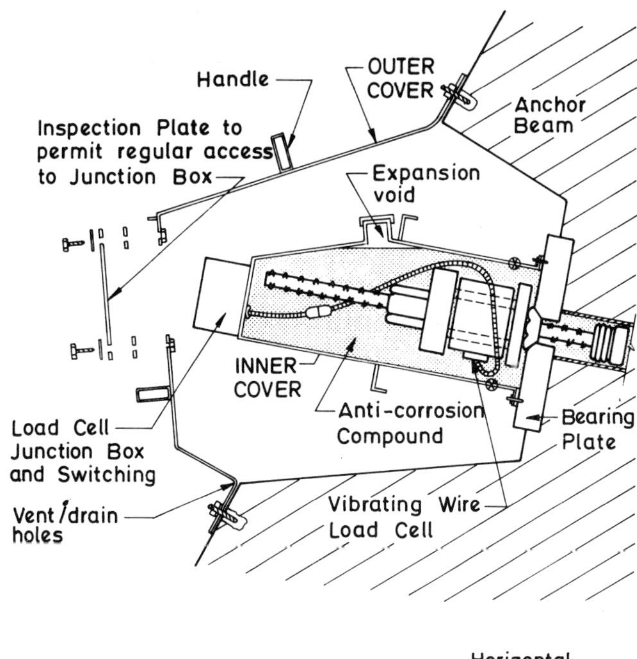

Fig. 7 Head detail of a load-monitored anchor.

standarised, prior knowledge of anchor hole depth is not required. System components can be fabricated in bulk prior to drilling and hole depths can be slightly increased, if appropriate, to match multiple unit length. There is, therefore, the potential for time saving during construction;

o later destressing and restressing of installed anchors are relatively simple to accomodate in the design of the structure, since long tendon jacking allowances at the anchor head (as associated with strand anchors) are not required.

The main disadvantage with bar anchors is that they require coupling and this is a potential weak link in the system. Close supervision and cleanliness requirements were, therefore, imposed on these operations and a photographic record was made of every coupling. Standard bar lengths were 12m, but this was reduced to 6m to aid handling in the difficult site conditions.

To comply with monitoring requirements, 12 of the bar anchors have permanently installed vibrating wire load cells. The detail of one of these installations is shown in Figure 7 and the

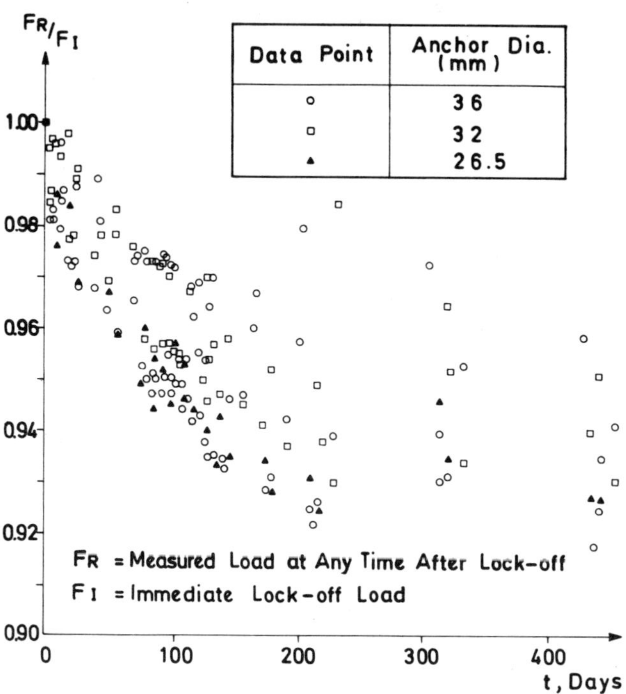

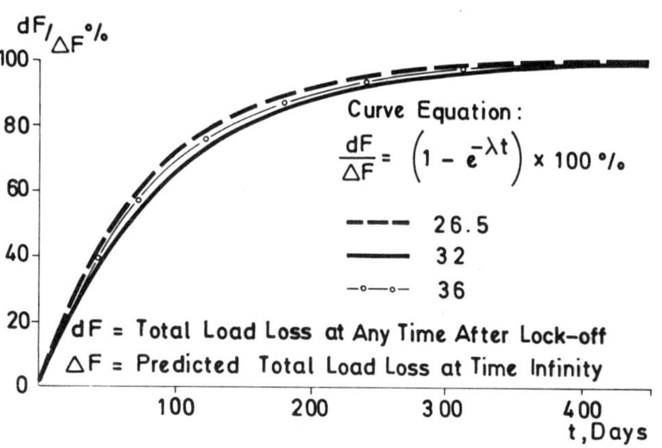

Anchor Dia. (mm)	ΔF (KN)	λ ($\times 10^{-2}$)	$\left(\frac{\Delta F}{F_I}\right)_{av}$ %
26.5	30.9	1.255	8 ± 1
32	32.9	1.058	6 ± 2
36	44.0	1.245	7 ± 3

Fig. 8 Load monitoring results and appraisal.

results of the first years monitoring are presented in Figure 8. From Figure 7 it can be seen that particular attention has been paid to anti-corrosion protection measures at the anchor head. All the bar anchors at the site have a similar, but smaller, inner cover so that a check on the load in any anchor may be undertaken. The data points in Figure 8 have be normalised with respect to the initial accepted load of each anchor, to take account of different working loads. All the anchors showed a marked initial load loss followed by a gradual reduction in the rate of load loss. This is attributed to (in order of significance) :

- o settlement of the anchor beams into the completely to moderately weathered tuff;
- o relaxation of the anchor bar steel.

After the first year, the total load loss in the monitored anchors comprised 6.5% ± 1.2% of the accepted load, which in all case equates to about 0.02% decrease in the free length. The data shown in Figure 8 have also been subjected to exponential decay analysis, as shown in the figure. Based on the regression line and assumming no other factors come into play e.g. corrosion, mechanical failure, creep, this analysis indicates that at time infinity the load in the monitored anchors would still be greater than the design load.

During excavation of the lower rock slope, in the northern part of the site a 25m long feature oriented approximately parallel to the overlying weak zone was uncovered. Detailed investigation of this 'fractured zone' was subsequently undertaken and this showed that unlike the weak zone, it did not contain a significant silt and clay proportion. Rather it comprised a rough and wavy, very closely to closely jointed, moderately weathered feature up to 0.3m thick. However, based on an orientation of the fracture zone derived from boreholes and trace mapping on the excavated face, known groundwater conditions and basic friction values plus measured field roughness, two dimensional stability analysis generated factors of safety less than unity for what was actually a stable field situation. Consequently, the design approach in this area adopted the concept of an internal body force aiding stability, to take account of unknowns such as continuity of the zone, three-dimensional effects and shear strength variation.

This design approach is summarised by the flow chart shown in Figure 9. Using this method it was established that an additional external restraining force of approximately 24,000kN was required to give a factor of safety of 1.4 for 1 in 10 year groundwater conditions, with a similar degree of restraint being implied internally. As this stabilising force was considerably greater than that required on the upper level, permanent multi-strand anchors were used at this location.

Altogether 12 multi-strand anchors were installed with working loads of 2000kN and 10m free lengths. No permanent load cells have been included on these anchors but the loads in 2 anchors have been regularly monitored by lift-off tests. In the first year of monitoring a total load loss of approximately 190kN was recorded (8% of the initially accepted load), which corresponds to a 0.04% i.e. 4mm decrease in the free length.

CONCLUSION

This paper has discussed a particular problem encountered during site formation in the volcanic rocks of Hong Kong. By their very nature of deposition and placement, volcanic rocks tend to be extremely variable and this situation is exacerbated when the effects of tropical weathering are superimposed. As was the case with this site, this variability will often not be fully revealed until excavation takes place. It is, therefore, extremely important that the design process should be extended to the field. Critical assumptions of the design should be communicated to site staff and a system of on-going review established. Even when such a system is working, the weather is one factor that cannot be controlled. It is fortunate that on this particular occassion no-one was killed or injured and in retrospect the landslide represented the best, if somewhat expensive, investigation of the weak layer. It must, however, never be forgotten that there is risk associated with any major excavation and this risk is compounded when the excavation is undertaken in an urban environment.

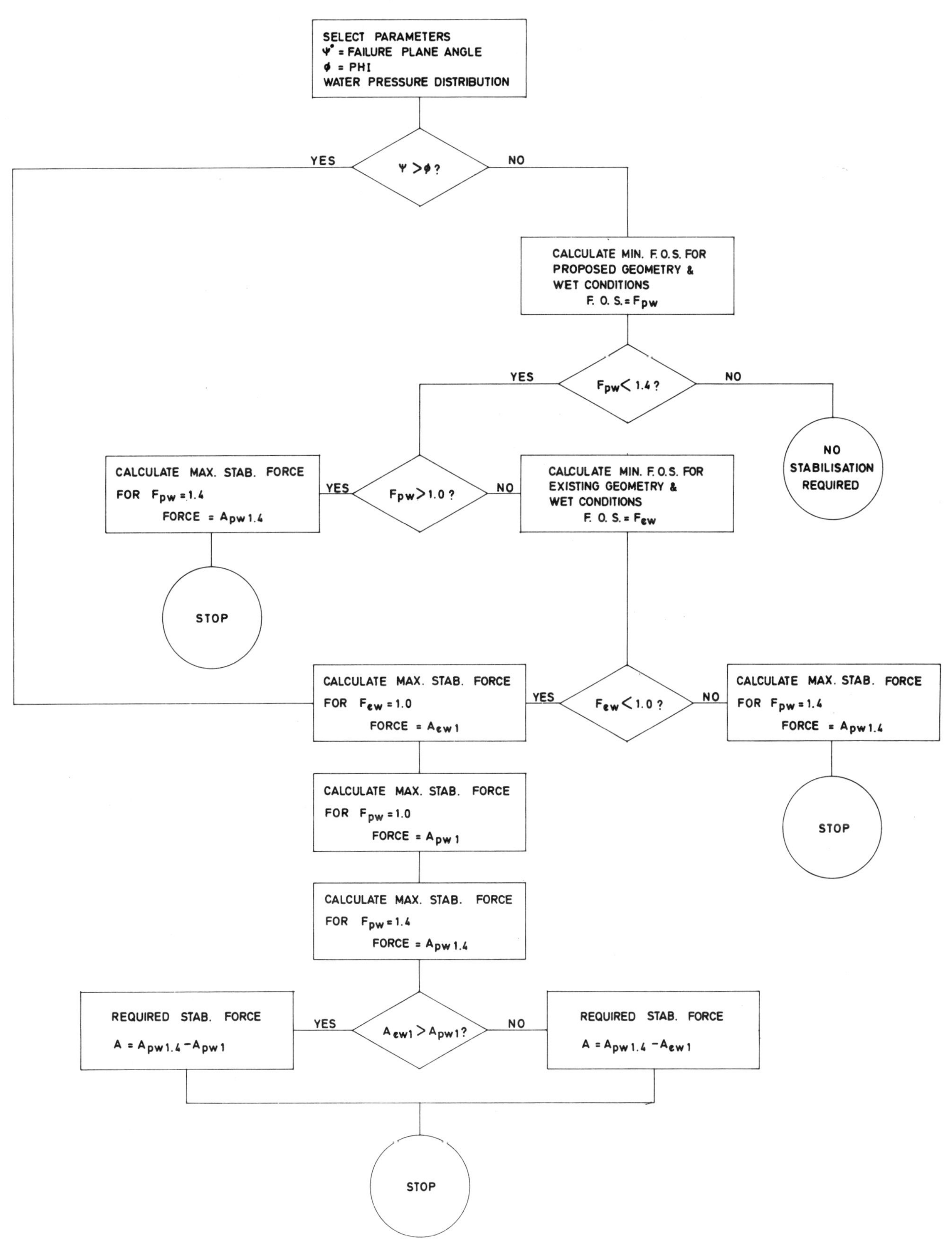

Fig. 9 Flow chart for design approach based on the internal body force concept.

ACKNOWLEDGEMENTS

This paper is published with the kind permission of Kalat Co. Ltd. Thanks are due to Raymond Kan and Norman Leung who were involved in the design and implementation of the works and in particular to Philip Lam who spent many hours over a calculator in the office and clambering about the site. The loan of probing equipment and the in situ shear box by the Geotechnical Control Office, Hong Kong is gratefully acknowledged. Thanks also to Anita Tam for typing the manuscript and to Andrew Cheng for preparing the figures.

REFERENCES

(1) ALLEN, P.M. and STEPHENS, E.A. Report on the Geological Survey of Hong Kong. Hong Kong Government Press, 1971, 107p + 2 maps.

(2) Barton, N.R. Estimation of In-situ shear strength from back analysis of failed rock slopes, Symposium of International Society of Rock Mechanics, Nancy 1971, 11-27.

(3) BRAND, E.W., PHILLIPSON, H.B., BORRIE, G.W. and CLOVER, A.W. In situ direct shear tests on Hong Kong residual soils. Proceedings of the International Symposium on Soil and Rock Investigations by In Situ Testing, Paris, 1983, vol. 2, pp 13-17 (discussion, vol.3, pp 55-56)

(4) BRAND, E.W. BORRIE, G.W. & SHEN, J.M. Field Measurements in Hong Kong Residual Soils. Proceedings of the International Symposium on Field Measurements in Geomechanics, Zurich vo. 1, pp 639-648.

(5) BRITISH STANDARDS INSTITUTION. DD81. Draft for Development-Ground Anchorages. British Standards Institution, London. 1982.

(6) ENDICOTT, L.J. Analysis of piezometer data and rainfall records to determine groundwater conditions, Hong Kong Engineer, 1982, vol. 10, no. 9, pp 53-56.

(7) GEOTECHNICAL CONTROL OFFICE. Geotechnical Manual for Slopes (First Edition). Hong Kong Government Printer, 1979, 242 p. (reprinted with corrections 1981).

(8) GEOTECHNICAL CONTROL OFFICE. Geotechnical Manual for Slopes (Second Edition). Hong Kong Government Printer, 1984, 295 p.

(9) GEOTECHNICAL CONTROL OFFICE. GCO Publication no. 3/84 : Model Specification for Pre-stressed Ground Anchors. Hong Kong Government Printer, 1984, 127p.

(10) HOEK, E. and BRAY, J. Rock Slope Engineering (Third Edition) Institution of Mining and Metallurgy, London, 358p.

(11) PETERSON, P. and KWONG, H. A design rainstorm profile for Hong Kong. Royal Observatory, Hong Kong Technical Note no. 58, 1981, 34p.

(12) ROYAL OBSERVATORY. Monthly Weather Summary December 1982. Royal Observatory, Hong Kong, 1983.

(13) VAIL, A.J. and HOLMES, D.G. A retaining wall in Hong Kong. Proceedings of the Seventh Southeast Asian Geotechnical Conference, Hong Kong, November 1982, pp 557-570.

Geotechnical designs for the construction of Fortress Hill Station, Island Line, Hong Kong

J. Costello B.SC., M.SC., C.ENG., M.I.C.E., M.I.H.E., M.H.K.I.E., F.G.S.
L.M. Mak B.SC., M.PHIL., C.ENG., M.I.C.E., M.H.K.I.E., F.G.S.
M.T.S. Chan B.SC. (GEOL. ENG.)
Freeman Fox (Far East), Ltd., Hong Kong

SYNOPSIS

Construction of the Fortress Hill Station, Island Line for the Mass Transit Railway Corporation (MTRC) involved excavation down 65 metres through soil and rock. The site is surrounded on three sides by multi-storey buildings. In order to achieve an open excavation of around 4,500 square metres, a system of temporary soil and rock anchors was used. The main rock excavation was carried out using 280 rock anchors with rock bolting and shotcreting to support the sides of the excavation as works proceeded. Extensive geological mapping was carried out during the excavation, and the anchoring system continuously modified to suit the exposed rock conditions. As the station box was constructed, the anchors were progressively de-stressed and the lateral loads from the ground and adjacent building foundations were transferred to the completed structure.

SITE LOCATION AND PROPOSED DEVELOPMENT

The site is located in the North Point area of Hong Kong Island (Fig. 1). It is bounded to the north by King's Road, which is the major eastern road link along the northern side of the Island. To the east, it is presently adjoined by a high-rise commercial development, with the excavation supported by tied back caisson walls as retaining structure. To the south, the site is bounded by the playground of Clementi Middle School and to the west by a multi-storey residential building known as 'Fortress Garden' (see site formation plan in Fig. 2). The area was originally a steep hillside, rising some 25m above King's Road to the south and west. The station itself is located some 15m below the street level of the adjacent King's Road, and comprises two levels, for a plant room and the main station concourse, with an escalator link down to the running tunnels and platforms beneath the station box.

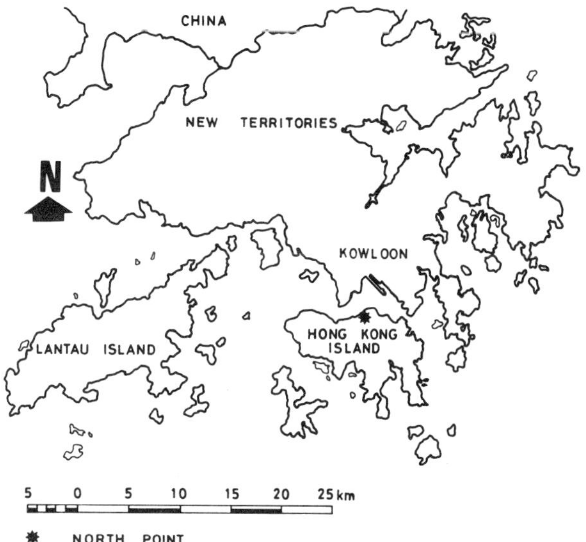

Fig. 1 Site location

Above ground level, there will be a commercial and residential development, of around 36 storeys. The reinforced concrete screen walls of the building will replace the temporary retaining structures, and the station box will form the foundation for the building. These developments are shown in typical sections in Fig. 3.

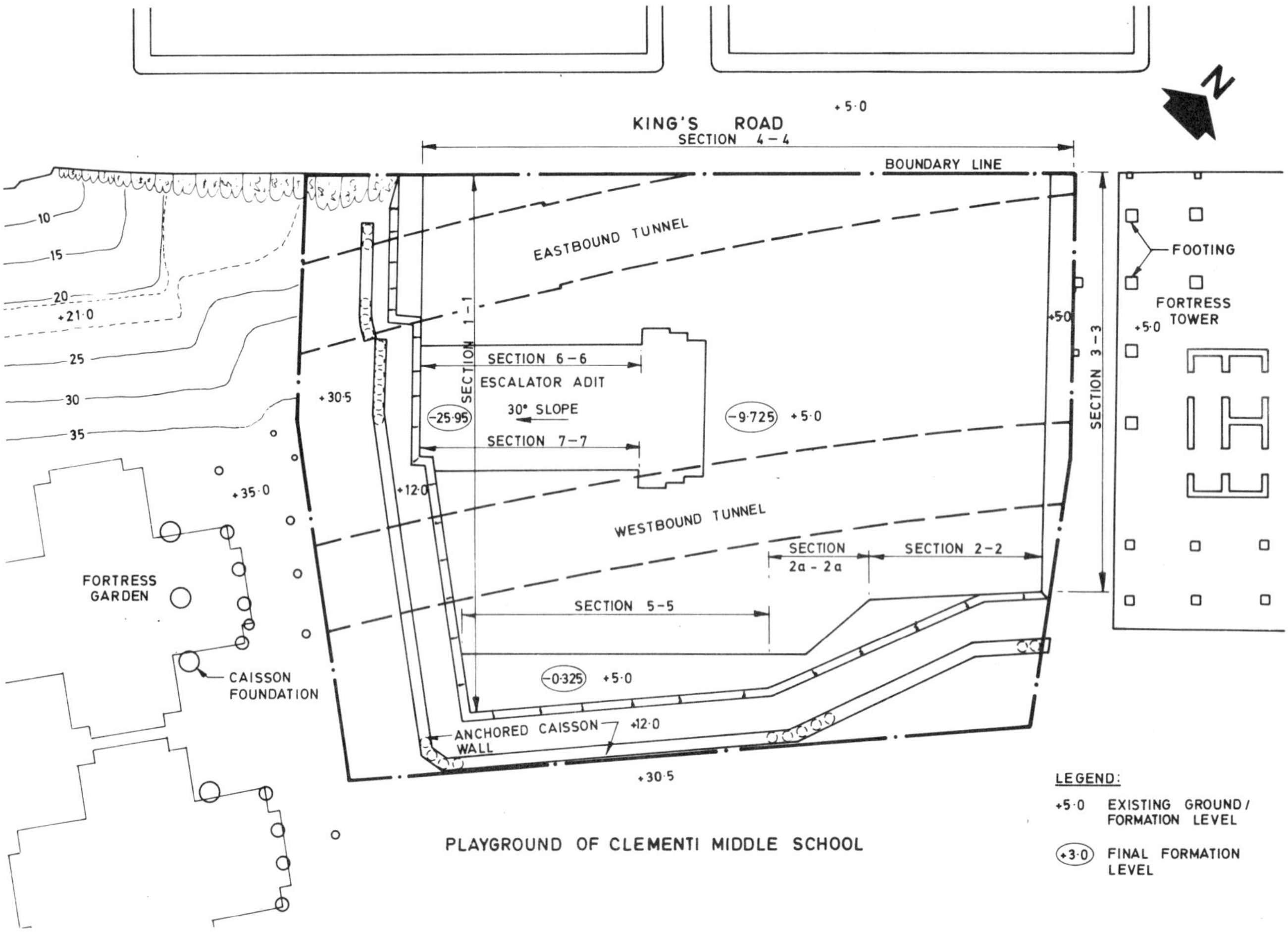

Fig. 2 Site formation plan

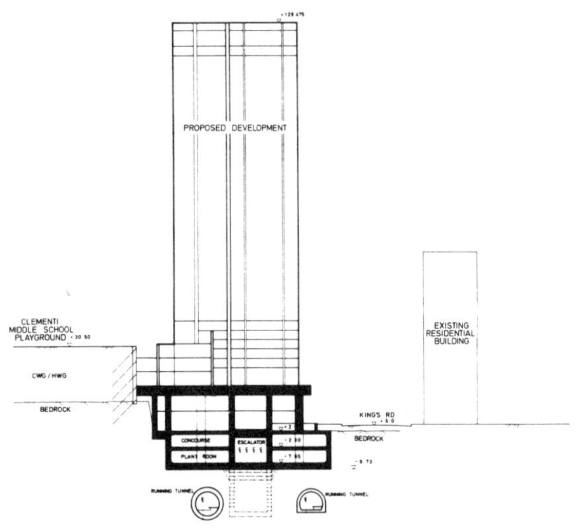

(a) Southeast - northwest

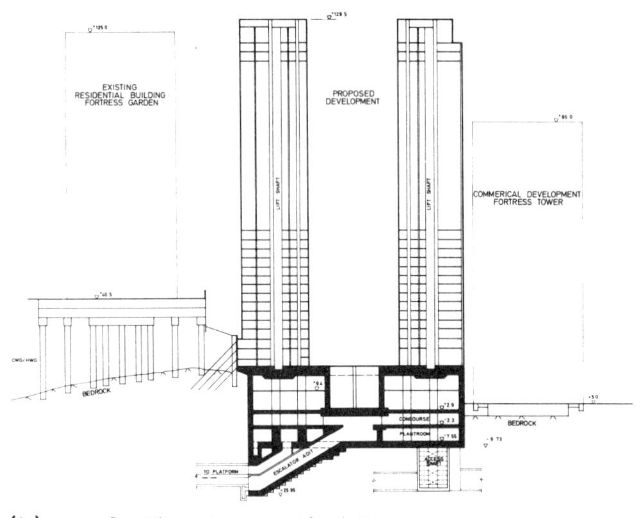

(b) Southwest - northeast

Fig. 3 Sections

CONSTRUCTION SEQUENCE

The construction of the station concourse was divided into two separate contracts. In the Advanced Contract Works which had been designed by others, the hill slope within the site was lowered to be level with King's Road (+5.00mP.D.) with the cutting on the south and west sides supported by temporary anchored caisson walls.

In the Main Contract, the ground was further lowered to a general level of -9.7mP.D. (see Plate 1) for the construction of a two-tiered box structure comprising the plant room and the concourse for the Fortress Hill Station. Local excavation was carried out to the maximum depth of some 31m (-25.95mP.D.) for the construction of the escalator adit, connecting to the two running tunnels beneath the station concourse. The excavation was supported by means of temporary rock anchors which were then de-stressed as the station box was constructed (see Plate 2) and became the permanent support for the rock slopes around the site. The construction sequence is illustrated in Fig. 4.

Plate 1 General view of excavation (southwest face)

GROUND CONDITIONS

Limited advance information on ground conditions was obtained from fifteen boreholes which were drilled within the site and adjacent areas. Seven of these boreholes were drilled for adjacent development in 1973, 1976 and 1978. The remaining boreholes were carried out under the instruction of MTRC in 1981.

Ground conditions as revealed by these boreholes were generally as follows :-

- Fill - comprised basically very loose to medium dense yellow-brown silty SAND with gravel. The thickness ranged from 0.8m to 11.0m.

Plate 2 Slope stabilisation using rock anchors

- Completely Weathered Granite (CWG) - comprised loose to very dense pink, yellow, white and brown silty coarse SAND with gravel. The thickness ranged from 0.8m to 14.8m.

- Highly Weathered Granite (HWG) - comprised very dense pink yellow, white and brown gravelly silty SAND with thickness ranging from 0.2m to 3m.

- Granitic Bedrock - comprised moderately strong to strong, mottled pink, light grey and dark brown, fine to coarse grained, moderately to slightly weathered Granite. The bedrock was found to be highly fractured in several boreholes, which might have been induced by local minor shearing movements.

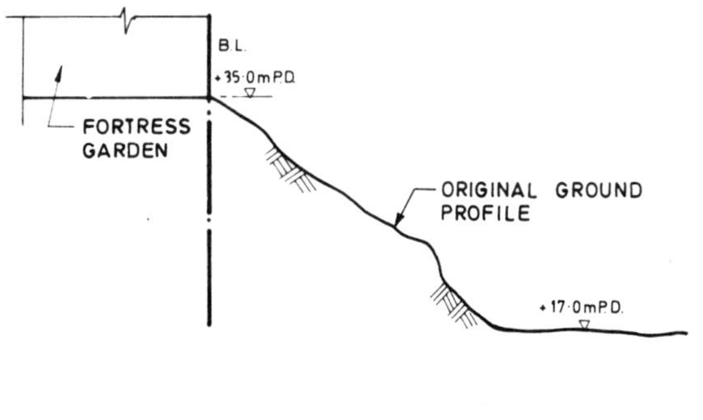

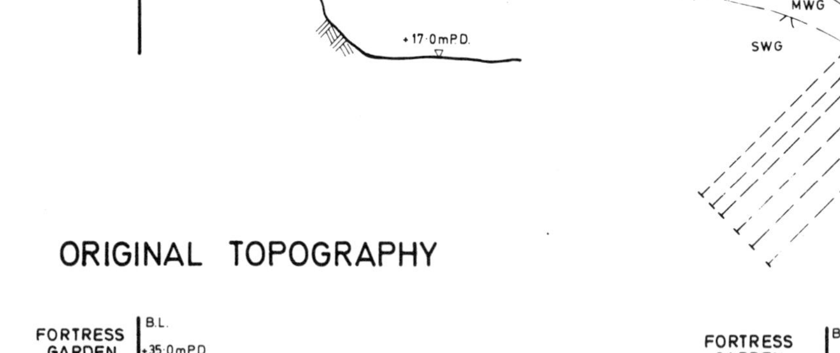

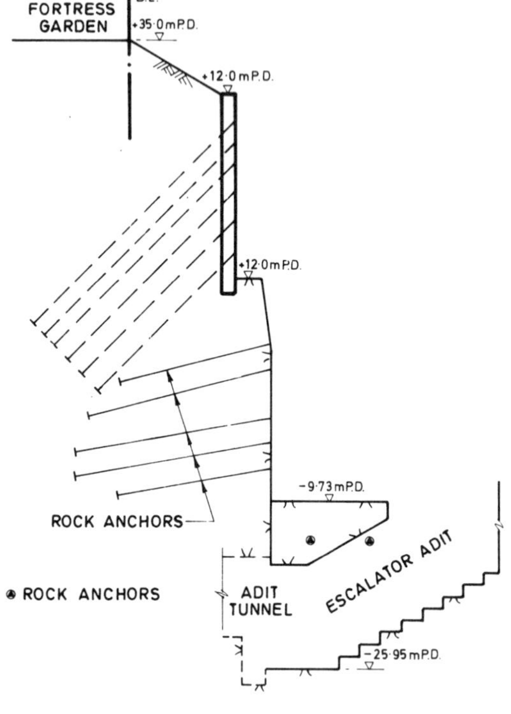

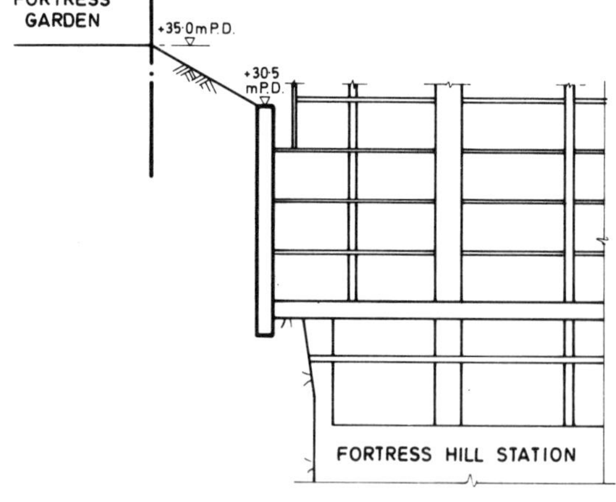

STAGE 1 – ADVANCE CONTRACT
TO INSTALL ANCHORED CAISSON WALL & EXCAVATE TO +5·0mP.D.

STAGE 2 – MAIN CONTRACT
TO EXCAVATE TO -9·73mP.D. WITH LOCALLY TO -25·95mP.D. & INSTALL ROCK ANCHORS

STAGE 3 – TO CONSTRUCT STATION STRUCTURE

STAGE 4 – TO CONSTRUCT SUPERSTRUCTURE

Fig. 4 Construction sequence

Based on limited piezometer readings, groundwater was anticipated to be located within HWG and close to the Bedrock.

As an illustration, a geological cross-section is shown in Stage 1 of Fig. 4. It can be seen that in the Advanced Contract, the anchored caisson wall was installed through soil and founded on bedrock. The bulk excavation in front the caisson wall was first in soil and then in rock. The main excavation for the station box was carried out in rock.

GEOTECHNICAL CONSIDERATIONS

Design parameters

The following soil parameters were adopted for the preliminary design, based on laboratory testing.

Soil Type	Bulk Density (kN/m^3)	Cohesion (kPa)	Friction Angle (Degree)
FILL	20.5	0	36
CWG	20.5	15	37
HWG	21.0	20	42

Table 1. Design soil parameters

Although these parameters are considered adequate and appropriate, they were based on the test results of only seven soil samples. Additional undisturbed samples were taken within and during the Advanced Contract caisson excavation, tests on which were later made to give further considerations of the recommended soil parameters.

No laboratory tests were carried out on rock samples in advance of the excavation works. However, reference was made to the Data presented in papers by Hencher & Richards[1] and Richards & Cowland[2], for Hong Kong Government sponsored research work into the frictional properties of joints in Granite near to the site.

Based on the suggested rock parameters, the following values were subsequently adopted in the design as given in Table 2.

Bulk Density	=	26 kN/m^3
Cohesion	=	0 kPa
Basic Friction Angle of a Smooth Surface	=	40 degrees
Roughness Angle	=	10 degrees

Table 2. Design rock parameters

Hydrostatic pressure

Ground-water was observed to stand at depths from +15.61 to +16.24mPD, within HWG layer. To take into account of fluctuation in ground-water level during wet/dry seasons, the design storm ground-water level was taken to be at +18.50mPD.

Due to limited piezometer installation, it was difficult to establish the piezometric head profile in the Granitic rock. A full hydrostatic head was therefore considered in the preliminary design of the basement walls and base slab. Subsequently, a drainage system was installed at the base slab level to release the hydrostatic pressure on the slab.

GEOTECHNICAL DESIGNS

Foundation design

The station would be founded on a 2m thick concrete slab with its underside level at -9.73mPD. The maximum bearing pressure would be between 3,000 and 4,000 kPa. Based on borehole information, the basement slab would be placed on slightly weathered granite with total core recovery in excess of 95%. Therefore, according to the Hong Kong Government Building (Construction) Regulations, the allowable bearing capacity could be designed to 5,000 kPa with adequate factor of safety against general shear failure.

Permanent retaining walls

The same parameters tabulated in Table 1 were used for the wall design. In addition, the wall friction between wall and soil behind was taken as zero. The permanent retaining wall is designed to resist the full hydrostatic pressure, soil pressure at rest and possible surcharge loads.

The bedrock profiles along the wall were taken from available borehole information. To cater for the irregularity of the bedrock profile, the design bedrock profile was assumed to be some metres below to ensure a safe design. Modifications to the design bedrock profile were made when further information on the actual bedrock condition was obtained during the caisson works in the Advanced Contract.

Basement walls

The basement walls were cast against granitic rock. The rock pressure on the basement walls depends greatly on the joint pattern. In view of limited rock joint information, calculation of rock pressure on the wall was carried out using various angles of possible failure plane, from which the maximum rock pressure was adopted in the design. The wall pressure was considered under two conditions. i.e. the temporary condition when the temporary anchors supporting the caisson walls had not yet been destressed but the hydrostatic pressure relief system was then in operation and the permanent condition when the temporary anchors had been destressed but the hydrostatic pressure relief system was then discarded. The results of calculation on wall pressure are briefly summarised in Table 3. For locations of sections, refer to the Site Formation Layout Plan in Fig. 2.

Temporary rock anchors and rock bolts

During excavation,temporary rock anchors and rock bolts were installed to stabilise the

Section No.	Top Elevation (mPD)	Basement Wall Pressure (kPa) Temporary Condition	Permanent Condition
1 - 1	+35.00 sloping down to +30.00	103.4	114 - 330
2 - 2	+38.50	129.0	144 - 360
3 - 3	+ 5.00	84.0	45 - 192
4 - 4	+ 5.00	60.0	18 - 165
5 - 5	+30.50	110.0	112 - 330

Table 3. Results of calculation on basement wall pressure

vertical cut rock slope against possible potential local and overall instability. Based on limited information, the rock anchors were initially designed to resist the theoretical worst possible slip failure, taking in account full hydrostatic pressure, vertical component from the anchored caisson wall, and overburden pressure and surcharge where necessary[3]. The requirements for temporary rock anchors and rock bolts were revised subsequent to joint mapping carried out during excavation works for the Advanced Contract and the Main Contract. The revision of rock anchors was also taken into account the construction sequence proposed by the main contractor.

Drainage system

In order for the station box to be stable under potential groundwater flotation forces when the overhead development was incomplete, a pressure relief system was installed beneath the base slab, permitting a thinner and more flexible slab than would have otherwise been necessary. The system comprised a series of interlocking channels with no fine concrete infilling under the slab, designed to intercept the seepage water and riser pipes to dissipate the pressure through the slab. This system could then be sealed off once the combined building and station weights were sufficient to maintain stability. Details of the pressure relief system are given in Fig. 5. Preliminary design of the base slab was carried out using conservative uplift pressure, and this was refined as the rock mass and groundwater condition was exposed in the excavation.

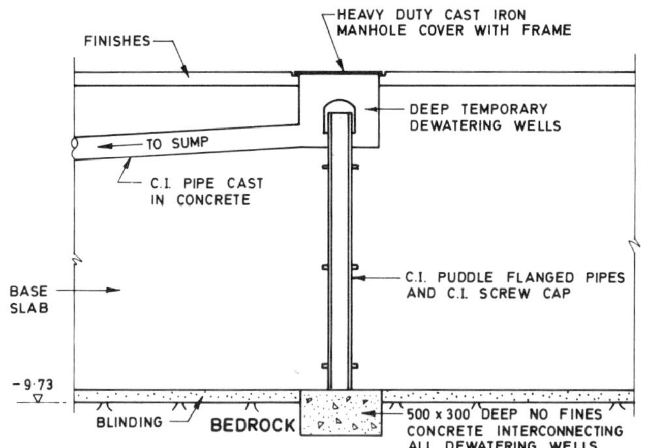

Fig. 5 Details of temporary dewatering well

DESIGN DEVELOPMENT AND MODIFICATION

The key element to the successful and economical use of the rock anchoring system was the philosophy of relatively conservative preliminary design, site inspection and analysis, followed by design refinement as excavation work proceeded. The anchor system gave an unrestricted working area within the excavation, presenting the contractor the minimum of obstruction to his excavation and construction operations, and had inherently more flexibility to cater for varying site conditions than a strutting system.

Design of both temporary and permanent works was only finalised as excavation work proceeded, and the actual rock and soil conditions were revealed. For the temporary works design, which essentially covered the rock anchors, extensive engineering geological mapping was carried out on the exposed final rock face. This required the services of an experienced engineering geologist being available to log the rock face immediately after blasting and scaling, to carry out a rapid kinematic stability assessment of the measured joint orientations and conditions exposed, and based on redesign/calculation, subsequent modification of the proposed anchoring arrangement as appropriate. Numbers and loads of anchors were adjusted, and in some cases anchors were deleted altogether (see Table 4), and the contractor was given immediate instruction through MTRC site staffs so that the

Section No.	Dip of Assumed Critical Joint Plane (Theoretical)	Dip of Critical Joint Plane Adopted in Design (Measured)	Original Anchor Force (Theoretical)	Revised Anchor Force (As Constructed)
1 - 1	65° to 70°	80°	5468 kN/m (134)	3440 kN/m (119)
2a - 2a	64° to 65°	77°	4417 kN/m (29)	2810 kN/m (25)
2 - 2	64° to 65°	52°	4230 kN/m (25)	3173 kN/m (14)
3 - 3	77°	instability unlikely	1898 kN/m (46)	739 kN/m (26)
4 - 4	69°	instability unliekly	1355 kN/m (72)	678 kN/m (40)
5 - 5	62.5°	85°	4057 kN/m (75)	2450 kN/m (52)
6 - 6	75°	85°	13000 kN (6)	4230 kN (2)
7 - 7	75°	85°	13000 kN (6)	4230 kN (2)
			(393)	(280)

Note : Nos. in ( ) indicate no. of rock anchors

Table 4. Summary of rock anchor design

installation could be made as soon as possible after exposure. Any local rock bolting or scaling requirements were included in these recommendations.

The exposed slope comprised typical Hong Kong Granite. In general, the rock was in slightly weathered and fresh state, with chloritised and slightly iron stained joint planes. The joints were moderately widely to widely spaced with some isolated areas of close spacing. The joint surfaces were mostly undular and curved in shape. Persistence of joint sets commonly exceeded 10m. The summary of the major joint orientations is shown in Fig. 6, whereas two examples of kinematic stability analyses of rock joints are indicated in Fig. 7. As the project proceeded, a programme of excavation sequences was developed, enabling the excavation, mapping, analysis, design refinement and anchor installation operations to proceed on a regular basis, accepted by all parties concerned.

The rock quality data was supplemented by examination of the rock tunnels beneath the site as they progressed, along with the access shaft, the geologist thus being able to interpolate geological data on discontinuity persistence and the continuity of major structural features through the site. The overall geological picture was therefore continuously revised and updated as work proceeded both above and below the station box. Throughout this operation, the total number of anchors was reduced from the 393 originally estimated to 280, representing a saving of 30% over the original preliminary design. In addition, the design of the permanent works were refined to reflect the actual conditions in the rock mass, resulting in an optimum design.

In the same way, the exposed ground conditions were constantly inspected, and the preliminary data on the design soil parameters were updated by means of additional sampling and testing as

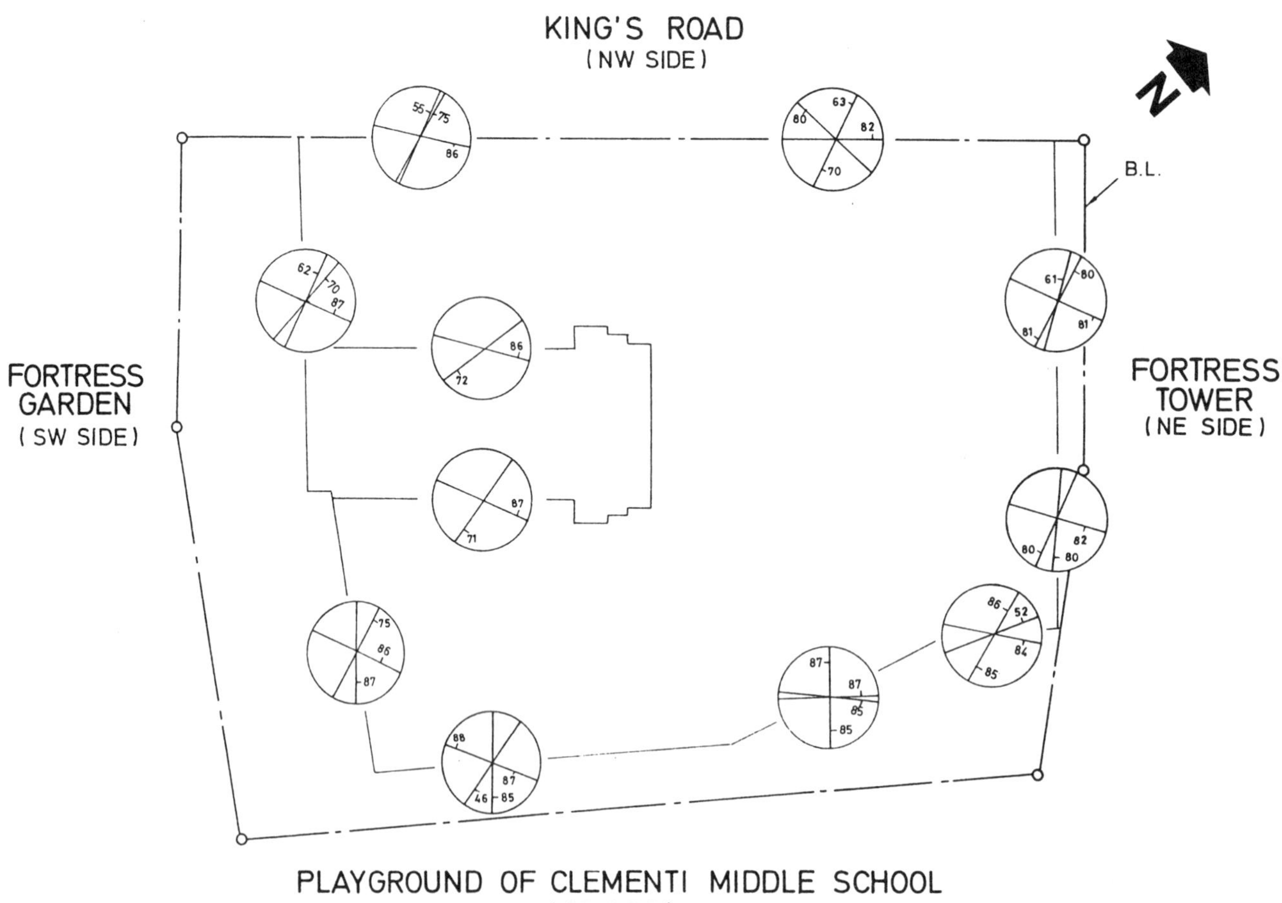

Fig. 6 Summary of major joint orientations

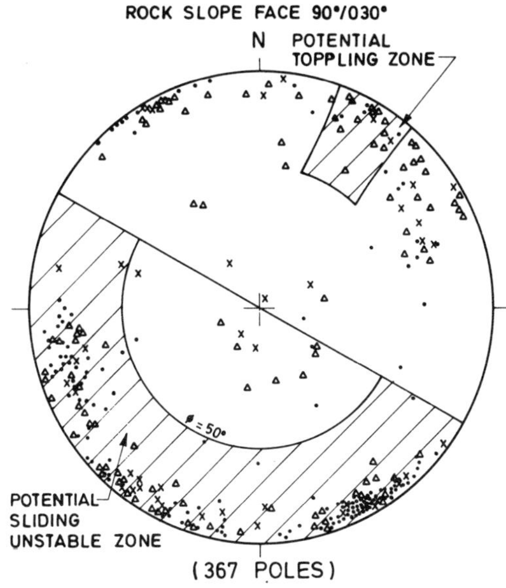

(a) Southern half of section 1 - 1

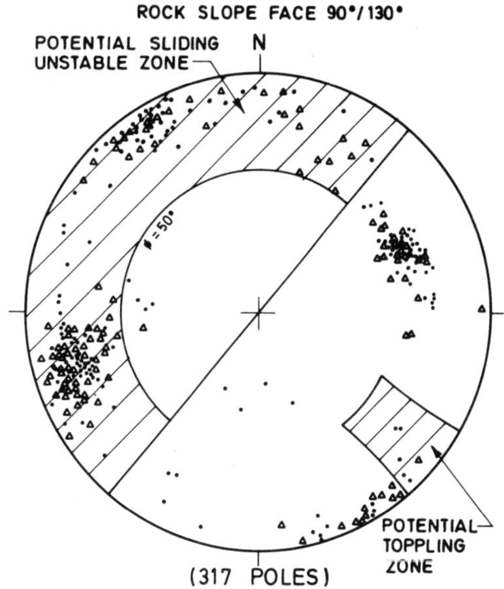

(b) Western half of section 4 - 4

LEGEND:

× DATA COLLECTED AT +12 TO +5mP.D.

△ DATA COLLECTED AT +5·0 TO −2·0mP.D.

• DATA COLLECTED AT −2·0 TO −9·7mP.D.

Fig. 7 Examples of kinematic stability analysis

the advanced earthworks proceeded. The actual bedrock level was found generally higher than those assumed in the preliminary design, with the Fill layer encountered to be deeper and extended further. A zone of Residual Soil was also identified below the Fill material. Concurrently, it was determined not possible to obtain additional undisturbed samples by rotary drilling method carried out within the caisson wall area since the ground had been disturbed by grouting and there was the risk of hitting the anchors. Instead, block samples were taken in various soil strata using hand trimmed method in front of the caissons for testing purpose.

Based on the results from the additional laboratory testing, the soil parameters adopted for the design of the permanent retaining wall were revised as shown in Table 5. Compared with those given in Table 1, the value of cohesion is lowered by 34 to 40% and friction angle by 3%.

As a result of these changes in design soil parameters and ground profiles, the lateral soil pressures on the proposed permanent retaining wall also revised. Calculation showed that the

Soil Type	Bulk Density (kN/m^3)	Cohesion (kPa)	Friction Angle (Degree)
FILL	20.5	0	35
RESIDUAL SOIL	20.0	12	30
CWG	20.5	10	36
HWG	21.0	12	42

Table 5. Revised design soil parameters

estimated lateral forces acting on the walls were 13 to 18% less than those previously recommended in the preliminary design.

CONCLUSIONS

The method of design and construction used for this project was only possible through the development of a good working relationship between the designer's geologist, the Resident Engineer's staff and the Contractor, with the active co-operation of the Government's checking engineers. It is therefore necessary to establish beforehand the general philosophy of the operation, so that all parties are aware that a cyclic sequence is required, otherwise the contractor's working procedure could be disrupted, with subsequent claims for delay. In particular, the contractor must be made aware of the need for allowing in his programme for the necessary inspection and analysis work and statutory approvals procedures. Providing that the system is understood by everyone at the outset, there are significant advantages to be gained in terms of overall programme from the overlapping of the design and construction phases, in addition to the optimisation of costs. The use of a variable anchoring system on this project proved to be a major factor in the timely completion of this major rock engineering exercise in a dense, urban area with minimal effect on the surrounding environment.

ACKNOWLEDGEMENTS

The authors wish to thank the Mass Transit Railway Corporation for permission to publish this paper and to W.K. Chan and S.W. Lee, the Project Director and Engineer respectively, for their guidances. Also special thanks are due to Windy Ng and May Chung for typing the manuscript and Andrew Cheng for preparing the figures.

REFERENCES

1. Hencher S.R. and Richards L.R. The Basic Frictional Resistance of Sheeting Joints in Hong Kong Granite. Hong Kong Engineer, February, 1982, p.21-25.

2. Richards L.R. and Cowland J.W. The Effect of Surface Roughness on the Field Shear Strength of Sheeting Joints in Hong Kong Granite. Hong Kong Engineer, October, 1982, p.39-43.

3. Hoek E.and Bray J.W. Rock Slope Engineering, 3rd Edition. Institution of Mining and Metallurgy, London, 1981.

Characteristics of rocks and their excavation in Makkah tunnels

Mahmoud Ali Darwish B.SC., ENG., M.S., PH.D.
Muhammad Hanif B.SC., ENG., PH.D.
Mining Engineering Department, King Abdulaziz University, Jeddah, Saudi Arabia

SYNOPSIS

The Holy city of Makkah is situated in the Pre-Cambrian mountainous region of western Arabian Shield. In order to meet the unique communications requirements, particularly during the pilgrimage season, a number of tunnels for traffic, pedestrians and utilities have been constructed in the city of Makkah and the nearby Hadj sites as a part of Saudi Government's effort for national development and providing maximum possible facilities to the pilgrims. The work on some tunnels is in progress and others are planned.

The Pre-Cambrian rocks comprised of granites and gneisses are generally competent. The rocks are, however, jointed and intruded by dykes of younger rocks resulting in weak interfaces which may cause localized ground control problems.

A brief description of important rock types and their geological and mechanical characteristics is presented. The design criteria and procedures of rock excavation and tunnel support and reinforcement are discussed. The need for using feed-back from construction and operational experience as well as local research effort in the updating of design criteria is emphasized.

INTRODUCTION

The Holy city of Makkah is at one hour's drive from the Red Sea port city of Jeddah in Saudi Arabia. The city area is comprised of a series of mountains and valleys with the Holy Masjid Al-Haram being situated in a central valley. Millions of Hadj pilgrims temporarily live in the city and perform religious rites in the city centre and nearby sites. This involves a unique traffic problem. Any number of surface roads in the city would not have solved the enormous problem of traffic congestion. A number of pedestrain and traffic tunnels were, therefore,constructed over the past decade and more are planned to be built. The main pedestrian tunnels connect the Masjid Al-Haram in the city centre with the sub-urban Muna valley where pilgrims stay for a few days during the Hadj. Some pedestrian tunnels are designed to shorten the walking distances between sections of the Muna valley. A few utility tunnels have been constructed near the Masjid Al-Haram to accommodate a large number of toilets and baths. The traffic tunnels are systematically designed to remove traffic congestion and reduce travelling distances. A number of tunnels form parts of ring roads around the city centre or of major radial roads. Some provide vehicular access to Muna from different parts of the city; Fig. 1 shows some of these tunnels. The authorities responsible for the construction of tunnels are Ministry of Communications, Ministry of Housing and Makkah Municipality. The master plan is prepared and updated by a coordinated effort, but for an individual project, the respective authority appoints one of the consulting engineers by tender who then carries out the design work. Diamond drilling for site investigation is carried out by a contractor to the requirements of the consultant who prepares the contract specifications and other documents. The contract is then awarded to a contractor and is supervised by the consultant.

Some 25 tunnels, totalling a length of more than 30 Km have already been commissioned making

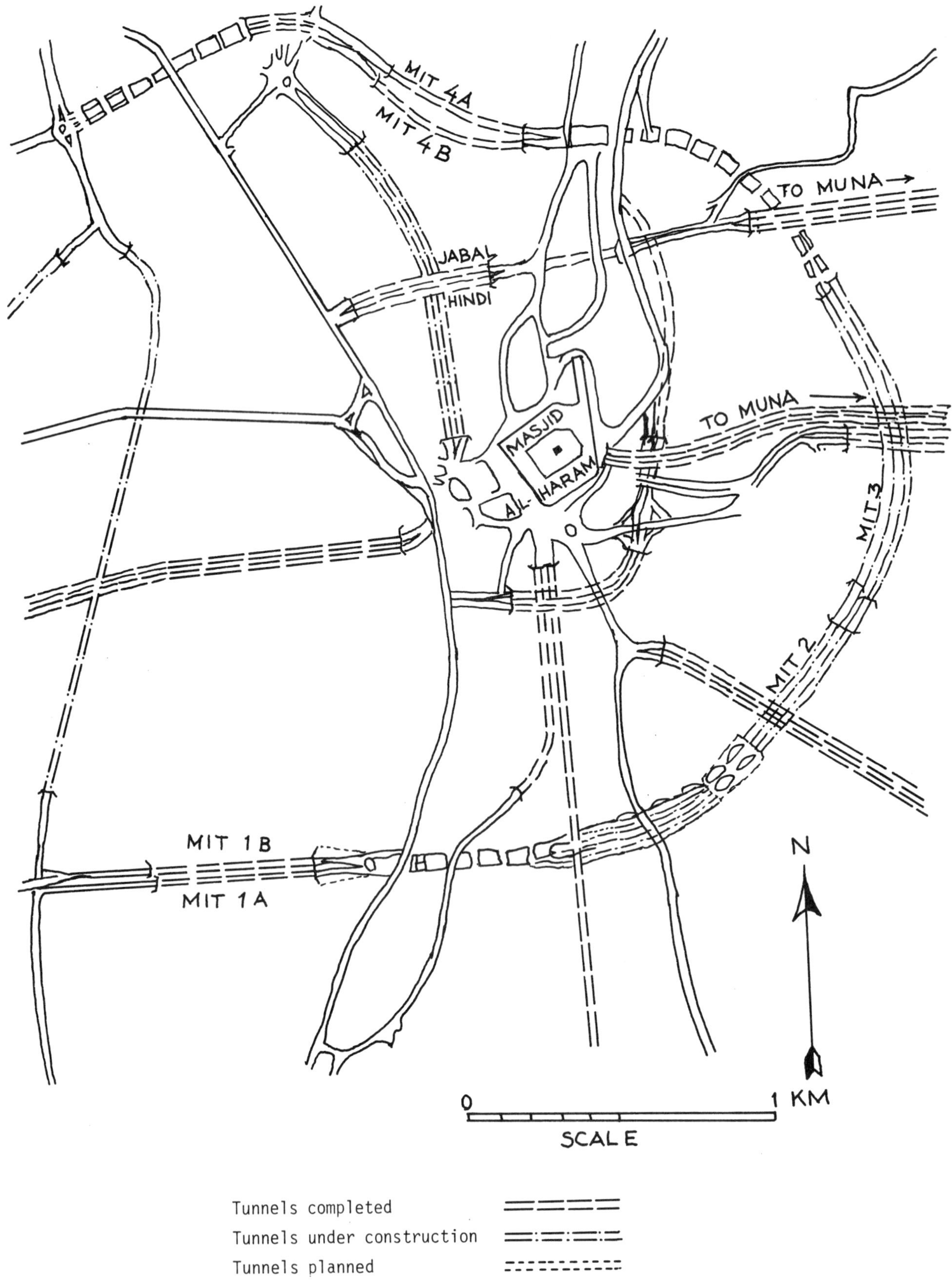

Fig. 1 A system of ring roads and other major roads in the city centre of Makkah Al-Mukarramah, showing some of the many tunnels in the city.

the traffic in the city compatible with the modern 8-lane and 6-lane expressways which connect Makkah with Jeddah and the Holy city of Madina. The traffic tunnels have 2 or 3 lanes and handle one-way traffic. Most tunnels are 14 m wide and 9 m or more in height.

Makkah has an arid climate with average summer temperature at 30-40^{0}C and average winter temperature at 20-30^{0}C. The mean annual rainfall which usually occurs between November and April, varies from 150 mm to 250 mm. Natural vegetation is sparce with isolated bushes. Artificially irrigated road-side trees and plants are abundant. Topography is varied mainly due to faulting and erosion with ridges and valleys. Many locations have scattered boulders, with debris composed of loosened rock and screes.

CHARACTERISTICS OF MAJOR ROCK TYPES

The tunnels in and around Makkah are driven in the mountain chain of the Pre-Cambrian belt of the Arabian Shield. The main igneous and metamorphic rocks exposed in the area are migmatite and quartz diorite which are locally intruded by younger, felsite and dolerite dykes. A brief description of the main rock types is as follows:

Migmatites;

The term migmatite is used here to denote metamorphic rocks (mostly gneisses) as distinct from quartz diorite. It is generally light gray to creamy, when fresh and yellowish brown, where weathered. The grain size is generally medium to coarse. It has a granoblastic texture and consists of feldspar (plagioclase), quartz,biotite and hornblende with some epidote and chlorite. The plagioclases are slightly altered to sericite. The weathering is mostly pronounced at the surface of natural exposures. The degree of weathering may be classified as slight to moderate, becoming generally insignificant at depth.

The mass engineering properties of migmatites largely depend on the discontinuities and the effect of weathering. Joints give rise to a blocky or columnar structure to the rock mass. The rocks show well developed joints at least in three directions. Close to the surface of natural exposures, relatively steeply dipping sets are separated at distances of 5 to 50 cm and gently dipping joints occur at an average separation of 75 cm, thus giving rise to blocks of small to medium size. The strength of the rock is assessed as strong where fresh and moderately strong where weathered. The primary porosity and permeability is very low. Due to joints, etc, the secondary, fracture induced, permeability may increase, however,it may be regarded as very low at the general tunnel levels.

Quartz diorite:

This rock type has a gray colour, holocrystalline lepidiomorphic texture and consists of medium to coarse crystals of plagioclase, quartz and hornblende and small amounts of biotite, chlorite and epidote. It is slightly metamorphosed and may locally contain hornblende phenocrysts.

Quartz diorite shows widely to very widely separated joints especially in three directions. The separation of two steeply dipping sets generally averages between 20 and 60 cm, whereas gently dipping joints are typically 1 metre apart, thus giving rise to medium sized blocks. Weathering of the quartz diorite is generally restricted to the upper levels of natural exposures. Except along shear zones, the rock may be taken as fresh at the tunnel elevations. The assessed strength of the rock is strong to very strong. The primary porosity and permeability is very low, and the latter is not significantly increased at the general tunnel elevations.

Felsite:

The felsite is a light coloured quartz and feldspar rich rock and includes quartz veins. It occurs as dykes within quartz diorite or migmatite. Felsite is whitish to light gray when fresh and yellowish gray when weathered.

Dolerite:

Dolerite is generally observed in the form of dykes of varying thicknesses forming a dyke swarm and are well-exposed in the tunnels. The average thickness is about 30 to 50 cm. Dolerite is dark green where fresh and dark yellowish brown when weathered. It is fine grained, holocrystalline and slightly metamorphosed being in the greenschist facies. Although they show slight to moderate weathering close to the surface, the rocks are essentially fresh after a few meters below the surface. Dolerite dykes generally show well-defined NE - SW orientation dipping 70^{0}-80^{0} in SE direction. The boundary between the dykes and country rock is always sharp and very smooth and contains a film of altered material along the contact surfaces. The strength of the fresh rock is assessed as moderately strong to strong and the rock mass is practically impermeable.

Overburden

The overburden consists of gravel and boulder sized blocks embedded in a silty and sandy matrix in an unsorted, angular and loose condition. The thickness of overburden generally varies from 0 to 2 m. It may be regarded as not of direct relevance to the tunnelling point of view.

SITE INVESTIGATION

Detailed site investigation was a rule in the early stages of tunnelling in Makkah, comprising detailed geological mapping, joint survey and analysis, diamond drilling and core logging, laboratory and in-situ rock testing and measurement of deformation and strains in tunnels during construction. The emphasis on detailed site investigation has been gradually reduced due to virtual absence of any serious ground control problems, insignificant deformation and relatively little variation in the exposed ground conditions.

Mechanical properties of rocks

Typical mechanical properties of the important rock types are as follows:

Table I gives the Schmidt hammer test results and estimated values of uniaxial compressive strength conducted in one of the tunnels.

Estimated average values of uniaxial compressive strength, Su, were derived from Brock and Franklin[2], point load strength index, I_s. I_s is given by $I_s = P/D^2$ where, P = Applied point load at failure and D = Distance between the loading points. Su is then, given by: $Su = 24\ I_s$. The results are given in Table II.

Laboratory tests were performed on core samples from diamond drill holes drilled for site investigation for MIT 1 and MIT 4 tunnels(Fig.1) and the results for uniaxial compressive strength, Modulus of Elasticity, Poisson's Ratio, natural density and point load index are summarized in

Table I Schmidt Hammer test results

Rock type	Schmidt Hardness (R)	Estimated Compressive Strength (MN/m^2)	Remarks
Quartz diorite	50.4	203	Fresh, fine grained, occasional staining.
Quartz diorite	49.0	180	Faintly weathered, coarse to medium grained.
Quartz diorite	48.5	176	Faintly weathered, coarse grained.
Migmatite	59.3	268	Medium to coarse grained fresh.
Felsite	48.1	172	Slightly weathered, 30 cm thick dyke.
Dolerite	41.9	119	Slightly weathered, fine grained, 50 cm thick dyke.

Table II Point load test results

Rock Type	Estimated Compressive Strength (MN/m^2)
Quartz diorite	201
Quartz diorite porphyry	125
Migmatite	220
Dolerite	200
Felsite	173

Direct shear strength test

Direct shear test was performed on saw-cut,fine grained quartz diorite specimens and on a altered (chloritized) joint surface between quartz diorite walls. A retest was performed on the cast specimens soaked in water for 5 days. The result was a lowering of both cohesion and angle of friction. Fig. (2) shows the difference between dry and saturated-drained conditions. The results of direct shear test are given in Table IV.

Table III Laboratory Test Results

Sample No.	Rock type	Unixial Compressive Strength (MN/m^2)	Modulus of Elasticity (GN/m^2)	Poisson's Ratio	Natural Density (Tonne/ m^3)	Point-load Index (MN/m^2)	Estimated Compressive Strength, Su (MN/m^2)
1.	Coarse grained quartz diorite	171	74.4	0.14	2.70	7.0	168
2.	Quartz diorite and dolerite contact	137	83.3	0.21	2.89	6.1	146
3.	Coarse grained quartz diorite	157	64.9	0.21	2.84	6.3	151
4.	Medium grained quartz diorite	150	111.0	0.14	2.97	6.2	149
5.	Fine to medium grained felsite	122	96.1	0.19	2.58	-	-

Table IV Shear Strength of Some Joint Surfaces

Specimen description	Cohesion (KN/m^2)	Friction Angle (Degrees)
1. Saw-cut quartz diorite	40	28
2. Altered(chloritized), undulating, slickensided joint surface.		
a) Dry	188	32
b) Saturated drained	130	29

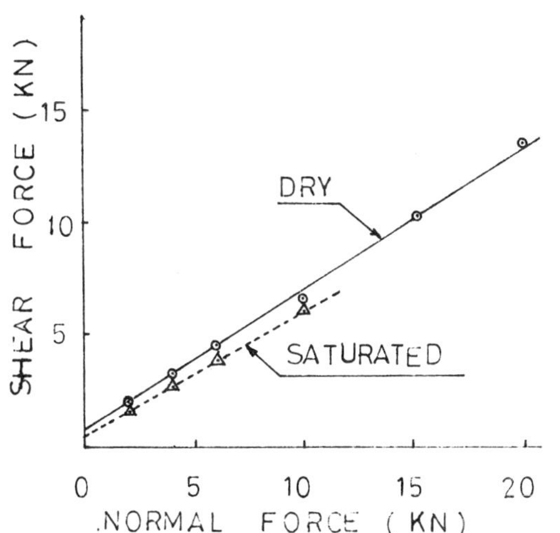

Fig.2 Direct shear test on altered joint

DESIGN CRITERIA OF TUNNELS

The depth of tunnels below the surface is relatively limited and rarely exceeds 100 meters. The rocks are generally strong and fresh. Ground water is rarely encountered. It is, therefore, not difficult to conform to the geometrical requirements of size and shape. The general shape is comprised of a semicircular arch roof with relatively short vertical walls. This shape is advantageous from the point of view of ventilation, floor space utilization and roof stability.

The ground support and reinforcement is designed on the basis of Rock Mass Quality (RMQ). The sides and roof of all tunnels are invariably shotcreted to a thickness of 50 mm as a measure of temporary support. The shotcrete may be reinforced, if so required, by wire mesh. In most cases shotcrete provides a satisfactory temporary support. This is followed by 500 mm thick concrete lining in the roof and sides with intervening 2mm thick P.V.C. Water proof membrane, where required by the Engineer. The membrane is protected against rupture by the shotcrete projections with a felt cushion.

25 mm diameter and 4 m long rock bolts are provided where required in accordance with the RMQ rating of the rock mass. The various types of rock bolts used according to the characteristics of the rock are mechanically anchored rock bolts, untensioned dowel bars, perfo bolts and resin anchored rock bolts. Near the portal and where the ground cover is less than twice the tunnel diameter Bernold sheeting with 200 mm x 200 mm H-section steel arches are employed. Steel arches are also used locally where ground control problems arise. The voids behind the sheeting are filled up by concrete before the usual 500 mm thick concrete lining is poured in. Six parameters are used to derive the Rock Mass Quality

rating, given by

Q = (RQD/Jn)(Jr/Ja)(Jw/SRF)

where:

RQD is Rock Quality Designation, graded from 10 to 100.

Jn is Joint Set Number, graded from 0.5 to 20. A massive rock is assigned the value of 0.5 to 1.0 while crushed rock is rated at 20. Rock masses with one, two, three and four joints sets are assigned Jn values of 2, 4, 9 and 15, respectively.

Jr is Joint Roughness Number. It varies from 4 for discontinuous joints to 0.5 for slickensided, planar joints.

Ja is Joint Alteration Number. This varies from 1.0 or slightly less for unaltered joint surfaces to 20 for extremely altered clay conditions.

Jw is Joint Water Reduction Factor, which is 1.0 for dry excavations or minor inflow to 0.05 for exceptionally high inflow or water pressure.

SRF is Stress Reduction Factor. This factor is based on an assessment of the likely stress relief due to deformation. It ranges from 1.0 to 20 and its rating depends on the presence of weakness zones, nature of the stress field in competent rock and/or anticipated plastic deformation.

The detailed rock mass description ratings for each of the above six parameters are given in the form of Tables by Barton, et al.[3]. The average values of these parameters for Makkah rock masses are as shown in Table V.

The mean rating of Rock Mass Quality, Q, for Makkah rock masses, lies in the vicinity of 20. Table VI, is used for determining the requirements of support and reinforcement which does not include 500 mm thick concrete lining of the roof and sides provided as the permanant support.

Most rock masses in Makkah fall in class 2 while some belong to class 3. However, at the portal and wherever the height above the crown is less than twice the tunnel diameter, Bernold sheeting with steel arches and concrete backfill is employed. This type of support is also commonly employed where localized ground control problems are experienced.

The problems of rock burst or of plastic deformation are non-existent. However, local

Table V Rating of RMQ parameters for Makkah rocks

Parameters	General Range	Average for Makkah Rock masses	Remarks
RQD	10-100	75-90	RQD rating is "Good"
Jn	0.5-20	6-9	Two sets + random joints to three joint sets
Jr	4 - 0.5	1-5	Slickensided, undulating to rough or irregular planar
Ja	0.75-20	1.0	Unaltered or stained
Jw	1.0-0.05	1.0	Dry excavations
SRF	0.5-20	1.0	Medium stress,elastic material

Table VI Supports for different RMQ ratings

Cl. No.	RMQ Rating	Support requirements for: Roof	Walls
1.	Q>40	Shotcrete 50mm.	Shotcrete 50 mm.
2.	40≥Q>10	Shotcrete 50mm, 25 mm dia. rock bolts as required.	Shotcrete 50 mm.
3.	10≥Q>4	Shotcrete 50mm with wirenetting as required. 25mm dia X 4m rock bolts @ 2m centres	Shotcrete as for roof, Rock bolts, 25mm dia X 4 m as required.
4.	4≥Q>1	Shotcrete 25mm with reinforcement, 25 dia X 4m rock bolts @ 1.5m centres.	Shotcrete as for roof, 25mm dia X 4m rock bolts @ 2m centres.
5.	1≥Q>0.4	50mm shotcrete with steel mesh. 25mm dia X 4m bolts @ 1m. centres.	Shotcrete as for roof, 25mm dia X 4m bolts @ 1.5m centres.
6.	0.4>Q	Bernold sheeting with steel arches and concrete backfill.	Same as roof.

ground control problems are some times caused by

the presence of water and the adverse orientation of dolerite dykes and joints with respect to the tunnel geometry as, for example, illustrated in Fig. 3.

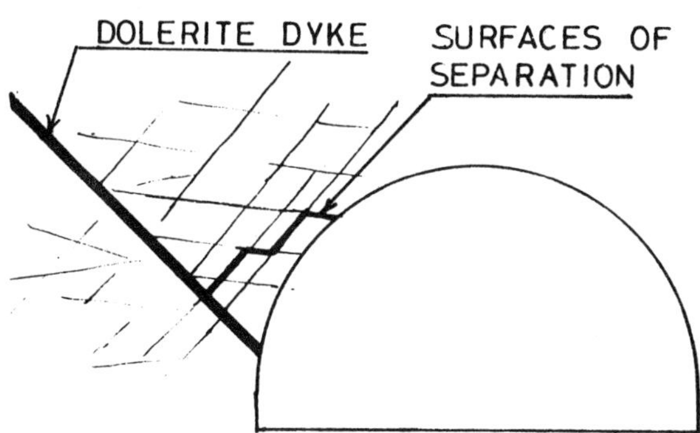

Fig 3 A localized failure due to shear movement along a dolerite dyke.

Since the tunnels are in the built up urban area, some small valleys and depressions have been backfilled and buildings constructed on them. In one of the Jabal Hindi tunnels for example, such ground was intercepted with a multistorey building above it. The tunnel was supported with steel arches as close to the face as possible and the face was advanced by first excavating a small diameter and then gradually enlarging in a number of stages, avoiding explosives wherever possible. A lack of water and adequate consolidation of well-graded rock/soil fill helped in the successful operation.

STABILITY ANALYSIS

An simple analysis was made to assess the validity of the RMQ design criterion assuming the depth of a tunnel below the surface to be 100 meters. The average values of measured rock parameters were:

Insitu density, ρ = 2.8 tonnes/m^3
Poisson's Ratio, ν = 0.18
Compressive strength, C_0 = 147 MN/m^2
Point load index = 6.4 MN/m^2
Cohesion, c = 130 KN/m^2
Angle of interval friction, $\phi = 29^0$

The virgin vertical stress was worked out as,

σ_v = 2.75 MN/m^2

Assuming that the horizontal stress is caused by Poisson's effect only, then, it is given by

$$\sigma_h = \frac{\nu}{1-\nu}\sigma_v = 0.61 \text{ MN/m}^2$$

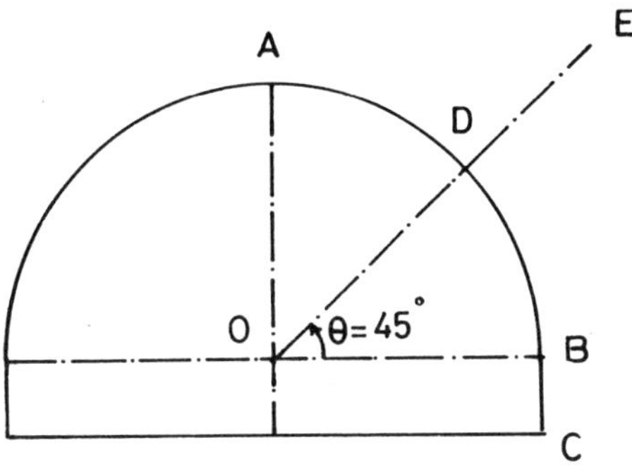

Fig. 4

With reference to Fig. 4, the radial or vertical stress at point A = 0

Tangential stress was worked out as = -0.92 MN/m^2

At point B, the radial or horizontal stress = 0 and tangential or vertical stress = 6.42 MN/m^2.

This vertical stress is only 4.3% of the uniaxial compressive strength of the rock. The effect of joints and size difference would be to reduce the margin, but the rock should be quite competent in resisting such compressive stresses.

As a second check, the induced horizontal stress in the abutment due to compression

$$= \frac{\nu}{1-\nu} \times 6.42 = 1.41 \text{ MN/m}^2$$

This is only 22 % of the point load index. Hence there would be little possibility of production of new near vertical tensile cracks in the abutments.

The scale Model tests by Hanif[4] indicated that the first signs of failure in sections identical to Makkah tunnels were the near vertical curved cracks in the abutments, with convexity outwords.

The tangential tensile stress at the crown of 0.92 MN/m^2 is only 14% of the point load index. The tendency of the tensile failure in the crown region would be countered by the arching action developed due to the elastic deformation in the form of vertical closure of the roof.

Shear stress

The shear stress at points along DE (Fig. 4) would vary with distance from point D. The shear stress in horizontal or vertical directions at points along DE with distances from D of zero, 1 m, 2 m, 3 m, 8 m and 24 m would be zero, 0.76, 1.13, 1.16,

1.47 and 1.19 MN/m^2, respectively.

At 2 m from D, where the shear stress is 1.13 MN/m^2, the radial stress, σ_r = 0.57 MN/m^2 and tangential stress σ_θ = 2.75 MN/m^2

Normal stress perpendicular to the direction of maximum shear = 2.29 MN/m^2

Max shear stress = 1.13 MN/m^2

Strength of a critically oriented, undulating, slickensided joint $= c + \sigma_n \tan \phi$

$= 0.13 + 2.29 \times \tan 29^0$

$= 1.40$ MN/m^2

This value is very close to the shear stress of 1.13 MN/m^2. Should a joint have less cohesion and angle of friction, as can happen in the case of dolerite dykes, especially if wet, localized failures can take place. It follows that the designing of Makkah tunnels on the basis of R.M.Q. is good in principle. However, the support and reinforcement requirements should be continuously reassessed on the basis of a systematic investigation, analysis of data and feed-back from experience.

ROCK EXCAVATION

Very hard rock, large tunnel cross-sections and relatively short lengths make tunnel broing machines impracticable and uneconomical at Makkah. Similarly, drilling and blasting would have obvious technical and economic advantages over machine ripping.

Tunnel construction starts with excavation for portals and slopes are then stabilized by shotcreting, with or without rock bolting. A special care is taken of the possible wedge failures. Fig. 5 shows a typical Makkah tunnel portal. Fig. 6 gives a simplified tunnel cross section.

It is a common practice in Makkah area to start a tunnel as a pilot, advance it by 10 to 35 metres and then, enlarge it to the full cross-section. This practice is advantageous not only for rock excavation efficiency but also for the flexibility of operation when the tunnel requires rock bolting as well as arch erection and concrete backfilling. This practice which also helps in limiting the size of a blast is not expected to cause a damage to the nearby structures. A burn cut is commonly employed. The holes are usually 45 mm in diameter, while 70 mm and 102 mm diameter

Fig. 5 A tunnel portal in Makkah Al-Mukarramah

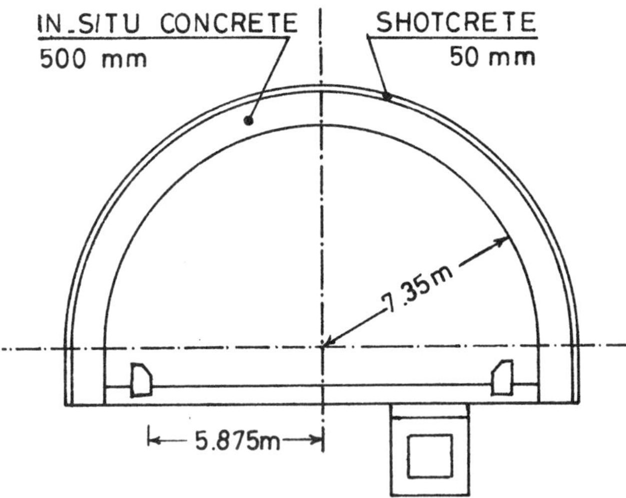

Fig.6 A typical tunnel cross section

holes are also employed. The depth of holes is 3.8 metres which gives an advance of 3.4 metres per round. A typical full face round of 157 blast holes in shown in Fig. 7a. Fig 7b shows the central holes comprised of, cut, loosening and some of the stoping holes using millisecond delays. The other holes are blasted using 30 second delays. With the exception of the two larger diameter empty holes, all others are 45 mm in diameter. 40 x 200 mm dynamite B cartridges are used in the bottom of the holes. The remainder consists of Prillit. The area of the face is 127 m^2. For a face advance of 3.4 m, the total explosives consumption is 660.36 kg. The consumption of explosives per cubic metre of rock

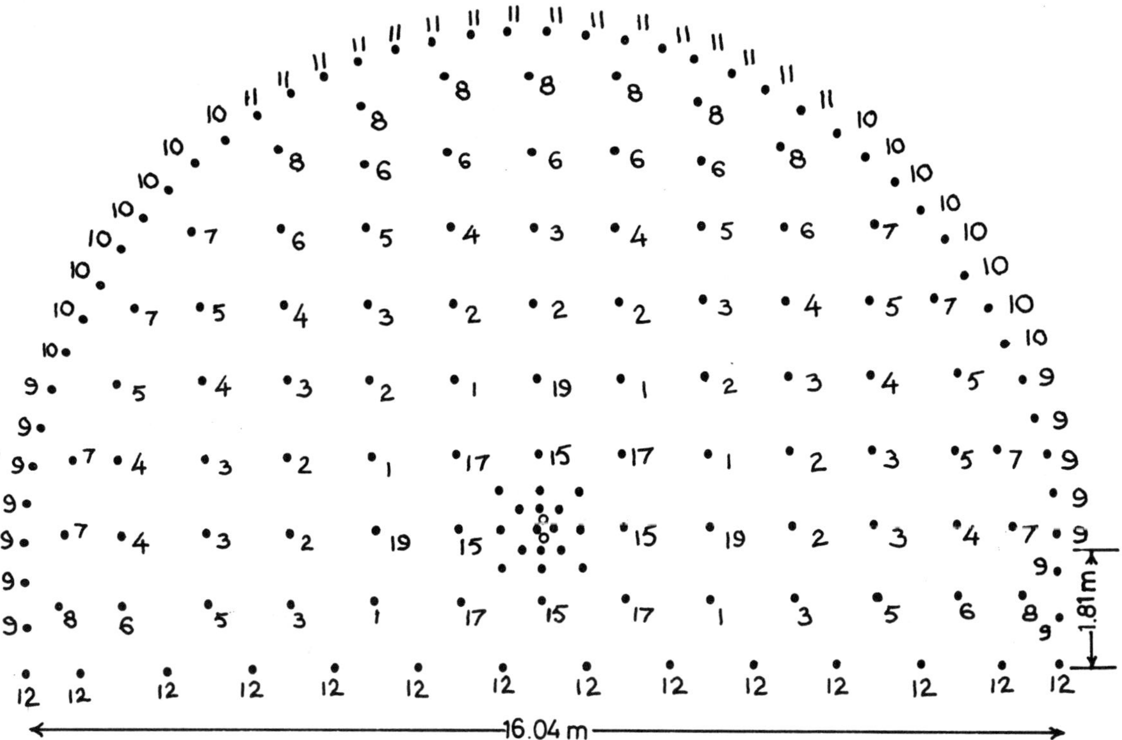

Length of holes = 3.8 m. Diameter of holes = 45 mm. Total weight of explosives = 660 Kg. Consumption of explosives = 1.53 Kg/m^3.

Fig.7a Full face blasting round

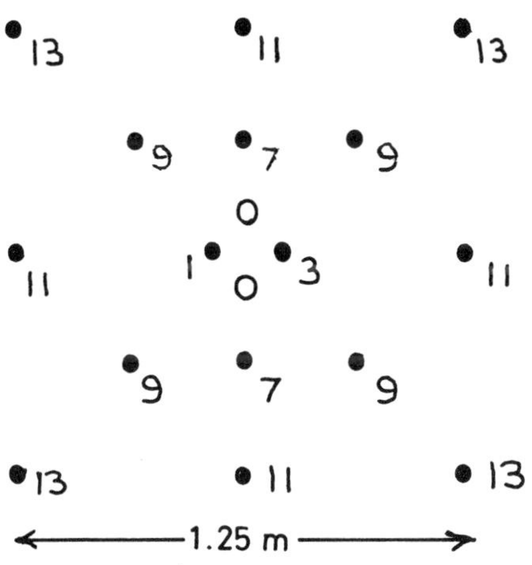

Fig. 7b Centre of ignitious burn cut.

blasted is 1.53 Kg. This falls within the range of 0.9 Kg/m^3 for large diameter tunnels to 3.6Kg/m^3 for small diameter drives. In practice, the consumption was later reduced to approximately 1.0 Kg/m^3 of rock blasted.

With all the tunnelling activity in the city, no blasting or other damage to the buildings has occured. An adequate precaution is taken to guard against fly rock, dust, fumes and excessive vibrations. Where necessary, the size of a blast is reduced. On occasions, the advance fer round has been halved to maintain acceptable public relations, even though it would not be technically necessary to do so.

Fresh air is provided at the rate of 600 m^3 per minute for adequate ventilation during construction. The muck is removed by front-end loaders and dumpt trucks of various makes and capacities but 14 m^3 trucks are successful.

The hard rock mined out from the tunnels is mostly crushed for use as concrete or asphalt aggregate. This also has an environmental advantage of reducing the rock dumps and the number of quarries.

CONCLUSIONS

The tunnels in Makkah are constructed as permanent structures and are excellent from the functional and architectural points of view. Finishing, lighting, ventilation and safety features are built to the highest standards. However, it would be advisable to continuously update the design criteria on the basis of feed-back from the construction and operational phases.

The criterion of rock mass quality for the

design of support and reinforcement for tunnels, although based on arbitrary quantification of essentially qualitative and descriptive engineering geological parameters, appears to be broadly acceptable as a preliminary guideline. The more specific design, however, should include a systematic analysis of joint survey data in relation to the tunnel geometry, in order to outline the possible locations and modes of failures. The stability analysis should, then, be carried out using the shear strength of joints and other relevant parameters. This should be followed by the design of reinforcement and support to counter the tendencies of instability.

Specific field trials may be undertaken of various possible support and reinforcement systems. Physical scale models and computer assisted mathematical models may also be used for the purpose. The research work may cover other aspects of tunnel construction too, such as blasting, materials handling, ventilation and safety measures.

The authors are continuing the study of the behavior of discontinuities in Makkah rocks and intend to start work on the physical scale model tests for the tunnels.

References

1. Khan, Mir Ayoob Ali, "A city of tunnels", The Saudi Gazette, Features, 1984.
2. Broch, E. and Franklin, J.A., "The point load strength test", Int. J. Rock Mechanics & Mining Sciences, vol. 9, p 669-679, 1972.
3. Barton, N., Lien, R. and Lunde, J., "Engineering classification of rock masses for the design of tunnel support", Norwegian Geotechnical Institute, Publication, 106, 1974.
4. Hanif, M. "Support and Deformation of Underground Openings", Ph.D. thesis 1974, Department of Mining and Mineral Sciences, University of Leeds, U.K.
5. Personal Communications and King Abdulaziz University, Mining Engineering Department students' training reports.

Stabilization of existing rock faces in urban areas of Hong Kong

B.I. Dubin B.SC.(ENG.), M.SC.(ENG.), P. ENG., M.E.I.C.
A.T. Watkins C.ENG., M.I.P.E.N.Z., M.H.K.I.E., M.I.C.E.
D.C.H. Chang B.SC.(ENG.), M. ENG., C. ENG., M.I.C.E., M.H.K.I.E.
Geotechnical Control Office, Hong Kong

SYNOPSIS

The current approach to investigation, design and construction of stabilization works to existing rock cut slopes in Hong Kong is presented. Several hundred rock cuttings of varying age are located within the urban areas of Hong Kong, the great majority having been constructed prior to the advent of modern geotechnical design standards. Occasional failures causing casualties and property damage require that these sub-standard slopes be upgraded. The Geotechnical Control Office of the Hong Kong Government has for a number of years been engaged in a programme of Landslip Preventive Measures (LPM) works to improve the stability of slopes in general. As a small but important part of this programme the stabilization of existing urban rock slopes has been included.

These works are distinct from many of the large scale rock slope works undertaken in conjunction with extensive infrastructure and housing development projects, in that a staged approach to investigation and design has not always been possible.

The urban rock slope works carried out under the LPM programme are normally implemented using simple investigation and design methods and the construction techniques generally comprise a combination of manual and small plant operations.

The practical aspects of these works are discussed including the influence of a number of common constraints, including the lack of works space, the necessity to maintain public safety during construction, and the need to minimise disruption to public transport and services and the environment. Stabilization treatment is usually limited to the surficial problems of the faces themselves relying on observational techniques, together with a study of the historical performance of the slopes.

Generalized design methods are discussed together with some aspects of the construction process.

INTRODUCTION

A general review of rock slope engineering in Hong Kong has been previously made by Brand, Hencher and Youdan[1], however, this paper will concentrate on more specific aspects of remedial and preventive works undertaken on formed rock faces within the urban area. In particular the works carried out on slopes which have been included in the Hong Kong Government's Landslip Preventive Measures (LPM) programme are discussed and the methods currently in use for investigation, design and implementation of remedial works to rock faces are presented.

Hong Kong is situated on the southern coastline of Kwangtung province of China and has a land area of just over 1 000 km². The terrain is very hilly typically with upper slopes steeper than 35°, midslopes in the range of 25-30° and footslopes of about 15°. Much of the original Kowloon peninsula has been flattened to provide filling for reclamation. Most of the land area below an elevation of about 5 mPD is reclaimed land.

The population of more than 5½ million is concentrated largely in the urban areas of Hong Kong and Kowloon. Fig. 1 shows areas where the

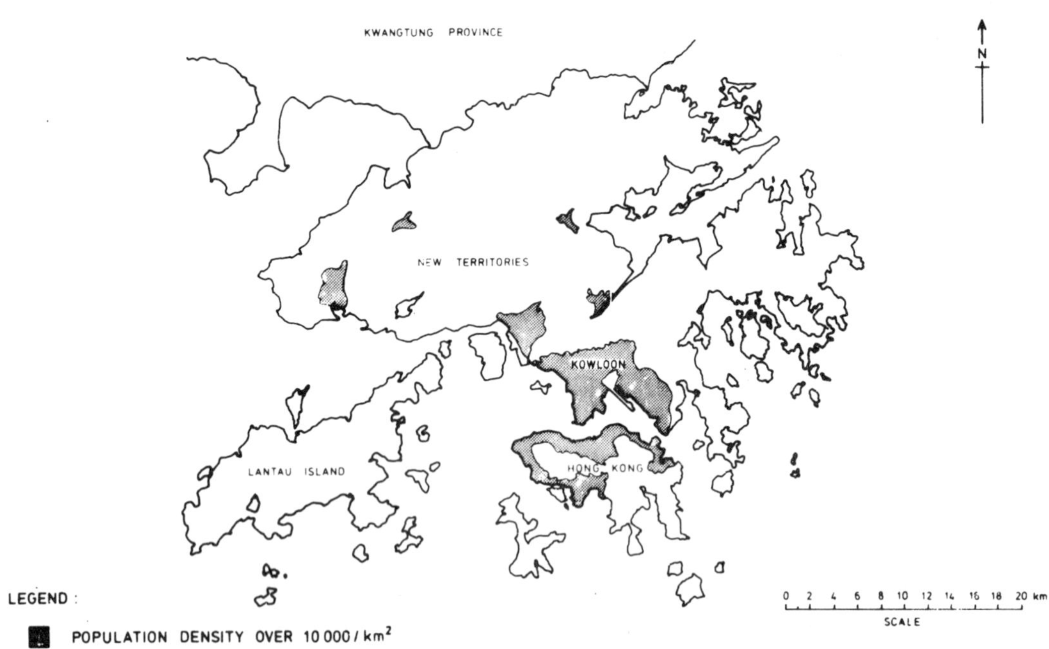

Fig. 1 Population density of Hong Kong

population density is greater than 10 000 persons/km². Some of this urban area is also coincidentally among the steepest terrain in the territory.

Landslides (including rockfalls) are a continual problem in Hong Kong and have over the years caused significant loss of life and property damage. The great majority of Hong Kong's landslides are directly associated with major rainstorm events and it has been suggested[2] that 70 mm/hr is the threshold value of intensity above which landslides are most likely to occur. Hong Kong is an area of low seismic activity and there is no positive record of landslides as a result of earthquake forces.

Since the end of the 2nd World War, Hong Kong has seen an unprecedented expansion, in terms of economy and population. This has caused enormous demand for useable land, resulting in extensive development on steep hillsides and many large slopes have been formed. Rockslopes 30 m high and up to 90° inclination are not uncommon. Prior to 1978 there were relatively few rock slopes constructed under modern engineering control, and faces formed by techniques more in line with quarrying resulted. Indeed, a number of former quarry sites have in fact become sites for development. From attempting to research old standards, it has become apparent that, during the period when many existing rock slopes were formed, no statutory design standards existed. Quarry faces were cut near vertically, and the accepted rule of thumb for rock cuts undertaken for road works and site formation was 1 Horizontal in 10 Vertical (84°).

Most of the rock slopes with which the LPM programme is presently concerned, were apparently formed by Government during the 1950s and 1960s. Gardiner[3] gives contemporaneous guidelines for the use of explosives for such work.

Much of the degradation of these old slopes results from blast damage from the original construction. Generally, pre-splitting and smooth face techniques do not appear to have been used and Gardiner[3] makes no mention of them. Scaling of loose materials (i.e. removal by prising with hand tools) likewise appears to have been limited. One of Gardiner's[3] typical examples is of a 20 shot blast in Hong Kong Granite, fired from 1.5 in (38 mm) diameter holes 20 ft (6 m) deep at 6 ft (2 m) spacing and burden. He calculates 10.65 lb (4.8 kg) of explosive per hole (based on a powder factor of 5 tons/lb). While this is not a large blast by mining standards, on examining the potential for blast damage, the following is found : For an explosive detonated in a rock mass, from Hoek and Bray[4], it is known that $V = k(r/\sqrt{w})^{\beta}$ where V = particle velocity (in/sec), $r/\sqrt{w}$ is the scaled distance at a particular point (radial distance and weight of explosive) and k and β are site specific constants. Using Gardiner's[3] example values, and assuming values of k and β at 60 and -1.5 (typical values from Hoek and Bray[4]), a particle velocity 6 ft (2 m) from each detonation of about 22 in/sec (0.6 m/sec) is calculated. This is very close to the threshold value for rockcracking. It is therefore not surprising that blast damage is so widespread on Hong Kong rock slopes (Fig. 2) and this has accelerated degradation due to the erosive effects of rainfall, and the prising effects of vegetation.

Fig. 2 Joint separation as a result of blast damage

Since 1978, Government has progressively developed standards for design and for construction. Powell and Irfan[5] have proposed design guidelines for use in those projects where new slopes are being formed or where new constructions are being undertaken close to existing slopes. Government has also implemented a Landslip Preventive Measures programme to carry out remedial and preventive works to slopes which threaten existing structures. The history of this programme has been reviewed by Bowler and Phillipson[6]. Initially the programme concentrated on fill slopes but in more recent years works on cut slopes including rock slopes have been undertaken.

Geological conditions

The geology of Hong Kong has been described by various reporters.

Until recently, the detailed geological descriptions of the rocks and rock formations given by Allen and Stephens[7] have formed the major reference source on this subject. However, work currently being undertaken by the geological survey section of GCO is superseding much of Allen and Stephens'[7] work.

The predominant rock types are igneous, of granitic and volcanic origin. Fig. 3 is a simplified geological map showing the distribution of the major rock types. It is interesting to note that granite predominates in much of the developed areas. The granitic rocks are younger than the volcanics and one of the main differences between the two from an engineering point of view is the joint pattern. Granite joint spacing is typically 0.5-2 m while the joint spacing in the volcanics is generally in the range 0.05-0.2 m. Large sheeting joints often occur in the near-surface granites.

Another difference between granitic and volcanic materials is in the depth of residual material formed as a result of weathering, which overlies the parent rock. Details of the typical weathering profile of these deposits are given by Ruxton and Berry[8] and are reviewed in the Geotechnical Manual for Slopes[9]. Mantles of residual material up to 60 m thick can overlie the granites while the mantle overlying the volcanics is normally up to about 20 m thick. This residual mantle has been classified based on degree of decomposition of the parent rock as follows :

Grade VI	Residual soil
Grade V	Completely decomposed rock
Grade IV	Highly decomposed rock
Grade III	Moderately decomposed rock
Grade II	Slightly decomposed rock
Grade I	Fresh rock

A more detailed discussion of the characteristics of the various weathering grades is given in the Geotechnical Manual[9] and by Hencher and Martin[10].

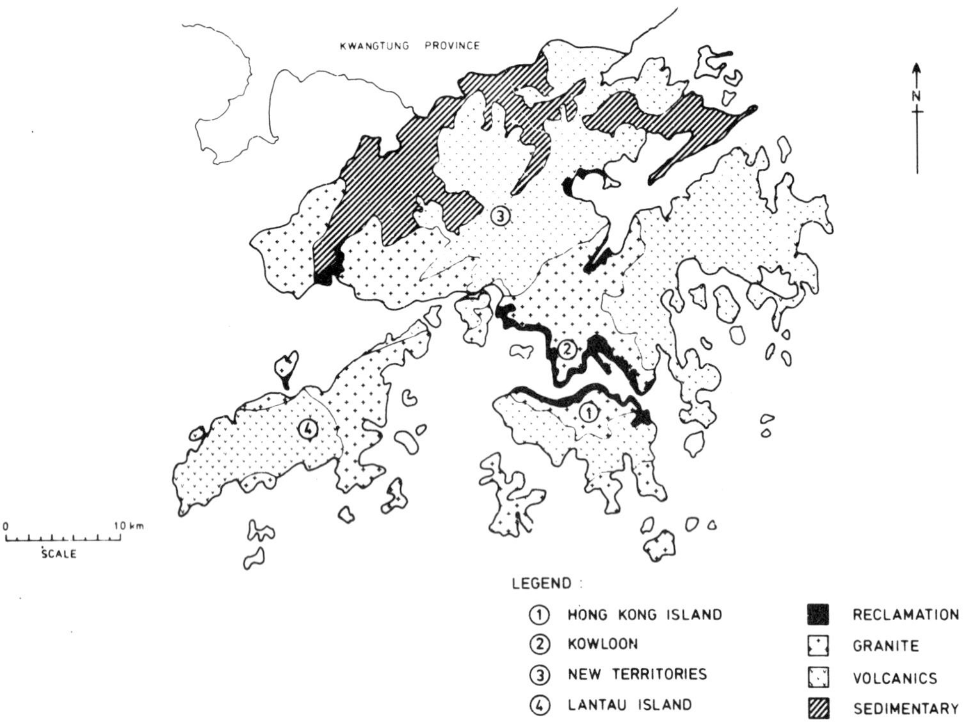

Fig. 3 Simplified geology of Hong Kong

The zones of highly and completely weathered rock commonly contain corestones of less weathered material. Material of Grades IV, V & VI are generally treated as soils in engineering analysis although the presence of relict joints is important. Details of the recommended procedures for analysis are given in the Geotechnical Manual[9].

The engineering properties of the rock materials (Grades I to III) are likewise often affected by the presence of more highly weathered zones containing soil material. Discussions of material characteristics and design procedures are given in the Geotechnical Manual[9] as well as by Hencher and Richards[11], Harris[12] and Richards and Cowland[13], to name but a few.

Constraints

The LPM programme is concerned with carrying out preventive works to slopes which pose a threat to life. Such slopes are commonly positioned very close to inhabited structures and in many cases very limited space is available to carry out investigation or to construct remedial works. In extreme cases it is impossible to even find a vantage point from which more than a small portion of the slope can be observed.

Preventive and remedial works to existing rock faces therefore must take such physical constraints into consideration and just as importantly, public safety must be preserved at all stages of the works. In a mine or a quarry slope, a small rockfall during or after slope construction would normally have insignificant consequences. The same rockfall in a slope above a crowded bus shelter in urban Hong Kong could constitute a major disaster.

Economic and environmental constraints are important also. Hong Kong has traffic densities among the highest in the world and accordingly roads must not be closed during remedial works to the slopes which threaten them. In Hong Kong's crowded conditions, environmental constraints are evident and the noise, dust, access restrictions and general public nuisance caused by LPM works must be kept to a minimum.

Such constraints can influence not only the form the eventual remedial works take, but also the investigation, design and construction methods necessary to achieve them.

INVESTIGATION TECHNIQUES

Background

The techniques used in the investigation of a rock slope can generally be divided into qualitative and quantitative types, although a number of methods encompass both categories. The economics of a particular project, including the time scale required to carry out a particular technique will normally influence the type of work undertaken. For example, there is little point in undertaking sophisticated analysis of what appears, to an experienced engineer, to be an unstable block. Where the cost of investigation would approach or exceed the cost of remedial works, far better, if possible, to eliminate the hazard by simply removing the block.

In most cases however, it is desirable to proceed logically, with a properly staged investigation, giving consideration to all techniques that are available. Unfortunately, many of the most valuable techniques cannot be applied to a particular rock face until scaffolding has been erected, and obscuring vegetation and surface coverings have been removed. As such works often increase the risk of a failure, it is common practise to adopt a design/build approach to rock slope stabilization works, with the major investigation and design work often being undertaken after site work has commenced. While this approach can cause contractual problems, in that it is difficult to define the exact scope of the works until after the contractor is on site, such problems are not insurmountable, and the design/build philosophy has yielded generally satisfactory results in the context of the LPM programme.

Nonetheless, prior to commencing field investigation or construction, it is obvious that considerable background information can be obtained for a particular slope.

Desk study

Generally, a rock slope investigation commences with a desk study. Background geology is examined through the references listed previously. It must be noted that the geological survey work currently being carried out by GCO has pinpointed a number of anomalies in the earlier work by Allen and Stephens[7]. It is therefore important to note that information obtained from background studies may not reflect the actual site conditions, and must be confirmed through fieldwork.

If available, construction information should be checked. It is possible in the context of LPM works to use the existing governmental record systems to good effect to retrieve information from other offices and departments. Possible sources of information are listed in the Geotechnical Manual[9] and can include Buildings Ordinance Office records (details of neighbouring buildings including foundations and loads, old site investigation data and inspection reports) and Highways Office records (which may include old profiles, construction plans and specifications). Existing GCO airphoto reports and area geotechnical studies such as the GCO GAS reports[14] should be checked, although as these studies are often large-scale, more site-specific information is usually required. Detailed airphoto analysis may be called for, with emphasis on geology, geological structure, history of development and previous instability. Information obtainable from GCO's Geotechnical Information Unit can include records of previous failures, previous site investigations, consultants reports and field sheets, records of previous governmental interest in the site and monitoring (both groundwater and ground movement) in the area.

Site visits

Site visits are undertaken at an early stage of the investigation to confirm geological information obtained by desk study. Normally abundant exposure can be found, but as noted this exposure is usually heavily obscured by vegetation and rubbish and it is often physically difficult to stand back and examine a slope.

Conventional site investigation

The techniques for a conventional site investigation in Hong Kong have been described by others[15, 9]. As many rock slopes are associated with soil slopes, techniques such as rotary drilling and soil sampling as well as rock coring, are of great value. Conventional instrumentation including standpipe piezometers to monitor groundwater levels and telltales to indicate localized ground movements are often installed. Special mention should be made of the impression packer device, which permits measurment of the orientation of structural discontinuities in a borehole. Details of the procedure are given in the Geotechnical Manual[9] and by Hinds[16], however, the machinery must be used with care if useful information is to be gained, and interpretation should be carried out by well experienced personnel.

Such techniques can normally be carried out regardless of the time scale of a particular project. It is not necessary to wait for the contractor to clear the site, although it can be useful.

The scale of typical rock slope remedial works under the LPM programme should be contrasted against a typical mine slope. LPM slopes are normally of the order of tens of metres in extent, both in height and length, while a mine slope can be hundreds of metres high and thousands of metres in length. Due to economies of scale which apply to the larger mine slope, the use of expensive techniques is more easily justified. However, economies in an LPM investigation must be tempered by an awareness of the consequences of a small failure. The failure of a one cubic metre block, which would be inconsequential in a mine slope, could have severe consequences, in human terms, in an urban rock slope. Therefore, the LPM approach is more conservative in design and execution.

Rock structure discontinuity surveys

Joint surveys, are the basis of all quantitative approaches to rock slope stability problems. The principles and methods are well explained by Hoek and Bray[4], Richards[17] and others[9, 12]. However, the comparatively small size of LPM sites, together with the degree of surface cover and obstructions often makes obtaining a statistically valid sample difficult. For example, old failure scars are normally covered with chunam (a plaster like material made from cement, lime and soil), which must be stripped off before measurements can be taken (Fig. 4).

As has been pointed out by Brand, Hencher and Youdan[1], the statistical approach often eliminates dangerous single joints. It is necessary to examine all joint data for evidence of these rogues. Identification by numbering of each joint measured in the field can be useful in this regard, although often such markings are destroyed by construction activity.

In LPM works, the results of joint surveys and kinematic stability analysis are used to pinpoint major instabilities and to afford the design engineer some idea of the most likely modes of failure when detailing works on site. Individual areas of concern are dealt with in greater detail, by means of more closely focused investigation.

Rock classification systems

Hong Kong's own rock classification system in terms of weathering grades, has been discussed previously. Simplified mapping and classification systems for use in Hong Kong have been described by others[18].

For particular problems, it may be useful to apply one of the more rigorous systems commonly used in tunnel analysis such as the Norwegian Geotechnical Institute (NGI) system[19] or the Rock Structure Rating (RSR) system[20]. It may be possible to correlate ratings from such systems with support requirements for a rock slope. A system recently proposed for tunnel classification in Hong Kong[21] could perhaps be expanded for this purpose.

Qualitative techniques

Although many conventional site investigation techniques, such as chunam stripping, can be considered as both quantitative and qualitative, a number of techniques in use on LPM sites are not widely disseminated in the literature. These include methods such as the pneumatic penetration probe, which is both quantitative and qualitative, and observational techniques including oblique terrestial stereo photography.

Pneumatic penetration probes

'P-Probes' have long been used as the basis of acceptance or otherwise of a rock-socketed caisson. An air-hammer is used to drill a hole from the base of a foundation caisson and the time for an arbitrary penetration distance is logged. Shorter penetration times are generally indicative of weak layers, although in very weak materials the reverse can be true. Discon-

Fig. 4 Obscuring effect of vegetation and chunam before and after stripping

tinuities can be located by means of a nail driven at right angles through a light wooden pole which when dragged up the inside surface of the hole will 'hang-up' on an open joint.

Such probes, often referred to as 'star drilling' can be of use in LPM works to indicate zones of weakness within the rock mass. It is relatively simple for an experienced operator to distinguish between soil, weathered zones associated with structural discontinuities and fresh rock. Fig. 5 is a typical plot of a P-Probe, showing the interpretation. Such methods must be used with caution, as they can be inaccurate.

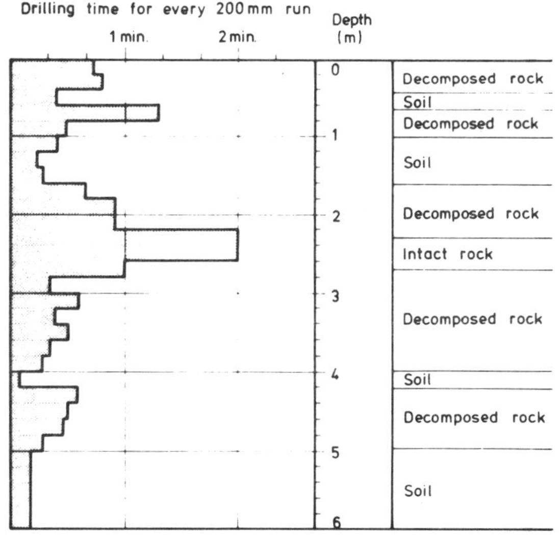

Fig. 5 Typical penetration probe result

Observational techniques

Typically, slopes are obscured by vegetation, surface protection materials and debris. The obvious first step in detailing specific remedial works is to erect scaffolding and remove the obscuring features. In some cases, a vantage point at a convenient distance from which the entire slope can be observed can be located. Binoculars can then be employed to examine and map as much of the face as possible. However, this is not always the case. Fig. 6 shows a slope where such a vantage point was not available.

Fig. 6 Problems caused by proximity of threatened structures

Scaffolding permits a close up examination of the face. Obviously, safety techniques are important when working on scaffolding and safety lines and a safety harness are essential equipment. Normally in Hong Kong, the scaffolding is constructed of bamboo. Bamboo has the advantage over conventional construction scaffolding in that it is very flexible, can be laid to suit the alignment of the face and tends to yield by cracking rather than by catastrophic failure. The flexibility can tend to be disconcerting at first to the engineering geologist carrying out detailed mapping operations.

Photographic techniques

Good color photography of the slope is useful. For easy reference, it is helpful to mark chainages and elevations directly on the scaffolding or slope face prior to photographing the slope. In addition, the scaffolding provides a grid which helps to locate and identify features on the face.

As an alternative to actually marking required works on the face, works can be detailed on transparent overlays to a photo and issued to the contractor. Fig. 7 shows an example of a photo overlay detailing stabilization works. Additional overlays can be used when checking the contractors setting out. Color xerox technology makes it possible to reproduce photo-overlays with sufficient clarity for contractual purposes.

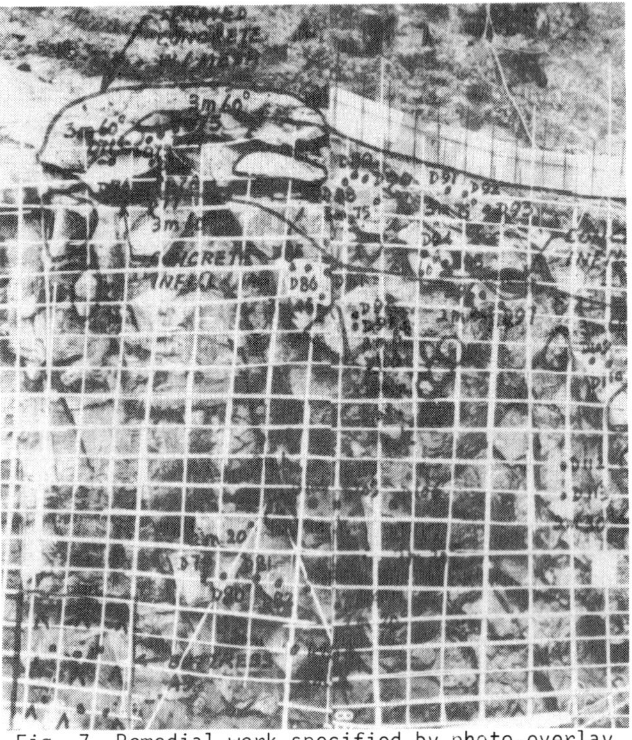
Fig. 7 Remedial work specified by photo-overlay

The use of photographic overlays for engineering geological mapping is discussed by Starr and Finn[22] and Starr, Stiles and Nisbet[23].

Oblique terrestrial stereo photography

Terrestrial photogrammetry has long been used in open-pit mines for survey purposes. Applicable techniques are discussed by Ross-Brown and Atkinson[24] and Hoek & Bray[4].

In addition to the methods described in the above references, a hand held camera can be used to produce stereophotographs acceptable for design purposes. Williams[25] describes methods by which accurate maps and stereopairs from which measurements can be taken can be produced using a hand held camera. Stereopairs useful for reference purposes when preparing engineering geology maps and photo overlays can be taken quite simply using a hand held camera and taking successive photos of the areas of interest, using a parallax movement of 3 - 5% of the focal distance and 100% overlap. Fig. 8 is an example of such a stereopairs. While these are of little use for quantitative purposes, good stereopairs can reduce the need for site visits to check outstanding features, resulting in a considerable saving of the engineer's time. Further, the ability to conveniently provide a realistic 3-dimensional visual representation of a particular problem, greatly facilitates consultation with other engineers and with the statutory authorities.

Fig. 8 shows the same area as that contained in the overlay at Fig. 7. The contractor's markings for the works detailed at Fig. 7 can be seen on the face.

Rock slope stability analysis

In many cases where quantitative techniques have pinpointed possible failure modes, techniques exist for rigorous analysis. Discussions of wedge failure and plane failure mechanisms together with analytical methods are given by Hoek and Bray[4] and by Golder Associates[26]. Such techniques are useful, but can be time consuming. Often in practical work they require the use of so many assumptions that such analyses must be approached with caution.

The argument can be made that detailed

Fig. 8 Stereopair showing proposed remedial work

numerical analysis for a rock slope that has been standing without a major failure for many years may be considered unnecessary, as the slope has stood the test of time. This view is obviously not altogether true, as degradation processes due to the prising effects of vegetation, erosion and human activities (such as leakage from underground services or increased loads due to building activity) can encourage an instability to develop over time.

In any case, the more sophisticated rock mechanics approach should be considered, along with the simpler design tools, and in the case of major rock slope stability problems its application is essential, to achieve a safe, economical solution.

DESIGN METHODS AND CONSTRUCTION TECHNIQUES

Hong Kong, in contrast to other technologically advanced societies, is a labor intensive community. Designs for rock slope remedial works must take this into account. Fookes, Sweeney, Manby and Martin[27] point out the necessity of taking local economics into account as well as presenting an overview of a number of low-cost, technologically simple techniques for slope stabilization.

It is interesting to contrast a typical Hong Kong technique - two men perched on bamboo scaffolding using a hand held air hammer to drill a hole for a rockbolt, against a mine slope in Canada, where one man on a large crawler drill might pattern bolt an entire bench face, or against a road slope in Austria where a derrick and jackleg may be used.

This difference in approach can be defined as an 'economy of culture'. While it is possible to obtain the equipment necessary to carry out works the same way as in Canada, or Austria, it is cheaper and easier to use the tried and tested methods which local contractors are familiar with and design techniques must take this factor into account.

Designs must be as maintenance free as possible. With the large numbers of slopes in Hong Kong, the limited maintenance resources must be kept free, as much as possible, to deal with problems. Therefore, rock slope designs should make use of durable materials, shotcrete is used rather than chunam for example. If possible, self-maintaining coverings are applied. Vegetation is preferred, if feasible, both for it's self-maintaining qualities and aesthetics. Self-flushing surface drainage channels as specified in the Geotechnical Manual[9] should be used. Trash racks and downspouts should be designed to avoid clogging. This philosophy tends to raise costs (for example, mesh reinforced shotcrete is approximately 5 times the cost of chunam but lasts considerably longer), however, it greatly reduces the need for maintenance on rock slopes, where access is often difficult.

Temporary support problems must be catered to

and designs must attempt to ensure stability at all stages of construction.

A number of typical stabilization techniques which could be chosen for use at a particular site are illustrated in Fig. 9 and can include the following :

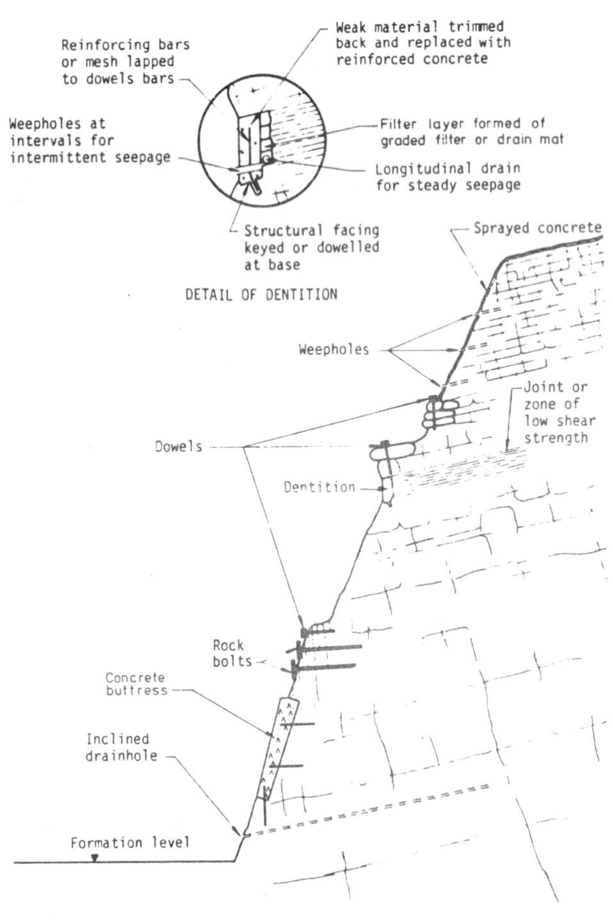

Fig. 9 Typical preventive works to rock faces

Removal methods

Scaling

As mentioned previously most existing rock faces are heavily obscured by vegetation. When rock slope stabilisation work commences on site, the vegetation, top soil, old surface protection and other obscuring features must be stripped off. Minor rock scaling using hand tools is often undertaken. Usually, the rock face is washed down by means of a water jet driven by compressed air to obtain a clean surface for detailed inspection by the engineer. Scaling must be carried out carefully in order not to affect the stability of the remainder of the face. As these works are normally carried out from bamboo scaffolding it is important that experienced workmen are employed. Proper safety techniques for men working on scaffolding must be employed and care must be taken to keep damage to the scaffolding from falling rock fragments to a minimum.

As it is necessary to avoid risk to the public at all times the majority of slopes require protective rock trap safety fencing to be erected prior to the commencement of work. Normally, a safety fence would consist of wire mesh and plastic tarpaulins supported by steel members (Fig. 10). Such an installation is, of course, not capable of stopping large blocks, but will successfully retain small fragments and soil material.

Fig. 10 Temporary safety fence 6.5 m high

Since most rock slopes have overlying soil slopes, preliminary scaling can also include removal of vegetation and soft surficial material from the soil slope. Such overlying soil slopes are normally treated to achieve an improved factor of safety, applicable to the particular risk as discussed in the Geotechnical Manual[9].

Rock excavation

Following the preliminary scaling and further inspection the designer may order further excavation of large rock masses. Scaling and removal of unstable blocks is preferred where possible, as it is a simple, maintenance free solution to rock slope problems. Potentially unstable blocks of rock are normally removed using hand-held powered equipment such as pneumatic hammers or drills. Occasionally, hydraulic splitters or proprietary silent non-explosive demolition agents are employed. Use of explosives is generally not feasible due to the proximity of inhabited structures. In the event that scaling and removal of rock blocks could endanger buildings or roadways adjoining the slope, alternative means of treatment or support must be considered. Safety aspects must be stressed, as removal of larger rock masses normally results in considerable damage to the scaffolding. The designer should also consider the economics of scaling/removal versus the support option.

Support methods

Detailed design of methods of providing localized support to portions of rock faces are given by Fookes and Sweeney[28], as well as by Starr, Stiles and Nisbet[23]. Methods commonly used in remedial works to rock slopes in Hong Kong, as part of LPM programmes are discussed below.

Buttresses

The design of a typical concrete buttress is illustrated in Fig. 11. Buttresses may be used to support overhanging rock blocks or heavily fissured sub-vertical rock faces and are normally designed as gravity retaining structures. Buttresses are normally connected to the rock mass by means of grouted, untensioned steel dowels. Steel reinforcement, either designed or nominal is incorporated as appropriate. Where buttresses are designed to take shear forces, then the potential sliding mass together with the buttress itself should be considered as the sliding mass, for design purposes.

It is essential that drainage is provided behind buttresses to prevent the build up of pore water pressure. Typically, this drainage takes the form of a proprietary drainage material placed along the joints and connected to weep-holes or other positive drainage.

It is important that buttresses designed to support overhanging rock masses have a direct contact with the mass above. In practice, this generally means that the buttress will protrude

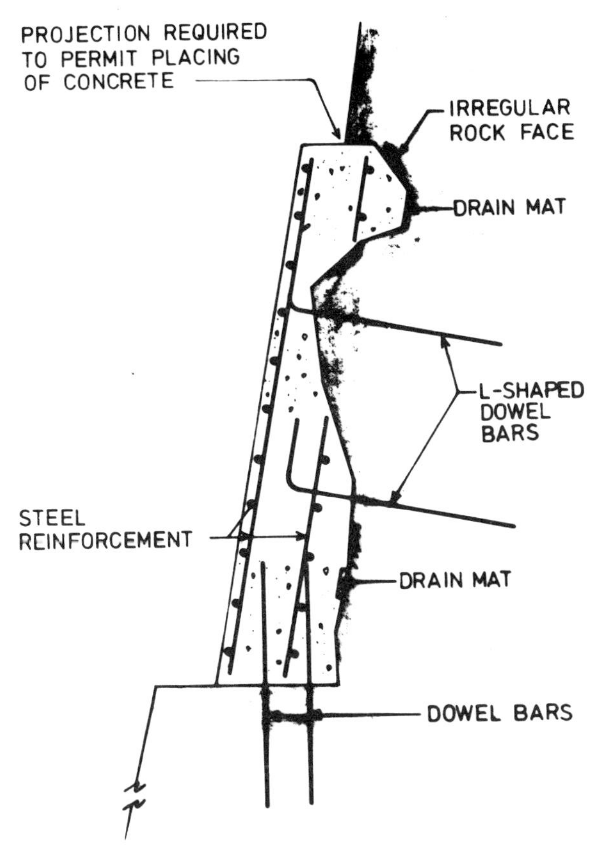

Fig. 11 Typical buttress

about 200 mm beyond the overhang to permit the concrete to be placed. A patterned surface finish to the concrete is often provided to make the finished buttress visually acceptable.

Masonry buttresses have been employed in the past. They have the advantage that the primary materials required (rock fragments) are normally readily available on site. However, due to the difficulty in providing reinforcement or in bonding in dowels, their use is not favoured.

Dentition

Concrete, which may be dowelled into the rock mass, is often used to seal joints or to fill relatively small voids and to support small overhanging blocks. Positive drainage is normally provided behind dentition.

Dowels

Dowels are steel bars mainly designed to provide shear resistance to potential sliding blocks.

Although very little research work appears to have been done on dowels, Sage[29] and Bjurstrom[30] give design methods which have been successfully used in LPM works. Lang[31] provides a brief discussion of the difference in behaviour of untensioned dowels versus tensioned rock bolts.

Bjurstrom[30] recommends that for dowels crossing rock joints at angles greater than 45°, the dowel effect can be calculated from :

$$T_{dowel} = d^2 \times \frac{2}{3} (\sigma s \cdot \sigma c)^{\frac{1}{2}}$$

where

T_{dowel} = shear resistance due to dowel effect (MN)

d = effect diameter of bar (m)

σs = yield stress of dowel steel (MPa)

σc = uniaxial compressive strength of the rock (MPa)

For dowels crossing rock joints at angles less than 45°, he recommends

$$T_{dowel} = T (\cos \alpha + \sin \alpha \tan \phi)$$

where T = tension force in the dowel due to shear displacement (the lesser of the yield stress in the dowel steel or the steel/grout/rock bond.

α = bar angle relative to the joint

ϕ = angle of internal friction of the joint surface.

Littlejohn and Bruce[32] give values of grout/rock bond which may be used in practice. For dowels in fresh Hong Kong granite, σc is generally taken as 100 mPa. This is a mid-range value for 'high-strength' rock as defined by the Canadian Manual on Foundation Engineering[33].

For Hong Kong conditions, the Geotechnical Manual[9] recommends a reduction in diameter of 4 mm to allow for corrosion and that the maximum working stress should not exceed 50% of the ultimate steel strength. As well, it is recommended that a minimum of 6 mm grout cover is required on the bond length and that centralisers should be used to ensure an even annulus of grout. Normally, in LPM works, dowels are installed in holes drilled using hand-held powered equipment. This limits the maximum hole diameter to about 50 mm, which provides ample cover for the 32 mm diameter dowel bar which is typically used.

In carrying out installation of large numbers of dowels, it is often useful to prepare design charts. Fig. 12 is an example of such a chart prepared to assist in design of dowels to fix unstable blocks resting on a sliding plane.

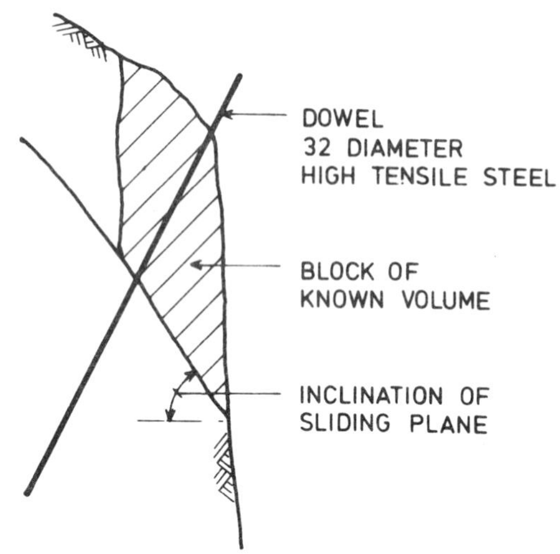

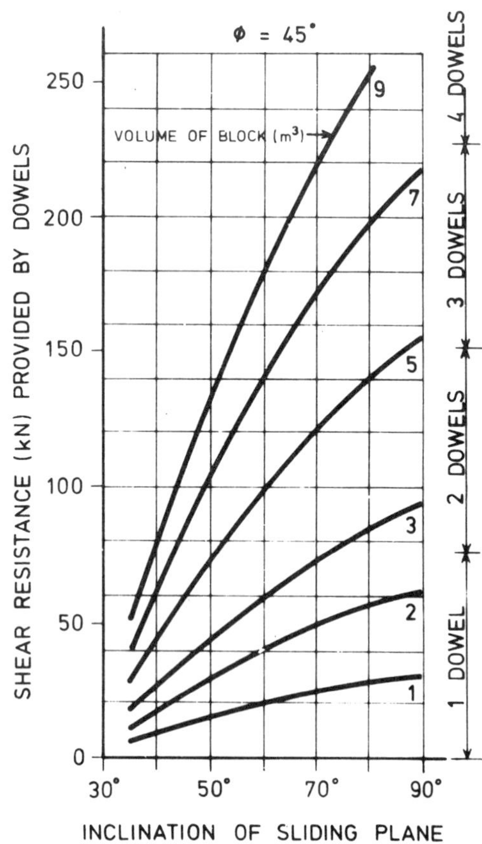

Fig. 12 Typical dowel design chart

Rock bolts

Bolts are tensioned steel bars designed to increase the normal stress of unstable blocks of rock and hence increase the mobilised shear strength. Design guideline for bolts in Hong Kong are given in the Geotechnical Manual[9].

Detailed design procedures can be obtained from Lang[31], Hoek and Bray[4] and Littlejohn and Bruce[32].

The tension load required in a rock bolt to prevent sliding of a block on an inclined plane can be calculated from the expression :

$$F = \frac{CA + (W\cos\psi - U - V\sin\psi + T\cos\theta)\tan\phi}{W\sin\psi + V\cos\psi - T\sin\theta}$$

where C = cohesion on joint

A = area of block in contact with joint plane

T = bolt load

W = weight of block

V = horizontal force from water in tension crack

U = uplift force

θ = angle between bolt and the normal to the sliding plane

ψ = inclination of sliding plane

ϕ = angle of friction on the plane

F = factor of safety

Detailed recommendations for bolt design are given in the references noted previously. In Hong Kong, typically the grout/rock bond is assumed to be of the order of 1 N/mm², although in all cases pull-out tests should be carried out to verify whatever figure is used. The bond length should be designed with a factor of safety of 2 against pull out. From a corrosion protection point of view, bolts should be designed to a higher standard than dowels, as discussed in the Geotechnical Manual[9]. Fig. 13 shows a typical bolt design which has been used successfully in LPM works.

Prestressed ground anchors

In LPM works, prestressed ground anchors would only be considered in exceptional cases due to the well documented difficulty in achieving sufficient corrosion protection, the requirements for long term maintenance and monitoring and the fact that the necessary anchor tendon length often must extend beyond the site boundary. Design procedures and installation guidelines are set out by Littlejohn and Bruce[32]. The Hong Kong Government has recently produced a model specification covering technical aspects[34].

Alternative support means

A number of other means of support have been used on occasion in LPM works, particularly for temporary support. For example, cable lashings anchored to grouted untensioned dowels have been used to provide temporary support to an unstable block during construction of permanent support. Cable lashings and similar methods (steel cable netting for example) have the advantage that they can be applied with minimum risk to workmen, who may otherwise be forced to work under or on a potential sliding mass. A discussion of these and similar methods is presented by Piteau and Peckover[35].

Surface and subsurface drainage

The provisions of surface and subsurface drainage measures are discussed in detail in Chapters 4 and 8 of the Geotechnical Manual[9]. The main purpose of these drainages measures is to improve slope stability by reducing infiltration and by lowering or controlling the groundwater level. U-channels are usually provided at the rock slope crest and toe to intersect surface runoff. All surface runoff should be collected through stepped U-channel or downspouts, provided with catchpits or baffle wall at junctions, and conducted to nearby storm water drainage services. Fig. 14 shows a design for a simple drain which has been used to good effect to control the groundwater level. In many cases, these and similar drains can be installed in holes drilled using hand-held equipment, eliminating the need to provide access for a drill rig.

In some Hong Kong slopes, the existence of disused air-raid tunnels underlying the slope is helpful, in that these tunnels may serve as drainage galleries and can be very effective in controlling the groundwater level.

Surface protection

Vegetative and rigid surface protection measures are discussed in detail in Chapters 8 and 9 of the Geotechnical Manual[9]. Pneumatically applied concrete with steel mesh reinforcement (shotcrete)

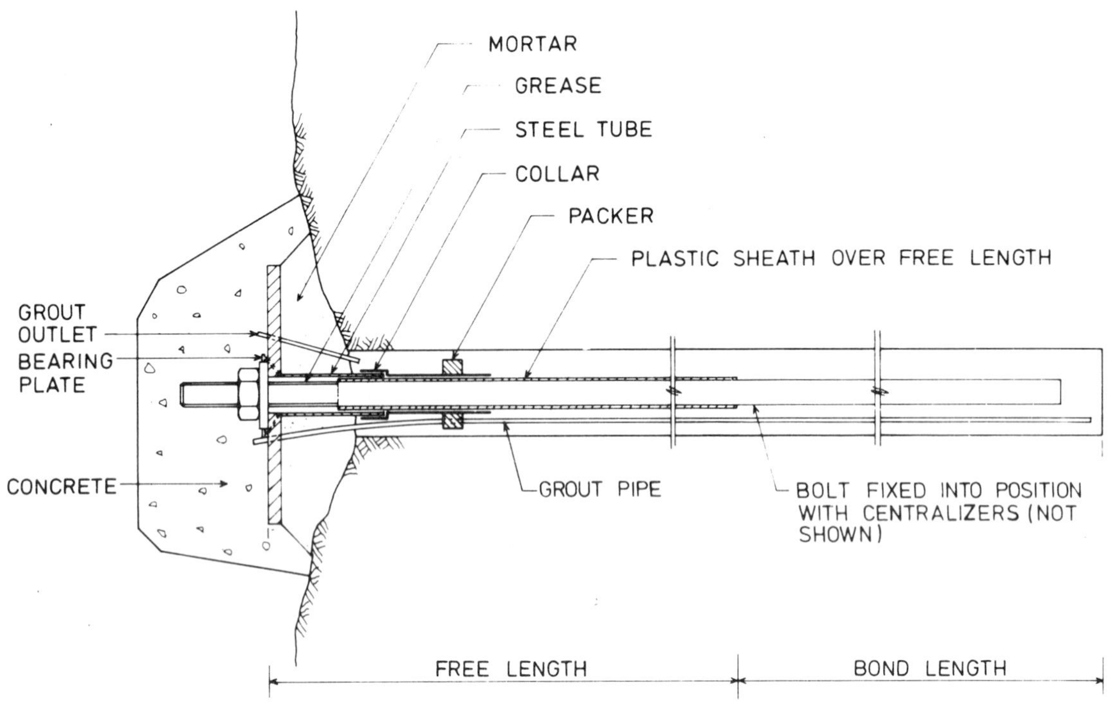

Fig. 13 Typical rock bolt detail

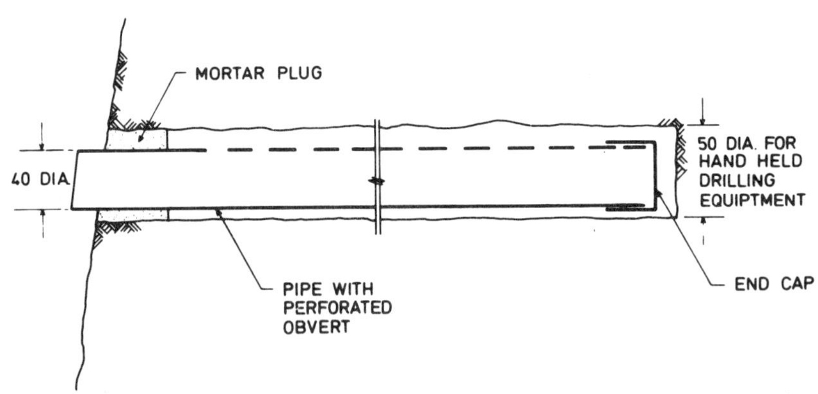

Fig. 14 Typical raking drain detail

is the most commonly used method for LPM works on rock slope surfaces. In addition to reducing infiltration, it can help to provide reinforcement to shattered zones or to distribute loads from bolt reinforcement. It is important that drainage holes or proprietary drainage mats are provided at rock joints to prevent pore water pressure build-up.

A number of alternative surface protection systems have been utilized in recent years, mainly in areas where aesthetic considerations have been judged to be important. These include the use of permanent mesh netting, as well as proprietary systems such as the On method for vegetation, and the Grasscrete and Meshbolt system.

Permanent mesh netting

Mesh netting for protection against minor spalling and rock falls is a relatively new technique in Hong Kong. Such mesh installed recently on LPM sites consists of PVC coated 2.7 mm diameter galvanized wire woven into triple-twist (3 half turns) hexagonal mesh of 80 x 100 mm, weighing approximately 1.7 kg/m². Laboratory tests carried out by the manufacturer indicate that the material may have a life expectancy of over 40 years in tropical atmospheric conditions[36].

Permanent netting is anchored at the top edge of a slope by galvanized rock dowels and cables and is held in close proximity to the rock face by 500 mm long galvanized bolts, installed at a vertical spacing of 3 m and a horizontal spacing of 1.5 m. On LPM slopes where permanent mesh netting has been installed, vegetation has established itself on the netting surface to achieve an attractive appearance. Permanent mesh netting is useful to protect against minor rock falls from a fractured rock surface but has no effect on overall slope stability.

The On method for vegetation

This proprietary method has made it possible to vegetate a rock mass or other non-soil surfaces. This method utilizes urea formaldehyde resin as a base with peat moss and bark compost as a supplement mixed together to form a plant cultivation base. The cultivation base is mixed with cement binder and selected plant seeds and is sprayed onto the slope surface using a wet process similar to hydroseeding. The On method has been tried on a sub-vertical rock face in the Mid-levels area of Hong Kong and has been growing well on a surface of Grade II-III rock.

Grasscrete and Meshbolt system

This is a proprietary combined rigid and vegetative treatment applicable in areas where shotcrete would normally be used. Inexpensive plastic formers are pinned to the slope surface, mesh reinforcement is fixed by grouted bolts, and the slope face is then shotcreted. Following the shotcreting operation the exposed tops of the formers are broken out and the voids filled with a mixture of soil and seed. The manufacturer claims that the system combines the advantages of shotcrete with the pleasing appearance of vegetation. This system has been used on some LPM slopes as an experimental alternative to the conventional methods (Fig. 15).

CONCLUSIONS

The authors have attempted to provide a detailed review of the methods currently in use for carrying out remedial works to rock slopes in the urban areas of Hong Kong, in the context of the Hong Kong Government's Landslip Preventive Measures programme. Work to rock faces carried out under this programme is unique for a number of reasons, including the large number of constraints of a physical, environmental, public safety and economic nature which must be catered for. Although the slopes commonly treated under the LPM programme are small, when compared to a mine slope, the consequences of a small failure on an urban rock slope can be very severe. Thus a more conservative approach to remedial works design is taken. Investigation and design

Fig. 15 Grasscrete surface protection

Plastic formers in place prior to application of spray concrete.

Finished surface with early grass growth.

techniques are varied and in many cases cannot be actioned until after works on site have commenced. Although this 'design-build' philosophy can cause minor contractual problems and certainly requires a higher level of engineering supervision than the more conventional approaches, it works reasonably well in the context of Hong Kong's peculiar culture and congested physical environment. Investigation techniques are used to highlight potential failure mechanisms, although in general, failures of large rock masses are uncommon. A large part of the investigation and design involves visual assessment, and small-scale detailing of remedial works to individual problem areas. The importance of maintaining a systematic approach through investigation, design and construction cannot be overstated.

ACKNOWLEDGEMENTS

The authors gratefully acknowledge the permission of the Director of Engineering Development and the Principal Government Geotechnical Engineer of the Hong Kong Government for the publication of this paper.

REFERENCES

1. Brand, E.W., Hencher, S.R. and Youdan, D.G. Rock slope engineering in Hong Kong. Proceedings of the Fifth International Rock Mechanics Congress, Melbourne, vol. 1, 1983, p. C17-C24.

2. Brand, E.W., Premchitt, J. and Phillipson, H.B. Relationship between rainfall and landslides in Hong Kong. Proceedings of the Fourth International Symposium on Landslides, Toronto, vol. 1, 1984, p. 377-384.

3. Gardiner, T.L. Blasting in Hong Kong; a Guide for Users of Explosives. Government Press, Hong Kong, 1965, 88 pp.

4. Hoek, E. and Bray, W.J. Rock Slope Engineering. (3rd ed.), Institution of Mining and Metallurgy, London, 1981, 358 pp.

5. Powell, G.E., and Irfan, T.Y. Investigation and design of rock slope remedial works for differing risks. Proceedings of Conference on Rock Engineering and Excavation in an Urban Environment, Hong Kong, 1986, in press.

6. Bowler, R.A. and Phillipson, H.B. Landslip preventive measures - a review of construction. Hong Kong Engineer, vol. 10, no. 10, 1982, p. 13-31.

7. Allen, P.M. and Stephens, E.A. Report on the Geological Survey of Hong Kong. Hong Kong Government Press, 1971, 116 pp. plus 2 maps.

8. Ruxton, B.P. and Berry, L. Weathering of granite and associated erosional features in Hong Kong. Bulletin of the Geological Society of America, vol. 68, 1957, p. 1263-1291 (plus 1 plate).

9. Geotechnical Control Office. Geotechnical Manual for Slopes. (2nd ed.), Geotechnical Control Office, Hong Kong, 1984, 295 pp.

10. Hencher, S.R. and Martin, R.P. The description and classification of weathered rocks in Hong Kong for engineering purposes. Proceedings of the Seventh South-east Asian Geotechnical Conference, Hong Kong, vol. 2, 1982, p. 167-168).

11. Hencher, S.R. and Richards, L.R. The basic frictional resistance of sheeting joints in Hong Kong granite. Hong Kong Engineer, vol. 11, no. 2, 1982, p. 21-25.

12. Harris, R. Stereographic projections for the engineering of rock. Hong Kong Engineer, vol. 10, no. 8, 1982, p. 37-48.

13. Richards, L.R. and Cowland, J.W. The effect of surface roughness on the field shear strength of sheeting joints in Hong Kong granite. Hong Kong Engineer, vol. 10, no. 10, 1982, p. 39-43.

14. Brand, E.W., Burnett, A.D. and Styles, K.A. The Geotechnical Area Studies Programme in Hong Kong. Proceedings of the Seventh Southeast Asian Geotechnical Conference, Hong Kong, vol. 1, 1982, p. 107-123.

15. Brand, E.W. and Phillipson, H.B. Site investigation and geotechnical engineering practice in Hong Kong. Geotechnical Engineer, vol. 15, 1984, p. 97-153.

16. Hinds, D.V. A method of taking an impression of a borehole wall. Imperial College Rock Mechanics Research Report, no. 28, 1974, 10 pp.

17. Richards, L.R., Leg, G.M.M. and Whittle R.A. Appraisal of stability conditions in rock slopes. Foundation Engineering in Difficult Ground, Newnes - Butterworths, London, 1978, p. 449-512.

18. Gamon, T.I. and Finn, R.P. Simplified descriptive scheme and classification system for

the logging of cut slope faces. Proceedings of the 20th Conference of the Engineering Group of the Geological Society of London, Guildford, 1984, p. 238-248.
19. Barton, N., Lien. R. and Lunde J. Analysis of rock mass quality and support practice in tunnelling and a guide to estimating support requirement, Norwegian Geotechnical Institute Report, no. 54206. Oslo, 1974.
20. Jacobs and Associates. Ground support prediction model (RSR concept). US Bureau of Mines Technical Report, no. 125, Washington D.C., 1974.
21. McFeat-Smith, I., Turner, V.D. and Bracegirdle, P. Tunnelling conditions in Hong Kong. Hong Kong Engineer, vol. 13, no. 6, 1985, p. 13-25.
22. Starr, D.C. and Finn, P.S. Practical aspects of rock slope stability assessment in Hong Kong. Hong Kong Engineer, vol. 7, no. 10, 1979, p. 49-56.
23. Starr, D.C., Stiles, A.P. and Nisbet, R.M. Rock slope stability and remedial measures for a residential development at Tsuen Wan, Hong Kong. Quarterly Journal of Engineering Geology, vol. 14, 1981, p. 175-193.
24. Ross-Brown, D.M. and Atkinson, K.B. Terrestrial photogrammetry in open pits : 1 - description and use of the photo-theodelite in mine surveying. Transactions of the Institution of Mining and Metallurgy, vol. 81, 1972, p. A205-A213.
25. Williams J.C.C. Simple Photogrammetry. Academic Press, London, 1969, 211 pp.
26. Golder Associates. Rock slope design review Report to Principal Government Highway Engineer, Hong Kong, 1974.
27. Fookes, P.G., Sweeney M., Manby C.N.D. and Martin R.P. Geological and geotechnical engineering aspects of low cost roads in mountainous terrain. Engineering Geology, vol. 21, no. 1/2, 1985, p. 1-152.
28. Fookes, P.G. and Sweeney M. Stabilisation and control of local falls and degrading rock slopes. Quarterly Journal of Engineering Geology, vol. 9, 1976, p. 37-55.
29. Sage, R. Pit Slope Manual, Chapter 6 - Mechanical Support. CANMET (Canada Centre for Mineral and Energy Technology), CANMET Report 77-3, 1977, 111 pp.
30. Bjurstrom S. Shear strength of hard rock joints reinforced by grouted untensioned bolts. Proceedings of the third International Rock Mechanics Congress, Denver, Colorado vol. 2B, 1974, p. 1194-1199.
31. Lang. T.A. Rock reinforcement. Bulletin of the Association of Engineering Geologists, vol. IX no. 3, 1972, p. 215-239.
32. Littlejohn G.S. and Bruce D.A. Rock anchors - state-of-the-art. Ground Engineering, vol. 8, nos. 3, 4, 5, 6, 1975, p. 25-32, p. 41-48, p. 34-45, p. 36-45, vol. 9, nos. 2, 3, 4, 1976, p. 20-29, p. 55-66, p. 33-44.
33. National Research Council of Canada. Canadian Manual on Foundation Engineering. (Draft for Public Comment). Associate Committee on the National Building Code, National Research Council of Canada, Ottawa, 1975, 318 pp.
34. Brian-Boys, K.C. and Howells, D.J. Model Specification for Prestressed Ground Anchors. Geotechnical Control Office, Hong Kong, 1984, 86 pp.
35. Piteau, D.R. and Peckover, F.L. Engineering of rock slopes. Landslides Analysis and Control, Transportation Research Board, National Academy of Sciences, Special Report no. 176, Washington D.C., Chapter 6, 1978, p. 192-227.
36. (Private Communication) CCL Systems. Lifespan of Maccaferri galvanized and PVC coated galvanized mesh, 1985.

Rock-socketed caissons for retention of an urban road

D.R. Greenway
G.E. Powell
Geotechnical Control Office, Engineering Development Department, Hong Kong
G.S. Bell
Highways Office, Engineering Development Department, Hong Kong

SYNOPSIS

The design and construction of rock socketed caissons for retention of a roadway on a steep sideslope is described. A retaining structure 9 m high and 100 m long was formed on a very restricted site by cantilevering caissons from granitic bedrock. The methods used to assess the lateral capacity of the rock socket are described, which included logging of actual geological conditions encountered during excavation. Construction details for the hand excavated caissons are described, including the safety measures employed to protect workmen.

INTRODUCTION

Tai Hang Road, an existing two lane road in urban Hong Kong, required widening to three lanes but the work was constrained by multistory buildings on the uphill side (Figure 1). This meant that the road must encroach upon steep marginally stable slopes on the downhill side. Slope conditions downhill of the roadway varied greatly along the 0.5 km project length and necessitated the use of a variety of slope stabilizing measures.

Along one 100 m segment of the roadway the downhill slope consisted of a steep 8 to 9 m high cut slope in decomposed granitic soil overlooking a private building (Figure 2). The boundary of the private lot was located approximately mid-height on this slope. Provision of three traffic lanes and footpaths required that the road widening extend up to the lot boundary, and a retaining structure was therefore needed. Further, two lanes were to remain open to traffic at all times, compensatory landscaping was to be provided for any trees removed during construction, and the retaining structure must not limit development alternatives on adjacent sites.

Rock socketed caissons 2 m in diameter were chosen for the retaining structure in order to satisfy these constraints. The hand excavated caissons were cantilevered from granitic bedrock which existed at a depth of approximately 10 m below the roadway (Figure 3). Alternative solutions utilizing gravity or tied-back structures were rejected primarily because the related temporary works would have restricted traffic during construction. Also, numerous utility services buried within the existing roadway would have made the installation of deadman anchors difficult. Rock anchors were rejected as they would have extended into or beneath private developments on the uphill side of the roadway, and access to the anchor heads for long term monitoring could not be guaranteed.

Fig. 1 Tai Hang Road, April 1984

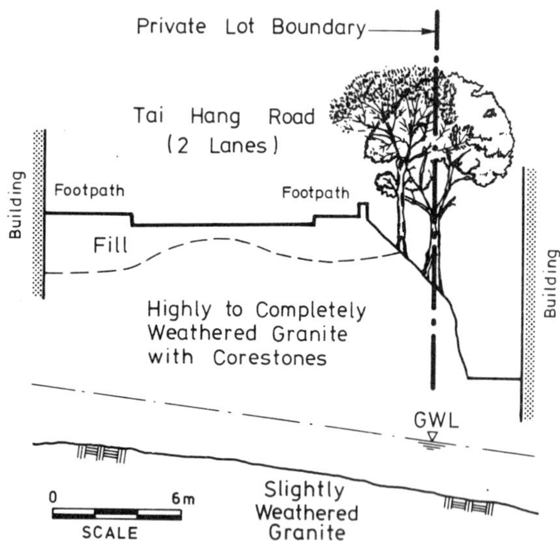

Fig. 2 Cross Section Before Widening

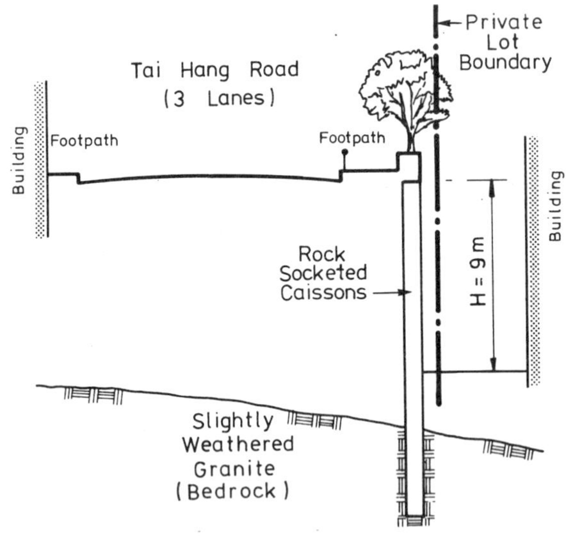

Fig. 3 Final Configuration

Hand dug caissons, which are commonly used in Hong Kong for building foundations, were therefore considered the most feasible solution in this segment. The caisson wall was constructed between January 1984 and October 1985 at a cost of HK$3 million.

SITE INVESTIGATION

Several drill holes were sunk along the caisson wall alignment to augment existing subsurface data. Field mapping of the accessible portions of the existing cutslope was also undertaken, and a joint survey of rock outcrops was made.

The site investigation showed that granite bedrock extended throughout this area at a depth generally not exceeding 11.5 m, overlain by weathered granite containing corestones of up to 2 m diameter within a dense silty sand matrix. Bedrock was moderately to widely jointed strong, pinkish grey medium to coarse grained granite with rough, stained sheeting joints dipping at approximately 10° in a downslope direction, spaced at 1.5 to 2 m intervals. Several sets of subvertical joints were also observed. There was generally a sharp transition from highly/completely weathered granite to the underlying slightly weathered granite. A shear zone several metres in thickness crossed the retaining wall alignment at one location, striking nearly perpendicular to the roadway. The highest observed ground water level was 2 m above bedrock.

DESIGN APPROACH

The lateral earth pressures acting on the wall are strongly influenced by the deflections of the structure. The rock socket will severely limit deformations in the soil immediately above the socket on both the active and passive sides. It is unlikely that adequate deformations will occur to mobilize active pressure, and the at-rest pressure was therefore adopted. A value of Ko = 0.5 was used[5]. On the passive side the thin soil layer above the rock socket is not likely to be significantly deformed and can not be relied on for any passive resistance at working loads.

Water pressures were included in the design, based on piezometric results and anticipated water rises, despite the provision of a drainage layer between (and behind) the caissons. It was anticipated that difficulties would be experienced in the placement of the drainage layer.

Alternative caisson diameters and spacings were evaluated. For construction reasons, it was decided that only caissons 1.5 m diameter or larger would be considered, as hand excavation of the rock socket might be difficult at smaller diameters.

Wall deflections were not considered critical in this case, as ground movements up to 50 mm could be tolerated by the buried utility services within the roadway.

ROCK SOCKET CAPACITY

A significant amount of work has been done on the

assessment of laterally loaded piles embedded in soil, including analyses for loads applied at the head of the pile[3, 10], loads induced in the pile by unsymmetrical surcharges[4], and loads induced by soil movements such as landslides[8, 9]. Methods for the design of cantilever walls in soil are similarly well documented[2]. A useful summary of commonly used design methods for cantilever and propped walls has been given by Burland, Potts & Walsh[1].

While the assessment of rock socketed piles is less well documented, the basic principles used for piles embedded in soil may be applied, albeit with certain adaptations based on rock mechanics. In particular, failure of the rock socket by movement along discontinuities must be considered as well as passive failure of the intact rock mass.

Joint Controlled Failures

Joints intersecting the rock socket that are oriented so that movement is kinematically possible could lead to failure affecting an individual caisson or a group of caissons (Figure 4). Planar or wedge failure geometries may be possible, and analyses of these follows established rock mechanics techniques[7].

Because of the strength and stiffness of the caisson, only failure modes which involve joints at or immediately below the caisson need to be considered. A preliminary analysis of planar failures was undertaken (Figure 5), using a total frictional resistance for sheeting joints in slightly decomposed granite of 56° [6, 11]. As can be seen in Figure 5, joints dipping upslope (into the slope) from 5° to 35° would significantly reduce the capacity of the rock socket. The orientation of critical discontinuities of this type are delineated on Figure 6. For the hypothetical case of a joint dipping upslope at 20°, the computed rock socket capacities for a range of socket depths is shown on Figure 7. Also shown on Figure 7 are the socket capacities required for limiting stability against overturning or sliding.

During site investigation, no persistent discontinuities were noted that would form kinematically possible wedge or planar failures. As the major sheeting joints dipped downslope, any sheeting joints that might intersect the

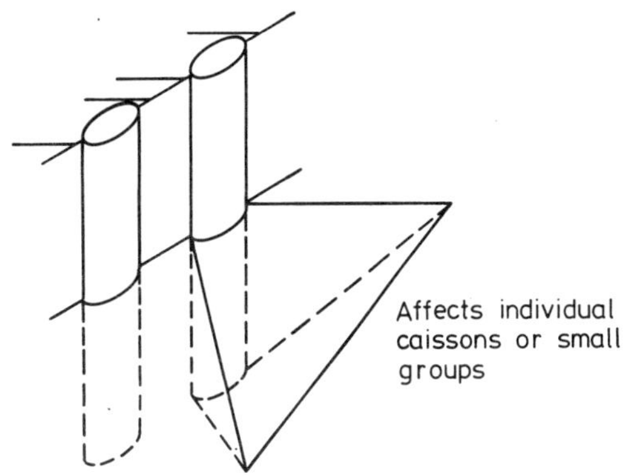

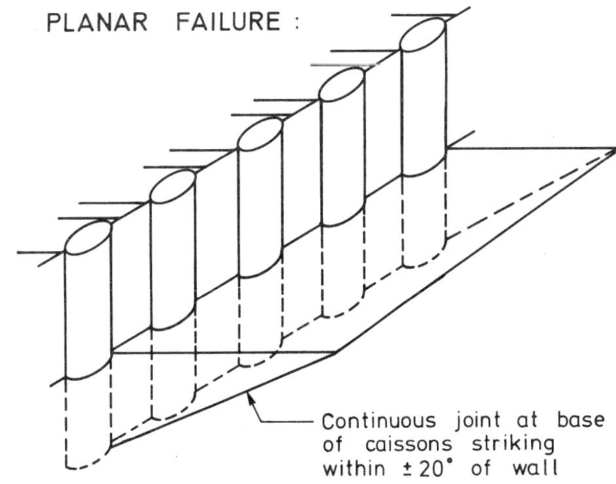

Fig. 4 Possible Joint Controlled Failure Modes

socket were not expected to daylight in the passive zone. Nevertheless, because of the severity of consequence of adversely oriented persistent discontinuities, the caissons were logged during construction.

Passive Failure in Rock Mass

The ultimate passive resistance of a rock socket when adverse jointing is not present is analagous to the lateral resistance of a pile embedded in soil. The deformations that the caisson would undergo before passive failure of the rock mass are likely to be significant, even in fairly brittle rock. The approach of applying a safety factor to the ultimate capacity in order to limit deflections is therefore appropriate for rock sockets.

Determination of the lateral capacity of a jointed rock mass is not exact, so a conservative value should be adopted. An allowable lateral bearing capacity of 2 MPa was adopted in this case, which was considered highly conservative in light of the high unconfined compressive

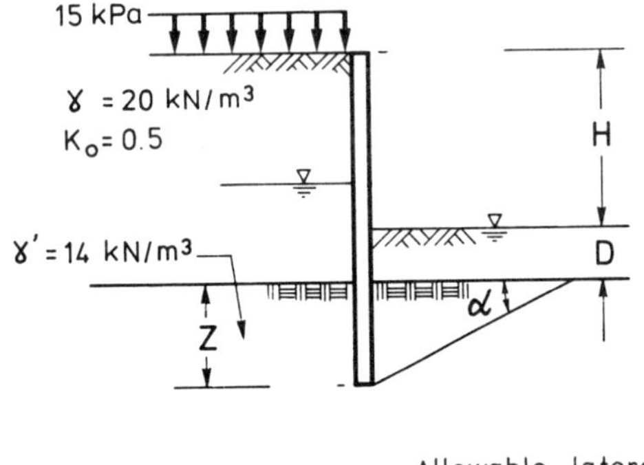

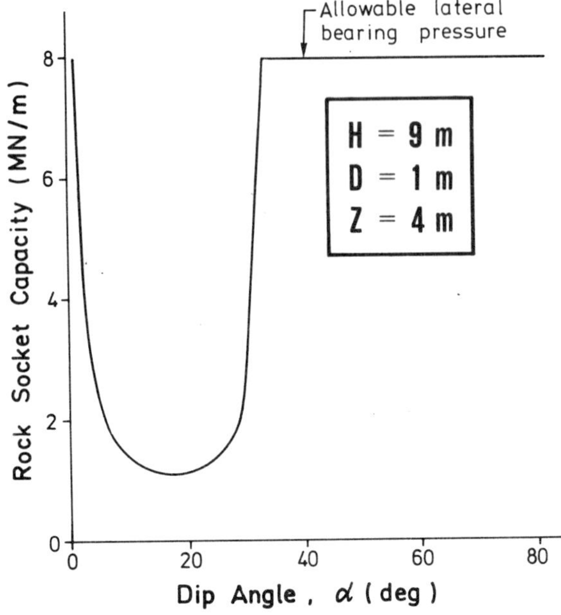

Fig. 5 Preliminary Analysis of Planar Failures

strength of this rock type (100 MPa or greater). Rock socket depths of 1.7 to 4.4 m were computed for contiguous caissons and caissons at two diameter centres, respectively.

CAISSON STRUCTURAL DESIGN

Structural design of the geotechnically feasible caisson arrangements was undertaken to check that adequate shear and moment resistance could be provided, and that deflections were not excessive.

The stress distribution within the rock socket has a significant effect on the location and magnitude of the maximum bending moments within the cantilever caisson. Several assumptions and their influences on shear forces and bending moments are illustrated in Figure 8. The assumption of full fixity at the rock level

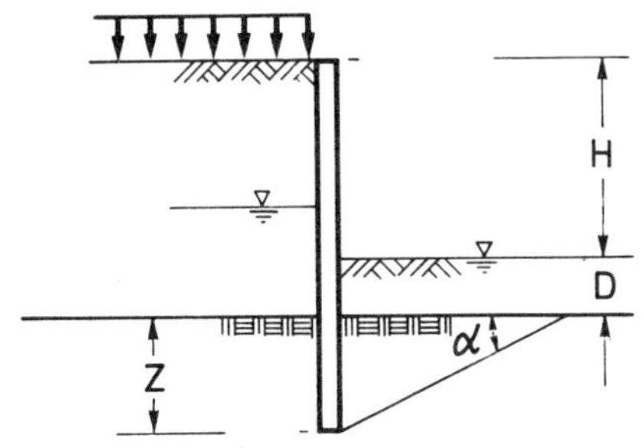

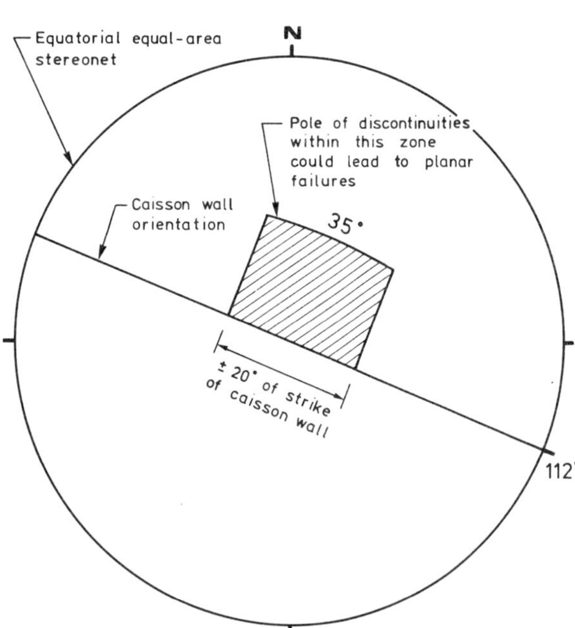

Fig. 6 Orientation of Critical Discontinuities for Planar Failure

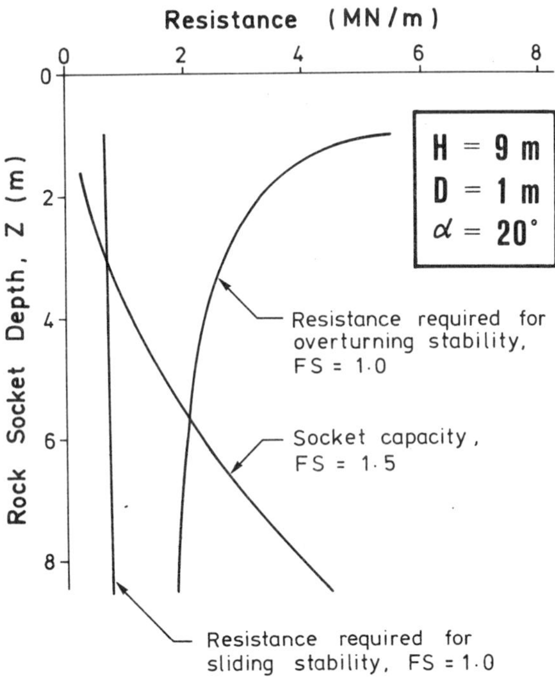

Fig. 7 Rock Socket Capacities when α = 20°

is not conservative, and the maximum bending moment will occur within the socket. For simplicity, the structural calculations were based on a uniform stress distribution in the socket, although the actual distribution is likely to be more complex. As discussed later, instruments have been installed in one caisson to ascertain the insitu conditions.

To cater for expected variations at the site, the reinforcing steel requirement was calculated for three bedrock depths (9, 10, 11.5 m). For ease of placing the reinforcement and compacting the concrete, vertical reinforcing steel in the caisson was designed to be held by a single series of circular reinforcing links.

Approximate deflections of the caissons were computed from elastic theory assuming the point

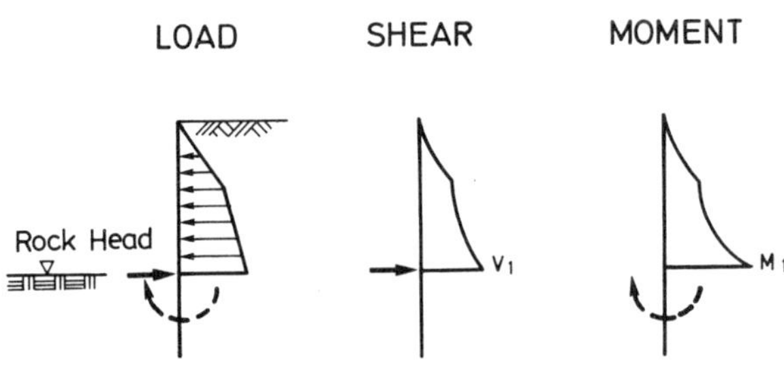

(a) Concentrated Forces in Rock Socket

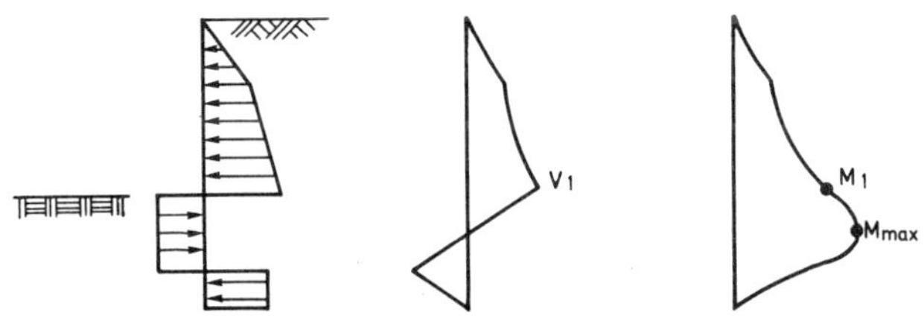

(b) Uniform Distribution with Depth

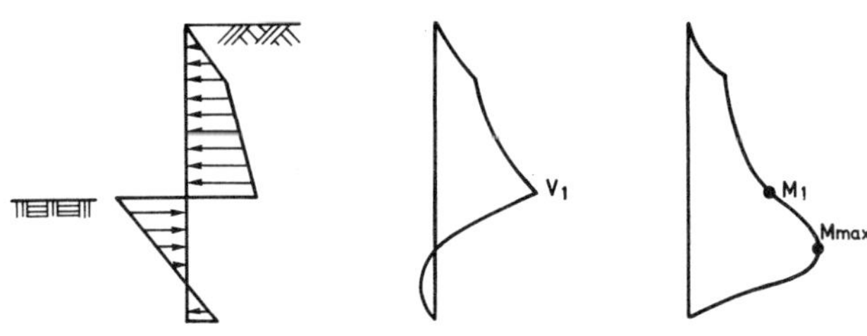

(c) Triangular Distribution with Depth

Fig. 8 Distribution of Load, Shear and Moment for Various Assumptions

of fixity was at rock head. To provide adequate moment capacity in the socket it was considered that the rock socket should have a minimum depth of a least 1.5 caisson diameters.

WALL DESIGN SUMMARY

Economic considerations governed the final retaining wall design, and it was found that 2 m caissons spaced at 4 m centres were marginally cheaper than other alternatives considered. For this alternative, deflections of the wall were anticipated to be less than 20 mm and adequate provision for drainage could be provided between caissons. The rock socket depths and other salient features of the final design are shown in Table 1.

Table 1 - Wall Design Summary

Caisson Type	Rock Socket Depth (m)	Total Length (m)	Number Anticipated
1	5.5	17.0	9
2	4.5	14.5	12
3	4.0	13.0	3

An economic study was also made of replacing the upper 9 m of each caisson with a concrete panel wall, and it initially appeared that this would be cheaper than full height caissons. As extensive temporary cut slopes would be created in this option, considerable inspection and monitoring would be required during construction, as well as temporary bracing. It was decided that the construction contract should be put out to tender with full height caissons specified, but tenderers were given the option of submitting an alternative design utilizing a combination caisson/concrete panel wall. Particulars for the alternative design were given in an appendix to the contract Particular Specification. However, no tenderer took up the option of submitting an alternative proposal.

WALL CONSTRUCTION

The caissons were entirely excavated by hand, as blasting could not be permitted in this conjested area. Each caisson was excavated by a two person crew, consisting of a male excavator and a female labourer; the latter remained above ground and operated the hoist. Once excavation of a caisson commenced it was very unusual for the crew to change.

Excavation in weathered granite could be accomplished with pick and shovel, and the soil removed by bucket from electrically operated overhead hoists. Concrete linings were provided in the weathered zone, consisting of cast insitu unreinforced concrete rings approximately 0.9 m high. When boulders were not encountered, it was possible to excavate and cast one ring every second working day.

Excavation in bedrock or boulders was done with a pneumatic drill utilizing 0.7 m drilling rods. It took an average of 6 working days to excavate a 0.7 m interval in bedrock. To limit dust created by the drilling, a pool of water was maintained in the base of the excavation. No lining was provided when the excavation was in bedrock.

During excavation in bedrock the caisson shaft was inspected at 0.7 m intervals and a geological log was prepared (Figure 9). The orientation of major discontinuities was recorded, along with details of any joint filling, seepage, or other unusual features. The inspection was routinely carried out by site staff, who compared observed joint orientations with the guidelines shown in Figure 6. Any questionable or unusual features found in the proposed rock socket were inspected by a geotechnical engineer.

The actual geological conditions encountered were generally as anticipated. No adverse jointing was found in any of the caissons. Several caissons in the vicinity of the shear zone had to be lengthened to a depth of 18 to 19 m, however, which increased the main reinforcing steel up to 63% over the Type 1 caissons shown on Table 1.

Alternate caissons were first excavated while traffic continued to run on the lane immediately adjacent. Traffic was diverted out of the adjacent lane to allow placement of the reinforcing steel and concrete in these caissons. The intervening caissons were then completed. With the approval of the adjacent lot owner, a mechanical excavator was used to remove the soil slope in front of the caissons

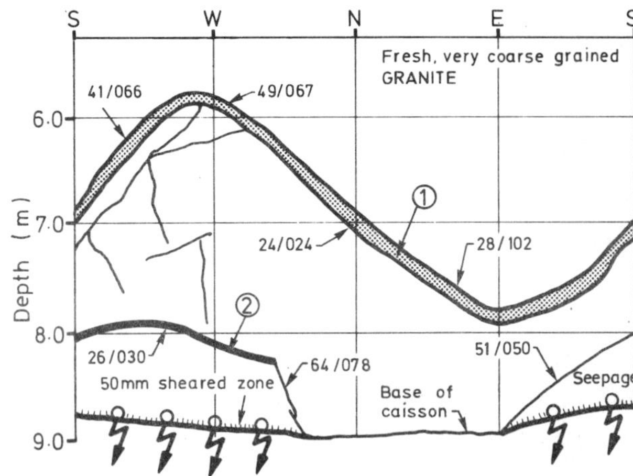

Fig. 9 Portion of Typical Geological Log of Rock Socket

(Figure 10). A temporary cutslope was formed between caissons (Figure 11) and concrete infill panels were then constructed. The infill panels were constructed in 1.8 m lifts and drainage aggregates were placed behind them. The infill panels incorporated galvanised dowel bars that had been cast into the caissons (Figure 11). An architectural facing will finally be applied to the wall in some locations.

During concreting of the caissons, difficulty was initially encountered in compacting the concrete forming the cover to the reinforcing steel within the rock socket. The contractor subsequently switched from grade 30/20 concrete with 50 mm slump to grade 30/14 concrete with 100 mm slump, and the problem was then overcome.

SAFETY

Safety is an important consideration during hand dug caisson construction. Suitable fencing and upstands were required around the top edge of the excavations, both to prevent people falling in and to prevent falling material from endangering workmen in the shaft. Ventilation of the shaft was also required, as was pumping the shaft below the water table, providing ready access to the shaft, and maintaining communication with the workman in the shaft at all times. The workmen were required to wear safety helments, and ear plugs were encouraged during pneumatic drilling.

Fig. 10 Excavation in Front of Caissons, April 1985

The workers were tempted to overload the electric hoists and their supporting frames when removing split blocks of rock from the caisson, and this was rigidly controlled.

TESTING AND INSTRUMENTATION

In addition to the usual concrete quality control, provision was made to nondestructively test the structural integrity of the caisson by sonic techniques[12]. Four 50 mm diameter PVC tubes were installed in each caisson for the full depth. These tubes were located at the quarter points of the circular reinforcing steel. when anomalies were indicated during sonic testing, the Engineer could then require coring of the caisson. The sonic testing was unsuccessful however, as the first two caissons sonically tested indicated that concrete irregularities were present that were not substantiated by subsequent coring. It can only be assumed that poor bonding between the PVC tubes and the concrete caused the sonic anomalies. This problem might have been prevented if the PVC tubes had been roughened on the outer surface, or if steel tubes had instead been used.

One caisson was instrumented to investigate the stress distribution in the rock socket. A series of five arrays of vibrating wire strain gauges were placed within the socket and readings were taken at intervals during and after construction. An inclinometer was also installed to monitor lateral movements of the caisson, and rotation of the caisson head was measured with a tiltmeter. Results of the monitoring programme will be reported in a later paper.

ACKNOWLEDGEMENTS

Permission of the Director of Engineering Development to publish this paper is gratefully acknowledged.

References

1. Burland J.B., Potts D.M. & Walsh N.M. The overall stability of free and propped embedded cantilever retaining walls. Ground Engineering, July 1981, p. 23-38.
2. CIRIA. A Comparison of Quay Wall Design Methods. Report No. 54, Construction Industry Research and Information Association, London, 1974, 125 pp.
3. CIRIA. Design of Laterally-Loaded Piles. Report No. 103, Construction Industry Research and Information Association, London, 1984, 86 pp.
4. DeBeer E.E. & Wallays M. Forces induced in piles by unsymmetrical surcharges on the soil

Fig. 11 Excavation Between Caissons for Infill Panel, April 1985

around piles. Proceedings, Vth European Conference on Soil Mechanics and Foundation Engineering, Madrid, 1972, Vol. 1, p. 325-332.

5. GCO. Guide to Retaining Wall Design. Geoguide 1, Geotechnical Control Office, Hong Kong, 1982, 153 pp.

6. Hencher S.R. & Richards L.R. The basic frictional resistance of sheeting joints in Hong Kong granite. Hong Kong Engineer, Vol. 11, No. 2, February 1982, p. 21-25.

7. Hoek E. & Bray W.J. Rock Slope Engineering. Third Edition, Institution of Mining and Metallurgy, London, 1981, 358 pp.

8. Ito T., Matsui T. & Hong W.P. Design method for stabilizing piles against landslide - one row of piles. Soils and Foundations, Vol. 21, No. 1, March 1981, p. 21-37.

9. Morgenstern N.R. The analysis of wall supports to stabilize slopes. In Application of Walls to Landslide Control Problems, R.B. Reeves (editor), American Society of Civil Engineers, April 1982, p. 19-29.

10. Poulos H.G. & Davis E.H. Pile Foundation Analysis and Design. John Wiley, New York, 1980, 397 pp.

11. Richards L.R. & Cowland J.W. The effect of surface roughness on field shear strength of sheeting joints in Hong Kong granite. Hong Kong Engineer, Vol. 10, No. 10, October 1982, p. 39-43.

12. Robertson S.A. Integrity and dynamic testing of deep foundations in S.E. Asia 1979-82. Proceedings, 7th Southeast Asian Geotechnical Conference, Hong Kong, 1982, p. 403-421.

Segment-shaped blasting in Baishan hydroelectric station excavations, China

Guo Zongyan
Zhou Yukun
Water Resources and Hydropower Construction Administration, Ministry of Water Resources and Electric Power, Beijing, China

SYNOSIS

The new technique of ' segment-shaped blasting' was first attempted in the downstream dyke excavation of the tailrace tunnel and in the rock excavation of the dam base in the Baishan hydroelectric station, Jilin Province, China. Total installed generating capacity will be 1500 MW and three units (3 x 300 MW) were in operation in 1984.

The new method of 'segment-shaped blasting' permitted rapid construction and ensured the minimum environmental effect.

Baishan hydroelectric station

Baishan hydroelectric station is located 39 km from the downstream Hongshi hydroelectric station and 250 km from the downstream Fengman hydroelectric station, Jilin Province, China, on the upper reaches of the Second Songhua River, and covers a catchment area of 19 000 km^2. The long-term discharge average of the river is 235 m^3/s. In Baishan district the weather is very cold in winter (lowest temperature, -44.5°C; highest temperature, 38.8°C; long-term average, 4.3°C). Annual rainfall averages about 800 mm.

The valley configuration of the Baishan dam site assumes an asymmetrical 'V' with a bottom width of 90-120 m and a slope of 1:2-1:1.6. The bedrock of the site is presinian age mixed rock with a saturated compressive strength of 800-1250 kg/cm^2.

The main purpose of this large-scale hydroelectric station is to generate electric power, with its attendant effects. The normal water level is 413 m and the reservoir capacity corresponds to 5310 000 000 m^3. Total installed generating capacity will be 1500 MW (five units). The construction period of the station was divided into two stages. The first stage is a underground powerhouse on the right-hand bank of the river with three units and was completed at the end of 1984. Total installed first-stage generating capacity is 900 MW and the annual energy generation will be 2003 000 000 kWh. The second stage is a surface powerhouse on the left-hand bank of the river with two units (2 x 300 MW).

The Baishan hydroelectric station comprises, essentially, a three-centred gravity arch dam, an underground powerhouse and a surface powerhouse (Fig. 1). The dam has a maximum height of 149.5 m, a crest length of 676.5 m and a crest elevation of 423.5 m. The underground powerhouse (121.5 m long, 25 m wide and 54.25 m high) is the largest in China. The lengths of three headrace tunnels are 329.03, 279.53 and 252.09 m, their excavation diameters being 7.5-8.6 m. The three tailrace tunnels are 250.07, 222.21 and 196.88 m long - of horseshoe cross-section. There are 37 underground chambers in the underground system. In the first-stage excavation 2140 000 m^3 of concrete and 15500 t of steel structure were used.

Use of 'segment-shaped blasting' in dam base excavation

For the excavation of the Baishan hydroelectric station dam base the construction site was narrow, the construction quantity involved in the project was large and the construction period was short so the 'segment-shaped blasting' method was adopted (Fig. 2) to increase excavation speed, to decrease the environmental effects of drilling the blast holes and to improve the quality of the dam base excavation. The construction sequence consisted of the excavation of a pilot tunnel before the water at the dam base was diverted.

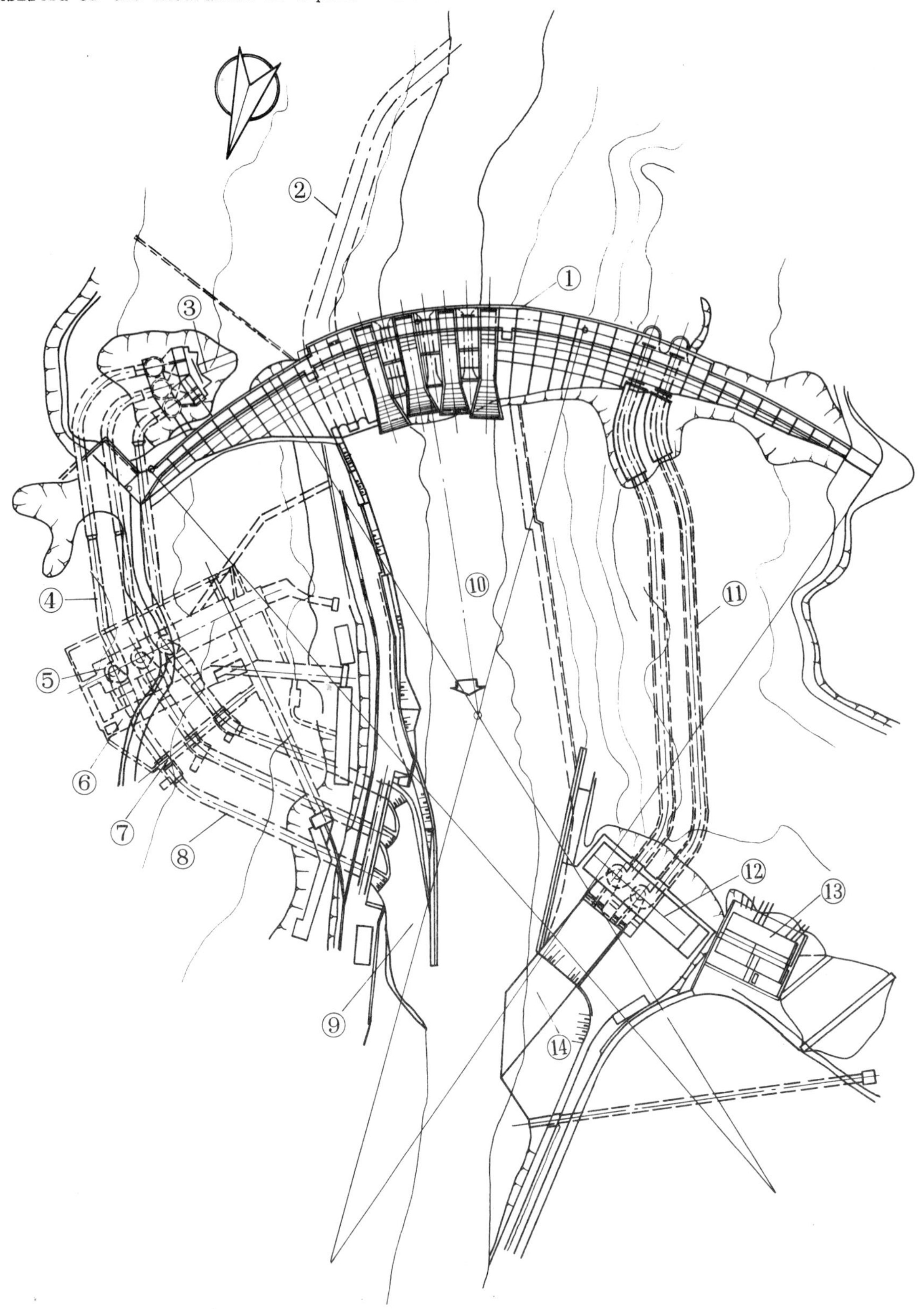

Fig. 1 Plan of construction layout, Baishan hydropower station: 1, arch dam; 2, diversion tunnel; 3, intake of underground powerhouse; 4, headrace tunnel; 5, main powerhouse; 6, main transformer house; 7, downstream chamber gate; 8, tailrace tunnel; 9, tailrace canal; 10, Second Songhua River; 11, headrace tunnel for second stage; 12, surface powerhouse; 13, switch yard; 14, tailrace canal for second stage

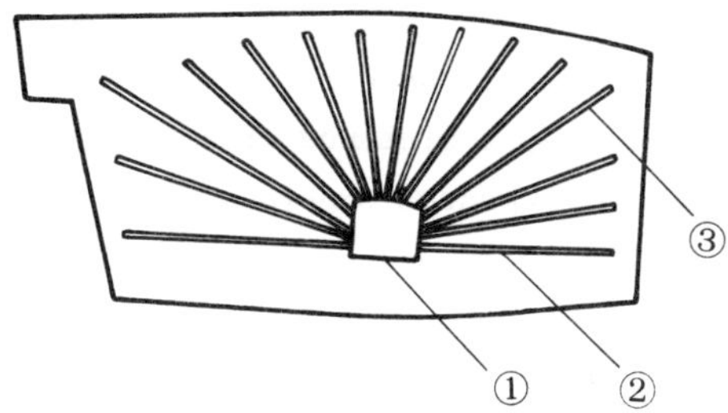

Fig. 2 Cross-section of layout of segment-shaped hole in dam base, Baishan hydropower station: 1, pilot; 2, presplit hole; 3, segment-shaped hole

Then through the pilot tunnel the holes were drilled to increase the excavation to the final size in successive segments. The segments were blasted as soon as the water at the dam base had been diverted into the diversion tunnel.

This new construction method not only enabled the dam base to be drilled before river was diverted and the dam base to be blasted as soon as the river diversion had been completed but also ensured the minimum environmental effect during drilling stage.

Selection of parameters for 'segment-shaped hole blasting' in dam base excavation

The dimension of the pilot tunnel must be as small as possible, but excavation costs are high as it must meet the operational needs of drilling or other excavating equipment. In accordance with sizes of drilling machines and other excavating equipment a pilot tunnel with 2.5 m wide and 3.0 m high was fixed. The results of a site test led to the following parameters: drill-hole diameters, 65 and 110 mm, respectively, and 7-17 m deep; distance between holes, 2.0-2.5 m; distance between hole ends, 2.5 m. To protect the dam base and to decrease the intensity of ground vibration the presplit holes were drilled on both bottom sides of the pilot with a diameter of 32 mm and were as long as the 'segment-shaped hole'.

The diameter of the charged cartridge is 60 mm and was used for the drill-hole with the diameter of 65 mm. The charged density of explosive is 2.4 kg/m. As the diameter of the drill-hole is 110 mm, the diameter of the charged cartridge will be 100 mm and its charged density of explosive is 7 kg/m. In fact, 'segment-shaped blasting' in the dam base excavation was divided into two parts - a dam base part and an intake part. The charged explosive quantity, which is instantaneously detonating for the dam base part, is 300-350 kg and for the intake part is 600-700 kg. The specific unit charge of 'segment-shaped blasting' in the dam base excavation is 0.45-0.53 kg/m^3.

Intensity of ground vibration in 'segment-shaped blasting'

During the excavation of 'segment-shaped blasting' in the dam base model RPSI-66 and RZSI instruments were used to measure the intensity of the ground vibration. The results showed a vibration acceleration of 1.016-0.011 g, a vibration velocity of 10.65-0.13 cm/s, a principal vibration frequency of 20-70 Hz and a vibration duration of 0.15-0.35 s (distance from measurement point to the focal point of blasting, 52-558 m; elevation difference, within 30 m).

The measurement results show that the intensity of the ground vibration is mainly dependent on explosive charging, distance from measurement point to the focus of blasting and blasting method, such factors as rock characteristics and topographic conditions also influencing the intensity of the ground vibration. For the Baishan hydroelectric station, if we treat the whole dispersive charges of 'segment-shaped blasting' as one concentrated charge in the geometrical centre of the 'segment-shaped blasting' and analyse all the measurement data, we obtain:

$$a_1 = 2.5 \left(\frac{Q^{1/3}}{R}\right)^{1.81} \qquad 0.0506 \leq \rho \leq 0.652$$

$$a_2 = 2.9 \left(\frac{Q^{1/3}}{R}\right)^{1.81} \qquad 0.0652 \leq \rho \leq 0.652$$

where a_1 is radial acceleration, a_2 is vertical acceleration, Q is instantaneously detonating charge, kg, and R is distance from measurement point to the blasting focus, m.

$$\rho = \frac{Q^{1/3}}{R}$$

For particle vibration velocity

$$V_1 = 12.8 \left(\frac{Q^{1/3}}{R}\right)^{1.76} \qquad 0.0652 \leq \rho \leq 0.807$$

$$V_2 = 8.9 \left(\frac{Q^{1/3}}{R}\right)^{1.49} \qquad 0.0652 \leq \rho \leq 0.807$$

where V_1 is radial velocity of particle vibration,

cm/s, and V_2 is vertical velocity of particle vibration, cm/s. The relationship between acceleration and the parameter ρ is shown in Fig. 3.

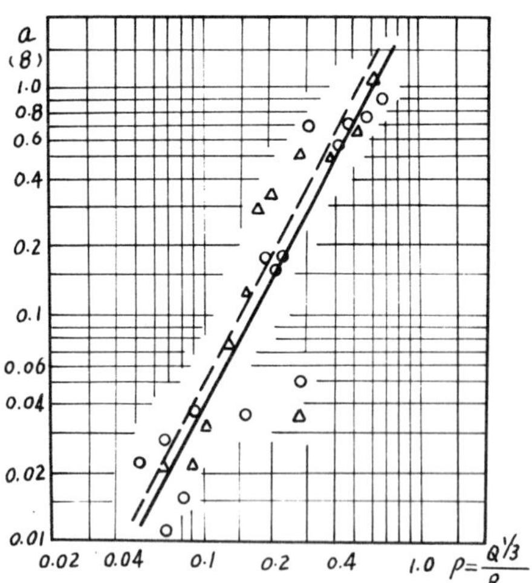

Fig. 3 Relationship between a and ρ (circle, radial component; triangle, vertical component)

The main characteristic of 'segment-shaped blasting' is to adopt the dispersive charges. Because of the mutual disturbance effect of vibration waves between each 'segment-shaped blasting' hole the intensity of ground vibration is lower - as low as one-fortieth to one-fifth of that of massive blasting or underwater rock plug blasting.

Use of 'segment-shaped blasting' in tailrace tunnel dyke excavation of underground station

There are three tailrace tunnels in the underground section of the Baishan hydroelectric station. The bottom elevation of the outlets of the tailrace tunnels is 279 m, but the water level outside the dyke is 290-292 m in the normal flood period. To excavate tailrace tunnels and outlets before filling the reservoir a concrete coffer dam was built around the outlets of the tailrace tunnels, four concrete support walls reinforcing the coffer dam from the inside.

The reservoir began to fill in November, 1982. To ensure safety during the flood period and to have the first unit in operation in 1983 the rock dyke and the coffer dam needed to be blasted within two to three months and the tailrace canal to be excavated to the designed elevation of not more than 282 m.

Besides the normal features of rock dyke excavation, such as large construction quantity and short construction period, also noteworthy was the very short distance from the blasting focus to the permanent concrete structures (2-30 m). The 'segment-shaped blasting' method was adopted to decrease the intensity of ground vibration and to increase the speed of rock dyke excavation. The excavation sequence of the rock dyke comprised, first, the excavation of an access from inside the rock dyke; second, a pilot tunnel was excavated along the axis of the rock dyke before the water outside the rock dyke was diverted. Finally, through the pilot tunnel the 'segment-shaped' holes and the horizontal cut holes were drilled, the horizontal cut holes and the 'segment-shaped' holes then being blasted as soon as the water had been diverted outside the rock dyke. The horizontal cut holes were designed to excavate two pilot trenches on both sides of the rock dyke for use as two free surfaces for 'segment-shaped blasting'.

Selection of parameters for 'segment-shaped blasting' in rock dyke

To avoid the poor geological conditions and to locate the pilot tunnel as close as possible to the rock dyke it (2.5 m wide x 3.0 m high) was located 1.25 m away from outside of the concrete coffer dam, which was covered with rock 4.6-5.3 m thick. To drill the horizontal cut holes of the initial trenches the height was increased to 4.5 m at both ends and the axis was moved to the portal side just below the concrete coffer dam (Fig. 4).

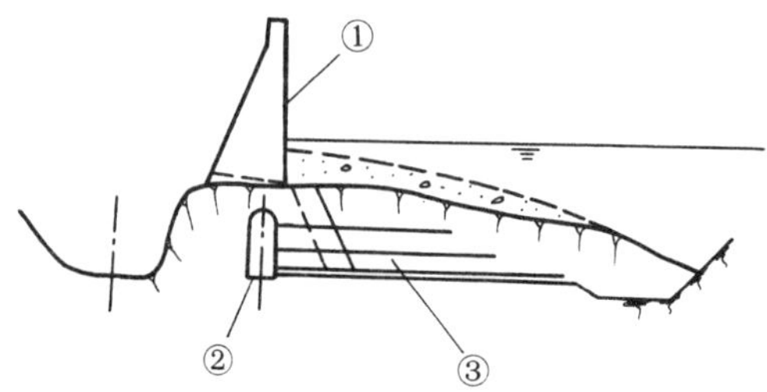

Fig. 4 Profile of the initial trench blasting in tailrace portal, Baishan hydropower station: 1, coffer dam; 2, pilot; 3, horizontal hole

The following parameters were adopted for the trench blasting: drill-hole diameter, 110 mm; minimum burden (distance between each horizontal cut hole line), 1.5-2.0 m; distance between holes (distance between each line of vertical holes), 2.5 m; specific unit charge, 0.6-0.7 kg/m^3; diameter of charged cartridge, 70-80 mm; and maximum instantaneously detonating charge, 401.5 kg. Additional drill-holes (diameter, 60 mm) were adopted between each hole on the bottom row to ensure the quality of the horizontal cut hole blasting. The total number of drill-holes was 48 and the maximum depth was 27 m.

In the pilot the other holes were blasted gradually from both trenches line by line. The lines of the holes were parallel and 'segment-shaped'. The parameters for the 'segment-shaped' holes were: drill-hole diameter, 60 mm; minimum burden (distance between each line), 2 m; distance between each end of holes (maximum distance between each hole in same line), 2.5 m; specific unit charge, 0.5-0.6 kg/m^3; diameter of charged cartridge, 56 mm; maximum instantaneously detonating charge, 248 kg (Fig. 5). Detailed information on the parameters for 'segment-shaped blasting' in the rock dyke is given in Table 1.

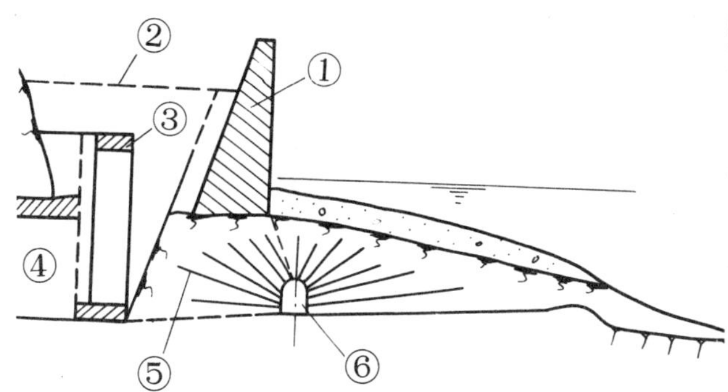

Fig. 5 Profile of dyke blasting in tailrace tunnel portal, Baishan hydropower station: 1, concrete coffer dam; 2, support wall; 3, concrete at gate pier; 4, pilot; 5, segment-shaped hole

Summary

The use of the 'segment-shaped blasting' method at Baishan rendered the environment unaffected by drilling in the 'segment-shaped' holes. Two months to compare 'segment-shaped blasting' with common blasting construction time was saved (37 500 man-days). The ballast piles were concentrated and the size of ballast was suitable for immediate transportation. After blasting, the blasted area was neat and flat, the direction of and distance from flying stones being predetermined and the nearby permanent concrete structures remaining safe.

Table 1 Parameters for 'segment-shaped blasting' in rock dyke of tailrace tunnel

	1	2	3	4	5	6
Blasting date	16/12/82	30/12/82	2/1/83	6/1/83	11/1/83	14/1/83
Blasting location	Downstream initial trench	Upstream initial trench	10 lines of 'segment-shaped' holes	12 lines of 'segment-shaped' holes	8 lines of 'segment-shaped' holes	8 lines of 'segment-shaped' holes
Drill-hole diameter, mm	110.60	110.60	60	60	60	60
Holes	28	20	76	100	64	67
Diameter of charged cartridge, mm	80.56	80.56	56	56	56	56
Charge, kg	1382	532	1830	2298	1721	1164
Blast volume, m^3	2000	840	rock, 12400; ballast, 5500			
Specific unit charge, kg/m^3	0.691	0.633	0.566	0.566	0.566	0.566
Delay no	5	4	10	12	8	8
Maximum instantaneously detonating charge, kg	401.5	185	219	248	235	149

Design considerations for underground blasts in an urban environment

T.N. Hagan B.ENG., PH.D., M.AUS.I.M.M.
Golder Associates Pty., Ltd., Melbourne, Australia

SYNOPSIS

High cost-effectiveness (measured in terms of the overall cost and period of excavation) and acceptable vibrations and flyrock can both be achieved provided that sufficient experienced effort is put into blast designs and, subsequently, into the implementation of these designs. Blast parameters considered in this paper include

(a) the diameter, alignment and length of blastholes,
(b) the shape and condition of faces to which charges shoot,
(c) the void volume into which blasted rock can expand,
(d) burden distance, effective subdrilling, stemming and both the size and shape of the blast block and, very importantly,
(e) the allocation of delay detonators.

Whilst most attention is paid to tunnelling, some consideration is also given to shaft sinking and the excavation of caverns/chambers by horizontal benching and downhole benching.

NOTATION

B - drilled burden distance

B_e - effective burden distance

d - blasthole diameter

INTRODUCTION

In any underground blasting operation, explosion energy should be utilised such that

(a) the fragmentation, looseness and profile of the muckpile allow the overall cost and period of excavation to be minimised, and
(b) overbreak does not cause a significant increase in excavated rock volumes and/or the costs of supporting or lining.

As operations proceed from rural to urban areas, this high demand for cost-effective blasting results is undiminished but the need to control ground vibrations and, in some cases, noise, airblast and flyrock increases, thus necessitating the application of greater expertise and care in both the design and execution of blasts.

UNDESIRABLE SIDE EFFECTS OF BLASTING

Any damage or complaint which results from blasting is caused by ground vibration, air vibration and/or flyrock. Of these undesirable side effects, ground vibration is usually the most difficult to control in underground blasts.

The ground vibration at a given point increases with the amount of explosion energy liberated within a given time interval. Less well recognised is the fact that ground vibration increases by a factor as high as about five when a free face (i.e., a rock/air interface to which charges shoot) that is reasonably near and extensive is replaced by one that is remote and of small effective area.

Air vibrations and flyrock can become problems where the distance between the blast location and the nearest tunnel portal or shaft collar decreases below certain critical values. These problems arise when a significant amount of explosion energy is vented to the atmosphere. Such venting occurs most readily

(a) where blastholes are unstemmed (as is commonly the case in tunnelling and horizontal benching) or have inadequate stemming columns,
(b) where the burden distance is too small, and/or

(c) where one charge fails to detach its burden from the rock mass before a dependent charge detonates.

DESIGNING COST-EFFECTIVE, ENVIRONMENTALLY-ACCEPTABLE BLASTS

Drilling

When drilling is not designed or carried out properly, blasts are unable to provide the required muckpiles and/or side effects. Optimum drilling is a prerequisite of optimum blasting.

Blasthole diameter

The blasthole diameter (d) should be as large as is practicable provided that any associated increase in charge weight does not result in excessive ground vibrations. When excavating the lower part of a cavern by downhole benching, for example, there are strong economic incentives to increase d, whilst there may be equally strong or stronger environmental pressures to restrict vibrations by limiting charge weight and, hence, d.

Blasthole alignment

Ideally, each blasthole should be parallel to the face to which it shoots. In burn cut rounds, the tunnel face is normal to the blastholes. Because the earliest firing charges are aware of the existence of only the (relatively poor) faces provided by the relief (i.e., uncharged) hole(s), ground vibrations from burn cut charges are

(a) appreciably higher than those produced by wedge cut, fan cut and drag cut charges, and

(b) considerably higher than those produced by charges in normal bench-type blasts.

When tunnelling with wedge cuts, blastholes are usually aligned so that the apical angle in the first wedge lies in the 55^0-65^0 range. As this angle increases, initial break out is achieved more easily and, hence, with lower ground vibrations. Unfortunately, wedge cut rounds produce significantly more airblast and appreciably more flyrock than burn cut rounds.

When tunnelling with fan cuts or drag cuts, the earliest firing charges are usually drilled at an angle of about 40^0 to the face. As this angle increases, the initial charges are tighter and, therefore, find it more difficult to break and eject their quota of the burden; for this reason, greater angles result in

(a) higher ground vibrations and

(b) greater "rifling" and, hence, noise and airblast.

Where downhole benches have heights up to about 3m and blasts shoot to free faces, satisfactory muckpiles and vibration levels can often be obtained by using vertical blastholes. As bench height increases, vertical front-row blastholes become progressively overburdened at bench floor level. Therefore, the replacement of vertical by inclined blastholes

(a) maintains toe burdens at their design values, thereby controlling both muckpile characteristics and ground vibrations, and

(b) increases the uniformity of the front-row burden distance from top to bottom of the face, thus reducing the ease with which explosion gases can vent prematurely to the atmosphere.

As the blasthole's angle to the vertical increases from 0^0 to about 25^0, fragmentation usually increases as a result of the reduction in premature venting of gases to atmosphere from alongside the tops of front-row charges. In the undesirable but common situation in which downholes are fired into a buffer of broken rock (created by the previous blast), and especially as the bench height increases beyond about 3m, angled blastholes are a prerequisite of acceptable ground vibrations.

Blasthole length

In tunnelling, horizontal benching and shaft sinking, there is a considerable economic incentive to maximise the mean advance per round. However, the mean advance per round must be compatible with the blasthole length and the maximum charge weight per delay period, the latter being dictated by ground vibration limits.

If he is unduly optimistic and/or fails to recognise the practical limitations (e.g., drilling errors) of his drill-and-blast system, an operator may, for example, drill 4m long blastholes but achieve a mean advance of only 3m. In this case, there would be 25% unproductive drilling and, of comparable importance, relatively high ground vibrations as a result of

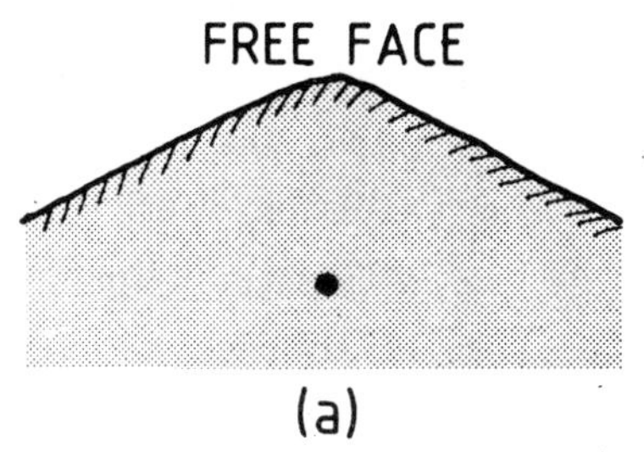

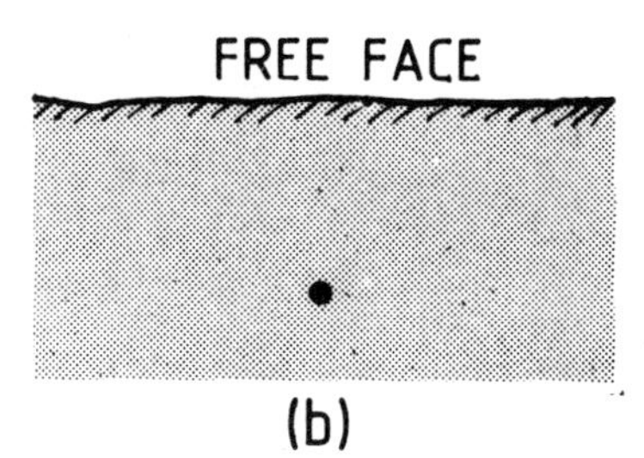

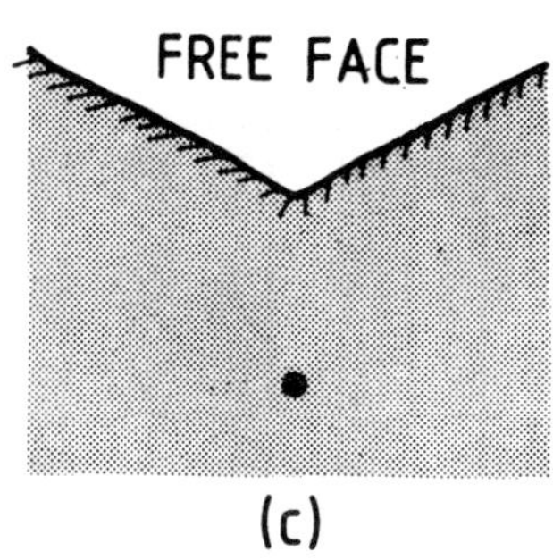

Fig. 1 Free faces which are (a) highly satisfactory (b) adequate and (c) unsuitable.

(a) the relatively long and heavy charges and, more especially,
(b) the effective overconfinement of the bottom 1m length of each charge.

There would be lower ground vibrations and probably lower costs and a greater advance rate if 3m long blastholes were drilled to pull a mean distance of 2.7m. Unlike full-face rounds in tunnelling and shaft sinking, horizontal bench blasts have very satisfactory free faces and, therefore, can have longer blastholes and heavier charges for a given ground vibration limit.

Where a downhole bench has a height exceeding about 8m, lowest overall costs are usually achieved by splitting this into two benches of equal height. If attempts were to be made to blast a single (say 10m high) bench, the small cost savings that might be realised in drilling and blasting would usually be outweighed by extra costs incurred as a result of increased ground vibrations and more overbreak.

Blast geometry

Shape and condition of face(s)

Ground vibrations are higher where the effective face
(a) subtends a small angle at the blasthole (in plan view),
(b) has not been cracked by a previous blast, and/or
(c) is choked by previously broken rock.

The angle subtended at a blasthole by that part of the face which is reasonably near should be as large as possible. Ideally, each charge should shoot to a concave face and all points on the face should be equidistant from the charge. Although this aim can never be fully realised, there is a strong incentive to obtain faces like that shown in Fig. 1a rather than those shown in Figs. 1b and 1c.

In downhole benching operations, blasting is facilitated by irregularities in the face and by cracks in the burden rock created by the previous blast. Where smooth unfractured faces exist, fragmentation and rock displacement are achieved with greater difficulty, and ground vibrations are higher, especially where the face is also convex relative to the charge.

When blasting in a burn cut round commences, ground vibrations are high for reasons which include the following.
(a) Even where a relief hole has a diameter as large as 200mm, it provides a free face which has a very restricted area and an unfavourable shape.
(b) The rock immediately around a relief hole contains few if any cracks created by previous blasts, especially near the (all-important) base of each blasthole.

This explains why the earliest firing charge is very often responsible for the peak of the ground vibrations produced. Because a relief hole represents a relatively poor face, burn cuts should be designed so that each of the early-firing charges can shoot to at least two equidistant relief holes. Ground vibrations decrease with an increase in the ease with which a charge can break and displace its burden rock and, hence, with
(a) a decrease in the distance between the charge and the relief holes, and
(b) an increase in the combined cross-sectional area of the relief holes.

Where ground vibrations are to be minimised, blasts should be fired to a free face rather than to a buffer of rock broken by a previous blast.

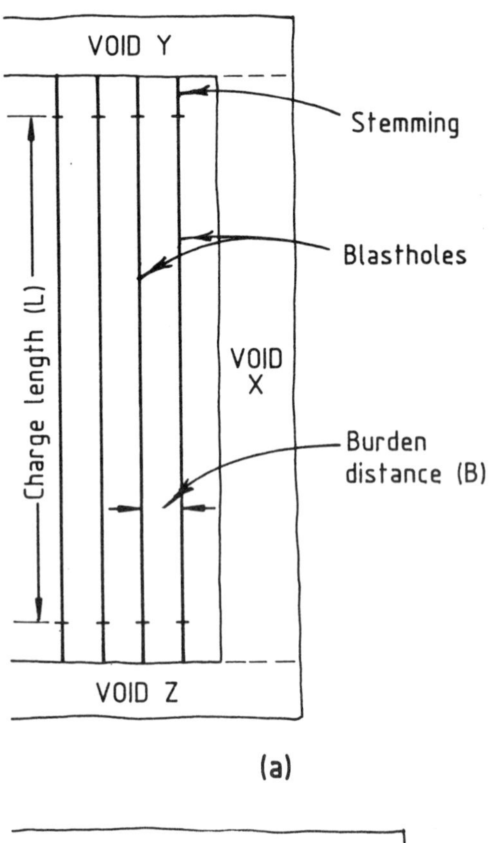

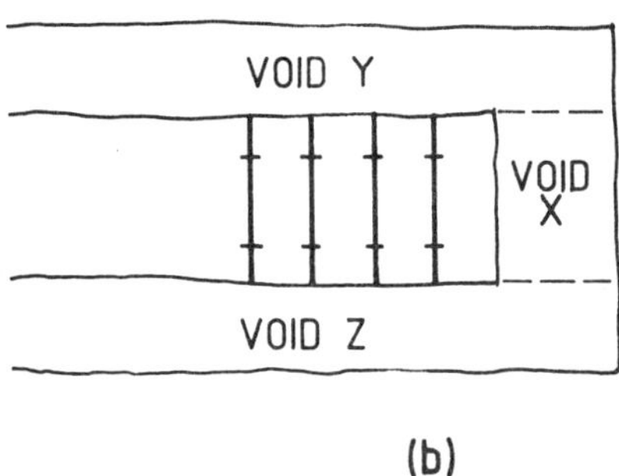

Fig. 2 The effect of charge length : burden distance ratio on the contribution of voids Y and Z to available expansion volume.

If the suppression of air vibrations and flyrock is the first priority, on the other hand, the presence of a buffer of broken rock can be beneficial.

Available expansion volume

When broken, rock expands. As the expansion volume decreases below about 15% of the volume of the solid rock in a blast, movement of the broken burden is increasingly suppressed, with a consequential increase in ground vibrations. A good example of the influence that expansion volume can exert on ground vibrations is provided by presplitting. For the instantaneous firing of a given charge weight, blasts which shoot to a good effective free face produce ground vibrations which are 15-25% of those produced by a presplit blast. In the latter operation, the burden rock fails to move and the explosion energy which would have appeared as kinetic energy in a free-face blast is now manifested as ground vibration energy.

In the interests of overall cost-effectiveness, as well as controlling ground vibrations, the following expansion volumes should be provided:

(a) ≥ 15% in burn cuts, and

(b) ≥ 25% when shooting to a raise, slot or expansion chamber.

When assessing the suitability of the expansion volume for a blast, operators should consider

(a) the principal direction of movement of the blasted burden(s), and

(b) the charge length: burden distance (L:B) ratio.

Where L/B is large, all of the desired expansion volume should be provided by the void which is in front of the blasthole. In Fig. 2a, for example, the burden rock is displaced principally to the right; for this reason, the explosion gases are aware of and appreciate the existence of void X, but are scarcely aware of the presence of voids Y and Z. In Fig. 2b, on the other hand, the small L/B value enables the explosion gases to displace rock upwards and downwards as well as laterally and, therefore, the available expansion volume is approximately equal to the sum of the volumes of voids X, Y and Z.

Burden distance

The effective burden distance (B_e = the burden distance at the instant of detonation of the charge) should be neither too large nor too small. If B_e is so great that the explosion gases find it difficult to break and displace the rock in a direction which is essentially normal to the blasthole's axis, the explosion energy saved through the reduction in rock movement appears as an increase in ground vibrations.

Where B_e is appreciably less than its optimum value, strain wave fracturing occurs so rapidly in front of the charges that much of the explosive's gas expansion energy is lost as airblast, noise and excessive kinetic energy of rock displacement before it is able to contribute fully to fragmentation. Needless to say, any increase in the amount of air vibrations and/or flyrock must be provided at the expense of fragmentation and muckpile looseness.

In multi-row bench blasts, it is most important that the burden distance for front-row charges is not excessive. If front-row charges fail to detach their burden from the rock mass by the time second-row charges detonate, progressive relief of burden will never be achieved, fragmentation and muckpile looseness will be reduced, and ground vibrations will increase.

Effective subdrilling

The effective subdrilling (i.e., the length of charge beyond the depth broken by a blast) should be the minimum distance for which good blasting results are obtained. Ground vibrations decrease with effective subdrilling.

In full-face tunnel rounds which have been well designed and executed, the depth of pull:blasthole length ratio is about 0.93 or more. As this ratio decreases, ground vibrations increase.

In downhole benching operations, the effective subdrilling needed to break cleanly to bench floor level is usually

(a) about 8d for blasts which fire to a free face, and
(b) up to about 15d for blasts which fire to a buffer of broken rock.

Where the effective subdrilling increases beyond these values, each additional section of the subdrill charge provides less energy for breakage and movement of the rock, and more of the energy liberated below bench floor level is manifested as ground vibrations.

Type and length of stemming

If neighbours are close to blasts at a tunnel portal, all blastholes in the initial rounds should be stemmed. Good stemming maintains high gas pressures within blastholes for longer periods of time. The greater stemming effectiveness achieved by using longer stemming columns and/or coarser more angular stemming materials increases the amount of effective work performed per unit weight of charge. Only well stemmed charges can deliver their full performance potential and cause minimal air vibrations.

Stemming columns shorter than about 18d usually cause appreciable air vibrations. Wherever possible, the length of the stemming column (L_s) should be greater than 20d. L_s should not be less than B_e.

Size and shape of bench blasts

In the interests of productivity, there is an incentive to fire as many rows of blastholes as possible in each bench blast. Fragmentation generally improves with an increase in the number of rows. Unfortunately, however, ground vibrations tend to increase with the number of rows; this is because progressive relief of burden is achieved with greater difficulty towards the end of a blast having many rows. Where a downhole bench blast has too many rows, back-row charges do not see an effective free face. The ground vibrations created by such (overconfined) back-row charges are appreciably greater than those for charges which can displace their burden rock forwards with reasonable ease. Where movement is less than a certain critical amount, fragmentation and especially muckpile looseness are reduced. Hence the need to ensure that the length:width ratio of the blast block encourages forward displacement which, in turn, promotes looseness and easy low-cost digging and gives low ground vibrations.

Delay allocation

The blaster's degree of control over the timing and, more particularly, sequence of detonations in a blast is high. For the following reasons, however, this control is never absolute.

(a) All delay detonators exhibit **scatter** (i.e., delay time variability).
(b) In large full-face rounds in tunnels and shafts, the range of available delay detonators may be smaller than that actually desired, necessitating the use of sub-optimum delay intervals between the detonations of dependent charges.

Blastholes should detonate in the sequence and with the timing which maximise the successive development of free faces which are sufficiently

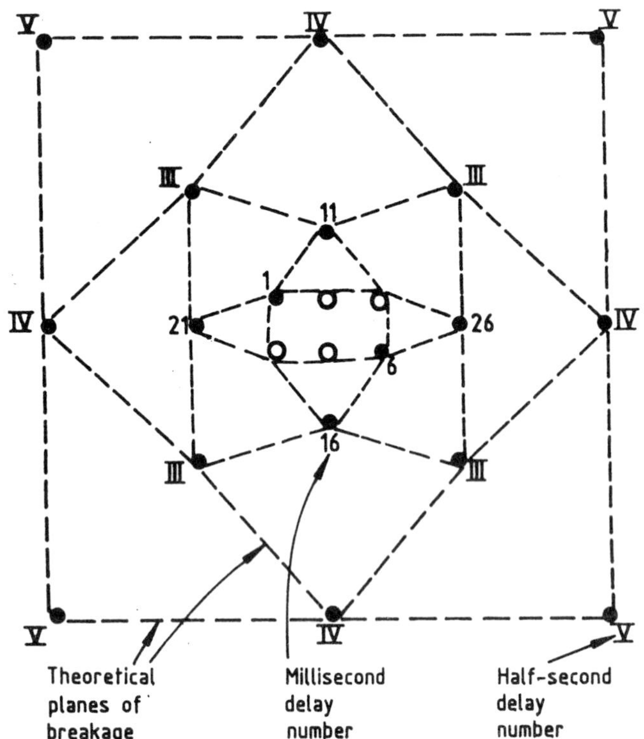

Fig. 3 Theoretical planes of breakage in and around a burn cut.

near and as extensive as possible. When allocating delay numbers in the initial design of a blast, the operator should construct lines of breakage as shown in Figs. 3 and 5. By doing this, any instances of poor sequencing and/or timing are exposed and superior delay allocations can then be made.

Every effort should be made to use the full range(s) of delays rather than several detonators of the same few delay numbers. Ideally, the total explosion energy in a blast should be liberated at a constant rate and over the longest practicable time period.

The absolute scatter for half-second (or long-period) delays is usually about five to twenty times that for millisecond delays. When using two or more detonators of the same delay number in a blast, therefore, the probability of experiencing superimposed ground vibrations is less for half-second delays than for millisecond delays. Half-second delays also give less over-break, reduced throw and a higher shorter muckpile but, unfortunately, slightly poorer fragmentation.

Delay allocations in burn cut rounds

In the burn cut itself, the earliest firing charge should be that which is closest to the relief hole(s). The sequence of detonations for all other charges in the cut should be consistent with the order of increasing free face effectiveness (see Fig. 3). Because the earliest firing part of a burn cut round has the most restricted free face and the smallest swell volume, it is often responsible for creating the peak of the ground vibration trace, even though the charge weights on later-firing delays may be somewhat greater than that on the earliest firing delay. For this reason, there is a special need to minimise the weights of explosive fired on the earliest-firing delays. The first two charges in a burn cut should never be fired on the same delay number (even of the half-second delay series) because of the possibility of negligible scatter and, hence,

(a) minimal progressive relief of burden, and
(b) superposition of ground vibrations.

Where ground vibration control is an important feature of the work, it may become necessary to reduce the charge weight per delay

(a) by restricting the blasthole length (and advance per round) and/or
(b) by shooting the cut and adjacent charges in one blast and the remaining charges in the round in a second blast.

In rounds with pulls of about 3m and more, the period taken for the earliest firing charge to completely eject its broken burden rock from the cut is typically 100ms or more. It follows, then, that the delay between consecutive detonations should exceed 100ms if the probability of **choking** (i.e., one charge shooting to a face which is at least partially choked by rock fragments created by an earlier firing charge) is to be minimised. As the time interval between the detonations of consecutive charges decreases below about 100ms, the probability of achieving progressive relief of burden decreases and, as a consequence, there is an increasing risk of **choking** and a **frozen** cut. As the intercharge delay increases between about 100ms and 200ms, the following benefits are obtained.

(a) The cut and, hence, entire round have a higher probability of pulling to the full drilled depth.

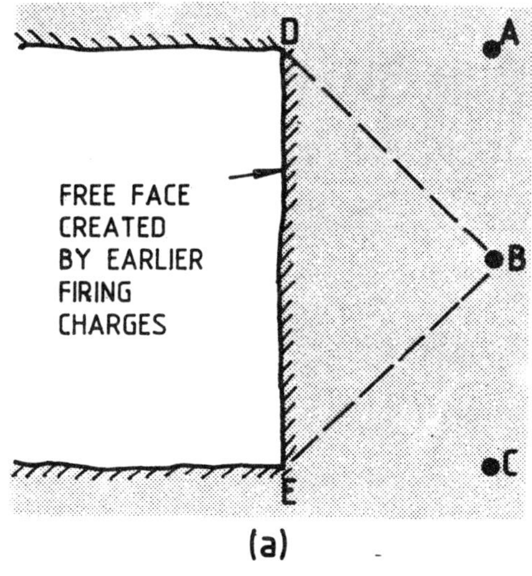

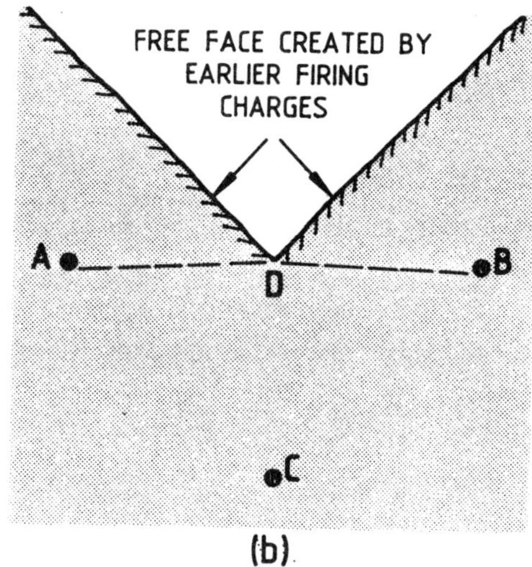

Fig. 4 Common situations in which charges with dissimilar free faces are fired on the same delay number.

(b) The muckpile has a higher and shorter profile with less flyrock.

(c) The tightness of the blast is less and, hence, ground vibrations are lower.

Where burn cut rounds are fired in tunnels with small cross sections, it is beneficial and often possible to use widely spaced delay numbers of the millisecond range in the cut and half-second delays outside the cut. In tunnels with large cross sections, the need for greater numbers of delay intervals tends to necessitate the use of intervals less than 100ms between detonations in the cut. In tunnels with very large cross sections, ground vibration limits may force the contractor to fire each round in two blasts of approximately equal size.

As blastholes become remote from the cut, it is often possible to drill them on a square or rectangular grid. Where this situation arises, one often sees poor sequencing which results from

(a) a desire to simplify the blast design and, hence, facilitate delay allocation at the face,

(b) inadequate consideration of the effects of scatter, and/or

(c) an inadequate number of delays for the face area to be blasted.

In the common situations shown in Fig. 4, for example, blastholes A, B and C are often initiated by detonators having the same delay number. Because of scatter, each charge has an equal probability of firing before the other two.

In Fig. 4a, it is quite clear that the free face for charge B is about twice as extensive as that for A or C. It follows, then, that lowest ground vibrations and best fragmentation result when positive steps are taken to ensure

(a) that charge B fires first, and

(b) that face DBE is fully developed before charges A and C detonate.

In Fig. 4b, the free faces for A and B are closer and more extensive than that for C. Therefore, A and B should detonate and create face ADB before C detonates.

Application of the recommended detonation sequences and timings to the cases shown in Fig. 4 allows larger blasthole patterns and/or lighter charges to be used. When A or C in Fig. 4a or C in Fig. 4b fires first, this charge sees an inadequate free face; because such charges experience difficulty in breaking out of their tight corner positions (especially C in Fig. 4b), a relatively large percentage of their energy is manifested as ground vibrations.

Delay allocations in angled-cut rounds

For a given charge weight per delay, the ground vibrations created by angled cuts are lower than those produced by burn cuts. In operations which have severe vibration limits, therefore, it may be prudent to select an angled cut rather than a burn cut, even though drilling and blasting efficiency for the latter is significantly greater.

Angled cuts are not as tight as burn cuts. In

an angled cut, each charge is more aware of the presence of the initial tunnel face and, largely for this reason, the cut can be pulled using a lower charge weight per cubic metre of rock. The time intervals required for angled cut blastholes to detach their burdens from the rock mass are less than those for burn cut blastholes. Because secondary free faces are developed more rapidly in angled cuts, inter-charge delays can be shorter than those used in burn cuts. For this reason and because angled cut rounds require fewer blastholes, a given range of delay detonators can be applied more effectively; the mean inter-charge delay is greater and, therefore, the probability of suffering inadequate progressive relief of burden and/or superposition of vibration waves is less.

In wedge cut rounds, optimum fragmentation and displacement are achieved when all blastholes in a given wedge are initiated by detonators of the same delay number. Where it is important that ground vibrations be minimised, it becomes necessary to accept a somewhat lower blasting efficiency which results from firing each charge on a separate delay number.

Although they exhibit lower drilling and blasting efficiencies than burn cuts or wedge cuts, fan cuts offer the best means of minimising ground vibrations. This is due to the operator's unparallelled ability

(a) to maintain a suitably small angle between each blasthole and the free face to which it shoots, and
(b) to fire blasts on a one-blasthole-per delay basis.

Delay allocations in benching

In benching operations, one should use a delay detonator down each blasthole rather than a surface delay system. Down-the-hole delays

(a) allow the use of longer inter-charge delays without experiencing cut-offs (i.e., misfires caused by blastholes which have been damaged/dislocated by earlier firing charges), and
(b) provide scatter (Where two or more charges on the same delay number are connected together by a detonating cord trunkline, they detonate almost simultaneously and, therefore, produce highly reinforced ground vibrations.)

Fig. 5 shows a very satisfactory allocation of millisecond delays (1-30 range) and half-second delays (1-10 range) for a five-row blast in a horizontal bench. Suitable features of this blast include

(a) the double delays between most adjacent charges in each row (thereby promoting the successive development of biplanar free faces),
(b) the large inter-row delay periods, and
(c) shooting two charges per delay on those detonators which exhibit the greatest scatter (i.e., the later-firing half-second delays).

In downhole benching, there are economic incentives to increase the length (measured along the tunnel or cavern) and, hence, size of blasts. Unless the delay allocations for larger blasts are well designed and implemented, however, higher ground vibrations and more overbreak will result. In general, the delay allocation should be such that

(a) the entire range of millisecond delays is used, and
(b) the number of charges on each delay is minimum and as constant as possible.

Whilst it is preferable to shoot <u>any</u> bench-type blast to a free face rather than into a buffer of rock broken by a previous blast, it is <u>especially important</u> to observe this rule when firing blasts with a length:width ratio greater than 1. The ground vibrations created by buffer blasts are appreciably greater than those produced by free-face blasts.

CONCLUSIONS

Good fragmentation and a loose, suitably profiled muckpile together with acceptable vibrations and flyrock may seem to be unattainable since, prima facie, these desired results of blasts appear to be incompatible. However, when sufficient <u>experienced</u> effort is put into blast design and, subsequently, into the implementation of that design, it is surprising how successfully these quite diverse requirements can be achieved within a blast. The high degree of compatibility of cost-effective muckpiles, low vibrations and minimal flyrock is explained by the fact that these three results of blasts are obtained

(a) by avoiding large blasthole diameters and

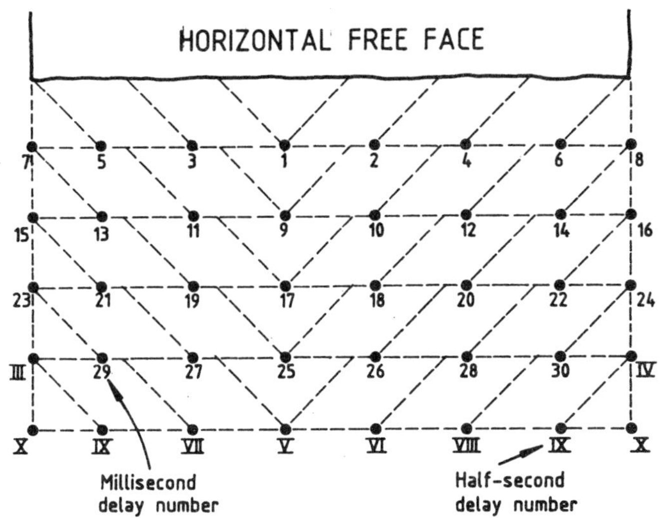

Fig. 5 Horizontal bench blast using millisecond delay and half-second delay detonators.

correspondingly heavy charges,

(b) by ensuring (through the selection of blast-hole pattern and delay allocation) that each charge sees a free face that is reasonably near and extensive,

(c) by ensuring that explosion gases are forced to expend almost all of their energy in breaking and displacing the rock before they emerge into the atmosphere, and

(d) by using the most extensive range(s) of delay detonators to minimise the superposition of vibrations from individual charges.

Rockbolting for excavations in sensitive urban areas

Luo Bangzhao
Changsha Institute of Mining Research, Changsha, China

SYNOPSIS

In tunnelling or excavation rock engineering in sensitive urban areas control of the displacement of the surrounding rock in tunnels and excavations to protect the surface constructions is vital. It is difficult to control the displacement by use of conventional concrete placement or concrete segmental ring but it is feasible considerably to lessen the displacement if bolts and shotcrete /wire mesh are installed and applied after excavation. By use of wedge-pipe bolts or quick-prestress pipe bolts with a mortar developed by the author and others, the displacement could be reduced significantly. The capacity, design and requirements of rockbolting in urban areas are considered, as well as the safe excavation depth for ground surface. The operational strength of several types of bolt and their recommended rate of anchorage depth are also discussed. The structural principle, features, in-situ testing and effects of wedge-pipe bolts and prestress pipe bolts with mortar are described.

INTRODUCTION

World-wide, cities of more than 1 million inhabitants may well number some 400 within the next 20 - 25 years. Owing to the limitation of ground surface, many urban projects such as metros, tunnels, caverns of various types, etc., have to be excavated underground. Thus, stresses in the surrounding strata are certain to be redistributed, and the surrounding strata will be deformed, displaced and also destroyed.

Excavations in the urban underground environment are usually shallow, ranging from a few metres to tens of metres below the ground surface. The rock stress in situ at a depth of 110m of overburden is only 30 kg/cm^2, as measured by the gravity stress field. This is favourable for the construction and protection of underground excavations but would have an effect, directly and indirectly, on surface buildings and the natural environment, causing them to deform and fracture or, more severely, to collapse and be destroyed. The economic loss caused by destruction of structures may equal, or even exceed the construction cost of the corresponding section of the underground excavation. Therefore, the stability of both the underground excavations and the surface buildings must be considered in tunnelling or excavating under cities.

Deformation and destruction of underground excavations occur when the ground pressure exceeds the maximum loading of supports. However, deformation and destruction of surface buildings are caused by settlement and deformation of ground surface resulting from the displacement of surrounding strata. Therefore, the key to supporting municipal underground excavations lies in consolidating the surrounding strata, which can control both the ground pressure and the displacement of surrounding strata.

There are two methods of constructing the underground excavations: the conventional method and the NATM.

1. The Conventional Method. Structures such as

steel supports, concrete placement or bricking rings are constructed within the underground excavations.

2. The NATM. The structures are built directly in/on the rock mass, by inserting bolts into the surrounding strata, shotcreting on and fixing wiremesh tightly on the rock surface.

Conventional supports cannot be set up early, and so cannot control the weakening of the surrounding strata; only if the surrounding strata are displaced to a certain extent do they passively support the destroyed surrounding strata. Therefore, conventional supports are not good effective in controlling the displacement of the surrounding strata.

With the NATM method, the surrounding strata can be consolidated and reinforced with bolts and shotcrete, so as to lessen considerably the displacement and also maintain or increase the strength of surrounding strata. Bolts and shotcrete as well as the surrounding strata support the ground pressure. Thus, the loading capacity of the surrounding strata can be improved, stability of both strata and underground excavations is also ensured. The NATM is indicated for supporting underground excavations in the urban environment.

ANALYSIS OF BOLTING CAPACITY

Stress-Displacement curve for surrounding strata

After excavation, the elastic-plastic solutions of stress and displacement from deformation of the periphery of the circular excavations (Fig.1) can be expressed as follows.

1. Radius of plastic zone

$$R_p = R_0 \left[(1 - \sin\phi) \frac{P_0 + C \cot\phi}{P_i + C \cot\phi} \right]^{\frac{1 - \sin\phi}{2\sin\phi}} \quad (1)$$

2. Pressure of peripheral deformation for excavations

$$P_i = (P_0 + C \cot\phi)(1 - \sin\phi) \times \left(\frac{R_0}{R_p} \right)^{\frac{2\sin\phi}{1-\sin\phi}} - C \cot\phi \quad (1')$$

where P_i == pressure acting on the supports in the process of deformation of surrounding strata

P_0 == initial stress of rock mass

C == cohesive force of rock mass

ϕ == angle of internal friction of rock mass

R_0 == radius of excavations

R_p == radius of plastic zone

3. Radial displacement of excavational periphery

$$U = \frac{1 + \mu}{E} \sin\phi \, (P_0 + C \cot\phi) R_0 \times \left[(1-\sin\phi) \frac{P_0 + C \cot\phi}{P_i + C \cot\phi} \right]^{\frac{1-\sin\phi}{\sin\phi}} \quad (2)$$

where μ == Poisson's ratio for rock mass

E == deformation modulus for rock mass

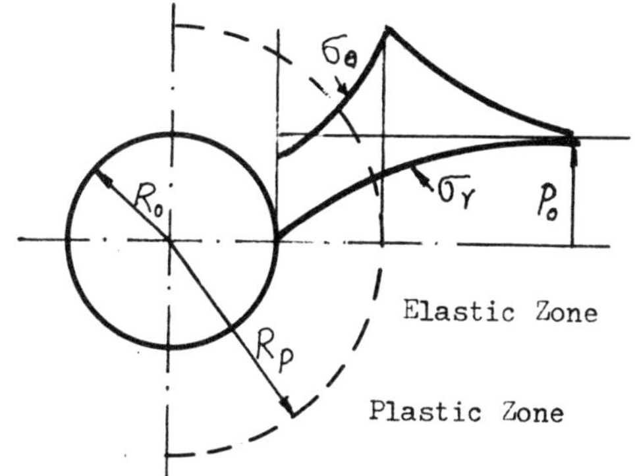

Fig.1 Elastic-plastic zone and stress distribution around circular excavations

The surrounding strata of underground excavations in the urban environment are mostly Tertiary and Quaternary, and therefore are soft, friable and instable. To take a marl for an example: if $C = 1$ or 5 kg/cm^2, $\phi = 20^\circ$ or 30°, $\mu = 0.35$ and $E = 5 \times 10^4$ kg/cm^2, given $P_0 = 30$ kg/cm^2 and $R_0 = 2.5$m, then the radial deformation U of the excavational periphery can be obtained from Equation (2), and the curve of P_i against U can also be plotted as shown in Fig.2.

Characteristic curve for bolting

The bolting rigidity (K_b) can be derived from the following expression:

$$\frac{1}{K_b} = \frac{ab}{F_0} \left(\frac{L}{SE_b} + Q \right) \quad (3)$$

where a,b == distance apart and spacing of bolts,

L == length of bolt

S == sectional area of bolt

E_b == elastic modulus of bolt material

Q == anchorage index of bolt, i.e. the displacemental increment with a tensional increment per ton

The maximum bolting resistance (P_{max}) can be computed from the following expression:

$$P_{max} = \frac{T}{ab} \qquad (4)$$

where T == maximum anchorage force (strength) of bolt

The characteristic curve of bolting can be expressed by the following equation:

$$U_i = U_{io} + \frac{P_i R_o}{K_b}$$

$$= U_{io} + P_i ab\left(\frac{L}{SE_b} + Q\right) \qquad (5)$$

where U_i == total displacement of surrounding strata with bolt anchorage

U_{io} == initial displacement of surrounding strata before bolting

Example. The size and performance of the mechanical bolts is L = 2m, D = 2.5cm, E_b = 2.1 x 10^6 kg/cm^2, Q = 0.13 cm/t, T = 14t, a = b = 1m and U_{io} = 2.0cm. Substituting these values in Equation (5),

$$U_i = 2.0 + 1.5P_i$$

and Equation (4) becomes

$$P_{max} = 1.4 \text{ kg/cm}^2$$

The characteristic curve of bolting is then plotted as shown in Fig.2.

From analysis of Equation (2) and Fig.2, the following conclusions may be reached.

(1) The displacement of the surrounding strata is proportional to the radius (R_o) of excavations, and inversely proportional to the deformation modulus of the surrounding strata.

(2) The displacement is proportional to the initial stress (P_o). If the friction angle of the surrounding strata is ϕ 30°, then $\frac{1-\sin\phi}{\sin\phi} > 1$, the displacement increases more rapidly and it will be proportional to a high power of P_o .

(3) If sufficient support is given to raise the value of ϕ to 20° - 30° and of C to 1 - 5 kg/cm^2, then stability of the rock mass can be expected. In general, bolt anchorage with shotcrete will be adequate.

From analysis of Equation (5) and Fig.2, the following conclusions may be reached.

(1) Bolts must be installed early -- that is to say, the value of U_{io} in Equation (5) should be as low as possible. Only, thus will the bolts properly serve the function of anchoring and controlling the displacement of surrounding strata. If the bolts are installed when the peri-

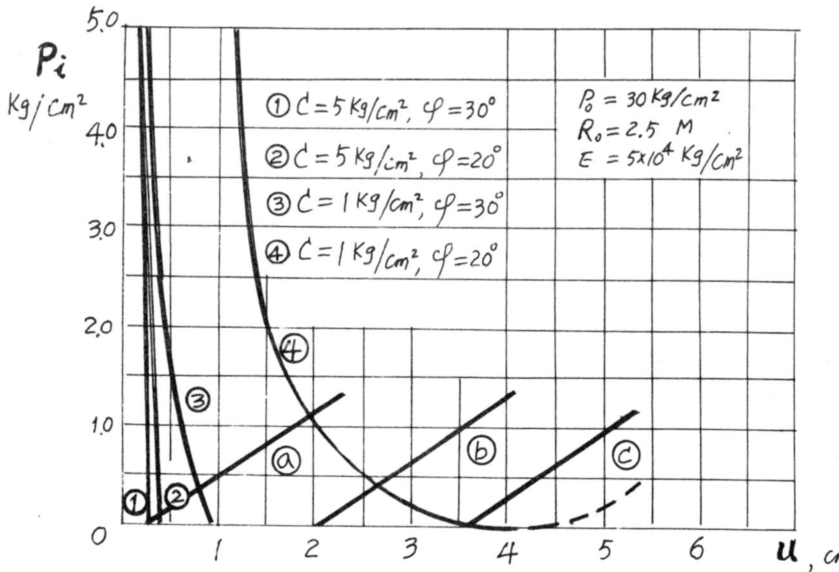

Fig.2 Interaction curve of pressure of host rock with bolting

pheral displacement of the excavation is only several millimetres, all of them have an anchorage effect on strata 1, 2, 3 and 4, as shown in Fig.2(a); if installed when the displacement is 2cm, the bolts have effect only on stratum 4, as shown in Fig.2(b); if installed at 3.5cm displacement, bolts have no effect on these strata, as shown in Fig.2(c).

(2) As can clearly be seen from Equation (5), in order to reduce the displacement of surrounding strata, it is necessary to shorten the distance apart and spacing of bolts, to decrease the anchorage index and to strengthen the bolt rigidity (S, E_b). To decrease the anchorage index, the structural form of bolts should be improved.

To strengthen the bolting resistance, the anchorage strength of the bolt must be high as far as possible, as obtained from Equation (4) and Fig.2.

However, non-yielding bolts of high rigidity may be destroyed if they are installed too early and their anchorage resistance is less than the deformation pressure of the surrounding strata. That is to say, to control the displacement of the surrounding strata, it is better to use those bolts which have strong rigidity, high anchorage force and high yield strength, so as to ensure the stability of the surrounding strata. The wedge-pipe bolts developed by the author and others have the above functions at the same time.

SAFE EXCAVATING DEPTH FOR GROUND SURFACE

After excavation of a circular cavern, the radius of the plastic zone can be computed from Equation (1).

As mentioned above, bolting with shotcrete can provide an anchorage resistance of 1 - 2 kg/cm^2, if P_i is taken as 1. By substituting the values C, ¢, P_o for strata and the radius R_o of excavations into Equation (1), the radius of the plastic zone can be calculated. Again, taking the safety factor as 2, it can be judged whether the plastic zone would have an influence on the ground surface; if not, the ground surface can be considered to be safe.

Example. For a soil stratum, $\phi = 10^o$, C = 1 kg/cm^2 and P_o = 5.4 kg/cm^2 (approximating the gravity stress for overburden with 20m thickness). Substitute these values into Equation (1). The values R_p is then equal to 2.11 R_o ; if R_o = 5m, then R_p = 10.5m. Taking the safety factor as 2, then the radius of the plastic zone could be considered to be 21m. It is obvious that the excavation has no unfavourable influence on the ground surface. If the ground conditions and other factor deteriorate (for instance, as a result of the values of C and ϕ for ground being much lower, or the excavation depth more shallow, or the influence of underground water and geological structure), excavating would have an unfavourable influence on the surface. In order to improve this unfavourable situation, it is necessary to adopt measures such as installing the bolts sooner and with much higher anchorage resistance, or setting up steel supports, spraying the shotcrete, concrete placement or pre-anchorage bolts.

Example. A tunnel conveying water to Tianjin in China

The tunnel passed through very poor strata (mylonite, breccia and sometimes exposed fault gouge) and intersecting faults with 27m depth of overburden. It is 11.38 km in length, 7.1m wide and 8.3m in height. Owing to the use of the techniques of smooth blasting and early bolting with shotcrete during excavating, the surrounding strata have been supported and are stable, and collapse did not occur. The construction method is as follows.

(1) The upper and lower benches were excavated in sections, and the upper bench was pre-excavated about 5m.

(2) The pre-anchorage bolts were installed in advance at the arch of tunnel before blasting. They are 1.5 - 2.0m in length, are spaced 40 - 50 cm apart and are of 15° - 25° elevation angle.

(3) The ' shotcrete-bolts-wire mesh-shotcrete ' process was used for consolidating the tunnel. In the fault zone, shotcrete was sprayed immediately after blasting; the radial bolts were installed after 1 - 2 rounds of blasting, their length being 2 - 3m with 0.8 - 1m spacing; the wire meshes (¢6 and ¢12 steel wire, mesh size 20cm) were put up after 3 - 4 rounds of blasting. The mesh, radial bolts and pre-anchorage bolts were connected with point welding, and attached to the rock surface as firmly as possible.

Finally, a layer of shotcrete of 10 - 25cm thickness was sprayed. After excavation of the lower bench, both sides of the wall were treated similarly with shotcrete and bolts.

(4) Measurements show that the total deformation in a section of bolts anchored with chotcrete was only 2mm, the maximum deformation being 4mm. However, collapse occurred in one or two sections of the tunnel, owing to late bolting with shotcrete (shotcreting 10 days after excavating), insofficient depth of bolting and with poor geological conditions. Even then, success was achieved by the use of bolts with shotcrete to treat the collapsed section.

BOLTING PRINCIPLES AND DESIGN

For urban underground excavations with bolting, important factors are the functions of hanging action, compound beam action and compound arch (ring) action. Sometimes there may be combined action of the three functions.

(1) Hanging Action

With regard to bedded or blocky strata with faults and fissures, if the roof of the excavation is not circular or full-arched, the bolt design should be based on the hanging function. In this design, the anchorage force (strength) of a conventional point-anchored bolt is often used as the hanging force. However, it should be noted that such a design method is not suitable for friction bolts such as ' Split-Sets ' and ' Swellex ' or full-length grouted bolts with binding force of 10 kg/cm^2, because the permissible (working) anchorage force of these bolts is much less than their full-length anchorage force.

The conditions of roof stability are shown in Fig.3. If the roof stratum anchored with the friction bolts is split into an upper and a lower segment by faulting and weak rock, the upper can be considered as stable stratum and the lower as unstable. In this case, the balanced condition of roof stability can be expressed as follows:

$$L_2 W \geqslant (L - L_2) ab\gamma$$

or

$$\frac{L_2}{L} \geqslant \frac{ab\gamma}{W + ab\gamma} \qquad (6)$$

where L_2 == upper segment of bolts which provided the hanging force

L == full length of bolts

a,b == distance apart and spacing for bolting, respectively

γ == unit weight of rock

W == anchorage force per unit length of bolts

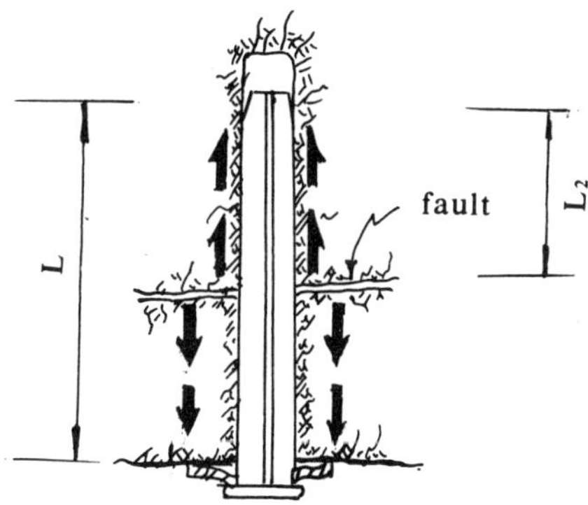

Fig.3 Analysis of stability of full-length anchored bolts

Obviously, the anchorage force in length L_2 is the permissible anchorage force of fully anchored bolts. If a=b=1m, γ=2.6t/m^3, the full-length anchored force of ' Split-Sets ' = 4-6t, i.e. W=4 t/m (on the basis of a bolt 1.5m in length); substitution into Equation (6) gives $\frac{L_2}{L} \geqslant 40\%$. That is to say, the permissible anchorage force is only 40% (1.6 - 2.4t) of the full-length anchored force when they are considered as support for the roof. Therefore, it will be dangerous to use bolts which do not have a firm top anchor (i.e. no 'root') in large-span excavations and in drifts liable to collapse. The permissible (working) anchorage force of some bolts is shown in Table 1.

(2) Compound Beam Action

If bolts are used to consolidate the strata as a compound beam, both the tensile strength of stratum and the bolting parameters must be considered in the design of bolting. Let the roof be a fixed beam with horizontal seam; let its thickness and span be h, b respectively; let the length and spacing of bolts be L, a, respectively; and let the even load on the beam be P. Then the tensile stress at the bottom of the beam is $\sigma = \frac{6M}{h^2}$; but at the middle of the beam span,

$$\sigma = \frac{pb^2}{24} \text{ . Thus, } \sigma = \frac{6M}{h^2} = \frac{0.25pb^2}{h^2}$$

Table 1 Permissible anchorage force and recommended rate of anchor depth for some types of bolts used for roof reinforcement

Index \ Bolt Type	Mechanical Point-anchored Bolt	Full Cement- or Resin-bonded Bolt	'Swellex' Bolt Atlas Copco (Sweden)	'Split-Sets' Ingersoll-Rand	Wedge-pipe Bolt
binding force or pressure on hole wall kg/cm^2		cement grout 18 - 30 resin 100 - 150	8 - 10	4 - 8	split segment: 8
W, t/m	point-anchored force for 0.1 - 0.2m in length, 5 - 10t	$<$ 20	10 - 20	4	top-anchored force for 0.15m in length is about 12t
loss rate of anchorage depth, $\frac{L_2}{L}$ %	$<$ 10	12	21	40	$<$ 10
recommended rate of anchorage depth, $\frac{L - L_2}{L}$ %	90	88	79	60	90
permissible anchor force during reinforcing roof, t	5 - 12	8 - 12 (to bolt broken)	8 - 10(bolt broken at 10t)	1.2 - 2.4	12

If the tensile strength at the bottom of roof stratum is σ_1, then

$$\sigma_1 > \frac{0.25pb^2}{h^2}$$

Again consider three coefficients: the safety coefficient K, ψ (as shown in Table 2) for reducing the coefficient of inertial moment and the safety coefficient of creep ξ (taking 1.2). Then

$$h > 0.5b\sqrt{\frac{\xi KP}{\psi \sigma_1}}$$

Table 2

Number of layers of strata in compound beam	1	2	3	4
ψ	1	0.75	0.7	0.65

The length of bolts can be calculated from h, i.e. the full length of bolts $L = L_m + h + L_\Delta$ (L_Δ expresses the segment length of bolts exposed outside the strata, and L_m the length of the top anchor).

The spacing and diameter of bolts should be calculated on the basis of the shear force which the bolts in the compound beam can bear.

Maximum shear force in the compound beam

$$Q = \frac{1}{2} pab$$

Maximum shear stress

$$\tau_{max} = \frac{3}{2}\frac{Q}{ah} = \frac{3}{4}\frac{ph}{h}$$

Total shear stress in the compound beam with a half-span

$$\tau = \frac{1}{4} ab \frac{3}{4}\frac{ph}{h}$$

Shear strength provided by the bolts within the range of the compound beam with a half-span

$$Q_a = \frac{1}{4}\pi d^2 \tau_a \frac{b}{2a} = \frac{\pi}{8a} bd^2 \tau_a$$

(where d == diameter of bolts;
τ_a == shear strength of bolts material)

Let the safety factor be K. Then

$$\frac{\pi}{8a} bd^2 \tau_a \geq K \frac{3}{16} \frac{pab}{h}$$

Thus

$$a \leq d \sqrt{\frac{2\pi h \tau_a}{3Kpb}}$$

(3) Compound Arch (Ring) Action

Given the spacing (a) and length (L) of bolts, and assuming that the pressure lines at both ends of each bolt anchored in the compound arch (ring) are at 45° to the bolt, then the effective thickness (t) of the compound arch, the radius (γ) of the centre line and the outer radius (γ') (Fig.4) can be derived. If radial pressure P occurs in the outer arch, then the tangential pressure stress (σ_t) exerting on the ring of arch can be approximately expressed as

$$\sigma_t \approx \frac{P\gamma'}{t}$$

If the monoaxial compressive strength of the rock mass is expressed as σ_c, then there will be an increase in the tangential stress which the rock mass reinforced by bolts and in triaxial stress can bear. From Coulomb-Mohr calculations, the result should be

$$\sigma_t = \sigma_c + \frac{1 + \sin\phi}{1 - \sin\phi}\sigma_2$$

Thus

$$P = \frac{t}{\gamma'}(\sigma_c + \frac{1 + \sin\phi}{1 - \sin\phi}\sigma_2)$$

where σ_2 == radial pressure stress provided by bolts. If the prestress of each bolt is expressed as N, then

$$\sigma_2 = \frac{N}{a^2}$$

The acceptable parameters of bolts anchoring the compound arch can be determined from the above equations.

The bolt design should allow for the fact that the bolts vary in quality, and that the anchorage capacity of the bolts will decreased with time, owing to corrosion or creeping. Thus the safety coefficient should be taken as 1.5 - 2 (i.e. 2 for non-grouted bolts and 1.5 for grouted bolts).

In designing the bolts, reference should constantly be made to previous results. As summarised by Lang many years ago, the following empirical rules can be laid down.

Minimum bolt length

(a) Twice the bolt spacing.

(b) Three times the width of critical and potentially unstable rock blocks, defined by average joint spacing in the rock mass.

(c) For spans of less than 6m, bolt length of one-half the span.
For spans of 18 - 30m, bolt length of one-quarter of the span in the roof.
For excavations higher than 18m, sidewall bolts one-fifth of the wall height.

Maximum bolt spacing

(a) One-half the bolt length.

(b) One and a half times the width of critical and potentially unstable rock blocks, defined by the average joint spacing in the rock mass.

(c) When weld mesh or chain-link mesh is to be used, bolt spacing of more than 2m makes attachment of the mesh difficult (but not impossible).

BOLTING FOR UNDERGROUND URBAN EXCAVATIONS

Bolts and bolting for underground urban excavations are characterised as follows.

(1) Early and Close to Face

As shown in Fig.2, bolts must be installed, so as to provide the supporting resistance early, and to consolidate the surrounding rock and control its displacement.

The results of theoretical calculations demonstrate that the peripheral displacement of surrounding rock will be essentially approximate to its maximum when the face is advanced about one diameter (d) of circular tunnel. If the bolting could be installed within the range of the length (one diameter), the initial strength of rock mass could be maintained and its loading

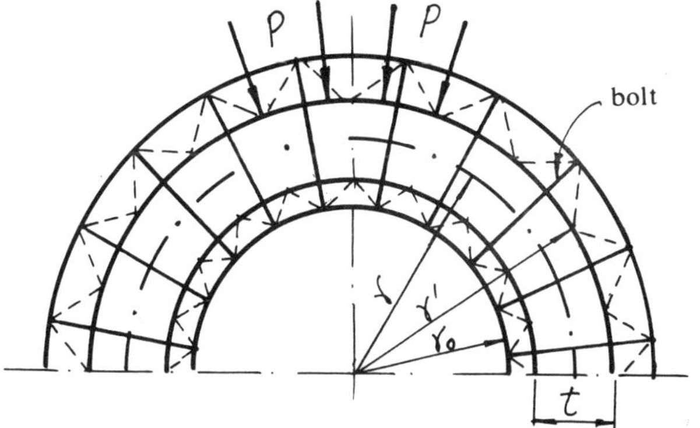

Fig.4 Compound arch (ring) with bolting

capacity be also enough.

(2) Blastproof

The bolts should be blastproof, because they must be close to the face. Further, the components of bolts, including the anchor in the borehole and the other parts exposed away from the rockface and their anchorage action, should not be damaged by blasting in tunnelling or excavating, or attenuated by the displacement of strata. Point-anchored bolts available at present are very vulnerable to blasting damage.

(3) Greater Adaptability to Strata

The strata passed through by shallow excavations vary widely, especially strata where soft and crushed rock predominate. In this case, the bolts must be adapted to such strata not only with better and moderate hardness, but also with weakness and crumbliness.

(4) Higher Rigidity

To lessen the displacement of surrounding strata as much as possible, the rigidity of bolts is required to be high, particularly the shearing rigidity, so as to strengthen the loading capacity of strata anchored with bolts.

To strengthen the shearing rigidity of bolts, it is best to replace the steel pipe by a rebar of the same cross-sectional area, because the former is much larger than the later in diameter, and the cross-section modulus is in cubic proportion to the diameter.

In addition, the prestress on the bolts should be increased up to a certain magnitude when the bolts are installed. In this way, the artificial loss of rigidity of bolts can also be avoided to a certain extent.

(5) Rust-proofing

Underground urban excavations are usually permanent. Thus, rust-proofing of bolts must be considered. An economical and feasible method of rust-proofing is to grout into holes or spray the shotcrete on the rock surface.

(6) Prestress

Photoelastic tests have shown that the stratum close to the surface of the surrounding rock is in tension (tensile stress zone) on condition that the axial prestressed bolts, but the inner strata is subjected to pressure (compressive stress zone), they are in the state of triaxial stress and then stable. Therefore, bolts with axial prestress give good results in consolidating strata.

Besides the bolts having axial prestress, if the prepressure can be exerted by bolts on the circle of boreholes, then the bolts would be more effective because they cause the surrounding rock be in the state of three-dimensional stress; wedge-pipe bolts developed by various authors offer these advantages, as shown in Fig.8. In addition, both'Split-Sets' and 'Swellex', developed by Dr J.J. Scott (USA) and Atlas Copco (Sweden), respectively, exert the prepressure on the circle of boreholes also. Tunnelling and excavating practice has shown that the capacity anchored by these bolts to the surface stratum of surrounding rock are much higher than that of the bolts often used at present; meanwhile, the surface stratum and the bolts have not been destroyed even where the bolts are close to the face. Unfortunately, 'Split-Sets' and 'Swellex' bolts are not so good in consolidation capacity to the entire anchorage layer, owing to the fact that they do not possess the axial prestress and adequate permissible anchorage force.

(7) High Anchorage Force

The bolts should have high anchorage force, so as to render the whole anchored stratum stable and reliable. In addition, it can be seen from Fig.2 that the bolts installed early must have much higher anchorage force, to balance the deformational pressure incompletely released and to effectively control the displacement and relaxation of surrounding rock.

No such bolts exist at present, which fully satisfy the above requirements. On the basis of these, the author's recommendation is that prestress-grouted bolts, wedge-pipe bolts and 'Swellex' bolts be used in urban underground excavations. With regard to excavations of 3 - 4 m span, 'Split-Set' bolts can be adopted as temporary support. In better ground conditions, grouted bolts (SN), mechanical bolts or resin bolts can be used also, but should be installed early, particularly the SN bolts. In addition, for a specific excavational section requiring special consolidation, the long anchor cable for ground preconsolidation can be used (Fig.5).

WEDGE-PIPE BOLT AND QUICK-PRESTRESS GROUTED BOLT

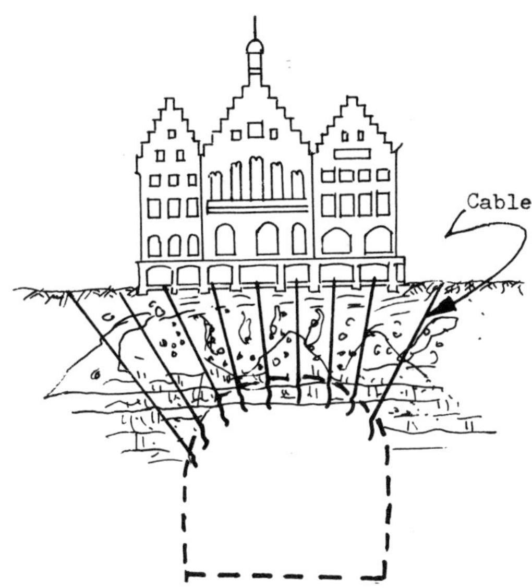

Fig.5 Preconsolidation of underground excavations with long anchor cable

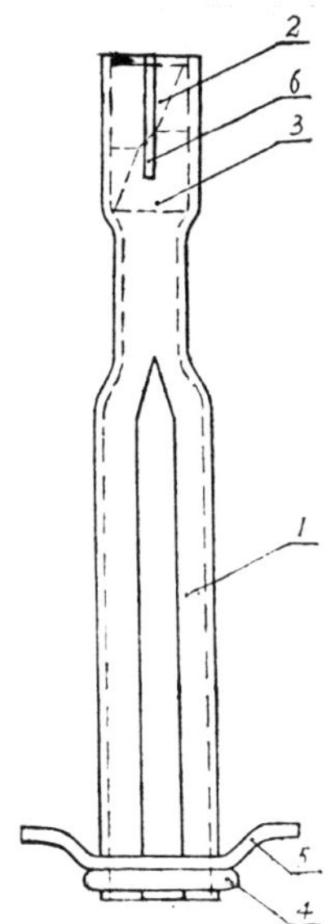

Fig.6 Structure of W-P bolt: 1. reducer pipe; 2. upper wedge; 3. lower wedge; 4. retaining ring; 5. washer plate; 6. straight slot.

Wedge-pipe Bolt

(1) Structure and Installation

The patented wedge-pipe bolt (W-P hereinafter) consists of a reducer pipe (1) with a slot along its partial length, an upper wedge (2), a lower wedge (3), a retaining ring (4) and a washer plate (5). The upper wedge and the retaining ring are welded on the inner and outer wall of the pipe respectively (Fig.6). The installation of W-P can be completed with an ordinary drill and steel rod. First, the segment split pipe is installed (Fig.7a), then the lower wedge is impacted by a long rod until the wedge segment expands to press against the rock wall of the borehole (Fig.7b). It takes only 1 - 2 min to install a W-P bolt.

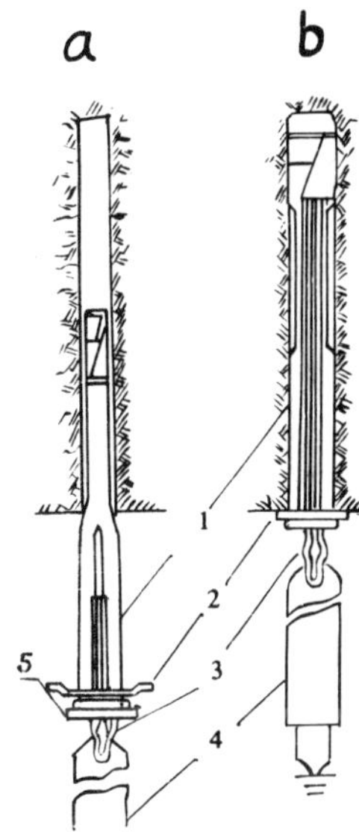

Fig.7 Installation of W-P bolt: 1. bolt; 2.plate; 3. rod; 4. stoper or drifter; 5. small ring.

Results of research have shown that the W-P bolt fulfils five functions, as follows (Fig.8).

(1) The anchorage force from the lower split segment of bolt can be up to 5 - 7t.

(2) The prepressure that the plate exerts on the rock surface is 5 - 7t.

(3) The anchorage force from the top-anchored segment can reach 12 - 14t.

(4) The pretensile force of the pipe body is 3 - 4t.

(5) The capacity to control the dislocation or displacement of surrounding rock is considerable. The forces on ground by the W-P are shown in Fig.8.

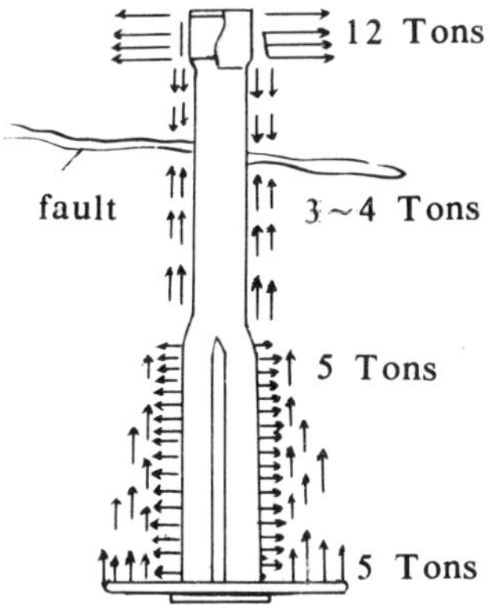

Fig.8 Forces on ground by the W-P

Structure and Advantages of the Wedge

The wedge structure of the W-P bolt is as shown in Fig.9. It is simple, conveniently installed and more adaptable to the borehole diameters. The breakage of rocks of the borehole wall near the anchor caused by the rod is a prerequisite for the wedging and expanding of slit-wedge bolts. Thus, if rock is too hard or too soft, the bolt is not good in anchorage. As for the W-P, its lower wedge is gradually wedged and expanded under the protection of the pipe wall and the impacting of the drill, respectively; so long as the impacting force is powerful enough, it can be wedged into; so long as the surrounding rock is not very soft and reacts sufficiently on the pipe wall, the wedge will be able to expand and firmly press on the surrounding rock. This is the reason why the wedge of W-P is widely adaptable to the surrounding strata. The results of calculation and pulling tests have shown that the wedge of W-P is quite suitable for rock with 300 - 2000 kg/cm^2 of strength for single-axis pressure.

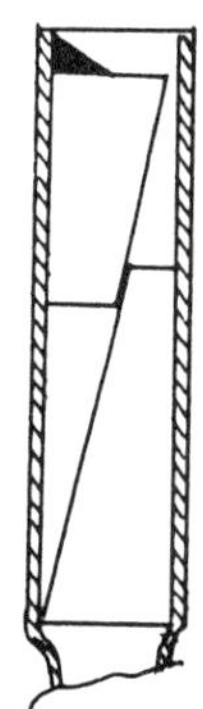

Fig.9 Wedge structure of W-P

Calculation for Anchorage Force

(a) Anchorage Force of Wedge (top anchor)

On the basis of static balance, the anchorage force of wedge can be calculated from analysing the pipe wall and hole wall after the wedge is expanded, as shown in Fig.10.

Let F_1 be the squeezing force exerted on the pipe wall by the borehole wall at the upper wedge;

F_2 be the squeezing force exerted on the pipe wall by the borehole wall at the lower wedge;

λ_1 be the frictional coefficient between steel and steel; and

λ_2 be the frictional coefficient between steel and rock.

Then $$F_1 = F_2$$

$$F = F_1\lambda_2 + F_2\lambda_2 = 2F_2\lambda_2 \qquad (7)$$

By reference to Fig.11, to analyse the forced state of the upper and lower wedges, we obtain

$$F_p = F_2\left(\frac{\lambda_1\cos\alpha + \sin\alpha}{\cos\alpha - \lambda_1\sin\alpha} + \lambda_1\right) \qquad (8)$$

It is known from Equation (8)that the force needed for installing the lower wedge is a function of the positive pressure F_1 or F_2 exerted on the rock at the top anchor, the wedge angles being α and λ_1

Take $\tan\alpha = 0.1$

Thus $\sin\alpha = 0.1$, $\cos\alpha = 1$

Take again $\lambda_2 = 0.35$, $\lambda_1 = 0.15$

Substitute these into Equations (7) and (8),

Thus $$F = 1.8F_p \qquad (9)$$

It is known from Equation (9) that the wedge structure of W-P can enlarge the thrusting force required for installation by a factor of about 1.8. This is due to the fact that the resistance acting in installing the bolt is λ_1, but the anchorage acting is λ_2 ; λ_2 is three times greater as λ_1.

Example. Results in field tests have shown that the total maximum thrusting force (F_p) was 7t when the W-P was hammered with a YSP-45 model drill.

Thus $F = 1.8 \times F_p = 1.8 \times 7 = 12.6$t

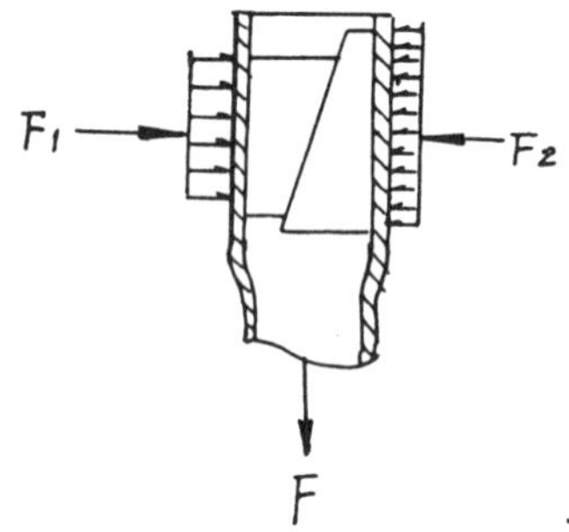

Fig.10 Anchorage of the wedge (top anchor) of W-P

(b) Anchorage Force of Split Segment

The results from in-situ pulling tests have shown that the anchorage force of the split segment of W-P can be up to 5 - 7t if the difference in diameter is about 2mm.

The author has calculated the stress near the borehole by the stress function method of elastic mechanics, and has also conducted experiments in photoelasticity. The results have demonstrated that the radius of the ' pear-like pressure zone for each 'Split-Set' bolt is only 0.2m and that the zones cannot be connected to one another, owing to the fact that radial stress of 'Split-Sets' exerted by the pipe wall on the borehole wall is low, only 4 - 8 kg/cm^2. If several tons of pre-tensioning force is exerted on the axial direction of the bolts, then the 'pear-like pressure zones' for bolts with spacing of 1m are connected to one another; W-P bolts are quite so with it (Fig.12).

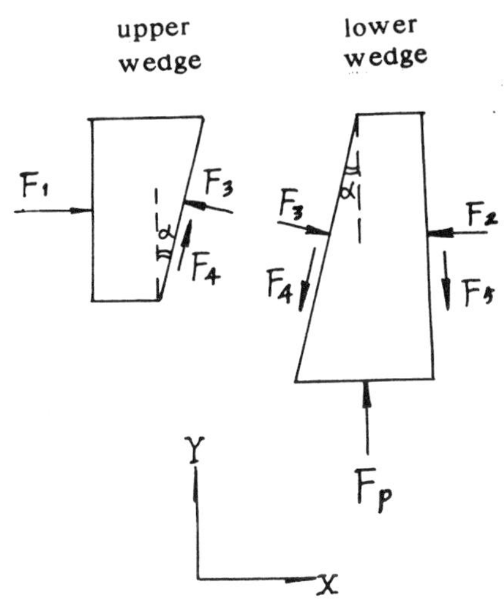

Fig.11 Analysis for anchorage force of the wedge of W-P

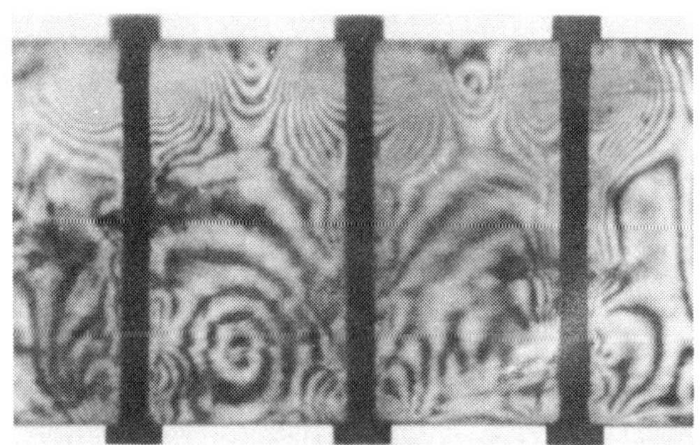

Fig.12 Connection of 'pear-like pressure zone' of W-P

(c) Pre-tensioning Force on the Upper Segment

Results of pull-strain experiments have shown that the upper segment of W-P can be accompaniedly obtained 3 - 4t pretensile force in the process of that the top anchor is installed by impacting the lower wedge with the hand-hammer, while with the driller is much higher, and even the washer plate has been pulled to be hollow upside-down (the elastic limit is 4t), as shown in Fig.13.

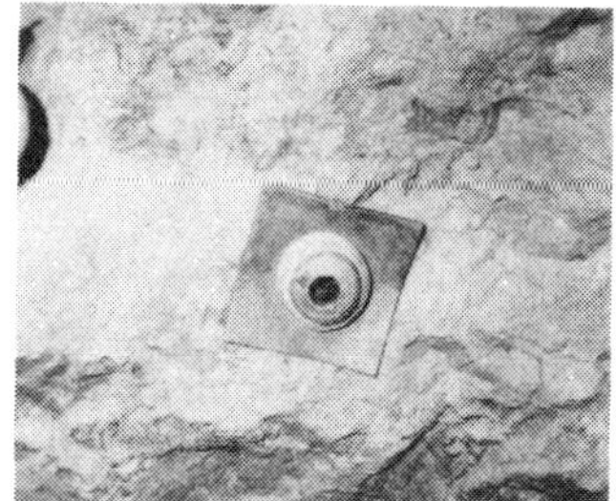

Fig.13 Washer plate photo after installing W-P bolts only with top anchor

(d) Pulling Tests In situ

The results of tests are as shown in Table 3 and Fig.14. W-P bolts have already been used in mine stopes, drifts, tunnels and caverns, with outstanding success.

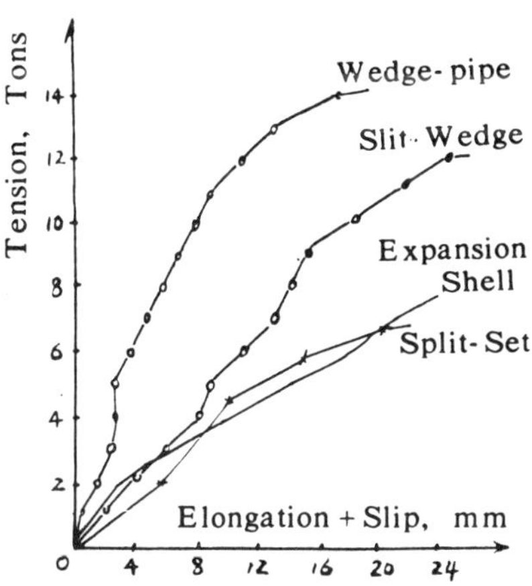

Fig.14 Pull force-displacement curves

Quick-prestress Grouted Bolt

The support provided by prestress grouted bolts is best, but it is the most expensive and the bolts are difficult to install. Quick-prestress grouted bolts developed by the author and patented are not only effective, but also cheap and convenient to install. In fact, they

Table 3 Results of Pulling Tests In-situ

Test Location	Bolt Type	Rock Condition	Average Anchorage Force, t	Notes
Xikuang-shan	W - P 'Split-Sets'	Changlong Group shale f = 4 - 6	13 5.8	
	W - P slit-wedge (¢25)	weathered shale	11.8 12	
	W - P slit-wedge (¢25)	weathered and weak shale	8 1.6	
	W - P slit-wedge (¢25)	siliceous limestone f = 18	11 7	8 bits for drilling a hole of 7.4m
	W - P slit-wedge (¢25)	Changlong Group shale	14 11.7	
Xiangxi	W - P 'Split-Sets' slit-wedge (¢25) slit-wedge (¢22) expansion shell	deep red slate f = 4 - 8	11.3 5.3 13.2 9.4 6.8	
	grouted rebar		13.2	¢16 spiral steel, anchorage force for 28 days
Mouding	W - P	settled sand rock (copper orebody) f = 15 - 17	12.14	

are W-P bolts without the split segment, as shown in Fig.15. The sequence of installation is as follows.

(1) Install (impact) the top anchor. It is obvious from the above analysis that the impacting imparts about 12t anchorage force to the top anchor and simultaneously 3 - 4t pre-tensioning force to the pipe body.

(2) Connect the pipe joint at the end of the bolt, and grout until the mortar is squeezed out from the space between the washer plate and rock face.

Obviously, installing this prestress-grouted bolt does not need those sequences of both particular pre-tensioning and erecting the exhaust guidance, owing to the fact that the ring space between the outer wall of the bolt in the borehole and the borehole wall is the natural exhaust hole. The mortar flows from the bolt pipe, and into the ring space through the top hole for displacing mortar, then out from the gap between the washer plate and rock face. Therefore, the bolts are quick in installation and cheap, and can be used as both temporary and permanent support.

CONCLUSION

(1) To lessen the unfavourable effect on

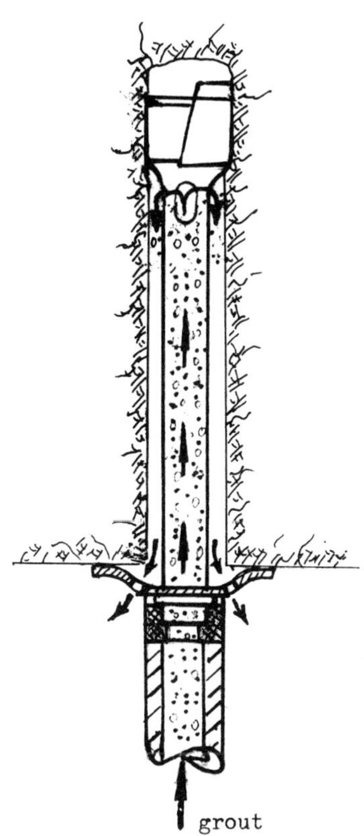

Fig.15 Quick pre-tension bolt with grouting

surface buildings as much as possible, the NATM is suitable for support of excavations in the sensitive area under an urban environment.

(2) As support of urban underground excavations, the bolts need to be installed early and close to the face, so as to consolidate the surrounding rock and lessen the displacement of surface ground. The bolts must be strong in rigidity, blast- and shockproof, prestressed and with enough high anchorage force, and preferably also have yield properties.

(3) Common friction bolts or full-length binding bolts with less than 10 kg/cm^2 bind force must be checked and calculated repeatedly when used in underground excavations with a large span, as their allowable anchorage force is much lower than that of their full length.

(4) W-P bolts combine and develop the merits of both classic mechanical anchor bolts and modern 'Split-Set' bolts, and overcome their shortcomings. Their anchorage force is about 12t. They can be yieldable and have more anchorage functions. They are safer and reliable for support; they are quick and convenient to install; it is easy to check their installation quality; they are blast- and shockproof; they are suitable for use in a wide range of strata; they are not strict in the requirements for hole depth, removal and storage, unlike 'Split-Set' bolts, which require a strict hole diameter. They represent a great advance in reliability of rockbolting, and are suitable to use as a temporary support in urban underground excavations and also as a permanent support after spraying shotcrete or grouting.

(5) Wedge-pipe prestress grouted bolts are quick-prestress grouted bolts suitable for use as a quick permanent support in urban underground excavations.

References

1. Qu Zhaoqi (1983). The Excavations for Water Power Station (in Chinese).

2. E. Hoek and E. T. Brown (1980). Underground Excavations in Rock . IMM, London.

3. R. Schach, K. Garshol and A. M. Heltzen (1979). Rock Bolting . Norwegian Institute of Rock Blasting Techniques, Norwegian.

Drill and blast tunnel excavations for the Buffalo Rapid Transit System, U.S.A.

D.S. McAllister B.SC., C.ENG., M.I.C.E., P. ENG.
Mott, Hay & Anderson, London, England; Transit & Tunnel Consultants, Inc., Buffalo, New York, U.S.A.
C.R. Nelson PH.D., P.E.
Charles R. Nelson & Associates, Inc., Minneapolis, Minnesota, U.S.A.
R.J. Strassburg M.SC., P.E., M.A.S.C.E.,
Hatch Associates Consultants, Inc., Transit & Tunnel Consultants, Inc., Buffalo, New York, U.S.A.

SYNOPSIS

Recent construction of a rapid transit line in Buffalo, New York, has demonstrated that rock excavation at comparatively shallow depths by conventional drilling-and-blasting can still be successfully carried out in an urban setting, in this present environment-conscious age, with the right planning, controls, and public relations effort. Drill-and-blast tunnel excavations for four stations are described together with similar excavations for arched and flat roofed-structures up to 14.7 m wide for crossovers, turnouts and storage tunnels, in some instances with as little as 5.5 m rock cover.

Excavation procedures and rock support systems are described where 7.3 m wide inclined escalator shafts were driven between parallel 8.5 m wide platform tunnels to leave rock pillars of only 2.3 m thickness. Platform tunnel excavations at one station approached within 10 m of the foundations of a three story building. Details are given of the necessarily restricted blasting at this location, together with vibration monitoring of the blasts measured at street level and within the building.

The history of rock movements is provided for a section of 13.7 m wide turnout structure with 7.5 m rock cover where an unexpected combination of severely weathered, deep vertical joints resulted in gross rock movements and settlement at the surface in order of 60 mm. The support system installed and monitored, and the successful completion of the excavation is described.

OVERALL PROJECT

May 1985 saw start of revenue operations over 8 km of the new light rail transit line in the city of Buffalo, which with its population of approximately one million is located at the east end of Lake Erie in New York State. Construction on this $525 Million project was started in 1979, and the full 10.3 km length of this first construction phase is scheduled to be opened in September 1986. It is expected to attract an average daily ridership of 45,500.

From a Yards and Shops facility located just south of the city centre, the line runs northwards on the surface through a pedestrian mall being constructed along Main Street, the city's principal street, as part of the transit project. This 1.6 km-long surface section extends through the central business, shopping and theatre districts. Descending on a 6% grade, the line continues just beneath Main Street for 2.8 km in a reinforced concrete double-box structure constructed in cut-and-cover through soft ground. It then descends on a 4% grade in a widening cut-and-cover box structure to the rock tunnel portal, from where the separate inbound and outbound circular running tunnels continue to follow Main Street for 5.6 km to the temporary terminus at South Campus Station.

Six closely-spaced stations are located on the surface section. Three stations are included in the cut-and-cover section. The rock tunnel section contains four stations constructed in tunnel and a fifth station originally planned to be in tunnel but revised to a cut-and-cover structure during preliminary engineering, on assessment of the subsurface exploration programme. Overall project cost statistics have been included in Table I.

Table I. Overall Project Cost Statistics

Rock Tunnel Section	Bid Costs ($ Million)	Present Costs ($ Million)
South Running Tunnels (3.24 km)	38.95	39.66
North Running Tunnels (2.32 km)	35.38	34.24
Delavan Station	16.25	18.50
Humboldt Station (Incl. Storage Facility)	21.58	21.55
Amherst Station (Cut & Cover)	9.20	10.24
Tonawanda Turnout	3.29	4.09
LaSalle Station	13.54	16.51
South Campus Station (Incl. Crossover)	28.19	29.15
Subtotal	166.38	173.94
Other Construction & Purchase		
Yards & Shops	11.63	12.31
Surface & Mall (1.6 km)	27.39	29.30
Cut & Cover Line (2.8 km)	54.93	54.55
Cut & Cover Stations (3 No.)	30.15	31.47
System Wide	79.78	85.15
Vehicles (27 Tokyu LRV)	21.83	21.69
Artwork	1.03	1.03
Total	393.12	409.44

Note: Design, Construction Management, Insurances, Land Purchase, etc. make up the balance of the $525 Million. With most contracts complete, "Present Costs" are expected to be close to final costs.

GEOLOGIC SETTING

The rock tunnels are located in limestones, dolomites and shales of the Lower Ordovician to Devonian Age. The specific formations encountered by the excavations in order of increasing depth are the Onondaga Limestone, Akron Dolostone, Bertie Dolostone and Camillus Shale. The strata are relatively undeformed with near-horizontal bedding, generally dipping to the south at approximately 7.6 m/km. The overburden of recent glacial deposits is generally less than three metres thick, except at a few isolated locations.

The vertical alignment was set to place the running tunnels, the principal station tunnels, and the other major tunnelled structures almost entirely within the Bertie formation. The 15 to 18 m thick Bertie formation is typically a grey, massive, thin-to-medium bedded, fine-grained dolomite with occasional bands of dark grey shale. The predominant discontinuities are joints and partings along the near-horizontal bedding planes with joint spacings of a few centimetres to a metre or so. Most other joints are vertical, with major orthogonal sets oriented approximately N75^{o}E and N40^{o}W, and with a minor set striking approximately N5^{o}E. The majority of the tunnels are aligned N40^{o}E. Average spacing of the vertical joints varies from approximately 1.8 m to 5.5 m. Generally the joints were found to be tight with wavy surfaces and were only slightly weathered. Occasional open joints with rough weathered surfaces and clay fillings were encountered, particularly when nearer to the bedrock surface. During the subsurface exploration programme, uniaxial rock strengths were determined by unconfined compression tests, supplemented by more extensive point load testing. Strengths as high as 345 MPa (50,000 psi) were identified in the Bertie, with average strengths being obtained of around 176 MPa (25,500 psi). Approximately 75% of the tests measured strengths greater than 138 MPa (20,000 psi).

The Onondaga Limestone formation, approximately 3 to 6 m thick, typically forms the cap rock in the area. Overlying the general tunnel horizon, it was only encountered in the inclined station escalator shafts and other shaft excavations. It is a coarse-textured massively-bedded, slightly fossiliferous limestone with occasional thin shale partings, and contains a high proportion of chert nodules. The Akron formation, approximately 2 m thick, lies beneath the Onondaga and immediately over the Bertie formation and was again only encountered in shaft construction. It is a fine-grained massive dolostone with occasional small solution cavities. Limited testing indicated the strength of the Onondaga and Akron to be approximately of the same order as the Bertie.

The Camillus formation underlying the Bertie was encountered on very few occasions in the tunnel excavations, being deliberately avoided when the vertical alignment was established, as far as pratical. Over 100 metres thick, it is a dolomite shale varying from a thin-bedded shale to a massive mudstone with layers and seams of gypsum and anhydrite. The upper strata contained zones of severely weathered and solutioned rock, commonly about 6 m thick, extending for hundreds of metres along the tunnel alignment. These zones were found to yield large quantities of water containing hydrogen sulphide concentrations in the range of 1 to 8 ppm. Average compressive strength of the unweathered Camillus Shale was less than half that of the Bertie.

In the subsurface exploration programme for the running tunnel contracts, 102 mm (4 inch) diameter borings were drilled to an average depth of about 28 m, and were spaced an average of 60 m apart along the 5.6 km of the rock tunnel section. Additional boreholes were drilled at station locations during the later design programme for those structures. Packer tests were performed, and multi-level piezometers were installed in selected boreholes.

Very low core recoveries were generally obtained from the lengths of borehole in the solutioned zones of the Camillus Shale. To provide more information, a 0.9 m diameter exploratory shaft was drilled 19 m through the fractured and solutioned Camillus Shale at the Amherst Station site. The shaft could not be dewatered by the available 0.032 cu. m/sec (500 gpm) pump, and a diver was engaged who, although unable to visually inspect the walls because of turbidity, was able to conduct a tactile survey, and in fact located many voids into which he could extend his entire arm. A full scale pumping test was conducted at this shaft, and at a 0.5 m diameter shaft drilled 35 m deep at a second zone of solutioned Camillus some 3 km to the south. These tests proved the hydraulic connection between the different solutioned zones and also with the groundwater in the overlying Bertie formation. Transmissibilities in the solutioned zones were calculated to be typically 800 sq. m/day to 1500 sq. m/day.[1]

Following discovery of the extensive solutioned zones in the Camillus Shale, the running tunnels were generally set to clear that formation. As a result, some lengths of running tunnel and some station excavations were brought

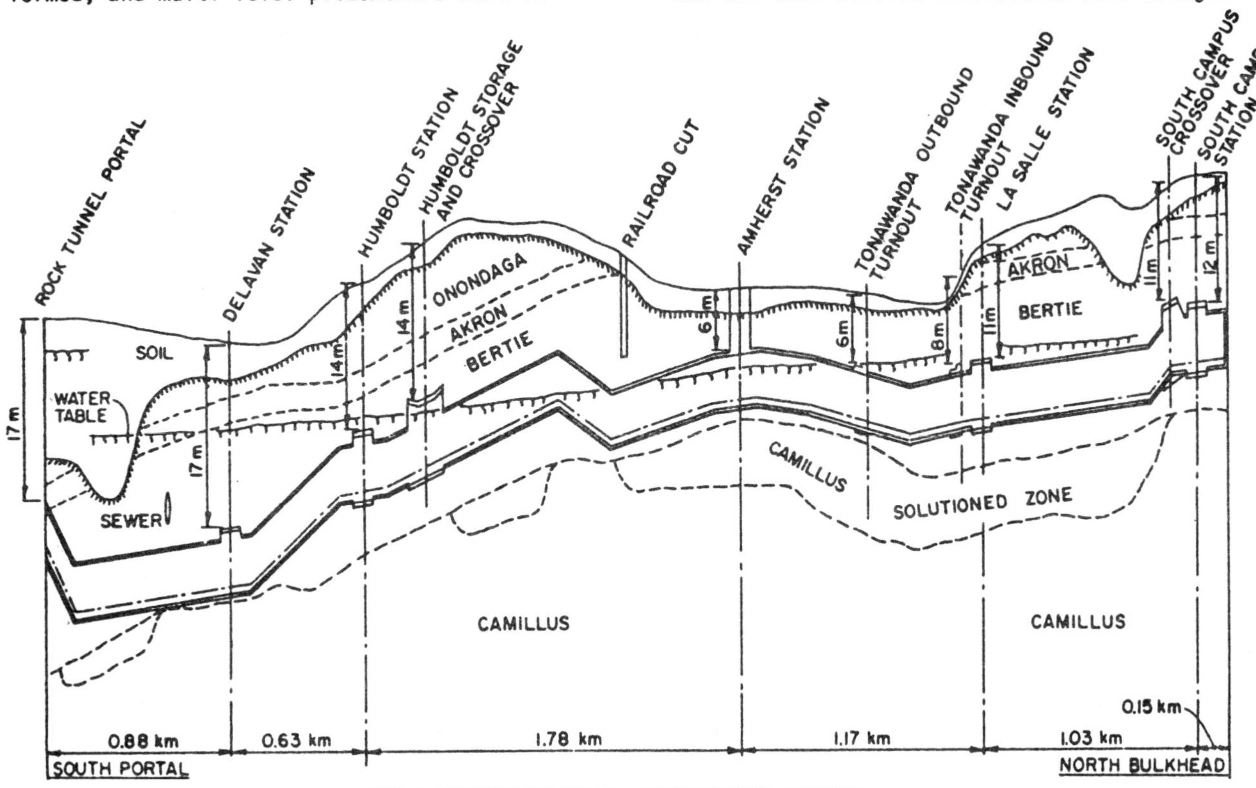

FIG. I PROFILE ON 5.64 km ROCK TUNNEL SECTION

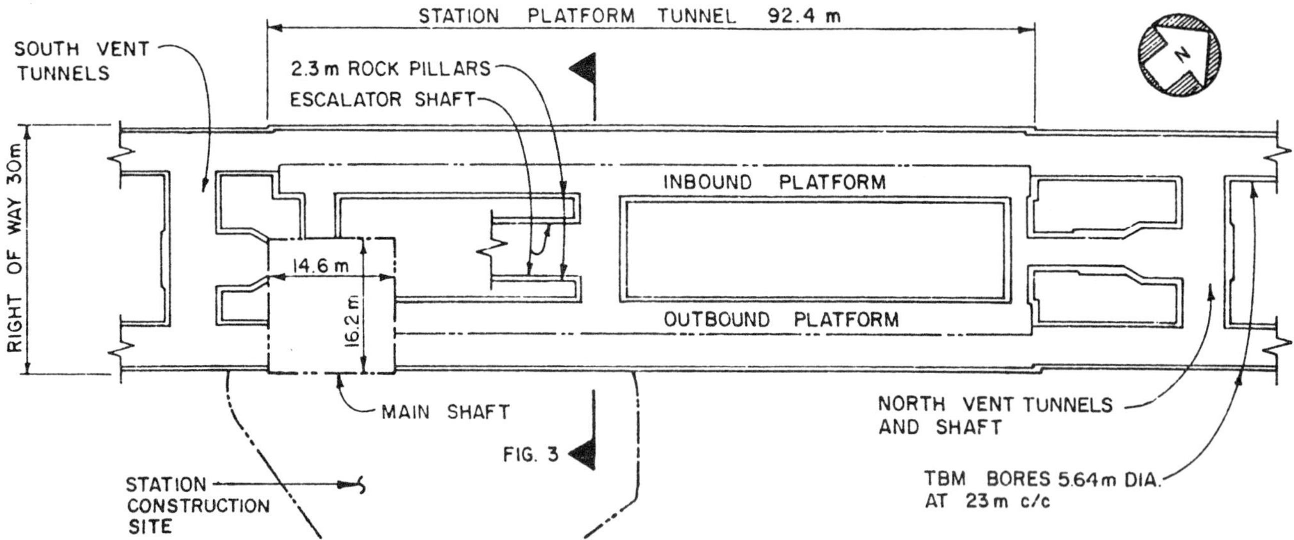

FIG. 2 PLAN OF DELAVAN STATION TUNNELS

closer to top of rock than would otherwise have been the case. At the Amherst Station site where the Camillus approached within 12 m of the surface, there was no alternative but to revise the station from a tunnelled to a cut-and-cover structure, regardless of environmental impact. Layout and location of the tunnelled Tonawanda Turnout, and location of the tunnelled storage and crossing tunnel facility also had to be fundamentally revised. Because of the unavoidably close proximity of the solutioned rock zones to the running tunnel invert, the somewhat unusual decision for a rock tunnel project was taken, requiring the contractors to lower the water table by means of a deep well dewatering system to at least 0.6 m below tunnel invert for most of the alignment.

From the pumping tests, the estimated maximum discharge for deep wells spaced along the 5.6 km alignment was 0.84 cu. m/sec (13,300 gpm). Special pipelines had to be included in the contracts to carry this discharge to a nearby stream because of the inadequacy of the city sewer system in this area. During construction, after a year of pumping from more than a dozen wells, the total discharge rate stabilized at about 0.27 cu. m/sec (4,300 gpm). Flows increased periodically as tunnels were excavated through areas of local recharge.

Because it was possible to establish the alignment in the generally high quality rock of the dewatered Bertie formation, full-face, hard-rock tunnel boring machines (TBMs) were able to be employed for the running tunnel excavations.[2] These 5.64 m (18.5 ft.) diam. machine excavations were continued as pilot tunnels through the locations of all the larger planned tunnels for station platforms, storage, crossing and turnout structures. Use of TBMs was specified in the running tunnel contract documents. However, the high strength of the rock precluded similar specification of mechanical excavation methods for enlargement of the TBM excavations in the station construction contracts, such as use of roadheaders, although this had been an initial hope in the effort to limit construction noise and vibration in this urban environment.

All the rock tunnels, apart from the running tunnels, were thus excavated in the conventional manner by drilling-and-blasting. The profile of Figure 1 provides data on the proximity of the principal drill-and-blast excavations to the surface.

STATION EXCAVATIONS

The 5.6 km length of Main Street designated for rock tunnelling is bordered by major city hospitals, schools, a University Campus, and other public institutions, as well as private homes and businesses. It was a governing requirement for this segment of the alignment that construction should be accomplished with a minimum of disturbance to the general public, and with minimum disruption to traffic and utilities. Specification of tunnel boring machines using off-street access shafts achieved this for the running tunnels. To

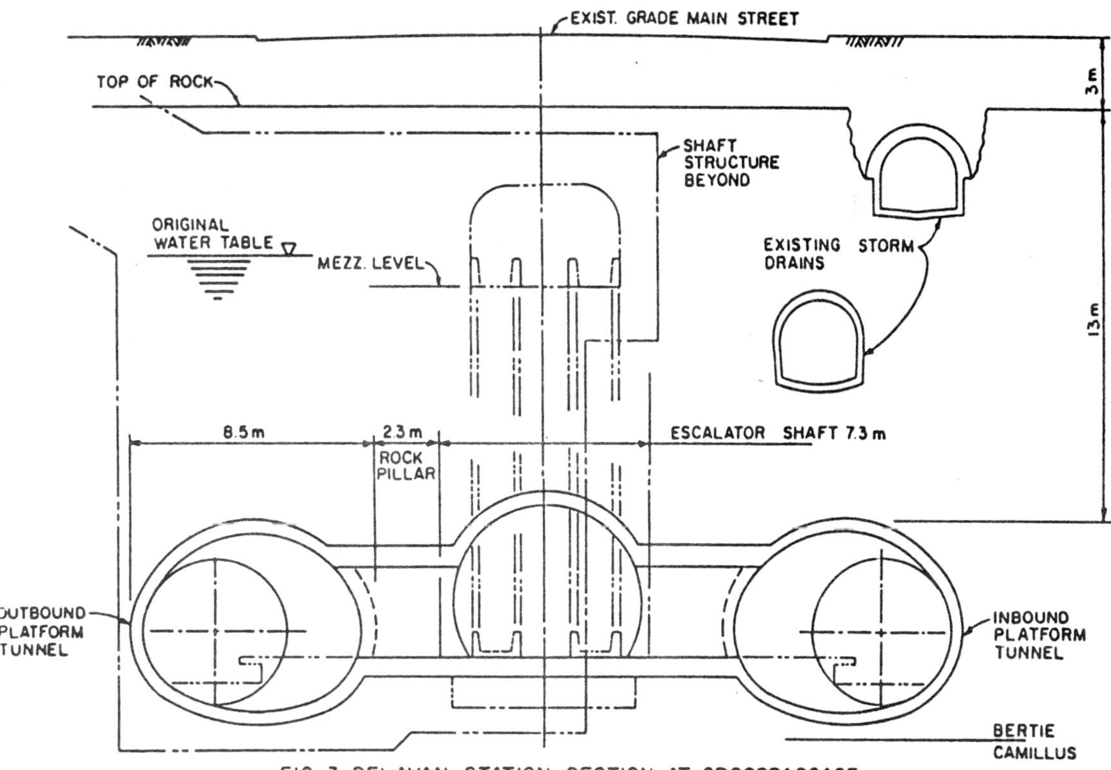

FIG. 3 DELAVAN STATION SECTION AT CROSSPASSAGE

minimize surface disruption at station locations, excavations for the stations and other necessary transit structures were to be carried out by tunnelling to the maximum extent practical.

The basic layout and construction sequence developed for the underground portions of the tunnelled stations were very similar for each station. The TBM-driven running tunnels were positioned at 23 m centres through the stations, placing the tunnels close to the limits of the 30 m wide Main Street right-of-way, and maximizing the available construction space between the tunnels. These running tunnels were excavated through the station areas from off-street work-shafts generally remote from the stations, and were concrete-lined to the station limits prior to start of station excavation.

A shaft measuring approximately 16 m by 14 m was located over one of the existing tunnel excavations, and immediately adjacent to the principal offstreet construction worksite which would eventually carry the necessary surface structures and bus interchange facilities. This shaft provided the main working access to track level for the station contractor, and was sized to accommodate the large number of service rooms required, and an elevator and emergency stairway connection to platform level.

Working through this shaft, and through a crosspassage to the adjacent tunnel, the existing TBM bores were widened towards the centre of the street to provide separate inbound and outbound platform tunnels. From a mezzanine level situated in the shaft just below street level, a single inclined escalator shaft was tunnelled to connect with a crosspassage driven between platform tunnels, to provide the main passenger access to the two platforms. Structures to provide for emergency ventilation and draught-relief for the tunnels and for heat exhaust from the platform areas were tunnelled separately at either end of the station, although as much of these structures as possible were also accommodated in the main shaft.

Housing in the shaft those service rooms required close to the trackway (train control, communication, etc.), largely avoided creation of the narrow rock pillars which would have been inevitable if the rooms had been accommodated in tunnels located between the platform tunnels. It also allowed a degree of flexibility during final design which would have been impossible with a tunnelled-room arrangement, as it was relatively easy to revise the shaft size to suit the changing room requirements. Narrow pillars were still created adjacent to the central escalator shaft, and on occasion, though less critically,adjacent to the main shafts. Figures 2 and 3 illustrate the Delavan Station excavations.

For the majority of the tunnelling, the Station Contract Documents placed no abnormal restrictions on the Contractor. Blasting was only allowed between the hours of 7 a.m. and 7 p.m., except for emergency work; the peak particle velocity of blast induced motion was limited to 50.8 mm/sec (2 ins. per sec.) at a distance of 60 m from the excavation, or at the nearest structures if closer; overpressure (air-bourne vibration) from the blasting was limited to 103.4 Pa (0.015 psi) at a distance of 60 m from the workshaft, or at the nearest structure; each shot was required to be monitored using instruments of the four-channel oscillagraph type to measure ground-bourne vibrations in three mutually perpendicular directions, and the air-bourne overpressures, and to record the results on strip charts; controlled blasting techniques were required to be used throughout the excavations.

The horizontally-bedded, generally high quality Bertie dolomite was very favorable for tunnelling. Minimum mandatory levels of primary support were specified for the excavations, consisting of different patterns of untensioned rockbolts, resin-encapsulated for their full length. These were required to be installed within three feet of the face of the excavation, within eight hours of blasting, and prior to the next blast. With full responsibility for safety, the Contractor was at liberty to install any additional bolts he considered necessary, for which he was reimbursed.

However, special requirements were placed on the Contractors at the narrow pillar locations where the proximity of platform tunnel and escalator shaft required rock pillars a maximum of 2.3 m wide at their narrowest point, without considering overbreak. These requirements included special supports and excavation sequences. Figure 4 shows a typical pillar.

To limit overbreak and minimize blasting damage at the thin pillars, smooth-wall blasting was specified with certain restrictions. Spacing of perimeter holes was not to exceed 450 mm; each loaded perimeter hole was to be loaded full-length with a light distributed charge (maximum 0.37 kg/m.); each apparently intact rock stratum of 300 mm or more thickness was to have at least one hole located therein; holes were not to be placed along discontinuities or through weathered zones; the blasting pattern was to be designed so that a burden not less than 600 mm nor more than 900 mm was left against the line of perimeter holes for removal with the final blasts; adjacent perimeter holes were to be fired on the same delay wherever possible; and only distributed charges were to be used within 1.5 m of the perimeter. Where overbreak extended more than 300 mm beyond the external neat lines of the tunnel structure,the Engineer could require filling with shotcrete to the structural neat line, or alternatively, filling with concrete and backfill grout. Additional rockbolting might also be required. All such remedial work was to the Contractor's account. It was also a requirement that trial blasts be conducted prior to reaching the narrow pillar areas, to demonstrate that this blast pattern left an adequate burden for the final trim blasting, and did not result in excessive overbreak.

Completion of excavation for both station platform tunnels was required past the pillar areas, prior to any excavation for crosspassage or escalator shaft. This was essentially a slashing operation, with blasting vibration being relieved by the existing TBM bores. On completion, the distance between the parallel station platform tunnel excavations was approximately 12 m. One face of each pillar was now exposed. The contract provided for cleaning out and grouting of any open joints found in the pillars, but in the event no such work was required. Patterns of short resin-encapsulated rockbolts were now installed through the pillars at 900 mm centres along the tunnels. These were also used to hold sections of rolled steel channel to the pillar face in the vertical plane. The bolts were angled to intersect the horizontal bedding plane joints. Any void between channel and rock surface was fully packed with shotcrete or concrete to ensure a full-thickness pillar, with steel channels able to assist containment of the pillar during later blasting for the escalator shaft.

A 3 m wide pilot tunnel was now excavated on

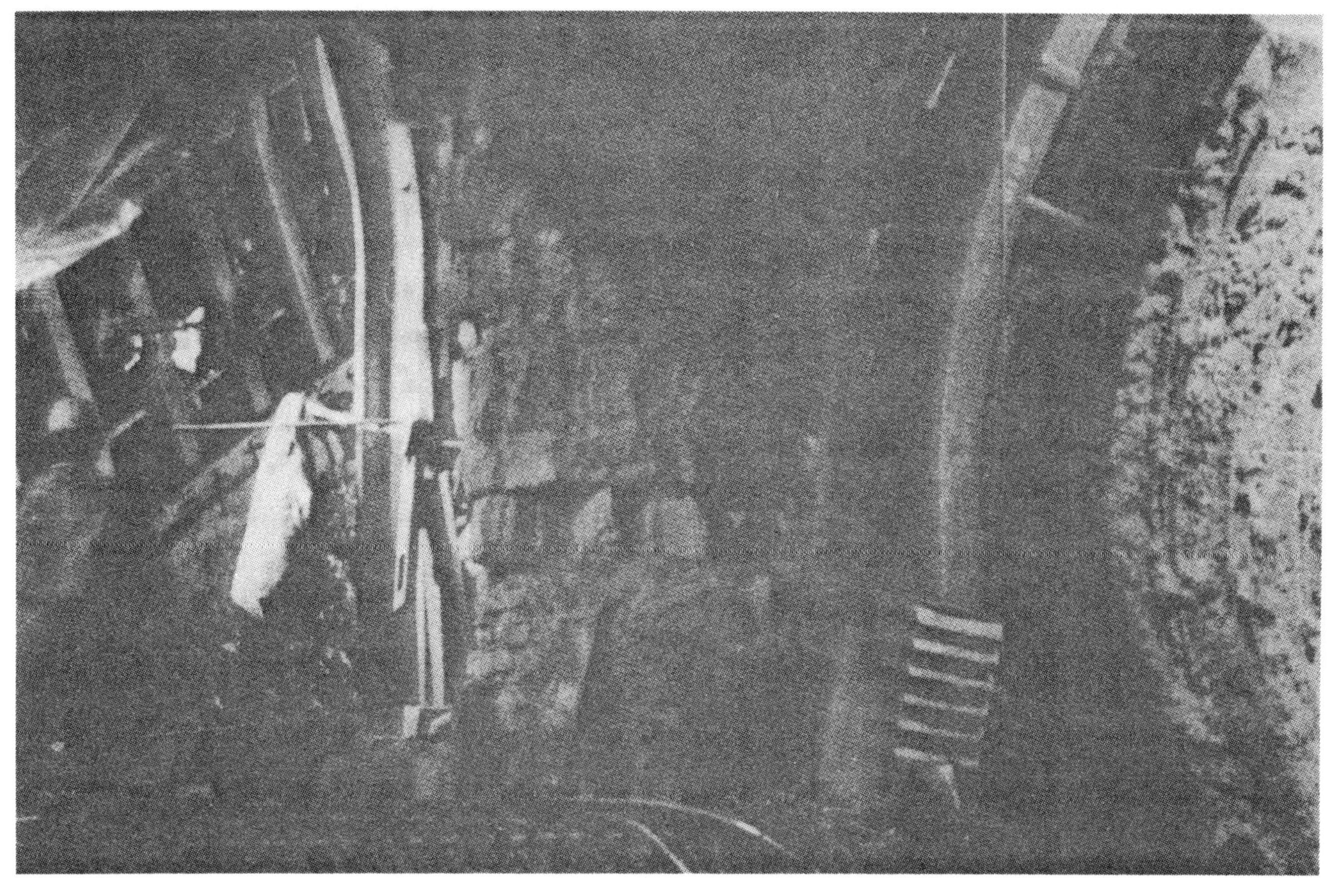

FIGURE 4. ROCK PILLAR 2.3 m WIDE AT LASALLE STATION

At left, 7.3 m wide escalator shaft excavation with steel rib supports; At right, 8.5 m wide platform tunnel excavation with steel channels, shotcrete, and steel rib supports.

the centreline of the crosspassage from one platform tunnel to the other, to its full height of 4.25 m, and was then widened to its full 5.8 m width by a smooth wall slashing operation. Prior to start of escalator shaft excavation, structural steel arch supports and lintel girders were installed at both intersections between the crosspassage and platform tunnels, and were closely blocked-off the rock. This steelwork was to form the major part of the reinforcement to the final concrete lining at the intersections, but was also designed to serve as additional primary protection to the excavations in this sensitive area. It had been required to be held on site throughout excavation of the platform tunnels for possible use anywhere in these tunnels if rock conditions warranted, to be replaced as used. In the event, such a situation never arose.

Excavation of the inclined escalator shaft was started by driving a 3 m wide pilot heading on the shaft centreline, from the track-level crosspassage to the main shaft at mezzanine level. The heading extended from tunnel crown level to a point 5.5 m below the crown, where the width of the final excavation was to be reduced for the narrower machine pit. Resin encapsulated rock bolts angled ahead of the advancing face were installed between each round. The Contractor was allowed choice of direction for the escalator excavations, however, all four contractors elected to drive in the uphill direction which had advantages for spoil removal and in containing blast overpressures. During excavation of this pilot heading, where confinement of the blasting would induce maximum lateral thrust on the pillar areas, the pillars were still approximately 4.4 m wide at their narrowest points.

The escalator shaft was now widened by a careful smooth wall slashing operation to peel off the final 2.1 m of rock against the pillars. Advance of the trim blasting was limited to increments of 1.2 m along the shaft. Rockbolts were installed between each advance. Steel wall

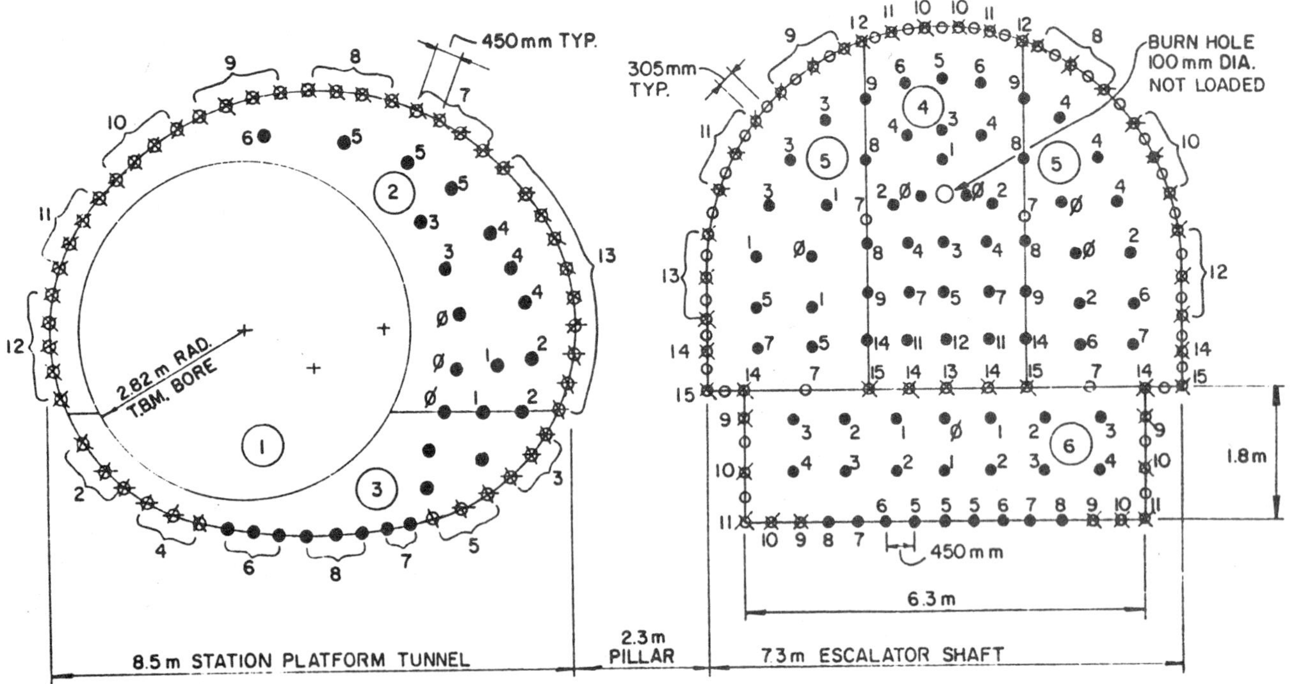

AS USED AT SOUTH CAMPUS STATION

● PRODUCTION HOLE USING 0.224 kg PER 305 mm STICK (0.91 kg PER m)

⊠ TRIM HOLE USING 0.372 kg PER m COIL

○ UNLOADED

(3) (5) etc. EXCAVATION SEQUENCE

5, 7, 8 etc. DELAY SEQUENCE

FIG. 5 EXCAVATION SEQUENCE & BLASTING PATTERNS @ NARROW PILLAR LOCATION

plates were bolted along the foot of the shaft walls and steel arch ribs were erected within 3.6 m of the face of the completed full-width excavation. Steel arch ribs and lintel girder supports were also provided for the intersection between the crosspassage and the escalator shaft. As at the adjacent platform tunnel intersections, this steelwork was designed to be used as additional primary support for the excavations, although in this instance, timing of its erection was left to the Contractor's judgement. This steelwork and the arch ribs in the escalator shaft all formed a substantial part of the reinforcement to the final concrete linings.

The pillars were now at a maximum of 2.3 m thick at their narrowest point, though overbreak on occasion reduced widths locally to just under 2 m. The final stage of escalator shaft excavation was mining the machinery pit below wall plate level. It was specified that a central wedge be removed prior to excavating to full width, again with 1.2 m advances along the walls. Figures 5 and 6 illustrate excavation sequence, typical blast patterns, and typical supports for escalator shaft and station platform tunnel excavation.

Prior to start of station excavation, multiple position borehole extensometers were installed from the surface over the escalator crosspassage intersections, and at other critical locations, to monitor vertical rock movements at approximately 600 mm intervals from tunnel crown to top of rock. At the first three stations to be constructed, inclinometers were also installed in boreholes driven from the surface vertically through the narrow rock pillars to approximately 10 m below tunnel invert level, to measure horizontal movements. The borehole instruments were monitored throughout progress of the different excavations, as were surface settlement points.

As might be expected, settlement at the ground surface was neglible at the deeper stations, Delavan, Humboldt and South Campus, where rock quality was good (R.Q.D. generally greater than 75%),and depths from top of rock to tunnel crowns ranged from 10 to 14 m, exceeding the 8.5 m width of the platform tunnel excavations.

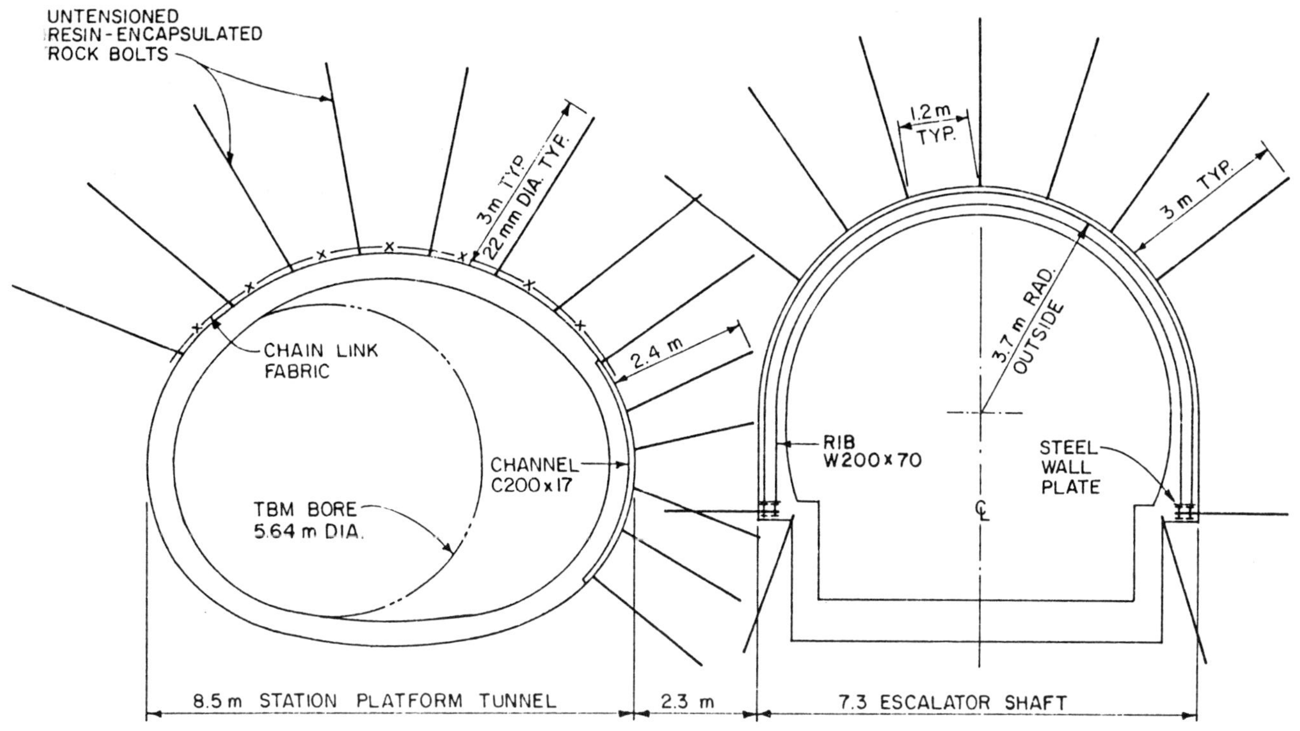

FIG. 6 PRIMARY SUPPORTS AT ESCALATOR SHAFT NARROW PILLAR

Readings taken were less than the accuracy of measurement. The extensometers indicated up to 3 mm of settlement at anchors 1 m above the excavations, and up to 1 mm at anchors 3 m above the excavations. The inclinometers recorded horizontal movements towards the station platform tunnels of up to 8 mm during widening of those tunnels from the initial TBM bores. Movements of up to 5 mm were recorded towards the crosspassage excavations as these were accomplished. After the rockbolts and steel channels had been installed in the platform tunnels, additional horizontal movements were not noted during the final excavation of the escalator shafts which formed the narrow pillars, although a number of inclinometers were damaged during rockbolt installation. For this reason, inclinometers were dispensed with at the last station constructed, LaSalle, in favour of additional extensometers.

At LaSalle Station, the rock cover over the platform tunnels was 9.5 m at the north end of the station, tapering down to 7.5 m at the south end, less than the width of the tunnel excavation. Not only was the rock cover less than at the other stations, but the rock forming this cover was more thinly-bedded and more weathered, with R.Q.D. values for the cores mostly in the 25% to 50% range. Recoveries were over 90%. The rock through the tunnel horizon itself, including the pillar area, was of generally good quality with R.Q.D. values over 75%.

In the area of the narrow pillars at the escalator shaft, there was 8 m rock cover over the 8.5 m wide platform tunnel excavations. The maximum settlement recorded by any of the five extensometers in this area at anchors set 1 m above the tunnel crowns was 3.5 mm relative to the ground surface. However, surface settlement measured at the five extensometer heads varied from 12 to 16 mm. No apparent damage was sustained by utilities or adjacent buildings.

Plans of the two largest structures constructed in the station contracts are provided in Figures 7 and 8. These are the Humboldt emergency storage and crossover facility, and the South Campus double crossover. A photograph of the completed Humboldt facility is included as Figure 9 to illustrate the size of the structure. Both structures were built entirely in tunnel, working from nearby station shafts and through the running tunnel excavations. There was 15 m of rock cover over the widest part of the Humboldt excavation, which measured 14.7 m. The

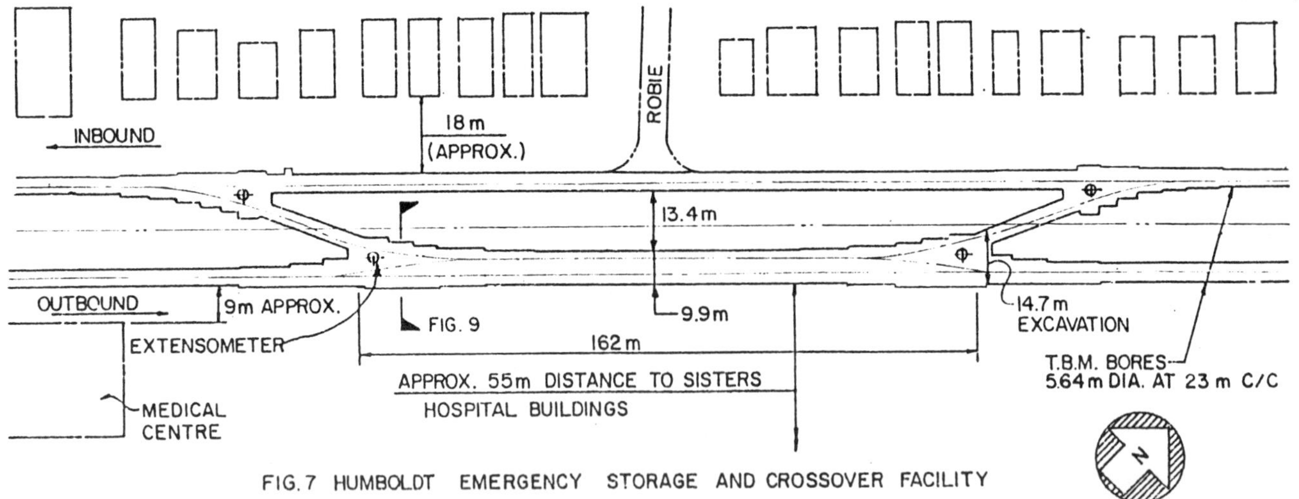

FIG. 7 HUMBOLDT EMERGENCY STORAGE AND CROSSOVER FACILITY

widest excavations at the South Campus crossover measured 12.2 m, and had rock cover of 10 m.

Surface settlements were negligible. The greatest rock movements at these two large structures took place over one of the South Campus double crossover excavations, inadvertently over-excavated by the contractor, where a settlement of 12 mm was measured 1 m above the excavation, and of 8 mm at a point 3 m above the excavation. The lack of problems encountered during construction of these larger structures was attributed in large measure to the fact that considerable flexibility was allowed in choosing their location, so that geotechnical considerations could be given full weight. This was not the case for the Tonawanda turnout structures, the two other large tunnels on the project, which had to be accommodated in the limited thickness of reasonably sound rock overlying the Camillus Shale solutioned zone south of LaSalle Station.

TONAWANDA TURNOUT CONSTRUCTION

In planning the Buffalo LRRT project, potential routes were examined for future extensions. One such route, known as the Tonawandas Corridor, would connect the initial 10.3 km mainline with the cities of Tonawanda and North Tonawanda approximately 9.0 km to the north. This route was recognized as being particularly cost effective because it could be built at grade on an existing railroad right-of-way. During preliminary engineering, tunnelled turnout structures were planned on the Main Street line, with short steeply-graded running tunnels continuing to the surface for connection with the Tonawanda right-

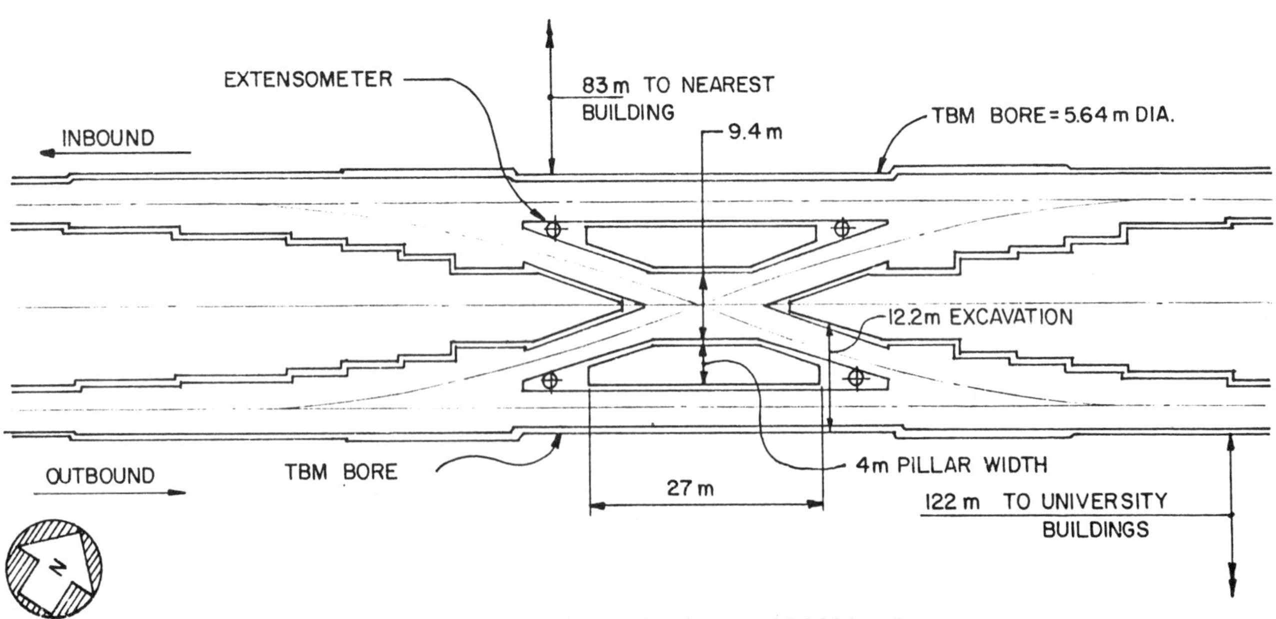

FIG. 8 SOUTH CAMPUS DOUBLE CROSSOVER

FIGURE 9. LOOKING NORTH AT HUMBOLDT EMERGENCY STORAGE TUNNEL

of-way. These were to be built in the first construction phase as part of the LaSalle Station contract, so that the Main Street line operations would not be impacted by future construction for the extensions. However, budgetary restrictions resulted in elimination of all Tonawanda Turnout work from the original 1978 capital grant application, and the Main Street TBM-excavated running tunnels were concrete-lined throughout the turnout areas during the running tunnel contracts. When limited funds became available in 1983, it was decided to let a separate construction contract to build just the tunnelled turnout structures themselves, with the barest minimum of stub tunnel lengths necessary to protect the Main Street line operations during future completion of the tunnels to the surface.

The original layout for the Tonawanda Turnout, which had involved the outbound Tonawanda tunnel passing beneath the inbound Main Street tunnel, had been an early casualty during preliminary engineering, when the Camillus Shale was found to be fractured and solutioned close to the surface through this area. A variety of alternative layouts had been prepared at that time, in preliminary form, all involving either shallow tunnelling or cut-and-cover construction. When belated funding finally became available, a shallow tunnel scheme was selected which incorporated plans for the future outbound Tonawanda line to use an existing rail bridge to cross over both Main Street and the tunnels beneath. However, it was realized that the shallow rock cover available for the large turnout structures was far from ideal. With the fractured Camillus Shale only 0.6 m beneath the outbound turnout structure, the cover of thinnly-bedded limestone over that structure was only 5.5 m. Last minute right-of-way difficulties placed the inbound turnout structure only 6.5 m beneath a railroad bridge abutment.

Fifteen borings taken during the preliminary and design phase were within or near the locations proposed for the turnouts. In addition, geologic mapping had been carried out through the running tunnel excavations. Twenty-five instrumentation boreholes were drilled within or near the inbound turnout limits, with eight inclinometers and thirteen extensometers installed to monitor rock movements near the railroad bridge as the TBM passed underneath the bridge abutment. These instruments showed the maximum horizontal closure of the tunnel walls to be about 12 mm, with a maximum vertical movement at the tunnel crown of less than 3 mm. As would be expected, vertical movements were greatest in the anchors nearest the tunnel crown, with shallower anchors showing less movement. Monitoring of the instrument sensor heads, carried out by optical survey, indicated that there was no appreciable surface subsidence due to running tunnel excavation.

The turnout structures, which were excavated using conventional drill and blast mining methods, are approximately 76 m long. The maximum unsupported width of mined excavation was 14.7 m with a maximum height of 7.1 m from excavated invert to tunnel crown. Two types of reinforced concrete liner were used in the turnout structure. For excavated widths of up to 10.4 m, a circular arch with a varying thickness at the crown was utilized. For the larger excavated widths at the north end of the turnouts, a flat roof structure with arched sidewalls was supported by columns and a deep girder on the tunnel centreline. Design liner thickness varied from 305 mm to 610 mm at the tunnel sidewalls, with crown thickness at 610 mm in the flat roof section and varying from 305 mm to 1300 mm in the arched-roof area. 30.5 m long stub tunnels were constructed at the north ends of the turnouts.

These circular arch reinforced concrete tunnels have an inside diameter of 5.2 m, with a design liner thickness of 305 mm. Mandatory 22 mm diameter fully-encapsulated, resin-anchored rock bolts were specified to be installed immediately behind the advancing work face. Bolts were typically spaced 1.2 m on centres, both longitudinally and transversely, with the number and length of bolts increasing with the tunnel width.

Figure 10 shows the general location of the Tonawanda Turnout. A plan and profile of the inbound turnout structure, along with typical liner configurations, are shown in Figure 11.

Inbound Turnout Construction

As can be seen from Figure 10, the drill and blast operations at both turnouts were carried out in well populated areas. At the inbound turnout, commercial buildings housing restaurants and a car repair shop were located within 21 m of the mining operations. A gold refinery, using vibration sensitive instruments and equipment to process precious metals, was located 50 m to the south. The tunnel enlargement was carried out directly beneath an abutment to a railroad bridge carrying occasional freight traffic.

In general, overburden varied from 0.6 to 1.5 m of fill material. Borings along the west curbline of Main Street indicated a fill depth of 2.4 to 3.0 m along a trench blasted and backfilled during utility construction. Main Street was locally depressed at the railroad bridge

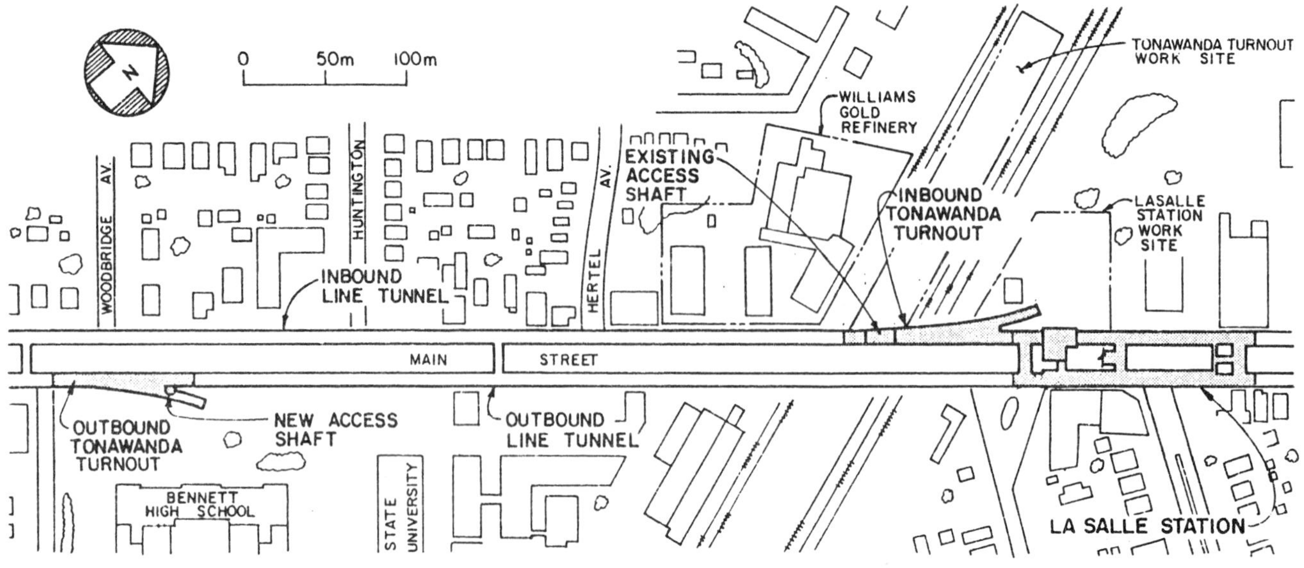

FIG. 10 TONAWANDA TURNOUT AND LASALLE STATION LOCATION

location, and this underpass construction had created an irregular rock surface with rock cover varying from 6.0 to 10.5 m above the tunnel crown. RQDs ranged from 61% to 89% between the tunnel invert and crown, with a trend of decreasing RQDs from the tunnel crown to top of rock. Rock quality was rated "fair" to "good". The boring logs indicated that discontinuities were mostly horizontal, closed to slightly open. Joints were generally slightly to moderately weathered, but occasional open and severely weathered joints were found throughout the strata. The logs also noted high angle and vertical discontinuities, slightly to moderately weathered.

Access to the south end of the inbound turnout was provided through a vertical shaft excavated during running tunnel construction several years earlier. The contract let for construction of the turnouts included a crosspassage between the mainline tunnels immediately north of the inbound turnout. This crosspassage would later be widened during construction of the adjacent LaSalle Station, for use as a tunnel ventilation shaft.

The Contractor began the inbound tunnel enlargement by removing the existing tunnel liner. Short holes drilled at an angle with respect to the longitudinal axis were shot with light loads to peel the liner away from the surrounding rock, and excavate the east invert to its final configuration. A heading was driven perpendicularly to the mainline tunnel at the north end of the turnout to provide two working faces, one for the stub tunnel to the north, and one for the turnout enlargement to the south. The Contractor then

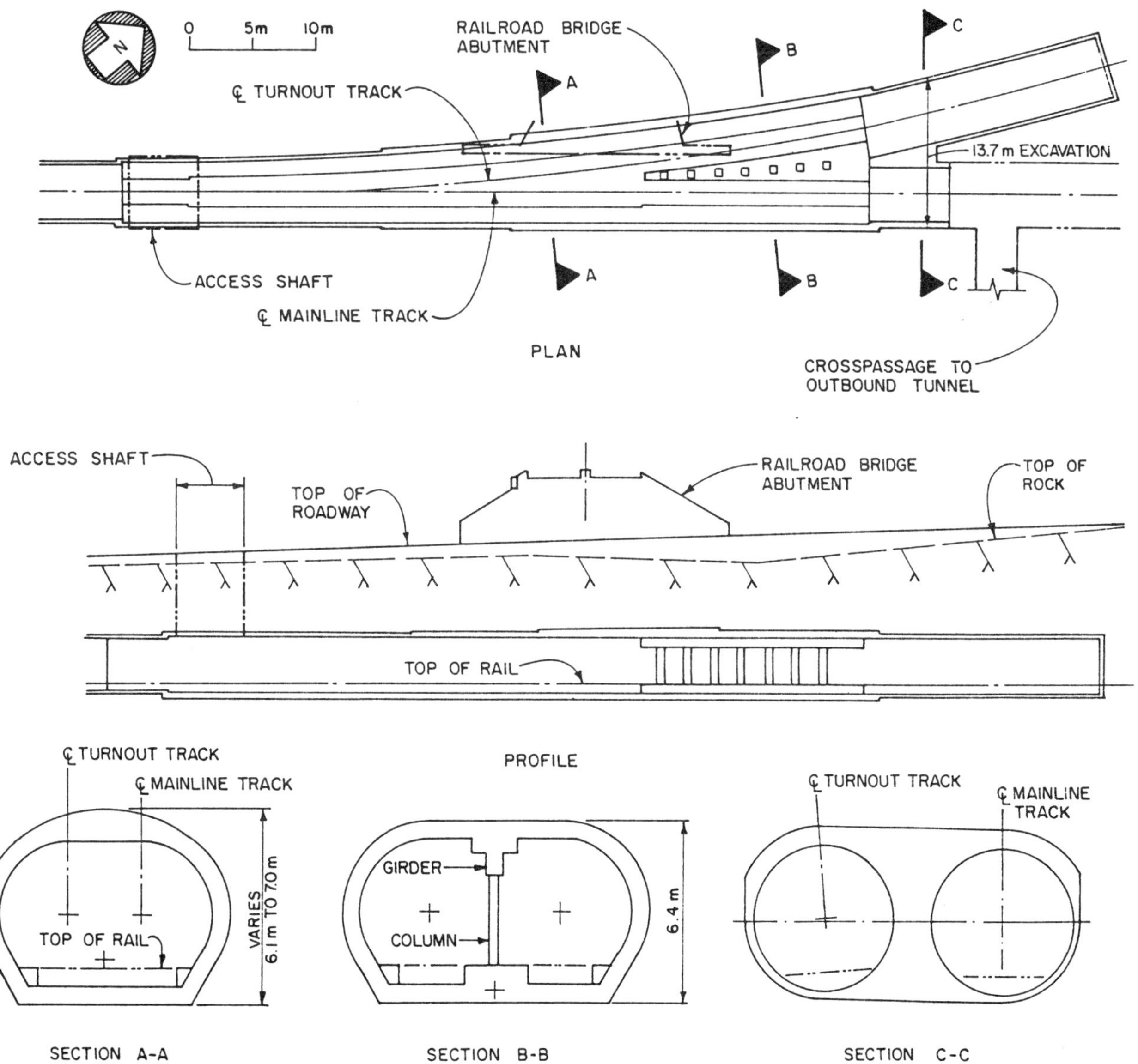

FIG. 11 TONAWANDA INBOUND TURNOUT

began simultaneously excavating a pilot heading for the stub tunnel, driving the turnout enlargement from both the north and south ends, and excavating the crosspassage at the north end of the turnout. As the tunnel enlargement proceeded, blasting vibration and overpressure were monitored for each blast. Portable instruments were set up at varying locations around the construction site. Peak particle velocity was held well within the 50 mm/sec specification requirement. A maximum velocity of 19 mm/sec was measured at the bridge abutment and occurred during the side slashing of the turnout enlargement approximately 21 m to the north.

When excavation was well advanced beneath the bridge abutment, optical survey of surface settlement points indicated that both the bridge abutment and the adjacent surface of Main Street were settling. On November 9, 1983, it was realized that the south end of the bridge abutment had settled by 5 mm, the north end of the bridge by 14 mm, and that the road surface overlying the widest part of the excavation had settled a full 56 mm. In addition, extensometers indicated that the roof of the excavation had settled up to 9 mm relative to the road surface. Close examination of the exposed tunnel crown revealed little sign of distress in the rock forming the roof of the excavation, and much re-checking of the surface settlement points was required before all parties could be persuaded that there was a problem.

However, at a site meeting on November 15, 1983, all excavation in the inbound turnout was halted when settlement of the bridge abutment was determined to be 6 mm and 18 mm at the south and north ends respectively. At the same time, maximum surface settlement in the roadway was measured at 65 mm, and with extensometers indicating up to 9 mm settlement of the excavation roof relative to the surface, this would imply that total settlement of the excavation roof had reached 74 mm. Analysis of the data indicated that for the two weeks prior to stopping blasting, the approximately 7.5 m thick rock mass forming the roof of the excavation had been settling at an average rate of 2 mm per day. All settlement ceased within 24 hours of the halt to blasting, during which period an additional 3 mm settlement was accumulated.

As shown in Figure 12, a series of near-vertical joints, running roughly parallel to one another from the west side of the excavation to the crosspassage north of the turnout, at an acute angle to the line of the excavation, appeared to isolate a large mass of rock. It was realized that this rock mass was forced by the angle of the joints to span a distance of 18 to 26 m rather than the 13.7 m maximum excavated width (measured perpendicular to the centreline of the mainline track) considered during the turnout design. It was also noted that the contractor had experinced large areas of overbreak along the west wall of the turnout enlargement, extending from the invert to the crown. This overbreak occurred along a series of steps along several near-vertical and deeply-weathered joints that angled up to the west away from the excavation.

Study of surface outcrops and of the tunnel exposures had shown that it was most uncommon for a vertical joint to extend continuously for any great depth through the different rock strata. The vertical joints would normally stop at some more pronounced horizontal fracture, sometimes to continue offset by a metre or so to one side. It was considered that this was probably still the case at the turnout location, but that deep weathering of a number of such parallel vertical joints, continuously stepped all the way to the surface, had allowed the gross movement of the tunnel roof with its long skew span. A further contributing factor may have been delay in installation of the rock bolts. Because of his excavation sequence and equipment arrangement, the contractor often did not install rock bolts until 24 hours after exposing the tunnel crown, a period well in excess of the specified 8 hour limit.

At the time excavation was halted, the turnout enlargement had proceeded from the north and south ends to a point north of the bridge abutment overhead. A rock protrusion approximately 6 m long extending from the west tunnel wall to the tunnel centreline still remained to be excavated. This protrusion formed a seat to the southwest end of the settling rock mass, and it's removal would significantly increase the unsupported length of the section of tunnel roof

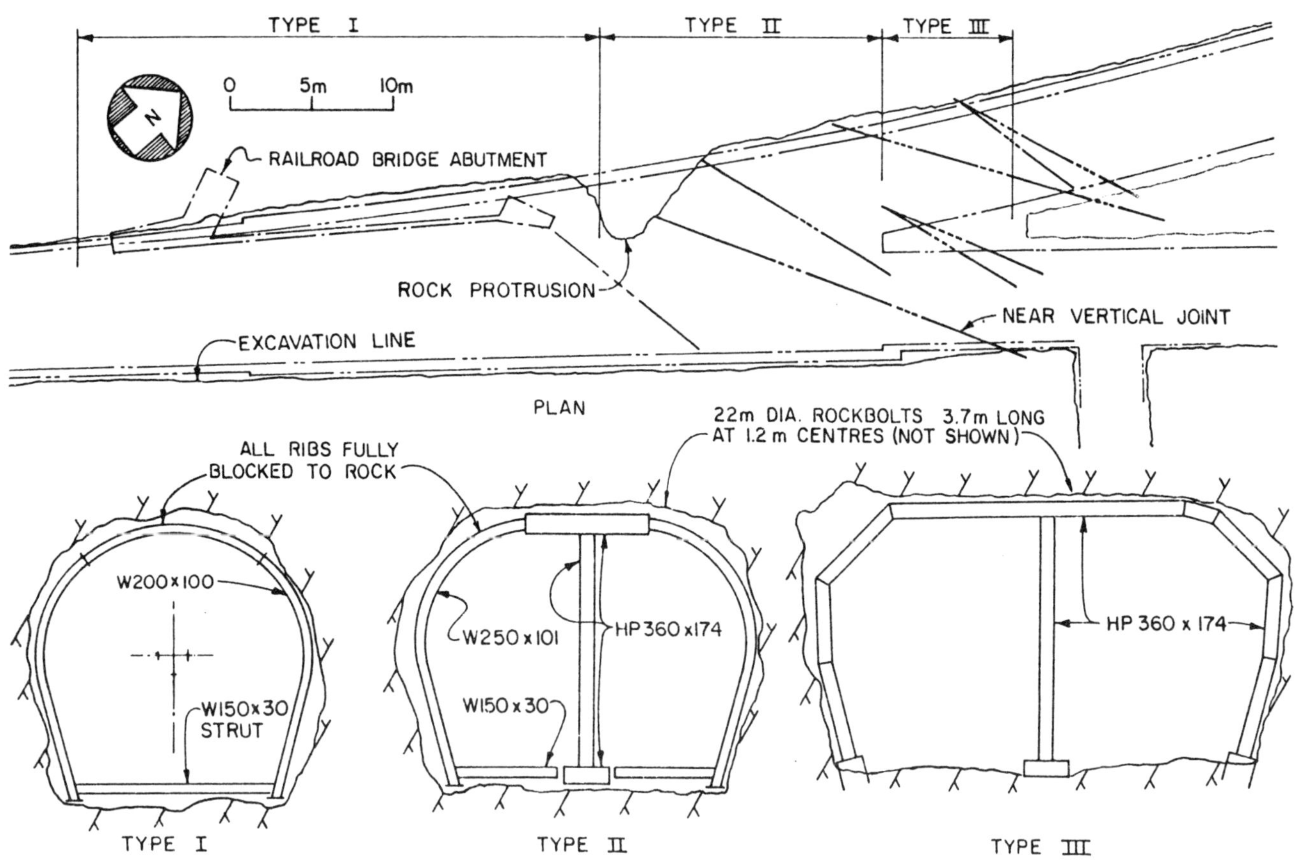

FIG. 12 TONAWANDA INBOUND TURNOUT - EXCAVATION AND PRIMARY SUPPORTS

between the parallel vertical joints. The decision was made to install steel rib supports from a point south of the bridge abutment to the rock nosing at the north end of the turnout enlargement.

To provide immediate support to the large flat roof at the north end of the turnout, the Contractor fabricated T-shaped supports from HP 360 x 174 pile sections stockpiled in his yard. The arm of the T was cut to length and butt plates welded to the ends. It was then hung from rock bolts close to the tunnel roof. A 1.2 m long x 0.9 m wide x 50 mm thick base plate was then bolted and grouted to the tunnel invert beneath. The wide base assembly was used to distribute the vertical load to prevent the column from punching into the weak underlying Camillus Shale. A column welded to a short base section was installed on top of the plate and the header beam was lowered and bolted to the top of the column. Four of the T supports were installed in the concrete nosing between the mainline and turnout tunnels at the north end of the turnout, with six additional T sections installed and eventually encased in the concrete columns in the flat roof portion of the turnout enlargement. For the four sets in the concrete nosing, side posts fabricated in a rough arch shape were bolted to the butt plates on either side of the header beams, and were anchored at their bases with rock bolts and grout. The remaining six T sections were completed by installing W 250 x 101 ribs rolled to match the radius of the concrete tunnel liner. For the arch shaped excavation underneath the bridge abutment, twenty-six W 200 x 100 ribs rolled to match the radius of the concrete tunnel liner were installed at spacings varying from 0.9 to 1.5 m. Because the rock at the base of the west tunnel wall was very thinnly bedded, it was felt that the Contractor would experience difficulty in pinning the bases of the ribs back with rock bolts. For this reason, invert struts were installed to provide lateral restraint to the bases of the ribs. All of the rib and T sections were fully blocked to rock with either shotcrete or with cement grout. Figure 12 shows the various types of rib supports.

Shotcrete was used extensively throughout the

remedial work. It was used initially to strengthen the rock protrusion that was temporarily providing support to the southwest end of the settling rock mass. This additionally protected the workmen from falling blocks, and also provided a means of monitoring any movement of the protrusion which would become immediately apparent by cracking of the shotcrete. No such cracking was observed, and the protrusion remained in place until excavation was resumed. The arms of the first T-sections installed were blocked to rock with high strength grout pumped into plywood forms. Once the shotcrete equipment and crew were mobilized, shotcrete was used almost exclusively to block the T-sections and rib sets. Shotcrete was also used to strengthen the rock nosing between the mainline and stub tunnels at the north end of the turnout, damaged by the stub tunnel blasting operations.

As many rib supports and T-sections as possible were erected and blocked prior to resuming rock excavation. The rock protrusion was excavated mechanically using a back hoe mounted hydraulic ram. Once the enlargement excavation was complete, the remaining supports were installed and blocked. As the excavation of the stub tunnel and crosspassage proceeded, loads on two of the columns at the north end of the turnout were monitored by wire strain gages installed on their webs. Upon completion of the stub tunnel and crosspassage excavation, the strain gages indicated that the columns had been loaded to approximately 1300 kN, which represents about 50% of their design capacity in axial compression. The stub tunnel and crosspassage excavations were completed with very little additional settlement of the tunnel crown, bridge abutment, or the surface above the excavation. All excavation and tunnel lining was completed without further difficulty.

Outbound Turnout Construction

As shown in Figure 10, the outbound turnout was also situated in a built-up area, with many small commercial and residential properties in close proximity to the excavation. A city school building was located within 40 m of the construction site, and public and private utility services pass directly over the turnout structure. Once again, great care was taken to control and monitor blasting vibrations and overpressures. A maximum peak particle velocity of 28 mm/sec was recorded at the school building, occuring during side slashing of the tunnel enlargement.

Access to the running tunnels was provided initially through the vertical shaft at the south end of the inbound turnout. It was thought during design that the crosspassage at the north end of the inbound turnout, if driven at the start of construction, could be used for access to the outbound turnout area. Rather than move materials, equipment and personnel through the existing shaft, the inbound turnout and crosspassage, and then 470 m down the outbound mainline tunnel, the contractor elected to sink an access shaft at the north end of the outbound turnout.

After sinking the shaft and punching into the mainline tunnel, the existing concrete lining was removed and the west invert excavated. The bulk of the tunnel enlargement was then excavated by side slashing into the mainline tunnel, working from the shaft to the south end of the turnout. 2.4 m long rounds with 20 kg per delay were employed, with smooth wall blasting techniques used in all perimeter holes around the final excavated shape. The short stub tunnel was then driven to the north. Throughout these excavations, timely installation of the mandatory rock bolts was rigorously enforced.

With only 5.5 m of rock cover over the tunnel excavation, which reached 14.7 m at its widest point, the outbound Tonawanda turnout was in fact the shallowest of the major tunnel structures on the project. It should be realized, however, that at its widest point the roof was supported to the north by the rock nosing between the diverging tunnels, and that the width of the excavation narrowed rapidly to the south as the tracks merged. As some settlement was considered possible, tunnelling at this depth would not have been attempted if it had not been possible to locate the excavation clear of existing buildings.

The deeply-weathered, near-vertical joints which caused the difficulties at the inbound turnout were not found at this location. However, the borehole logs did indicate thinly-bedded rock overlying the tunnels, with some

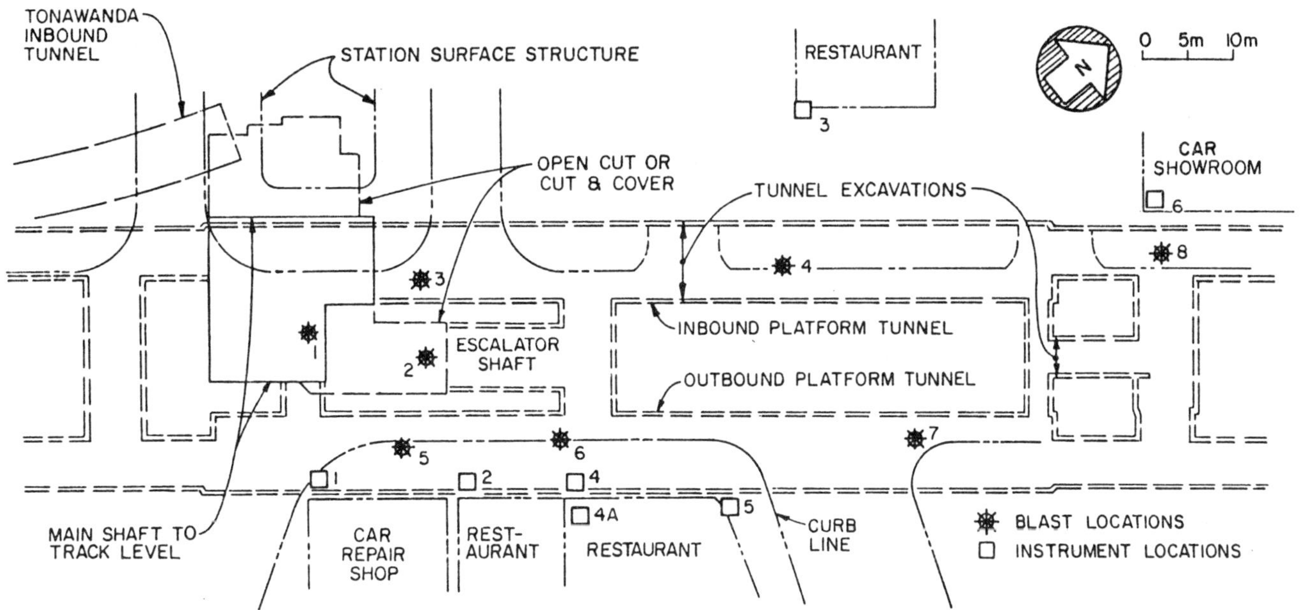

FIG. 13 LA SALLE STATION: INSTRUMENT & BLAST LOCATIONS

RQDs in the 60% range, and with weathering along many of the horizontal bedding joints. Surface settlements were mostly in the 10 mm range, but a maximum surface settlement of 27 mm was recorded over the widest span. Maximum settlement of the tunnel crown relative to the road surface at this location was indicated by an extensometer to be 3 mm, to give a maximum settlement of 30 mm at tunnel level. No damage was sustained by any utilities or structures as a result of this settlement, and all excavation and tunnel lining at the outbound turnout was completed without serious difficulty.

LASALLE STATION CONSTRUCTION

With total cover from road surface to crown of station platform tunnel excavation at one point only 9.4 m thick, LaSalle was the shallowest of the four tunnelled stations. With commercial and residential properties lining Main Street through the area, it thus provided one of the greatest challenges with respect to control of blasting vibration.

As shown in Figure 13, the station surface structures were situated on the west side of Main Street, with commercial and residential properties in close proximity to the construction site. Buildings housing a car repair shop and several restaurants were located on the east side of Main Street, within 1.0 m horizontally and 10.0 m vertically of the outbound platform tunnel. At the north end of the station, a car sales showroom fronted by large plate glass windows (removed during blasting) was immediately adjacent to the north vent shaft excavation. To the south, the inbound Tonawanda Turnout enlargement was still being concreted at the start of the LaSalle Station contract, and normal operations continued at the vibration sensitive gold refinery.

The general sequence of excavation and specified restrictions on blasting were the same as at the other tunnelled stations. As was the case at all the stations, vibration due to blasting was monitored at structures adjacent to the mining operations. Some of the monitoring locations are shown on Figure 13. A summary of maximum peak particle velocities recorded at these locations appears in Table 2.

The Contractor began his mining operation by sinking the 17.6 m by 12.6 m main shaft to the tunnel level. As the shaft excavation proceeded, vibration levels were monitored on the sidewalk fronting the buildings on the east side of Main Street. A seismograph was also set up on the tunnel invert in the Tonawanda Turnout stub tunnel, to measure vibration levels on the newly cast concrete tunnel liner. The existing pavement and overburden were stripped to top of rock and the weathered layers of cap rock were removed by mechanical means. Perimeter holes on 300 mm centres were then drilled for the full depth of the shaft using track mounted percussion drills. These holes were then filled with pea gravel to the bottom of the first lift level. 2.4 m deep production holes were drilled individually for each lift. The explosive charge in the first round at any lift level was held to 2.2 kg per delay and 46.2 kg total, with four 102 mm burn holes used for relief. Subsequent shots in the lifts, with a maximum charge of 6.6 kg per delay and 185 kg total, were designed to detonate into the relief space provided by areas excavated by previous rounds.

After completing the shaft, a 4.4 m wide by 5.4 m high crosspassage was driven from the shaft to the outbound TBM tunnel. No difficulty was experienced in limiting vibration at the adjacent structures to the specified 50 mm/sec peak particle velocity during excavation of the shaft and crosspassage. Upon gaining access to the outbound line tunnel, the Contractor prepared to slash the platform tunnel enlargement in both the north and south directions. To limit overbreak and damage to the rock adjacent to the excavation, smooth wall blasting techniques were employed for each blast. Perimeter holes were loaded with light distributed charges, and as many as possible were shot simultaneously to remove the final burden. The detonation sequence was designed to push succeeding layers of rock into the relief space provided by the existing TBM excavation. Typical blasting patterns used for the outbound tunnel enlargement are shown in Figure 14.

Due to the close proximity of the buildings on the east side of Main Street, difficulty in limiting blasting vibrations to the specified maximum levels was experienced immediately. As the tunnel enlargement blasting proceeded to the north, attempts to reduce vibration by altering the blast design met with only limited success. Even though the amount of explosive per delay was reduced by decreasing the length of the round and increasing the number of delays per round, vibrations continued to be significantly above the contractural limits. It was theorized that the tolerance on the timing of the non-electric blasting caps, as reported by Winzer,[3] was causing near-simultaneous detonation of holes on closely spaced delays. To overcome this problem, subsequent blasts in this area were initiated with electric blasting caps and an electric blasting machine. It was also considered that the existing rock bolts that had been installed during TBM excavation increased rock confinement at the tunnel crown. This artificially-created confinement would tend to increase blasting induced ground motion in the surrounding rock. To reduce this effect, additional (unloaded) line

Table 2. LaSalle Station: Peak Particle Velocities at Varying Distances from the Blasts.

Shot No.	Blast Description	Instrument Location	Distance From Blast (Metres)	Peak Particle Velocity (mm/sec.)
1	Main Shast: Production Shot	1	17	22
2	Escalator Shaft: Production Shot	2	15	32
3	I/B Platform Tunnel: Production Shot	3	47	13
4	I/B Platform Tunnel: Production Shot	3	17	24
5	O/B Platform Tunnel: Production Shot	2	14	83
6	O/B Platform Tunnel: Production Shot (Sidewalk)	4	5	75
6	O/B Platform Tunnel: Production Shot	4A	5	30
7	O/B Platform Tunnel: Production Shot	5	24	27
8	North Vent Shaft: Burnhole Raise	6	5	83

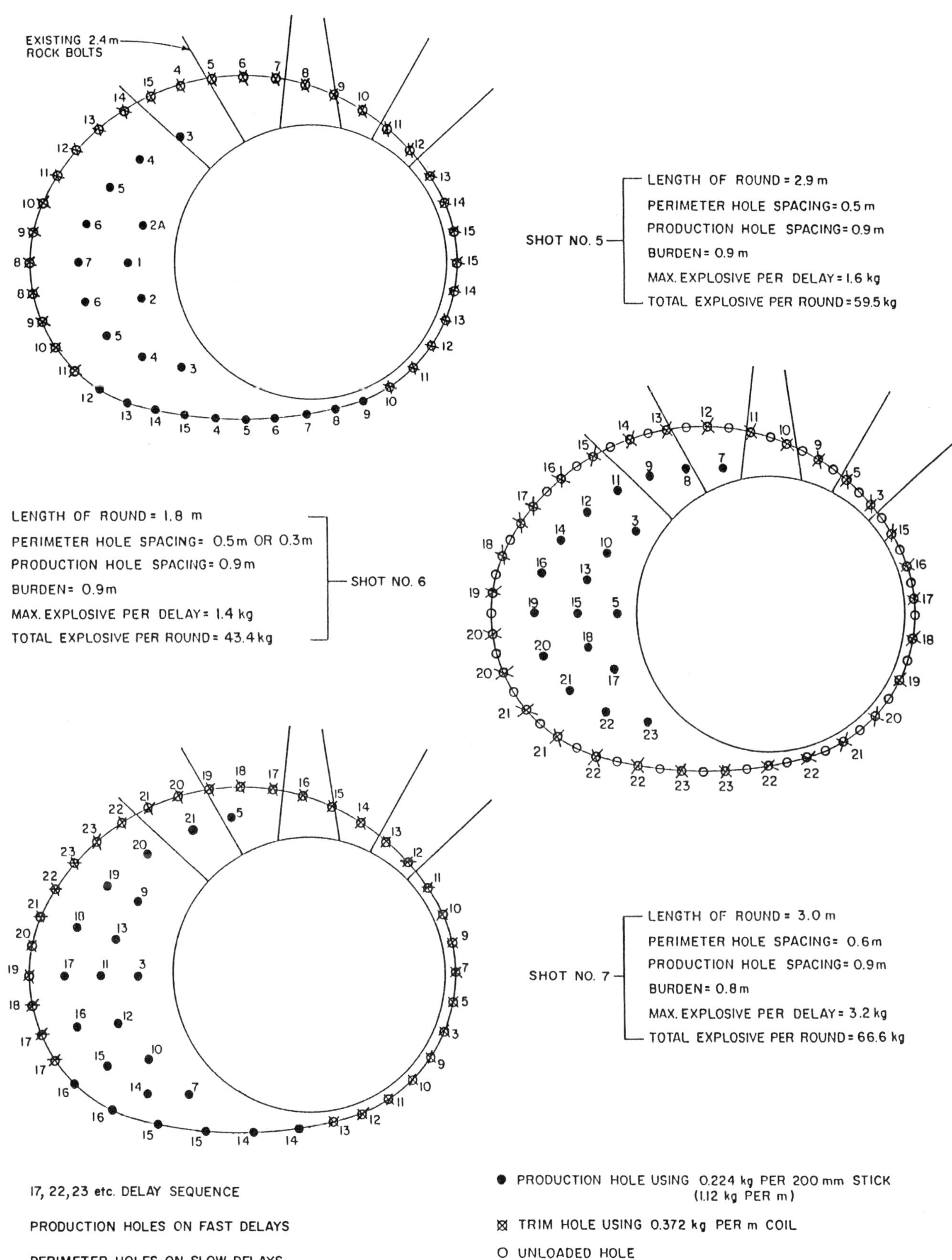

FIG. 14 LA SALLE STATION BLASTING PATTERNS

holes were drilled in the area of the rock bolts. Also, the length of each round was further reduced.

Although the above measures reduced the vibrations of subsequent shots, peak particle velocities remained significantly above the 50 mm/sec limit. An examination of the seismograph records indicated that the buildings adjacent to the monitoring stations were being subjected to ground motions at high frequencies. A careful inspection of the masonry and steel frame structures proved that they had already tolerated vibrations in excess of the limits specified without ill effects and in all probability would continue to do so. To avoid further reductions in blasting productivity, it was decided to continue with the current blast design and carefully monitor the buildings. A seismograph was used to monitor vibrations on the roof of one of the buildings for each blast. In addition, frequent damage inspections of the building were made. Blasting continued at a reduced level until the distance from the working face to the buildings was sufficient to limit vibrations to the specified level. A plot of measured vibration levels versus distance from the monitoring point fronting the northern-most building is shown in Figure 15. This clearly shows the marked reduction in peak particle velocity with the small increase in distance from 6 to 9 m.

Completion of the remaining LaSalle Station excavations was without incident, and brought rock excavation on this present phase of the Buffalo transit project to a successful conclusion.

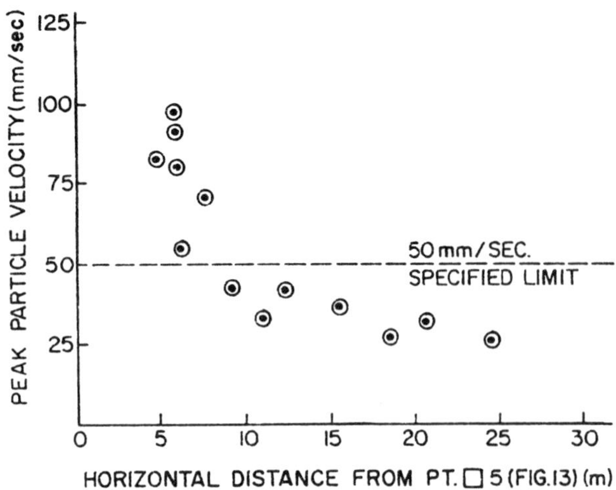

FIG. 15 BLASTING VIBRATION LEVELS AT LA SALLE

Acknowledgement

The Authors wish to thank the Niagara Frontier Transportation Authority for permission to publish this paper.

Owner: Niagara Frontier Transportation Authority.

Designers & Construction Managers for the rock tunnels: Transit & Tunnel Consultants, Inc. (T&TC), a joint venture of Mott, Hay & Anderson, London, and Hatch Associates, Buffalo.

Special Consultants to T&TC for rock mechanics and blasting technology: Charles R. Nelson and Associates, Minneapolis, Minnesota.

General Soils Consultant to the owner: Goldberg-Zoino & Associates of New York, Buffalo.

References

1. J.D. Guertin and R.F. Flanagan, "Effect of artesian aquifer on feasibility of Buffalo LRRT project", Proceedings of Tunneling '82, England.

2. J.G. Ball, K.W. Torpey, D.S. McAllister & R.F. Flanagan, "Rock Tunneling for a rail transit system, Buffalo, New York", RETC Proceedings 1981, San Francisco.

3. S.R. Winzer, Martin Marietta Laboratories, "The firing times of MS delay blasting caps and their effect on blasting performance", prepared for the National Science Foundation, Washington, D.C., June 12, 1978.

Use of probabilistic stability analysis and cautious blast design for an urban excavation

A. McCracken B.SC., C.ENG., M.I.C.E., M.I.M.M., M.A.E.G.
G.A. Jones PR.ENG., M.A., F.S.A.I.C.E.
Steffen, Robertson and Kirsten, Inc., Consulting Geotechnical Engineers, Johannesburg, South Africa.

SYNOPSIS

The siting of a large city hospital in a disused quarry in urban surroundings required a number of problems related to geotechnics to be addressed. The architects' plans included: utilising the existing quarry face as a natural feature of the site; excavating a deep slot through a quartzite ridge to allow connecting access to existing hospital facilities; and excavating a multi-level car park in the back of the quarry and flank of the ridge. Investigations were carried out to determine the geometry of the quarry floor and to obtain data for an assessment of excavatability, stability, blasting design and rock support requirements.

The systematic data processing and stability analyses undertaken utilised a suite of Steffen, Robertson and Kirsten (SRK) computer programs, JOINTS, which handles the raw data and evaluates stability by probabilistic methods. The acceptability of the operative stability of the natural face and excavations in terms of public utility and safety during construction was assessed. The level and nature of the support for acceptable stability during and after the construction and excavation phases was determined on this basis.

The urban environment, in particular the proximity to existing hospital facilities, the close engineering tolerances for construction and stability considerations, required that strictly controlled blasting practices should be adopted. A cautious blasting design, based on the geotechnical conditions and site constraints, was prepared utilising the SRK program BLASTS. The blast design criteria and final design are discussed in some detail together with the excavation procedures, support requirements and scheduling, and the supervision requirements of work of this nature.

The site of the proposed Pretoria Academic Hospital is situated only two kilometres from the centre of the capital. The new building will be some 10 levels high and will mainly be situated in the excavations of an old quartzite quarry. Existing hospital buildings will be separated from the new structure by a prominent quartzite ridge. An areal view of the site is shown in Fig. 1.

The architects have attempted to integrate the building aesthetically with the natural features of the site. In this regard, it is planned that the new and existing buildings will be linked by an elevated corridorway. This corridorway passes through the crest of the ridge and requires an excavation 80 metres long by 7 metres wide and a maximum of 20 metres high.

The existing quarry face, which is up to 40 metres high, is to be retained as a natural slope to the rear of the main hospital block, separated from the building by an access road and walkway. Parking facilities will be made available by excavating a portion of the back of the quarry and the flank of the quartzite

Fig. 1 Aerial view of proposed hospital site

ridge. Figs. 2 and 3 diagrammatically illustrate the proposed building plan with these aspects highlighted.

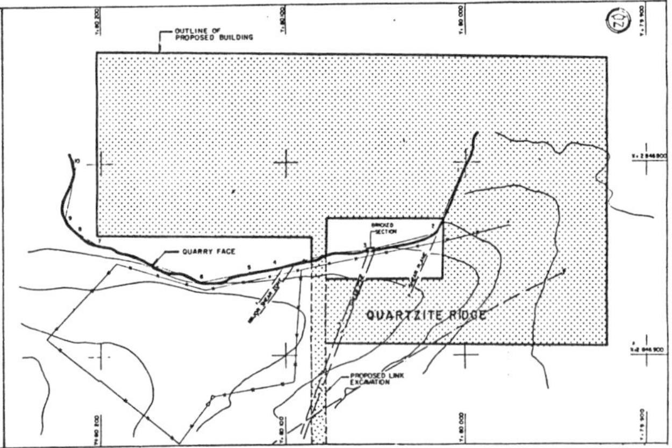

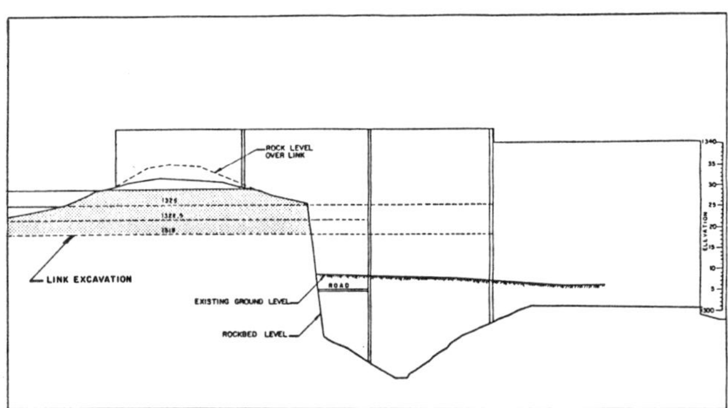

Figs. 2 & 3 Proposed building plans

The nature of the site regarding final utilisation and public liability obviously requires that the stability of the existing and excavated faces is ensured absolutely. In addition to stability requirements, the urban nature of the site and its environmental sensitivity necessitated that strict excavation procedures be developed.

POTENTIAL ROCK RELATED PROBLEMS

Potential rock engineering problems for the proposed hospital are connected with quarry geometry, the existing quarry face and the proposed excavations for the link corridor and car park.

Quarry geometry

The geometry of the quarry floor was largely unknown due to the lack of plans and it being used as a dump site. To determine the bedrock profile, a gravimetric survey was conducted. A programme of borehole drilling further identified the fill materials and quarry floor elevations. These provided the basis of the foundation investigation for the hospital buildings and infrastructure.

Existing quarry face

The quarry face is steeply dipping to sub-vertical with minor local overhangs. Moderately good conventional blasting methods were used to quarry the quartzite. The face is generally unbenched and therefore blocks of ground falling from the slope would not be caught. The stability analysis and resulting support of the face had to be designed to an acceptable level of risk due to the free vehicular and pedestrian access immediately below the slope face.

Planned excavations

Initially, a tunnel was considered for the pedestrian access through the ridge to the existing hospital buildings. This option was rejected in favour of an open cut, or slot, due to the limited thickness of roof cover which would remain. The rock mechanics problems related to the link slot were in respect of blasting, overbreak and stability.

Blasting in an urban environment places constraints on the operations. Flyrock and vibration levels must be controlled whilst still achieving satisfactory fragmentation. Noise must also be suppressed.

Overbreak and blast damage of the final cut faces had to be minimised due to the engineering requirements for construction. In the slot, the final faces had to be smooth blasted to a close distance behind the finished wall line to minimise making up with concrete and to maintain the inherent strength of the rock. In the excavations for the car park, it was intended that the rock faces would form the final interior walls.

Stability of the excavations, and especially of the link slot, had to be ensured during excavation and construction phases and in the long term. The support design also had to prevent convergence of the slot walls due to horizontal stresses released by the excavation.

ROCK MASS CONDITIONS

A programme of detailed structural mapping by line and cell[1] methods to collect rock joint information was conducted. This covered the foot of the entire quarry face and the exposures on the flanks and crest of the ridge. Together with the information from exploration drilling, this formed the major source of data. In situ and laboratory testing provided material strength properties.

General geology

Fig. 4 shows a geological section through the link slot. The major excavations will be cut in the quartzites of the Daspoort Formation. These are generally slightly weathered to fresh fine to medium grained thickly bedded, moderate to wide bedded very strong rock quartzites. Highly to completely weathered laminated and closely fractured very weak rock shales underlie the quartzites. A decomposed diabase sill locally underlies the quartzite and intrudes the shales. This material was recovered as a firm to stiff sandy clay.

The excavations in the shales and diabase

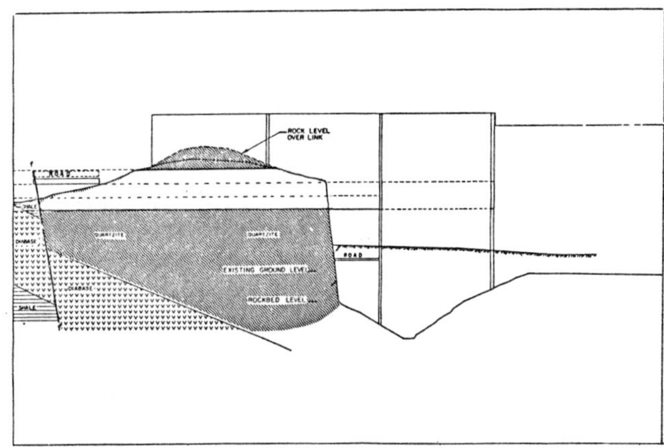

Fig. 4 Geological cross section

will be shallow. A stable slope angle was the only input required to excavation design in these materials.

Rock strength

Intact strength

Intact rock strength from point load testing yielded an equivalent mean uniaxial compressive strength of approximately 250 MPa, ie extremely high strength, with a range of 30 MPa to greater than 300 MPa. Such consistent high strengths indicate that failure would be discontinuity controlled.

Joint-compressive strength

Schmidt hammer tests yielded a distribution of joint compressive strength as follows: mean:100 MPa; range:46-209 MPa; standard deviation 35 MPa.

Joint shear strength

The program JOINT2 determined the peak shear strengths of the discontinuities based on the method of Barton and Choubey[2]. Values were obtained using effective normal stresses of 50, 100 and 200 kPa, this being approximately equivalent to vertical thicknesses of rock of 2, 4 and 8 metres. This was modelling the shear strength components acting near the surface of the face rather than for a deep-seated failure. Table 1 shows the resulting distribution of shear strength parameters.

Geological structure

Major structural features and rock mass fabric were identified from the mapping.

Major features

A northeast-southwest trending fault bounds the southern flank of the ridge. This down throws the quartzite against the shale. Major shear zones trending NNE-SSW are present and obvious

Normal Stress	Shear Strength Parameters				Units
		Mean	Range	SD	
50 kPa	τ	102	36 -318	51	kPa
	ϕ	55	35,3-64,1	7,6	degrees
	c	28	1 -251	43	kPa
100 kPa	τ	174	72 -408	56	kPa
	ϕ	52,7	35,4-64,1	7,1	degrees
	c	37	2 -215	38	kPa
200 kPa	τ	302	140 -600	94	kPa
	ϕ	50,1	34,1 -60,6	6,3	degrees
	c	57	5 -240	44	kPa

TABLE 1 : Discontinuity shear strengths

in the quarry face. Two such features, 3 metres and 5 metres wide, appear to have caused some stability problems in the face. Sandy and silty gouge are associated with the shears. These features can be seen in Fig. 5.

Fig. 5 View of quarry face showing shear zones

The major significance of the larger of the shear zones was its proximity and orientation relative to the proposed excavation for the link corridor.

The dominant discontinuity in the quartzites is the bedding. The strike and dip continuity of the beds is in excess of 50 metres. This, together with the northerly dip and east-west configuration of the quarry face made planar failure of blocks of ground very possible. Longitudinal and cross joints were also persistent within the bedding units and, together, defined a blocky rock mass. Shear joints subparallel to the major features were also present. Stereoplots from the program JOINT4 for the joint data are shown in Fig. 6 indicating the joint sets as described above.

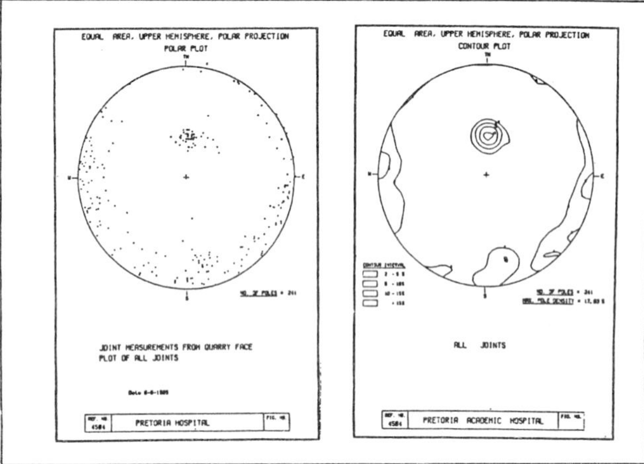

Fig. 6 Stereoplots of mapped discontinuities

DISCONTINUITY DATA PROCESSING

The discontinuity data were handled using the suite of computer programs developed at SRK called JOINTS[3]. These utilise a probabilistic approach which numerically models the rock mass properties for subsequent input to specific analytical and design programs. Through these programs the engineering properties are evaluated in terms of the statistical parameters of their conditions of structure, strength and hydrogeology. By using a probabilistic approach, the inherent variability of the rock mass is assessed which then provides a rational basis on which to undertake engineering design.

The programs which comprise the joint analysis system and their interrelationships are shown in Fig. 7. The programs used in this project are shown shaded. Brief descriptions of these programs are given where relevant.

STABILITY ANALYSES

The stability analyses undertaken comprised kinematic analysis, deterministic analysis and probabilistic analysis.

Kinematic analysis

Kinematic analysis established the possible modes of failure with respect to the rock mass structure and the geometry of the proposed excavations and existing quarry face. Fig. 8

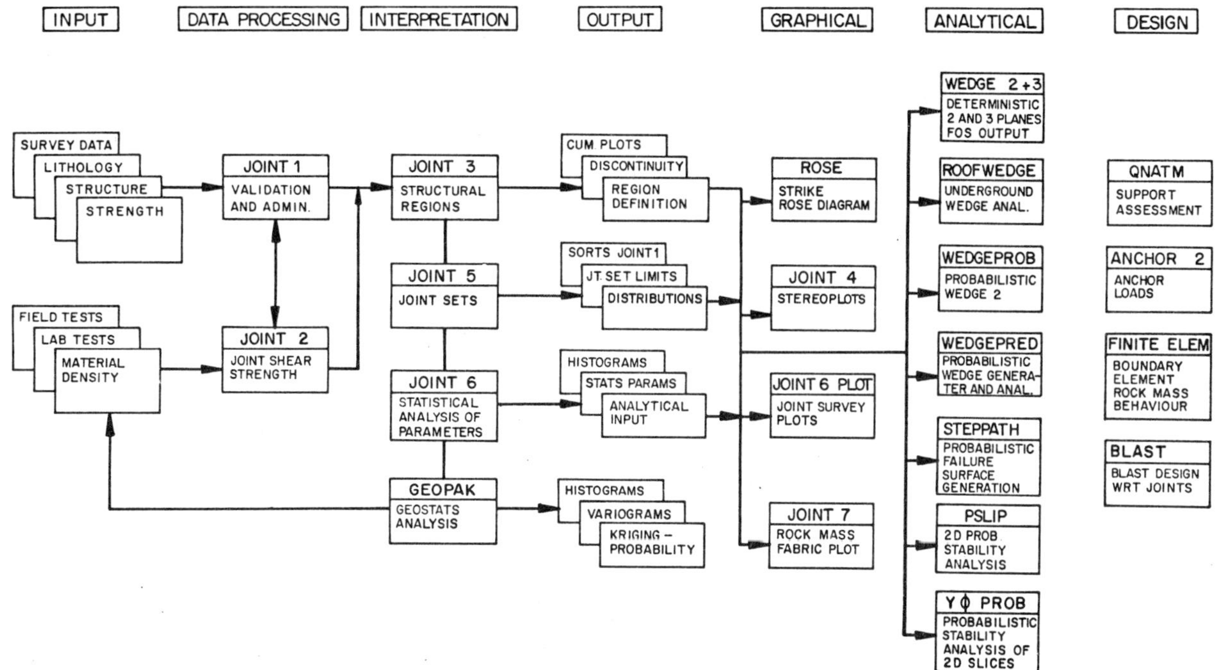

Fig. 7 Joint analysis system flowchart

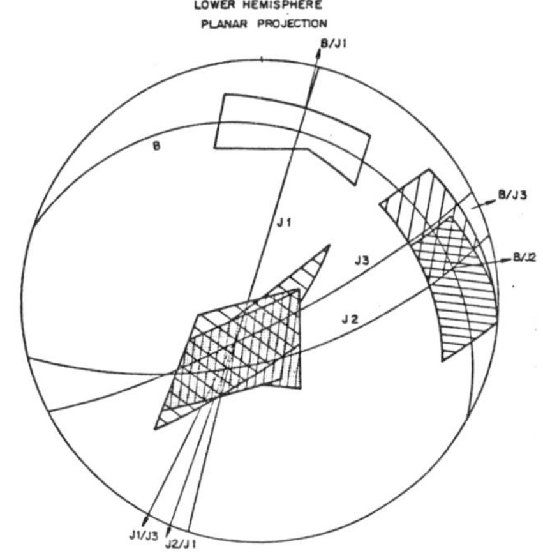

Fig. 8 Kinematic stability analysis

is a stereographic representation of the kinematics of potential failure wedges and planes. The mean plus and minus one standard deviation of dip and dip direction for each discontinuity set was used. The planar projection of the mean orientation of each discontinuity set, together with the areas of intersection and directions of sliding of failure wedges, are shown.

Due to the shape of the proposed building, the excavation geometries are largely limited to north-south cuts, ie east and west facing walls such as the corridor slot and major parking garage cuts; and east-west cuts; limited to the north facing slopes of the main quarry face and minor parking cuts. The slopes were taken as being vertical, as will be the case in the proposed rock excavations and as is the worst case for the quarry face. Kinematically possible failures were defined as follows:

North facing cuts and existing quarry face:

Bedding plane failures

East facing cuts:

Wedge failures - Bedding and joint set 2; Bedding and joint set J1 and J2; J1 and J3

West facing cuts:

Wedge failures involving J1 and J2 and/or J2 and J3 joints

Deterministic analysis

Bedding plane failure was initially analysed using conventional limiting equilibrium techniques to determine a factor of safety. The stability of a block is the ratio of the resisting and disturbing forces.

Assuming zero cohesion and no water pressures, the factor of safety is a function of the bedding angle, α, and the friction angle, $\emptyset$, in the equation :

$$FOS = \frac{R}{D} = \frac{W\cos\alpha\tan\emptyset}{W\sin\alpha}$$

In classical deterministic analyses, unique values of the input parameters, usually the mean or a more conservative figure, are used. Using the mean values for the quartzite rock mass, the central factor of safety against bedding plane failure of a 2 metre thick block was calculated at 2.53. Assuming a disturbing force of the mean plus one standard deviation and a resistivity force of the mean minus one standard deviation this still gave a factor of safety of 1.45. These figures would be considered acceptable in most engineering calculations. These unique figures, however, do not take account of the actual variability of disturbing and resisting forces. Certain lower bound conditions of friction angle and upper bound values of bedding angle may occur together and lead to failure. In the situation where free public access is allowed below a slope face, the probability of occurrence of this situation must be determined and the acceptability of this risk assessed.

Probabilistic analysis

To quantify the risk of failure, probabilistic analyses were undertaken for bedding plane failure and the wedge failures indicated to be kinematically possible.

Bedding plane failure

The bedding plane failure can be regarded as a classic capacity demand type model[4]. The shearing resistance of the failure surfaces is the capacity and the disturbing forces due to the weight of the potential failure block and the angle of the plane on which it rests is the demand.

From the histograms for dip angle and peak friction angle standard normal distributions of the resisting and disturbing forces as shown in Fig. 9 were derived. The probability of failure, ie the probability that the demand exceeds the capacity, is a function of the amount by which the upper tail of the demand distribution, D, exceeds the lower tail of the capacity distribution, C. The value of the probability of failure, P(f), is a function of the statistical parameters defining the two distributions. For the quarry face:

$$P(f) = 0.0034 \text{ or } 0.34\%$$

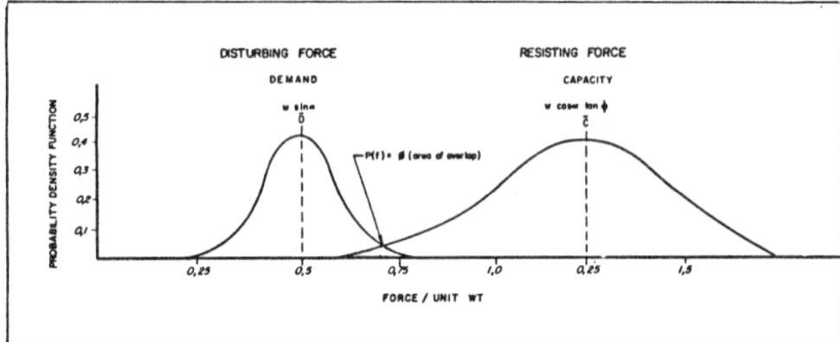

Fig. 9 Capacity - Demand Model for plane failure

The acceptability of a particular value of probability of failure depends upon the function, service life, public access, size and monitoring of the excavation. In the case of the quarry face being associated with the hospital access road, the strictest criteria have to be applied. Table 2, compiled from the works of various authors[5,6,7], presents the comparative significance of probabilities of failure for various excavations. It shows that for unmonitored permanent slopes with free public access, probabilities of failure of less than 0.5% have been shown to be acceptable. Thus, the derived probability of bedding plane failure of 0.34% was acceptable.

Prob of Failure (%)	Design criteria for acceptable probability of failure		
	Serviceable Life	Public Access	Surveillance Required
50-100	Effectively zero	Forbidden	Ineffective
20-50	Very very short term (temporary open-pit mines - untenable risk of failure in temporary civil works)	Forcibly prevented	Continuous sophisticated monitoring
10-20	Very short term (quasi-temporary slopes in open-pit mines - undesirable risk of failure in quasi-temporary civil works)	Actively prevente	Continuous monitoring with instruments
5-10	Short term (semi-temporary slopes in open-pit mines, quarries or civil works)	Prevented	Continuous simple monitoring
1,5-5	Medium term (semi-permanent slopes)	Discouraged	Conscious superficial monitoring
0,5-1,5	Long term (quasi-permanent slopes)	Allowed	Incidental superficial monitoring
Less than 0,5	Very long term (permanent slopes)	Free	No monitoring required

TABLE 2 : Significance of Probability of Failure in excavation design

Wedge failure

The computer program WEDGEPRED was used to determine the probabilities of wedge failure of the various combinations of joint sets and excavation geometry which would affect the faces. In this program, the relevant parameters of two discontinuity sets are sampled randomly from the distributions determined from programs JOINT2 and JOINT6. These are then used to model the rock mass behind a specified slope face. Potential failure wedges are identified and analysed deterministically for stability. A probability density function of factor of safety is obtained from which the probability of failure is derived. In addition, the predicted number of failure wedges is given together with weights and geometries for the individual wedges. The results of this analysis for the quarry face and the link excavation are given in Table 3.

Face	Failure Wedge	Prob of Failure P(f) %	Expected Number of failures	Vol of Failures m^3	Avr Height of Wedge m
Quarry face 250m long	$B/J_1, J_2$ or J_3	0,1	(Friction angle exceeds dip)		
	J_1/J_2	0,0	(Intersection angle steeper than slope)		
	J_1/J_3	1,0	1	3,2	9,5
Slot, West face 60m long	$B/J_1, J_2$ or J_3	0,0	(Friction angle exceeds dip)		
	J_1/J_3	1,0	2	2,1	1,4
	J_1/J_2	0,0	(Generally dipping into face)		
Slot, East face 60m long	$B/J_1, J_2$ or J_3	0,0	(Friction angle exceeds dip)		
	J_1/J_3	3,0	6	5,4	2,1
	J_1/J_2	2,0	3	4,1	2,3

TABLE 3 : Probability values and numbers and volumes of wedge failure

The values of probability of failure and numbers and volumes of failures are within acceptable limits during the excavation and construction phase. Some support measures were required, however.

EXCAVATABILITY ASSESSMENT

The excavatability of the quartzites and shales was assessed using the method of Kirsten[8]. This determines an excavation index, N, based on rock mass parameters where:

$$N = M_s \cdot \frac{RQD}{J_n} \cdot J_s \frac{J_r}{J_a}$$

and M_s = rock mass strength number
J_n = joint set number
J_s = ground structure number
J_r = joint roughness number
J_a = joint alteration number

The values of the parameters were based on information from exploratory drilling and field assessments, with the results as shown in Table 4.

Material	Excavatability Mean	Index (N) Range	Class	Excavation by
Weathered shales	9	5- 20	2	Power tools/easy ripping to hard ripping
Quartzites	1525	250-20000	4	Very hard to extremely hard ripping/blasting

TABLE 4 : Excavatability assessment

In the urban environment, it would be an advantage to be able to rip the rock and thereby reduce the requirement for blasting. Although it was possible for some of the quartzites to be ripped, the stability requirements of the excavation had to be recognised. Ripping would lead to loosening of blocks of ground. This would cause stability problems and may have led to over-excavation of the cuttings. Drilling and blasting alone were therefore recommended for the excavation of the quartzites.

BLAST DESIGN CRITERIA

Blasting in an urban environment for civil engineering excavations places constraints regarding noise levels, vibrations and fly rock. In addition, the requirements of the excavation must be considered with special regard to stability considerations and limiting overbreak. The excavations for the corridor slot required most careful consideration.

Both stability and urban blasting criteria can be met through cautious blasting practices. Presplitting or, alternatively, smoothwall or trim blasting, which can be more effective, was recommended to force the final excavation walls. The increased cost of controlled blasting techniques over that of conventional methods is generally more than redressed by reduced support and overbreak make-up costs. The primary aim of the blasting was to maintain the integrity of the rock mass behind the final face as far as possible and especially with respect to the shear zones. Fragmentation was a secondary consideration regarding the size of the excavation and given the urban environment.

In terms of particle velocities, presplitting can, in fact, cause more damage behind the proposed final face than conventional blasting. Controlled trim blasting is therefore advocated.

The blast design utilising the program BLASTS has been compiled by SRK. This utilises the geotechnical parameters of intact rock strength, Young's Modulus, joint orientation, spacing, continuity and condition, and hydrogeology as input, together with excavation dimensions and requirements, and explosive-type parameters and outputs a blast design in terms of bench height, burden, spacing, hole size, powder factor, delays, decoupling, sub drill, preferred free face and initiation sequence. A flow diagram of the input and output of BLASTS is shown in Fig. 10.

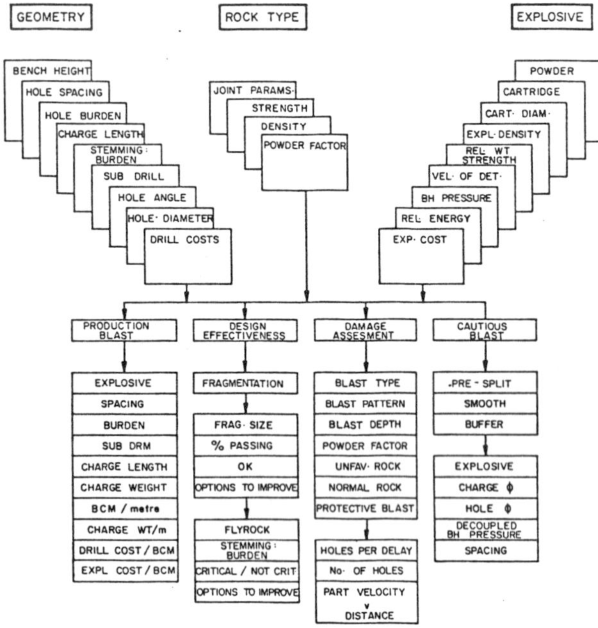

Fig. 10 Blast design program BLASTS - Flowchart

```
==========================================================================
BLAST -- BLASTING DESIGN PACKAGE -- REV 1.10  --  GEMCOM (PTY) LTD.

NAME OF PROJECT : SLOT
==========================================================================
Geometry   record number                                                 2

Geometry description                                          SLOT
Bench height                     (metres)                            4.000
Hole spacing                     (metres)                            1.367
Hole burden                      (metres)                            1.367
Spacing : Burden ratio                                               1.000
Charge length above grade        (metres)                            2.633
Charge length : Bench height ratio                                    .658
Stemming : Burden ratio                                              1.400
Sub-drill : Burden ratio                                              .300
Hole angle                       (degrees)                          90.000
Hole diameter                    (mm)                               50.000
Drilling accuracy                (cms/m)                             2.500
Drilling costs                   (Rand/m)                            5.000
==========================================================================
Rock type  record number                                      .          2

Rock type name                                               QTZ SLOT
Joint friction angle             (Phi + i    degrees)               38.000
Rock factor                                                         12.000
RQD                              (0% - 100% )                       85.000
Compressive strength             (MPa)                             200.000
Density                          (Tons/bcm)                          2.650
Discretionary vibration constant  (try 750 )                       750.000
==========================================================================
Explosives record number                                                 2

Explosives name                                              SINEX SLOT
Cartridge (0)    or   Powder (1)                                         0
Cartridge diameter               (mm)                               38.000
Explosives density               (gm/cm**3)                          1.370
Relative weight strength                                           124.000
Borehole pressure                (MPa)                            5600.000
Explosives cost                  (Rands/kg)                          3.000
==========================================================================


 2.      Smooth blast

SINEX SLOT
Required charge diameter         (mm)                               22.582
Hole diameter                    (mm)                               50.000
Decoupled borehole pressure      (MPa)                             831.207
Hole spacing                     (m)                                 1.089

This spacing should be less than 2 * critical joint spacing
==========================================================================
BLAST TYPE         1) If blasting to free face
                   2) If box cut/no free face
                   3) Pre-split                                          1

BLAST PATTERN      1) Square
                   2) Chevron                                            1

BLAST DEPTH        1) Single row
                   2) 2-3 drilled burdens
                   3) 4-5 drilled burdens                       .
                   4) ) 5 drilled burdens                                2


                                                                         1

 ENGINEERS DISCRETION
1)  Slope designed on unfavourable structure
2)  Slope designed on normal rock mass
3)  Protection of surface or underground structure                       2
NUMBER OF HOLES PER DELAY                                                1
DISTANCE FROM BLAST (m)                                             25.000
MAXIMUM ESTIMATED GROUND VELOCITY (mm/sec) IS :                     36.425
DISTANCE FROM BLAST (m)                                             10.000
MAXIMUM ESTIMATED GROUND VELOCITY (mm/sec) IS :                    140.427
DISTANCE FROM BLAST (m)                                              5.000
MAXIMUM ESTIMATED GROUND VELOCITY (mm/sec) IS :                    318.985
DISTANCE FROM BLAST (m)                                              2.500
MAXIMUM ESTIMATED GROUND VELOCITY (mm/sec) IS :                    860.783
```

Fig. 11 Cautious blast design for slot excavation

The blast design

Part of the output from BLASTS for the link slot cautious blast design is presented in Fig. 11.

The contractor appointed to undertake the drilling and blasting is required to submit his proposed blast design to the engineer for geotechnical approval prior to undertaking the work. The drilling equipment must be commensurate with the excavation requirements in that it should have the ability to drill blastholes so that continuous vertical sidewalls may be formed.

EXCAVATION SUPPORT REQUIREMENTS

Link excavation

The support assessment applied mainly to the link excavation. Support requirements were calculated for the quartzites alone. The sideslopes of the excavation through the shales were to be cut at 40` which would preclude failure. The Norweign Geotechnical Institute's Q System was used to classify the quality of the rock mass and thereby determine the support requirements.

Rock mass qualities were determined for the following: in the quartzites was determined for zones, from surface to 5 m below ground level; quartzites below 5 m; and shear zones. The results are shown in Table 5. To determine the support, an excavation support ratio, ESR, of 1.6 was applied due to the temporary nature of the link cut.

Zone	Q Value	Rockmass Quality	Support
Quartzite 0-5m	6,8	Fair	Untensioned grouted dowels, 1 - 2m spacing
Quartzite 5m	15,2	Good	Generally no support required, local spot bolts
Shear zones	0,15	Extremely poor	Tensioned grouted dowels, 1m spacing, mesh reinforced, 30mm to 50mm shotcrete

TABLE 5 : Rock mass qualities and support requirements

Link excavation portal

The section of the quarry face where the link will exit, ie the link portal, required particular attention regarding support. Potential stability problems due to blast damage were particularly significant due to the greater degree of freedom of the rock mass. Support required to be installed prior to excavation of the link corridor. Tensioned and grouted rock bolts, 5 m long, at 1.5 m centres around the perimeter of the excavation, were recommended (Fig 12).

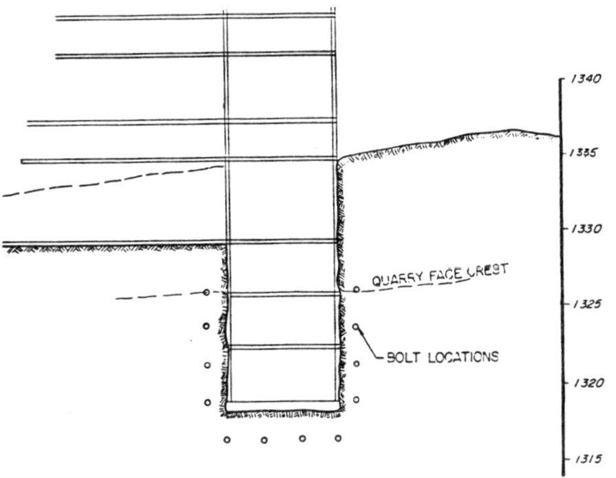

Fig. 12 Slot portal presupport

Support proximate to a major shear

It was recommended that the major shear zones delinated by the investigation should be avoided when finalising the link corridor position in order to preclude particularly severe adverse ground conditions. It may, however, prove unavoidable to locate the slot other than proximate to a major sub parallel shear (Fig 2). In this case, it is recommended that the affected wall be supported by tensioned and grouted cable anchors which would be anchored in competent strata beyond the shear zone.

EXCAVATION PROCEDURES AND GEOTECHNICAL SUPERVISION

Benched drill and blast operations are required to excavate the quartzites. The free faces for blasting should be north facing, ie where possible a southerly direction of excavation advance was advised.

The excavation of the slot should be completed prior to the construction of the floor decks.

Support requirements should be identified and installed timeously as excavation proceeds to allow ease of access and ensure safety as the cut is deepend. A suitably qualified engineering geologist responsible for these duties should be available during the works. He should also monitor the results of blasting with respect to overbreak and instability relating to poor excavation practices. A detailed record of blasting effects, support recommended, support installed and geotechnical conditions should be maintained.

CONCLUSIONS

A probabilistic approach to the stability analysis quantified the risk of failure and showed that this was acceptable in terms of the function of the excavation. Such a quantification would not have been possible using a classical approach. The BLASTS program designed on acceptable cautious blast so that damage to the rockmass may be minimised, stability maintained and construction costs reduced. Collectively, the investigations and analysis undertaken have ensured a high level of confidence regarding the successful implementation of the excavation and construction phase.

ACKNOWLEDGEMENTS

The authors wish to express their appreciation to the architects department of the Pretoria Academic Hospital Structural Engineering Consortium and Transvaal Provincial Administration for their assistance and input during the project and for their permission to publish the paper.

REFERENCES

1 Call, R D, Savely, J P and Nicholas, D E. Estimation of joint set characteristics from surface mapping data. Proceedings of the 17th U S Symposium on Rock Mechanics. Utah, 1976.

2 Barton, N and Choubey, V. The shear strength of rock joints in theory and practice. Rock Mechanics, Vol 10 pp 1-54, 1977.

3 Steffen, O K H, McCracken, A and Haines, A. Probabilistic analysis of discontinuity data for excavation design. Proceedings of the International Symposium on Fundamentals of Rock Joints. Bjorkliden, Sweden, 1985.

4 Harr, M E. Mechanics of particulate media - A probabilistic approach. McGraw-Hill Inc, 1977.

5 Kirsten, H A D. Significance of probability of failure in slope engineering. The Civil Engineer in South Africa. Vol 25 No 1, 1983.

6 Steffen, O K H. Monitoring of rock slopes. Proceedings of the Conference of Southern African Surveyors. Johannesburg, 1982.

7 Yuceman, M S and Tang, W H. Long term stability of soil slopes - a reliability approach. Proceedings of the 2nd International Conference on Application of Statistics and Probability to Soil and Structural Engineering. Aachen, W Germany, 1975.

8 Kirsten, H A D. A classification system for excavation in natural materials. The Civil Engineer in South Africa, Vol 24 No 2, 1982.

9 Page, C H. User documentation for cautious blast design program BLASTS. Steffen, Robertson and Kirsten, Johannesburg, internal document, 1985.

Seismic refraction surveying in urban areas

P.W. McDowell B.SC., M.PHIL., C.ENG., F.I.M.M., M.I.GEOL., F.G.S.
Portsmouth Polytechnic, Portsmouth, England

SYNOPSIS

Seismic refraction surveying can be very useful for the sub-surface investigation of urban areas where open or underground excavations are proposed. A common application is the mapping of the rockhead surface beneath a cover of superficial deposits, particularly where buried channels are expected. Determining the depth of weathering is also possible, but usually calculations are less accurate in this case. Seismic refraction surveying can also be used to map step-faults which bring different rock types into juxta-position and fracture-zones within an otherwise homogeneous rockmass.

The urban environment, unfortunately, has several characteristics which can severely limit the use, or accuracy, of the seismic refraction method. Probably the greatest problem is achieving an adequate signal-to-noise ratio, as there is usually a high background noise level and explosives are generally prohibited. Paved surfaces usually produce a velocity inversion situation, which introduces errors into the depth calculations. Buried pipes and trenches can confuse the results and interpretation, unless suitable field procedures are adopted.

An example is provided of a successful seismic refraction survey for a sewage pipeline near Runcorn in the United Kingdom. The seismic spreads were set out on the concrete paved hard shoulders of the M56 motorway, and a large dropping-weight system was used as an energy source. Analysis of the amplitudes of first arrival events enabled 'bedrock' refractions to be differentiated from direct waves through the concrete pavement. Despite the limitation of a velocity inversion situation, the seismic profiles clearly mapped the irregular erosional surface of the Keuper Marl, beneath the superficial deposits, and enabled the pipe-line to be re-routed completely within Keuper Marl without a change of invert level of the tunnel.

INTRODUCTION

A pre-requisite for rock engineering and excavation is a ground investigation to ascertain the nature and distribution of rocks, rock mass discontinuities and groundwater. Usually, the most cost-effective approach is to combine direct methods, such as drilling and trenching, with geophysical investigations. Seismic refraction is a geophysical method which is commonly used to map the rock-head surface beneath a cover of superficial deposits, particularly where buried channels are expected. The extent of weathering can also be assessed by this method, although the accuracy of depth determinations, in this case, is lessened by the relatively small increase in seismic wave velocity with depth. Seismic refraction surveying can also be used to locate faults which bring different rocks into lateral juxtaposition; an early example of this being the successful mapping of the extent of upfaulted blocks of quartzite at the site of a government building in South Africa. (Gough D I 1953)[2]. Steeply dipping shear and fracture zones, in an otherwise homogeneous rock mass, are usually more readily located by the seismic refraction method than by vertical drilling, because the relevant portions of the direct and refracted ray-paths are essentially horizontal. An interesting example is the mapping of both the low velocity zone produced by the Glen Lia Fault within the Foyer's Granite of Scotland and the associated thickening of the overlying lower velocity Boulder Clay (Cratchley, C R et al. 1972) [1].

The urban environment, unfortunately, has several characteristics which can severely limit the use or accuracy of the seismic refraction method. There is, usually, a high level of background noise; both transient, such as from moving traffic, and continuous, as from operating machinery. The survey time is greatly increased, if the operator has to wait for quiet spells, and it may be necessary to work at night-time. Near-surface ground conditions are often complicated by the presence of landfill, backfilled trenches and ancient foundations, which can produce confusing seismic records. Paved surfaces often produce a velocity inversion situation, which also introduces errors into the depth calculations.

These problems are considered in the context of recent technological developments and a case study is provided to illustrate how some of the limitations to the use of the seismic refraction method in urban environments can be minimised.

SIGNAL-TO-NOISE RATIO

Without the use of explosives, the achievement of an adequate signal-to-noise ratio has, until recently, been difficult in urban areas. Apart from the high background noise level already mentioned, the surface layer of superficial deposits, or fill, severely attenuates the seismic waves as they travel between the source and receivers. Explosives cannot usually be used at, or near, the ground surface and, if required, should be placed within boreholes at the base of the surface layer. (Linehan, D et al. 1962)[3]. The need for explosives has been lessened by the development of alternative energy sources and a significant improvement of seismograph systems. Where the depth of investigation exceeds 100 m, the project may justify the use of continuous vibration sources and the digital processing of reflected and refracted events, as in the Chicago Deep Tunnel Project. (Mossman, R W et al. 1972)[4]. For shallow depth investigations, dropping weight systems are usually adequate especially when used in conjunction with signal enhancement seismographs for seismic refraction surveys. Other alternatives to explosives are air-guns, gas-guns and a system which fires a shot-gun cartridge into the ground. The advantage of portability is offset by the high frequency of these sources, relative to dropping weight systems, and the inability to use these sources directly on paved surfaces.

The signal-enhancement procedure requires a number of repeated impacts at the same location, so that seismic signals, having the same arrival times and wavelengths, will reinforce each other; whereas the addition of random extraneous vibrations will not increase the background noise level to the same extent. A controlled and consistent mechanical energy source, consisting of a large metal block falling onto a metal plate, well coupled with the ground, is usually necessary to produce a strong signal with a clear onset from one or two repeats of the procedure. Small dropping weight sources and sledge hammers are unlikely to be adequate in urban areas, even when used with signal enhancement seismographs. The mechanical source should be designed to: minimise ground vibrations during the free fall of the mass; incorporate the facility to synchronise with the seismograph and record the instant energy is imparted to the ground. A coupling plate may not be necessary on hard road surfaces and could be a disadvantage in this case.

Careful consideration should also be given to the placement of geophones. Loose fill, topsoil and dry non-cohesive soils can result in severe signal attentuation and distortion, whereas wet clay provides excellent coupling. The geophones may need to be planted in firm ground at the base of shallow-dug pits or, in extreme cases, clamped within boreholes. On paved surfaces, perhaps surprisingly, better results may be obtained by placing the geophones in a mound of wet clay, rather than by cementing the sensors directly to the hard surface. An array of geophones linked together in series will produce a stronger signal than a single geophone and can, if spaced correctly, filter out background noise and surface wave events.

SHALLOW IRREGULARITIES

The interpretation of seismic refraction results can be difficult in urban areas where near-surface irregularities, both natural and manmade, are common. There can be a considerable scatter of data on the time-distance graphs, with anomalously early arrivals being produced by old foundations and with late arrivals resulting from deep pockets of fill. These can be incorrectly interpreted as variations in the depth to refractors of interest. A small increase in the thickness of a surface layer of fill of low velocity can, for example, be interpreted as a significant increase in the depth to bedrock. This ambiguity can be resolved by the use of energy points along the seismic spread, and off-set from the ends of the spread, as well as at each end. Additional seismic spreads, with small length and close geophone spacing, may also be required to resolve the near surface variations.

Paved surfaces constitute a problem for seismic refraction surveying, in urban areas, when they overlie fill or superficial deposits of lower velocity. This is a classic velocity inversion situation with the low velocity material not being represented on the time-distance graph. Depths to seismic refractors can be grossly overestimated, unless this situation is recognised and calculations are carried out using correct values for the low velocity layer. These can be obtained from alternative methods, such as borehole logging.

CASE HISTORY

The geophysical survey for the Flood Brook Storm Water Sewer at Runcorn, in the United Kingdom, illustrates some of the advantages and limitations of seismic refraction surveying in urban areas. Along the original route of the tunnel for the sewer, (Figure 1), construction work had been interrupted by difficult ground conditions, in the form of saturated sands and gravels, approximately 100 m to the south of the M56 motorway. Boreholes confirmed that these superficial deposits were infilling a deep channel cut into the underlying Keuper Marl (Figure 2) and further site investigations, incorporating seismic refraction surveying, were instigated to provide a map of the Keuper Marl surface in the vicinity of the motorway.

Seismic traverses were set out along the hard shoulders of the M56 motorway, close to a railway line and a housing estate (Figure 1). The problem of signal-to-noise ratio was overcome by carrying out the surveying in the middle of the night and using a large dropping weight energy source. This consisted of a 180 Kg mass falling approximately 1.5 m onto a 90 Kg coupling plate on the ground surface. A clearer signal may have been obtained by dropping the weight directly onto the hard shoulder, but this was not attempted because of the risk of damage to the road surface. Using a seismograph that indicates the background noise level, it was possible to operate during lulls in traffic movements and obtain good seismic records.

A more difficult problem, at this site, was the distortion of the seismic results by high amplitude direct compressional waves through the concrete pavement. Effectively, a single layer time-distance graph was produced until this event was overtaken by critically refracted rays from the surface of the Keuper Marl (Figure 3). The unusual step-out, in this case, is the result of rapid attentuation of the direct waves through the thin layer of concrete; whereas recorded events from the thick Keuper Marl layer maintained their amplitude relatively well over the length of the seismic spread. No refracted energy was recorded from the superficial deposits immediately beneath the pavement, because they formed an intermediate low-velocity layer. An indication of the velocity for deeper superficial deposits was obtained from a study of second arrival events. It was then necessary to obtain confirmation of velocity values for the superficial deposits, independently, in order to calculate depths to the Keuper Marl surface. Traverse 4, which was on natural ground adjacent to the motorway (Figure 1), yielded velocity values between 360 and 660 m/s for sands and gravels above the water-table and 1400-1600 m/s for saturated sand and gravel. Although the use of similar velocities, where not recorded, could lead to errors in interpretation, the calculated depths to the Keuper Marl, for geophone stations at borehole locations, were very accurate.

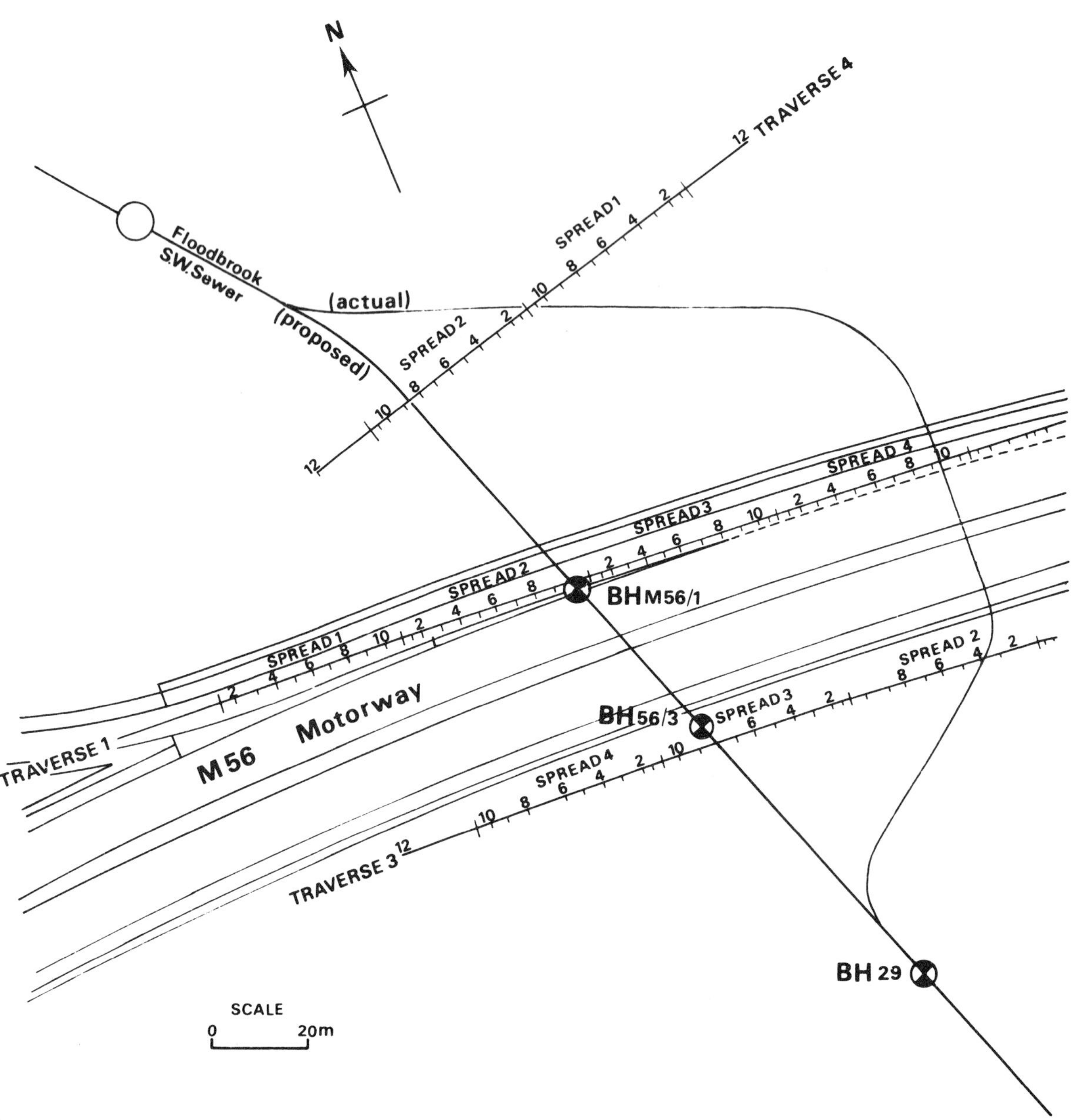

Figure 1: Flood Brook Storm Water Sewer Scheme.
Layout of seismic traverses.

Horizontal Scale
0
20m
Vertical Scale
0
10m
BH19
FILL
GWL
SLIP ROAD
BH29
FILL
BH56/3
FILL
PROPOSED M56 MOTORWAY
M56/1
S A N D
A N D
GWL
G R A V E L
GWL
TUNNEL LEVEL
K E U P E R M A R L

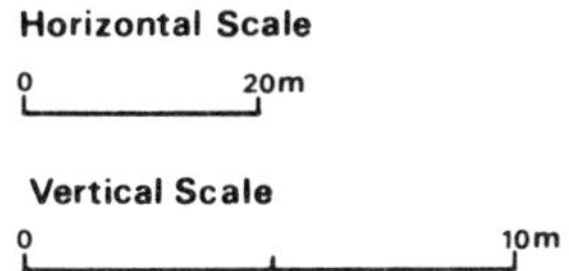

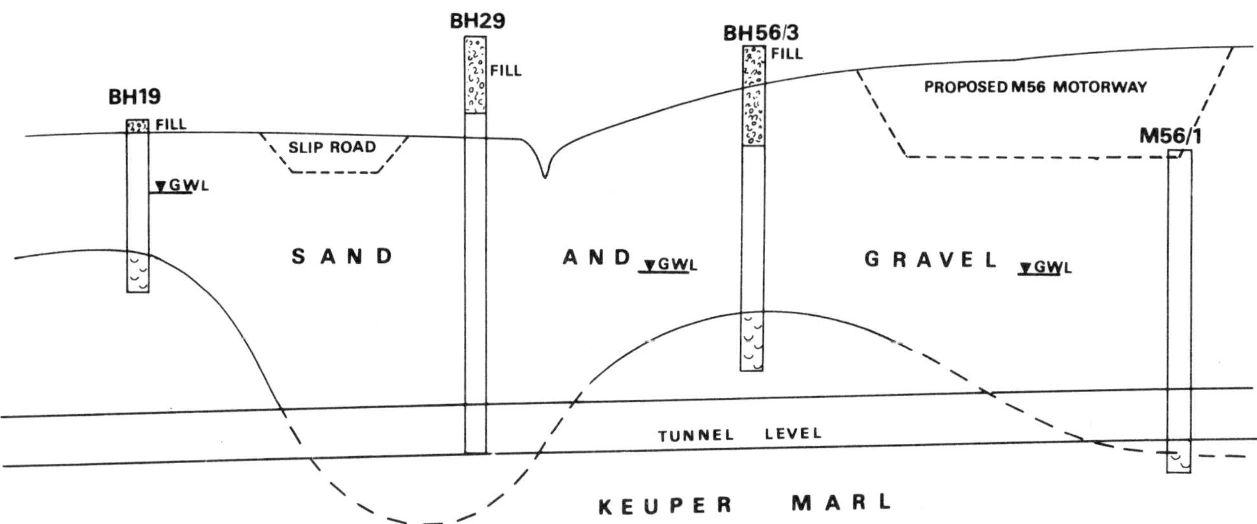

Figure 2: Profile along centre line of proposed tunnel route.

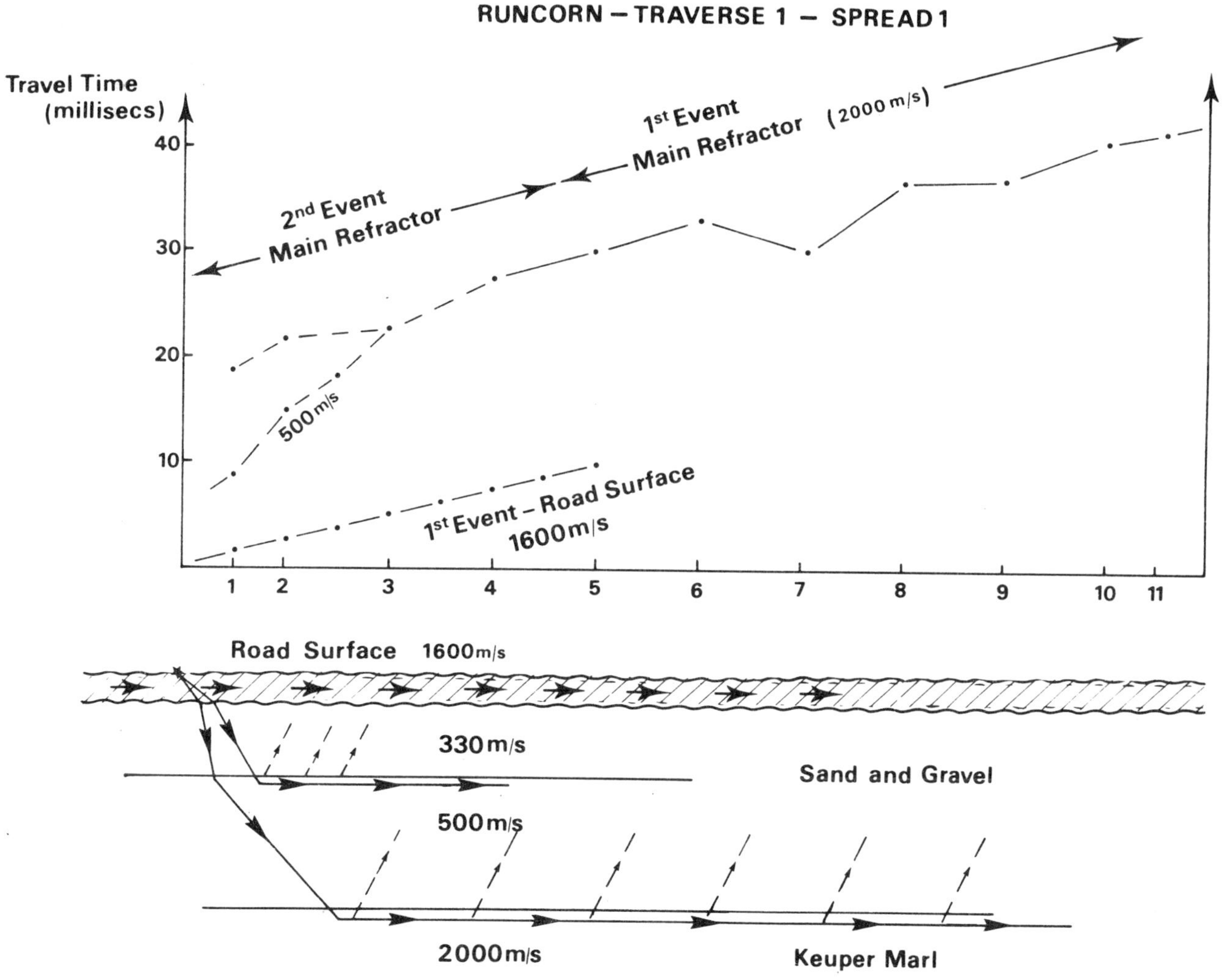

RUNCORN – TRAVERSE 1 – SPREAD 1
Travel Time (millisecs)
40
30
20
10
2nd Event Main Refractor
1st Event Main Refractor (2000 m/s)
500 m/s
1st Event – Road Surface 1600 m/s
1
2
3
4
5
6
7
8
9
10
11
Road Surface 1600 m/s
330 m/s
Sand and Gravel
500 m/s
2000 m/s
Keuper Marl

Figure 3: Typical ray-path diagram and t-d graph for seismic refraction survey along M56 Motorway

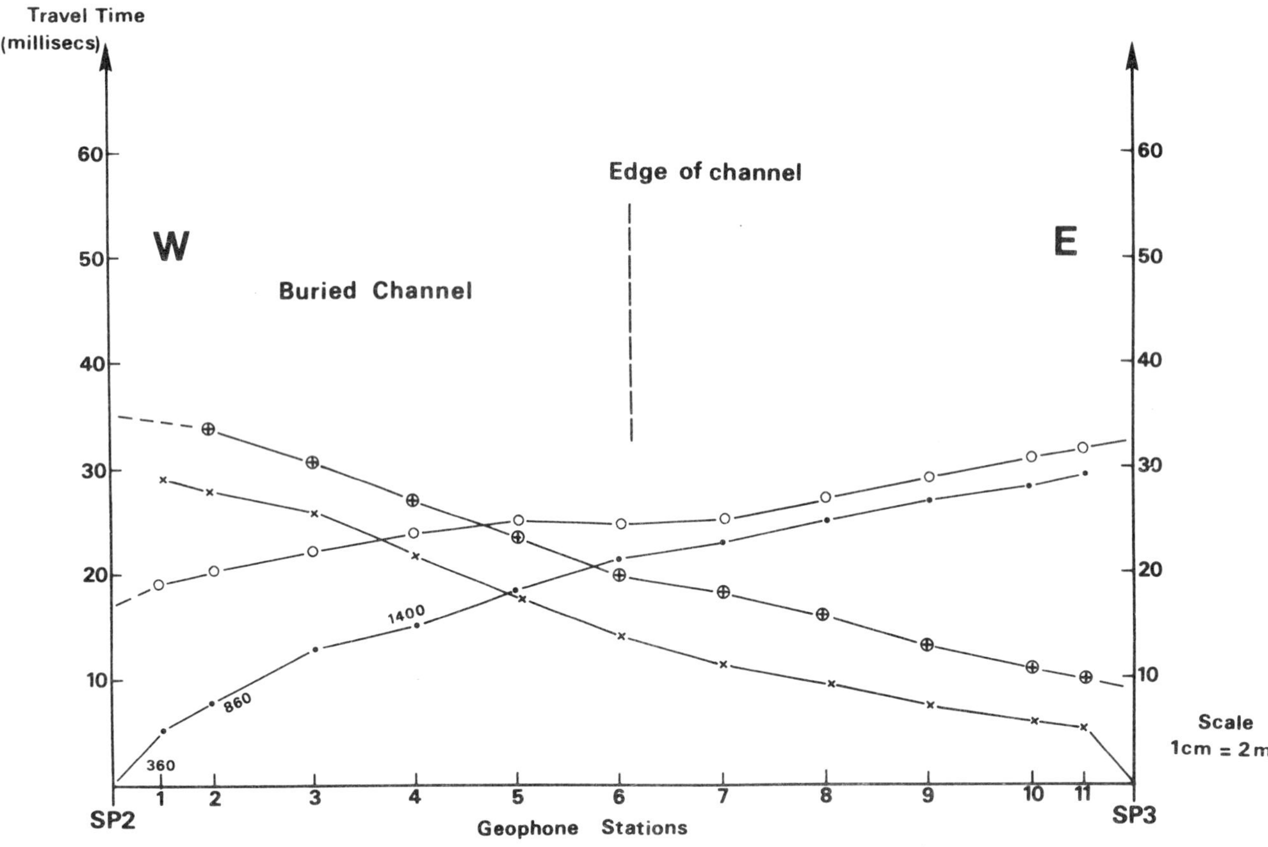

Figure 4: Characteristic t-d graph over edge of buried channel.

Despite the site difficulties, the seismic refraction profiles enabled the buried channel to be mapped adequately for this particular project. It was shown that the original tunnel route was in the worst possible position, relative to the buried channel, and the depth to the Keuper Marl decreased both to the east and west of the original route in the vicinity of the M56 motorway. The eastern edge of the channel was shown up very clearly by the time-distance graph of Spread 4/Traverse 1 (Figure 4) with the edge corresponding to the change in apparent velocity values for the Keuper Marl. With confirmation by further drilling, the tunnel was realigned for excavation completely within the Keuper Marl horizon, as shown in Figure 1.

CONCLUSIONS

Seismic refraction surveys can be carried out successfully in urban areas where rock engineering and excavation is proposed. The many difficulties produced by the urban environment can be minimised by careful planning of the surveys by experienced geophysicists and the use of recently developed energy sources and seismographs. Accuracy in depth determinations may be reduced but, used in conjunction with boreholes, the seismic results can be valuable both for mapping variations in ground conditions and providing an independent assessment of rock mass quality.

References

1. Cratchley C R, Grainger P, McCann D M and Smith D I. "Some Applications of Geophysical Techniques in Engineering Geology with Special Reference to the Foyer's Hydroelectric Scheme." 24th International Geological Congress (Section 13 Engineering Geology) 1972, pp. 163-175.

2. Gough D I. "The Investigation of Foundations by the Seismic Method". Transactions

South African Institution of Civil Engineers, Volume 3, No. 2, 1953, pp. 2-11.

3. Linehan D and Murphy V J. "Engineering Seismology Applications in Metropolitan Areas". Geophysics, Volume 27, No. 2, 1962, pp. 213-220.

4. Mossman R W, Heim G E and Dalton F E. "Seismic Exploration in the Urban Environment". 24th International Geological Congress (Section 13, Engineering Geology) 1972, pp. 183-190.

Application of seismic surveying, orientated drilling and rock classification for site investigation of rock tunnels

Ian McFeat-Smith B.SC., PH.D., C.ENG., M.I.C.E., M.A.S.C.E.
Charles Haswell & Partners (Far East), Hong Kong
G.K. Nieuwenhuijs B.SC., M.SC.
W.C. Lai
Electronic & Geophysical Services, Ltd., Hong Kong

SYNOPSIS

This paper outlines the merits and limitations of various site investigation techniques for rock tunnels at portals, areas of low cover, urban areas and for sub-aqueous crossings. A model is presented which includes the integrated use of seismic refraction surveying, the drilling of vertical and orientated boreholes and collation of the data obtained in terms of a simple rock classification system.

This approach has been tried and tested in Hong Kong's sedimentary and igneous rocks on a number of tunnel projects involving the excavation of some 60 kms of rock tunnels from 20 portals and shafts. Emphasis is given to the use of geophysical methods to locate rockhead and zones of potentially difficult tunnelling ground.

INTRODUCTION

Ground conditions play a predominant role in the excavation of tunnels. In rock tunnelling for example the presence of zones of soft ground requires the installation of heavy temporary support. As a result of delays incurred the cost of encountering soft ground can be as high as six times as that for rock. In anisotropic rock where the geology varies substantially over short distances it can be impracticable constantly to change the method of support, hence the most expensive support system may have to be installed throughout.

Variations in rock hardness influence selection of the form of excavation, either by drill and blast or boring machine. They affect the progress rates gained by such techniques and the running costs incurred. The latter include the cost of filling overbreak with concrete and the rate of consumption of drill bits or cutting tools.

Comprehensive site investigations are therefore required for all driven tunnel projects to allow accurate costing and planning. These must be well balanced according to the local geological conditions. In Hong Kong's igneous terrain where returns from aerial photography and geological mapping can be limited and access for drilling cannot always be obtained, a seismic refraction survey may well provide the most cost-effective solution for locating zones of adverse tunnelling ground.

As demonstrated previously[1] high delays to tunnelling operations can occur at tunnel portals. This arises as Contractors, still in their learning curves, experience some of the worst conditions likely to be encountered in the tunnel route. A close site investigation is therefore required at portals and other areas of low cover.

Some form of categorisation is necessary to link the data collated from such investigations to the tunnelling operations to be carried out.

This need not be interpretative if a simple system is used. It is logical to relate this to the support characteristics of the rock mass as this is the greatest time (and hence cost) dependent tunnelling operation that is affected by variations in geology.

PORTALS AND AREAS OF LOW COVER

As a result of the rapid pace of development of engineering projects in Hong Kong decisions made during the preliminary studies become irreversible. Hence much emphasis is placed in these circumstances on the need to reach the most practical solutions as early as possible. Tunnel alignments and portal positions are selected primarily according to engineering and land requirements. There may be little time available to examine alternative portals by drilling. Where some latitude exists, portals are therefore often located according to the topography. Geologically favourable positions can be found at the head of spurs although a check should be made for evidence of major landslip zones and for thick boulder colluvial deposits. A brief seismic refraction survey over a grid-patterned area by the hammer method can provide invaluable data at this critical stage.

As outlined in previous work[2], portals in Hong Kong are commonly sited at the break of slope between the flat urban/cultivated areas and the hilly terrain in the more rural areas. The Aberdeen Road Tunnel[3] and Western Aqueduct[4] Projects provide examples of this. As a result of the high cost of land and the need to minimise the environmental impact of projects in these areas, the extent of the works at portals is often fixed irrespective of the local geological conditions. Thus in some cases onerous portal formation works may have to be undertaken.

Drilling and Testing

Drilling at portal areas is carried out primarily to define the geology of the area to be excavated in terms of soil and rock profiles and to measure the properties of each unit by in-situ and laboratory testing. Emphasis is given in Hong Kong to the need to establish accurately the position of rockhead, in view of the erratically weathered nature of the local igneous rocks, and to examine the condition of the rock mass beyond this. It is similarly important to determine water levels and the overall hydrogeology of the portal area for stability analysis due to the need to cut steep slopes (to minimise land usage) and considering the instability of some of the local soils.

At areas of low cover where penetrative weathering, faulting and other adverse conditions may be encountered site investigations are carried out for the following:

- Design changes
- Estimation of quantities
- Determination of construction difficulties
- Estimation of water inflows
- Selection of preferred methods of construction

Investigations at these areas include the use of orientated boreholes and seismic surveys. Inclined boreholes are normally drilled in the direction of the tunnels at an angle of 45 degrees[5]. Their value is considerable particularly for proving fault zones and because they give a degree of lateral coverage not provided by vertical drilling .

Location of Rockhead by Seismic Refraction Surveying

Seismic refraction surveying provides an important site investigation tool for tunnel projects. The survey is used to produce a continuous simplified map of the acoustic velocity structure of the subsurface.

The basic method for seismic refraction surveying is well established[6]. It requires measurement of the travel times of acoustic waves from a sound source to a set of positions with known co-ordinates. This produces a graph of the travel times as a function of distance from the sound source. The data is interpreted to give information about the velocity variations with depth.

The normal layout of a seismic refraction "spread" is to lay the geophones (sound receivers) along a straight line and position the sound source at either end and, if possible, also in the middle. The maximum length of the spread will depend upon the strength of the sound source compared with any background noise such as nearby traffic or even wind. The maximum depth at which information is obtained will be approximately one third of the spread length. The resolution with which the different horizons are determined will depend upon the number of geophones used. The accuracy of this method for depth determination is about $\pm$ 20% but is improved to about $\pm$ 10% when the data is correlated with borehole information.

A strong acoustic velocity contrast will usually exist between the overburden soils and rockhead. The refraction method is therefore well suited to determine rockhead levels. Interpretation of the records is based upon a simplified model of the subsurface, namely that of a layered earth with the different layers having an increasing velocity as a function of depth. As a result the method may not detect certain geological features which can therefore give rise to misinterpretation in the final analysis. These features are summarised as follows:

- Extreme changes in bedrock topography
- Isolated boulders
- Zones of completely to highly weathered rock lying below rockhead.
- Thin layers of unweathered material.

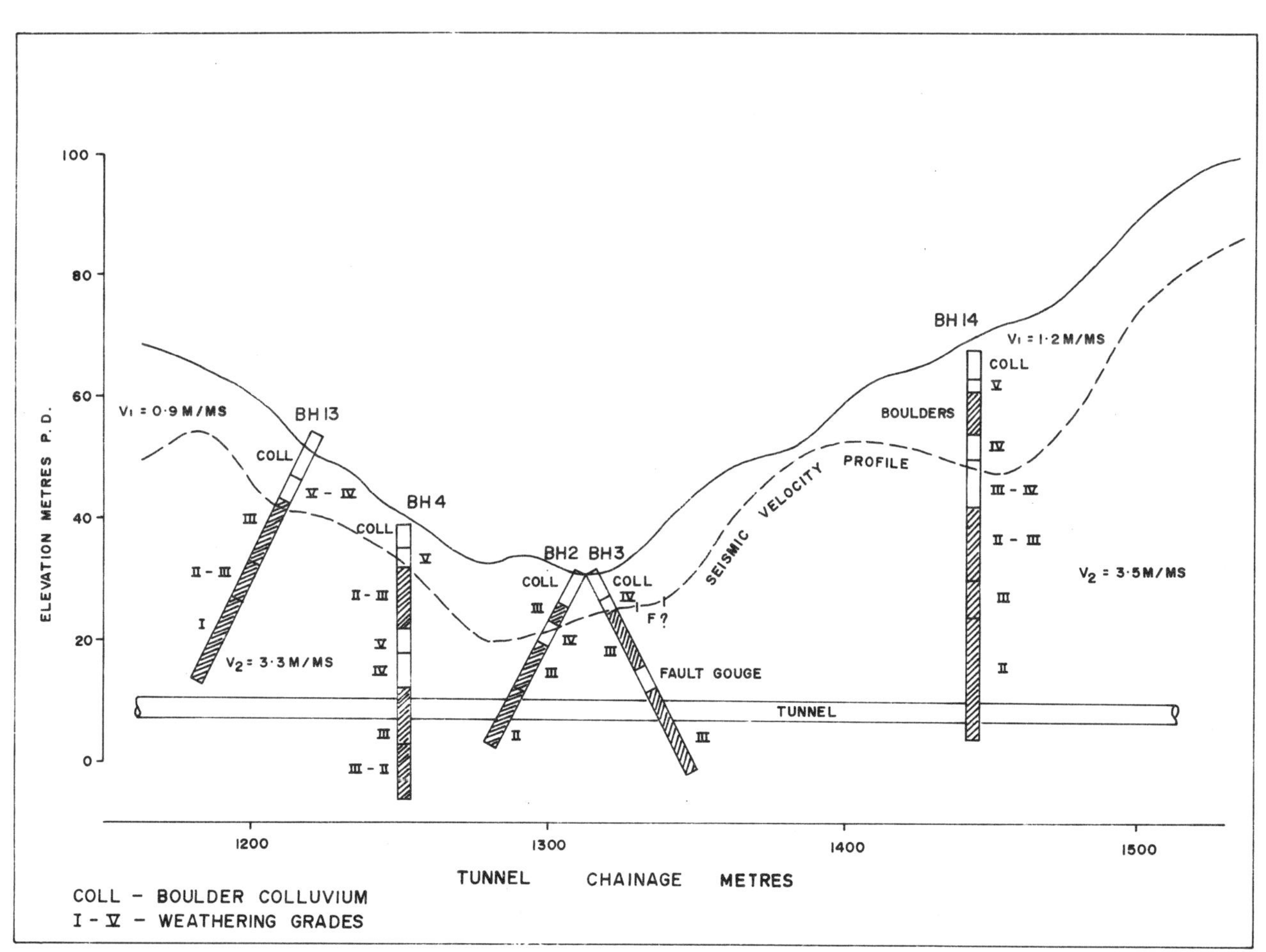

FIGURE 1 CASE HISTORY SHOWING INVESTIGATION OF A VALLEY CROSSING BY SEISMIC REFRACTION SURVEY AND ORIENTATED DRILLING

An example of both the value of seismic refraction surveying for locating rockhead levels and some of the factors that must be taken into account when using this technique is presented in Fig. 1. In this case a survey was conducted above the alignment for a proposed tunnel in Hong Kong at a valley crossing. The terrain was composed of coarse volcanic tuff of the Repulse Bay Formation, overlain by boulder colluvial deposits.

The survey was conducted in advance of drilling, hence the boreholes were used to check the predicted velocity profile. As illustrated in Fig. 1, good correlation was obtained between the profile and the interface between weathering zones III - IV which is generally acceptable as rockhead for engineering purposes. In addition, the survey indicated the presence of a minor fault which was later proved by the drilling of BH3.

An important anomaly was also found in this crossing. At BH4 the grade II - III rock had acted as the main refracting boundary obscuring the presence of an 11.8 m deep zone of completely to highly weathered volcanic tuff. Conversely, at BH14 the rockhead had acted as the main refracting zone in spite of the presence of boulders at a higher level.

Location of Discontinuity Zones

Seismic velocities of rock material commonly fall within the range 300 - 7000 metres/second[7,8]. The velocity is influenced by many factors and is not a diagnostic for a particular rock type. Borehole correlation is normally required to assign rock types to the different velocity zones detected by a survey. Although exceptions exist, the velocities shown in Table 1 generally increase from left to right.

Empirical relationships found between weathering grades and seismic velocities of the igneous rocks around Hong Kong are given in Table 2.

Table 1

Sedimentary rock	- igneous rock
Acid igneous rock	- basic igneous rock
Unconsolidated sediments	- consolidated sediments
Dry soil	- water saturated soil
Porous rock	- Solid rock
Fractured or cracked rock	- unfractured rock
Weathered rock	- fresh rock

Table 2

Seismic Velocity	Weathering Grade	Description
0.3 - 0.7 m/ms	VI - V	Soil or completely weathered rock
0.8 - 3.0 m/ms	IV	Highly weathered rock
3.0 - 4.0 m/ms	III	Moderately weathered rock
4.0+ m/ms	II - I	Sightly weathered or fresh rock

Fracturing and faulting are known to lower the velocity of rock masses even where no increase in weathering is present. Measurements at rockhead can therefore be used to identify the presence of adverse tunnelling conditions in terms of increased weathering, fracturing and faulting, although differentiation between these features by seismic methods alone is not always possible.

The ability of a survey to detect such zones depends upon:

- The width of the feature
- The relative velocities of the adverse zone and adjacent rock
- The accuracy of measurement of the energy arrival time

If the latter is 1 millisecond and the rock velocity is 4.0 m/ms then a zone with a velocity of 2.0 m/ms should be detectable if it is more than 15 metres wide; if the adverse zone has a velocity of 1.0 m/ms the zone should be more than 5 metres wide.

It follows that an adverse zone must either be relatively wide or contain sufficiently low velocity material to be identifiable by seismic means. In most cases the geophysicist will receive only indications of the presence of low velocity zones. The minor fault zone located in the previous case history and later proven by borehole BH3 to be 5 - 7m wide provides a fair example of this. In surveys for several tunnel projects under construction, zones of widths in the order of 3 - 40m have been located in the

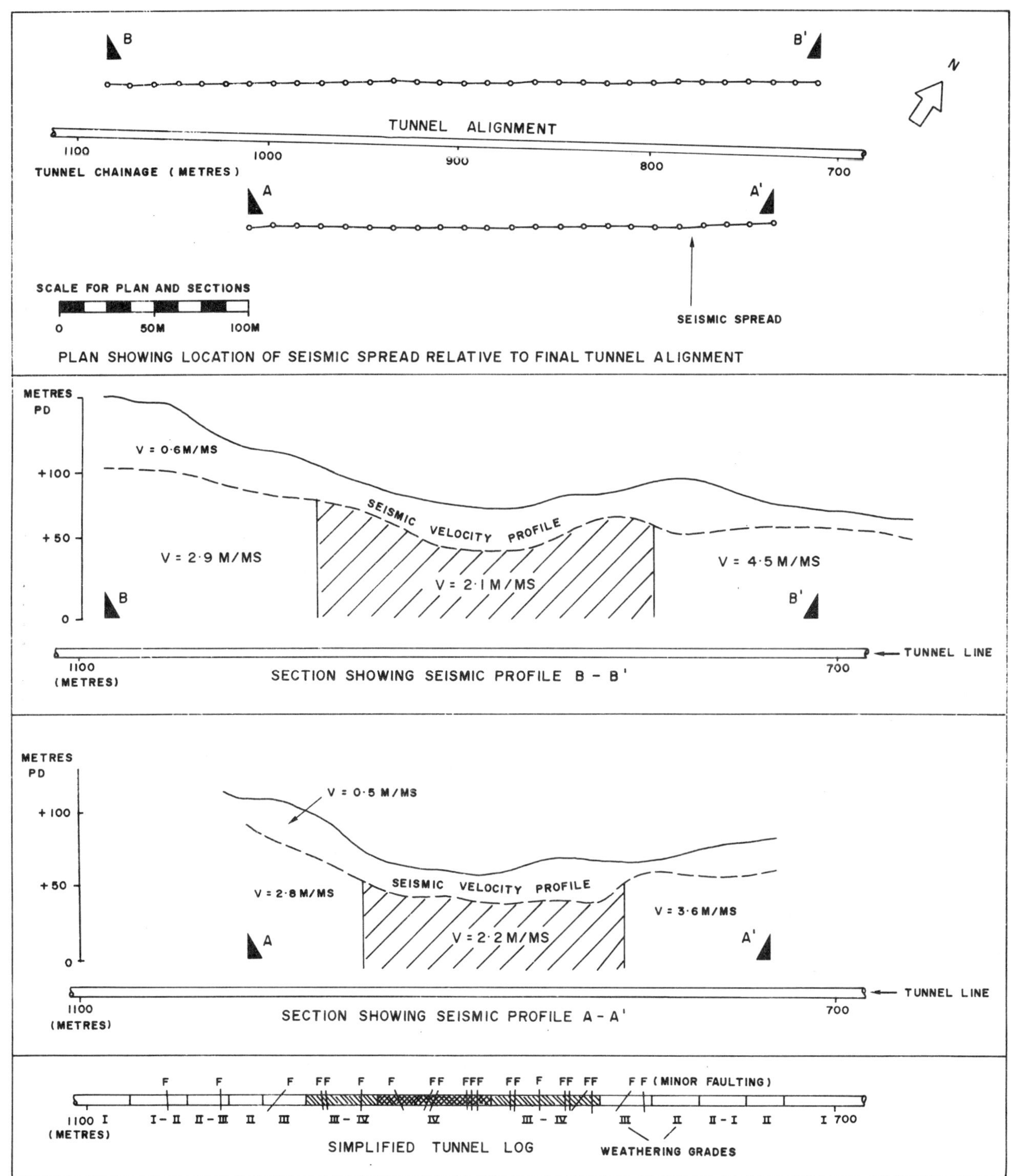

FIGURE 2 CASE HISTORY SHOWING LOCATION OF MAJOR DISCONTINUITY BY SEISMIC REFRACTION SURVEY

local volcanic tuffs at a frequency of about 1 - 3 per km of route length.

There is no certainty that all of these anomalies will be present as significant discontinuity features at tunnel level. However, experience in the local rock types has led the investigators to attempt to check conditions within each zone by the drilling of inclined boreholes whenever possible. Where this is not practicable, due often to tight investigation programming or lack of access, such zones are assumed to be adverse when planning and costing the project.

The following case history provides an account of how a very wide low velocity zone was detected by seismic refraction surveying for a tunnel in the New Territories where there was no reason from terrain evaluation methods to assume the presence of such features.

In this case two parallel surveys were conducted to locate the most favourable alignment for a rock tunnel along the side of a steep hill. As illustrated in Fig. 2 alignment A - A' offered less cover to the tunnel than B - B' but was preferred as it involved a shorter route length of tunnel. Both surveys identified the presence of an area of low velocity rock as shown. The zone was 130 m wide along alignment A-A' and 160m wide along alignment B- B'. The trend of the feature appeared to be perpendicular to the alignments.

There was some reason to doubt that the zone would exist at tunnel level. No evidence of faulting existed at the surface. It also seemed likely that if the feature were associated with weathering, then such weathering would pinch out with depth, particularly as the overburden cover to the tunnel was in the order of 80 - 180m. In the event, the alignment shown was favoured for engineering reasons and a lack of time to consider the matter further. Due allowances were made in the Bills of Quantities for tunnelling in adverse ground, in particular to allow for the installation of heavy temporary supports.

During excavation of the tunnels difficult conditions were encountered in the form of near continuous minor faulting and highly variable weathering, including zones of grade V. This required the installation of substantial quantities of steel arch ribs and lagging. As shown in Fig. 2 the total width of the feature proved to be 160m wide and its location correlated closely with that indicated by the seismic survey.

Horizontal Drilling

In order to locate rockhead at tunnel portals and to obtain the specific data required on rock quality in the area beyond, the authors favour the drilling of long horizontal boreholes. The value of horizontal drilling is substantial as each metre drilled relates directly to conditions within the tunnel.

The overall progress reported for six horizontal boreholes drilled to lengths of 150 - 220 m for tunnel projects in Hong Kong[9] can be sub-divided as follows:

Setting up - Erection and stabilization of the rigs were recognised as being important aspects and took 8 - 11 days.

Drilling - The average rates achieved in the local extremely strong, welded volcanic tuffs varied between 3.2 - 5.5 m/day averaging 4.6 m/day.

Grouting - Effective grouting was difficult to achieve because of the pressure from water outflows. Stage grouting was normally used and took about 6 days.

The major difficulty with drilling such boreholes is that they deviate from their intended path. Fig. 3 shows the displacements calculated from tropari orientation tests normally taken at 20 m intervals. The average displacements of the holes were 6, 10 and 15 m at drilled lengths of 100, 150 and 200 m. This

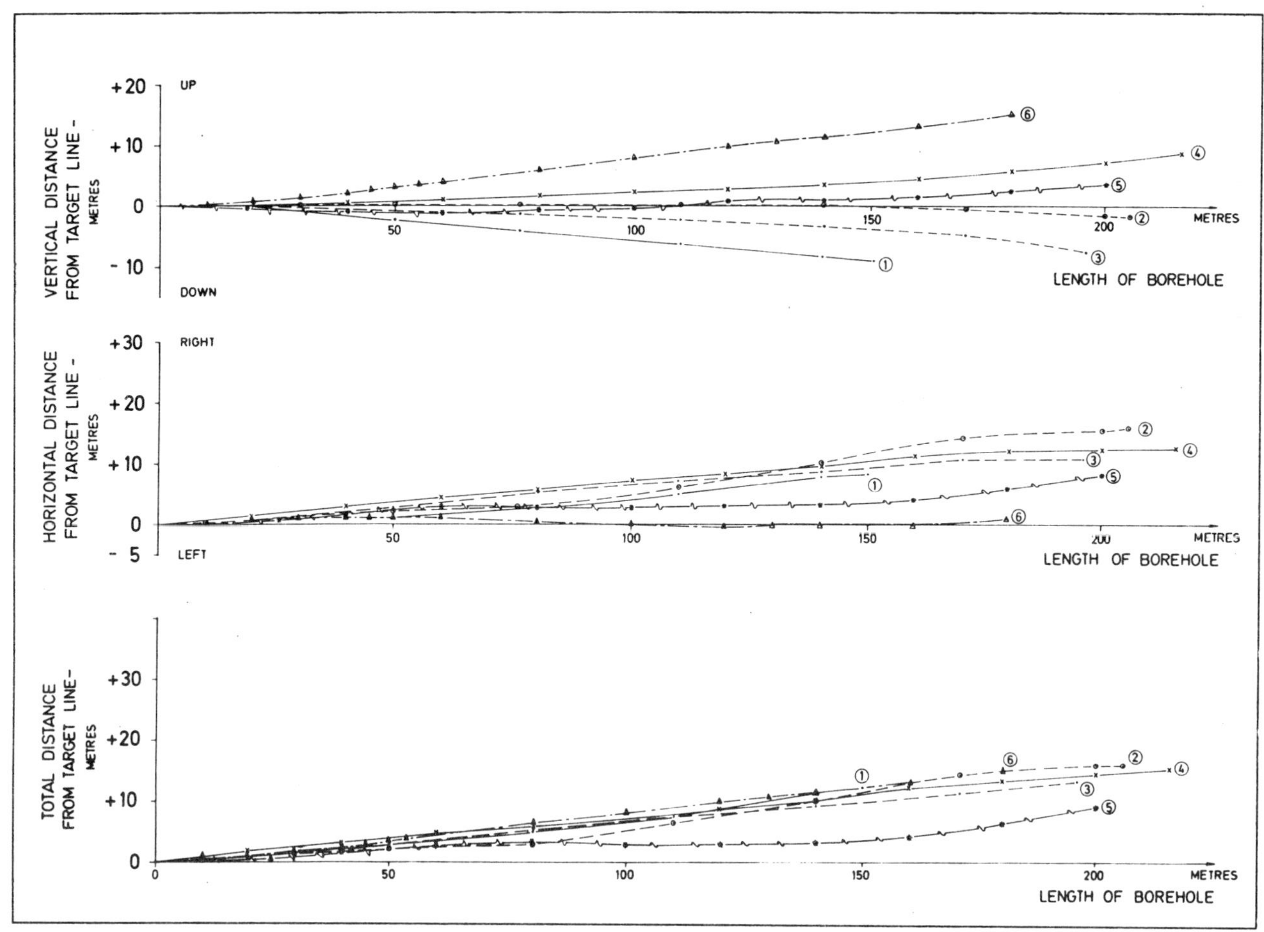

FIGURE 3 DEVIATION OF HORIZONTAL BOREHOLES FROM TARGET LINE

was either up or down but in all cases to the right.

In this case the displacements incurred were not considered to be significant relative to variations in the geology of the local volcanic rocks. As illustrated in Fig. 4, a high geological return close to the tunnel alignment was obtained from each borehole. This included measurement of water outflows which provides an estimate of the quantities of water expected in the tunnels.

URBAN AREAS

Tunnels driven under developed areas are commonly required for infrastructure facilities. These are often characterised by the need for easy access and hence are constructed as close to the surface as possible. The Hong Kong Mass Transit Railway (MTR) provides an example of such a project.

Tunnelling close below major buildings and roads, even in competent rock, creates particular problems and requires detailed investigation. This is often impeded by limited access and the need to minimise disruption in the city centre. In these situations investigations are carried out to obtain data on the following features:

- Areas of low rockhead or penetrative weathering
- Rock quality particularly where high loadings on the tunnel lining are envisaged
- Areas where settlement may occur

Even in rock tunnels where no loss of ground is expected, water inflows into the tunnels may drain the overlying soils and induce high settlements. Examples of this have recently occurred in Hong Kong.

Techniques used in these areas generally include the drilling of many short boreholes. Prelimary studies for the Hong Kong MTR Island Line included about 1500 vertical boreholes spaced at distances of 15-20 m in key problem areas.

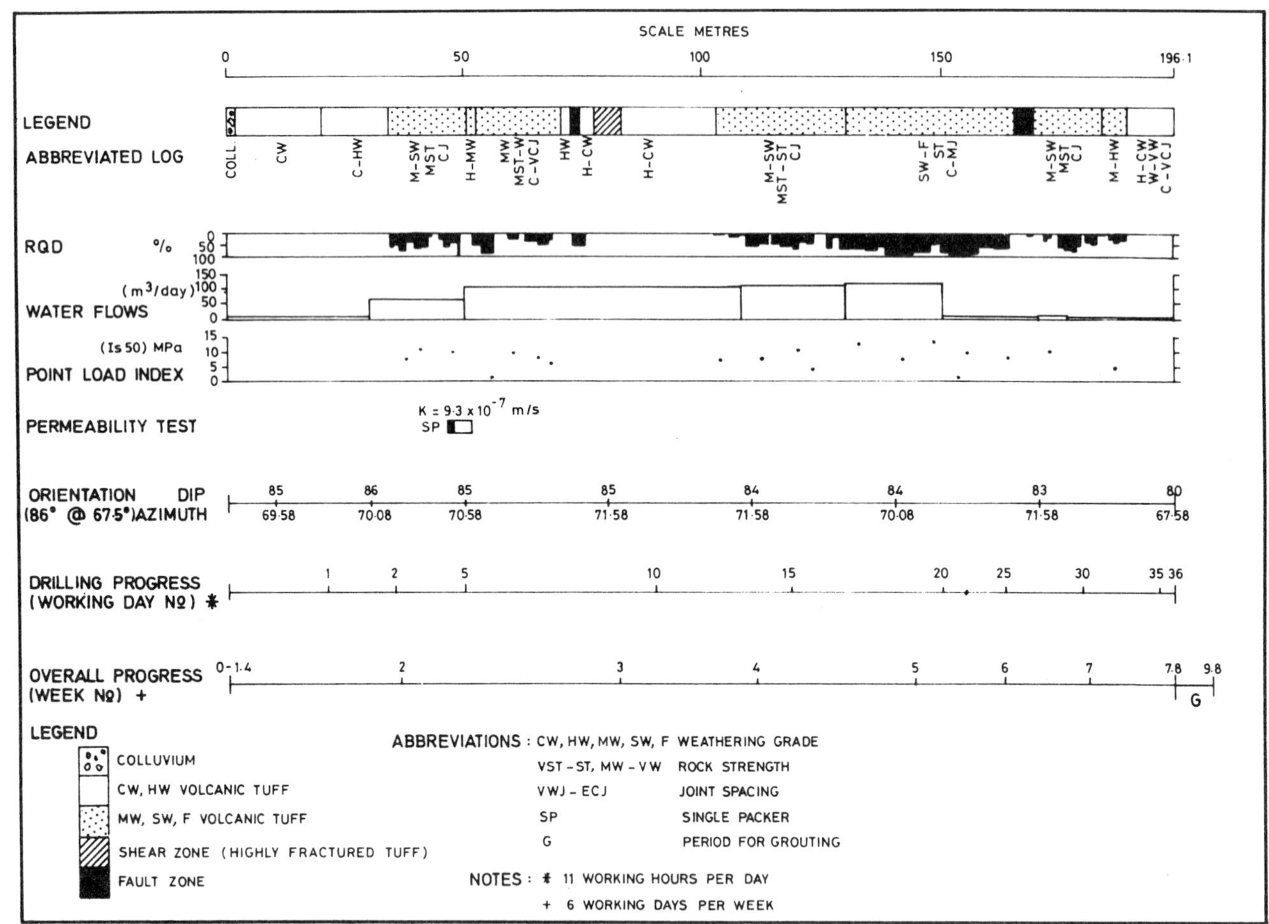

FIGURE 4 LOG FOR HORIZONTAL BOREHOLE SHOWING TYPICAL GEOLOGICAL RETURN AND DRILLING PROGRESS

Access to such tunnel systems is generally gained via shafts. Where portals and caverns are to be constructed or cover is substantial, orientated boreholes can be drilled. On the Island Line two long horizontal boreholes were drilled at the Lei Yue Mun Portals in Chai Wan whilst inclined boreholes were employed to investigate a cavern to accommodate Tai Koo Station.

Seismic surveying techniques can also be employed gainfully in developed areas as outlined in the following sections.

Refraction Surveys Using Hammers

The principal limitation to the accuracy of seismic surveying is the signal to noise ratio. Where necessary an increase in this ratio can be achieved by either increasing the signal or decreasing the noise. The latter can be effected by electronic filtering, which is normally only partly successful because the frequency bandwidths overlap to a large extent. Thus the signal may have to be increased to produce better quality results. In a normal refraction survey using explosives an acceptable signal to noise ratio can be obtained by increasing the amount of explosives used in every shot.

In developed areas such as Hong Kong, it is often found that the use of explosives is either not permitted or is governed by very strict controls. This lengthens the time taken for a survey to be executed and makes it more expensive. Other sound sources may therefore be necessary in urban areas.

Hammer impacts on a metal plate embedded in the ground can be used as a sound source for short spread lengths of up to 50 metres. A hammer survey has the advantages of being quick and cheap, and causes no damage to the environment.

In order to use the hammer as a sound source for longer spreads the technique of stacking has been developed. By adding the signals obtained from successive hammer blows electronically in a memory, the signal to noise ratio can be greatly improved. The signal from each hammer blow takes the same time to arrive at the receiver whereas the noise will be randomly distributed. With a sufficient number of records stacked the noise will be averaged out. The ratio will improve as the square root of the number of records stacked. With this technique reasonable records can be obtained for spread lengths of more than 100 metres in relatively noisy conditions.

Tomographic Acoustic Measurements

As outlined previously, where the layered earth model does not exist refraction surveys may not be capable of detecting variations in velocities. As an alternative, tomographic acoustic surveying may provide a more satisfactory solution.

Tomographic reconstruction techniques are used in the field of medical imaging. Algorithms, used for medical purposes, are applied in geophysics for the interpretation of crosshole, and borehole to surface measurements. The algorithms are used to calculate a two dimensional velocity grid in the area of interest from the travel times[10]. There are no assumptions to be made in the interpretation, thus the accuracy of locating adverse zones depends only upon the layout of the survey and the frequency and accuracy of measurements.

The technique can be used to study geological situations which are thought to be too complex to be resolved by the refraction method or too far below the surface to investigate accurately using surface measurements. Investigation boreholes can be used after a little preparation for this purpose. As illustrated in Fig. 5A borehole BH1 can be used to position the sound source at different levels, whilst acoustic measurements are made in BH2. It can be seen that the acoustic waves traverse the area of interest at widely differing angles allowing detailed assimulation of the shapes of any adverse features present within the rock mass.

As shown in Fig. 5B an advancing tunnel face can also be used for transmission purposes thereby providing information about conditions between the tunnel face and boreholes ahead of the face. Blasting for the tunnel excavation can be used as the sound source. As the tunnel face approaches the borehole the information gained becomes progressively more detailed.

MARINE, ESTUARY AND RIVER CROSSINGS

The need for sub-aqueous crossings is becoming increasingly apparent along the Coast of China where redevelopment of the infrastructure of major cities/ports and the intervening transportation corridors are urgently needed.

The advantages of tunnelling for these crossings, either by bored tunnel or immersed tube can be considerable in comparison with bridge crossings which offer no environmental protection.

Investigations of bored tunnels by drilling and testing can be expensive in view of the need for drilling platforms and difficulties imposed by tides and currents. The use of long orientated boreholes drilled from either bank is recommended whenever possible. In addition, shallow water seismic reflection surveying provides an extremely useful tool for such studies[11].

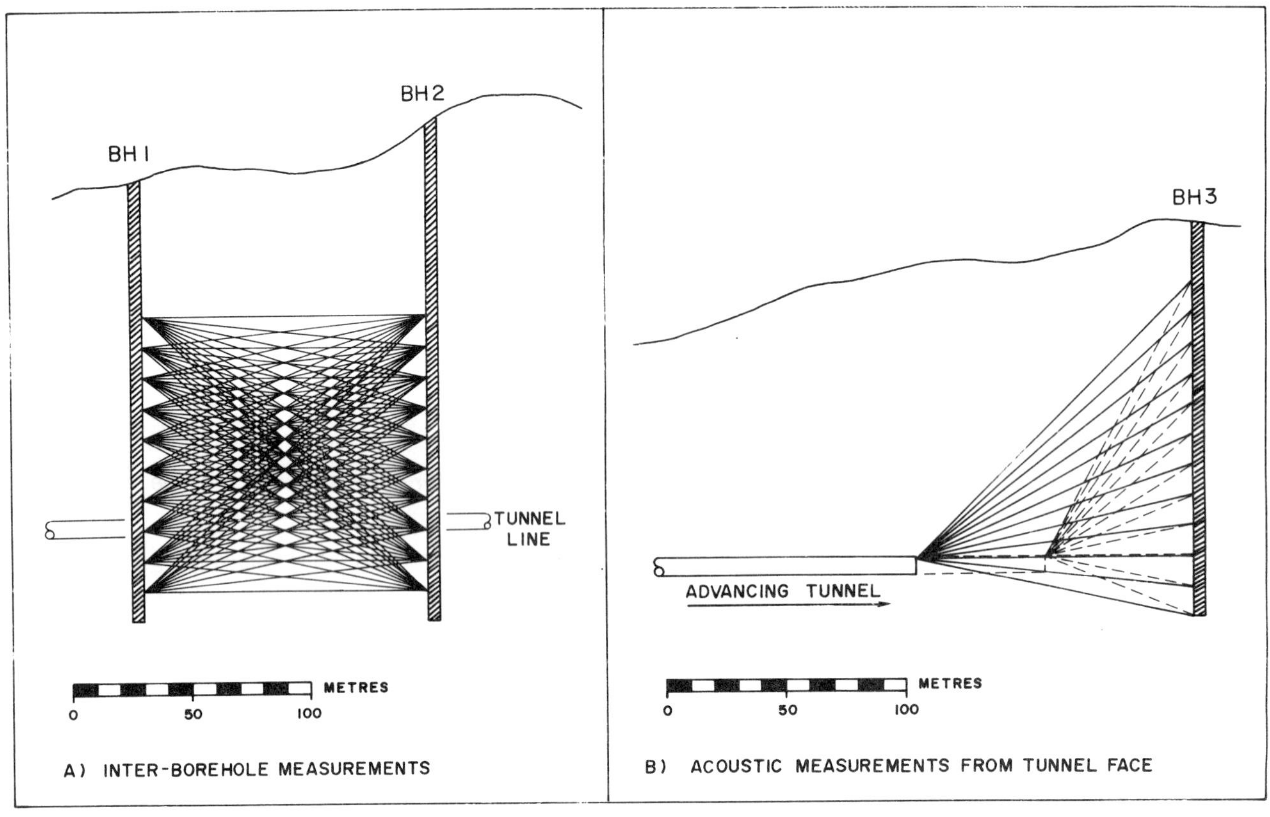

FIGURE 5 EXAMPLES OF ACOUSTIC TOMOGRAPHIC MEASUREMENTS

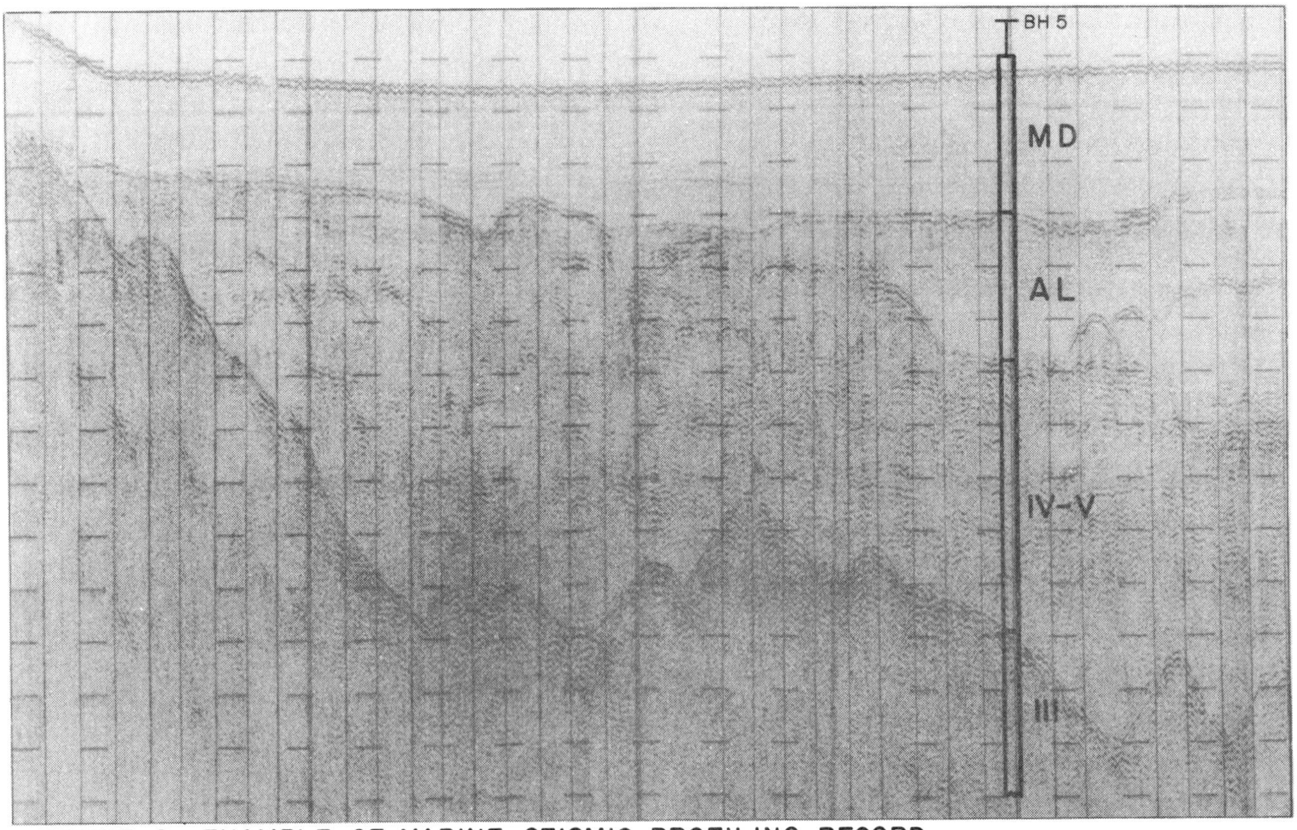

FIGURE 6 EXAMPLE OF MARINE SEISMIC PROFILING RECORD

MD = MARINE DEPOSITS
AL = ALLUVIUM
V - IV = DECOMPOSED ROCK
III - II - I = MODERATELY DECOMPOSED ROCK

For the latter both the sound source and the receiver are towed behind a small boat which carries the recording equipment. The sound source emits bursts of high frequency sound pulses at short intervals. The waves reflect off interfaces between media with different acoustic impedance. These reflections are received by the hydrophone receivers and, after suitable amplification and noise suppression, are recorded on either paper or magnetic tape, or both. By displaying the reflections from subsequent pulses adjacent to each other, a continous geological profile is produced as shown in Fig. 6. Here the horizontal axis is the boats position and the vertical axis is the two way reflection time.

The different horizons of interest are identified, traced and correlated with data from boreholes drilled in the area in order to convert the reflection times to depths.

Refinements to the technique include electronic position fixing systems; computerised navigation systems to control and record the boats navigation; and filters which will eliminate the effects of swell from the records. All of these refinements will increase the speed with which a survey is executed as well as the accuracy of the results.

This approach normally gives excellent results. Factors which can adversely affect the quality of the record are:

- Solid waste deposited on the sea bed and organic material trapped within marine sediments. These can provide a strong reflector of seismic energy obscuring reflections from deeper layers. The presence of solid waste is becoming an increasing problem in urban surroundings.
- Variable conditions such as erratically weathered rock or reworking of the sea bed by dredging operations.
- Narrow river crossings. It is found that under these circumstances multiple reflections from the river bed can confuse the record.

When the record quality is very poor the seismic refraction can be used to locate rockhead. The technique is basically the same as that used on land but is more difficult to execute in water-covered areas.

ROCK CLASSIFICATION

A significant limitation of the CSIR[12] and NGI[13] rock classification systems is that the input data required is normally far in excess of the sample obtained from site investigations. In addition, the number of parameters used is substantial (at least six) and, with particular reference to the NGI system, errors obtained when averaging the input parameters tend to be compounded. For these reasons such systems are most suitably applied at excavation faces where the rock mass can be observed directly. A simpler approach is recommended for the purpose of classifying conditions monitored from site investigations.

Fig. 7 shows a system developed specifically for use in massive igneous rocks such as those in Hong Kong. The approach, which has been outlined previously[1], assumes that the rock types are of a generally strong nature in their unweathered condition. Classification is made of the materials present in the tunnel crown, in accordance with the primary function of the system for predicting temporary support requirements. The application of the system can be extended for prediction of geologically related imponderables such as the overbreak and advance rates gained by different methods of excavation.

CONCLUSIONS

Emphasis has been placed in this paper on the use of various site investigation techniques for obtaining data of practical value for costing and planning tunnel projects.

Variations in the use of geophysical techniques such as seismic refraction, reflection and tomographic surveys have been presented. These are supported with case histories from projects in Hong Kong to illustrate the benefits to be gained from these cost-effective methods. The conclusion to be drawn here is that geophysics can provide us with the basic framework of the geology of the sub-surface and much of the detail but that drilling is always required to clarify the picture and, in particular, for investigation of key areas where difficult engineering works are to be undertaken. The potential of the seismic refraction method to locate zones of adverse ground where no other evidence is available on the surface has been shown.

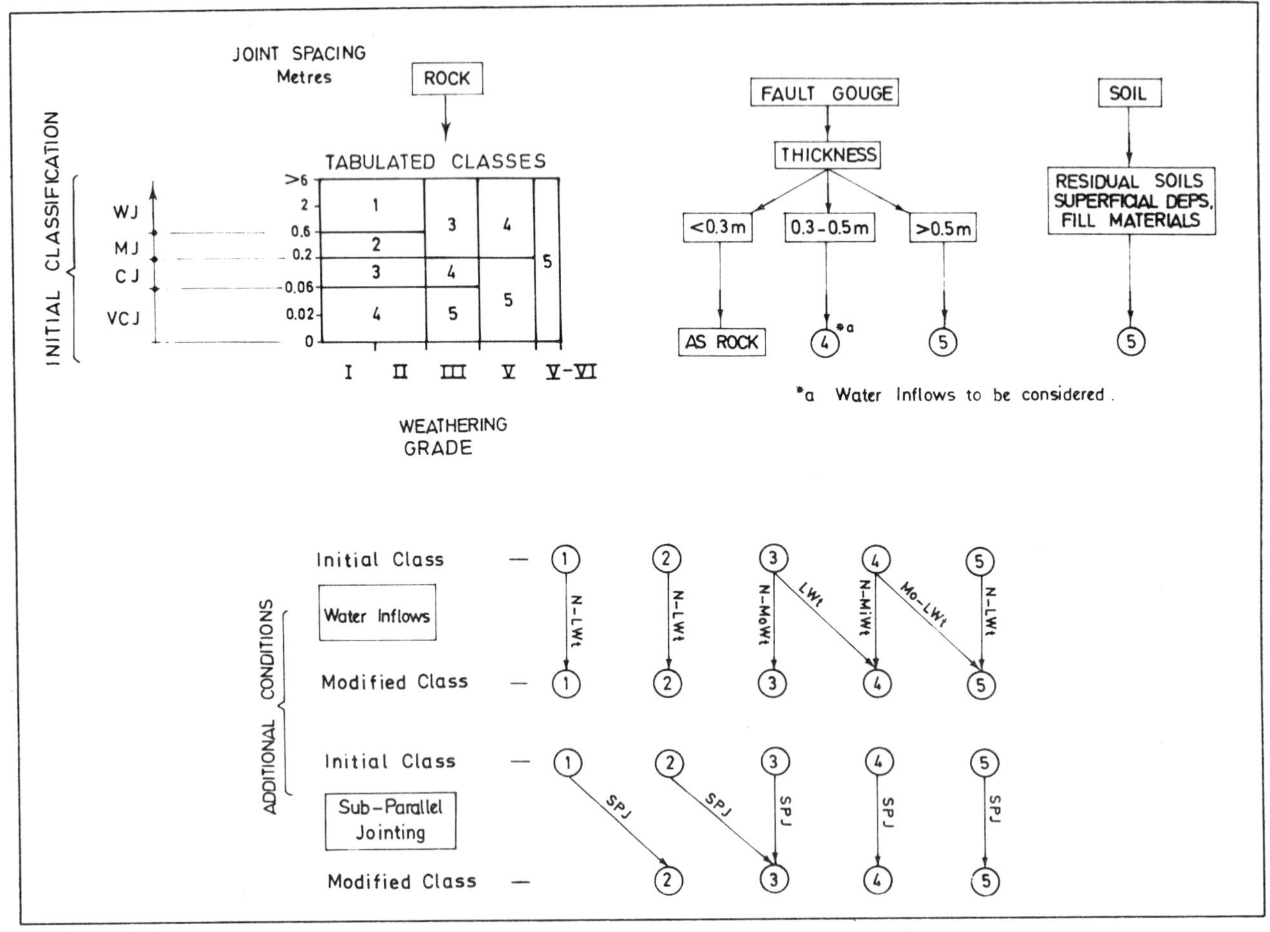

FIGURE 7. ROCK CLASSIFICATION SYSTEM FOR HONG KONG ROCKS

Notes for use of rock classification system

1. Determine material type present in tunnel arch and classify as ROCK, FAULT GOUGE or SOIL.

2. For classification of ROCK, monitor average conditions present in tunnel arch.

3. Use joint spacing and weathering conditions on table above to determine initial classification.

4. Check for effect of additional conditions, namely water inflows[14] and sub-parallel jointing.

5. For classification of FAULT GOUGE measure average thickness and classify as shown above.

An outline has been given of the geotechnical requirements for tunnel construction work at portals, areas of low cover, urban areas and for sub-aqueous crossings. The advantages of drilling inclined and long horizontal boreholes to investigate adverse conditions at these areas have been illustrated.

The classification system presented has an important application in tunnelling and provides a logical conclusion to a well executed site investigation. In addition, the same approach, used in conjunction with a comprehensive tunnel logging system[14], has been successfully employed for payment purposes in tunnel contracts in the United Kingdom[1].

References

1. McFeat-Smith I., Turner V.D. and Bracegirdle D.R. Tunnelling Conditions in in Hong Kong. Hong Kong Engineer, Vol. 13, No. 6, June 1985.

2. McFeat-Smith I. and Haswell C.K. Preliminary studies for tunnel projects in Hong Kong. Tunnelling '85 Symposium, IMM. Brighton, UK, March 1985.

3. Anon. Ground problems at Hong Kong's Aberdeen Tunnel. British Tunnelling Society Meeting. Tunnels & Tunnelling, May 1984.

4. McFeat-Smith, I. Geotechnical feasibility study and site investigation for the Western Aqueduct Tunnels, Hong Kong. Proceedings of the Seventh S.E. Asian Geotechnical Conference, Nov. 1982, p.171-187.

5. Coates D.J., Carter P.G. and McFeat-Smith I. Inclined drilling for the Kielder Tunnels. Quarterly Journal Engineering Geology, Vol 10, 1977, p.195-205.

6. Redpath, B.B. Seismic refraction exploration for engineering site investigations. National Technical Information Service, U.S. Department of Commerce, 1973.

7. Introduction to geophysical prospecting, Dobrin, M.B. 3rd ed. McGraw - Hill Book Company Inc, 1976.

8. Applied geophysics. Telford, W.M., Geldart, L.P., Sheriff, R.E. & Keys, D.A. Applied geophysics. Cambridge University Press, 1976.

9. McFeat-Smith, I. The drilling of long horizontal boreholes for site investigation purposes. Meeting on Geological Aspects of Site Investigations. Bulletin No. 2, Geological Society of Hong Kong, 1985, p.31-40.

10. Cosma, C. Determination of rock mass quality by the crosshole seismic method. Bulletin of the International Association of Engineering Geology, no. 26-27, 1983.

11. Ridley - Thomas, W.N. The application of engineering geophysical techniques to site investigations in Hong Kong. Proceedings of the Seventh South East Asian Geotechnical Conference, Nov. 1982. p.205-226.

12. Bieniawski, Z.T. Rock mass classification in rock engineering. Proceedings of Symposium on Exploration for Rock Engineering, Johannesburg, Vol. 1, 1976. p.107-117.

13. Barton, N. Recent experiences with the Q-system of tunnel support design. Proceedings of Symposium on Exploration for Rock Engineering, Johannesburg, Vol. 1, 1976. p.107-117.

14. McFeat-Smith, I. Logging tunnel geology. Tunnels and Tunnelling, May 1982.

Rock trap design for presplit rock slopes

N. Mak B.SC., M.SC., D.I.C., F.G.S.
D. Blomfield B.SC., C.ENG., M.I.C.E., M.H.K.I.E.
Ove Arup & Partners Hong Kong, Ltd., Hong Kong

SYNOPSIS

This paper reports a series of field trials to investigate the design of rock traps for presplit rock slopes up to 12m high formed adjacent to a residential development. The work involved releasing over 13,000 boulders on a variety of slopes and measuring their trajectories with simple and inexpensive equipment.

INTRODUCTION

The treatment of instability of rock slopes may be divided into :-

(a) the prevention of release of substantial quantities of rock along relatively continuous joint sets.

(b) the control of small individual blocks of rock after release.

The approach to the first type of instability is well documented and takes the form of draining the slope and/or providing structural support, such as rock anchors, dowels, buttresses etc. For the second type of instability structural support may often be provided by shotcrete or mesh but for reasons of cost or aesthetics rock traps are frequently adopted. Such traps usually comprise a system of drop zone and barrier. The design of rock traps is usually based on the data obtained by Ritchie[1] and design charts published by Fookes and Sweeny[2]. The work Ritchie carried out in 1963 was to establish ditch dimensions to contain rockfalls. The rock slopes examined were not necessarily presplit and were often natural slopes. The boulders under consideration were 'of all sizes' and presumably frequently larger than would be encountered on a presplit face in highly fractured rock. The use of this information and the Fookes and Sweeny charts for the design of rock traps was therefore considered likely to lead to a conservative design when applied to relatively smooth presplit rock faces.

THE INVESTIGATION

At the outset of the investigation several parameters were considered to influence the trajectories of boulders falling from a rock slope. They were the type, shape and size of the boulders, and the height, angle and profile of the slope. These factors interact with one another in the complex process of rock falling from a slope. The objective of the investigation was to examine the effect of each of the parameters.

The investigation was carried out on a 42 hectare site located on the south western shore of Port Shelter (Ngau Mei Hoi) in Hong Kong. This site is formed from volcanic rock in varying states of decomposition belonging to the Repulse Bay Formation. The rock forming these slopes is a strong to very strong, dark grey fine grained closely to moderately jointed fresh to slightly decomposed volcanic tuff, forming irregular shapes on degrading.

The site formation work involved the forming

of approximately 8000m^2 of rock cuttings. These rock cuttings were formed by presplit-blasting using a combination of Tovex and A.N.F.O. placed in 65mm diameter holes drilled to a maximum length of 15m at 0.5m spacing. The final face was achieved by manually trimming and scaling.

The investigation was carried out at thirteen locations on various rock slopes on the site. The slope height, H, slope angle, θ, and slope profile, P, for each of the slopes was measured and recorded. The slope profile was estimated by measuring offsets at 250mm increments along a reference line parallel to the slope face. A typical section is shown in Figure 1.

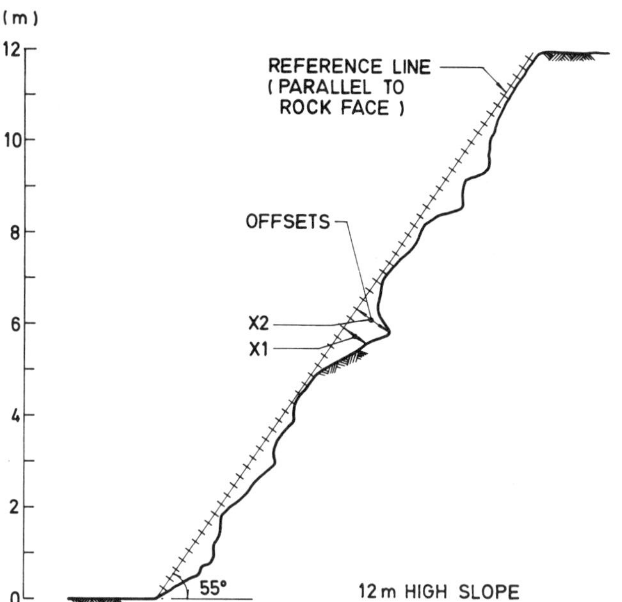

SLOPE PROFILE MEASUREMENTS

FIGURE 1

The ground at the toe of each slope examined was levelled and covered with a layer of compacted rock fill to provide similar energy absorption characteristics at every location. As the rock exposed in the cut faces was closely jointed and all major areas of instability treated by measures such as dowelling, bolting and scaling it was considered that any rockfalls would be of small blocks. It was therefore decided to carry out the investigation using 100mm to 300mm angular boulders collected from the various rock cuttings on the site.

The trajectory of each boulder was recorded in terms of the height, h, above ground level that the boulder attained at two fixed distances, D, from the toe of the slope, and the total distance travelled, d, measured from the toe of the slope. Figure 2 illustrates the principal parameters and conditions that were considered in the investigation and a test set up used for recording the trajectory of a boulder.

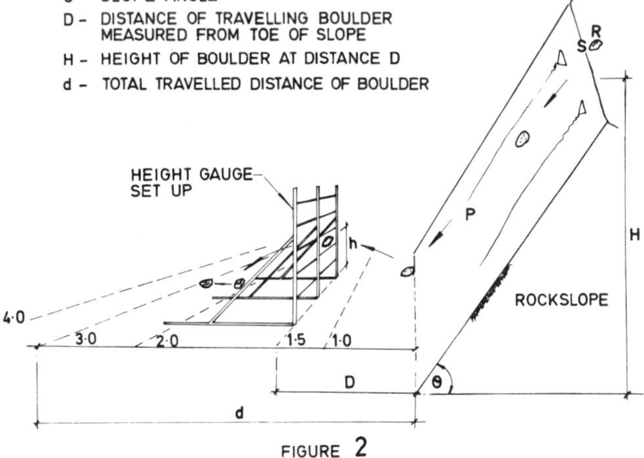

FIGURE 2

The height gauge comprised a series of five horizontal bamboo canes at 300mm intervals spanning between wooden stands at a fixed distance from the toe of the slope. A boulder passing between any two canes was recorded as passing below the upper cane.

Preliminary experiments indicated that within the range of parameters to be tested the height gauge should be positioned within 2.0 metres of the toe of the slope as beyond this distance most of the boulders were travelling at less than 300mm above ground level.

Therefore, all the tests were undertaken with two gauges located at 1.0 metre and 1.5 metres from the toe of the slope.

The measurement of the total travelled distance of a boulder was defined using a series of red lines painted on the ground, spaced at 1.0 metre intervals, starting from the toe of the slope. A boulder stopping between two reference lines was recorded as having stopped by the distance to the further line.

The boulders were initially divided into two groups: 100-200mm blocks and 200-300mm blocks. Preliminary tests with these two groups indicated that there was no significant difference in the trajectories of the boulders from each group and so the two groups were amalgamated. To achieve statistically representative results it was decided that 1000 boulders should be released from the crest of each slope.

PRESENTATION OF RESULTS

The results of thirteen experiments undertaken were categorized into five groups according to slope angle and height as follows :-

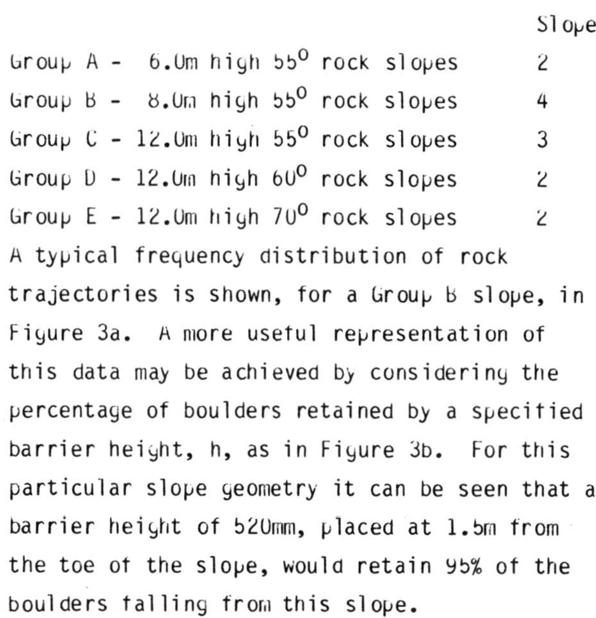

	No. of Slopes
Group A - 6.0m high 55^{0} rock slopes	2
Group B - 8.0m high 55^{0} rock slopes	4
Group C - 12.0m high 55^{0} rock slopes	3
Group D - 12.0m high 60^{0} rock slopes	2
Group E - 12.0m high 70^{0} rock slopes	2

A typical frequency distribution of rock trajectories is shown, for a Group B slope, in Figure 3a. A more useful representation of this data may be achieved by considering the percentage of boulders retained by a specified barrier height, h, as in Figure 3b. For this particular slope geometry it can be seen that a barrier height of 520mm, placed at 1.5m from the toe of the slope, would retain 95% of the boulders falling from this slope.

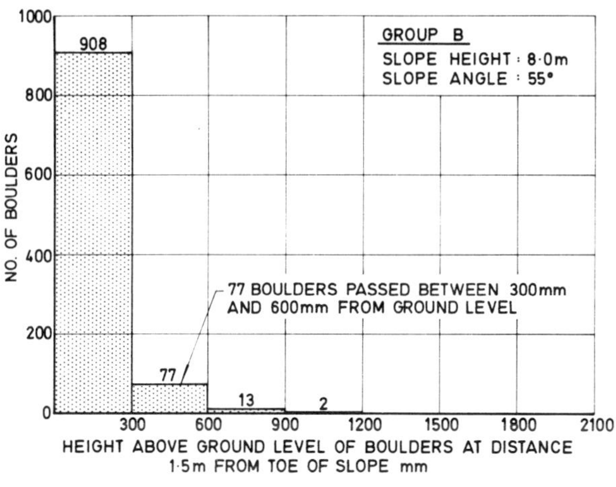

FIGURE 3a

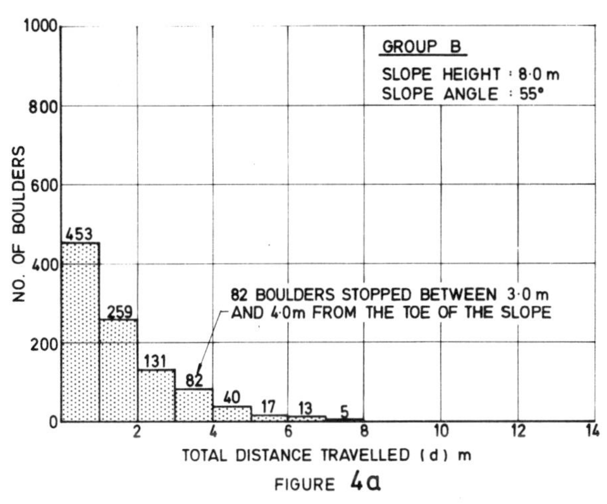

FIGURE 4a

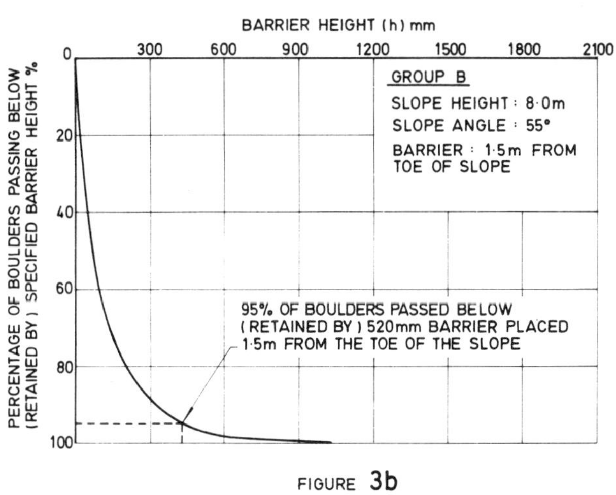

FIGURE 3b

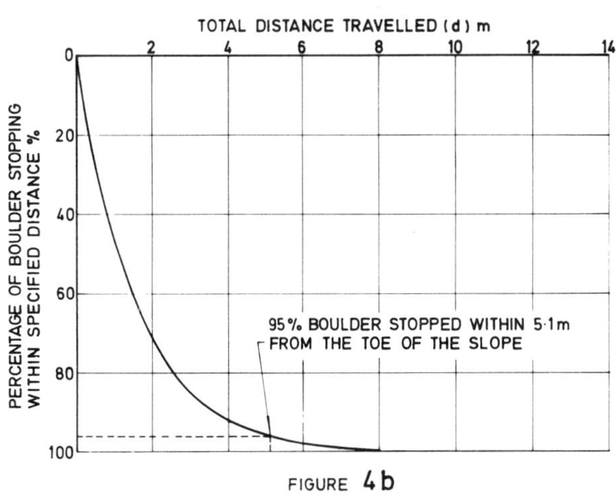

FIGURE 4b

A typical frequency distribution of travel distance of fallen boulders, d, on a Group B rock slope is shown in Figure 4a. As with Figure 3a and 3b it is considered more useful to present the data in the form of percentage of boulders which stop within a specified distance, d, as in Figure 4b. For this slope it can be seen that 95% of the boulders stopped within 5.1 metres of the toe of the slope. The results for Group B slopes are summarised in Figure 5.

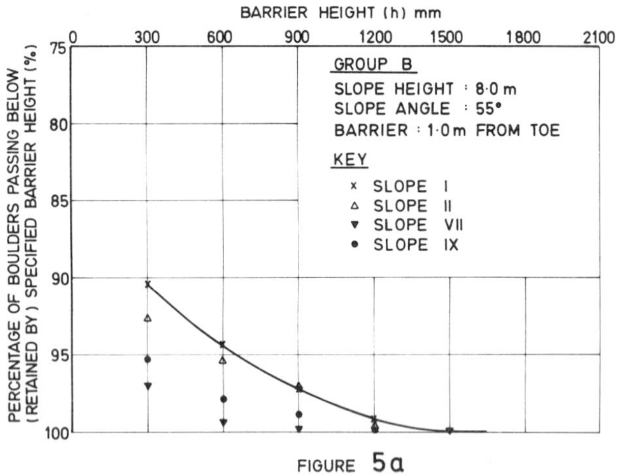

FIGURE 5a

To generalise the trajectory of a rockfall on presplit faces envelopes have been drawn in Figure 5a and 5b to represent the lower bound value of percentage of boulders retained by a specified barrier height placed at both 1m and 1.5m from the toe of the slope. An envelope has also been drawn on Figure 5c to give a lower bound value of the percentage of boulders stopping within a specified distance.

Similarily envelopes have been drawn from the results of the remaining groups and these are summarised in Figures 6 and 7.

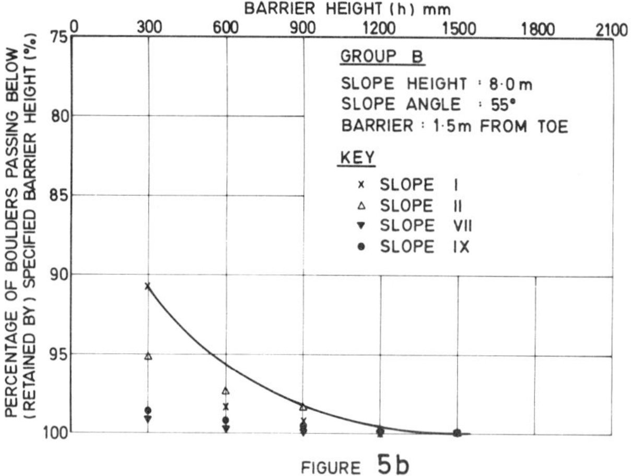

FIGURE 5b

INTERPRETATION OF RESULTS

There are several points of interest in the investigation. The boulders for each experiment were released from rest at the crest of the rock slope. This represents an upper limit on the potential energy for boulders released from the slope and consequently each result signifies an upper bound value on the parameters measured.

As Group B contained a large number of slopes of the same height and angle but of differing profiles this group may be used to examine the effects of slope roughness.

Figure 8 shows the profiles of slopes in this group, where Slope I has the most irregular face, and Slope VII has the smoothest slope face.

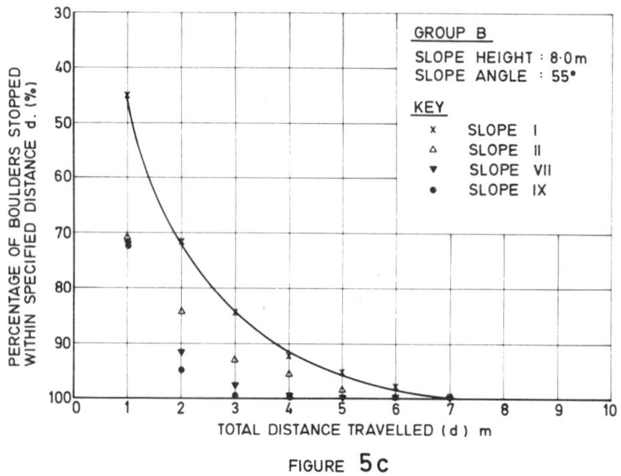

FIGURE 5c

For the design of rock traps, the results of most significance are those which represent the retention of a high percentage of the boulders. Referring to Figure 5, it can be seen that even under these extreme profile conditions, the percentage of boulders retained for a given barrier height is not sensitive to slope profile at the higher percentages. Therefore

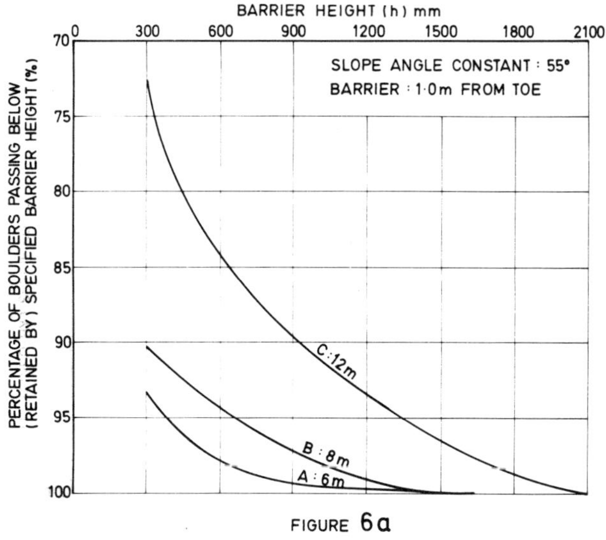

FIGURE 6a

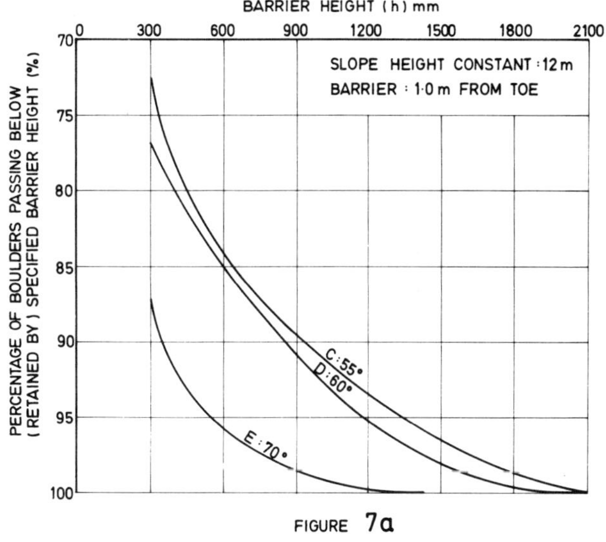

FIGURE 7a

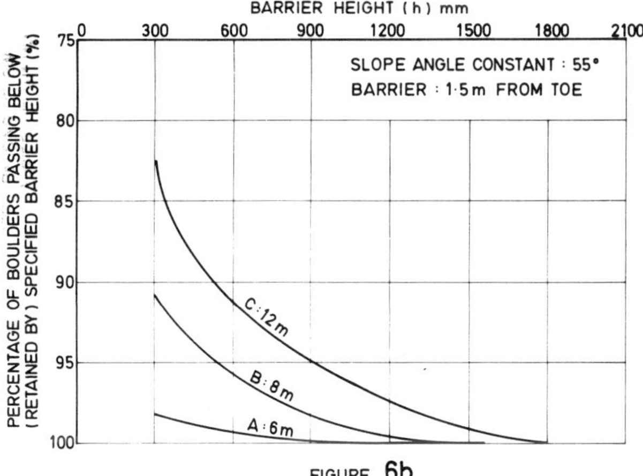

FIGURE 6b

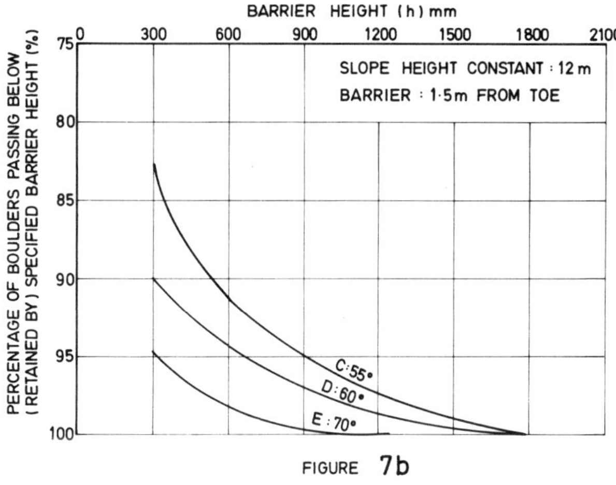

FIGURE 7b

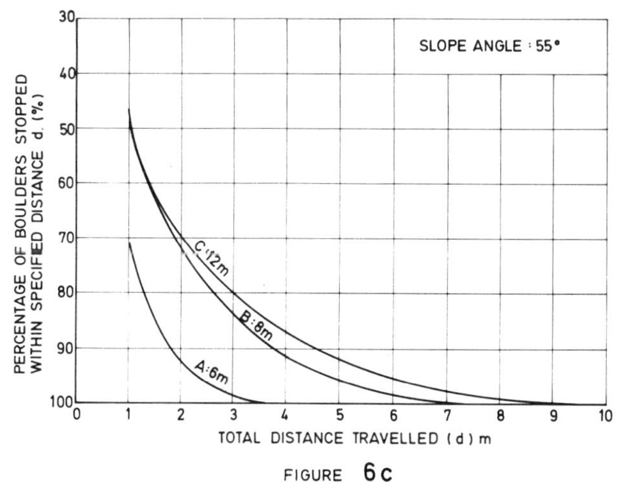

FIGURE 6c

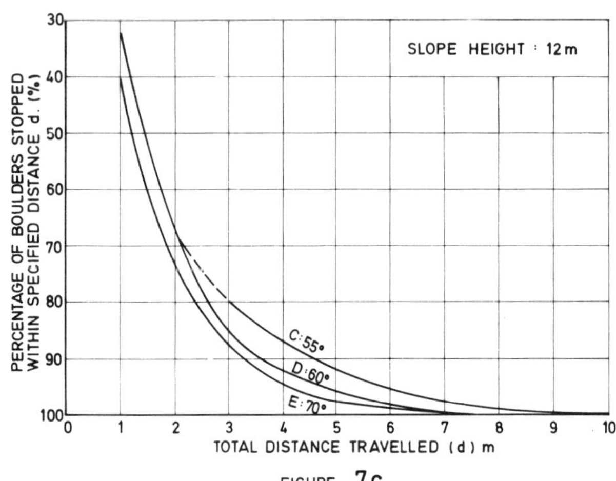

FIGURE 7c

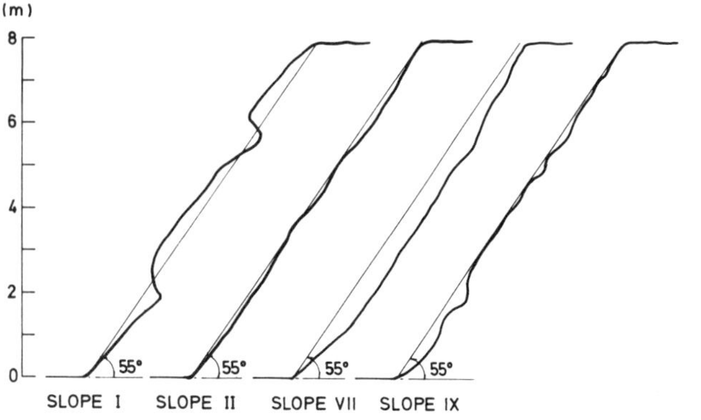

FIGURE 8

for design purposes within the conditions examined the influence of slope profile on boulder trajectory is negligible. Site observation did, however, indicate that at distances very close to the toe of the slope the trajectory of the boulder was influenced by the slope profile but this influence rapidly diminished with increasing distance from the toe of the slope.

The effect of slope height on the rock trajectories for a constant slope angle of 55^0 on presplit rock faces can be seen by comparing results for Group A, B and C (Figure 6). It is apparent that an increase in slope height results in a larger stopping distance and requiring a higher barrier for a given width of drop zone. For a given slope height the effect of slope angle on presplit slopes can be seen by comparing Groups C, D and E as shown in Figure 7. An increase in slope angle requires a smaller barrier height and results in a smaller stopping distance since the horizontal component of velocity and angular velocity of the boulders decrease with increase in slope angle (as described by Ritchie).

ROCK TRAP DESIGN

The investigation indicated that for recently exposed presplit rockfaces, with small angular boulders of between 100mm and 300mm, the size of the boulders had negligible influence on their trajectories. The angle and the height of the slope, however, were found to have a major influence on boulder trajectory and consequently on the design of a suitable rock trap. The influence of slope profiles, within the range studied was only minor.

The results presented in Figures 6 and 7 can therefore be used as simple charts for the design of rock traps for presplit rock faces with slope heights varying between 6.0m to 12.0m, and slope angle between 55^0 to 70^0.

It is recommended the design of rock traps should be at all time, aimd at the highest percentage retention of boulders practicable. An important factor affecting the choice of an appropriate percentage retention of boulders is the degree of risk involved in the area of consideration. A higher percentage should be aimed for if the rock trap is to comprise a drop zone and a barrier than if the trap is to consist of only a wide drop zone. The reason for this is that boulders not retained by the former system, can still have substantial kinetic energy and thereby pose a greater risk to life.

For a 12.0 metre high, 55^0 presplit rock slope, where 95% retention of boulders is considered appropriate, Figure 6 and 7 indicate that :-

a) a barrier height of 1350mm is required at 1.0 metre from the toe of the slope, or
b) a 925mm high barrier is required at 1.5 metre from toe of the slope.

For a rock trap without a barrier a 4.5m wide drop zone would be a viable alternative, retaining 90% of the boulders.

For comparison purposes the Fookes and Sweeny Chart based on Ritchie's work would give a drop zone of 4.5m wide and a barrier height of 1.75m for this slope geometry.

CONCLUSION AND RECOMMENDATIONS

In an urban environment it has often been the case that insufficient space has been available for the introduction of rock traps at the base of rock cuttings. This has lead to the extensive use of meshing and other slope protection measures such as shotcrete to prevent small rocks falling from the slope and endangering the public. These measures are often unsightly and expensive.

The simple but extensive test performed on the

site showed that for a presplit rock face the problem of small scale rockfall could be overcome by the use of a small rock trap.

The design chart presented in Figures 6 and 7 were used for establishing the size of rock trap for a practical range of presplit slopes where boulders upto a size of 300mm were considered to be potentially unstable. Whilst this work was for a particular site the method of obtaining the data and the conclusions reached may be of more general guidance to engineers carrying out similar works.

It is hoped that further trials will be undertaken so that these charts can be extended to provide useful guides on the design of rock traps for a wider range of rock slopes and boulder size. Such investigation should in particular identify the effect on boulder trajectory of providing intermediate benches in rock slopes.

ACKNOWLEDGEMENTS

The authors wish to express their gratitude to the Architectural Office of the Building Development Department of the Hong Kong Government for permission to publish the results of these trials, to Leighton Contractors PTY Limited for their assistance in carrying out the field work and to the staff of Ove Arup & Partners for their assistance in preparing this papar.

REFERENCES

1. Ritchie A.M. (1963) Evaluation of rockfall and its control. Highways Research Record 17, 13-28.
2. Fookes, P.G. & Sweeny M,(1976) Stabilisation and control of local rockfalls and degrading rock slopes. Q. JL. Engng Geol 1976 Vol.9 pp 27-55.

Design and excavation of stable slopes in hard rock, with particular reference to presplit blasting

G.D. Matheson B.SC.(HONS.), PH.D., F.G.S.
Transport and Road Research Laboratory, Livingston, Scotland

SYNOPSIS

Instability in hard rock can be divided into natural and induced types. The former results from a geometric imbalance between discontinuities in the rock mass and the land surface; the latter is induced into the rock mass during excavation. Excavations in hard rock should be designed to minimise both types of instability by optimising both the design slope and the excavation technique.

In an urban environment it may not be possible to select a design slope which minimises the natural instability. In such circumstances knowledge of the optimum design slope can be used to estimate the type and amount of remedial work.

The approach advocated involves stability assessment based on available data updated periodically during excavation, and the use of an excavation technique which does not adversely affect the final face. In this latter respect presplit blasting is recommended as a means of limiting the blast disturbance produced by bulk blasting - the main source of induced instability in most surface excavations in hard rock.

The paper suggests design and excavation methods which have particular relevance to urban environments and describes methods which can be used to collect, present and evaluate field data. A method of presplit blasting, used successfully in a wide variety of highway excavations in Scotland, is outlined and guidance given on important aspects.

INTRODUCTION

Land is usually at a premium in an urban environment and surface rock excavations must utilise the steepest possible side slopes to minimise land take. In addition full stability assessment is difficult to carry out because data on discontinuities are scarce. Consequently optimisation of design in terms of the geotechnical conditions is seldom possible and extensive remedial work is frequently necessary. Difficulties such as these are often cited as the reasons for adopting empirical designs and standard excavation techniques. The knowledge that instability in the final face can be rectified by remedial work appears to give reassurance to the design engineer. This is seldom justified and not enough forethought is given in many projects to devising logical procedures which could minimise disruption to the contract schedule, claims for unforeseen ground conditions and damage to the rock slope.

NATURAL INSTABILITY IN ROCK SLOPES

Instability in a rock slope generally arises from an imbalance between the geotechnical conditions within the rock mass and the surface morphology. In hard rock the situation simplifies to a geometric problem involving the orientation of planar weaknesses (discontinuities) in the rock mass relative to the slope of the ground surface. The discontinuities are believed to be stress induced and to result from one or many phases of regional stress acting in the brittle levels of

Plate 1 Natural instability resulting from coastal erosion in sandstone.

Locality: Old Man of Stoer, NW Scotland

Plate 2 Plane failure resulting from natural instability in a highway excavation in gneiss.

Locality: A894, Kylesku, Scotland

the earth's crust; they thus pre-date the land surface and have a strong influence in developing the landform. Very little, if any, change in the discontinuity geometry occurs immediately prior to the development of instability.

Although the geological trend is towards the elimination of topography, natural processes, such as river and coastal erosion, glaciation or extreme weathering, can increase topographic contrast and may activate, reactivate or accelerate instability in a rock mass. Man-made excavations for civil, mining or other engineering purposes, can be regarded as processes of rapid slope steepening. The effect on a rock mass is thus similar but the time scale much shorter. If geometric conditions are unfavourable then excavation can either trigger or accelerate failure by removing support and allowing movement of individual rock blocks to occur.

It is recognised that factors other than the geometry of the situation can have an extremely important role to play in any failure mechanism. The role of ground water, freeze-thaw cycles, and the frictional and morphological characteristics of the discontinuities are particularly important in this respect. Changes in ground water conditions and seismic vibration are also recognised as important trigger mechanisms. These factors are not considered in this paper.

Instability in this context is therefore defined as a state of imbalance between discontinuity bounded blocks constituting the rock mass and the ground surface.

Examples of natural instability are shown in Plates 1 and 2.

INDUCED INSTABILITY IN ROCK SLOPES

Induced instability can be caused by the excavation process or can arise from changes in the principal stresses resulting from the removal of rock. The former is generally

Plate 3 Blast induced instability in a highway cutting in gneiss.

Locality: A832, Gruinard, Scotland

Plate 4 Blast induced instability in a highway excavation in dolerite.

Locality: A90, Forth Road Bridge Approaches, Scotland

associated with the uncontrolled use of high explosives which can cause widespread damage to the design slope. Such damage is caused mainly by the dilation of existing fractures in the rock mass. Stress relief manifests itself through the formation of new fractures related to the land surface; they therefore post-date rock removal. Care should be attached to the identification of these latter fractures as it is very often possible to show that they have been formed by the blasting.

Damage to a design slope can be minimised by using carefully controlled techniques of excavation, designed to maintain as far as possible the natural conditions of stability in the rock mass. Presplit blasting is an example of such a technique.

Induced instability is defined as imbalance caused by the excavation process and is superimposed on any natural instability which exists.

Examples of induced instability are shown in Plates 3 and 4.

DESIGN OF ROCK SLOPES

Data collection

Site investigations for the design of slopes in hard rock must obtain factual information on the orientation and importance of discontinuities. Without such information there is no option, at least initially, but to resort to an empirical design based on experience in supposedly similar conditions. All reasonable effort must therefore be made through desk studies, field mapping, trial pits, drill holes and other techniques, to obtain relevant data. If data cannot be obtained from the immediate area of interest it may still be possible to extrapolate information from nearby locations. In urban environments relevant data are likely to be obtained from records of earlier excavation work and through study of exposed rock faces.

Methods of collecting and recording the data are well established[1] and will not be described in the present text. It is important however, to remember that it is the accuracy, relevance and representativeness of the data which are of prime concern rather than quantity alone.

The concept of a stability domain, a region of rock containing a definable pattern of discontinuities, is important and should be kept in mind during the collection of data.[2] The recognition of domains, perhaps related to local geology, considerably assists evaluation. Indiscriminate grouping of data can obscure discontinuity patterns and make subsequent evaluation meaningless.

Presentation of data

Field data relating to the orientation of discontinuities are difficult to analyse without graphical assistance. Current forms of presentation rely on types of stereographic projection in which the distribution of both poles and intersections of discontinuity planes are studied. The latter are usually plotted by using great circles. Recommendations and full instructions on a manual method of presentation are given elsewhere.[3] Where large numbers of discontinuity readings have to be processed, or weighting of the field data in terms of importance has to be carried out, then computer assistance is necessary. An example of a computer assisted stereographic projection is given in Fig. 1.

Computer techniques can also be used to aid presentation and assist evaluation of field data without recourse to stereographic projection. An example using simple frequency histograms is given in Fig. 2. This facilitates visual analysis of the data in terms of variation and grouping. However to those familiar with stereographic projection the outcome does not compare favourably with a stereoplot.

A full listing of original data should be considered mandatory in any presentation.

Kinematic evaluation

The potential for instability in a design slope can be evaluated from stereographic projections using the kinematic approach.[4-6] In this the geometric conditions satisfying common modes of failure are depicted on stereoplots of the data. The logic involved and a method of using separate stability overlays have been included in a previous publication[3] and will not be described further. In practice overlays can be combined and the potential for failure in each mode assessed by reference to a single stereoplot using poles for plane failure, intersections for wedge failure, and a combination of poles and intersections for toppling failure. In each mode the possibility of failure increases with increasing numbers of points, or higher contour levels, within the area of instability defined

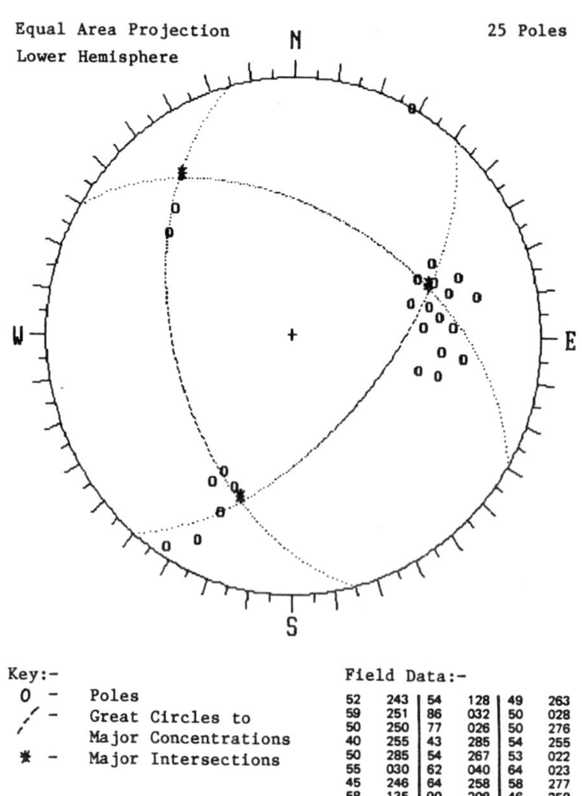

52	243	54	128	49	263
59	251	86	032	50	028
50	250	77	026	50	276
40	255	43	285	54	255
50	285	54	267	53	022
55	030	62	040	64	023
45	246	64	258	58	277
58	135	90	208	46	258
				43	266

Fig 1. Computer Assisted Stereographic Projection

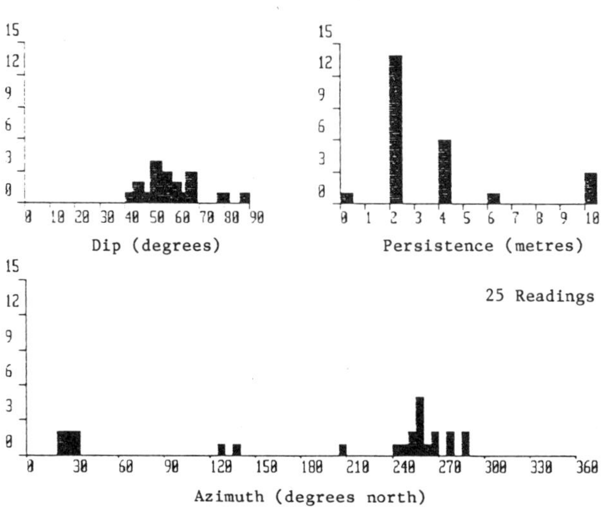

Fig 2. Frequency Histograms of Field Data

for that mode. The method, illustrated in Fig. 3, is simple and quick, and allows immediate visual assessment of the effect of changes in design; it thus permits optimisation of the design slope in terms of the local discontinuity geometry.

Design methods for urban environments

Design methods for rock slopes in urban environments should, as elsewhere, have the main objective of minimising the natural instability caused by the proposed slope steepening. In hard rock this entails optimisation of slope geometry in terms of discontinuity geometry and importance.

The engineer may however be forced to adopt a design which is not optimum; complicated and costly remedial work may then be required to strengthen the final face. Special methods of working may also be required to prevent failure and ensure safety during the excavation process. Knowledge of the type and extent of possible failures is consequently very important to the design process.

An awareness of the imbalance between optimum and proposed designs is a prerequisite to the efficient selection, design, and costing of remedial work. Even in situations where no variation in slope design is permitted it may still be important to establish how far the proposed design departs from the optimum. This knowledge allows both the method of working and the technique of excavation, factors which are usually controllable, to be carefully chosen to minimise remedial requirements. Such evaluations can be made using the techniques of stability assessment outlined in the present paper.

In situations where the acquisition of reliable data proves a problem then a process involving periodic collection, updating and evaluation of data is recommended as the excavation proceeds. For such a process to be successful a certain degree of flexibility in thought, design and contractual conditions is necessary. It also implies that a certain element of contingency planning has been included in the contract.

The procedure of continued observation during the excavation and regular updating of both data and stability assessment has much to commend it in all excavation projects. No site investigation in rock can be expected to reveal the exact conditions which will be encountered on the final face.

EXCAVATION OF HARD ROCK

Blast disturbance

Considerable damage can be caused to a rock slope by the uncontrolled use of high explosives. Field studies conducted on a number of projects in Scotland suggest that, at least in highway excavations, the damage induced is not one of

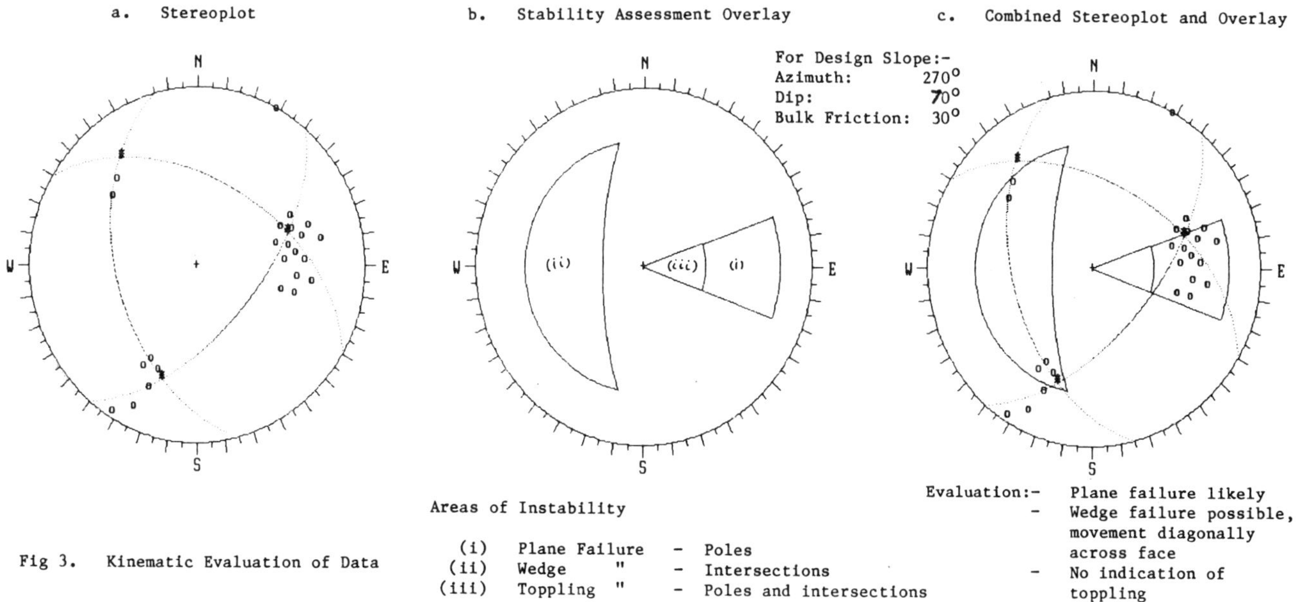

Fig 3. Kinematic Evaluation of Data

shattering of the rock but one of disturbance of the rock mass through the dilation of mainly pre-existing fractures. The same is likely to be true of most excavations for civil engineering purposes in urban environments.

The concept of blast disturbance is illustrated in Fig. 4. An explosive charge has been placed in a drill hole in a fractured rock mass. Before detonation the fractures are predominantly tight. Immediately after blasting the rapid expansion of gases from the explosive causes dilation of adjacent fractures in a zone surrounding the hole, and ground heave results. When the gas pressure subsides some fractures are unable to close fully leaving them in a dilated state. If the stresses produced by the blast exceed the strength of the rock then new fractures, usually produced through tensile failure, may be formed. In general these are found close to the drill hole, and seldom extend far into the surrounding rock.

The extent of blast disturbance associated with various methods of excavation has been

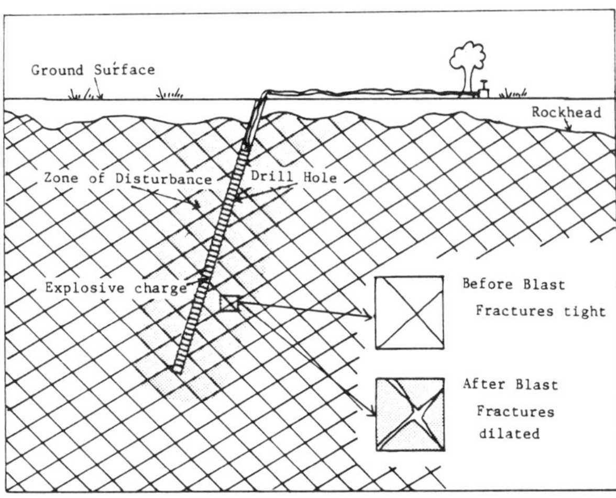

Fig 4. Blast Disturbance

established by direct measurement of dilation, and inferred through the use of seismic refraction techniques. The description of the principles and the methods used have been published previously.[2] Widths of disturbance inferred from the use of seismic refraction techniques in several excavations are given in Table 1. Disturbance was not detected in the natural slopes studied, in old faces excavated predominantly by hand, or in faces excavated by presplit blasting. Extensive disturbance was

Table 1 Widths of disturbance associated with various methods of excavation in Scotland

Excavation Type	Excavation Method	Approx Age (Yrs)	Rock Type	Inferred Width of Disturbance
Natural	-	10,000	Dolerite	Nil
	-	"	Basalt	"
	-	"	Gneiss/ Schist	"
Rail	Hand and Black Powder	100+	Dolerite	"
Quarry	Bulk	0-60	"	6m
	"	10	Dolerite & Sediments	8m
Highway	Bulk	20	Dolerite	6m
	"	5	Gniess/ Schist	2m+
	Smooth Blast	8	Basalt	4m
	Pre-split	7	"	Nil
	"	5	Gneiss/ Schist	"

however detected in rock slopes excavated by bulk, smooth and trim blasting, the first mentioned being associated with the greatest width of disturbance. The highest degree of instability, as determined visually, was associated with the greatest width of disturbance in all the excavations studied. The only faces excavated with high explosives in which disturbance was not detected seismically were those formed by successful presplit blasting.

Presplit blasting

Theory

Presplit blasting is a form of controlled blasting designed to minimise damage in the final face. The method involves the drilling of closely spaced holes along the proposed design slope and charging these relatively lightly. The presplit charges are detonated simultaneously thus forming a discontinuity along the design slope. Bulk blasting techniques are then used to fragment and loosen the rock in the excavation area. The presence of the presplit discontinuity limits disturbance caused by bulk blasting and minimises induced instability in the design slope. The method is shown diagrammatically in Fig. 5.

At present there is not universal agreement regarding the theoretical basis for presplit blasting. Two theories are in current use: in one the presplit fractures are formed during the short-lived shockwave or dynamic phase

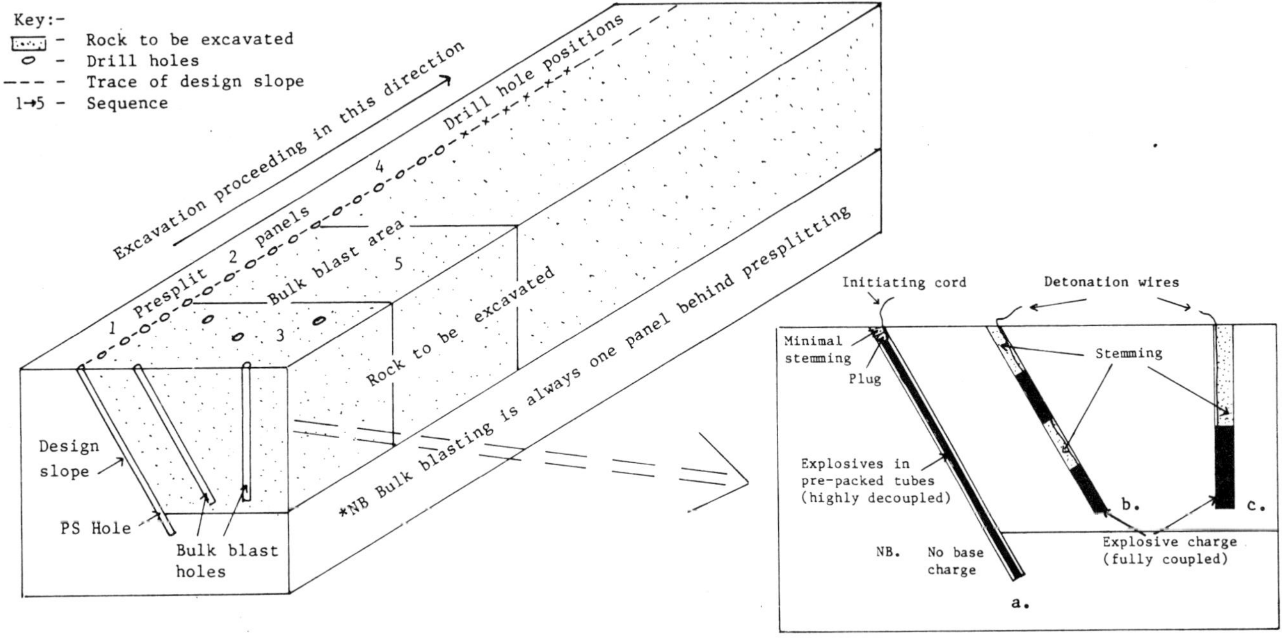

Fig 5. Presplit and Associated Bulk Blasting

Details of Charging a. Presplit holes
b. & c. Bulk blast holes

immediately following detonation of the explosive charges; in the other it is sustained gas pressure during a quasi-static phase which results in propagation of fractures between drill holes.

Laboratory and field studies of presplitting[7] indicate that the second of the two theories more satisfactorily explains the results. In this theory the dynamic component plays a relatively minor role, its effect being to form radial fractures which cut the drill hole wall - if none already exist. The quasi-static component plays the most important role, and sustained gas pressure during this phase is held responsible for the formation of a directional stress field and the preferred propagation of fractures between drill holes. A presplit discontinuity is thus formed. The basic mechanism is summarised in Fig. 6.

Geological and geotechnical conditions

Experience has been gained from presplitting sedimentary, metamorphic and igneous rock types, mainly in Scotland. Faces have been successfully formed in Devonian sandstones, basalts and andesites, in Precambrian schists and gneisses and in widely different ages of granitic and basic intrusives. The geotechnical conditions

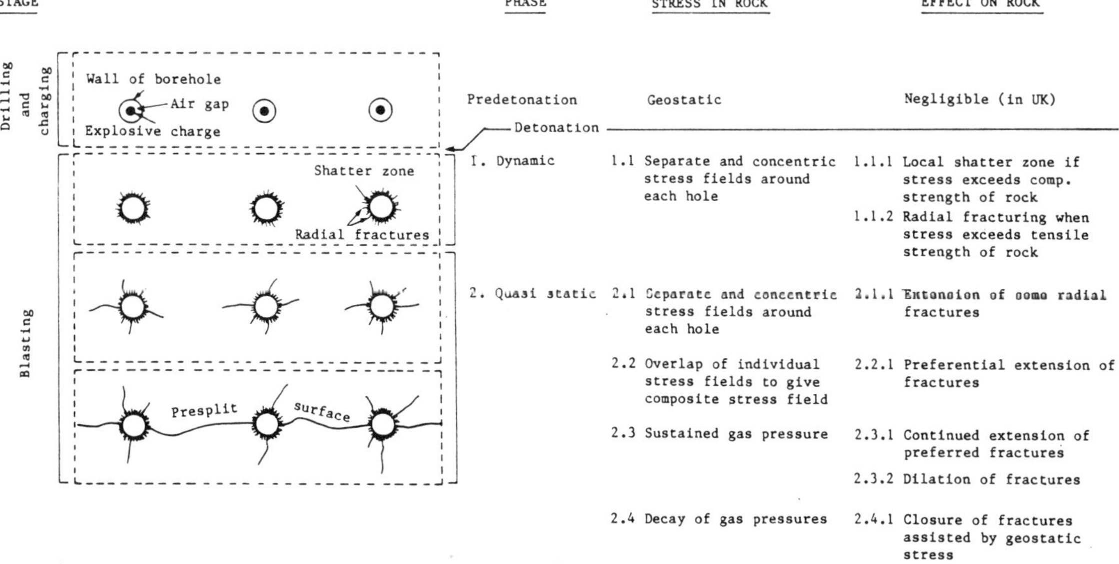

Fig 6. Basic Mechanism of Presplit Blasting in Surface Excavations

Plate 5 Face formed by presplit blasting in andesitic lava

Locality: M90, Moncrieffe Hill, Scotland

Plate 6 Face formed by presplit blasting in phyllitic schist

Locality: A82, near Inverbeg, Scotland

have been equally diverse and have ranged from intensely fractured phyllitic schists to blockily fractured granitic intrusives. In all rock types a number of prominent discontinuity sets have usually been identifiable although structural domains have tended to cover relatively small areas. This latter aspect reflects the complexity of the local geology.

Success in forming a presplit face has been found to relate neither to the petrological type nor to the discontinuity intensity but to the relative orientation of important discontinuities to the proposed presplit plane. Discontinuities at moderate to high angles have little effect on the formation of the presplit plane. As the angle approaches 15 to 20 degrees some overbreak can occur and at angles of less than 15 degrees important discontinuities tend to take over completely and form the final face. Knowledge of the existence of such planes is therefore of vital importance. If they are favourably orientated and important enough they can be used as natural presplit planes; contract presplitting can then be dispensed with and the natural planes used to limit blast disturbance.

Although fracture intensity itself has not proved to be a critical factor, highly fractured or easily weathered rocks may give rise to a different problem. The presplit face may become unstable as a result of physical and/or chemical weathering. Masonry, concrete or shotcrete can be used to protect parts of the face likely to be affected by such weathering.

Examples of stable faces formed by presplit blasting are given in Plates 5 and 6.

Drilling

Accurate setting up and close monitoring during the initial stages of drilling have been found to be of vital importance to successful presplitting. A simple drill orientation device (DOD) has been developed (Plate 7).[8] Its use has considerably increased the accuracy of drilling presplit holes (Table 2).

In theory any rig capable of accurate drilling to the diameter and depth required can be used. In practice down-the-hole hammers have proved to be consistently more accurate than drifter rigs.

Plate 7 Drill orientation device for increasing accuracy in drilling

This could arise from differences in rod stiffness and the location of the percussive mechanism - less flexure is likely in the case of down-the-hole hammers.

Accuracy of drilling ensures even charging and becomes increasingly important as the height of the proposed presplit face increases. It is therefore relevant to define accuracy in terms of linear rather than angular deviations. This allows the maximum height of a face to be governed by the expertise of the driller and the equipment used. Deviations, measured at the base of the presplit face have been defined as acceptable up to one-third and one-fifth of drill hole spacing for in-plane and from-plane distances respectively. Such accuracy should be well within the capabilities of a competent driller using a down-the-hole hammer to depths of 15 metres. Avoidance of large variations in accuracy between adjacent holes is however more important than absolute accuracy.

In general presplit holes should be drilled in one lift whenever possible. Small benches are undesirable as near surface disturbance by presplit and bulk blasting usually causes loss of the bench edge. This leaves a bevel which can project falling rock out from the face. Where benches are required they should be a minimum of 4m wide.

Hole diameter, hole spacing and explosive charging have proved to be fundamental, dependent variables which should not be varied unnecessarily. The limits of experience for holes drilled in successful presplit contracts in Scotland are given in Table 3.

Table 2 Accuracies in drilling presplit holes

A. Using traditional methods of rig alignment

ROCK TYPE: Banded Gneiss

	Cut 1	Cut 2	Cut 3
FACE HEIGHTS (m):	10	12	12
DIP (degrees):-			
No of measurements:	221	1242	51
Mean:	70.5	71.2	73.5
S.D.:	2.8	3.1	4.6
Range:	62 - 86	60 - 90	64 - 82
Design:	71.6	71.6	71.6 (3 in 1)
AZIMUTH (degrees north):-			
No of measurements:	151	584	51
Mean:	217	217	227
S.D.:	13.5	15.5	17.6
Range:	181 - 300	175 - 255	200 - 282
Design:	235	211 - 222 (curving face)	202

B. Using the DOD

ROCK TYPE: Interbedded greywacke-shale

FACE HEIGHT (m):	15
DIP (degrees):-	
No of measurements:	66
Mean:	69.9
S.D.:	1.0
Range:	67.5 - 72.3
Design:	70.0
AZIMUTH (degrees north):-	
No of measurements:	66
Mean:	- 111.3
S.D.:	2.5
Range:	-106.6 - 117.5
Design:	120

Table 3 Limits of experience of drilling on successful presplit contracts in Scotland (to July 1985)

DIAMETER:	60 to 100mm	Usually 100mm
DIP:	50 to 90 degrees	Optimum angle chosen
AZIMUTH:	Right angles to road centre line	Chosen for setting-out convenience
SPACING:	0.6 to 1.0m	Ten times hole diameter
DEPTH:	4 to 28m	Depends on local geology
FACE LIFT:	3 to 27m	Usually 10-15m

Explosives and charging

The type, charging and detonation of the explosives used in presplitting must be carefully controlled. The dynamic component is deliberately subdued by using explosives with a relatively low velocity of detonation (VOD) and highly decoupling the charges. VODs are typically between 2,300 and 3,000 metres per second and decoupling ratios (the diameter of the charge to the diameter of the hole) of approximately 1:4 are in common use. The characteristics of the U.K. explosives used in the majority of contracts are given in Table 4.

Table 4 Characteristics of explosives package for presplitting and commercially available in the UK

Brand Name	Trimobel	Gelamex - DC1	Gelamex - DC2
Manufacturer	Nobel's*	ECP**	ECP
Explosive			
Type	Semi-gelatine	High strength amm. gelignite	High strength amm. gelignite
V.O.D. (m/s)	2400	3000 (approx.)	3000 (approx.)
Density	1.15	1.4	1.4
Weight strength	87% BG	80% BG	80% BG
Bulk strength	65% BG	78% BG	78% BG
Packaging			
Type	Polythene tube (linkable)	Polythene tube (linkable)	Polythene tube (linkable)
Tube diam. (O.D.)	25 mm	30 mm	30 mm
Tube length	750 mm	381 mm	381 mm
Charge	Continuous	Individual, separated	Individual, separated
Weight expl./tube	320 gm	95 gm	190 gm
Weight expl./metre	426 gm	250 gm	500 gm
Linear strength (100% BG base)	371 gm/m	200 gm/m	400 gm/m
Method of initiation	Detonating cord	Detonating cord	Detonating cord

* Nobel's Explosives Co. Ltd., Nobel House, Stevenston, Ayrshire, KA20 3LN.
**Explosives and Chemical Products Ltd., Commonwealth House, 1-19 New Oxford Street, London, WC1A.

Poor results from earlier projects suggest that the use of PETN based explosives in the form of single or multistrand cords should be avoided for all but initiation purposes.

Slight increases in the intensity of fracturing resulting from the dynamic component may occur where the explosive has been in contact with the borehole wall. With the low charge weights used, this is proving to be of little importance. In already discontinuous and crystalline rock the effort of ensuring that charges are centralised in the hole by using "feathered" collars[4] or other types of spacers is not considered worthwhile. The quasi-static gas pressure can be expected to distribute itself uniformly even when the charge is not central.

Early techniques of charging were to tape portions of a high gelatine explosive at regular intervals on an initiating cord. However pre-packed explosives in the form of separate or continuous charges contained in linkable tubes are now in common use. Originally these were made of waxed cardboard and more recently of polythene. Omega-shaped flexible clips made from tubular plastic have now become available. The detonating cord is placed in the back of the clip and run down the entire length. Prepacked explosive cartridges are then snapped into the clip at the required intervals and the assembled charge train lowered into the hole. Variation of the explosive charging is easily obtained by changing the explosive charge type, weight or spacing.

All four types of explosive charging (Fig. 7) have proved successful. The use of the linkable tubes and clips has proved particularly useful in ensuring that each hole is charged to the base; with detonating cord alone obstructions can cause the charge train to hang in the hole.

In early presplit contracts it was standard practice to use a relatively heavy charge at the base of each charge train. The purpose of this was twofold: to weight the end of the detonating cord and ensure that it reached the base of the hole, and to obtain clean breakout at the toe of the intended face. This practice was discontinued because of the damage induced into the face at the critical position for face stability. The technique of drilling presplit holes at least one metre below formation level and using even charging throughout has now become standard practice. If desirable, any toes remaining can be removed by "pop" blasting after mucking out.

Stemming has traditionally always been used in the top 2 metres of presplit holes. Presplit fractures cannot however be expected to form

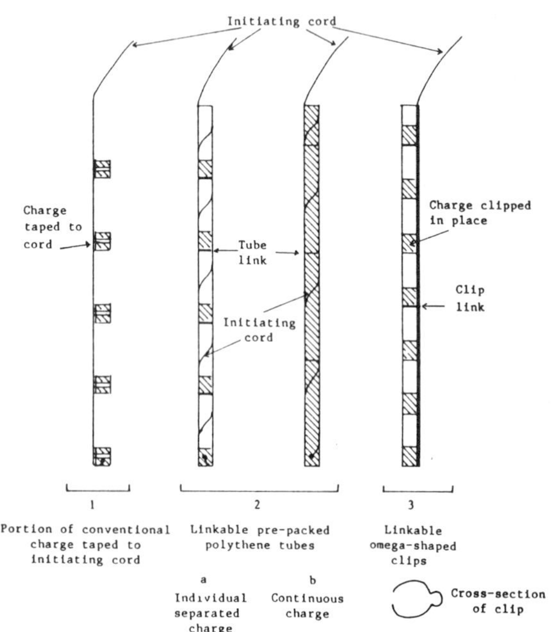

Fig 7. Charging Techniques and Packaging for Presplitting

between uncharged, stemmed sections of drill holes. In addition excess gas cannot escape and can be expected to dilate existing fractures thereby causing ground heave. Stemming should therefore be kept to a minimum. This latter practice however, tends to increase the amount of fly rock and the intensity of the air blast. The practice of using minimum stemming in the form of a plastic bag filled with heather (or scrub) to hold up 0.5 to 1.0 m. of granular material at the top of the hole has been found to reduce this to acceptable levels. In areas where air blast and fly rock have to be kept to a very low level then matting, held in place by chain or wire mesh, can be used.

Detonation

According to the theory of presplitting being followed here, the main requirement is that the gas pressure develops simultaneously in each presplit hole. Both the techniques of attaching "instantaneous" electrical detonators to the down line of each charge train and connecting each down line to a main initiating line (itself detonated electrically) have been used successfully. The latter method considerably reduces the possibility of misfires and is to be recommended; the configuration of down line and main line does not seem to be important.

Associated bulk blasting

Bulk blasting is designed to fragment rock in the excavation area and assist removal by mechanical means. In doing so it induces considerable local disturbance. Clearly there must be a limit to both the intensity and proximity of such blasting before disturbance manages to penetrate the presplit discontinuity and enter the design slope. This has been found to vary with the explosive type and charging as well as the rock conditions. Field experience indicates a lower limit for proximity of two metres for uncontrolled blasting.

To avoid having to calculate and monitor the proximity of bulk blast holes to the presplit plane it is easier to rake adjacent bulk holes parallel to the presplit holes. Decking or other methods of distributing the charges in the bulk holes also promotes even fragmentation up to the presplit plane. Increasing the number of holes and reducing the charge weights is also recommended. Care should be taken to ensure that the disturbance from the bulk blast holes cannot penetrate under or around the presplit plane.

Mucking out to the presplit face must be carried out carefully to avoid mechanical damage to the design face. Use of plant buckets in a downward sweep has proved best; upward movement can cause substantial lift. It is advisable for remedial work to be done as mucking out proceeds, and not left until the whole face is excavated and the higher parts of the face become inaccessible.

Levels of vibration

Maximum seismic vibration might be expected from blasting carried out in a confined rock mass. However in the presplit technique described here the charges are usually highly decoupled and there is therefore a high impedence contrast between explosive, air and rock. A reduction in resultant peak particle velocity over that expected from the weight of the explosive alone is consequently experienced. A comparison of the different site laws using conventional and decoupled charges from one site[9] is given in Fig. 8. Vibration levels 3 to 6 times below

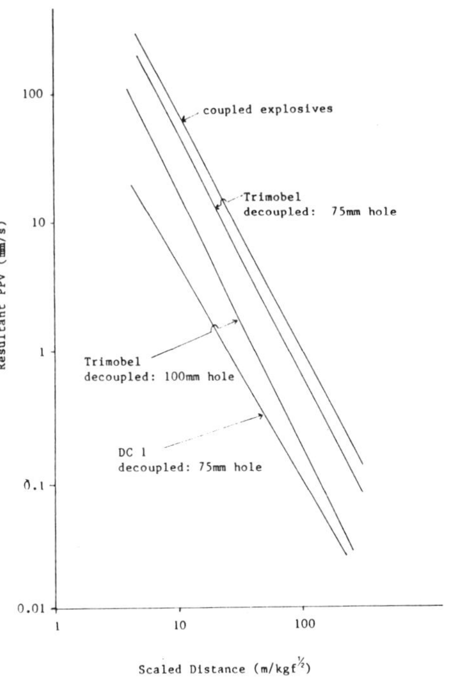

Fig 8. 'Site Law' Regression Lines

that predicted have been measured on other sites where 1:4 decoupling ratios were used.

Water has the effect of coupling the explosive charge to the host rock. Increases in the level of vibration in proportion to the amount of water in the drill holes can therefore be expected.

Studies of vibrations from bulk blasts travelling through presplit planes have not shown any reduction in vibration level, indicating that the rock on either side of the plane is acoustically coupled.[10] Standard methods (resultant peak particle velocity related to scaled distance) of predicting vibration levels are therefore applicable - providing that intervening fractures have not dilated.

Threshold limits of between 12.5 and 50mm/sec. for resultant peak particle velocities are in common use. In some instances (near hospital sites) the threshold was set as low as 3mm/sec. It is strongly recommended that reasonable limits are set individually for each site by establishing the site laws by trial blasts and paying attention to local circumstances. Guidance on vibration limits and on carrying out trial blasts is given by New.[11]

The maximum size of presplit panel which can be detonated at any one time depends on the vibration limits set. The minimum panel size recommended is 6 adjacent holes; the optimum is 12. Where delays are used it is important to allow vibrations from one blast to subside before another is initiated; enhancement may otherwise occur.

Specification, monitoring and control

Blasting is a destructive process and presplitting cannot normally be repeated if unsuccessful. The specification recommended is a combination of "method" and "end product" types. Its effectiveness depends on carefully defining those aspects critical to successful presplitting and permitting contractual choice for the others. Aspects found to be vital were drilling accuracy, charging, detonation, and the proximity of bulk charges to the presplit plane. Optimum conditions, based on field experience, are given in Table 5.

A trial presplit is included as standard in each contract. This allows evaluation of the contractor's options and ability to meet specifications. The procedural flowchart, blasting record sheet and suggested specification requirements used in most recent contracts have been published.[7]

Excavation methods for urban environments

Excavation methods used in urban environments should be chosen to minimise induced instability in the final design slope. Presplit blasting, in conjunction with bulk blasting, should be considered in hard rock where there are no important natural discontinuities which can be exploited. If there are important natural

Table 5 Optimum conditions specified for presplitting in highway contracts in Scotland

DRILLING TOLERANCE:	In-plane	Hole spacing/3
	From-plane	Hole spacing/5
EXPLOSIVES AND CHARGING:		
LINEAR STRENGTH:	200 - 400 gm/m (BG)	Adjust to rock conditions
DECOUPLING RATIO:	1 : 4	1 : 1 if hole filled with water
STEMMING:	0.5m minimum	keep as short as practical
BURDEN	9m minimum	may be possible down to 6m.
PANEL SIZE:	Minimum 6 holes	usually 12
DETONATION:		
PRESPLIT PANEL:	Simultaneous (0.0 delay)	Panels can be delay linked
BULK TO PRESPLIT:	Minimum 50ms delay	No upper limit
DISTANCE OF NEAREST BULK CHARGE TO PRESPLIT PLANE:		2.0m minimum

discontinuities orientated approximately parallel to the proposed design slope then these may be able to be used as natural presplit fractures and presplitting omitted.

The technique of presplitting found most successful in highway projects in Scotland has relied on the theory that it is the quasi-static component of the presplit blast, and not the dynamic one, which is of primary importance in the formation of the presplit discontinuity.

Specification is vital and the combination of a method and end product type would appear to be appropriate. It is important to ensure compliance and it is advisable to draw up strict methods of monitoring and control.

Vibration levels from presplit and associated bulk blasting may be important in an urban environment. These can be predicted from the site laws which should be established individually for each excavation proposed.

ACKNOWLEDGEMENTS

The work described in this paper forms part of the programme of the Transport and Road Research Laboratory and the paper is published by permission of the Director. The co-operation of the Scottish Development Department and their Consultants and Contractors is gratefully acknowledged.

REFERENCES

1. Anon (1977). The description of rock masses for engineering purposes. Quarterly Journal Engineering Geology, 10, pp 356 - 388.

2. Matheson, G D (1985). The stability of excavated slopes exposing rock. Symposium on Failures in Earthworks. Institution of Civil Engineers Paper 17, March 1985, London.

3. Matheson, G D (1983). Rock stability assessment in preliminary investigations - graphical methods. Department of the Environment Department of Transport, Transport and Road Research Laboratory Report LR 1039: Transport and Road Research Laboratory.

4. Hoek, E and Bray, J W (1971). Rock slope engineering. Institute of Mining and Metallurgy London.

5. Goodman, R E (1976). Methods of engineering geology. West, New York.

6. Goodman, R E (1980). Introduction to rock mechanics. Wiley, New York.

7. Matheson, G D (1983). Presplit blasting in highway rock excavation. Department of the Environment Department of Transport. Transport and Road Research Laboratory Report LR 1094: Transport and Road Research Laboratory.

8. Matheson, G D (1984). A device for measuring drill rod and drill hole orientations. Department of the Environment Department of Transport. Transport and Road Research Laboratory Report SR 817: Transport and Road Research Laboratory.

9. New, B M (1984). Explosively-induced vibration in civil engineering construction. Ph.D Thesis (Unpublished), University of Durham.

10. Devine J F, R H Beck, A V C Meyer, and W I Duvall (1965). Vibration levels transmitted across a presplit fracture plane. United States Department of the Interior, Bureau of Mines Report RI 6695.

11. New, B M (1985). Ground vibration caused by civil engineering works. Department of Transport. Transport and Road Research Laboratory Research Report RR 53 (In Press).

Rock support prediction in the deep basement of Causeway Bay East Concourse for the Mass Transit Railway, Hong Kong

C.R. Matson B.SC., M.ENG.SC., F.G.S., M.A.I.M.M.
Maunsell Geotechnical Services, Hong Kong
H.H. Choy B.SC., P.E. (B.C.), P.G. (ALB.)
Formerly, Maunsell Geotechnical Services, Hong Kong (now, Geotechnical Control Office, Hong Kong)
A.M. Gibson B.E., M.I.P.E.N.Z.
Mass Transit Railway Corporation, Hong Kong

SYNOPSIS

The predicted temporary rock support was significantly affected by the perception of geology and support parameters assumed at various stages in the construction of Causeway Bay East Concourse for the Mass Transit Railway Corporation. The 35 metre deep basement was built top down through the soft upper layers, and bottom up through the granite bedrock. The site is located in a densely populated urban area of Hong Kong.

A nominal allowance of 100 rockbolts was included in the contract document for temporary rock support.

Typically there was 18 metres of bed rock to be excavated prior to the "bottom up" construction. To improve progress the Contractor proposed that 40% of the rock be mined from headings while "top down" construction through the soft section was proceeding. At this stage the geological data indicated that 2,300 bolts were needed for the total area of the two rock faces on the most critical south wall.

This was reduced to the support of individual wedges and planes as geological data was gathered. Groundwater pressure models and factors of safety were varied in stability analyses. A factor of safety of 1.2 for a drained groundwater model was finally adopted. Reductions in the rock bolt support were also achieved by assuming an increased angle of friction to 50° for tight, clean joints. On this basis it was expected that 240 rockbolts would be required for the two critical faces.

The bolts actually installed in the south wall totalled 171 with half being for local minor instability. Even less would have been installed if immediate designation of rock support had not been necessary. Back analysis indicated that 35 bolts would have stabilized the actual wedges on the south wall.

INTRODUCTION

Construction of the Causeway Bay East (CAB(E)) Railway Concourse for the Hong Kong Mass Transit Railway Corporation (MTRC) commenced in May 1982. The site was typical of the other stations on the Island Line in being built in a densely populated urban location.

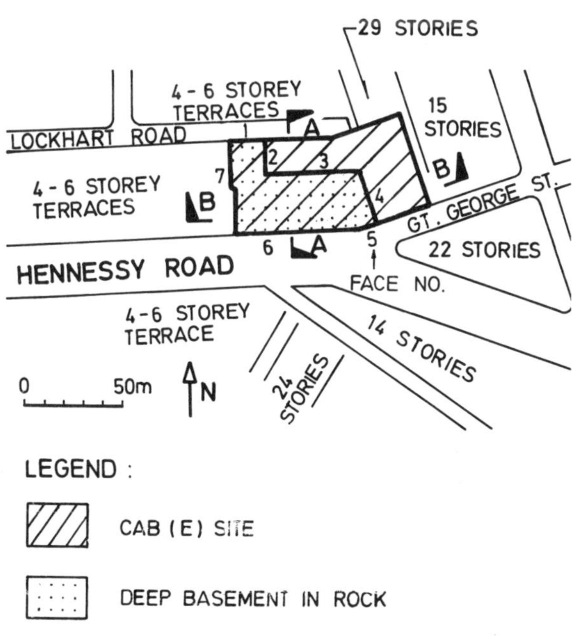

Fig. 1 Site plan

Figure 1 shows the CAB(E) site in relation to the surrounding buildings and roads. The 4 to 6 storey terrace buildings generally have shallow timber pile foundations. The foundations of the multi-storey blocks are generally deep concrete piles founded in decomposed granite. Hennessy Road is a main traffic artery on Hong Kong Island and is underlain by many services.

The 35 metre deep basement for the Concourse was built by "top down" construction from the ground level of +4m P.D. through the soft upper layers of alluvial soils and decomposed granite using diaphragm and hand dug caisson walls, and bottom up through the granite bedrock. As is shown on Figures 2 the upper portion of the site was excavated in soft ground to generally just above the upper platform slab (U.P.S.) level of -14m P.D. The deep basement section was excavated in rock below the U.P.S. down to -31m P.D.

The method of construction allowed the above ground structure to be built at the same time as the railway concourse by forming the main columns prior to bulk excavation of the basement. At the same time as the columns were under construction, the diaphragm and hand dug caisson cut off walls were being built. "Top down" construction, by the casting of floor slabs and excavating below each slab in succession permitted the permanent works to support the perimeter wall in excavating to the base level of the U.P.S. Below this level rock was excavated to the bottom of the basement at -31m P.D. with "bottom up" construction of the remaining station levels from the lower platform slab (L.P.S.) to the U.P.S. On completion of excavation there was typically 18 metres of rock with 17 metres exposed beneath the U.P.S.

With the surrounding buildings, roads and utilities, any failure of the rock side walls could not be tolerated. Failures of such wedges of rock would undermine the diaphragm and caisson cut off walls, allow ingress of groundwater and the overlying soil (Figure 3).

This paper will confine itself to a discussion of the predicted rock bolt support of the rock faces below the U.P.S. at various

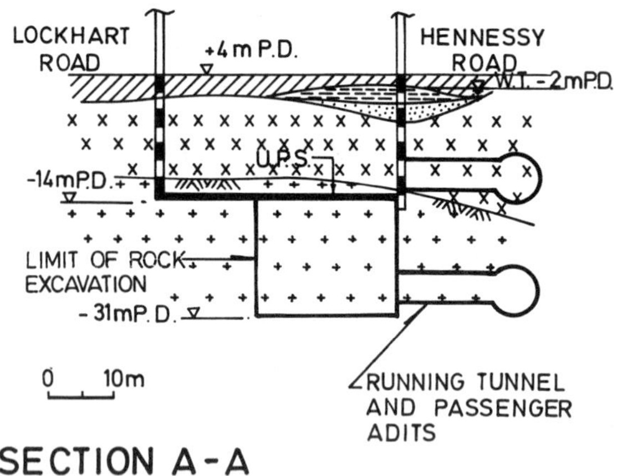

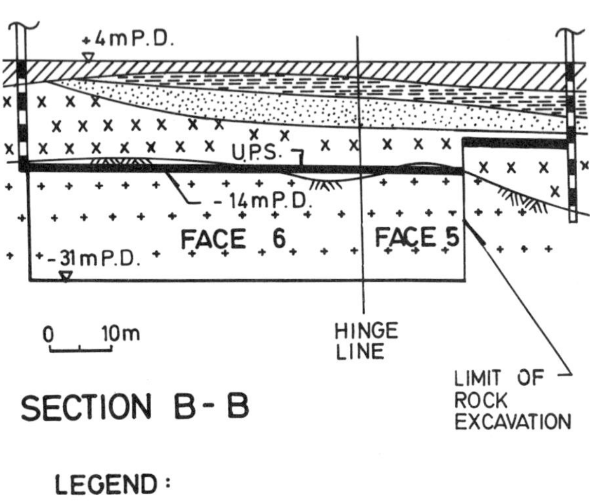

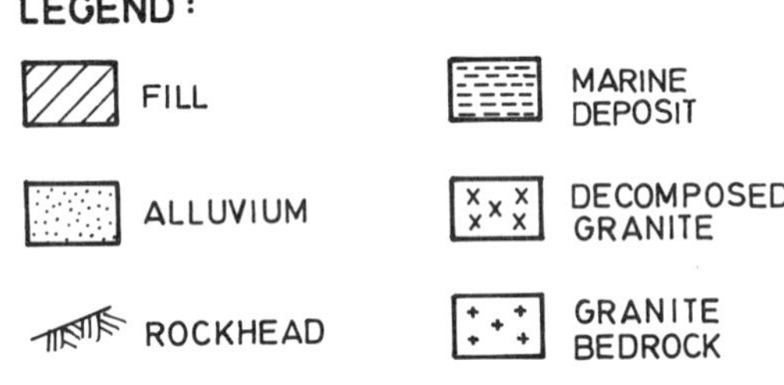

Fig. 2 Geological sections

stages from design to completion of the excavation. In particular the two southern faces, Faces 5 and 6 (Figures 1, 2 and 3), are dealt with in detail as the other five faces did not have potential major instability. The envisaged temporary support varied as more geological information became available and as construction developed. For clarity, construction and support will be discussed in five stages. These are :

Stage I - Design and Documentation (Pre-May 1982)

Stage II - Rock Mining Proposal (November 1982)

Stage III - Early Rock Mining (April 1983)

Stage IV - Top Down Rock Excavation (September 1983)

Stage V - As Built (January 1984).

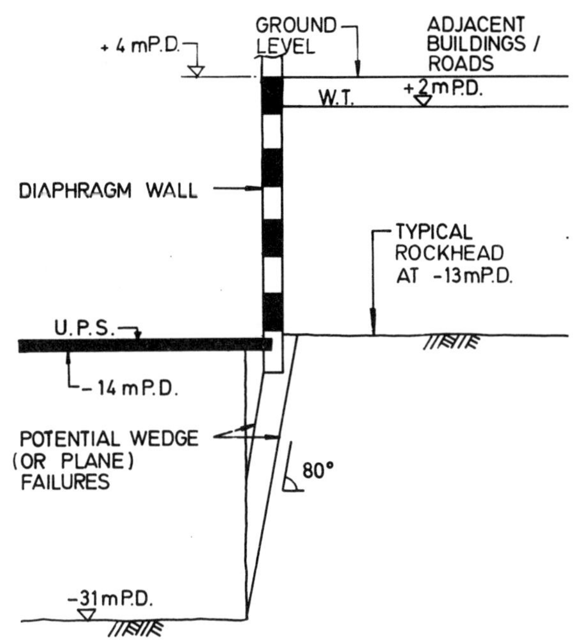

Fig. 3 Failure mode of rock side wall

STAGE I - DESIGN AND DOCUMENTATION (PRE MAY 1982)

Site investigation and geology

At the design stage seven vertical boreholes were drilled around the perimeter of the site. The boreholes were drilled at least 5 metres into slightly weathered granite with three holes being extended to the bottom of expected excavation. The deepest hole was drilled to -31mP.D. These holes were essentially used for assessing the soil profile and properties, the broad position of rockhead and determining the soundness of the granite bedrock.

The bedrock was generally described as strong, grey and pinkish grey slightly weathered granite. This applied for the vast majority of the rock to be excavated between the U.P.S. to below the L.P.S., though in several holes moderately and highly weathered granite occurred below the U.P.S. Jointing was described as varying from closely to moderately widely spaced, with references to 30°, 60°, and near horizontal joints and occasional vertical joints. The RQD's varied between 25% and 75% in the upper 5 metres of rock including the moderately weathered zone. Below this they generally varied from 50% to 100% and were often greater than 75%.

Construction

Having completed construction down to the U.P.S. it was envisaged that rock would be completely excavated from beneath the U.P.S. at -14m P.D. to the bottom of the excavation of -31m P.D. A shoulder of rock was left so that the diaphragm and caisson walls would not be undermined. Thus, unless jointing was unfavourable, or the walls had terminated within weathered rock, underpinning of the walls would not be required. In the event of unfavourably orientated joints then support would be provided.

Support

The contract was arranged such that the Engineer designated the temporary support of unstable blocks of rock. It was envisaged that rock bolt support of the side walls would be installed from the floor of the excavation as unfavourably orientated joints and potentially unstable blocks were identified. Problems may have arisen however with large potential plane failures which may not have been identified until excavation had exposed the toe. In this situation, support installation well above the then current level of excavation may have been necessary.

To provide the temporary rock support a nominal allowance of 100 rockbolts was included in the contract document for the seven rock faces (Figure 1). The rock was perceived as basically sound with instability resulting from randomly unfavourable and not very persistent joints.

The permanent support of the lower rock section was to be similar to the upper soil section where the floor slabs were struts for structural walls. The presence of any rock bolts was ignored in assessing loads for the

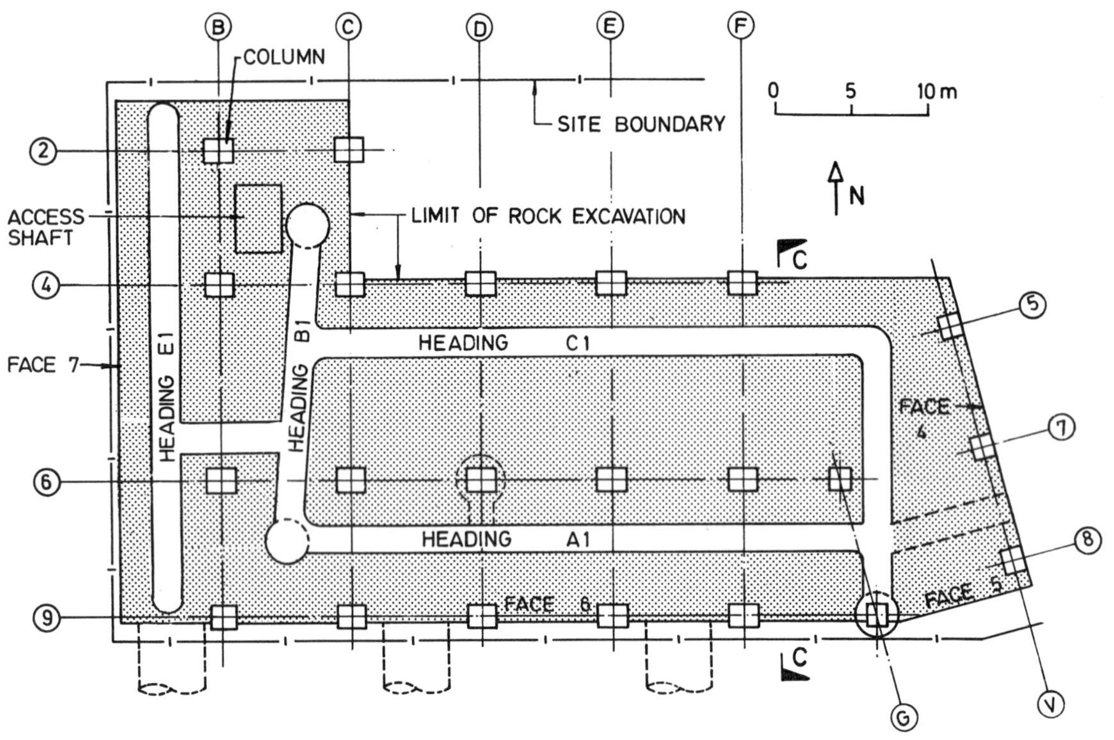

Fig. 4 Underground mining - plan of upper headings

permanent works. In the soil sections the structural walls were the diaphragm and caisson walls while in the rock section the walls were to be cast against the exposed rock. The permanent support is not considered any further in this discussion.

STAGE II - ROCK MINING PROPOSAL (NOVEMBER 1982)

Progress of the works

Three months into the Contract, construction of the perimeter walls, although approximately 75% complete, was some 5 weeks behind programme due to problems with boulders, utilities and running marine sands. In addition excavation below the ground floor level in the east of the site, had encountered a very soft decomposed granite fill in which it appeared unlikely that programmed excavation rates could be met. Excavation of caissons for column construction in the centre of the site had shown rock levels up to 10 metres higher than expected. These factors, coupled with the fact that the programmed rate for the bulk rock excavation ($1800m^3$ per week) was already approximately double the figure achieved in similar circumstances elsewhere in Hong Kong, led to the conclusion that some other means of improving rates of progress would have to be implemented.

To speed up rates of progress the Contractor proposed that some $9000m^3$ (or 40% of the rock) be removed by underground mining at the same time as excavation was proceeding through the soft upper layers. Figures 4 and 5 show the Contractor's proposed method. Two metre wide headings were driven at both upper and lower levels. The lower levels were designed to operate as ore passes during excavation of the main caverns. Following completion of the caverns, the crown and rib pillars were to be removed by top down excavation.

Site investigation and geology

Drilling at 4 metre centres had been undertaken along the length of the perimeter wall to determine the depth to bedrock. An additional four holes were drilled within the site again to define rockhead. Core orientation was not undertaken on any of these holes.

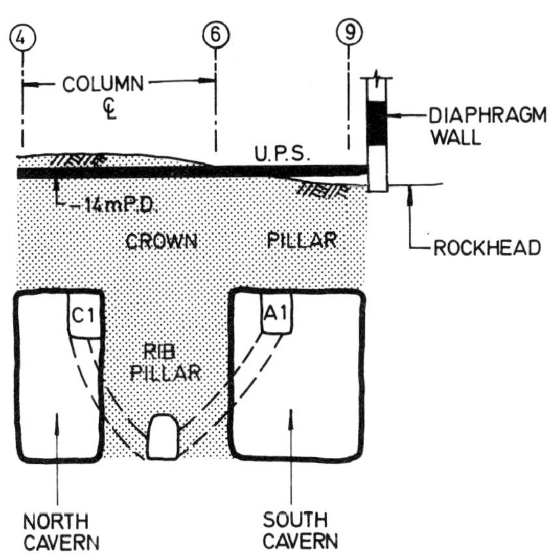

Fig. 5 Section of underground mining

The information obtained from this drilling was invaluable for defining rockhead, and hence the crown levels for the headings. However very little useful information was obtained for defining the rockbolt support of the sidewalls.

Support

The Contractor's proposed method did not produce any major changes in the side wall support to be designed by the Engineer. As the caverns were to be excavated out to the final side wall profile, the Engineer's support would be installed during cavern excavation. In the area of the crown pillar the Engineer's support would obviously not be installed until after removal of the crown. On excavation, the loose rock left after blasting of the crown would provide a working platform for installation of this support.

The presence of the crown pillar would effectively act as a stiff strut while the underground mining was in progress. The proposed method increased the number of potential modes of side wall failure. However they all depended on failures in the crown. The Contractor took full responsibility for the integrity of the crown.

STAGE III - EARLY ROCK MINING (APRIL 1983)

Progress of the works

Excavation of exploratory headings at the crown level of the proposed caverns was in progress with headings A1 and C1 now complete. Rates of progress on top down works had improved slightly as the decomposed granite soil in the west half of the site proved to be dry and easy to excavate.

Geology

The 2 metre diameter headings driven at -21m P.D. allowed a greater appreciation of the rock quality and discontinuity data prior to excavation than would have been available from

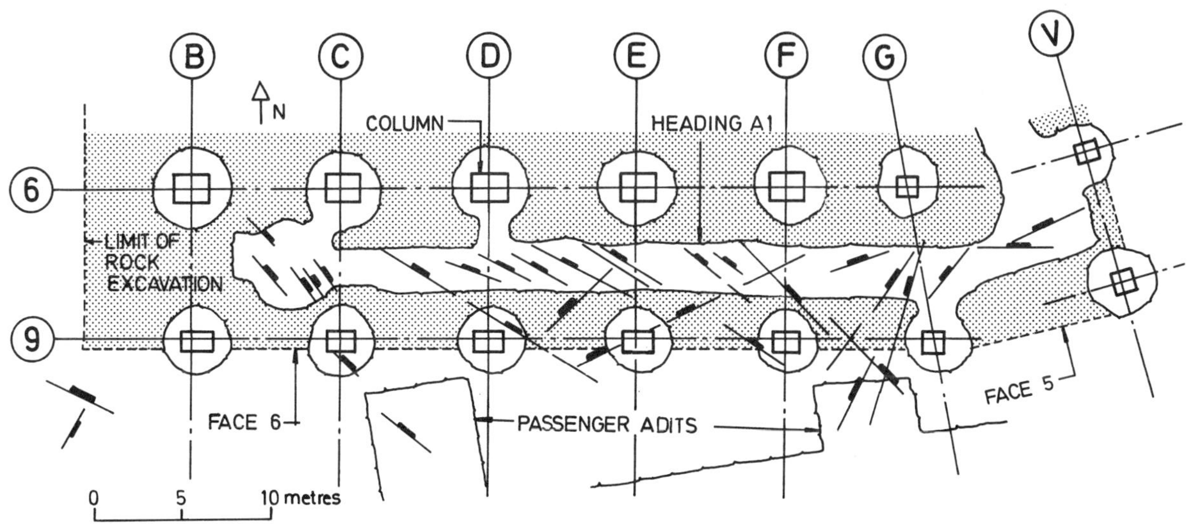

Fig. 6 Faces 5 & 6 showing jointing (Stage III)

the caisson excavations alone. The preliminary mapping (Figure 6) along headings A1 and C1 revealed that the majority of the discontinuities sampled were joints with a few shears. The discontinuity system comprised three sets of well-developed joints as shown in the stereo plot (Figure 7). J1 and J2 were a pair of conjugate, subvertical tectonic joint sets and J3 was a set of stress-release sheeting joints. The mapping indicated that joint sets occurred in a repetitive series, with one joint plane following behind another.

Stability analysis

Analysis of the stereoplot (Figure 7) showed that the stability of the southern faces (Nos. 5 and 6) was of concern. Face 5 had potential plane failures along J1 and Face 6 had potential wedge failures along the intersections of joint sets J1 and J2. With the repetitive nature of the parallel joints it became necessary to consider the need to support the entire areas of Faces 5 and 6.

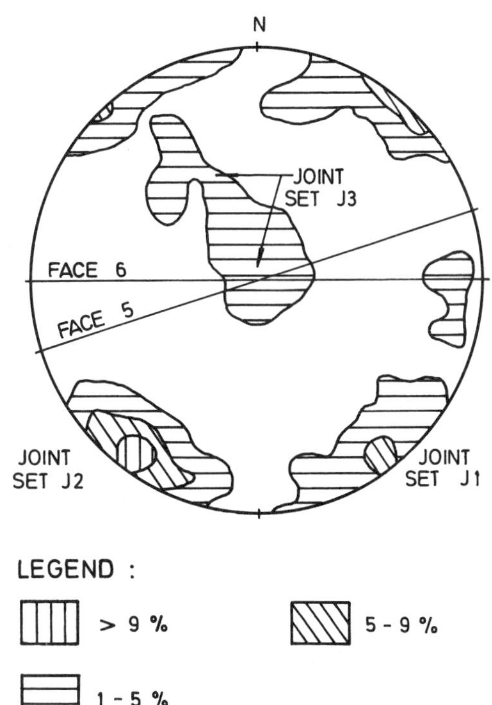

Fig. 7 Joint orientation (equal area polar plot)

Support

From the preliminary mapping and in view of the serious consequences of failure, the preliminary design assumed the joint sets were repetitive and would extend the full height of excavations. Such continuity had been observed at other MTR sites. Instability was therefore perceived to result from extensively developed and persistent joint sets.

An angle of friction of 40° was assumed. The external destabilizing forces acting on an unstable block were the weight of soil overlying the block, the load from the upper concourse including the weight of the diaphragm/caisson walls, the hydrostatic force and the traffic load from adjacent roads. These forces, other than the hydrostatic force, could be accurately estimated from known geometry. The hydrostatic force could not be estimated with such accuracy. Four models for the pore pressure distribution on the rock faces were considered as shown on Figure 8.

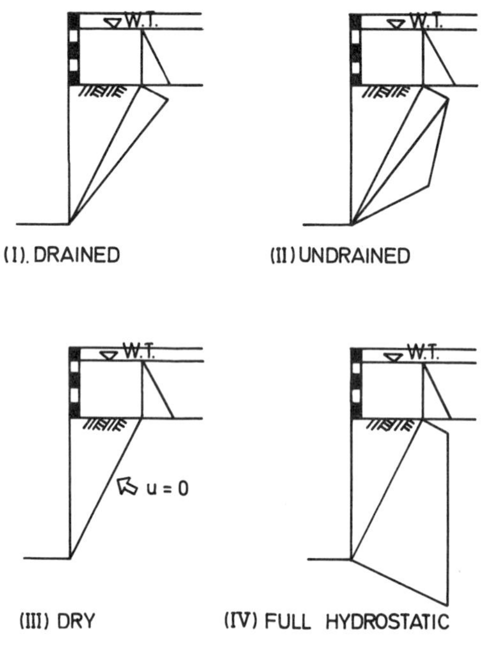

Fig. 8 Groundwater models

The hydrogeological conditions relating to the four models were:

a. Model I: drained model where hydrostatic pressure was released in a linear manner over the length of the joint.
b. Model II: undrained model where hydrostatic pressure was released only at the toe of the block. This is a

conservative model.

c. Model III: dry model with no hydrostatic pressure on the block. This model is applicable for a fully drawndown condition in the rock or when joints were fully grouted. The presence of the adjacent rail tunnels lent credibility to this model at least for the condition prior to lining the tunnels.

d. Model IV: full hydrostatic model where groundwater does not drain into the excavation. It corresponds to situations where the toe of the joint is blocked or where the rate of recharge of joints is faster than the rate of discharge at the toe. This model was not likely to occur during excavation unless all the rock faces were shotcreted and were without weepholes. It was considered to be an unnecessarily conservative model and was not used in analysis.

The total temporary stabilization forces acting on Faces 5 and 6 derived from the preliminary design using the three hydrostatic models were:

	Total Required Supporting Force (kN)		
Face No.	Model I	Model II	Model III
5	25,700	35,000	15,800
6	153,900	197,200	127,200

With an adopted factor of safety of 1, the preliminary support was evaluated for the conservative hydrostatic Model II. The analysis indicated that over 2,300 bolts of 100 kN working load would be required to secure Faces 5 and 6.

Rock support - alternatives

Further consideration was given to alternative means to reduce costs for the temporary rock support. These included grouting, strutting, rock anchors and dowels. Grouting and strutting were considered to be too expensive. Rock anchors were considered but discarded as the adjacent rail tunnels were within 5 to 10 metres including one under construction in compressed air. Dowells were also considered and although not as efficient as rock bolts in that greater numbers would be required, it was hoped that significant cost savings could be made.

Sensitivity analysis

Sensitivity analyses were carried out to assess changes in the required support force. The factors which were varied included the height of the rock excavation, dip angle of the joint, angle of friction, factor of safety and the hydrostatic conditions (models I, II or III). An example of the sensitivity analyses is illustrated in Figure 9 where the height of rock was 15 metres and a factor of safety of 1 was used.

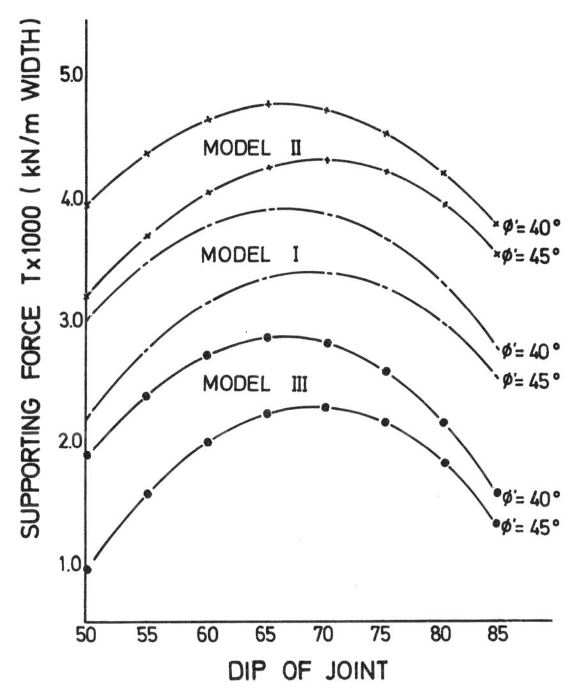

Fig. 9 Sensitivity analysis for support forces

The results of the sensitivity analyses can be summarized as follows:

a. An increase in the height of rock head by 3m, and therefore the size of rock wedge, would require a 10% increase in support.

b. An increase in the assumed angle of friction by 5° would result in a 10 to 20% decrease in support.

c. An increase in the assumed factor of

safety by 0.2 would require a 10 to 20% increase in the estimated support.

d. An increase in the hydrostatic forces from Model III to Model I and then to Model II increases the estimated support by approximately 50% for each step (Figure 9).

Clearly the hydrostatic force was the major destabilizing factor.

STAGE IV - TOP DOWN ROCK EXCAVATION (SEPTEMBER 1983)

Progress of the works

Excavation of rock by the mining method was abandoned by the Contractor in June 1983 following completion of all headings and commencement of the north cavern. Rates of top down construction had exceeded expectations assisted by good soft ground conditions. The high rock levels had been dealt with by over excavating the rock under each slab by 2 metres and casting the slab on a falsework support. This allowed easy access underneath and resulted in almost no programme loss. By September 1983 top down excavation had reached 2 metres below the U.P.S., and construction of the slab was in progress on falsework.

Geology

The geology at this stage was sufficiently well defined from mapping of the headings, the caissons and the adjacent rail tunnels to recognise that total support of Faces 5 and 6 would not be required.

Probable major wedges were predicted by projection of mapped discontinuities onto the final rock faces. The predicted wedges permitted the temporary support to be modified from a full face support to the support of specified areas. Four potentially unstable wedges were predicted on Face 6 (Figure 10). However, as expected, the potential plane failures on Face 5 could not be verified by mapping of the preliminary excavations.

Support

The support system to stabilise the four wedges identified on Face 6 was evaluated using a factor of safety of 1.4 with the dry Model III as well as a factor of safety of 1.0 for the drained Model I. This approach resulted from a consideration of the sensitivity analysis. During the inspection of the heading and caisson excavations, it was concluded that for the major clean joints with slight staining an angle of friction 50° (Barton, 1977) would be

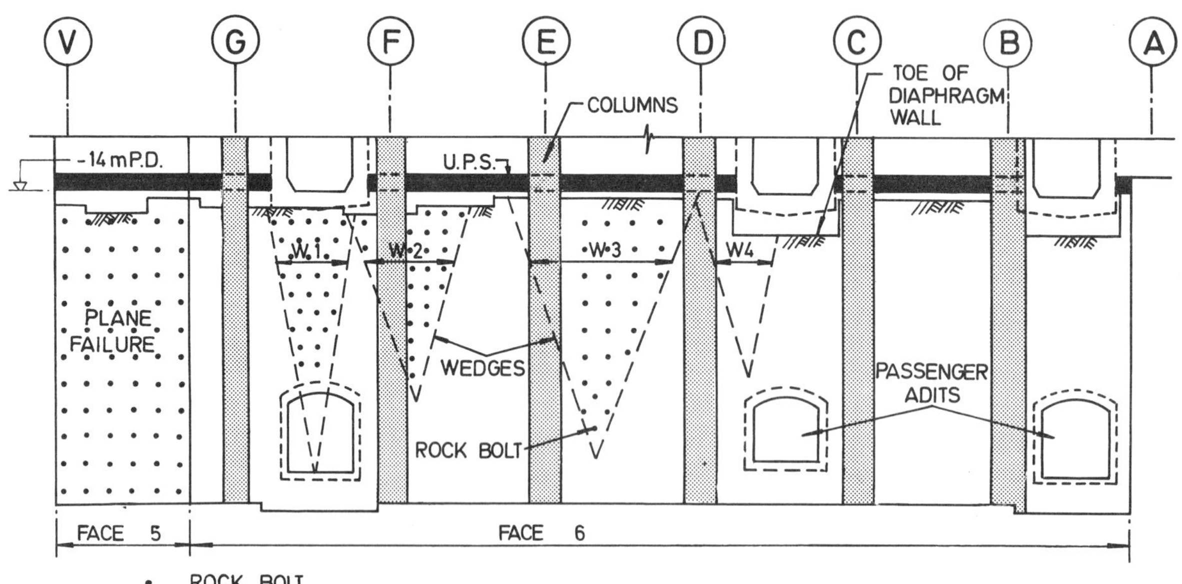

Fig. 10 Stage IV predicted instability and support

appropriate. However, the angle of friction for shears infilled with dense, decomposed granite was retained at 40° (Barton, 1974). The analyses suggested wedge 4 was stable and the other three wedges would need support. The potential plane failure of joint set J1 on Face 5 was also evaluated with the same design parameters. As its stability was marginal, support of the top half of the rock excavation, supplemented by mapping of major joints was recommended.

The number of rock bolts required for Faces 5 and 6 was 148 for hydrostatic Model I (Factor of Safety of 1.0) and 77 for dry Model III (Factor of Safety of 1.4) using 100 kN rock bolts installed at a 5° dip. For final design, the fully-drained hydrostatic Model I with a factor of safety 1.2 was adopted for temporary support (GCO, 1979). The number of rock bolts required was 240 for Faces 5 and 6. This number does not include for spot bolting of local instability.

Rock support - alternatives

Active support by bolting and passive support by dowelling were still under consideration for Faces 5 and 6. Dowelling would be effective only after some displacement of the rock mass. The peak friction angle adopted in the design might thus be reduced due to shearing of irregularities on joint surfaces. Furthermore, the dowel bars would need adequate anchorage to resist the high tensile forces from the hydrostatic pressure and would therefore be of similar length to rockbolts. For these reasons rockbolting was adopted.

Implementation of rock support measures

Excavation of rock below the U.P.S. proceeded rapidly. With the excavation equipment already proven and operating, the critical activity became identifying and installing the rock support. As the responsibility for design of rock support rested with the Engineer it was essential that a conservative and rapid system of identifying support measures be implemented in order to avoid claims for delay. Generally one or two faces were worked at a time with 2 metre heights being removed every 3 days on each face.

Immediately following a blast and while loose rock was being removed the face would be inspected. The already identified support measures for expected major plane or wedge failures were marked on the face for immediate installation. Written confirmation of these required support measures were issued as soon as possible following the physical marking on site. Locally unstable blocks were also marked for bolting or removal. Shotcrete, mesh or other support measures were also identified. The maximum time available for the support designation so as not to cause delays was 5 to 6 hours.

Following this designation, and generally while the support measures were being installed, the face would be mapped and the support requirements checked and amended where necessary. These amendments were often not available until after the excavation level had been lowered a further 2 metres. The revised support requirements were generally less than those expected prior to full exposure of the face and also less than was actually installed. In most cases it did not prove necessary to regain access to a face.

STAGE V - AS BUILT (JANUARY 1984)

Progress of the works

Excavation of the rock was completed in January 1984 approximately 5 weeks ahead of programme. Construction of the base slab, external walls and intermediate slabs was carried out bottom up using falsework. A further 4 weeks was saved in this period and the basic structure was completed 9 weeks ahead of the specified date.

Monitoring of the rock faces while exposed was confined to visual inspection and occasional bolt lift off tests to check residual tension. These tests indicated that serious loosening of the bolts was not occurring. Rock joints in the south face remained dry, probably due to the dewatering effects from the adjacent tunnels. Joints on other faces began leaking several days after exposure and continued with a steady flow until covered by concrete.

Geology

Detailed mapping of major discontinuities and analysis of wedges identified on site were carried out as the excavation proceeded and the side walls were exposed. This mapping confirmed the joint system recognised earlier, though it was found that only two major wedges were exposed at their predicted locations on Face 6 (Figure 11). No plane failures were identified on Face 5 due to the lack of persistence of joint J1 at this location.

Support

A total of 171 bolts was installed on Faces 5 and 6 with length varying from 2 metres to 5 metres (Figure 11). Over half of these were installed for local instability exposed during excavation. The remainder were installed as pattern bolts to support actual wedges, or as a precaution against having to regain access high up on the rock face. Back analysis indicated that 35 bolts would have secured the two major wedges. It is estimated that approximately one third of the total number of bolts on Faces 5 and 6, both spot and pattern, could have been eliminated with full time supervision by specialist staff and if time had allowed a full analysis.

For all 7 faces, a total of 467 rockbolts varying from 2 metres to 5 metres in length were installed.

DISCUSSION

Increased site investigation

If time had allowed a more detailed site investigation, including core orientation, at the design and documentation stage it is likely that the joint sets which affected Faces 5 and 6 would have been identified. However, predicting the continuity or persistence of individual joints within a set cannot be done from drillholes.

Both the presence of joint sets and long continuity of individual joints will increase the required support (assuming orientation is unfavourable). It is therefore reasonable to assume that with core orientation and more drillholes that the billed number of rockbolts would have been increased. If in addition to recognizing joint sets it had been assumed that the joints were persistent it is possible that the number of rockbolts would have been significantly over estimated in the original

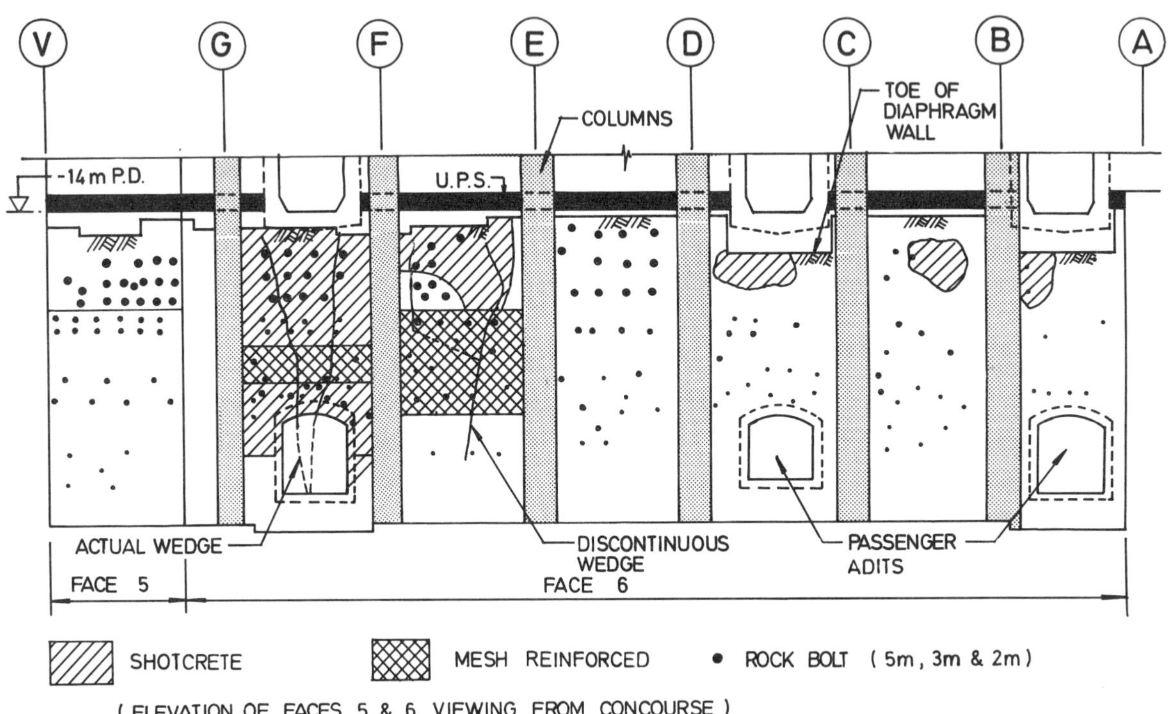

Fig. 11 Wedges and installed support Stage V

Bill of Quantities. It is likely in this event that the rate for rockbolts would have been lower and the temporary support actually installed would have been cheaper for the owner. However it is not necessarily the case that the overall job would therefore have been cheaper as a consequence.

Design parameters

At the design and documentation stage the rock support philosophy was essentially that, as the rock was sound, nominal support by spot bolting was appropriate. Design parameters for rock support were therefore not considered. In the early stages of rock excavation the presence of apparently repetitive joint sets forced a detailed assessment of design parameters as major instablity on Faces 5 and 6 was possible. Initially an angle of friction of 40°, with a conservative undrained groundwater model (Model II) was used with a factor of safety of 1. Finally an angle of friction of 50° with the drained water model (Model I) and a factor of safety of 1.2 were adopted for designing the temporary rock bolt support. The actual variations in design parameters are summarized in Table 1.

Geological data and interpretation

The predicted rock bolt support varied significantly at each of stages I, III and IV very largely as a result of available geological information and the interpretation of that information. The initial estimate (Stage I) of 100 rockbolts resulted from the perception that the granite bedrock was sound and that spot bolting would be required for not very persistent joints which were randomly orientated in an unfavourable direction.

The position at Stage III was very different in that available geological information indicated that two major, persistent and repetitive joint sets were going to control the stability of at least Faces 5 and 6. This assessment suggested a substantially increased construction risk requiring more support. It resulted in over a 20 fold increase in predicted temporary support requirements with a corresponding escalation in estimated cost.

At Stage IV the geology and specifically the jointing was better defined so that three large wedges and one potential plane failure (Face 5) were of concern. This predicted instability only required a 2 fold increase in support from the 100 number bolts at Stage I. It did not however take into account random instability.

The rockbolt support actually installed (Stage V) was for a combination of random spot bolting, for discontinuous and/or random joints, and pattern bolting for persistent joint sets. In the latter case, for some instances on Faces 5 and 6, the joint sets were discontinuous with depth and the pattern bolting was abandoned.

Increased technical supervision

Considerable information on the geology of the rock was obtained during excavation of headings. In the absence of this advance information the pressure on specialist staff to map, calculate and advise the required support measures would have been greater. In order to avoid delays to the works, the maximum time available for this activity would have been 5 to 6 hours. This would have required the geotechnical specialists to be resident on site and to work odd hours to suit site progress.

The added staff costs would have been compensated for by a reduction in the cost of the bolts installed. The main aim in very early designation of rock support was to permit rapid installation and continued excavation without delays. In the absence of full time on site specialist staff it was necessary to adopt a conservative approach to minimize the need to regain access to a face. The practice of full time supervision of rock excavation for economical support designation is well established in many places outside of Hong Kong.

CONCLUSIONS

1. Increased site investigation would very likely have resulted in reduced costs for rockbolt support of the rock faces.
2. The geology or perception of the geology is likely to have a very significant

Table 1 : Summary of temporary rock support

STAGE	ROCK SIDEWALL SUPPORT	DESIGN PARAMETERS	GEOLOGY	SUPPORT PHILOSOPHY
I: Design and documention (Pre May 1982)	100 rockbolts for all 7 rock faces	-	Sound rock, randomly unfavourable, not persistent joints	Nominal support, spot bolting
II: Rock Mining proposal (November 1982)	-	-	-	Essentially no change. Side wall support combined with cavern support
III: Early rock mining (April 1983)	2300 rockbolts for Faces 5 and 6 only	ϕ = 40° F. of S. = 1 Water Model II (undrained)	Extensively developed repetitive joint sets -wedge and plane failures over all of Faces 5 and 6	Full support, conservative undrained water Model (II) with marginal F. of S.
IV: Top down rock excavation (September 1983)	148 rockbolts (Face 6 + 1/2 Face 5) 77 Rockbolts (Face 6 +1/2 Face 5) 240 Rockbolts (Face 6 + 1/2 Face 5)	ϕ = 50° F. of S. = 1 Water Model I (Drained) ϕ = 50° F. of S. = 1.4 Water Model III (Dry) ϕ = 50° F.of S. = 1.2 Water Model I (Drained)	3 wedges on Face 6 and possible plane failure on Face 5 predicted from mapping	Varied F. of S. with water model I and F. of S. of 1.2 adopted. Top half of Face 5 to be fully supported
V: As built (January 1984)	171 rockbolts in Faces 5 & 6 (A total of 467 rockbolts were used for all 7 faces)	ϕ = 50° F. of S. =1.2 Water Model I (Drained)	1 full wedge and 1 part wedge confirmed. Random blocks	Half of 171 bolts for local instability (spot). Half pattern bolts. Ensure early installation to maintain progress and then confirm support.

effect on actual temporary support as well as the predicted support. Two extremes of the preceived geology at CAB(E) and which required estimation of temporary support were:

a) Randomly unfavourable and not very persistent joints requiring only spot bolting.

b) Extensively developed and persistent joint sets requiring pattern bolting or other support measures.

3. Varying the combinations of design parameters, while perhaps playing with numbers, will effect the extent and quantities of rock bolt support.

4. At CAB(E) manipulation of the design parameters resulted in significantly smaller changes in the perceived level of required support, than the gross changes resulting from geological appreciation.

5. At CAB(E) full time specialist supervision of rock support would have reduced the number of rockbolts installed as well as the cost.

ACKNOWLEDGEMENTS

The authors express their appreciation to the Mass Transit Railway Corporation for permission to publish this paper and make the point that the views expressed are those of the authors and not necessarily those of the Corporation. They also express their thanks to Marples Ridgway/LTA Joint Venture for their cooperation and assistance. In addition the authors wish to thank their many colleagues at the MTR and Maunsells, many of whom contributed to aspects of work discussed in this paper.

Client: Mass Transit Railway Corporation
Engineer: Maunsell Consultants Asia
Contractor: Marples/LTA JV
Project Management/Supervision:
Mass Transit Railway Corporation

REFERENCES

Hoek, E. and Bray, J.W. (1980). Rock Slope Engineering. (Third edition). Institution of Mining and Metallurgy, London, 358p.

Geotechnical Control Office (1979). Geotechnical Manual for Slopes. Hong Kong Government Printer, 227p.

Barton, N.R. and Choubey, V. (1977). The Shear Strength of Rock Joints in Theory and Practice. Rock Mechanics, Vol. 10/1-2, pp 1-54.

Barton, N.R. (1974). A Review of the Shear Strength of Filled Discontinuities in Rock. Norwegian Geotechnical Institute Publication No. 105, 38p.

Urban quarries in São Paulo city: operation, development, control and regulations

N.F. Midea
Instituto de Pesquisas Tecnológicas do Estado de São Paulo S/A, São Paulo, Brazil

SUMMARY

For several years environmental issues arising from quarrying activity in urban areas of Estado de São Paulo, Brazil - one of the largest concentrations of population in the world - have been under consideration.

Prevention of damage to the environment has come to be regarded as an essential element of planning, and there has been continuous research in this field. The results of laboratory investigations and field studies in rock blasting with explosives have been collected and submitted to relevant state organizations to assist in framing appropriate regulations to ensure safe and economical quarrying without damage to the human environment.

This paper provides a brief history of the problem, and reviews results obtained through co-operation between the private and the public sectors.

It summarises the experience of a research group of IPT (Instituto de Pesquisas Tecnológicas do Estado de São Paulo), concerned with the science and technology of rock masses, and including, apart from the author, Sector of Instrumentation of Rock Mass, directed by Civil Eng. Fernandes, A.C.C.; Sector of Rock Blasting, directed by Min.Eng. Fabiani, P.Abel, and Sector of Numerical Data Treatment, directed by Math. Furumoto, Shintaro.

INTRODUCTION

1. The Socio Economic Problem

São Paulo City and outskirts, with approximately 15 million inhabitants occupying 8000 km^2, faces a serious problem caused by mining activity (quarrying) in areas in which live an expanding population. Risks of damage and discomfort are present in the daily noise of the production blastings.

Efforts to find a solution to this problem led to the involvement of public adminstration bodies with private mining companies and community associations, among other sectors of society, and of IPT, which was in charge of the technological aspect only. Since IPT, through the Rock Mechanics Group, has, for at least 10 years, given technical assistance to a great number of mining and civil construction companies, a considerable body of data, here presented, has been generated.

Figure 1 shows one of the most intensively studied areas of Brazil - the administrative region of São Paulo City. However, the problem also arises in other major centres of population: Rio de Janeiro, Porto Alegre, Salvador and Belo Horizonte. The results found for São Paulo are expected to be extrapolated to other areas.

1.1 The conflict between quarry and environment: its causes

For the same reasons as those for which former civilizations were attracted by the advantages

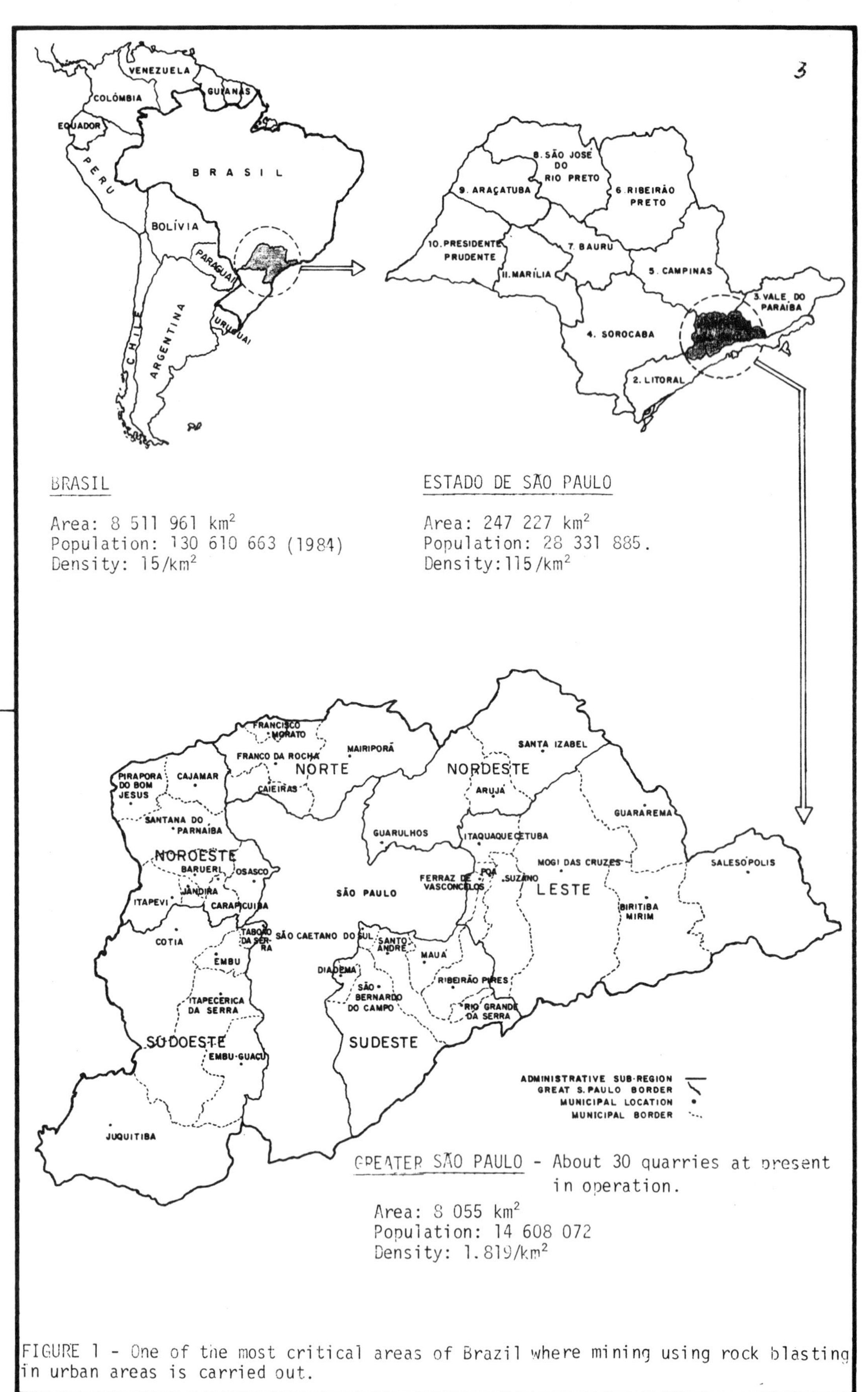

FIGURE 1 - One of the most critical areas of Brazil where mining using rock blasting in urban areas is carried out.

of proximity to such large rivers as the Nile, the Tigris and the Euphrates, the less favoured sections of present-day populations are attracted by the improved economic conditions and the infrastructure offered by the extractive industry. However, the expected benefits have not been forthcoming from the production, in populated areas, of crushed stone for the civil engineering industry. On the one hand, the investing producers aim to achieve large short-term profits by using techniques of rock fragmentation relying on unsuitable explosives and creating permanent danger of damage to structures and injury to people. On the other hand, communities are not always sufficiently honest to resist the temptation to stimulate and magnify the disorder caused by mining activities in order to gain illegal benefits by way of compensation.

These aspects indicate the real problem, derived, as has been said, from the unplanned approach of the population to existing quarries. On this basis the problem is being examined in order to find the real roots of the equation.

1.2 Problem equation

The problem should be solved under the following conditions:

- continuity of the economic mining process;
- indispensable and sufficient safety and comfort for the neighbourhood.

For ten years IPT kept in touch with the problem, providing technical assistance and consultant services for industrial, social, public and political bodies. The following factors were recognised:

1) Absence of specific regulations, with official norms and procedures, related to the main undesirable mechanical effects arising from the rock blasting technology.

2) Insufficient and inadequate public control, since public bodies have neither specialists nor the required equipment, apart from having to apply a subjective judgement in place of the missing specific regulations.

3) Technical feasibility of urban areas being mined to supply other industries economically without detriment to public welfare.

1.3 Proposed solution

During different stages of IPT studies, some facts to be considered emerged, especially such radical proposals as:

- closure of quarrying areas owing to very serious events such as the death of a person caught by a flying fragment, or to political pressures exerted by part of the local population.
- proposals to move the mining areas to new localitites, as was done in Estado do Rio de Janeiro.

Such solutions were proved to be impracticable from technical, economical and political points of view, and it was decided to make a direct attack on the problem according to the following flow-chart.

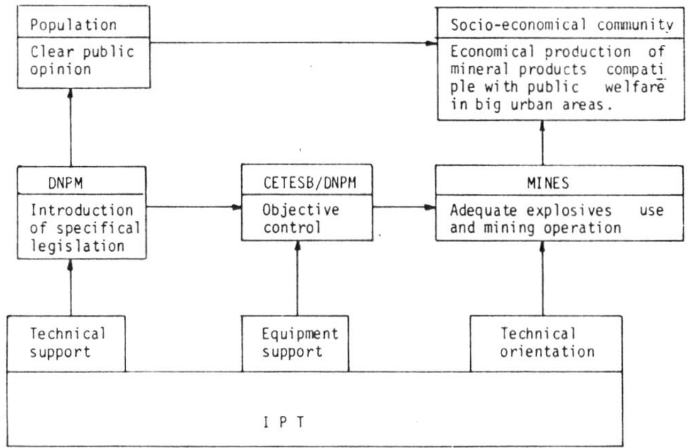

This integrated action was carefully studied and was finally agreed with the organizations most directly concerned:

- DNPM (Departamento Nacional de Produção Mineral: National Department of Mineral Production) is a federal organization with legislative and control functions in the mining area;
- CETESB (Companhia Estadual de Controle da Poluição e Meio Ambiente: State Company for Pollution and Environment Control);

- Prefeitura de São Paulo (Municipal Administration of São Paulo City);
- Private miners, producers of mainly quarried crushed stone for use in the civil construction industry;
- IPT, already described.

Some bureaucratic resistance from the organizations involved to the introduction of the proposed solutions was unavoidable.

2. Introduction of proposed solutions

The introduction of action according to the scheme outlined above had to take account of:

- technical aspects
- legal aspects of legislation and control
- sociopolitical aspects

2.1 Technical aspects - IPT functions

The most important mechanical effects bringing up risks of damage and discomfort to the adjacent population are identified as:

- flyrock
- ground vibrations
- air-shock

All IPT technical work was orientated to a double aim:

- acquisition of exact technical knowledge of the phenomenon through the study of causes of origin and intensity regulation, and
- definition of admissible values of energy transferred in the form of waves.

2.1.1. Flyrock

Secondary blasting to reduce the size of the biggest blocks resulting from bench blasting was the main source of rock fragments being propelled over the permissible boundaries. In general the number of secondary blasts is very high and they are usually produced without any control of geometry of drilling, stemming of quantity of explosive charge. The presence of a specialist to give technical assistance and make recommendations on the use of alternative mechanical technology for primary comminution is part of the solution. In relation to primary bench blasting, it is necessary to draw attention to certain risks arising from incorrect stemming, degree of confinement, presence of loose blocks on the top and face of the bench being blasted, drilling geometry (burden, spacing, diameter, length, inclination, etc.) and powder factor.

The ABNT (Brazilian Association of Technical Norms) is preparing a standard forbidding flyrock fragments beyond the boundaries of the mine property area.

2.1.2 Ground vibrations

The most controversial environmental problem caused by the use of explosives was that of ground vibrations resulting from the seismic energy released by blasting. Almost ignored by the mining industry until recently, this hazard has prompted IPT to undertake a series of research studies at existing large hydroelectric power stations such as Ilha Solteira, Jupiá, Agua Vermelha, Três Irmãos, Capivara, Porto Primavera, etc. (Parana Basin) and Tucurui (Tocantins River).

The need to excavate rock simultaneously with the building of concrete structures of gravity dams demanded strict observation of explosive charges (Q) detonated at a distance (D) from the concrete. It was necessary to establish a relationship between Q, D and the energy of the dynamic event, in the form of particle vibration velocity (Vp), for each locality. At present IPT define this relationship by statistical analysis of the independent variables Q,D and the dependent Vp. An equation of the form

$$Vp = a\ Q^{b} D^{c} \tag{1}$$

is obtained, where a, b, c are parameters derived from multiple regression and not pre-fixed, as in Devine, Langefors, etc., systems. Factors for blasting design, energy transmission properties of rocks and excavation geometry are included. Multiple correlation gives a better coefficient compared with regressions with fixed parameters.

Figure 2 is a typical representation of multiple correlation of Q, D, Vp. Data was obtained from rock excavations of large hydroelectric power stations: 791 sets of data from measurements in 11 different job sites.

Following the model of equation (1), the regression equation is

$$Vp = 85.30\ Q^{0.61} D^{-1.36} \tag{2}$$

The usefulness of this equation requires homogeneity in the propagation conditions,

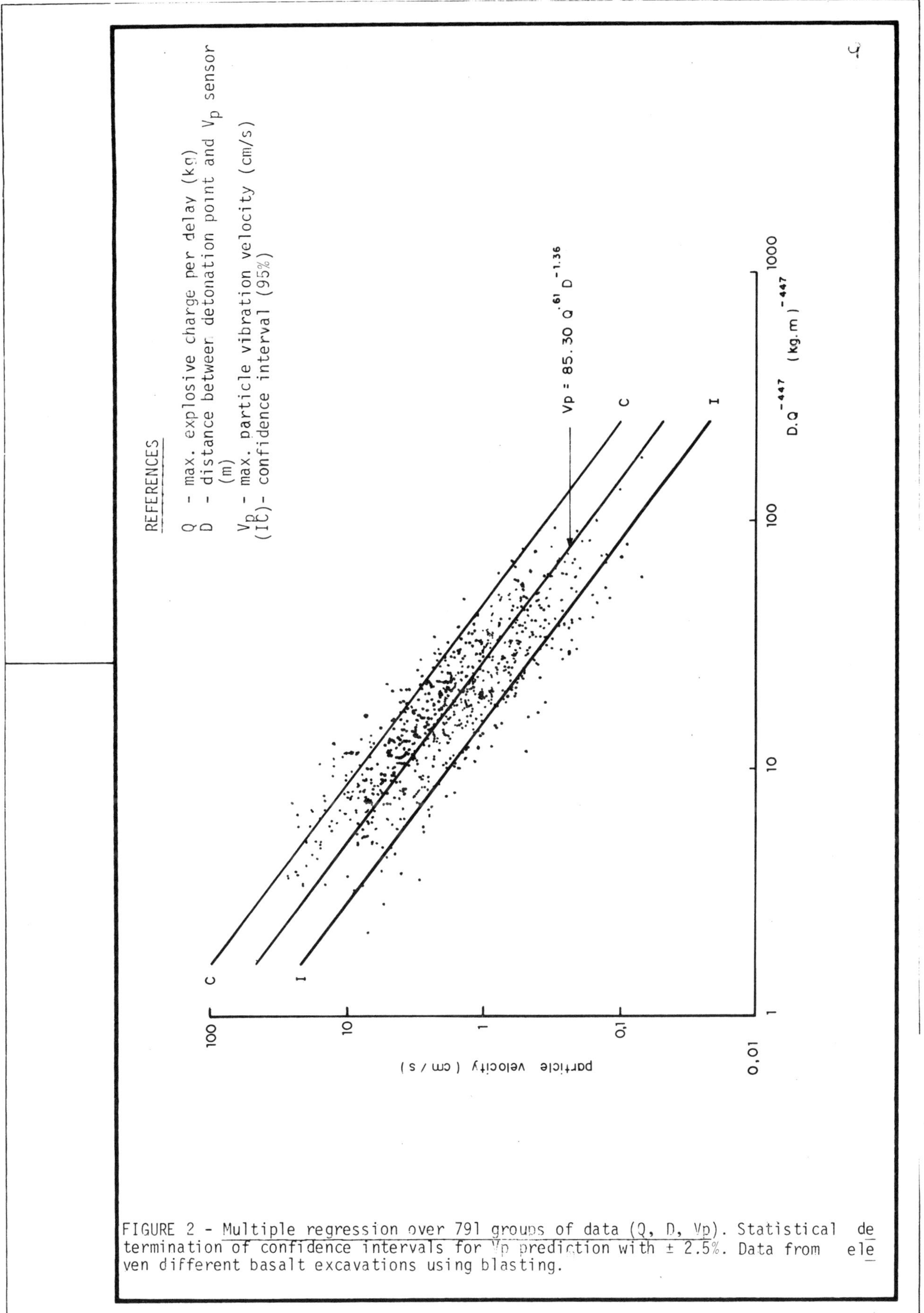

FIGURE 2 - Multiple regression over 791 groups of data (Q, D, Vp). Statistical determination of confidence intervals for Vp prediction with ± 2.5%. Data from eleven different basalt excavations using blasting.

weight of explosives, blasting accessories and blasting design. Besides this, a common variables domain of Q and D with a fair distribution will lead to a reasonable contribution in the statistical explanation of Vp.

In the case of urban mining, control of Q depends on typical structures associated with the area surrounding the quarry. However, controversy arising from the definition of admissible levels of vibration energy capable of being resisted by these structures must be expected. As construction of dwelling structures without quality control increases, so it becomes more difficult to arrive at the definition. Thus, to define the threshold value, the results of IPT work at problem sites were analysed and compared with the international values. The chosen value was Vp = 1.5 cm/s, which is not expected to cause problems, whatever the frequency class. Practical experience with such a threshold value promises satisfactory conditions of environmental safety and comfort.

SOME TYPICAL DATA: PRODUCTION OF CRUSHED STONE AND TECHNOLOGY IN ESTADO DE SÃO PAULO

- Annual production (mean for 1983 and 1984, considered critical years) = 12 x 10^6 m^3 /year
- Production per capita = 0.5 m^3/year
- Number of operational quarries 50

From Figure 3 one can see that largest values for Vp (mean 1.6 cm/s) occur inside quarry areas. For measurements outside the limits of such properties the mean value is 0.2 cm/s, which is higher than the median and mode of the statistical distribution of Vp, characterizing a skewed distribution in favour of safety.

It is seen from Figure 4 that 200 m can be considered an initial standard for distance to residential structures of the type usually associated with Brazilian quarries, and this finding is based on observation of several kinds of rock mass (gneisse, granite, limestone, basalt) and mining companies.

Occurence of damage

As is well known, the probability of structural damage caused by vibrations is a function of several factors, which makes estimating it difficult. The occurrence of damage depends not only on the vibrational characteristics (amplitude, velocity, frequency), but also on the characteristics of the structure (geometry, mechanical strength, natural frequency, etc.). For the solution of the problem for greater São Paulo, in a short-term study, the adoption of an initial experimental value was proposed, this value to be adjusted during different application stages.

Studies were recently carried out in an area where residential constructions afforded an opportunity to monitor cracked walls. Cracks were of unknown origin and of appreciable age. As a result, monitoring with a mechanical instrument (Tenso-test) showed that the effect of blasting vibrations on the initial amplitudes of cracks is of the order of 1/25 of the effect on amplitude produced by daily temperature variation. The type of instrument used does not permit the association of the amplitude with the vibrating frequency.

In the Butantã Administrative Region of the Municipal District of São Paulo City, 27 pre-existing cracks in 6 local residences, with a maximum width of a few millimetres (see photographic illustrations) were monitored.

- Primary blasting frequency twice a week, with a minimum distance of 250 m between the monitored houses and the bench being blasted, and maximum explosive charges per delay of 44 kg gave maximum particle vibration velocity of 1.15 cm/s measured on the balcony (1.20 to 1.50 m from the floor).
- Characteristic values of vibrating motion of cracks due to blasting:
  . maximum amplitude (measured) = 30x10^{-3}mm
  . frequency (estimated) $10 < f < 30$ Hz
- Characteristic value of oscillating motion of cracks due to daily temperature variations:
  . maximum amplitude (measured) = 800x10^{-3} mm
  . frequency f = 1 cycle per 24 h.

2.1.3. Air-shock

Parameters related to blasting rounds,

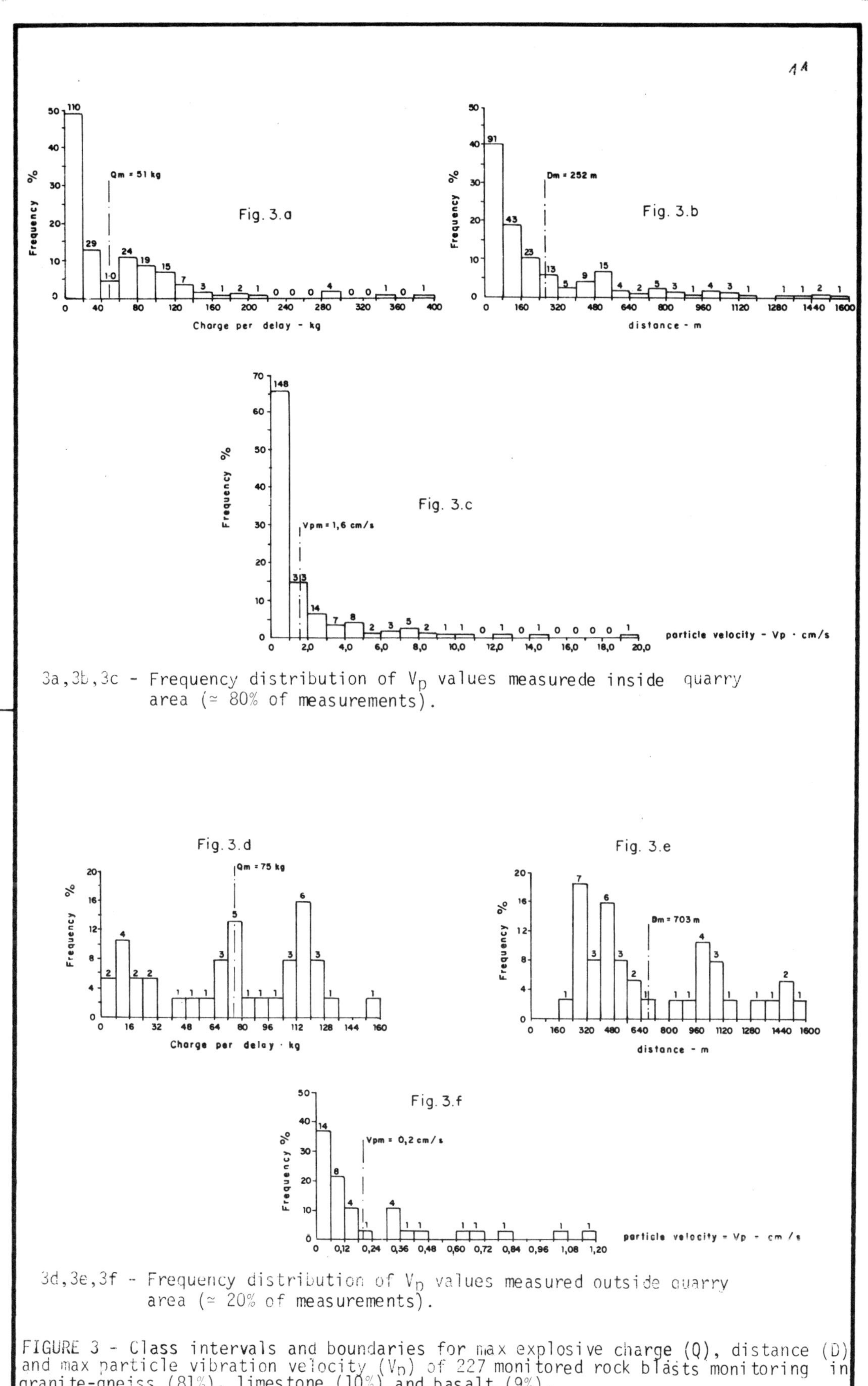

3a,3b,3c - Frequency distribution of V_p values measurede inside quarry area (≃ 80% of measurements).

3d,3e,3f - Frequency distribution of V_p values measured outside quarry area (≃ 20% of measurements).

FIGURE 3 - Class intervals and boundaries for max explosive charge (Q), distance (D) and max particle vibration velocity (V_p) of 227 monitored rock blasts monitoring in granite-gneiss (81%), limestone (10%) and basalt (9%).

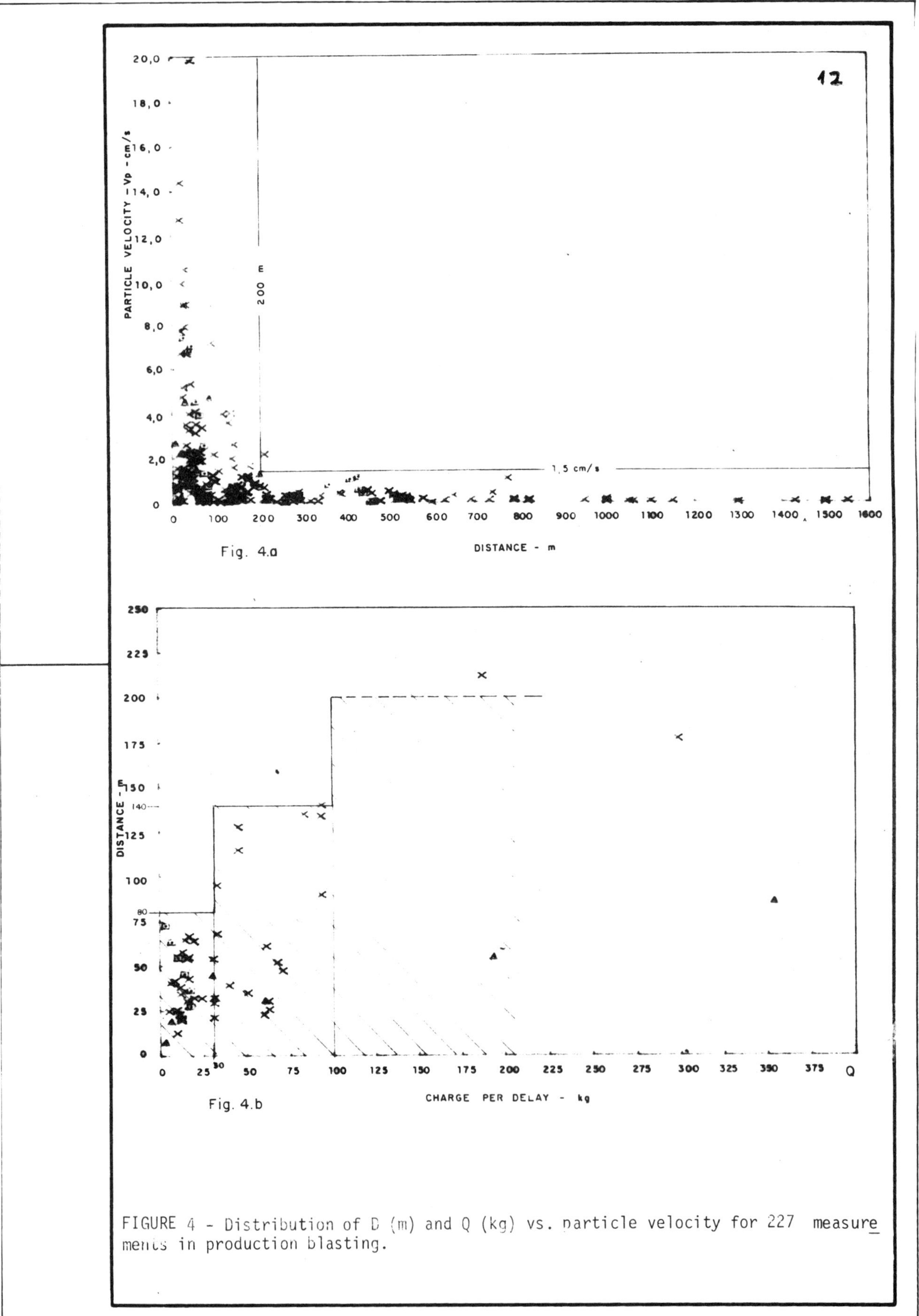

Fig. 4.a

Fig. 4.b

FIGURE 4 - Distribution of D (m) and Q (kg) vs. particle velocity for 227 measurements in production blasting.

explosives and accessories used, and psychological effects were dealt with in IPT technical recommendations to the quarrying industry in order to decrease air-shock discomfort. However, the climate of the region is of great complexity and nothing can be done about the wind speeds and thermal inversion. Thus, it becomes very difficult to establish variation criteria for air-shock propagation. The most important consideration is that the blasting be carried out under acceptable weather conditions, in geometrically and psychologically favourable conditions.

In the light of other local practical experience and the general international standards associated with discomfort and damage to fragile parts of structures (glass, windows, plaster, etc.), the maximum permissible value for air-shock pressure is 130 dB. Residents themselves stated that values below this limit are acceptable.

2.2 Legal aspects - ABNT functions

Present-day Brazilian legislation contains no regulations regarding adequate control of blasting operations in urban areas. There are federal laws forbidding the emission of polluted elements to the mining environment without stating the quantity and nature of the element, as is the case for flyrock, ground vibrations and air-shock. DNPM, the body in charge of this legislation, awaits the final ABNT decision on the standards.

ABNT Commission CE: 18.2.7 in charge of technical standard production, includes representatives of the main public and private bodies, and of federal and local state interested organizations, directly and indirectly involved with this work. ABNT expects to have finished its work by the end of this year, and DNPM will proceed to issue proposed federal legislation based on this work and an edict calling for control of mine operation in urban areas.

2.3 Control aspects

The provisions of this legislation could be enforced by a federal organization (DNPM) and by public organization at state and at municipal level by the regional administration of the relevant municipality. The administrative region of São Paulo City is shown in Figure 1 and the various administrative functions are shown in the flow-chart of Section 1.3. See also pictures.

3. Final considerations

We conclude that:

- it is necessary and important to major Brazilian urban centres to continue mining within their boundaries, to avoid the cost increase following moving the mining activity planned localities far away from the centres of consumption (higher costs of transport and other benefits introduced such as labour demand, logistic structure, etc.), on condition that safety and comfort in the environment be respected.
- there exists sufficient technical background to enable the required conditions of safety and comfort in the environment to be observed. It is merely a question of enthusiasm in considering the typical complexity of situations created by unplanned population growth.
- the above conclusions apply to the resolution of problems in Estado de São Paulo. It is believed that these conclusions could be extrapolated to other communitites if it is accepted that the same basic human needs and rights apply to all peoples.

ACKNOWLEDGEMENTS

The author wishes to acknowledge the technical cooperation of IPT colleagues in the production of this paper:

- Eng. Fabiani, P. Abel, text revision and English version.
- Geol. Cerello, Luiz, data for Figure 1.
- Roney Serpa and Luiz Bassi: mountings and drawings.

He also acknowledges the cooperation of Eng. Ary Simonetto from Administração Regional do Butantã, Prefeitura de São Paulo, who authorized the publication of technical data of IPT Technical Report No. 22 206 - May, 1985.

Typical location
of urban quarry

Air blast monitoring in
a residential area

Cracks probably induced
by vibration

BIBLIOGRAPHY

INSTITUTO de PESQUISAS TECNOLÓGICAS do ESTADO de SÃO PAULO - Análise de operações com explosivos na Pedreira Raposo Tavares e sua interferência na integridade do meio ambiente do Jardim Boa Vista, São Paulo. São Paulo, 1985 (Relatório oficial no. 22 206).

MIDEA, N.F. & DINIZ, M.J. - Avaliação dos efeitos provocados pelo uso de explosivos nas minerações em áreas urbanas. In: III SIMPÓSIO SOBRE NORMALIZAÇÃO DE CIMENTO, CONCRETO e AGREGADOS. São Paulo, Brasil, 1984 (inédito).

MIDEA, N.F.; DOZZI, L.F.S.; COELHO, L.F.M. - Avaliações estatísticas e considerações sobre controle de vibracoes devidos a explosivos durante a construção de usina hidroelétrica de Nova Avanhandava - (paper presented in the 2. Jornadas Argentinas de Ingenieria de Minas, San Juan, 1981, Argentina).

SAAD, J.C.; GUAPO, L.A.; FABIANI, P.A.A.; DOZZI, L.F.S. - Equação da previsão do nível de vibração oriunda de detonações em maciços basalticos da Bacia do Alto Paraná. (Paper presented in the Simpósio sobre a Geotecnia da Bacia do Alto Paraná. Rio de Janeiro, 1983).

Design and construction of high rock cuts for the Kornhill Development, Hong Kong

T.R.C. Muir B.SC., A.R.C.S.T., C.ENG., M.I.C.E.
Mass Transit Railway Corporation, Hong Kong
B.K. Smethurst B.SC.(ENG.), M.SC., C.ENG., M.I.C.E., M.I.M.M., M.A.S.C.E., F.G.S.
Formerly, Mass Transit Railway Corporation, Hong Kong (now, Balfour Beatty Construction (International Division), Oman)
R.P. Finn B.SC., M.SC., M.I.M.M., M.H.K.I.E., M.I.GEOL., F.G.S.
Freeman Fox and Partners (Far East), Hong Kong

SYNOPSIS

The setting and scope of the Kornhill Development are outlined with the design and construction of its high rock slopes discussed in detail. The site geology revealed by the various methods of investigation is stated and is highlighted where it influenced the choice of presplitting and support methods. The need for and establishment of a site mapping/design team to enable a tight construction programme to be met are described, together with the cycle from mapping through to completion of slope treatment. The versatility of shotcrete in slope treatment and drainage work is stressed.

SITE DESCRIPTION

The Kornhill Development is situated on the northern slopes of Mt. Parker at Quarry Bay on the northern shore of Hong Kong Island and comprises an 'S' shaped 22ha land-site for 44 no. 30-storey residential tower blocks for 30,000 persons, and 2 commercial centres. The development is an integral part of an overall scheme for the area, involving at-grade and elevated roads, underground railway tunnels and a station, a service reservoir and school sites. The principal aspect of this paper is the design and construction of rock slopes within and around the perimeter of the land-based works. The approximate location and the original topography, the final form of the site formation, and the site are shown in Figs. 1, 2 and Plate 1 respectively. 3.5Mcu.m. of rock and 1.5Mcu.m. of weathered granitic soil and superficial deposits (bank volume) had to be removed over a three year period from the land site at a rate of up to 10,000cu.m./day with original ground levels being lowered by up to 80m. The first main programme feature of the Project was the early formation of two large temporary excavations for entrances to an underground station in the north of the site and this included the formation of 9000sq.m. of temporary rock slope and 3000sq.m. of temporary soil slope. At the same time, excavation for the final development area proceeded from south to north, resulting ultimately in the removal of these two temporary excavations. 58,000sq.m. of permanent rock slope and 5000sq.m. of permanent soil slope were produced. The main perimeter rock slope was 800m long and up to 65m high. A typical section is shown in Fig. 3.

SITE INVESTIGATION

Before construction, the topography consisted of several north-north-east trending valleys with a subsidiary west-north westerly trend. Valley sides were steep, ranging from 30° in soils to vertical in rock. Between the valleys, the natural topography comprised three broad sub-rounded ridge crests, varying in level from +140 mPD, at the southern edge of the site to +80mPD at the northern edge. In the eastern half of the site, the original topography had been extensively altered by quarrying, and excavation and backfilling for structures. The pre-construction vegetation cover appeared to be thinner over an area underlain by coarse grained Hong Kong granite.

At the time of preliminary design, information was derived from several sources.

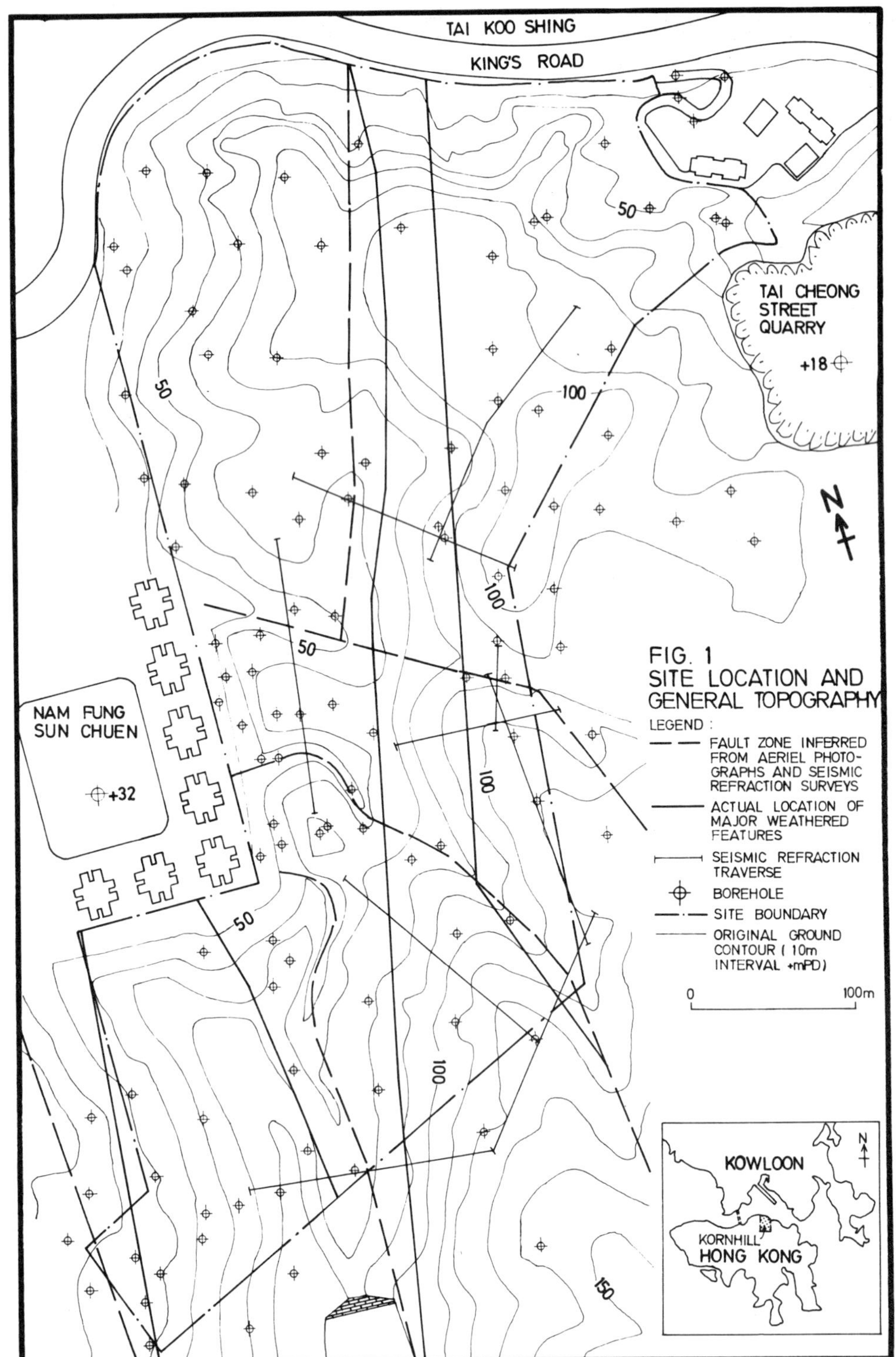

Information prior to 1970 comprised several series of aerial photographs dating back to 1949. These showed several prominent valley features and also gave an indication of excavation and filling around structures. Between 1970 and 1980, a total of 64 boreholes, 6 horizontal probeholes and 12 trial pits were sunk on adjacent sites. Between 1980 and 1981, 137 boreholes and 2 seismic refraction surveys were completed within the site boundary. This was supplemented by engineering geological surface mapping and discontinuity surveys during 1981. Boreholes were located to provide the maximum amount of information necessary to :

(i) design temporary rock and soil slopes, tunnels and shafts for the advanced works contracts,

(ii) indicate rock quality at proposed formation level,

(iii) permit reliable estimates of rock and

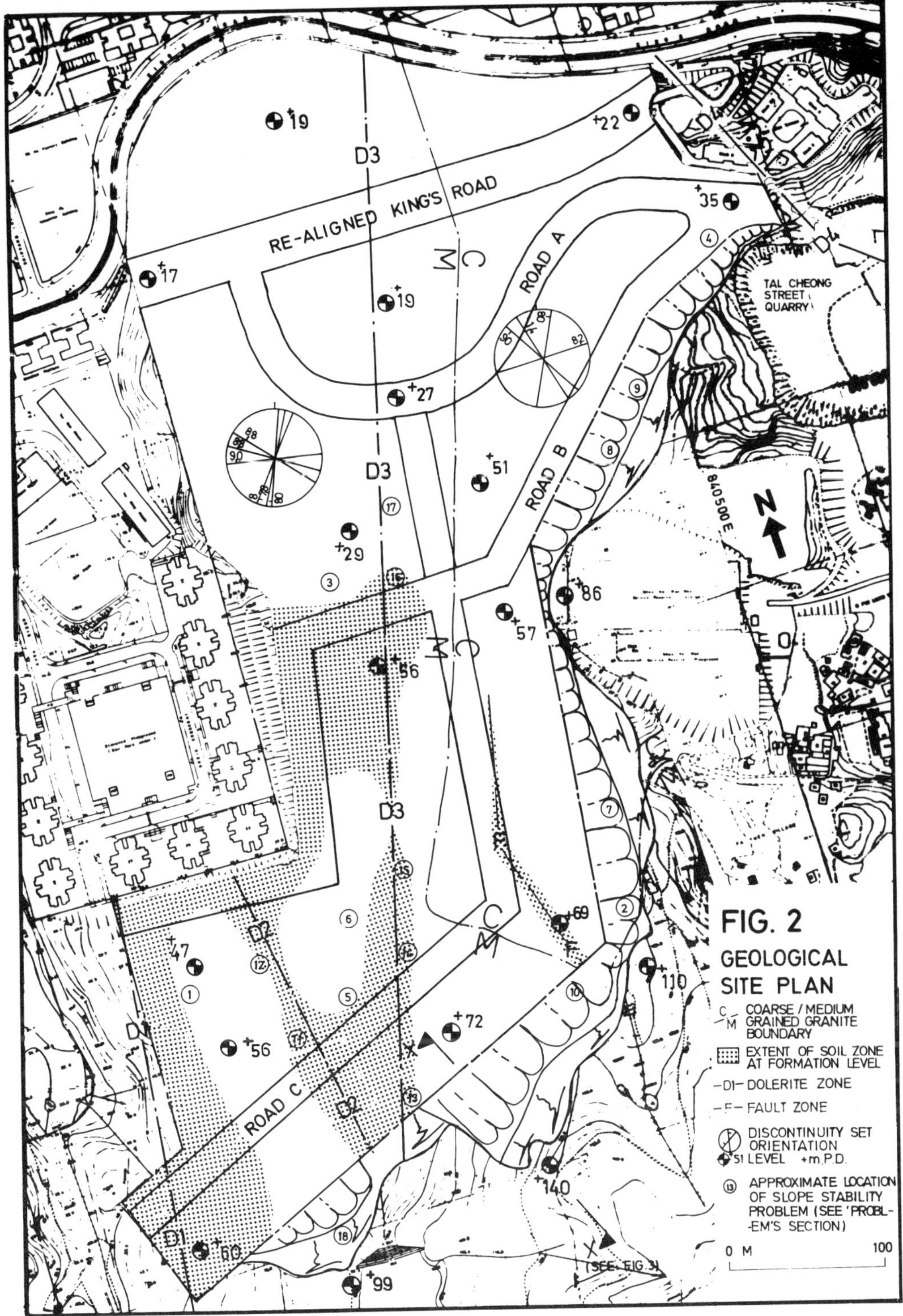

FIG. 2 GEOLOGICAL SITE PLAN

soil quantities,

(iv) design permanent rock and soil slopes.

The normal criterion for proving bedrock was 5m of continuous sound core. Boreholes around the location of the future high perimeter rock slopes were generally advanced to 2-3m below proposed slope toe level to detect adversely orientated discontinuities and to monitor groundwater levels before and during construction. The orientation of the discontinuities was originally measured using a Craelius core orienter, but the Borehole Impression Device gave more representative results (Ref. 3). Samples were taken of different types of joint conditions for direct shear box testing. These provided basic friction parameters which were compared with previous experience in the Hong Kong Granite. Piezometers or standpipes were provided in most

Plate 1 : General view of Kornhill Development from the north in mid-construction. The main perimeter rock slopes can be seen in the background, and partly-formed slopes and bulk excavation in the foreground

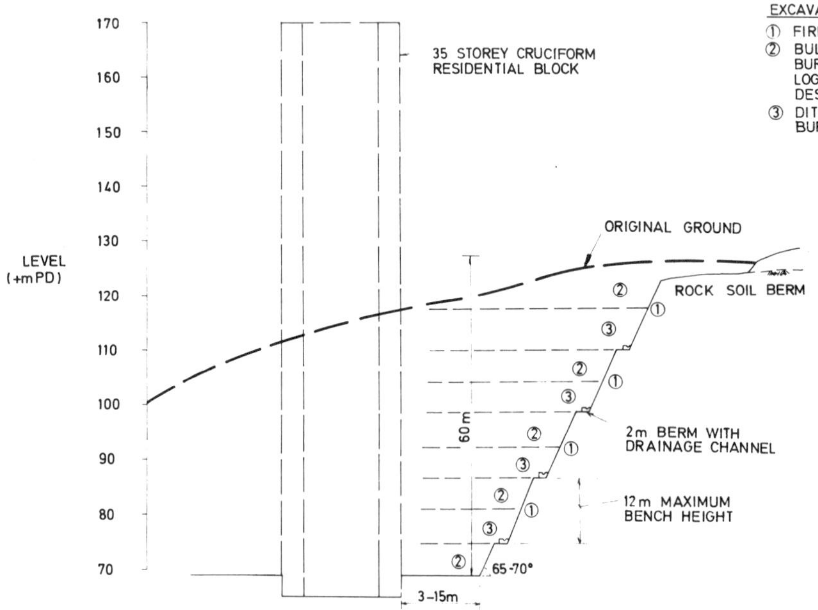

TYPICAL SECTION X-X AND EXCAVATION SEQUENCE FIG. 3

holes.

During the construction period, rapid probing using production drilling rigs was found extremely useful in re-defining rock-soil interfaces and limits of faulting. Boundaries were generally distinguished by examination of the air-flush returns and rates of penetration of the 76mm bit. Colour was also a useful criterion. This technique was much cheaper and quicker than cored drilling, and the equipment was readily available on site. Rock levels established by this method were sufficiently accurate to be used in re-designing the slopes where this proved necessary. Probing investigation at location (4) (see Fig. 2) is shown in plate 2. Hydraulic rigs were generally used as these had been specified in the contract on account of their lower noise and dust emissions, although better results could be obtained from the less powerful and more sensitive pneumatic rigs ('air-tracks').

SITE GEOLOGY

The site was located within the outcrop of the Hong Kong Granite. Two types of granite were present, a medium grained type underlying about 60% of the site, and a coarse grained type underlying 40%. The granite was affected by at least four dolerite dykes (zones D1-D4) and a suspected fault zone (zone F). The solid geology is shown in Fig. 2 and discussed in more detail elsewhere (Ref. 2).

The granite had been subjected to sub-tropical weathering and the weathering profile was similar to that illustrated in the Geotechnical Manual for Slopes (Ref. 1) with the exception that a corestone rich zone (Zone C) was only occasionally developed. The transition from soil to continuous rock occurred over a metre or less. Residual soils were typically 10m thick on ridge crests not affected by dyke zones, thinning on valley sides to nil in valley bottoms. There was slight evidence that residual soils were occasionally thicker over the medium grained granite which may have weathered preferentially in this area. However the greatest soil thickness (up to 40m) occurred along the three dolerite zones within the medium grained granite (D1-D3) and along the fault zone (F) within the coarse grained granite. Colluvium occurred in isolated patches on valley sides throughout the site, rarely exceeding 2m in thickness. Alluvium up to 10m thick was encountered in the major faulted valley on the western side of the site, and in isolated areas of the other valleys, Actual bedrock levels encountered on site compared well with those deduced from nearby site investigation data. The seismic refraction interface and ripping limit co-incided with the top of moderately weathered rather than the top of slightly weathered granite.

Plate 2 : Rapid probing to prove bedrock using hydraulic blasting rig.

However, seismic traverses were less successful in locating wide zones of deep weathering (zones D2, D3, F), possibly because of a high water table in these areas. The boundary between the two granite types was not usually deeply weathered but appeared to have a low seismic velocity.

The orientation of discontinuities in the Hong Kong Granite as a whole is discussed elsewhere (Ref. 4). It became clear early in construction that the dominant trends at Kornhill were a low angle and two vertical sets, all caused by cooling. The medium grained granite had additional sets parallel to its contact with the coarse grained granite (Ref. 2). These are shown on Fig. 2. Few major sheeting joints were encountered. The only clear tectonic joints were vertical sets for short distances either side of, and parallel to, dykes D1 to D4. There was no characteristic surface texture or mineralisation to enable different sets to be distinguished, but some sets clearly cross cut others, crossed the contact between the granites or were controlled by it.

Apart from minor erosion on ridge crests (especially around the site of former structures), there were few signs of pre-existing instability in the area, although the steep rock areas in some valleys may have been exposed by landslipping of thin soil cover. This may have been triggered by the water seepage visible at most of these faces.

ROCK SLOPE PRELIMINARY DESIGN

When the general layout of the development was established, rock slope heights were estimated from a bedrock contour plan produced from boreholes, outcrops and seismic profiles. Stability analyses were carried out using available discontinuity data and basic friction angles from direct shear box tests (cohesion and water pressure ignored), to determine slope angles for the various orientations to ensure overall stability. Optimum slope angles were all found to be in the range 65-70°. All temporary and permanent rock slopes had to be designed in accordance with standard procedures (Ref. 6) taking account of recommendations in the Geotechnical Manual for Slopes (Ref. 1). Maximum rock bench heights of 12m were adopted to balance pre-split drilling progress against deviation of longer holes, particularly since profiles were inclined at 65-70°. Level drainage berms were adopted between benches to facilitate future maintenance and landscaping. A berm width of 2m was chosen for these purposes. At the rock/soil interface, provision was made for a berm up to 5m wide to allow for anomalies in the depth of soil such that the height of soil cuts could be increased by up to 3m without affecting the designed slope toe line. This was adequate for 95% of the slopes. However, in two locations not directly covered in the site investigation (locations (7), (18), Fig. 2), rock levels were lower than expected and were re-defined by use of rapid probe holing.

Several proposed slopes crossed valleys which were suspected to contain fault zones. However, many boreholes in these valleys showed high rock levels. In view of the uncertainty on the extent of weathering of the rockmass at these locations, slopes were designed at 65-70°, with a facility for regrading or slope treatment to be added if necessary, once their exact location and condition had been exposed. This was shown as provisional on the working drawings.

Between the preliminary design stage and the construction stage, extensive advance works were undertaken. These involved 350,000cu.m. of excavation and included haul roads, shafts, two long 4m diameter tunnels in the central part of the site, and later Mass Transit Railway tunnels in the northern part. The exposures were used to draw-up a site specific descriptive and classification system, for standardising logging procedure. Rock mass and discontinuity data from the advance works supported assumptions made in the preliminary design.

Wet season groundwater levels from standpipes and piezometers located across the site tended to be within 2-3m of bedrock level. No excessive groundwater pressure was encountered. Water pressures were not analysed in the preliminary design and it was expected that post construction drawdown would occur. The long term water-table was expected to be well behind the finished slopes even with a 3m rise in groundwater level corresponding to a 1:200yr storm. This has proven to be the case on site. The main potential groundwater problem was the possible consequence of storm water ingress into toppling joints.

PRESPLITTING

The Contractor used a standard 76mm diameter drillbit for both presplitting and bulk blasting. Trial presplitting was carried out early in the contract as shown in Table 1. The adopted pattern (Table 1 and Fig. 4) produced generally satisfactory results. Pre-splitting in highly to slightly weathered granite is shown in plate 11. In a few places cracking between presplitting holes was incomplete at the gap locations between the individual Tovex cartridges. Removal of a large 'hang-up' is shown in plate 3. In some areas, infilled or open joints running roughly parallel to the original ground profile tended to open, causing loosening of blocks due to venting of gases through the discontinuities. Discontinuities running sub-parallel to the face occasionally caused 'peeling' of the profile (plate 4).

The single most important factor influencing the effectiveness of a presplit blast on this site was the accuracy of the hole alignment. Deviation in some areas was a source of concern in longer holes, particularly where it was in the direction of the dip of the slope. In shorter vertical presplit holes (6m), such deviation was not apparent and seemed linked to the high axial loads generated by hydraulic rigs, causing flexure of the longer drill stems and hence bit deflection. This could be improved by using down-the-hole hammers, where the axial load is acting immediately behind the bit and not at the top of the drillstem. An example of misalignment is shown in plate 5 and deviation in plate 8. Presplit slopes were set out by batter rails at 5 metre centres on the rock head, or in the case of subsequent benching operations, on the berm level. Each hole was positioned and corrected for level by dipping from a string wire stretched between the rails.

Where wide variations in existing ground levels occurred there were corresponding variations in presplit drilling depths. Early trials at standardising drilling depths and backfilling the holes to formation level before charging proved unsuccessful as cracks also propogated between holes within the backfilled sections, as shown in Fig. 5. Drilling schedules were therefore produced daily, giving the depth of each hole to be drilled by a particular rig. Methods to set the drill rig boom to the correct angle varied from a simple

Presplit Trial Panel	Area (m^2)	Hole Spacing (m)	Charge Vertical Spacing c/c(m)	Charge Concentration (Kg/m^2 of presplit surface)	Comments
A1	30	0.5	0.8	0.53	Total length of trial face was 30m , with six groups of six 76mm holes each 11-13m long. The trial was in closely jointed material. Charge pattern was based on 0.4m long, 25mm o Tovex 210 (confined velocity 4800 m/sec) sticks taped to detonating chord with a bottom charge of 2 sticks. No uncharged guideholes were used. Due to restrictions on air blast, detonating chord was not allowed above ground level and a detonator was used at the top of the detonating chord.
A2	30	0.5	1.2	0.37	
B1	45	0.75	0.7	0.4	
B2	45	0.75	1.0	0.27	
C1	60	1.0	0.7	0.3	
C2	60	1.0	1.0	0.2	
Adopted Pattern	-	0.6	1.0	0.33	This pattern was chosen as the best balance between satisfactory slope surface quality and drilling progress.

Table 1 Pre-splitting Characteristics

Plate 3 : Removal of large hang-up on rock face after pre-splitting and bulk-blasting.

Plate 4 : Effect of vertical discontinuities sub-parallel to the slope profile on the presplitting.

Plate 5 : Mis-alignment of pre-split drill-holes in widely jointed granite. The pre-split has however been successful.

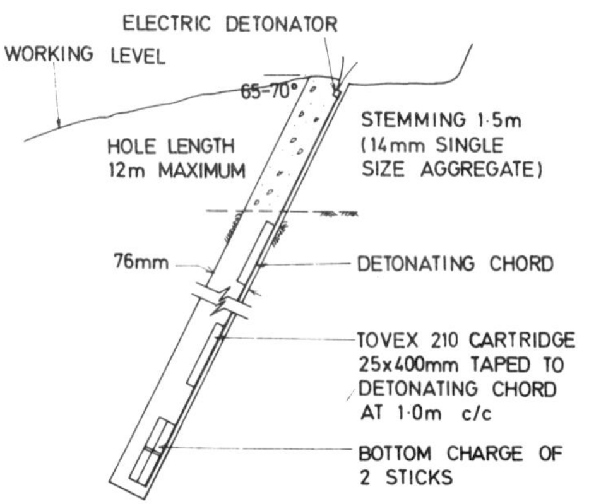

TYPICAL PRESPLIT CHARGE DISTRIBUTION FIG 4

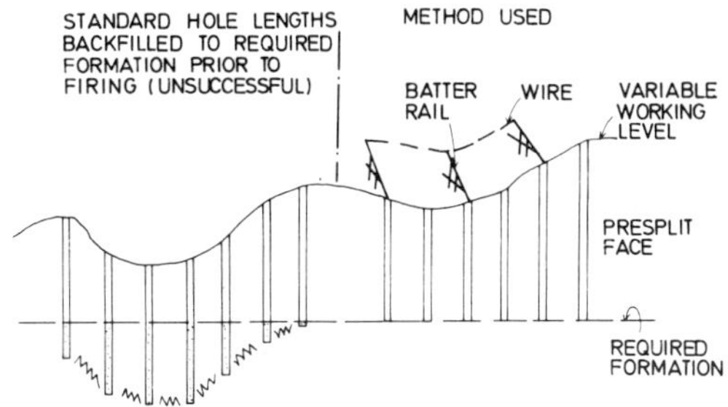

ELEVATION OF DIFFERENT PRESPLIT DRILLING PATTERNS USED AT KORNHILL DEVELOPMENT FIG. 5

template and spirit level, to a sophisticated optical inclinometer, fixed to the rig boom. Local drillers preferred the former method, with orientation being set by a plumb line arrangement using a bamboo tripod offset 3m in front of the presplit line. The accuracy of this method was approximately 10mm/m depth of presplit hole.

It is generally accepted that 'confined' presplit blasts produce peak particle velocities (ppv's) several times higher than bench bulk blast at the same distance with the same charge weight per delay. Presplit and bulk blasting were constrained in certain parts of the site by the presence of adjacent property, a service reservoir, water mains and adjacent existing quarry slopes of doubtful stability. It was felt that the high ppv's during presplitting would exceed the imposed limits, where applicable, or could cause instability in adjacent slopes. Whilst bench heights and charge weights could be reduced by increasing the number of delays for the bulk blasts, presplit holes had to be drilled to full depth and fired on the same delay to achieve optimum effect.

In situations where blasting is taking place in close proximity to installed rock support, there exists the possibility of damage, particularly to grout, due to ground vibrations. Whilst most specifications impose restrictions on blasting within a certain distance of support, such restrictions rarely relate to peak particle velocities. At Kornhill, four trial holes were drilled and grouted in a rock slope 48 hours before an adjacent presplit blast, and overcored after the blast. The grout appeared undamaged. Using vibration data from this somewhat crude trial and from elsewhere on this site, a limit on resultant PPV of 110mm/sec was chosen.

In the area of the site adjacent to the former Tai Cheong St. Quarry (see Fig. 2), it was necessary to form a 21m high pre-split face such that there was only 4m between the top of the presplit face and the vertical quarry face behind. In this and other similar cases it was decided to use chemical rock breaking compounds (CRB) which are packaged in powder form and mixed with water to produce a slurry which can be poured into drillholes. The resultant chemical reaction causes expansion which induces extremely high tensile stresses at the hole circumference, generating a crack between adjacent drillholes. Cracking was normally evident after 24 hours, with the maximum effect being achieved generally within three days. To prevent leakage of CRB slurry into open joints and potential instability of adjacent slopes, the holes were lined with polythene tubes, sealed at the bottom end. A typical CRB slope is shown in plate 6. The negative aspects of this method were cost (up to twenty times the cost of the explosive for the same presplit), lining the holes, and the delay of up to three days before the face could be exposed. At Kornhill it is considered that the advantages outweighed the disadvantages.

SLOPE TREATMENT DESIGN PROCESS

The programme for bulk excavation, slope forming and building work was very tight. It was essential to install slope support measures as the formation was lowered so that problems of access for installation were reduced, and building works could commence as final platform levels were achieved. It was intended that support would generally be installed at least one bench height (12m) above the working level at any given time to minimise the possibility of damage to the permanent slope works from blasting. In practice, the contractor generally installed support 15 to 25m above this level. This is illustrated in plate 7. The permanent site formation and slopes had to be approved by the Government Geotechnical Control Office soon

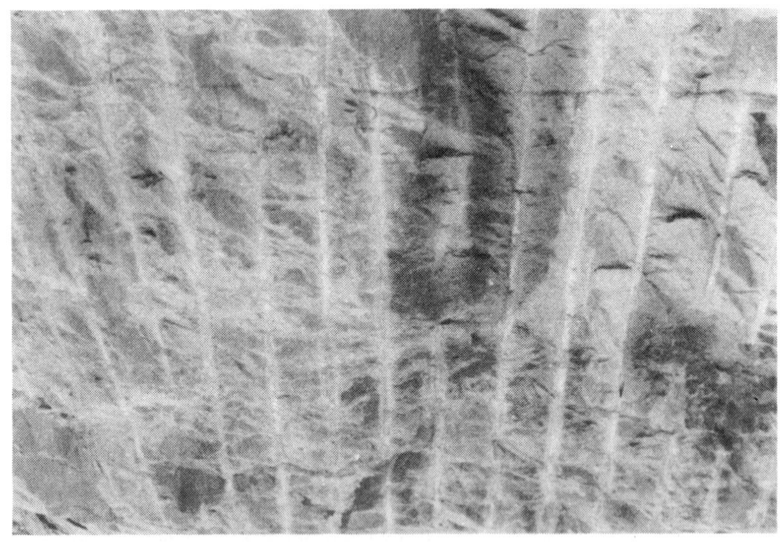

Plate 6 : Close-up of CRB pre-splitting. This should be compared with Plate 10. The result was successful in Highly weathered to fresh granite.

Plate 7 : Installation of slope treatment. The upper slopes have completed slope treatment (and drainage channels under construction), mid-level slopes are scaffolded prior to slope treatment, a newly formed slope is being trimmed by hand, and drilling rigs are preparing the next bulk blast at the base.

after completion.

This entailed submission of as-built engineering geological drawings (1:100) for all major rock slopes, photographs with overlays, stereographic projections demonstrating overall stability, and calculations justifying the slope treatment measures.

It was realised at an early stage that it would be necessary to set up a design team on site to minimize the turn-round time. Geological mapping of the slopes was critical and a separate team comprising a geologist and three technicians was included in the site supervision staff, to map cut slopes immediately after trimming. The design staff requirement was estimated from consideration of site formation drawings, mining plans and the programme. This indicated that 300sq.m./day of completed slope would require assessment in the first year of the project. This would be in addition to other site duties.

A design team of a geotechnical engineer, an engineering geologist, four assistant engineers and two technicians was used at the peak of site work. A slope treatment design manual was produced once potential failure modes and the applicability of slope treatment had been confirmed.

The availability and extent of the areas of slope to be mapped were entirely dependent on the contractors excavation cycle. The typical cycle is shown in Fig. 3. A newly formed face was expected to be divided into panels ideally 15m long by 6m high. However, the location and areas of face exposed varied greatly due to changes in the contractors working method and the complexity of the site. A system was established whereby discrete panels of the exposed face could be mapped and uniquely identified for future reference. From a mapping point of view, this was complicated by the fact that adjacent panels might not be exposed in sequence. Panels were randomly located soon after exposure, and controlled by survey stations painted on the face. These were subsequently located by an electronic distance measuring device. Approximately 700 such panels were numbered in order of exposure. Panel lengths varied from 5 to 50m, height from 2 to 16m and areas from 10 sq.m. to 200 sq.m. with an average of 90 sq.m. A key elevation showing typical panel layout is given in Fig. 6. The interaction of excavation, logging and design is shown in Fig. 7. Panels were defined by a horizontal, chainaged, painted reference line and vertical paint lines. The panel was drawn and photographed at 1:100 scale, and an objective discontinuity survey carried out along the reference line by two technicians with the data being checked by the geologist prior to issue to the design team. A typical panel is shown in plate 8, with its elevation sheet and a material description sheet in Fig. 8 and Ref. 4 respectively.

As the cut slope face was exposed during the site formation, the salient features were checked against the original geotechnical design assumptions, and where the slope was potentially unstable, slope treatment was considered. A summary of the original geotechnical design data was provided on site, with any additional data collected during the advance works. Areas of potential instability were marked on the logging sheets during inspection of the face by the design engineer.

The overall stability of rock slopes was assessed from analysis of aggregated discontinuity data from adjacent panels. Plots of up to 1000 discontinuities were produced by transmitting data from a site microcomputer to a

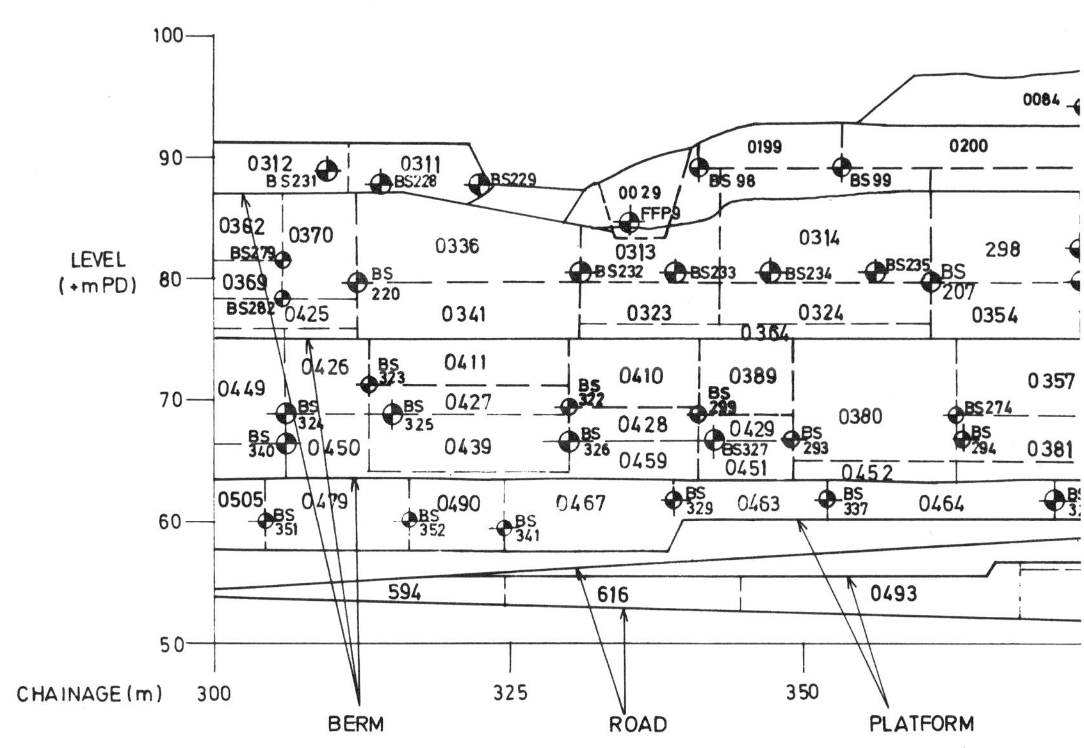

TYPICAL KEY ELEVATION FIG. 6

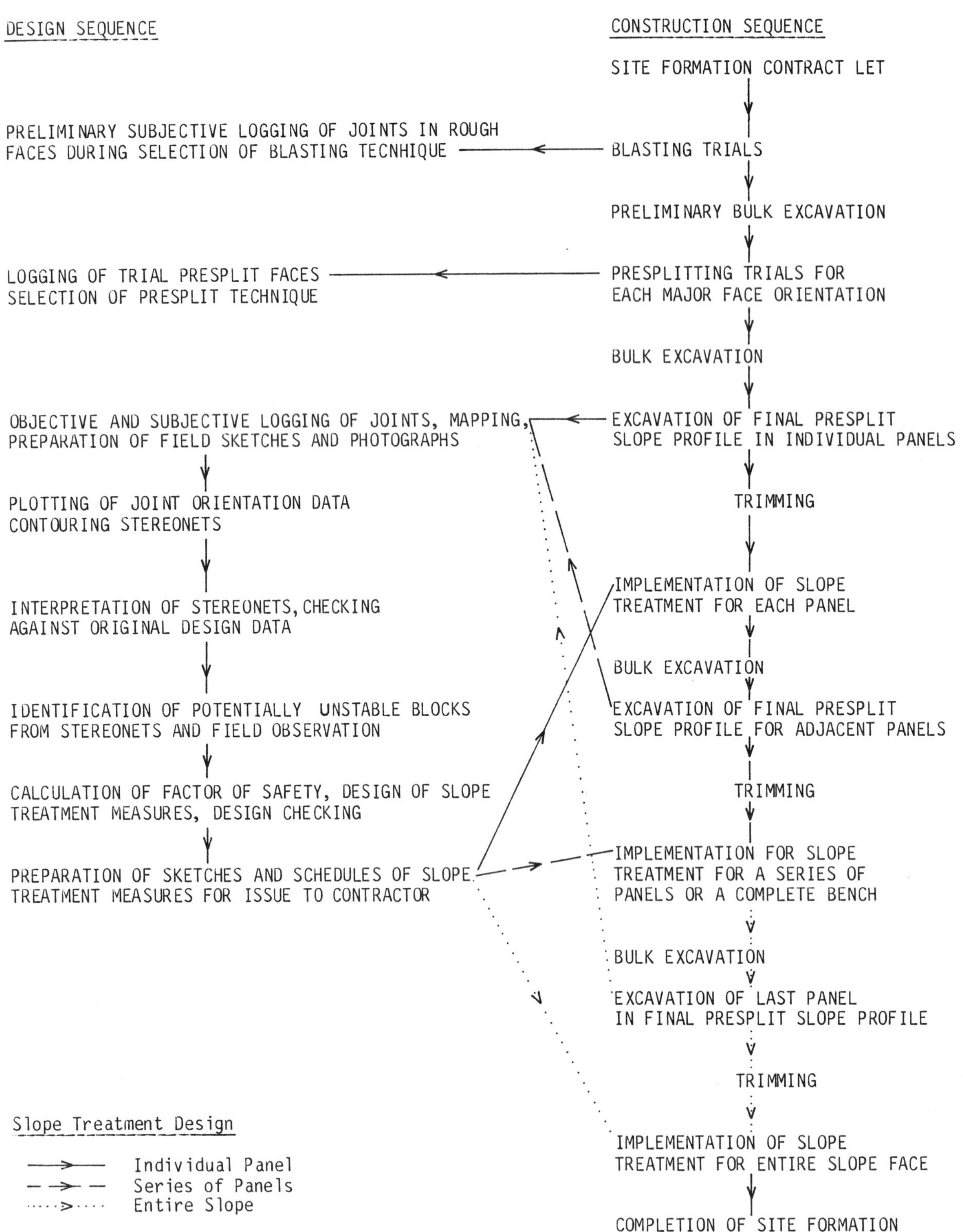

FLOW CHART FOR THE LOGGING OF ROCK FACES AND THE DESIGN OF SLOPE TREATMENT FIG. 7

Plate 8 : Typical logging panel. Note survey station 'AH1' on the left and deviation of pre-split drill holes.

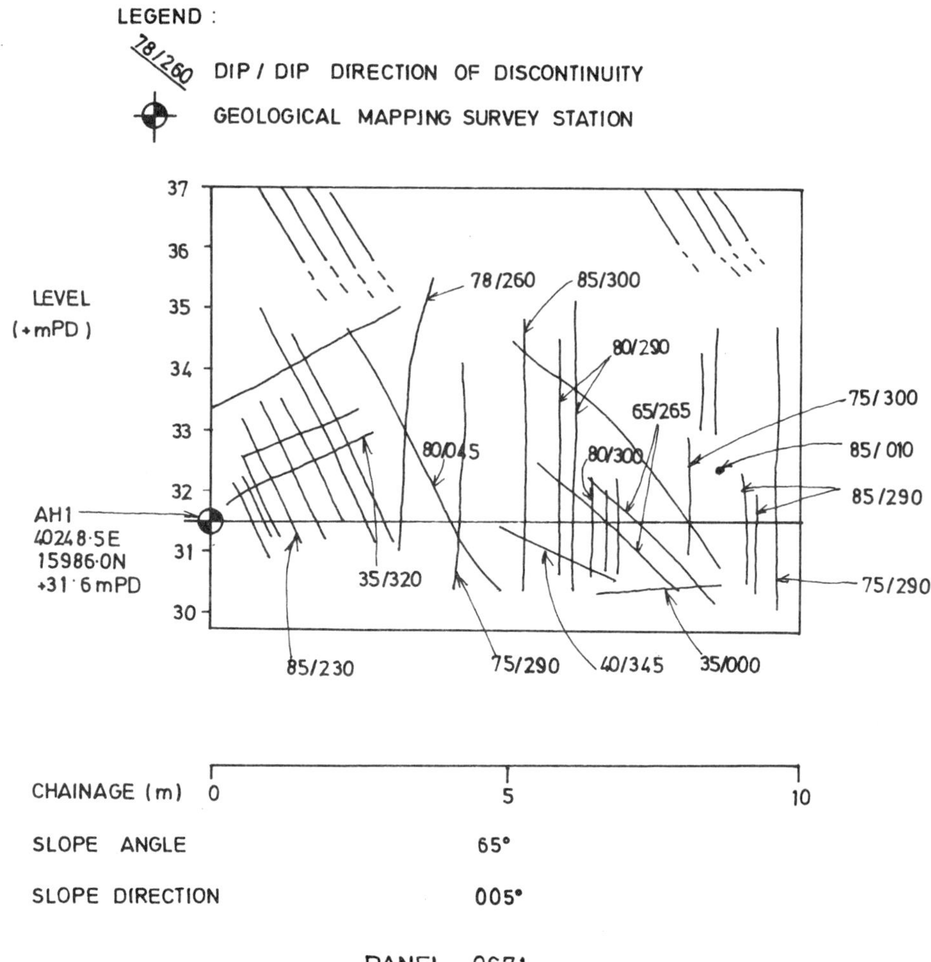

TYPICAL PANEL LOGGING ELEVATION FIG. 8

distant main frame computer via a telephone modem. Stability of entire faces could thus be checked rapidly. A total of approximately 50,000 data were measured and repeatedly plotted in various combinations, a task which would otherwise have been impracticable. Stereographic projections included both scatter and contour plots with an appropriate daylight envelope on each. Intersections of at least the three main concentrations were plotted on both diagrams, unless the second or third concentration had a very low percentage significance. Faults and shear zones were also plotted. Individual poles falling within potential planar or toppling zones (Ref. 6) were checked. If persistence was very low (say < 1m) and spacing extremely wide, then the slope was considered safe against overall failure. Similarly if apparent intersections fell within potential wedge zones (Ref. 6), the persistence and spacing of the sets and the actual existence of the line of intersection were checked from the panel data and by subjective site checks. If the factor of safety calculated by the design procedures (Refs. 1 & 6) was inadequate, appropriate stabilisation measures were employed as detailed in Table 2.

Discontinuity - controlled failure modes were not significant. Toppling was most common, but slope orientations limited this to four isolated instances, locations (1)-(4) in Fig. 2. Potential plane failures were only locally developed. Isolated wedge masses were found at locations, (8), (9) in Fig. 2. The rock mass was generally not closely or randomly jointed and rockfall was not possible. Wedges were usually bounded by one sub-vertical plane and were analysed as plane sliding parallel to the steeper plane. Where seepage was occurring on the slope face at half-height or above for the block in question, full head water pressure was assumed in analysis. Where seepage was occurring below half height or even below the block itself, half-head was used. Where adjacent standpipes or piezometers indicated a high standing groundwater level or piezometric level, this information was used in relation to the block under study, with an appropriate groundwater level increase due to 1:10 or 1:1000 year storm events. As the height of block increased, the choice of groundwater level became increasingly significant and had to be based on experience. Toppling blocks were considered more likely to develop 'full head' than sliding blocks. Shotcrete (as a joint sealant) and inclined drains were used to limit future possible groundwater conditions to those assumed. The use of post-tensioned bolts may have had a slight 'tightening' affect on the rock mass, but this was not assumed to render it impermeable. Piezometers outside the slope cut line continued to be monitored after completion of the site formation. Present or future additional destabilising factors were allowed for where appropriate including the influence of building loads (up to 5000kPa) or highway loading (15kPa) near slope crests.

After carrying out detailed stability calculations, the final form of slope treatment was agreed by the designer and site supervisory staff, to suit site constraints. A typical slope treatment elevation is shown in Fig. 9, for the same slope panel as in Fig. 8 and plate 8. Detailed design calculations and drawings were submitted to the Geotechnical Control Office on a regular basis for checking, and regular site inspections made by their representatives prior to approval.

Ideally slope treatment was intended to be instructed in each panel on a once only basis. However further instructions were necessary for a number of panels. The basic reasons were :

- exposure of large scale geological problems in lower or adjacent panels, necessitating additional works in previously exposed panels.
- damage due to secondary, non explosive excavation on faces (e.g. for drainage channels).
- changes in slope treatment at the request of the contractor.
- trimming and scaling were sometimes incomplete at the time of assessment and design. Consequently, amendments were required where blocks could not be trimmed, or the removal of unstable blocks revealed further problems.

A site-based team familiar with the design of the original slope treatment proved invaluable in allowing a rapid re-design. The originally estimated and actual rates of rock slope exposure are shown in Fig. 10. Also shown is the rate of submission of as-constructed reports for slopes to the Geotechnical Control Office which marks the end of the design process, and the reduction in design staff input with time.

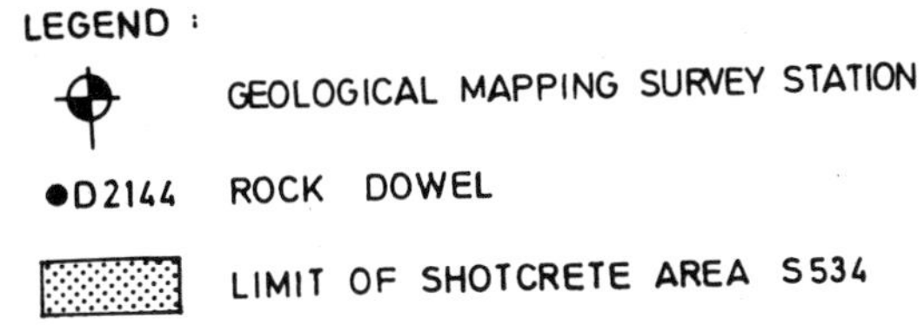

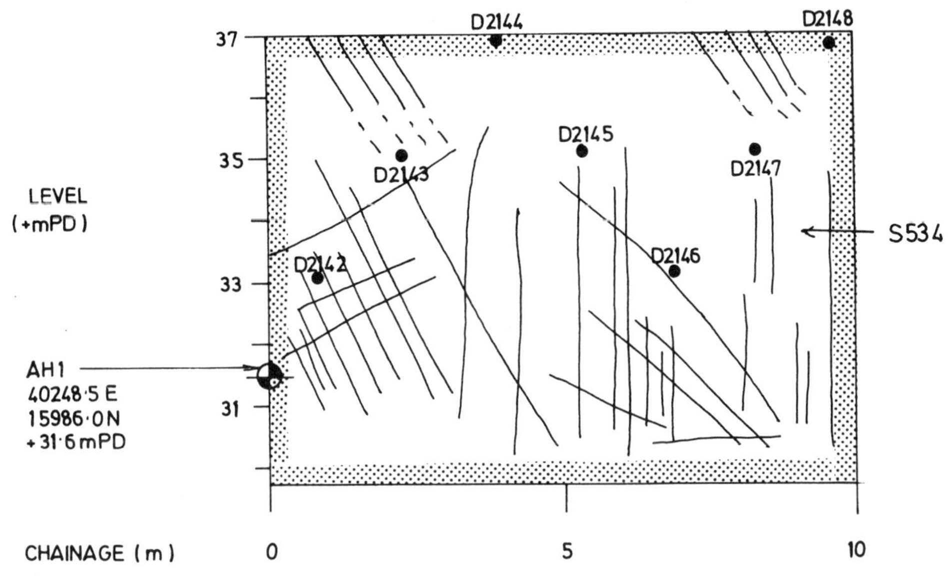

PANEL 0671

TYPICAL SLOPE TREATMENT ELEVATION FIG.9

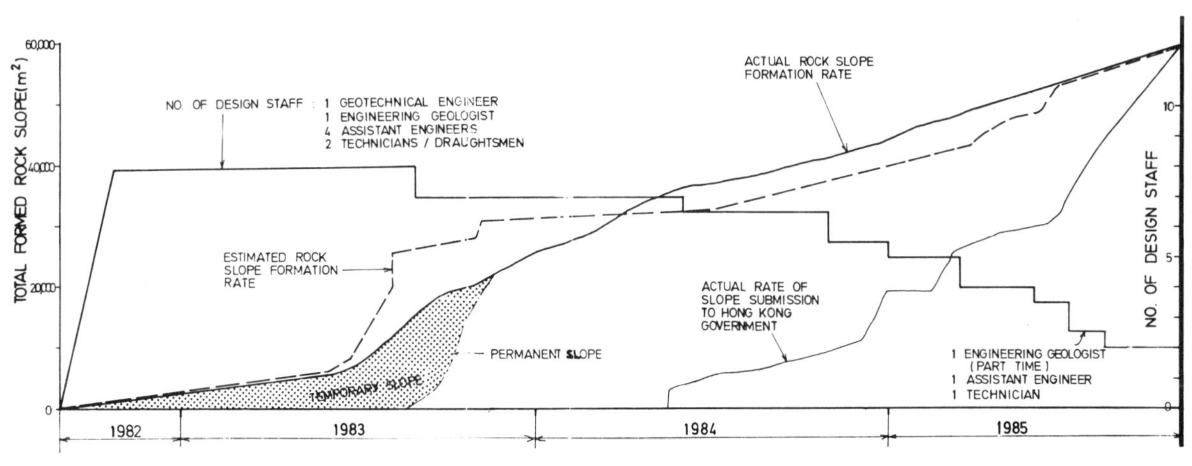

RATES OF ROCK SLOPE EXPOSURE / COMPLETION, AND DESIGN STAFFING FIG. 10

Adjacent panels were shown in the as-constructed reports combined in a single record elevation, showing geological boundaries, significant discontinuities (generally with persistence > 3m), and orientation data. A composite contoured stereoplot of all discontinuity data for the slope section was added to this drawing with an appropriate daylight envelope. Upon satisfactory completion of drainage, trimming, and slope treatment works, the as-built position of all permanent slope support was added.

Where large concentration of bolts were proposed, bamboo scaffolding was erected from the berm immediately below and hand-drilling used. Scaffolding is visible in plate 7. However, holes greater than 50mm diameter, or more than 4m long could not be hand-drilled. In such cases, installation was effected from a crane-slung cage, with a rotary percussive air-flush drill ('drifter') mounted on a horizontal axle within the cage. Bolt drilling from a gondola is shown in plate 9. Maximum crane reach was 28m. Approximately 75% of all rock slope support was installed from scaffolding as this had a less disruptive effect on other operations. The types and quantity of slope treatment used at Kornhill are shown in Table 2. The predominant use of 50mm shotcrete and 19mm diameter dowels reflect the fact that most unstable blocks at Kornhill were small, and occurrence of weathered zones in rock faces was the main problem.

Plate 9 : Bolt drilling from a gondola.

PROBLEMS

The main problems were related to deep weathering and high groundwater levels in fault zone 'F' and two of the dolerite dykes (D2, D3) crossing the site. The zones were extremely difficult to locate because they rarely corresponded with prominent topographic features, while some major features proved to be in sound rock. This is illustrated in Fig. 1.

Zone F was found to be 10m wide when excavation started. Probe holes indicated the zone was sub-vertical and would extend fully down a 40m high rock slope at location (10), Fig. 2. Groundwater levels in the zone were close to existing ground level. The relatively narrow geometry of the zone indicated that cohesion on the sides and base of the potential shallow non-circular failure mass would be significant. Cohesion was estimated from back analysis of a 100cu.m. failure of an existing slope elsewhere on site, and from block samples of relict joints collected from various parts of the site and tested by direct shear box. Parameters obtained were adequate. A system of inclined drains was used, with 150mm shotcrete over the whole zone to retain local relict joint blocks and prevent erosion and long term degradation of the soils. Groundwater levels were monitored by two piezometers. The original and stabilised appearances of the zone are shown in plates 10 and 11.

Dolerite zone D2 extended north-south for 200m across the western part of the site. At location (11) (Fig. 2), a 10m high rock slope was re-designed as a 45° soil slope. At location (12), a 7m high rock slope was re-designed to a 40° soil slope due to high groundwater levels.

Dolerite zone D3 extended north-south across the entire site for 600m. The width of the

Slope Treatment	Qty Used	Details/Max. Working Load	Application/ Comments
Trimming (hand bar)	58,000 m^2	-	Blocks less than 0.5 m^3 Relatively slow
Trimming (Krupp Hammer)	2,000 m^2	-	Blocks greater than 0.5 m^3. Rapid and much superior to hand trimming provided well supervised
Rockfall Fence	3,000 m^2	H-section steel upright and cross members, with 50-75mm horizontal planking	To retain $1m^3$ boulders on 45° slope or equivalent
Hanging Mesh	4,000 m^2	2mm wire, 50mm aperture lattice, usually fixed at top and bottom	To control and deflect (not support) falling boulders up to 1 m^3
Shotcrete (50mm)	9,200 m^2	25/10 grade, dry mix process. 50mm ø weepholes 1.5m c/c, 1.5m long. D83 mesh, 16mm ø fixing dowels 2m c/c 0.5m long. Mesh cover and spacing 50mm	- to protect soil slopes steeper than 50° or too dense to hydroseed, especially where access too difficult for no-fines concrete - falls of up to $0.5m^3$ in closely jointed/fractured/weathered rock - joint sealant to prevent infiltration
Shotcrete (100mm + 1 layer of mesh)	3,980 m^2		- falls of up to 1 m^3 (100mm + 1 layer) or 2 m^3 (150 + 2 layers) in closely jointed/fractured/ weathered rock.
Shotcrete (150mm + 2 layers of mesh)	622 m^2		- in conjunction with rock dowels/ bolts typically at 2 - 3m c/c in large closely jointed/fractured/ weathered rock masses.
No fines - concrete	20 m^3	20 mm max. aggregate size	- infilling of weathered joints, or deep depressions in soil slopes, provided access is easy
Buttress	190 m^3	30/20 grade, 50 mm ø weepholes at 1.5 m c/c	- to hold closely jointed/fractured masses or large blocks, on steeply dipping planes where movement is expected to be mainly vertical, provided there is a stable footing within 3-4 m of block. - fixing dowels used where buttress rests on inclined footing
Inclined drain	90 No	65 mm ø Netlon lining and filter fabric in 75mm ø hole, up to 10 m long	- used in conjunction with 50mm shot-crete for areas of high groundwater level, and/or where low water levels assumed in bolt/dowel design, especially for potential toppling failure
Rock dowel (19mm ø)	2,000 No.	Dywidag high tensile ribbed	- individual blocks requiring force 0-35kN (19mm ø) and 35-179kN (38mm ø)
Rock dowel (38mm ø)	150 No.	bar, max. W.L. 35kN (19 mm ø, in 75 mm ø hole), up to 8m long, torqued to 5% working load by hand wrench after grouting	- in conjunction with shotcrete + mesh at 2 - 3m c/c in large closely jointed/fractured/weathered rock masses
Rock bolt (22mm ø)	123 No.	Dywidag expansion shell type, upto 7m long in 50mm ø hole, max. W.L. k5KN, tensioned by hydraulic jacks and grouted.	- individual blocks requiring 35-95kN force - large masses divided into blocks of above size - no subsequent monitoring possible

TABLE 2 - KORNHILL DEVELOPMENT - SLOPE TREATMENT

Plate 10 : Weathered zone H (prior to stabilisation).

Plate 11 : Weathered zone H (after stabilisation).

weathered zone along dolerite D3 was found to vary rapidly as shown in Figs. 2, 11, where it was crossed by other valley features (themselves unfaulted), and formation levels were close to original ground. At location (1), (Fig. 2) a 20m high proposed rock slope was found to be completely weathered and was regraded to 40° with inclined drains to reduce high groundwater levels. At locations (14), (15) 10-13m high rock slopes were regraded to 45°. At location (16), rock levels at the position of an 8m high retaining wall above a 19m high rock slope were found to be lower than anticipated. Probe holes proved rockhead to be up to 9m lower than indicated by the nearest available drillholes and seismic traverses. A 70° slope in soil below the wall foundations would have been difficult to stabilise with an anchored retaining wall, and it was decided to lower the founding level of the original wall by 6m. At location (17) the condition of several critical rock slopes was proved to be unweathered by advance probing. Foundation design for tower blocks was also affected but is not considered here.

Several existing rock slopes adjacent to the Development boundary were to remain. Such slopes generally did not satisfy current standards for new rock slopes. Slopes within the site were stabilised, and it was essential to demonstrate that slopes outside it had not been adversely affected by excavation work. Blast vibrations near slopes adjacent to the site were reduced, even where not close to occupied property. Where this was impractical, CRB pre-splitting was used. As well as vibration monitoring, slope faces themselves were monitored by periodic readings of levels and co-ordinates, and checking across pre-existing tension cracks or similar. Such monitoring indicated negligible change in all cases.

SHOTCRETE TESTING AND USAGE

Previous research on quality, testing and potential failure modes of in-situ shotcrete were insufficient to enable it to be 'designed' as a structural concrete (Refs. 7, 8).

At Kornhill the 'dry mix' process was used successfully. The introduction of water at the

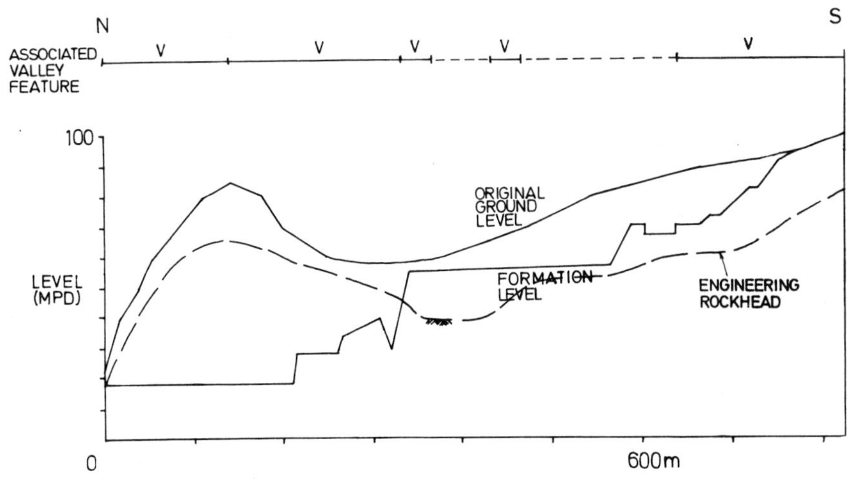

VARIATION IN WEATHERING ALONG DOLERITE D3 FROM N. TO S. FIG 11

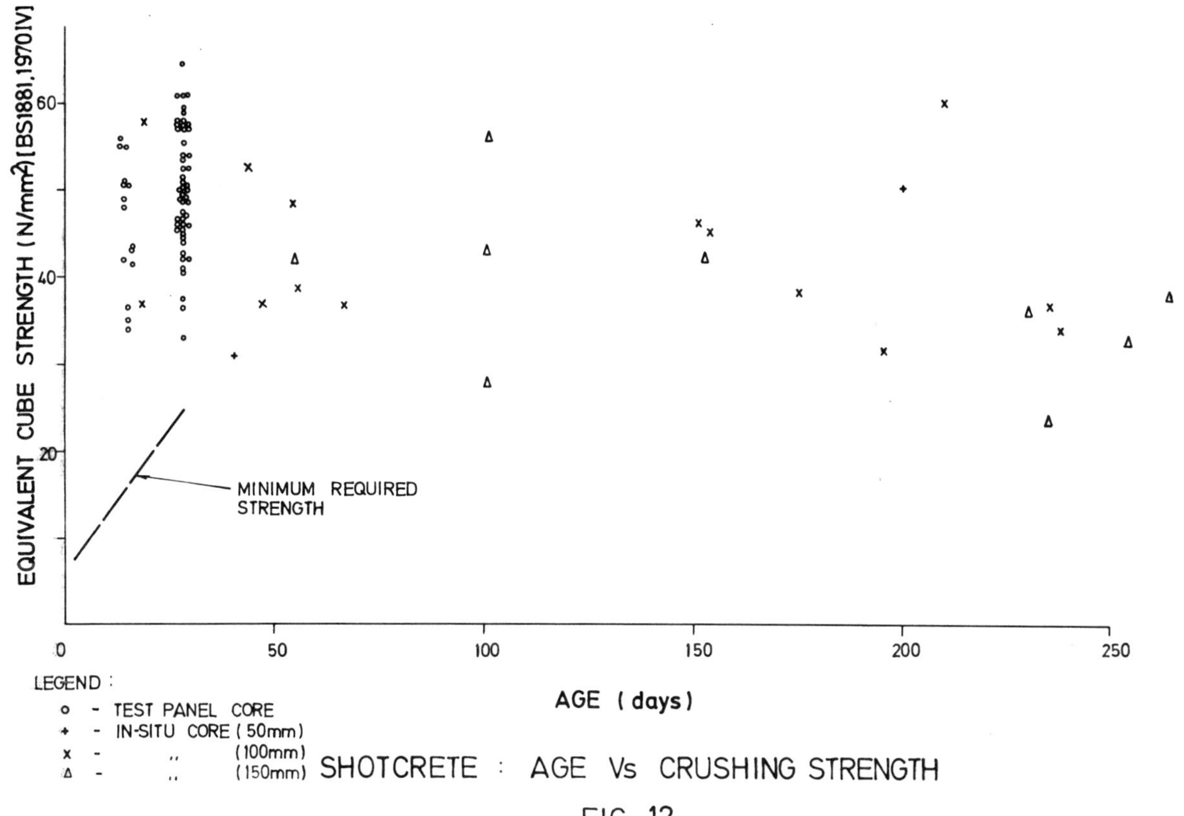

SHOTCRETE : AGE Vs CRUSHING STRENGTH

FIG. 12

nozzle did not appear to have an adverse effect. A test panel was made during each shotcrete pour. After several days exposure, six 100mm diameter cores (L/D ratio 1-2) were taken and cured until crushing, usually at 28 days. Rebound of in-situ shotcrete was generally 10-15% (mainly coarse aggregate) and no curing was carried out after placement. 100mm diameter cores (L/D 1-2) were taken at random locations and age throughout the whole site, and saturated immediately prior to crushing. Due to the difficulty of in-situ sampling, only 50 No. in-situ samples were taken compared with 400 test panel cores. Saturated densities of in-situ and test cores were similar (2.1-2.3Mg/cu.m.). Fig. 12 shows an age/strength plot for all in-situ samples against all test panels for the same shotcrete batches. In-situ strengths are generally above the specified minimum, but with scatter which may be largely due to variable curing conditions.

The scatter in 100 and 150mm shotcrete strength results may also be due to effects caused by steel mesh layers even after correction for the steel itself. There were few obvious defects in these samples. In this respect, it would be preferable to use tensile strength, pullout or Windsor probe tests. Schmidt Hammer or ultrasonic tests would be difficult to perform due to the rough surface texture of the shotcrete. Excess voids and segregation (especially behind mesh) were not significant in in-situ samples, and cover to mesh (50mm) and total thickness were generally achieved. These factors in conjunction with the reasonable densities and strengths obtained are considered to have produced a material durable enough for Hong Kong conditions.

However it was not practical because of the time that would be involved to carry out cement content, carbonation or chloride tests on in-situ samples. Bond between shotcrete and slope surface was good, provided faces were cleaned by high pressure airline and/or water jet prior to application. Overspraying of several panels was carried out several months after the initial application to increase the shotcrete thickness. The initial layer was thoroughly cleaned using water jets, and a bonding agent applied to avoid separation of the layers and subsequent coring indicated this was effective.

Surface collection drainage on rock slopes consisted of 300mm dia. raised concrete U-channels, connected via concrete lined stepped channels and catchpits into the development main drainage system. The total length of channel constructed was 5km and, in order to avoid delays in the excavation cycle due to the presence of cranes, concrete trucks and other equipment, it was decided to defer the completion of many of the channels until formation level had been reached.

With this approach it was not practical to use cranes for concreting after reaching formation level, as lifts of up to 55 metres were required. Therefore shotcrete was used, pumped up to berm level and sprayed inside a vertical wooden form and then vibrated. Collapsible steel internal moulds fixed to the correct grade within the form formed the invert and side walls.

DISCUSSION OF SLOPE ANGLES AND PROFILES

The discontinuity data obtained in the site investigation stage was sufficiently representative to ensure the formed slopes were stable against overall failure. The overall discontinuity pattern exposed in the final faces could have permitted steeper angles of up to 80° in some areas, however a more intensive and costly site investigation may not have produced such a clear picture. Steeper pre-split drill holes would have reduced deviations.

The contractor experienced some difficulty in forming berms to the required width and level. Where incorrectly formed an acceptable profile was achieved by secondary excavation or make-up in concrete. Wider berms and steeper slope angles would have been easier to form from an excavation viewpoint and could have reduced bulk excavation.

CONCLUSIONS

The design and construction of the high rock slopes at Kornhill is considered to have been implemented in an optimum fashion. Design data provided by the site investigation was largely adequate and problems not detected proved relatively minor.

Faults, dykes and other features may cause wide zones of deep preferential weathering in Hong Kong's sub-tropical climate. Regional orientations of such features should be borne in mind, and where they are encountered during site investigation (even if unweathered) it may be necessary to establish their location by further intensive investigation, particularly at the proposed position of high rock slopes or critical structures. The use of long inclined drillholes is preferred to vertical ones for this purpose.

Rapid probing by production drilling rigs proved to be an effective mid-contract investigation method.

The very tight programme for excavation and slope treatment was met by detailed organisation of logging and design of slope treatment. The procedure used may be applicable to all scales of project with appropriate changes. The proposed types of slope treatment were adequate for the problems encountered.

Shotcrete has proved to be a versatile technique, with the method specification producing material of consistent quality.

ACKNOWLEDGEMENTS

The authors are grateful to Headstar Ltd., Mass Transit Railway Corporation and Freeman Fox (Far East) Ltd. for permission to publish data in this paper. They also wish to thank those assisting in preparing and reviewing the text and figures.

Client - Mass Transit Railway Corporation/Headstar Limited Joint Venture

Employer - Mass Transit Railway Corporation

Consultant - Freeman Fox (Far East) Ltd.

Project Management/Site Supervision - Mass Transit Railway Corporation

Contractor - Aoki Corporation

REFERENCES

1. Geotechnical Control Office (Hong Kong Government) Geotechnical Manual for Slopes (1979, 1984) Government Press.

2. Finn R.P. and Gamon T.I. Differentation of different phases of the Hong Kong Granite by study of discontinuity patterns. Hong Kong Geological Society Newsletter (in press).

3. Gamon T.I. (1984) A comparison between the Core Orienter and the Borehole Impression Device. Conference on Site Investigation (BS5930) Surrey 1984. Geological Society of London (Engineering Group Regional Meeting).

4. Gamon T.I. and Finn R.P. A simplified Descriptive Scheme and Classification System for the logging of Cut Slope Faces. Conference on Site Investigation (BS 5930) Surrey 1984. Geological Society of London (Engineering Group Regional Meeting).

5. Gamon T.I. and Finn R.P. The Structure of the Hong Kong Granite. Hong Kong Geological Society Newsletter (March 1984).

6. Hoek E. and Bray J.W. Rock Slope Engineering 1977. Institution of Mining and Metallurgy.

7. Use of Shotcrete for Underground Structural Support. Proceeding of the Engineering Foundation Conference, Berwick Academy, South Berwick, Maine, (July 1973). American Society of Civil Engineers/American Concrete Institute, Publication SP-45.

8. Shotcrete for Ground Support. Proceedings of the Engineering Foundation Conference, Tidewater Inn, Easton, Maryland (October 1976). American Society of Civil Engineers/American Concrete Institute, Publication SP-54.

Problems and control of groundwater in fissures: a case history

Charles R. Nelson P.E.
Charles R. Nelson & Associates, Inc., Minneapolis, Minnesota, U.S.A.

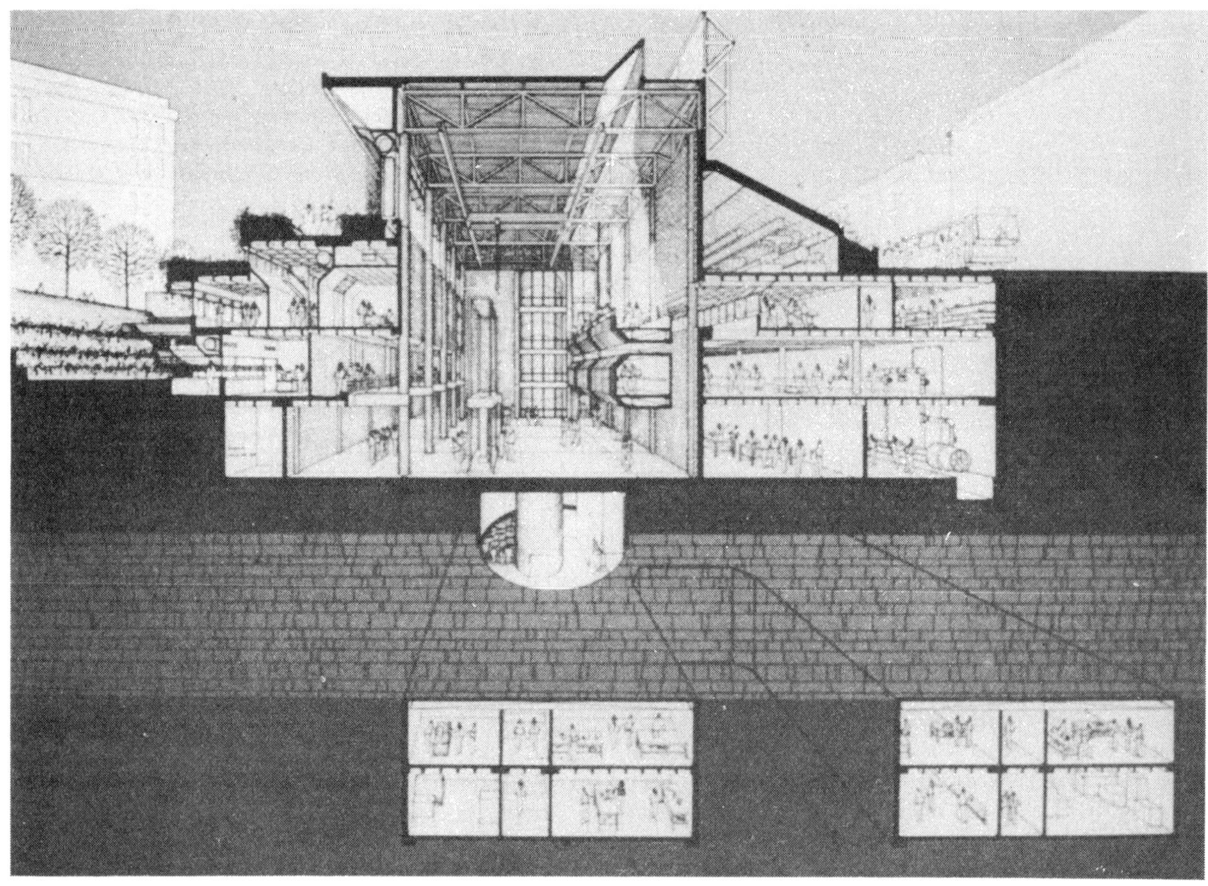

Figure 1 Building Cross Section

INTRODUCTION

The recently completed Civil and Mineral Engineering (CME) Building on the Minneapolis campus of the University of Minnesota USA comprises four upper floors of open cut construction to 12 m below grade and two lower floors of mined space 22 to 30 m below grade. A flat-lying 9 m thick layer of limestone separates the two parts of the structure. The mined space was excavated in a soft sandstone below the limestone roof. One reason for selecting a design concept that included mined space was to demonstrate that it could be economic to construct and that substantial energy savings might be achieved. This innovative building was selected as the National Outstanding Civil Engineering Achievement of 1983 by the American Society of Civil Engineers.

The limestone contained open horizontal fissures that were water bearing and tight vertical fissures producing minor amounts of water. Both sets caused water-related design and space use problems with the overhead vertical fissures being the most difficult to control. Some water entered the shotcrete layers and to some extent followed the rebar pattern. In addition, a few resin anchored rockbolts leaked water. The water entered the shotcrete layers and followed the rebar pattern to some extent.

GEOLOGY

The rocks at the CME site are horizontal-layered (Figure 2), with three major layers controlling the building's design geometry--soil cover, the Platteville limestone, and the St. Peter sandstone. Intervening shale layers also influenced parts of the design.

Figure 2 Geology

The unconsolidated soil layer, about 12 m thick, varies from clays to sands to boulders. The second layer, the strong Platteville limestone about 9 m thick, is self-supporting with light rockbolt reinforcement over roof spans of at least 15 m. The last major layer, the very weak but massive St. Peter sandstone, is about 45 m thick. The Glenwood shale, 1 m thick, with interbedded clay seams acts as an impervious layer below the limestone. The sandstone can be made self-supporting in vertical walls more than 9 m in height with only light reinforcement but is not stable in roof spans over about 3 to 6 m.

Given these basic properties, the most suitable geometry for the building was the location of major space in the upper soil and in the sandstone beneath the limestone dividing layer. The limestone would then serve as both a substantial foundation for the upper structure and as a roof over the mined space in sandstone. The cut-and-cover space contains classrooms, offices, and large laboratories, which experience major traffic flow or need surface access for deliveries. The mined space contains laboratories and office space having smaller needs for surface access.

Groundwater is perched in the soil and limestone by the 1 m thick shale between the limestone and sandstone. The main groundwater table is located in the sandstone 6 to 10 m below the limestone.

SHAFT-EXCAVATION AND SUPPORT

Two circular shafts 13 and 8 m diameters, one at either end of the structure, were designed for access to the mined space. Excavation was by blasting. The large diameter shaft is shown in Figure 6. A 2 m high collar of reinforced concrete was placed on the limestone around the rim of each shaft to prevent entry of surface water. Backfill was then placed outside the rim to provide a level work site adjacent to each shaft. The shafts were lined with 0.3 m thick reinforced shotcrete after being excavated full depth.

MINED SPACE-EXCAVATION AND SUPPORT

The CME Building mined space is roughly a rectangular, torus-shaped space, with a central pillar 14 m by 27 m forming the hole of the torus (Figure 3). The pillar is a central support column for the limestone roof having an 18-m clear span on all sides. The space laid out around the pillar has outer dimensions measuring approximately 53 m by 48 m, with a height of 8 m.

The sandstone was excavated using a crawler-type front-end loader having a custom bucket with teeth on its top edge as well as its bottom edge. The upper teeth were used to scrape the shale layer off the bottom of the limestone roof. Very light explosive charges were used first to loosen a small top heading in the sandstone. The loader was then able to excavate both the sandstone and most of the Glenwood shale.

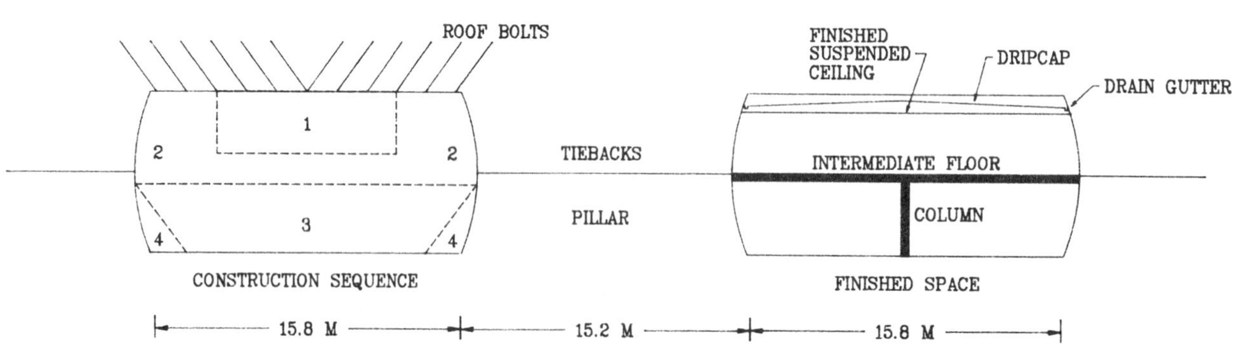

Figure 3 Construction Sequence and Space Finishing

Figure 4 Wall Reinforcement

The support method for the large, flat limestone roof was rockbolting. As the roof was exposed it was reinforced with 3 m long rockbolts placed 1.5 m on center and angled towards the end of the span. After the roof bolts were in place the roof was coated with two layers of shotcrete. The first layer, of 2.54 cm minimum thickness, was applied to seal any remaining shale and prevent its swelling and eventual spalling. Next, rebar was placed (12.4 mm bars at 0.5 m on center in one direction and 16 mm bars at 1.5 m on center in the other direction) and attached to the rockbolts. The rebar and rockbolts were then covered with a second 2.5 to 5 cm thick layer of shotcrete.

The total excavation proceeded to full width (18 m) and down to slightly more than 50% of the design height (8 m) before wall reinforcement was installed. This reinforcement consisted of 6 m long rebar tiebacks grouted into the sandstone wall at 3 m spacings. Each bolt was attached to a transfer plate destined to hold a beltbeam of poured concrete. This beam, which continued around all the walls at midheight, served multiple purposes (Figure 4). First, it distributed the wall tie forces on the sandstone. Second, it acted as a footing into which the upper wall shotcrete could anchor. Third, it provided a ledge upon which the intermediate floor could rest.

The sandstone walls were first stabilized by sodium silicate spray grouting and then overlaid with 15 cm of reinforced shotcrete to insure long-term stability.

The shotcrete for both shaft and cavern lining was placed in layers. First the vertical rebar was placed and covered by shotcrete, then the horizontal rebar was placed and covered by shotcrete.

HORIZONTAL FISSURES IN THE SHAFTS

Two horizontal fissures were encountered all around each shaft. These varied in width from tight to 5 cm around the circumference of each shaft. Water production of these fissures was about 1 lps for each shaft. A perforated plastic hose was recessed in the horizontal fissures and covered with shotcrete while being allowed to drain. After the full shotcrete thickness was applied the drain hose was pumped full of a portland cement-fly ash based grout (Figure 5). No measurable running water was evident on the shotcrete surface after these procedures but major areas over rebar locations were damp or wet. These damp areas show as dark areas on Figure 6.

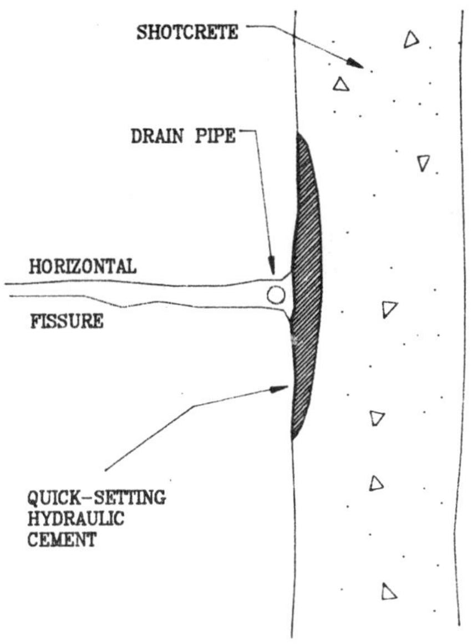

Figure 5 Horizontal Fissure Treatment

VERTICAL FISSURES IN THE ROOF

The fissures encountered on the underside of the roof were generally of one set or strike only. The total length of the fissures that leaked water was about 200 m in about 2000 m^2 of roof. These fissures were usually less than 1 mm wide and often appeared tight. The water would spray out with an estimated pressure head of 1 to 3 m.

These fissures were much too small to insert drain hose. For roof control purposes and anticipated construction difficulties a recess into the fissure for a drain was not considered feasible. A wide range of flexible covers were placed over the fissures and covered with shotcrete. None could be made to work. They all peeled off in seconds to hours.

Figure 6 Damp Shotcrete

The successful solution was to install rigid hat shaped metal troughs under the fissure, mechanically anchor the trough to the roof, provide drains at about 3m spacing and cover with shotcrete (Figure 7). The metal troughs were 30 cm wide, 5 cm deep and 3 m long. The roof was very flat and the fissures quite straight. For a rough roof or irregular fissures, much shorter troughs would be required. A drain pipe was located below each drain trough with frequent connections to the trough and cleanouts on the pipe. Total flow was estimated at 40 to 80 lpm.

DRIPCAP

The mined space was finished as office and research laboratory rooms. There were some leaks from the shotcrete covered roof and it was assumed that other leaks could develop over the life of the structure. A light metal drip cap was suspended below the roof to divert such leakage to gutters at the sides of the rooms (Figure 3).

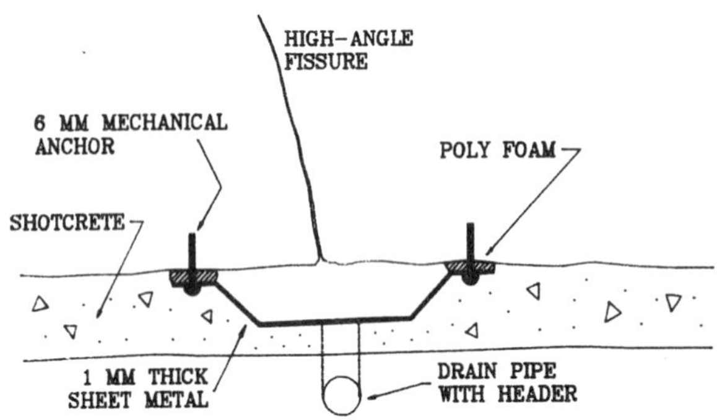

Figure 7 Vertical Fissure Treatment

Control of a large natural slope in a suburb of Vienna

R. Poisel DR.TECHN., DIPL.-ING.
W. Eppensteiner UNIV.-DOZ., DR.PHIL.
Department of Soil Engineering, Geology and Rock Engineering, Technical University of Vienna, Austria

SYNOPSIS

The region under consideration is a slope of an alpine, geologically young transverse valley, which follows mostly tectonically formed zones of weakness. The slope face is parallel to the bedding planes of the rock (dolomite), which dip 40° - 60°. In the valley bottom a river, a road and a row of houses lie beneath the 80 m high rock slope. After several rock falls the risks to the houses and their inhabitants from landslides and rock falls were to be investigated. First the slope, approximately 1 km long, was geologically mapped and, as a result of this mapping,the slope was divided into four zones of homogeneity. In each zone a typical profile was geodesically, geologically and geophysically mapped in detail. No signs of a possible landslide larger than 10 000 m³ could be found. The investigations showed that a rock tower of some 1000 m³, which rests on a bedding plane dipping to the valley and which is already cut off the rock slope, as well as some rock blocks are in danger of falling. Earthquakes, which are very likely because of the neighbourhood of the geologically young Vienna Basin, are an additional danger in this context. The geophysical investigations located fault zones in the upper crest parallel to the slope, which could not be found in the upper surface of the slope.

At the present time excavations are made to learn something about these fault zones. A programme of measurements for the zones in danger of falling is being developed, and measuring devices are to be installed.

INTRODUCTION

The region under consideration is a valley first having been developed only in suitable areas, but in the course of time having become more densely populated, thus now being built up throughout under a steep rock slope (see Fig.1; Beethoven composed his 'Missa Solemnis' in a lonely inn in 1824, now situated among other buildings).

Stabilizing workings to prevent rock falls have been carried through by the local mountain rescue team, completely satisfying the local population. Nevertheless, several rock falls have occurred recently and the population was afraid of a possible landslide. Thus the danger to the houses and their inhabitants by landslide and rock fall was to be investigated.

GEOMORPHOLOGICAL SITUATION

The "Gorge of Moedling" cuts through the margin of the Vienna Basin, which is a Tertiary trough. Hence, this valley is an alpine, geologically young, transverse valley, which was cut into the Tertiary erosion surface by the River Moedling. The valley mostly follows tectonically formed zones of weakness (see Fig.2). Whereas erosion has removed those zones in the area of the valley, they have

Fig.1 The "Gorge of Moedling"
region of profile 1; see rock tower in profile 2 in the centre, adits to the air raid tunnels on the right-hand margin of the photo

largely remained in the north and south areas, covered by detrital and soil formations and thus not always morphologically perceptible.

The northern valley side of the gorge is largely dominated by bedding planes generally pointing southwards. The minor formations are determined by the cutting of these bedding planes with parting planes tectonically laid out.

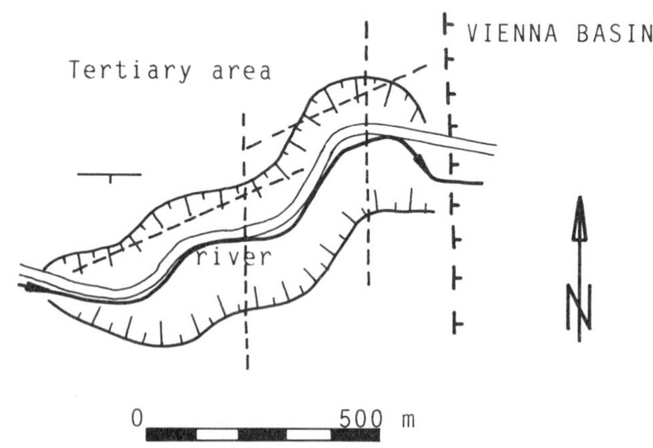

Fig.2 Lay-out of the "Gorge of Moedling"
margin of the Vienna Basin ("Thermenlinie")
bedding
faults
rock slope

GEOLOGICAL SITUATION

The whole area is built up by dolomite mostly in a fairly advanced stadium of dolomitization, but containing parts less dolomitized and lime-rich.

The rocks are irregularly bedded, but partly almost massive. The bedding joints represent continuous parting surfaces only to a minor extent. The bedding planes are normally not smooth but are very rough and adhere together.

Layers of marl and clay, possibly very disadvantageous to the stability of the slopes, have not been found anywhere.

The beds of the dolomite generally dip between 35° and 65° to the south. Where the valley direction of the gorge runs east-west, the beds dip parallel to the slope. Thus, the northern slope is much more endangered by mass movements than the southern slope. But where the valley diverges from that direction, the banks run into the slopes, thus creating a more favourable situation. Roots, especially those of pines, have often proved to represent a danger for rock blocks in steeply dipping bedding planes. On the one hand, they lift the blocks by the pressure of the growing roots and, on the other the roots and the accumulation of earth, soil etc. cause some accumulation of water. In the case of freezing, the pressure of the roots and of the ice gradually lift up the affected rocks from the underlying banks. This mechanism frequently causes the sliding or falling of rocks.

TECTONICS

Owing to its brittleness, as a result of tectonic loading at the margin of the Vienna Basin, the dolomite is heavily broken over large areas - in certain parts it is even crushed to the size of sand or silt. It is, however, sufficiently bound together by calcareous cement, especially in the regions with small size fractions.

Faults parallel to the western margin of the Vienna Basin, generally running north-south, cut through the slopes and also the adjoining areas of the Moedling Gorge (see Fig.2). The steep dipping of these faults varys rapidly between east and west.

Another system of steep faults runs approximately ENE-WSW, i.e. parallel to the valley respectively to the slopes, over major distances of the gorge. Both in the region of the faults, and especially where one fault cuts another, the dolomite is fractured heavily and partly mylonized.

GEOLOGICAL MAPPING

The approximately 1 km long slope was geologically mapped and was divided into four zones of homogeneity. In each zone a typical profile was first geodesically and then geologically mapped in detail (scale 1:200).

Profile 1

In the region of profile 1 the surface of the slope consists chiefly of the bedding planes of the dolomite. Only particular blocks of rock of a few cubic metres are undercut and, thus, in potential danger of falling.

Profile 2 (Fig.3)

Contrary to profile 1, in profile 2 not only are there some rock blocks of a few cubic metres in danger of falling, but there is also a rock tower of some 1000 m^3 that represents a real risk. This tower essentially rests upon undercut bedding planes and is separated from its hinterland by a vertical master joint 10 cm wide in its upper part. The rock situated on the mountainside of this joint is also necessarily undercut but is relatively stable at the present, as shown by the open joint. The complete slope near profile 2 as well as the areas north-east and south-west from it are dominated by the intersections of master joints and fault zones running north-east to south-west with fault zones generally running north to south and belonging to the system of the "Thermenlinie" parallel to the western margin of Vienna Basin.

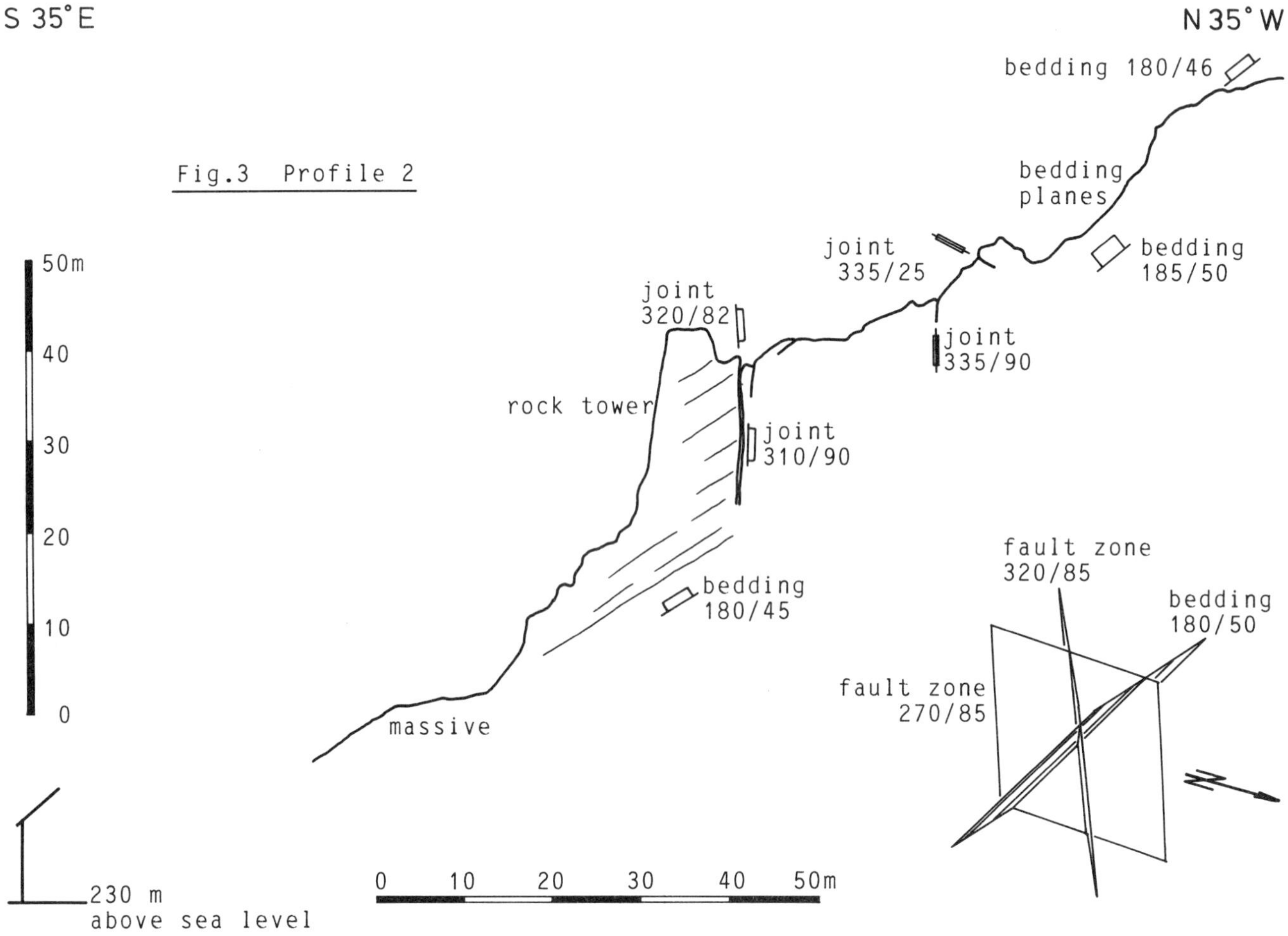

S 35°E N 35°W

bedding 180/46

Fig.3 Profile 2

bedding planes

joint 335/25

bedding 185/50

50m 40 30 20 10 0

joint 320/82

joint 335/90

rock tower

joint 310/90

fault zone 320/85

bedding 180/45

bedding 180/50

fault zone 270/85

massive

230 m above sea level

0 10 20 30 40 50m

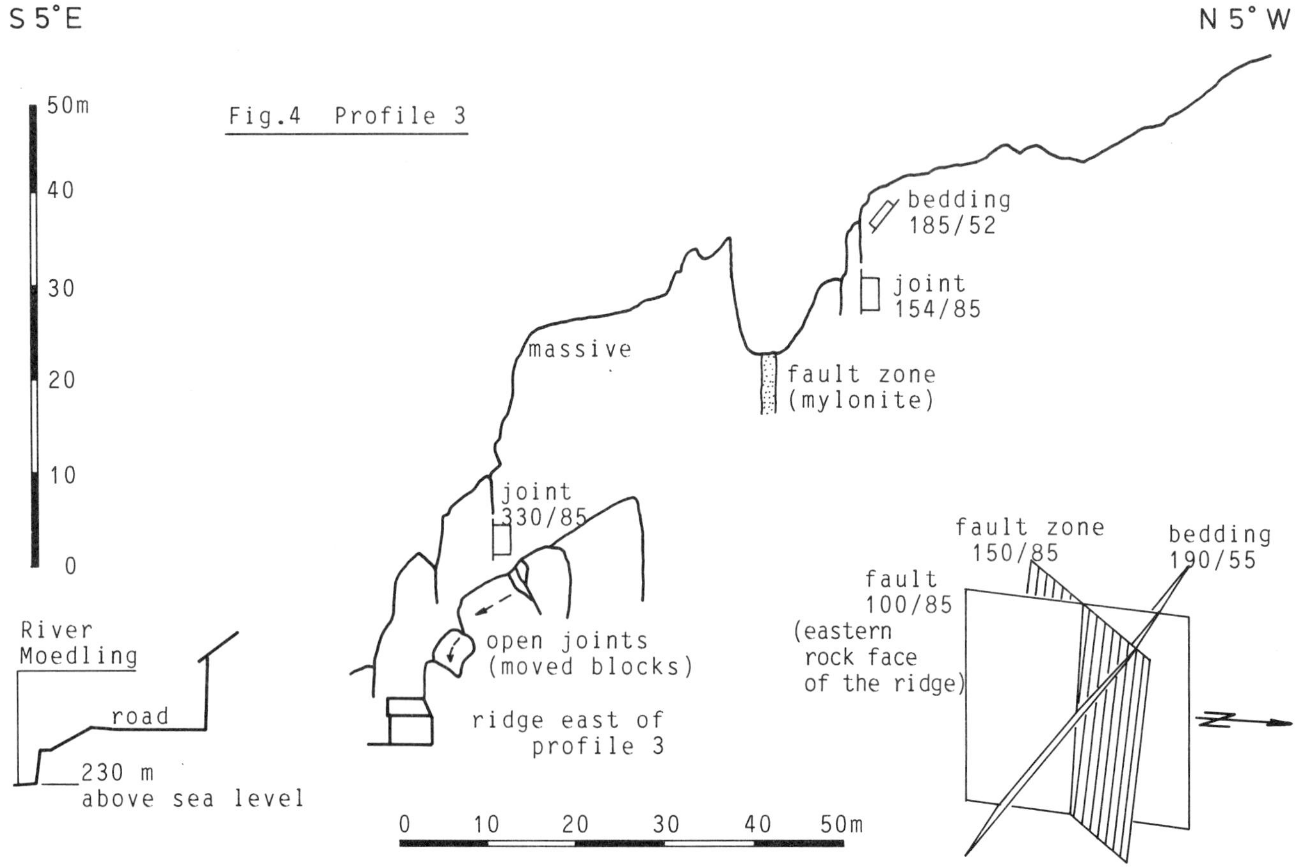

S 5°E N 5°W

50m 40 30 20 10 0

Fig.4 Profile 3

bedding 185/52

joint 154/85

massive

fault zone (mylonite)

joint 330/85

fault zone 150/85

bedding 190/55

fault 100/85 (eastern rock face of the ridge)

River Moedling

open joints (moved blocks)

road

ridge east of profile 3

230 m above sea level

0 10 20 30 40 50m

These intersections cause an intensive splitting of the dolomite. Thus there is an erosive wash-out deeply sweeping into the slope in this region.

Profile 3 (Fig.4)

In the region of this profile the eastern rock face of the ridge is formed by a fault running parallel to the "Thermenlinie". But there are no larger rock face formations on the western side of the ridge. The situation at the bottom of the rock face may be seen in Fig.5. There several blocks rest upon each other. In order to estimate the stability of this rock block system, measurements would be necessary. Only minor parts of the massiv tower adjoining mountainside are undercut. Furthermore the bedding planes only show a slight degree of separation. Mountainside the tower borders on an erosively washed out fault zone. The rock masses adjoining mountainside in the profile are massive, but have some vertical joints parallel to the valley.

Profile 4

This profile follows the bedding planes of the dolomite (dip angle approximately 60°) with the exception of some parts of minor dip. Several blocks in this profile of a few cubic metres, however, are undercut in regard to the general dip of the bedding planes. But in these areas the dolomite is relatively massive and without bedding joints. The terrace at the level of 270 m above sea level is partly overgrown with grass and bushes, partly covered with debris, and probably hides a steep zone of mylonite.

GEOPHYSICAL INVESTIGATIONS

For the geophysical investigations of the four profiles multiple spacings electronic logging, refraction seismic and infra-red geothermy have been used. Investigations by refraction seismics brought forth a depth of the rock floor at the bottom of the valley of 2 to 6 m under the overlay of loose rock. In the region of the base of profile 2, a zone of loose rock extending over 60 m has been found. As has been mentioned already in the geological description of profile 2, there is an erosive wash-out deeply sweeping into the slope in this region caused by the intensive jointing of the dolomite. The rock tower mentioned in the description of profile 2 has not yet been affected by this wash-out. On the plateau of profiles 1 and 2 and in the slope area of profiles 3 and 4 fault zones could be located by all three procedures concurrently. These zones could not be found on the surface of the slope and of the plateau.

Fig.5 Rock block system at the bottom of profile 3

CONCLUSION

The investigations have not indicated any impending landslide over 10 000 m³. No damage to the lining in the air raid tunnels in the area of profile 2 has been observed, which would suggest any movement of the rock slope.

There have, however, been found several blocks and block structures (< 10 000 m³) in danger of falling. One of them is the rock tower in profile 2. Earthquakes with an intensity of 7 - 8 MSK, very likely because of the neighbourhood of the geologically young Vienna Basin, are an additional danger in this context.

In order to observe the possible movements of the rock tower in profile 2, a technique using a convergence measuring instrument as used in tunneling has been suggested (see Fig.6). Thus reliable statements on the stability of the area may be made after having observed the movements of the rock tower in connection with the weather conditions (temperature and amount of precipitation) over an adequate span of time.

Additionally, it has been suggested that several excavations should be made in the upper area of the slope respectively in the Tertiary area in order to learn more about the fault zones located by the geophysical investigations.

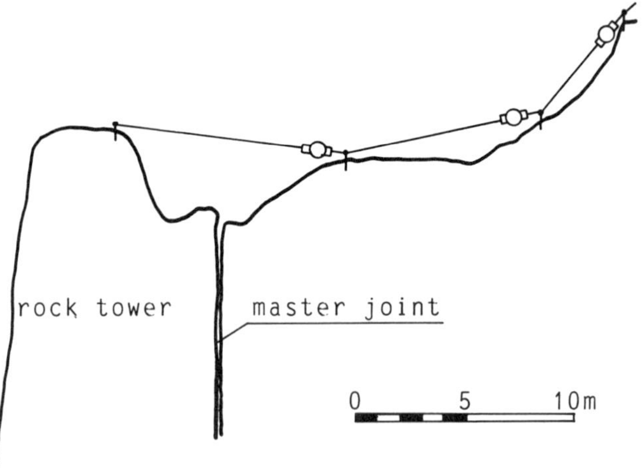

Fig.6 Measuring arrangement for the possible movements of the rock tower in profile 2

References

Drimmel, J.: Earthquakes in Lower Austria (in German). Wissenschaftliche Schriftenreihe Niederösterreich 51. Niederösterreichisches Pressehaus: Wien,1981.

Geological map of Vienna and its environs 1:75.000. Editor: Geologische Bundesanstalt Wien.

Grill, R., Küpper, H., et al.: Explanatory notes on the geological map of Vienna and its environs (in German). Geologische Bundesanstalt: Wien, 1954.

Geotechnical maps, risk management and vibration limits in tunnelling in Helsinki

J. Pöllä M.SC. (MIN.ENG.)
Y. Kähönen B.SC.
P. Raudasmaa M.SC.
Geotechnical Department of the City of Helsinki, Finland

SYNOPSIS

When tunnelling under the city there are mainly two types of difficulties: geological and environmental. Geological problems arise because it is very difficult to investigate the quality of rock under built-up areas, and therefore it is not easy to estimate the reinforcement measures and the need of grouting etc. One way to solve this problem is the use of rock classification and the estimated reinforcement measures.

Environmental problems are caused mainly by ground water, blasting vibrations and existing tunnels and caverns. To avoid financial losses and damages to the environment, a good map system is essential. The permissible values of ground vibrations must be reasonable to keep the environmental effects to the minimum and still allow an economical blasting work. Because it is very difficult to predict all the factors concerning ground vibrations, the well-thought principles of sharing the risks must be declared, too.

THE GEOTECHNICAL CONDITIONS IN HELSINKI

The geological structure of Helsinki can be divided into bedrock and overlying soil deposits. The solid rock is Precambrian, while the soil dates from the Pleistocene Age and has been deposited during and after the Ice Age.

The ground level varies between +62 m level and -15 m level. The face of the bedrock has been observed to lie at -70 m level at its deepest point. The differences in elevation are not great, but they can vary very rapidly from place to place. The bedrock consists mainly of granite, gneiss and amphibolite. Although it contains numerous cracks and zones of weakness, it is well-suited for tunnel construction. The soil cover contains mainly frictional soil constituents, such as sand, silt and moraine. Frictional soil areas together with rock outcrops cover about 55 % of the city area. Clay soils cover approximately 35 % of the Helsinki area, and the thickness of the clay course may be up to 40 m. The ground water table is usually only 1 - 3 m below ground level.

GEOTECHNICAL MAPS

The Geotechnical department has a system of geotechnical maps at scales of 1:500, 1:2,000 and 1:10,000. The maps are used primarily for city planning purposes, but they are very useful in the planning of rock tunnels, too, where geological and geotechnical data are needed for large areas.

The detailed geotechnical maps provide data on about 130,000 drilling points, and on the type and depth of the drilling on those points. The boundaries of geological layers have been drawn as contour lines on the basis of the drilling data. Both the drilling points and the lines have been drawn on separate plastic overlays. The desired combinations of data can therefore be obtained by copying these one on top of the other. These overlays are easy to amend, so the maps (1:500, 1:2,000) are continuously updated.

Foundation maps of ground water risk areas are also compiled. The maps are especially important in tunnel planning, since great risks are involved in tunnelling where building foundations may be damaged. There are, for example, about 100 buildings constructed on timber piles. The geotechnical maps also provide information on ground water levels. In areas with risk-prone timber pile foundations, the Geotechnical department has been monitoring the ground water level since 1972.

The location of existing rock tunnels and caverns can be seen on geotechnical maps, too. It is, of course, very important to know the location of these when planning new tunnels and caverns. The City Survey department updates the maps showing underground space in detail.

GEOTECHNICAL MAPS IN HARD ROCK TUNNEL PLANNING

In Helsinki there are more than 130 km of tunnels. The cross section is usually 10 - 20 sq.m. The subway system accounts for 10 km of tunnels the rest being for waterpipes, waste water, electricity and district heating etc. The city also has a considerable abount of rock caverns, such as oil storage facilities, water tanks, civilian air-raid shelters etc., totalling about 1.6 million cu.m.

Tunnel planning consists largely of searching for and circumventing zones of weakness, and adapting the rhythm of excavation to prevailing conditions. Investigations and planning go hand in hand. Prognoses of geotechnical conditions are first made with the aid of geotechnical maps and then elaborated with stage-by-stage investigations.

Geotechnical maps are indispensable in defining the position of a tunnel both horizontally and vertically. The maps are also needed when establishing suitable places for access tunnels and in giving information on the type of foundations for buildings. This also facilitates the calculation of ground water and vibration effects. One of the greatest risk factors in tunnelling is grouting, most of which is done in the areas of high ground water risk where the compression of soft clay layers and the decay of wooden piles can raise the cost of tunnelling considerably. The availability of good maps reduces the risk.

Although the geotechnical maps include a lot of information, they are not sufficient for detailed planning in every place. Often the level of bedrock surface is known with required accuracy. But there are areas, which call for more information. It is sometimes very difficult to carry out investigations due to the existing

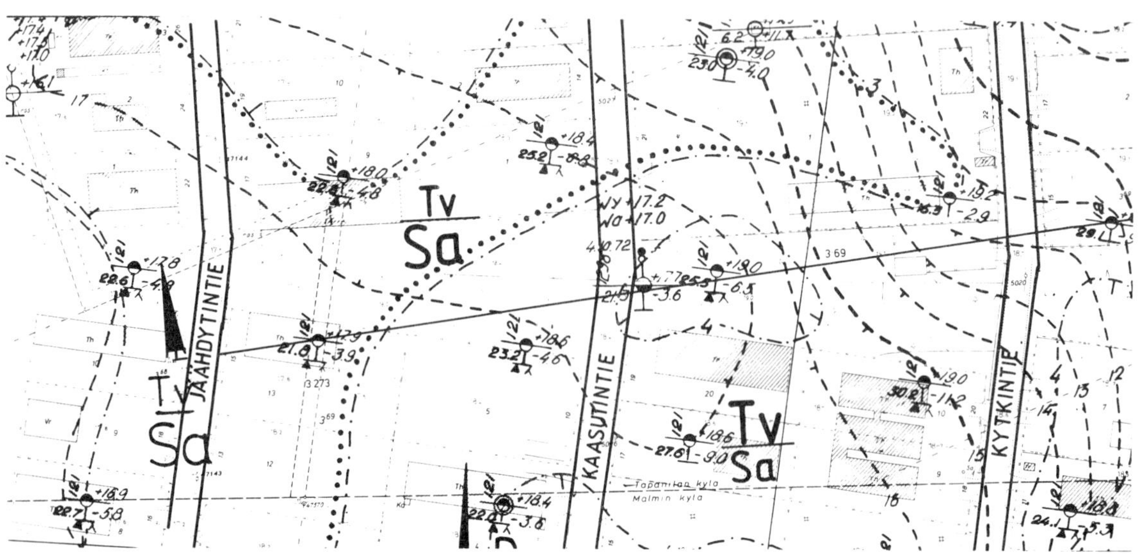

Fig. 1 Geotechnical soil map 1:2,000. On this map are shown the points of investigation, boundaries of different soil types and the clay layer depth contours. The final map is obtained by copying the different elements together. (Tv = peat, Sa = clay)

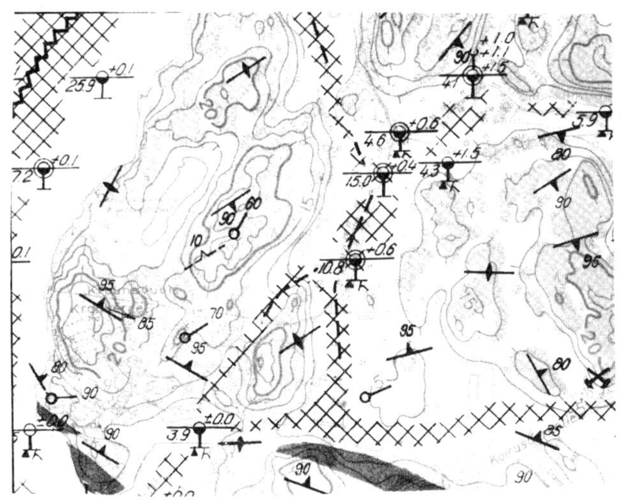

Fig. 2 Geotechnical bedrock map 1:10,000. This map shows the investigation points, geological symbols, weakness zones and tunnel spaces (original print in colour).

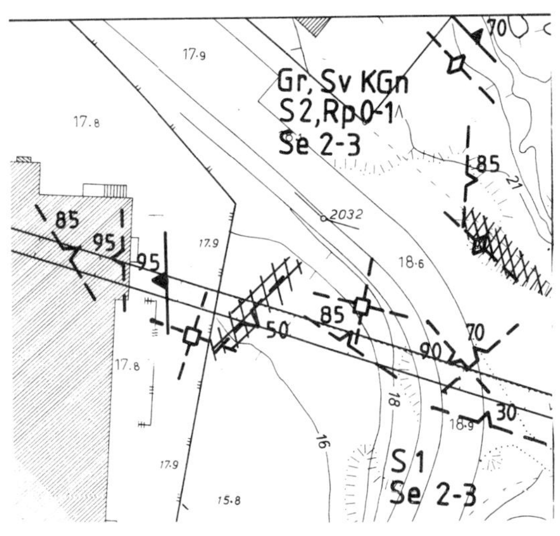

Fig.3 Detailed geological map based on the rock outcrops and from the underlying tunnel.

buildings and pipes and cables under the streets. That is why the Geotechnical department collects all the relevant data available concerning bedrock. This work covers even such areas where there are no tunnelling plans in sight.

The methods used for collecting the information are as follows:

- Close cooperation between authorities, who excavate canals for pipes and cables, so that the Geotechnical department can take the levels of bedrock from the excavations.
- Geotechnical mapping in all major excavations blasted for buildings to determine the quality of rock.
- All geological information is gathered from existing rock tunnels. Today, this mapping is performed during construction work and includes all rock tunnels whether they are owned by the city or private companies.
- The information on the foundations of existing buildings is available in the archives of the city authorities. Unfortunately the information on older buildings is not complete.

Finnish contractors say in a study conducted in Helsinki, that if the investigations are poor, they include in their tender a risk reserve, which is about 20 % of the tunnelling cost. In Helsinki, 3 - 5 % of the tunnelling costs generally consist of planning and bedrock investigations. It is possible to gain good economic and technical results with only a small investment in investigation and planning largely because geotechnical maps are available for the whole city area.

ROCK CLASSIFICATION FOR RISK-SHARING DURING CONSTRUCTION WORK

Because it is not economically feasible to find out the quality of rock in built-up areas, certain risks must be taken. The risks are mainly economic, seldom technical. During the past few years the zones of unexpectedly poor rock have resulted in long negotiations on "who pays and how much". To avoid this kind of situations we have tried to develop some rules. The very first of these were based on the geological classification and especially on the quality and number of joints. This proved to be too specific. Today we use the following methods depending on each case:

1. When the investigations are sufficient no classification will be used. The contractor pays all expenses caused by blasting and mucking the poor rock. Rock bolting and shot-

creting is divided into two stages: the one to be done during the excavation work and the other to be done after it. There are two price categories for bolting and shotcreting. The first price category covers reinforcement work interrupting blasting operations, and the second category is applied to reinforcement work after the excavation is completed. The first price category should also include all the expenses caused by starting the reinforcement work, i.e. costs of interrupting the drilling, blasting and mucking. The same system is used for grouting works. This means that if we estimate that there will be 1,000 sq.m of shotcreting during excavation work, the real amount being 1,300 sq.m, the city pays 300 sq.m at a given unit price, and if the real amount is smaller, the sum of contract will be decreased. The system refers to both bolting and grouting.

2. When the investigations are insufficient, the system for reinforcement work and grouting is the same as in the first case. To cover the expenses caused by poor rock the rock classification system will be used. The rock is divided into two groups. The first class rock is good or moderate and no extra is paid when blasting rock of this type. The second class rock interrupts blasting, and shotcreting or bolting must be carried out immediately after the round has been blasted. The estimated amounts of rock types are included in the contract and all the changes in the estimated amounts are handled as in case 1.

There are also slightly modified classifications. The main aspects for classification are that, first, they must refer to the subjects where the greatest financial risks are involved and, second, they must be as simple as possible to avoid misinterpretations. In regard to the whole project these methods have proved economically feasible. However, it demands experience to keep the differences between the estimated and final bolting, shotcreting and grouting work to the minimum.

The estimated measures for grouting and reinforcement are included in the papers given to the contractors before they make their tender. The unit price will be reasonable, because the contractor does not know whether the estimated amounts are too small or too great, and the sum of contract depends on the final measures. The price difference between first and second class rock is 20 - 35 % depending on each case.

VIBRATION LIMITS FOR BLASTING

In a modern city the devices, which are extremely sensitive for ground vibrations induced by blasting operations, represent a potential financial risk for the contractor.

When carrying out the ground vibration measurement, we in Finland usually pay attention to the vertical component.

The permissible values of ground vibrations are usually as follows (v = vibration velocity mm/s, a = acceleration in g):

- ordinary buildings founded on rock v = 50 mm/s and on soil v = 18 mm/s
- unlined rock caverns v = 70 mm/s, shotcrete-lined 110 mm/s
- shotcrete, 0 - 3 days old v = 10 mm/s, 3 - 7 days old v = 35 mm/s
- steel pipes in tunnels v = 50 mm/s
- relays a = 1 - 8 g depending on the importance of the relay. In the case of very important relays, there must be an electrician to watch the system.
- computers and similar devices a = 0.25 g.

When planning blasting where ground vibration problems occur, it is important to know the relationship between distance/charging and ground vibrations. In Finland the vibration velocity is calculated from the relationship

$$v = k\sqrt{(Q/R^{1.5})},$$

where Q is instantaneously detonating charge in kg, R is distance in m. The constant k describes the properties of soil and rock. The maximum value of k is 400 in hard intact granite. By distance increase the k-value decreases, because there are more and more zones in rock which are damping the vibration.

To find out whether the blasting work has damaged structures or devices, the buildings are inspected both before and after the blasting work. The range for inspection is 50 m for buildings and 80 - 100 m for computers and

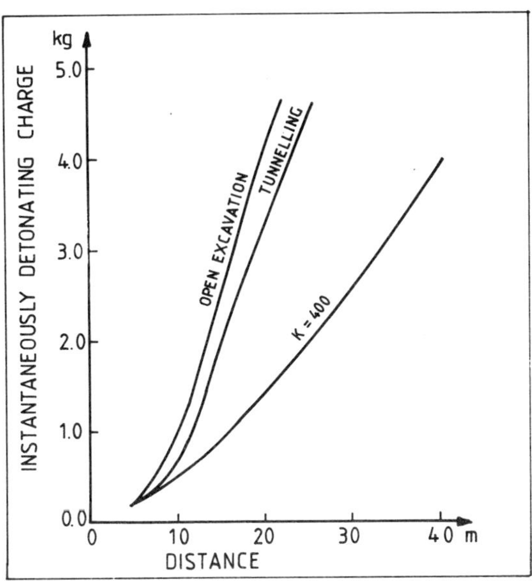

Fig. 4 Instantaneously detonating charge Q as a function of R. Values are measured in the Helsinki Metro. (v = 50 mm/s for buildings founded on rock and v = 18 mm/s for buildings founded on piles.)

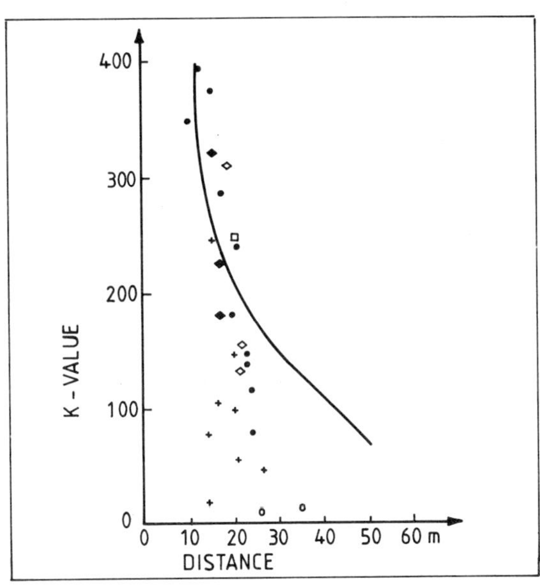

Fig. 5 The variation of k-value as a function of distance R according to the results obtained in the Helsinki Metro.

sensitive devices. When the tunnel goes deep under the city there will be no problems. The tunnel for discharging the purified waste water goes at the -70 - -80 m level and no problems occurred. The computers can easily be mounted on special shock absorbers and the contractor can blast rounds with 5 m advance. However, when the tunnel is closer to the ground level, there will be some severe problems. The planning of a service tunnel from the southern parts of the city to the north was very difficult because of a ground water risk area. The tunnel was routed off that area. Instead it was routed beneath a computer-center, because it was thought that the computers could easily be isolated by rubber-mats. When the tunnel reached the area, it was found out that there was a special computer-run going on in three shifts. It was not possible to perform any measures to protect the computers from the effects of ground vibrations. So the tunnel routing had to be changed to go through the ground water risk area. The water pressure was 0.6 MPa and 11 tons of cement was needed to keep the amount of leakage within the bounds of permitted values.

During the planning stage it is not reasonable to search all computers and devices along the long tunnel route. So nobody can say in advance how much explosives can be used to maximize the length of the round and still keep the vibrations under the given limits. This is why in such cases there are different unit prices for different amounts of allowed instantaneously detonating charges marked by Q. Usually there are three classes of unit prices, one for Q < 1 kg, second for Q = 1 - 2 kg and third for Q > 2 kg. The vibrations are controlled all the time and in that way the maximum allowable Q can be determined. Contractor runs no risk if he has given the right price. So we can say to the contractor, for example, that according to our estimates 200 m of the tunnel must be blasted with Q less than 1 kg, 500 m with Q = 1 - 2 kg and the rest with Q more than 2 kg. The contractor gives the unit prices and when the work is done the amounts are checked and the sum of contract is decreased or increased accordingly. It has been observed, that if the price for Q > 2 kg is 100, then the price for Q = 1 - 2 kg is 120 and the price for Q < 1 kg is 140.

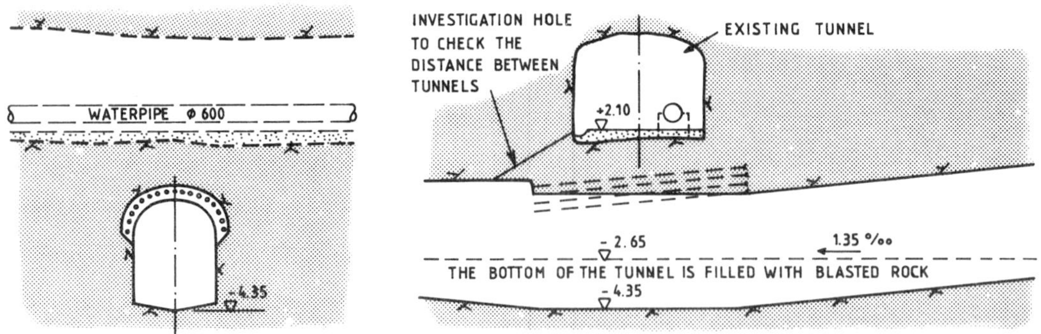

Fig. 6 A method to underpass existing tunnel. The rock between the two tunnels is reinforced by cement-grouting and pre-tensioned bolts. The bolts are Ø 20 mm L = 8,000 mm c/c 300 mm.

Because ground vibrations exceeding 10 - 20 mm/s make people feel unpleasant, it is important to inform beforehand those living in the immediate vicinity of the blasting work.

A METHOD TO UNDERPASS EXISTING TUNNEL

In Helsinki there are many kilometres of sewage tunnels. The flow in tunnels is based on gravity and the inclination of the tunnels is 0.1 - 0.2%. Because the level of the sewage tunnel is very accurate given by the levels in the treatment plants, it is sometimes unavoidable to go very close to the existing tunnels. Another solution is, of course, to go deep and build a pumping station, but this method is far more expensive.

The latest underpass was carried out in the Roihuvuori sewage tunnel in eastern Helsinki. In this case the vertical distance between the existing water-pipe tunnel and the sewage tunnel is about 2 m. The work was planned and performed as follows.

The existing tunnel was mapped to find out the best place for the crossing. The thickness of the loose macadam and rock lying on the bottom of the tunnel was investigated.

The sewage tunnel was inclined downwards so that the machines could operate in the tunnel. After the excavation was completed the bottom of the tunnel was filled to the right level.

The permitted vibration velocities were 50 mm/s for the water-pipe and 70 mm/s for the tunnel. The vibrations were measured all the time and although there were vibration velocities of over 100 mm/s for the pipe, it did not broke. The water-pipe tunnel was scaled 50 m to both directions from the crossing point. Also the pipe was protected against falling rock.

When the sewage tunnel came closer to the water-pipe tunnel investigation holes were drilled to find out the exact distance between the tunnels. When the distance was 4 m, a 0.5 m edge was blasted on the roof of the sewage tunnel. Then 17 holes were drilled to rock between tunnels for bolting. Before the bolts were installed the rock was consolidated by cement grouting. The grouting and bolting was carried out in two stages. In the first stage 9 holes were grouted and the pressure used was 0.2 MPa. After grouting pre-tensioned bolts were installed and the moment used for tensioning was 200 Nm. In the second stage grouting pressure was 0.5 MPa for 7 holes. The bolts were of one similar type. After bolting no blasting was allowed for 3 days.

The roof and the walls of the tunnel were shotcreted. The thickness of the layers was 70 mm and wire mesh 3.4 #100 mm was used.

References

1. City of Helsinki, Geotechnical department, The geotechnical soil, rock and foundation map sheets of Helsinki, 1955 - 1980

Slope remedial works in weathered rocks for differing risks

G.E. Powell B.E., M.E., DIP. ED., M.I.E.AUST.
T.Y. Irfan B.SC., PH.D., F.G.S.
Geotechnical Control Office, Hong Kong

SYNOPSIS

This paper describes a staged general approach in the investigation of the stability of slopes in weathered rocks and design of appropriate remedial works based on the risk to life category in the urban environment of Hong Kong. Particular emphasis is paid on discontinuity-controlled failures above the "rock-soil interface" and the subjective assessment of risk posed by small scale failures. A scheme used to geotechnically map the slope faces in weathered rocks is outlined. Three case histories are given to illustrate how stability investigation and design of remedial works are varied to suit the risk category of the slope.

INTRODUCTION

In Hong Kong a large number of rock and mixed soil/rock cuttings and remedial works are carried out as part of the development of an extensive infra-structure and housing programme to cope with a rapidly expanding economy and population. The standards for overall slope stability are regulated by the Geotechnical Control Office of Hong Kong Government. The Geotechnical Manual for Slopes[1] outlines the standards of safety required of slopes based on risk to life and property, and provides general guidance for slope design.

A number of deaths have occurred in Hong Kong through the fall of relatively small rock fragments, as well as through larger volumes of soil and rock demolishing squatter huts and buildings. To safeguard the public the Government requires systematic examination of potential instabilities in both rock and soil slopes. In a highly developed urban environment this necessarily means examination of both large and small scale instabilities.

The rocks in Hong Kong are often weathered to considerable depths. In a typical cutting the slope may consist of all rock, rock and soil mixtures, or all soil materials. The Manual recommends a classical limit equilibrium approach for soil (e.g. Janbu[2]). For rock, the methods recommended by Hoek and Bray[3] are commonly used. However, in weathered rocks, it is not always easy to quantify the soil-rock interface and rock failure modes, controlled by discontinuities, extend well within the weathered soil mass. There have been some large "soil slope" failures in Hong Kong which have been controlled by relict joints[4].

In this paper a staged approach based on the risk category of the slope is recommended in the investigation of the stability of slopes and design of remedial works with particular emphasis on subjective assessment of risk posed by small scale instabilities and discontinuity-controlled failures above the soil-rock interface. A scheme used to geotechnically map the slope faces in weathered rocks is presented.

GEOLOGICAL SETTING

The predominant rock types in Hong Kong are granitic rocks (granites and granodiorites) and acid volcanic rocks (fine and coarse tuffs and rhyolites, agglomerates and some ignimbrites). These rocks cover the majority of the Territory

(Fig. 1), with granites being the most common in areas of major urban developments. Metamorphic rocks are limited in extent but present special slope stability problems due to their complex geology and discontinuity characteristics.

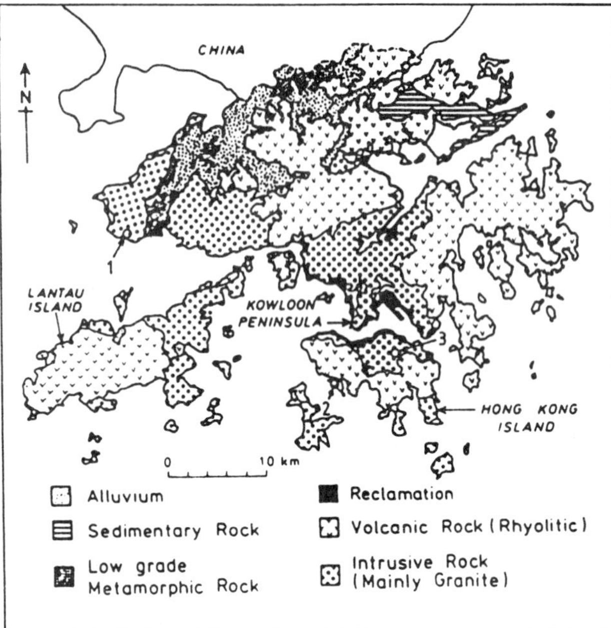

Fig. 1 A simplified Geological Map of the Territory of Hong Kong

The rocks in Hong Kong have weathered to considerable depths under the subtropical climatic regime which has prevailed in the region over a long period time.[5] The weathered soil mantle can be up to 60 m thick on the granitic rocks and up to 20 m thick over the volcanics. Large corestones are common in the soil mantle of coarse-grained granites and coarse tuffs but not in the finer grained equivalents. The soil mantle is generally underlain by a mixed soil and rock zone of highly and moderately weathered rock. The fresh rock may be at a considerable depth and may even not be met in normal engineering excavations.

The weathering profile is often complicated by faulting, hydrothermal alteration and dyke intrusion, with the transitional rock and soil zone being very thin or missing.[6] In the volcanics usually there is a gradual transition to rock at shallower depths, mainly due to the closely jointed nature of these rocks. However, in bedded sequences of volcanic, volcaniclastic and metamorphic rocks, bands or beds of more weathered material may alternate with fresher rock horizons as a result of difference in resistance to weathering of the various rocks.

SLOPE FORMATION TECHNIQUES

The majority of slopes formed in Hong Kong before 1979 were not the subject of geotechnical design and the rock faces were often rough blasted with little or no remedial works applied to them, leaving overhanging and fractured faces. For smaller excavations hand techniques were commonly used, and relatively smooth finishes were obtained using wedges, or continuous line percussion air drilling.

Since 1979 the Hong Kong Government has required finished rock faces to be either pre-split or a smooth face finish obtained by other means.[7] Local contractors are still relatively inexperienced in the use of pre-splitting techniques, and as the commonly used conditions of contract do not penalise the contractor for poor face finish, it is inevitable that on new faces there are still cases of overblasting and incorrectly applied pre-splitting techniques, particularly lack of directional control of the pre-split holes and charge selection. Pre-splitting minimises rock damage and the extent of follow up scaling and remedial measures, but has the disadvantage of making potentially unstable blocks harder to identify.

In some cases restricted site conditions preclude the use of pre-split blasting, either because of the necessity to limit vibrations on adjacent structures or because only a relatively thin sliver of rock is to be taken off the face. In these cases hand techniques or the use of mechanical equipment such as rock breakers are resorted to even at significant depths into the slightly weathered rock.

ASSESSMENT OF RISK FOR ROCK SLOPES

In Hong Kong, soil slope design is carried out to standards specified in the Geotechnical Manual for Slopes.[1] The Manual gives guidance on the recommended approach for stability assessment and specifies that slopes must have acceptable theoretical factors of safety related to the risk category of the slope. The required factors of safety for both new slopes and remedial and preventive works to the existing soil slopes are given in Table 1.

The variables in a rock slope analysis are slope geometry, discontinuity properties and orientation, and water pressure. The assessment of stability in rock is commonly carried out by subjective methods of precedent and experience, by statistical analytical techniques or better still by a combination of the two. Whilst it is possible to decrease the factors of safety used in designing remedial works as confidence in the variables increases this is generally not done in Hong Kong and the factors of safety used are those recommended by Hoek and Bray.[3]

It should be noted that an adequate factor of safety for the overall slope is not sufficient. Small scale instabilities, weathering and deterioration may create more problems than a single major discontinuity in the design of high risk slopes. This still, however, leaves an engineering judgment to be made on the minimum size of block which will be stabilised. The main factor in determining the extent of superficial stabilisation measures in a heavily populated urban area is the likelyhood of a rock fragment reaching an area where people will be present and unprotected, although maintenance, construction expedience and economic consideration play a role. Table 1 shows the authors adopted range of acceptable rock slope failure sizes with risk category, and gives examples of locations in each risk class for small scale rock failures.

ROCK MASS CHARACTERIZATION

Characterization of the rock at an early stage in any engineering project in zones or grades of similar geological, engineering and weathering characteristics is desirable. Engineering geological assessments of rock grade are cost effective and allow the information gained from a limited number of expensive or time consuming tests to be applied at other locations with similar rock mass characteristics.

Weathering has profound effects on the physical and mechanical properties of the rock material and discontinuities within a rock mass. Classification of weathering profile in terms of grades provides a simple starting point to group rock units with similar engineering properties which then can be used to describe and map the rocks for engineering purposes.

Weathering eventually converts rock to an engineering soil and, therefore, the weathering profile may be described in terms of three basic units : rock, rock and soil, and soil. A preferred weathering grade classification based on Dearman[8] and recommended by BS 5930[9] distinguishes 6 grades from fresh rock to soil (Table 2). The grades are easily recognizable and are related to engineering performance.[10] The scheme has been successfully applied in characterizing rocks for both foundations and slopes.[11]

A geological descriptive scheme of zones based on Ruxton and Berry[5] is commonly used in Hong Kong[1] with four zones based on the percentage of corestones present in the rock mass. However, this zonal scheme is not detailed enough to be applied to slopes in all types of weathered rocks, since zones of engineering

Table 1 Risk categories for soil and rock slopes

Risk category (Risk to life)	Factors of safety for cut slopes[1]		Typical examples of localities in each Risk Category (for small scale rock failures)	Typical rock slope height (m)	Tolerable volume of rockfall reaching risk area (m^3)
	New slopes	Existing slopes (Remedial and preventive works)			
High	1.4	1.2	Public waiting areas (bus stops, railway platforms etc.). Heavily used pedestrian areas. Light structures likely to be occupied during heavy rain.	over 15	0
Low	1.2	1.1	Densely used open space and recreational facilities. Occupied buildings. High speed roads.	over 7.5	Depends on engineering judgement and situation
Negligible	> 1.0	> 1.0	Country parks and lightly used open air recreation areas. Distributor roads.	less than 7.5	less than 50

Table 2 Scale of weathering grades of rock mass[9, 6]

Term	Grade		Description
Fresh	I		No visible sign of rock material weathering; perhaps slight discoloration on major discontinuity surfaces.
Slightly weathered	II		Discoloration indicates weathering of rock material and discontinuity surfaces. All the rock material may be discoloured by weathering.
Moderately weathered	III	III_i	Less than 10 percent of the rock material is decomposed and/or disintegrated to soil either as continuous or discontinuous seams.
		III_{ii}	10 to 50 percent of the rock material is decomposed and/or disintegrated to soil. Fresh or discoloured rock is present either as continuous framework or as corestones.
Highly weathered	IV		More than half of the rock material is decomposed and/or disintegrated to soil. Fresh or discoloured rock is present either as a discontinuous framework or as corestones.
Completely weathered	V		All rock material is decomposed and/or disintegrated to soil. The original mass structure is still largely intact.
Residual soil	VI		All rock material is converted to soil. The mass structure and material fabric are destroyed. There is a large change in volume, but the soil has not been significantly transported.

significance may be missed by being lumped together into one zone (Table 3). This is particularly important in the zone directly above the soil/rock interface where discontinuity-controlled failures are likely to occur.

Geotechnical mapping

It is a requirement of the Geotechnical Control Office that "large scale engineering geological drawings (marked-up transparent overlays to photographs or otherwise) showing all salient rock features and the dimensioned details of proposed support and drainage measures together with calculations to justify the design of all support measures and to confirm overall stability, are to be submitted for approval immediately after excavation of new rock slopes. The same requirement also applies to existing slopes which require remedial works.

The geotechnical mapping of many slope faces using the rock mass grading scheme set out above have been undertaken by the authors. The mapping was carried out by visually broadly delineating the rock mass grade boundaries and other features of engineering significance on overlays to colour photographs (Fig. 2). The classification of a rock mass into broadly recognisable zones or grades of similar engineering properties is also a first step in determining the mode and type of failure and the appropriate stability analysis.

MODES OF FAILURE IN WEATHERED ROCKS

The general stability of slopes in weathered rocks is governed by the mass properties such as structure, discontinuities, weathering profile, material properties such as fabric, grain size, state of weathering and alteration, strength and the groundwater. In Hong Kong, precipitation plays a predominant role, whilst temperature variations, dynamic forces of earthquakes and blasting, creep properties, weatherability and insitu stresses are considered to normally have only a secondary effect. In rocks failure modes are generally controlled by the orientation, pattern and properties of

Table 3 Failure modes and slope remedial measures in weathered rocks

ROCK MASS GRADE	SOIL %	WEATHERING GRADE OF ROCK MASS[6, 9]	ZONE[5]	TYPICAL PROFILE	FAILURE MODE/ROCK TYPE	DOMINANT FAILURE TYPE	SLOPE REMEDIAL MEASURES	
SOIL		WIV RESIDUAL SOIL	Zone A		Material controlled (in all rocks)	Erosion Non-circular Circular	Trimming to lower slope angle Drainage to lower groundwater Retaining structures Chunam/vegetation to reduce erosion	
	100	WV COMPLETELY WEATHERED	Zone B		Dominantly material controlled (occasional relict joint control in all rocks)			
SOIL & ROCK	90	WIV HIGHLY WEATHERED			Discontinuity/material controlled (strong cleavage/schistosity control in metamorphic rocks)	Non-circular Rockfalls Planar along discontinuity	Trimming to lower slope angle Drainage to lower groundwater Retaining structures Shotcrete/mesh	Protection methods e.g. intercepting slope ditches and catch nets
	50	WIIIii MODERATELY WEATHERED	Zone C		Discontinuity controlled (Joints in volcanic and granitic rocks, Bedding planes and joints in sedimentary rocks. Cleavage, schistosity and joints in metamorphic rocks)	Widely Jointed: Planar, Wedge, Toppling Closely Jointed: Rockfalls, Circular, Toppling, Planar Combination	Widely Jointed: Rock bolts/anchors, Dowels, Buttress, Dentition Closely Jointed: Shotcrete/mesh, Dowels, Buttress	
	10	WIIIi						
ROCK	0	WII SLIGHTLY WEATHERED	Zone D				Rock bolts Rock anchors Buttresses	
		WI FRESH ROCK						

Fig. 2 Engineering geological map of slope face

discontinuities whereas in weathered soil masses material properties exert greater influence in determining the failure mode (Table 3).

Joint controlled failures above soil-rock interface

The soil-rock interface is normally taken as the upper boundary of the Moderately Weathered Zone (BS 5930). Above this level the slope is considered to act as a soil slope and is analysed in accordance with traditional soil mechanics practice, assuming a perched water table at the interface. However, in the Highly and Completely Weathered Mass (grades W IV and W V) above the soil-rock interface relict joints will be present and may control the stability of the slope, and the assumption of a soil type failure mechanism may not always be conservative. A number of major joint-controlled failures in material above the soil-rock interface have occurred in Hong Kong (Table 4).

Investigation of joints in this zone prior to cutting is particularly difficult, even with sophisticated coring techniques, such as triple tube core barrel with air foam flush.[12] Only a relatively small section of joint is recovered. No indication of joint extent of waviness can be obtained and the usual techniques of core orientation, such as impression packers, are of little use. Whilst it may be possible from adjacent borings to identify similar joint types in terms of infill and general dip direction, to assume a continuous plane between these points would usually result in an unreasonably conservative design.

For the reasons given above mapping of the extent of the highly weathered zone should be carried out as cutting proceeds. These zones should be carefully inspected for persistent joints. A subjective approach is not recommended, as the joints are often similar in colour

Table 4 Joint controlled failures in highly and completely weathered rocks in Hong Kong

Location	Rock type	Failure type	Joint type	Reference
PEPCO slope, Tsing Yi Island	Completely weathered granite	Complex planar	Low angle hydrothermally altered cooling joint (?)	Internal Government Report (SPR 10/83)
Junk Bay Road	Completely weathered granite	Planar	Kaolin infilled tectonic joint	Internal Government Report (SPR 3/83)
Area 19, Tuen Mun	Completely to high weathered metasedimentary volcanics	Planar	Shear plane	Hunt[22], Koo[23]
Fat Kwong Street, Kowloon	Decomposed granite	Complex sliding block	-	O'Rorke[4]
Tuen Mun Highway, Tuen Mun	Completely weathered granodiorite	Successive toppling-sliding	Tectonic - sheeting joints	Internal Government Report (SPR /83)
Table Hill Service Reservoir, Sheung Shui	Completely to highly weathered phyllites and metasandstones	Complex planar and wedge	Clay infilled faults and schistocity	Personal observation

and texture to the weathered parent rock and are often masked by a thin layer of loosened material. Rather, a systematic inspection by an experienced engineering geologist (line survey, etc.) should be used.

Subvertical joints, generally kaolin infilled with manganese staining and showing signs of slickensides are widespread throughout the granitic and to some degree in volcanic weathered rock masses. These may provide a vertical release plane for planar and wedge failures and because of their tendency to open, possibly due to stress relief, a means for water to rapidly enter the slope and provide a driving force to the potential failure.

Failures below the soil-rock interface

In Hong Kong, most failures occur through the mechanisms of planar, wedge and toppling or combination, controlled by the orientation, pattern and properties of discontinuities[3] (Table 3). The authors are not aware of any rotational failures in highly jointed rocks in Hong Kong.

METHODS OF INVESTIGATION AND ANALYSIS ACCORDING TO RISK

The methods and extent of investigation to assess the stability and to design the remedial works for an existing or newly cut slope depend on the site conditions, geology and the risk category. The design of rock slopes involves both engineering and geology and in addition, a combination of the knowledge of precedent with the art of estimation and judgement.[13]

A staged approach is recommended to allow an initial assessment of instability and likely mode of failure, coupled with risk category to determine the extent of further investigation. Fig. 3 gives the various steps in assessing the stability of a slope and in the design of appropriate remedial works for a sound geotechnical investigation.

One of the most important aspects of rock slope analysis is systematic collection and presentation of geological data in such a way that it can be easily evaluated and incorporated into stability analysis. The geological and statistical aspects of data collection and

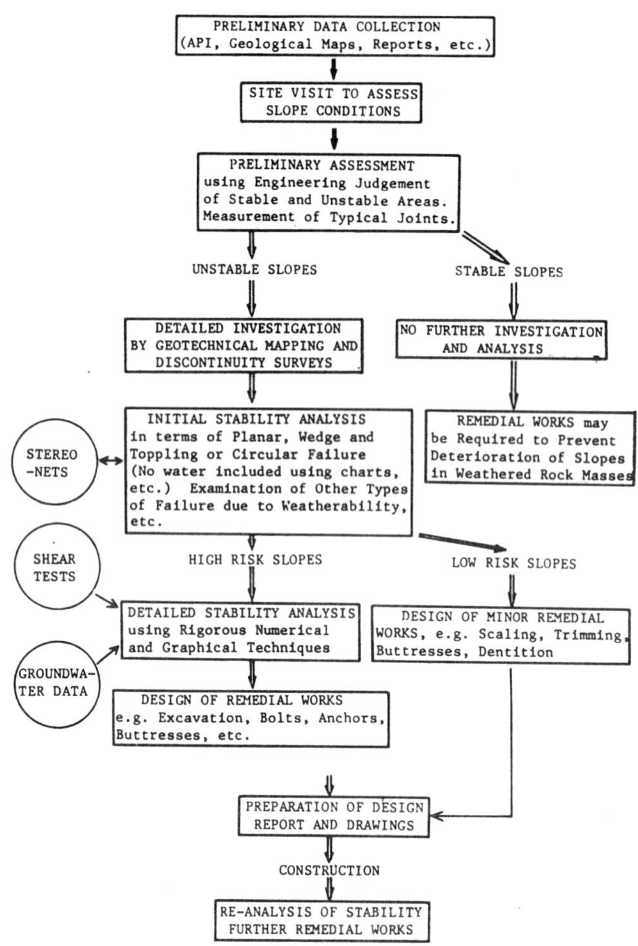

Fig. 3 Flowchart for rock slope stability assessment and remedial works

presentation are described in standard textbooks such as Hoek and Bray[3] and Goodman.[14] The description of properties of discontinuities is covered in detail by ISRM.[15] For critical joints which may cause large scale instability or for high risk slopes it is prudent to describe fully the properties of the joint surfaces. In Hong Kong special attention is paid to moderately dipping sheeting or stress release joints in granitic rocks where the joint surfaces may be more weathered than the rock itself with lower shear strength properties.[16]

The necessity of detailed analysis and the design of appropriate remedial works depend on the size of potential failure and the risk category of the slope. For low risk slopes or for minor instability problems it is adequate to carry out the design of minor remedial works such as scaling, trimming, dowelling, buttressing and dentition after initial stability analysis in terms of planar, wedge, toppling or circular failures. Other types of failure due to rapid weatherability, spalling, blasting effects may also be considered.

For high risk slopes detailed stability analysis of the unstable sections using numerical and graphical techniques incorporating shear test data and groundwater conditions are carried out to determine the factors of safety. Site investigation including coring, packer tests, sampling and installation of piezometers is undertaken to obtain more reliable information on the continuity of critical joints, groundwater conditions and shear strength of rock joints. The present site investigation practice in weathered rocks of Hong Kong is reviewed in detail by Brand and Phillipson.[17]

Finally, the drawings illustrating the slope, engineering geological details and the proposed remedial measures and the accompanying design report is prepared to be submitted to the necessary authority which is the Geotechnical Control Office in Hong Kong.

REMEDIAL AND PREVENTIVE MEASURES

Rock slopes are often designed for overall stability and allowance made for redesign and stabilisation of local areas. For an ecomomical design, about 10 percent of the slope area may require some form of treatment.[18] The stability of a rock slope may change considerably with time because of rock deterioration, strength reduction and change in groundwater conditions. For an ecomomical overall design the decrease in stability should also be taken into consideration and remedial measures and maintenance costs should be included in the original cost estimate of the project.

A useful review of remedial measures is given in Fookes and Sweeney[19] and a detailed treatment in Landslides - Analysis and Control[20] and Hoek and Bray.[3] An overview of remedial measures commonly used in Hong Kong has been given by Brand et al.[16] A simplified summary of most common remedial works applied to slopes in each grade of weathered rock mass is shown on Table 3. There are various alternatives in each case. The choice of an appropriate remedial measure depends on the scale of potential instability, the risk category of the slope and the cost. Costs in one environment or location or risk situation may be entirely different from those in another. Time, access, and space available for remedial works in urban areas are generally limited compared those of highway slopes in non-urban areas.

A programme of post-construction inspection, by both geotechnical and non-geotechnical staff, should be devised for both low and high risk slopes to assess the continuing effectiveness of the slope stabilisation works.

CASE HISTORIES

Slope A. The mixed soil and rock slope about 20 m high and 100 m long was formed in 1967 in widely jointed granite. The rock section at approximately 80° had a rough face containing numerous overhangs, loose blocks and a number of prominent differentially weathered sheeting joints with seepage. A new road and footpath was to be constructed at the base of the slope to provide access to a school. The expected high pedestrian usage required the slope to be categorised as high risk and no failure could be tolerated from the slope. Detailed survey and analysis of critical joints was carried out to ensure that all potential failures, minor and major, were considered.

The wide joint spacing allowed individual blocks to be stabilised economically using buttresses and dowels. Thorough scaling of small blocks on the rock face and trimming of highly to completely weathered soil section to a stable slope angle was carried out. Because of wide jointing of the rock mass shotcreting was not necessary. Deterioration was limited by plastering the upper slope and providing dentition with drainage to the infilled sheeting joints.

Slope B. An existing slope, 10 to 25 m in height and 600 m long was formed in closely jointed volcanics by rough blasting in 1972. The slope angle averaged 80° and a number of failures in the thin soil cover (highly to completely weathered rock and colluvium) at the crest had occurred since its formation.

A new road and footpath leading to an industrial estate was to be constructed beneath the slope. It was expected that pedestrian traffic would be light and so following an initial assessment of the face from a hydraulic platform, limited scaling of the face and trimming of the soil mantle was specified, together with a safety zone and catch fence at the base. The space available for the catch zone was less than that calculated from Ritchie's method.[21] Additional space was obtained by narrowing the footpath. It is expected that, rarely, fragments of rock may cross the catch fence, but the risk to the public of such an event is acceptably low.

Slope C. Slopes over 1.5 km in length with heights 20 to 60 m were recently excavated in weathered granite during the construction of two catchwater channels for a controlled tip site. The rock sections of the slopes, grades WI to WIII, were cut at empirical slope angles of 10 on 6 to 10 on 3 by bulk blasting and the final face by pre-split blasting. The soil sections, grades WIV and WV, were excavated at 1 on 1. Benches were provided, wherever rock conditions permitted, at about 7.5 to 10 m in both soil and rock. However, these were rather irregular or non-existent in the moderately to highly weathered granite zones and in closely fractured areas. Some sections of the slope were in a poor state with excessively shattered zones, oversteepened and potentially unstable areas, overhangs, etc. This was in part due to the geological conditions (state of weathering and fracturing of the rock, steeply dipping joints, etc.) but in some cases due to the blasting technique used which has not fully accounted for the varying rock conditions.

After the initial site visit, the risk to life was assessed to be negligible but the height of the slope put it in the low risk category along some sections where large potential instabilities were observed. It was decided that potential failures smaller than 50 m³ need not be stabilized. Volumes smaller than this entering the catchwater, about 8 m wide and 4 m deep, which would act as a "catch zone" were regarded as a maintenance problem. The geotechnical mapping and stability analyses were then programmed and carried out taking into consideration the risk situation of various sections of the slope (Fig. 3). The remedial works altough minor in nature but extensive, were designed for the rock sections of the slope after initial stability analysis which included the use of stereonets, simple calculations and failure charts. Scaling was minimised. Protective works in the form of chunaming/shotcreting were necessary in some moderately to highly weathered rock sections to prevent further deterioration of the slope. Major rock excavation was required in two areas due to the presence of a number of critical wedges and planes and the large volume of unstable mass involved to be stabilised.

ACKNOWLEDGEMENTS

This paper is published with the permission of the Director of Engineering Development Department of the Hong Kong Government.

References

1. Geotechnical Control Office 1984. Geotechnical Manual for Slopes (Second Edition). Geotechnical Control Office, 242 pp.

2. Janbu, N. 1954. Application of composite slip surface for stability analysis. Proceedings of European Conference on Stability of Earth Slopes, Stockholm, vol. 3, p. 43-49.

3. Hoek, E. and Bray, J.W. 1977. Rock Slope Engineering (2nd edition), Institution of Mining and Metallurgy, 402 pp.

4. O'Rorke, G.B. 1972. A cutting failure in Hong Kong granite. Proceedings of the Third Southeast Asian Conference on Soil Engineering, Hong Kong, pp 161-169.

5. Ruxton, B.P. and Berry, L. 1957. Weathering of granite and associated erosional features in Hong Kong. Bulletin of Geological Society of America, vol. 68, p. 1263-1292.

6. Irfan, T.Y. and Powell, G.E. 1985. Verification of founding depth of large diameter piles on granitic rock. Hong Kong Engineer, August 1985, p. 11-17.

7. Starr, D.C., Stiles, A.P. and Nisbet, R.M. 1981. Rock slope stability and remedial measures for a residential development at Tsuen Wan, Hong Kong. Quarterly Journal of Engineering Geology, vol. 14, p. 175-193.

8. Dearman, W.R. 1976. Weathering classification in the characterisation of rock : a

revision. Bulletin of the International Association of Engineering Geology, No. 13, p. 123-127.

9. British Standards Institution 1981. Code of Practice for Site Investigations, BS 5930 : 1981, British Standards Institution, London, 147 pp.

10. Bell, F.G. (ed) 1978. Foundation Engineering in Difficult Ground. Newnes-Butterworths, 598 pp.

11. Irfan, T.Y. and Powell G.E. 1985. Engineering geological investigations for foundations on a deeply weathered granitic rock in Hong Kong. Bulletin of International Association for Engineering Geology, in press.

12. Phillipson, H.B. and Chipp, P.N. 1982. Air foam sampling of residual soils in Hong Kong. Proceedings of the ASCE Speciality Conference on Engineering and Construction in Tropical and Residual Soils, Honolulu, p. 339-356.

13. Piteau, D.R. and Peckover, F.L. 1978. Engineering of rock slopes. In : Landslides - Analysis and Control (eds. R.L. Schuster and R.J. Krizek). Special Report 176 published by Transportation Research Board, National Academy of Sciences, Washington, p. 192-228.

14. Goodman, R.E. 1976. Methods of Geological Engineering. West Publishing Company.

15. International Society for Rock Mechanics 1978. Suggested methods for the quantitative description of discontinuites in rock masses. International Society for Rock Mechanics Committee on Laboratory Tests. International Journal of Rock Mechanics and Mining Sciences & Geomechanics Abstracts, vol. 15, no. 6, p. 319-368.

16. Brand, E.W., Hencher, S.R. and Youdan, D.G. 1983. Rock slope engineering in Hong Kong. Proceedings of the Fifth International Rock Mechanics Congress, Melbourne, vol. C, pp 17-24.

17. Brand, E.W. and Phillipson, H.B. 1984. Site investigation and geotechnical engineering practice in Hong Kong. Geotechnical Engineering, vol. 15, p. 87-153.

18. Richards, L.R., Leg, G.M.M. and Whittle, R.A. 1978. Appraisal of stability conditions in rock slopes. In : Foundation Engineering in Difficult Ground (ed. E.G. Bell), Newnes - Butterworths, p. 449-512.

19. Fookes, P.G. and Sweeney, M. 1976. Stabilisation and control of local falls and degrading rock slopes. Quarterly Journal of Engineering Geology, vol. 9, p. 37-55.

20. Schuster, R.L. and Krizek, R.J. (eds) 1978. Landslides - Analysis and Control. Transportation Research Board, National Academy of Sciences, Washington, Special Report 176, 243 pp.

21. Ritchie, A.M. 1963. Evaluation of rock-fall and its control. Highway Research Board, Highway Research Record 17, p. 13-28.

22. Hunt, T. 1982. Slope failures in colluvium overlying weak residual soils in Hong Kong. Proceedings of the ASCE Conference on Engineering and Construction in Tropical and Residual Soils, Honolulu, p. 443-462.

23. Koo, Y.C. 1982. Relict joints in completely decomposed volcanics in Hong Kong. Canadian Geotechnical Journal, vol. 19, p. 117-123.

Stability evaluation of some urban rock slopes in a transient groundwater regime

L.R. Richards B.E., M.SC., PH.D., D.I.C., N.Z.C.E., C.ENG., F.I.M.M. F.G.S.
Golder Associates, Maidenhead, England
J.W. Cowland B.SC., M.SC., D.I.C., C.ENG, M.H.K.I.E., M.I.C.E., F.G.S.
Geotechnical Control Office, Hong Kong

SYNOPSIS

A detailed study was commenced in 1979 to evaluate the stability of fourteen excavated rock slopes in a densely populated area of Hong Kong. The paper describes the field and laboratory investigations and the determination of field shear strengths for the discontinuities.

Failure mechanisms were defined for the slopes and analyses carried out to determine their stability conditions. The stability of one of the slopes was found to be particularly sensitive to groundwater levels. The groundwater monitoring programme was augmented and continued until the end of 1983. The piezometric data provided valuable information on transient groundwater rises along continuous rock joints under conditions of tropical rainstorms. Water pressure distributions measured on these joints are shown to be significantly different to those normally assumed for analytical purposes. The study demonstrated that the overall slopes had acceptable factors of safety.

INTRODUCTION

This paper describes a detailed investigation that was carried out to define the overall stability of several excavated rock slopes in the North Point area of Hong Kong. The study was carried out by Golder Associates on behalf of the Buildings Ordinance Office.

North Point is a prominent spur of land between Causeway Bay and Quarry Bay on Hong Kong Island. The study area was located near the tip of the spur where extensive building development had occurred in the 1960s and 1970s, resulting in many multi-storey residential buildings being placed in close proximity to large steep rock faces (see Figure 1). The need for a detailed stability study of the area became apparent with the observation of persistent sheeting joints dipping out of most of the slopes.

Figure 1
Example of rock slope in study area

Fourteen rock slopes were evaluated. They were typically 15 m to 30 m high and totalled about 600 m in length. The purpose of the investigation was to define the overall stability of rock slopes in the area. The main emphasis was on identification of major potential failures which could endanger the stability of adjacent buildings. Consideration of isolated rockfall problems, or remedial works, did not form part of the brief. A detailed investigation in the winter of 1979/80 laid the foundations for analysis of the overall stability of the most critical slopes using the groundwater measurements made in the wet seasons of 1980-1983.

GEOLOGY

Hong Kong consists primarily of a thick sequence of volcanic rocks into which a series of granites have been intruded. North Point is formed of Hong Kong Granite which is one of the largest of the intrusions. In the study area, this was found to be mainly medium to coarse grained and equigranular.

In common with most igneous and pyroclastic rocks in Hong Kong, the granite has weathered in situ. However, as expected from the location of the study area on a prominent ridge, the depth of weathering was not found to be great. Material of decomposition grades IV to VI was found only in a thin surficial layer and with a maximum thickness of 5 m.

A large number of sheeting joints was encountered in the rock slopes. The term "sheeting joint" is used here to describe rock discontinuities which have developed parallel, or sub-parallel, to an existing or pre-existing, ground surface. They are typically found in granites and gneisses. Their origin is generally considered to be due to stress changes arising from removal of overburden.

GROUND INVESTIGATION

Firstly, a review was made of the available data for the area. This included previous geotechnical work, routine investigations for foundation designs, and broader studies carried out for preliminary assessment of landslide risks over significant areas of the Territory. Extensive use was made of topographic maps and aerial photographs to provide information on original topography, development and construction problems. A typical cross-section through the study area is shown in Figure 2.

Geotechnical Mapping

Elevations of the slope faces were obtained by terrestial photogrammetry and these were used as base maps for geotechnical mapping. Contoured elevations of the slopes were obtained from the terrestial photogrammetry by plotting vertical slices in relation to an arbitrary datum plane behind the face. The close contour interval of 5 cm enabled very fine details to be identified on the elevations. Terrestial stereopairs were marked up in the field with geotechnical data and these were used for guidance in plotting geological features during the photogrammetric contouring.

The restricted space between buildings and slopes was often a critical factor in the photogrammetric work because of the need for unacceptably wide angle lenses or tilt. In one case, it was impossible to obtain proper photogrammetric control and a bamboo scaffold was erected over the slope to allow manual mapping and data collection (see Figure 3).

Discontinuity Surveys

The attitude, continuity and aperture of discontinuities were measured by window mapping at stations across the faces after visual

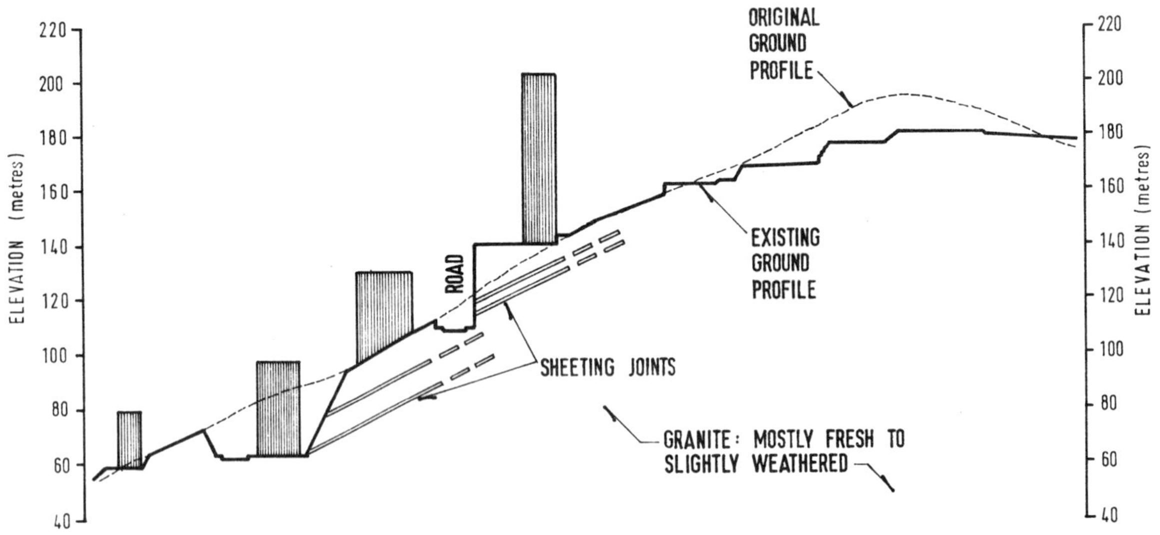

Figure 2

Typical section through study area

identification of structural domains. Access to the higher levels on some faces was obtained by the use of mobile platforms. In one case (see Figure 3) a bamboo scaffold was used for access.

The data obtained from these surveys were plotted up as scatter diagrams to identify the predominant sets and to enable kinematic analyses to be carried out. These joints sets are summarised in Figure 4. The most common joints are subvertical with a typical spacing of about 1m. The other joints are moderately low angle (about 30°) and have variable spacing. Sheeting joints fall within the above category.

Figure 3

Scaffolding of slope for access and mapping purposes

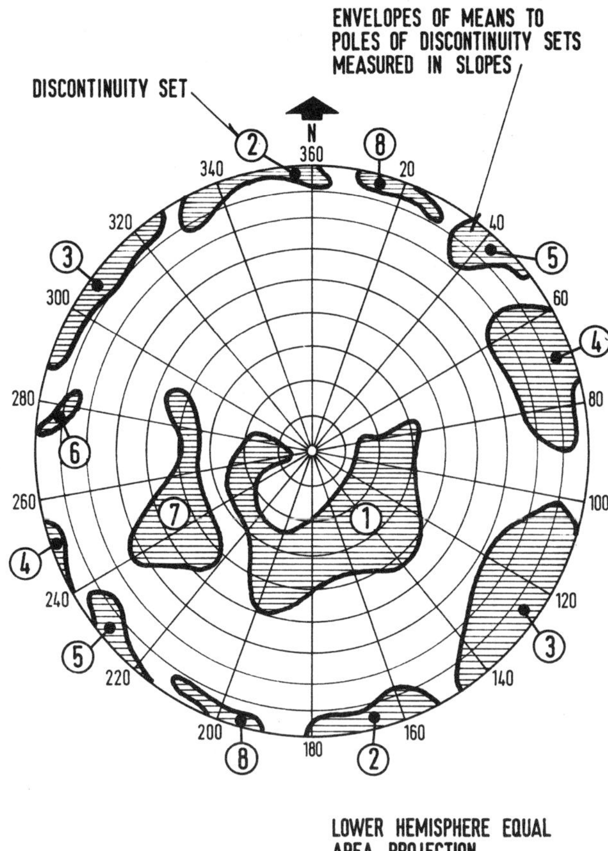

Figure 4

Stereographic projection of discontinuity sets

Borehole investigations

Although excavated faces provided good exposure, structural information was often inadequate because of the face orientation with respect to sheeting joint attitudes or because of blasting damage. Where required, additional information was obtained from borehole investigations.

In order to obtain maximum core recovery, triple-tube core-drilling at HQ size was specified. 23 holes were drilled with a total length of 340m. Water flush was not permissible because of its potentially adverse effect on slope stability. The use of air-foam allowed low air pressures to be maintained (about 450 kPa) and resulted in excellent core recovery being achieved. An example of 100% core recovery of a sheeting joint in slightly weathered granite containing a completely weathered infill is shown in Figure 5. More comprehensive details are provided in Phillipson and Chipp [1].

Figure 5

Example of high core recovery in weathered sheeting joint

On completion of coring, an impression packer survey was carried out wherever possible to orient the core and the discontinuities. In some holes, broken ground made it impossible to remove the packer without damaging the impression film. Orientation of the core allowed identification of those low angle joints which were dipping out of the slope. These were then correlated with data from other holes and with surface geological mapping. Figure 6 summarises for one section the borehole interceptions with outwardly-dipping, low-angle joints.

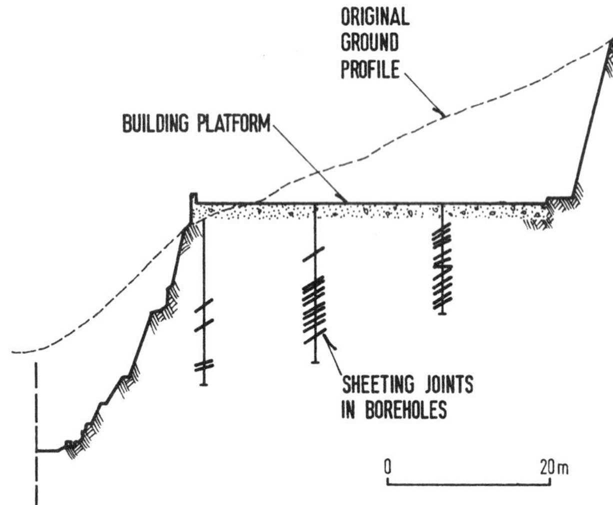

Figure 6

Section through slope showing sheeting joint orientations from borehole data

Laboratory tests

Tests were carried out on samples obtained from the borehole investigations. These consisted of routine index tests and a comprehensive programme of direct shear tests on sheeting joint samples at varying stages of weathering. The direct shear tests are fully described in Hencher and Richards [2].

The laboratory shear tests were carried out using normal stress levels appropriate to the slopes (typically 0 - 250 kPa). The basic friction angle of the sheeting joints was consistently found to be 40° regardless of the weathering grade of the rock material.

Roughness surveys

Roughness measurements were carried out wherever a sufficiently large expanse of sheeting joint was available. Circular plates of different diameters were fixed to a Clar compass in order to span irregularities of different amplitude and wave length. Measurements were made on a regular grid array. The paper by Richards and Cowland [3] provides a comprehensive account of these roughness surveys and the assessment of field shear strength for the sheeting joints. It was concluded that roughness angles of 16° and 8° were appropriate for overall slopes and minor failure mechanisms respectively. The field shear strengths were therefore $\tau = \sigma \tan 56°$ and $\tau = \sigma \tan 48°$ for the measured basic friction and roughness values. These values are of course applicable only to the slopes for which they were measured.

GROUNDWATER INVESTIGATIONS

Annual Rainfall Variation

The average annual rainfall in Hong Kong is 2.2 m. There is a distinct wet season from April to October with the rest of the year being relatively dry. Groundwater monitoring was carried out between 1980 and 1983. The total rainfall for these years was: 1980, 1.7 m; 1981, 1.6 m; 1982, 3.2 m; 1983, 2.9 m. The first two years of the study were thus significantly drier than normal and were followed by two wetter-than-average years.

Piezometers

Preliminary analyses showed that the stability of one of the slopes was particularly sensitive to groundwater levels. A comprehensive evaluation of the groundwater regime was carried out using both manual and automatic measurement of standpipe and pneumatic piezometers. This aspect of the work has previously been described by Cowland and Richards [4].

The rock slope in question is that shown on Figure 6. It is about 30 m high at an angle of 65° in slightly weathered granite. Recharge to the slope occurs from further up the hillside via the vertical and sheeting joint systems. Seepage was reported from the major sheeting joints in previous dry seasons.

Pneumatic piezometers were chosen as the primary instrumentation system because of their rapid response time. Standpipe piezometers were used as secondary, or backup, instrumentation. Monitoring was carried out by manual and automatic methods.

The first year's experience of the automatic monitoring of pneumatic piezometers with mechanical chart recorders proved to be unsatisfactory. During this first year, the development of an automatic standpipe monitoring system was being finalised (Pope, Weeks & Chipp [5]). This 'bubbler' system measures the back pressure at which bubbles escape from the end of a tube within the standpipe. The system appeared to offer significant advantages and was introduced into this study with considerable success.

Piezometers were installed at the locations shown in Figure 7. The piezometer tips were all located in sheeting joints. Measurements were taken once a day in the dry season, nine times a day in the wet season, and thirty six times a day (i.e. every forty minutes) during periods of heavy rainfall or following major storm warnings.

The base groundwater level was found to fluctuate by only by about 1m between wet and dry seasons. In addition, transient groundwater rises were measured during the wet season. A typical rise was about 5 m and the maximum rise recorded was 9.8 m. A typical duration was about 5 hours, with a range in durations from less than 80 minutes (once for a 7.6 m peak) to 30 hours. Some typical examples are shown in Figure 8.

It was thought initially that a simple relationship could be established between rainfall and groundwater response. Examination of the observed peaks showed that such a relationship was complex and not amenable to simple mathematical treatment. The measurements at 40 minute intervals showed that the sheeting joints experienced groundwater surges in times of heavy recharge. Despite the relatively short duration of the surges, there was often an interval of a few hours between surges being recorded at different piezometers in the slope (see Figure 8).

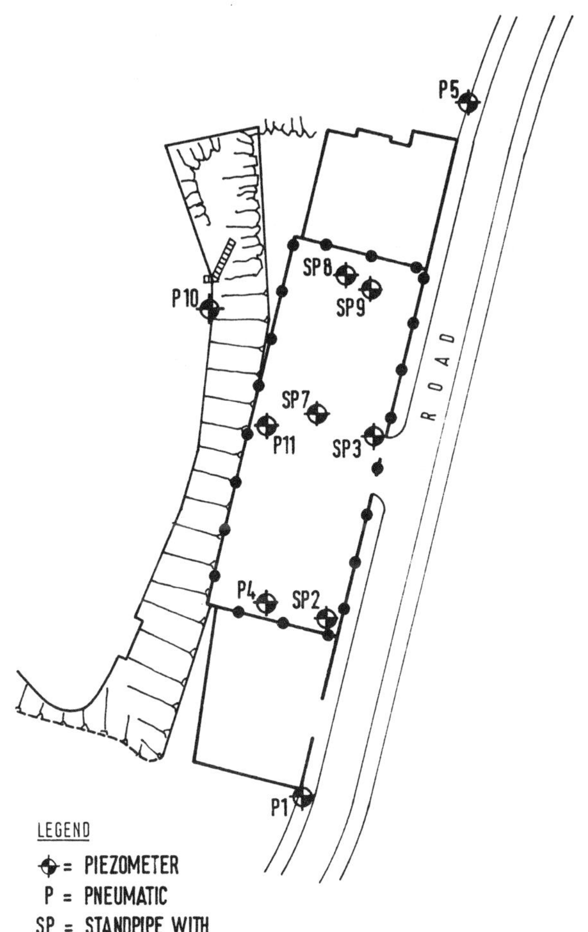

Figure 7

Plan of slope showing layout of piezometers

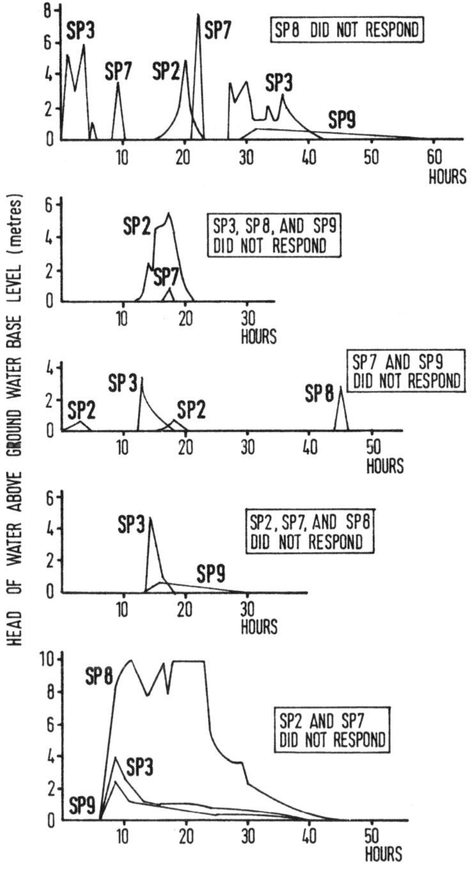

Figure 8

Example of groundwater surges following rainstorms

The most notable aspect of the surges was their apparent randomness when compared with piezometer locations for any given rainfall event. At no time during the study did all the automatically-monitored standpipes respond to the same rainstorm. Typically only two piezometers responded in any one day. Conversely, from the many peaks that had durations of less than 80 minutes it is evident that the absolute peaks were probably missed with the 40 minute reading interval.

It is clear from the measurements that the total surface area of the large sheeting joint was never subjected to large simultaneous transient groundwater rises. The total water pressure acting on these sheeting joint is therefore far less severe than would normally be assumed from the use of the concept of a rising groundwater table.

STABILITY CONSIDERATIONS

Failure modes

The slopes under consideration are in jointed rock masses where failure mechanisms are directly related to existing planes of weakness in the rock. Stability considerations have consisted of an examination of kinematic possibilities followed by limit equilibrium analyses of specific failure modes.

The most critical failure modes in these slopes are those governed by sheeting joints dipping out of the faces. Sensitivity analyses of planar sliding were carried out assuming a bilinear phreatic surface. Factor of safety equations for this situation are given in Stimpson [6]. These equations are not valid for the full range of possible geometries and were therefore revised for the purposes of these analyses. Wedge and toppling failures were analysed using standard methods. The detailed stability checks also incorporated loads due to adjacent buildings where appropriate. It is worth noting that the building loads frequently act as a stabilising influence on the low angle sheeting joints due to the added friction that would be mobilised on the joint.

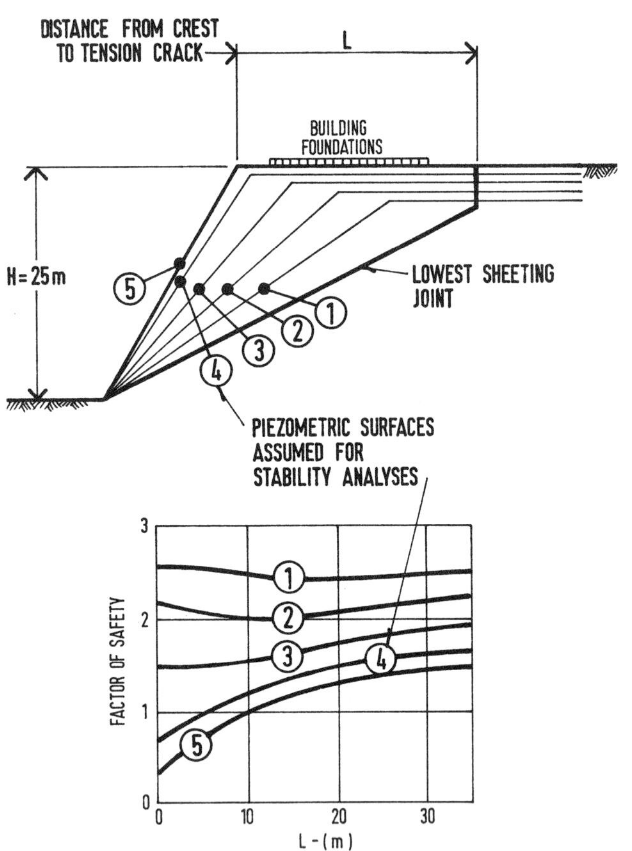

Figure 9

Sensitivity analyses for slope stability evaluation

Results of analyses

With the exception of one slope, the sensitivity analyses showed that an acceptable factor of overall safety was obtained for the probable range of slope parameters. In the case of the other slope, certain groundwater conditions could lead to unacceptable stability conditions as shown in Figure 9.

More extensive groundwater measurements, as shown in Figure 8, were made in this slope. For the most severe groundwater pressures observed during the period of monitoring, the overall factor of safety for the slope is about 3. This relatively high factor of safety is obtained because of the fact that transient groundwater rises were never observed to occur simultaneously along the whole surface of the sheeting joint. This produces less severe groundwater, and therefore stability, conditions than the usual assumption of an overall rise in groundwater.

CONCLUSIONS

A detailed investigation was carried out to define the overall stability of several excavated granite rock slopes containing sheeting joints dipping out of the slope. Specific stability conclusions were determined for each slope. The more interesting technical conclusions are as follows:

- The use of air-foam resulted in excellent core recovery being achieved. 100% core recovery was frequently attained even when a drill run passed through material of a variety of weathering grades.
- An impression packer provided good orientation data for low angle joints.
- The basic friction angle of the sheeting joints was found to be a consistent value, regardless of the weathering state of the rock joint.
- Measured joint roughness values were included as a component of the field shear strength of the sheeting joints.
- The relationship between rainfall and groundwater response was found to be complex and not amenable to simple mathematic treatment.
- Discrete groundwater surges were measured in sheeting joints in times of heavy recharge. These surges were typically only of a few hours duration, and automatic monitoring with reading intervals of a few minutes would be necessary to determine the absolute maximum level of a surge.
- Where transient groundwater rises were monitored, it was found that the total surface area of the large sheeting joints was never subjected to large simultaneous transient groundwater rises. The total

water pressure acting on these sheeting joints is therefore far less severe than would normally be assumed from the use of the concept of a rising groundwater table.

- Considerably more research is required to appreciate fully the mechanism governing transient groundwater rises in rock joints.

- All the slopes in the study area were found to have acceptable factors of safety with regard to overall stability.

Acknowledgements

The authors gratefully acknowledge the contribution of Mr G.E. Rawlings of Golder Associates who supervised a large part of the ground investigation. This paper is published by permission of the Director of Engineering Development of the Hong Kong Government.

REFERENCES

1. Phillipson, H.B. & Chipp, P.N. High quality core sampling - recent developments in Hong Kong. Hong Kong Engineer, Vol.9, No.4, 1981, pp. 9-15.

2. Hencher, S.R. & Richards, L.R. The basic frictional resistance of sheeting joints in Hong Kong Granite. Hong Kong Engineer, Vol.10, No.2, 1982, pp. 21-25.

3. Richards, L.R. & Cowland, J.W. The effect of surface roughness on the field shear strength of sheeting joints in Hong Kong Granite. Hong Kong Engineer, Vol.10, No.10, 1982, pp. 39-43.

4. Cowland, J.W. & Richards, L.R. Transient groundwater rises in sheeting joints in a Hong Kong granite slope. Hong Kong Engineer, Vol.13, No.2, 1985, pp. 27-32.

5. Pope, R.G., Weeks, R.C. & Chipp, P.N. Automatic recording of standpipe piezometers. Proceedings of the Seventh Southeast Asian Geotechnical Conference, Hong Kong, Vol.1, 1982, pp 77-89.

6. Stimpson, B. Simple equations for determining the factor of safety of a planar wedge under various groundwater conditions. Quarterly Journal of Engineering Geology, Vol.12, No.1, 1979, pp. 3-7.

Construction and blasting methods at Kornhill site formation works

M.C.F. Smith M.I.C.E., M.H.K.I.E.
Mass Transit Railway Corporation, Hong Kong
D.G. Morton B.SC., M.I.C.E., F.H.K.I.E.
Freeman Fox and Partners (Far East), Hong Kong

SYNOPSIS

A Mass Transit Railway Corporation associated development site is being formed as the Kornhill Development. The site is closely surrounded by buildings on three sides, including King's Road along the northern perimeter, and beneath the site is MTR Tai Koo Station. The site formation works consisted of the excavation of 5M cu.m. of material, of which 3.7M cu.m. was rock, to form various platforms with ancilliary works on which the Development buildings will be situated. Major cut slopes have been formed in both soft and rock material.

The works are essentially a quarrying operation in an urban area resulting in vigorous restraints.

The environmental factors and their effect, at various stages of the works, on methods of quarrying and transporting the resultant spoil are discussed, with an emphasis on blasting controls. As increased excavation production became necessary so a review of the blasting design was undertaken and a more comprehensive vibration monitoring programme set up, to check the effects on adjacent properties and the MTR Tai Koo Station cavern below the Kornhill site.

The monitoring results enabled empirical expressions to be established for blast design vibration predictions and the limiting vibration levels were increased in stages under controlled conditions for the Tai Koo Station cavern, without resulting in damage.

Guidelines for blast design in the Kornhill granite to avoid causing damage to surface buildings and underground structures are given.

INTRODUCTION

The Mass Transit Railway Corporation (MTR) was established by Government in 1975 to construct and operate a mass transit railway system. The Corporation is directed to operate on prudent commercial principles, and to minimise external funding for construction the Corporation entered into a number of property development joint ventures, in association with the railway construction, to develop airspace and take advantage of enhanced land values in the vicinity of the MTR.

The largest of these property developments is at Kornhill, adjacent to Tai Koo Station and associated with the recent Island Line construction. Tai Koo is an underground station constructed within a rock cavern beneath Kornhill and is an integrated station/development with direct links into the Kornhill development.

GEOLOGY

Hong Kong lies on the edge of an eroded South China mountain chain, composed of folded and metamorphosed volcanic and sedimentary rocks. In the central area of Hong Kong Island and Kowloon there is a younger intrusion of granite rock. The granitic and volcanic rocks are mainly crystalline and are often deeply weathered.

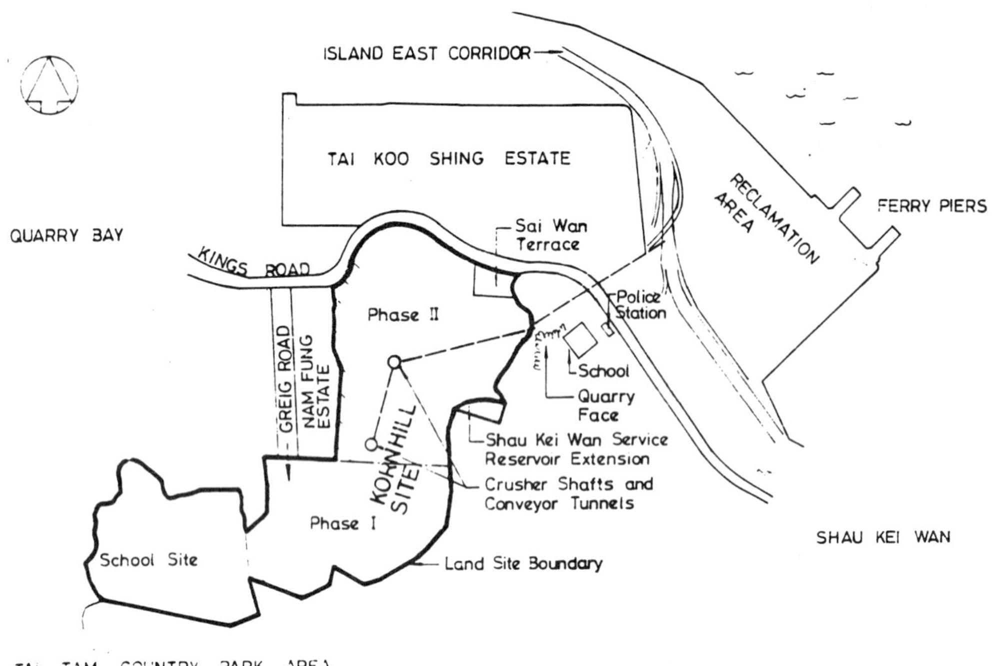

Fig. 1   Site Location Plan

The Kornhill rock mass is part of the granite intrusion which ranges from an un-weathered condition to various stages of decomposition, overlain by completely weathered granite. The completely weathered granite contains corestones and boulders with rock outcrops as well as surface vegetation. The Site contains significant faults. In the Phase I area (see Fig. 1) these are normal to the proposed excavation sequence and slope formation, being formed generally in a NE to SW line. The water table coincides with the top of the bedrock which is moderately weathered. The north eastern area of the site contains coarser grained granite, the change occurring at a major N-S discontinuity in the middle of the site. The bedrock is characterised by deep discontinuities or sub vertical seams of weathering below the base of the general weathered granite.

DEVELOPMENT

Government planning had zoned Kornhill as a development area in conjunction with straightening a section of King's Road; in addition the Sai Wan Ho area in front of the Tai Koo Shing Estate was designated for reclamation, thus allowing the Island Eastern Corridor highway to be built and providing further development land.

In 1980 Freeman Fox and Partners (Far East) were commissioned to determine the viability of excavating the hillside immediately to the south of Tai Koo Shing to produce platforms for a potential development of 30-35,000 residents. They demonstrated the feasibility and that the resultant production of some 12M tonnes (fill) could be partially accommodated in the proposed reclamation at Sai Wan Ho. It was therefore decided that the discrete but linked works of the three parties, Government, Developer and the MTR, should be co-ordinated and executed by the MTR who retained the services of Freeman Fox and Partners (Far East) for the detailed planning and design.

The development consists of 8,868 flats in 44 blocks with over 100,000 square metres of commercial, recreational and community facilities, all in a 17.2 hectare site bounded by Tai Tam Country Park to the south (with prominent features of Mt. Butler and Mt. Parker) and by the existing King's Road to the north,

Fig. 2 Model of Kornhill Development (adjacent buildings omitted)

which at the time, was the only east-west route along the eastern part of HK Island. In close proximity to the site on the west and north sides are commercial and residential buildings. Virtually within the site is Sai Wan Terrace, which had to be retained.

The Tai Koo Shing area is a neck between densely populated Shau Kei Wan and Quarry Bay, and consequently the King's Road was notorious for traffic congestion here as well as other locations further west. Access to the site from King's Road was therefore restricted, leaving the only access via Greig Road to the west of the site.

The Island Line Railway works completion date of mid 1985 was dictated by Government. This in turn fixed the Kornhill Development programme, in that the hillside had to be removed before MTR opening. The other major criterion was the early excavation of the areas in which the Tai Koo Station entrance and vent shafts had to be built. The planning of the excavation production was also complicated by the complex timing of filling requirements at Sai Wan Ho reclamation, to prevent future mud wave problems and to allow existing drainage services and culverts to be extended. Commercial requirements also influenced the programme, it being necessary to excavate the southern half (Phase I) of the Site to allow this part of the residential development to be completed first. Also zones containing platforms and cut slopes tended to have completion dates to suit building works rather than allowing the most efficient excavation plan.

All these factors meant that the granite shoulder of the Kornhill ridge, a volume of approx. 5M cu.m., had to be removed at consistent production rates of 40,000 cu.m. per week for 2½ years of a 3 year contract period. It was recognised that it would not be possible to move this material by road and the solution was to introduce primary crushing and move the material by conveyor underground. An underground system of conveyance was advantageous. By letting advance works contracts, shafts and tunnels were excavated for the installation of rock crushing equipment and material transportation.

MATERIALS HANDLING SYSTEM

It was decided to have two crushers and these were at fixed locations to suit the development programme and finished platforms.

The rock crushing system consisted of two identical 42" Krupp gyratory cone crushers which reduced the as-blasted rock from approx. 1 cu.m. size to 225mm down crushed material.

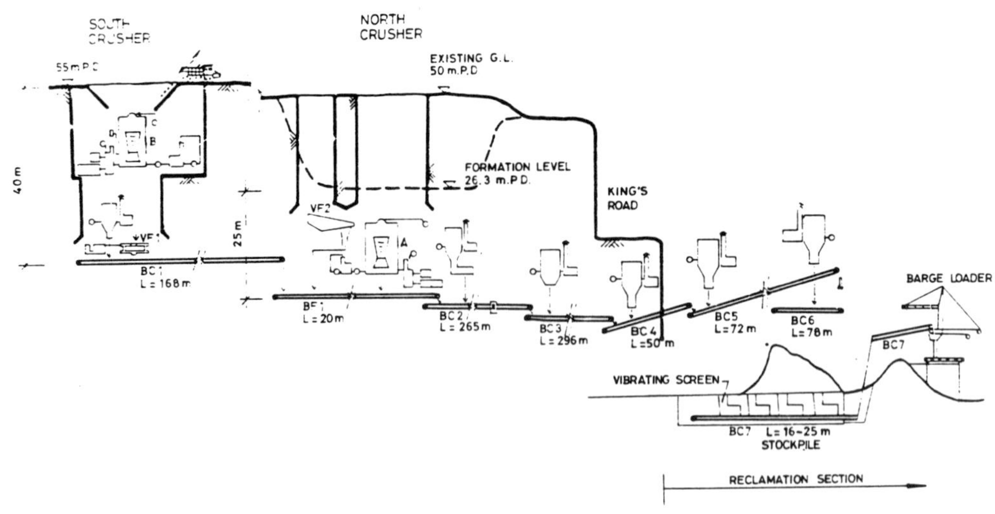

Fig. 3 Schematic Layout of Materials Handling System

One crusher was located at ground level (finished platform level) and was preceded by fixed grizzly bars to scalp off fine material. All material passed down a 40m deep drop shaft to a reciprocating plate feed (for regulation) to a conveyor belt 1,400mm wide. The second crusher (to the north) was located some 45m below original ground level to allow for a reduction in level to finished platform. Material was fed via a drop shaft and vibrating grizzly feed to the crusher and then discharged onto a conveyor belt 1,800mm wide, also carrying the discharged material from the first crusher.

The crushed rock was transported via one feeder belt and 6 conveyor belts with a total length of 930m, a stockpile reclaim tunnel with 4 vibrating feeders, and a conveyor belt leading to a barge loader capable of handling 3,000 tonnes of material per hour. The first three conveyors, 730m long, were inside a tunnel above the railway tunnels and beneath King's Road and Tai Koo Shing Road, and thence fed material onto a 3-belt conveyor system leading to a surge stockpile.

The control of dust was achieved by dust collectors at every material transfer location, by water sprays at the first crusher the stockpile and barge loader, and by rubber curtains at the stockpile. In the tunnel the dust collectors also provided a number of air changes per hour to ensure a good working environment. Excessive noise was controlled by installing the plant underground and carefully designing the transfer chutes to reduce rock impact on steel surfaces. The external conveyor flights were enclosed for additional safety and dust hazards, whilst the crushing/conveying system was provided with a fail safe interlocking solid state programme control system and an automatic weighbelt recording system.

CONSTRUCTION FACTORS

The Kornhill Site is in close proximity to residential and industrial areas as well as a major highway, making the Works essentially a quarrying operation in an urban area. Environmental restrictions were therefore imposed for the safety of the public, to minimise noise, the safety of utility services, the control of dust, the movement of vehicles and the use of explosives. In view of public sensitivity to the generation of surplus energy from blasting the latter had the most stringent controls. Additionally the use of explosives was subject to the requirements of the Mines Division of the Government's Labour Department. These controls governed the type of plant used

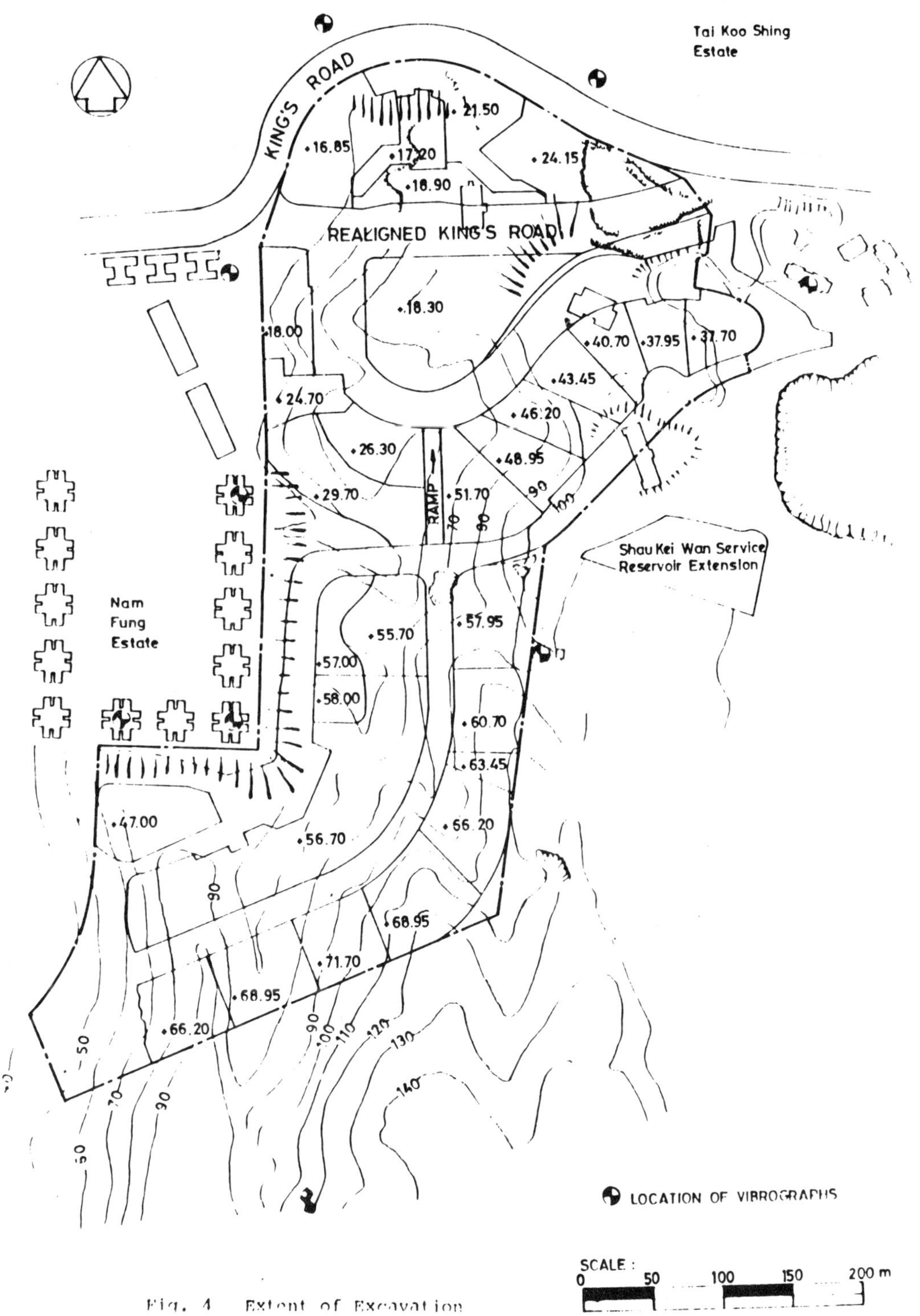

Fig. 4 Extent of Excavation

and affected the quarrying plan.

CONSTRUCTION METHODS

The Contract works commenced in July 1982. A rockfall fence was erected around most of the perimeter of the site. Excavation by non-explosive methods commenced in the Tai Koo Vent Shaft areas, followed by a roof-over blast screen to allow blasting to commence.

Originally it was envisaged that a bulk slurry form of explosive would be used for site blasting. However for bulk slurries the Mines Division requested the whole of the site to be covered with a roof-over screen, because of their concern over the higher energy causing flyrock, and escape from drill holes. Roof-over screening of the site would have made excavation/removal of material impractical and consequently the Contractor opted to use site mixed Ammonium Nitrate-Fuel Oil (ANFO) prills and cartridge water gel explosives. The latter were obtained from an explosives sub-depot at the Mt. Butler quarry, set up to supply explosives daily to MTR construction sites, and from the Mines Division controlled store at Stonecutters Island.

By the end of 1982 blasting operations had started in the south east section of the site. To achieve and maintain the programme excavation rate it was necessary to create sufficient operating benches for drilling, blasting and mucking away without mutual interference. In addition, the quarrying plan had to allow for properly maintained access routes to the crushers for stripping the overburden, and where practicable, for ripping the rock.

The choice of bench height was governed by the stability of the formation and the environmental restraints on the blasting. During the Contract planning stage, blast simulation by computer had been carried out using a blasting model and plotting theoretical crack patterns in three forms, the intersections of which gave the expected sizes of the rock fragments. Estimates of the explosive charge (powder factor in Kg/m^3 of rock) were determined based on assumptions of geology, physical properties of the rock, blast design and coupling of the explosive, to produce maximum rock fragment size of 1 cu.m., for which the crushing system had been designed. Vibration predictions were then checked based on a USBM derived expression and this limited the charge weights and indicated that 6m. bench heights would be suitable. During early trial blasts with ANFO, however, actual ground vibrations measured were less than predicted and the Contractor was therefore able to plan with increased bench heights, up to 12.5m but commonly 10m, thus increasing the production rate for much of the higher level excavation in the south and north central zones of the site.

It was also anticipated that approx. 0.5M cu.m. (bank) of the total material to be removed would be rippable. The seismic propagation velocities for the Kornhill granite were known to be around 4,000m/sec. for slightly weathered granite and for moderately/highly weathered granite 1,000m/sec. Ripper tractors of Cat. D8 or D9 size are capable of rippable production in rocks with seismic velocities of 1500-2000m/sec., but in the event, no large quantities were able to be ripped, mostly because of the varying level of the moderately weathered granite.

Haul roads were generally 10-12m wide and restricted in gradient to 10%, although this was not always achieved. Dust was controlled by water truck bowser sprays. The haul road network focussed into the two crusher locations from the benches and along the platforms. The site became segmented by the operations of overburden stripping, blasting areas, and haul roads and it therefore became necessary to surround each segment or zone by a blast protection screen and obtain blasting permits as the bedrock was exposed. Excavation was carried out in a general south to north direction and thus eventually the northern situated crusher became the most used. As the platforms were exposed so surface drainage works (rock channels and concrete culverts) were constructed. The exposed cut slopes were reinforced by bolts, dowels and shotcrete as required. At the completion of the Works some 800m of perimeter slopes upto 65m high had been formed, together with 500m of internal slopes upto 25m high.

CONSTRUCTION PROGRESS 1983

During the first half of 1983 the excavation programme was delayed. The delays occurred partly by the site activities causing interference, but largely because of the intermittent working of the crusher/conveyor system caused by electrical and mechanical breakdowns. Problems also occurred with excessive wet soft material and extraneous material fed into the system. The long learning phase by the Contractor of operating the crusher/conveyor system also added to the delays.

After obtaining Government approval, soft material was removed by road haulage and the resources and working hours were increased in the MTR Railway vent shaft areas to overcome these delays. Activities late at night were restricted to loading and removing blasted material to stockpile in the site. These measures enabled work in the Phase II area to commence on programme.

The performance of the crusher/conveyor system improved in the second half of 1983 but delays still occurred. Relocation of blast screens took a long time and congestion of the various activities caused inefficiency. An isolated flyrock incident occurred in September and this resulted in withdrawal of the blast permit for one week. The Contractor then adopted a very cautious approach and increased the stemming length but this led to poor fragmentation at the top of the free face and necessitated secondary breaking, which reduced daily output.

In October 1983 an agreement was negotiated with the Contractor to achieve the original Contract completion date, by increasing resources. By December 1983 the excavation of the Phase I area had been substantially completed and excavation in the Phase II area had commenced in earnest. The best weekly production through the crushing/conveying system up to that date was 120,500 tonnes.

CONSTRUCTION PLANT

By early 1984 the Contractor's plant consisted of:-

Atlas Copco hydraulic air flush drills type ROC712H
A Caterpillar D10 tractor/ripper
Caterpillar D9L tractor/rippers
Komatsu tractors
Caterpillar 245 face shovels
Caterpillar/Komatsu track and tyre loaders
Various backacters (some with hydraulic breakers)
Caterpillar 769 & 773 dump trucks (45 tonne)
Komatsu 32 tonne dump trucks
Numerous 18 tonne highway tippers

Most of the equipment was new for the Contract. This ensured that the noise emanating from the plant was contained within current practical limits. During plant maintenance particular attention was paid to the condition of exhaust systems, cooling fans and worn/noisy parts. Much of the heavy plant created noise levels in the range 80-100dBA at distances of 10m.

Typical noise levels, for aggregated plant activities, recorded in the buildings around the site ranged from 68dBA to 75dBA, with the dumping/crushing at the south crusher generally causing the higher noise levels at Nam Fung Estate. During blasts sometimes over 80dBA was recorded. Ambient noise at Sai Wan Terrace and Tai Koo Shing Estate was generally 60dBA during the day and evening.

Hydraulic percussive air flush drilling rigs were used to comply with the specification's noise and dust pollution criteria; also hydraulic rigs are more easily adjusted to suit ground conditions and have a better penetration rate than air powered rock drills. The rates of drilling and bench height became important factors in the new programme. During peak production on the 8m/10m benches the average rate of drilling per rig was 25m per hour, for 76mm dia. holes.

A typical 10m high bench explosives pattern was a staggered grid of blast holes 76mm diameter at 2.5m centres drilled at approx. 20°-10° angle back to the free face. Over the central sections of the site typically two blastholes containing at least 15Kg of explosive were connected to the same electrical explosive timing delay. This small diameter drilling at close spacing provided good fragmentation at economic unit costs. Subdrilling varied but was of the order of 1m. These measures assisted fragmentation at the toe of the bench.

The rock slope profiles were formed by

presplitting, as was the later foundation pocket perimeters. Generally this was done by loading 76mm diameter holes at 600mm centres with decoupled 25mm water gel cartridges. Presplit blasts are heavily confined (sink blasts) so the length of instantaneous detonation, the normal method, was restricted because of the vibration constraints. Portions of the presplit face were detonated with electric delays, using detonating cord for initiation. Presplit blasts had to be detonated at least 18m away from previously fixed rock slope reinforcement to avoid disturbance and general production bench blasts were not allowed within 12m of installed slope support reinforcement.

In order to increase the rate of excavation removal from the site, and to provide contingency removal if the crushing system broke down, a new exit haul road was formed in the north west corner of the site onto King's Road. The junction was controlled by traffic lights with the soft material removed via this haul road, averaging 15,000 cu.m. (bank) per week for sustained periods.

BLASTING CONSTRAINTS

Progress continued reasonably well into early 1984 but the programmed excavation production rate was still not being achieved. Then in April 1984 a serious flyrock incident occurred in which rock was projected several hundred metres causing damage. This had been preceded by several weeks delay due to a fire in an electrical control room that damaged the automatic programme control of the crusher/conveyor system. Drilling, bench heights, blast design and charge weights were then all reviewed with two objectives in mind, to minimise flyrock and to increase production.

Flyrock

The principal causes of flyrock are reduction in burden, inadequate stemming and out of sequence firing. It was therefore agreed to provide additional protection. The Contractor, with the approval of the Mines Division, introduced mobile screens 8m long x 15m high which were placed within 10m of the bench free face. Additionally, for areas within 150m of adjacent property, portable roof-over screens 3m high were placed over the blast area of the bench. All screens were frames of steel members covered by a steel mesh. Heavy duty blasting mats were also considered but in conjunction with the portable roof-over screens proved to be impractical. The Mines Division would not allow sole use of the blasting mats except in the middle zone of the site.

The blasting layouts were also revised so that there was always a space of 3 x burden distance between any two adjacent blast areas. The use of sequential timing blasting machines was proposed but not approved by Mines Division. 10mm crushed aggregate was used as the stemming column and the use of drill cuttings banned. This also reduced the dust arising from blasts. The priming charge was relocated from top to bottom priming, particularly where bagged ANFO was used when there was water in the blastholes. Finally all blastholes were to be drilled vertical.

Blast design

The major blasting design restraint contractually specified was that the quantity of explosive was to be restricted to ensure that the max. particle velocity did not exceed 25mm/sec. at the nearest occupied building. This restraint would obviously hinder excavation production as the rock level was reduced over Tai Koo Station and to a lesser extent around the perimeter of the site. It was therefore decided to see whether the particle velocity limit could safely be increased.

To predetermine the quantity of explosive to be blasted without damaging adjacent buildings, first the intensity of ground vibration must be predicted and second the level of allowable vibration without causing building damage must be known. Confidence of predictions and a safe limit of allowable vibrations on the adjacent buildings and Tai Koo Station had not been established but no damage had occurred to date.

There is no acceptable theoretical approach to determine ground vibrations from blasting in rock. Based on sinusoidal stress wave forms empirical expressions have been obtained, but it is usual to obtain field measurements and scale these to predict ground vibrations. If the various factors are considered dimensionally, then the distance from the detonation scales as the cube root of the charge size (or energy released). The US Bureau of Mines however has derived a useful propagation law from various quarrying vibration data,

$$V = K \frac{W^A}{R^B}$$

where V = peak particle velocity

R = distance

W = charge wt/delay

K is the ground transmission constant

A&B are constants related to the geology between blast & recorder

which could, for practical purposes, be expressed in square root scaling distance terms.

From numerous field measurements in all types of hard rock, the USBM expression can be typically expressed as

$$V = 1143 \left(\frac{R}{W^{\frac{1}{2}}} \right)^{-1.6}$$

where $\frac{R}{W^{\frac{1}{2}}}$ is known as the scaled distance

The Kornhill blasting design had been based on this expression for predicting peak particle velocities as a measurement of ground vibration. However statistical analyses of recorded vibrations by regression indicated that the empirical expression ranged from $V = 3802 \left(\frac{R}{W^{\frac{1}{2}}} \right)^{-2.05}$ on an area basis (Phase II area) to $V = 293 \left(\frac{R}{W^{\frac{1}{2}}} \right)^{-1.07}$ on a distance basis of $10m > R < 50m$, with low correlation coefficients.

It was estimated that production rates of at least 8,500 cu.m. (bank) per day were required down to the level +38 to achieve the completion date of mid 1985. This required nominal 9m benches with charges of 28kg per delay over the Tai Koo Station area. Initial predictions based on this, using an arbitrary curve in a conservative zone of prediction, indicated a low probability that vibrations of 50mm/sec would occur at the crown of Tai Koo Station.

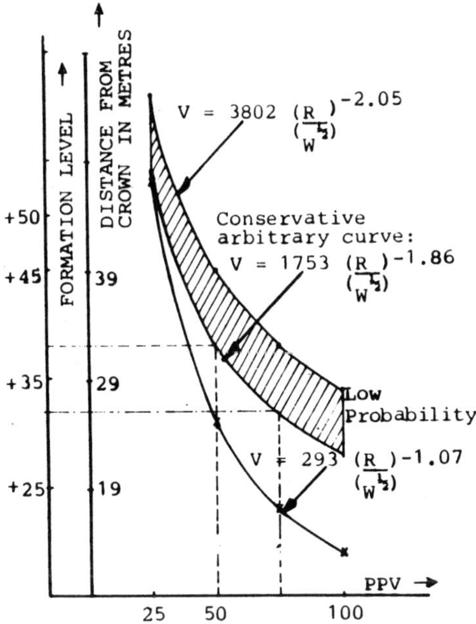

Fig. 5 Suggested Blasting Criteria

At that time the Phase II area was at levels of +60, +70 and +80 and the required production rates were being achieved, without reaching peak particle velocity predictions or the limiting ppv of 25mm/sec at Tai Koo and adjacent buildings. So additional monitoring stations were introduced into the Tai Koo cavern to carry out further research and establish the feasibility of allowing a higher limit of vibration and a means of confident prediction.

Because of the poor correlation in the derived expressions the results were examined more closely. This showed that when the scale factor $\frac{B}{A}$ from the trivariate expression $V = K \frac{W^A}{R^B}$ was greater than 0.6 or less than 0.2 the predictions were unrealistic. So all results were checked and only used when the scale factor $\frac{B}{A}$ was between 0.2 and 0.6. Regression lines were then produced for the various components of recorded vibrations and these were grouped into small scaled distance (6 to 15) blasting and large scaled (10-40) and for Tai Koo Station further subdivided into the areas west and east of the major fault.

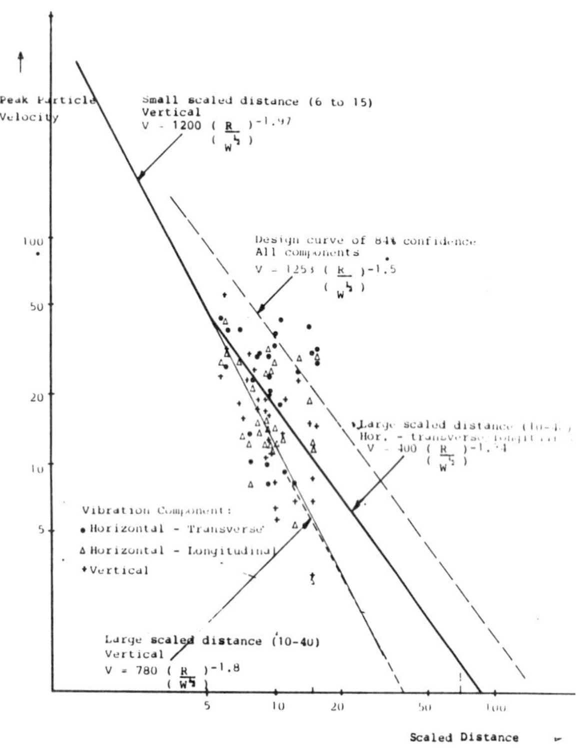

Fig. 6 Regression Lines for recorded vibrations at West end of Tai Koo Station

Fig. 6 shows a typical early plot for blasts at the west end of the Tai Koo Station. The design curve $V = 1253 \left(\frac{R}{W^{\frac{1}{2}}}\right)^{-1.5}$ was considered to be an 84% confidence line, based on a regression of $V = 790 \left(\frac{R}{W^{\frac{1}{2}}}\right)^{-1.5}$ for general control purposes. All the monitoring data when analysed produced regression lines within this design curve envelope, and in most cases actual measured vibrations were less than predicted.

Blast damage

Having established a system of analysis and usable expressions for the blast designs with suitable parameters for the site, it was necessary to determine the level of allowable vibration on adjacent structures.

Damage to a building occurs due to its movement as a reflection of the ground movement. Building response cannot be related directly to one factor of the ground movement. Typical natural frequencies of a building and its elements vary from 1Hz to 150Hz. It is therefore generally agreed that the best guide for assessing risk of damage is the maximum velocity of the ground vibration, the peak particle velocity. There are various published guides, based on direct measurement research, listing peak particle velocity limits for avoiding damage to various structures. The USBM's latest recommendations suggest the limits, for cosmetic damage only, to be 19mm/sec to 50mm/sec for frequencies ranging from 3Hz to 100Hz. (At a higher frequency a greater vibration level is permissible). However, a German DIN Standard proposes lower limits (of the order of 50%), whilst the Swedish limits are higher at 110mm/sec for frequencies above 30Hz, but only relate to buildings founded on hard rock. Other published material indicates that where the seismic propagation velocity is high, as in rock, then there is little risk of damage to buildings subjected to vibration velocities of 100mm/sec. Similarly, high vibration velocities have been recorded in concrete sections in underground caverns without detriment.

The buildings adjacent to the site were all concrete structures founded on rock, some piled, (except for the Sai Wan Terrace blocks) which, although containing visible cosmetic defects, were technically in good condition. Tai Koo Station was considered to be technically in a very good condition, with all the concrete lining works completed and past the curing stage. It would, however, contain architectural finishes and E&M railway equipment during the blasting programme, so the effects of vibration velocities on these and the concrete structure contained within the rock mass had to be assessed.

The excavation geometry, primary support and lining arrangement of the Station works effectively created an 11m deep constrained zone causing the rock arch to behave like a stiff beam, transferring load in shear to the abutments with little or no bending. The rock arch and concrete lining were assumed to act compositely when subjected to the vibration from proximity blasting. It is generally considered that damage to rock occurs when the ground vibration velocity is between 700-1000mm/sec for medium tensile strength rock of 10MPa such as granite. This damage threshold is usually within 2m beyond the excavated profile. However in the design of the rock arch a depth of disturbance of 5m below the final formation level was allowed, for the effects of blasting and weathering. So the potential problem in the Station was considered to be only when the blast-produced compression waves were reflected at the lining face and became tensile, creating a risk of spalling.

The installed equipment considered to be at risk were relays associated with signalling and computer equipment for the Automatic Fare Collection system. The standard MTR Specification required that E&M equipment be designed to continue functioning if subjected to an acceleration of 1g. An acceleration of 1g at 100Hz frequency is equivalent to a vibration velocity of 17mm/sec. However this equipment was in rooms in the middle of the Station and any vibration of the arch lining would be damped or well dissipated before reaching these rooms.

An assessment on the condition of a Water Authority reservoir was also necessary. The two part reservoir structure had a concrete base slab with mass concrete walls, alleged to be in good condition and founded in bedrock. The reservoir contained approx. 50,000 cu.m. of water and it was estimated that the natural frequency of the structure was 30-50Hz. From the vibrations already recorded it could be predicted that blast induced ground vibration frequency would be in the range 40-60Hz when the blasting was 50-65m distance. Theoretically for a ppv of 25mm/sec at a frequency of 40Hz, the initial amplitude of movement would be 0.1mm, unlikely to cause a defect. The specification had limited particle velocities to 13mm/sec at the reservoir, based on Water Authority/Mines Division advice and it was therefore proposed to base the blast design on limiting ppvs on the reservoir to 25mm/sec. However permission was not granted and the original restrictions were enforced. Fortunately this did not prove critical.

A more important area of major restraint was the King's Road perimeter area. On account of the steep faces and rock jointing of the slope fronting King's Road it was impractical to have high bench blasts. As well as the vibration restriction of permitted particle velocities of 25mm/sec on the utility services in King's Road, air blast effects had also to be considered.

Air blast is air pressure over and above atmospheric caused by the transmitted energy from the blast in the form of pressure waves. Published information on air blast indicates that it is generally less of a problem than ground vibration damage. The USBM has undertaken considerable research and has established restriction limits: poorly fixed windows may crack at 150dB, structural damage may occur at 180dB. They have recommended a maximum level of 134dB based on a minimal probability of superficial damage to residential structures.

Generally recorded air overpressures in quarrying operations are below 125dB. Results of monitoring at the reservoir indicated air overpressures less than 0.9 mbar ($\equiv$ 133 dB) except for two blasts, one being the flyrock incident in April 1984. Generally the average was around 0.4 mbar ($\equiv$ 123 dB). The Code of Practice on Wind Effects Hong Kong was also referred to, to check wind pressure design, wind being an air overpressure in the frequency range 0-2Hz. The design wind pressures are 1.2 KPa ($\equiv$ 146 dB) and 4.3 KPa (163 dB). Only a small number of results of air overpressure were recorded and good correlation for a regression analysis could not therefore be achieved. However it was established that an air overpressure limit of 134dB would not limit the charge weights when the ground vibration limits were considered.

Vibration monitoring

At Kornhill three types of vibrograph monitors were used, one fixed and two portable.

(a) VME Seistector 1400MP (Fixed)

Designed to be bolt-fixed within a structure, this model records the maximum of the peak particle velocity of the three component axes. The frequency response characteristics are linear up to 200Hz.

(b) VS1600 Sprengnether Seismograph (Portable)
This model automatically records, in a separate unit from the transducer, the three component velocities and air wave velocity on film and displays their peak values together with time. The frequency response is only linear upto 100Hz and therefore has a limited ppv applicability.

(c) LOG 1 Digital Seismograph (Portable)
This model, also with a separate recording unit, records all seismic components and sound waves and additionally notes the resultant ppv of the three components. The frequency response characteristics are linear up to 500Hz; this greater frequency range being useful for monitoring proximity blasting in Tai Koo Station.
The transducer was able to be bolted directly to the crown lining, which was an advantage as it was suspected that some rogue results may have been caused by dynamic characteristics of the support brackets that were needed for the other models.

Although results for each component of the ground motion were analysed, the expressions established for forecasting were based on the resultant ppv of the three components. For the models that did not record the resultant, the maximum directional ppv was used. The resultant ppv was never more than 10% greater than the maximum directional ppv.

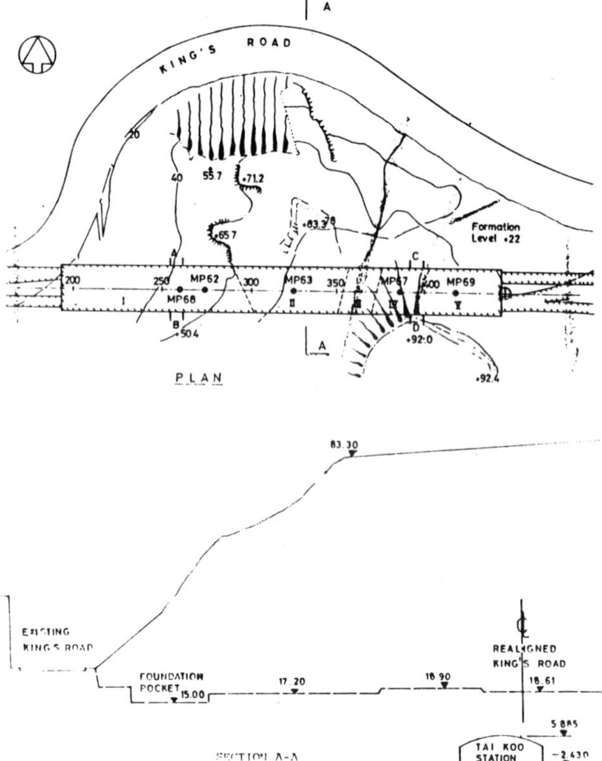

Fig. 7 Geological Zones and Vibration Monitoring Points in the Tai Koo Cavern

Fault Zones MP Monitoring Point

Geological Zones

Zone I Generally strong pinkish-grey granite generally medium to fine grain size. Very little penetrative weathering. Slabby rock conditions due to shallow dipping sheet joints. Least depth of cover after KOH excavation 11m.

Zone II As Zone I but generally massive with little sheeting joints.

Zone III Major fault zone with loose rock conditions.

Zone IV Generally strong pinkish-grey granite generally coarse to medium grained. The joints are moderately spaced, are predominately sub-vertical, with some extensive sheet jointing in places and some close fracturing.

Zone V As Zone IV but the sub-vertical joints are wider, major and extensively weathered. Depth of cover 16m after KOH excavation.

IMPROVED CONSTRUCTION PROGRESS 1984

Whilst additional vibration monitoring was in progress, excavation rates of 10,000 cu.m. (bank) daily were being achieved using 10-12m benches with charge weights of upto 54Kg/delay. The Phase II area at this time was generally at between +60 and +70 levels and recorded vibrations were all less than 25mm/sec. As the excavation progressed, so from the analyses of the vibration results, prediction curves were developed. These curves were then used to determine the charge weight for predicting a limiting peak particle velocity of 50mm/sec at the Tai Koo Station crown, and an 84% confidence design curve to a limit of 25mm/sec at adjacent buildings. This allowed sufficient size of blast volumes (nominal 9m benches) down to +42 such that excavation rates of 55,000 cu.m. (bank) per week could be regularly achieved. Similarly for the lower 6m benches and the smaller benches to formation level. 55,000 cu.m. (bank) per week had to be regularly achieved for 12 months to allow completion in June 1985.

At this stage a number of public meetings were arranged with residents/owners and Government at which information was given on the lack of progress and increased activity. Some activities were worked in the evening with every effort made to reduce the disturbance. Generally the Contractor arranged to blast daily at around 10:00 a.m. and 1:00 p.m. The local population were warned by sirens and gongs of impending blasts, and, for blasts near King's Road, pedestrians and traffic were halted clear of the site.

During the last half of 1984 the Contractor's average excavation rate was 56,500 cu.m. (bank) per week; the best week's production being 163,000 tonnes through the crusher/conveyor system.

The King's Road perimeter with a need for a more careful and slow mode of excavation limited the bench progression from south to north. Boulders and rock masses dipping towards King's Road were carefully broken up by hand held air breakers or non-explosive demolition agents (expansive chemicals), with blasting being prohibited within 10m of the face of the slope.

CONSTRUCTION PROGRESS 1985

On the north east boundary of the site is an old quarry with a steep, sometimes undercut, face. A soft rock trap area was built at the bottom of this quarry and steel mesh draped down the face. Despite the precaution of presplitting by expansive chemicals, during adjacent blasting some quarry face blocks peeled off but did not cause damage. In the same area a Contractor was building concrete bridgeworks and careful regulation of the blasts was need to ensure that the fresh concrete was not damaged.

During the final stages of the excavation the drilling/blasting was closely monitored to minimise sub-formation damage. Overdrilling and non-verticality of drillholes sometimes occurred, causing overbreak and uneven break. As backfilling of overlong holes was ineffective, surface inspection was used to assess any damage below formation.

Generally the bulk excavation was completed by the Railway Opening date of 31st May 1985. Some blasting did continue and trains were stopped either side of the Station, as a precautionary measure; passengers were informed but not cleared from the Station. The blast noise levels with the public present ranged from 110dBA to 117dBA and the associated vibrations were 10-25mm/sec ppv. No complaints were received, yet at these levels of disturbance complaints had been expected.

During the E&M fitting out and building finishes stage in the Station, production blasting caused noise levels up to 128dBA (associated with ppvs of 35-40mm/sec), yet, again, there were no real complaints. It is suggested that because they occurred regularly and people received advance warning, there was a higher degree of tolerance.

CONCLUSION

1. Potential damage and disturbance from large production quarrying can be minimised by proper planning and control during the duration of the works. The public are becoming more unwilling to accept nuisance and disturbance, and it is therefore essential to maintain good relations by informing and explaining to them what is going on, and by showing that real efforts are being made to restrict nuisance and disturbance.

2. It is essential that blast designs are governed by sensible vibration criteria to allow optimum economic excavation production. Air blast noise may be the severest restriction, but was not for the Kornhill site, partly due to the prevailing high ambient noise in the area.

3. For any quarry in an urban area, that uses explosives for excavation, it is important to establish guidelines for blast designs by a proper vibration monitoring programme. Only then can excavation production be safely optimised with minimum disturbance. The level of allowable vibration on all adjacent structures and services should be determined according to the state of the structure and then sensibly factored for blast designs.

4. In the absence of a monitoring programme the USBM typical expression can be used for safe blast designs in the Hong Kong granite. However for optimum blasting production in the Kornhill granite a blast design curve of $V = 790 \left(\frac{R}{W^{\frac{1}{2}}} \right)^{-1.5}$ can be used but with a suitable confidence limit. Table 1 is a comparison of the charge weights per delay allowed for different expressions used on the Kornhill site.

5. The following expressions best fitted the collated results of measured and assumed resultant ppvs from the vibrations recorded in Tai Koo Station:-

The Fault Zone

large scaled distance blasting

$$V = 235 \left(\frac{R}{W^{\frac{1}{2}}} \right)^{-0.87}$$

West of the Fault Zone

large scaled distance blasting

$$V = 300 \left(\frac{R}{W^{\frac{1}{2}}} \right)^{-1.2}$$

small scaled distance blasting

$$V = 735 \left(\frac{R}{W^{\frac{1}{2}}} \right)^{-1.5}$$

East of the Fault Zone

large scaled distance blasting

$$V = 420 \left(\frac{R}{W^{\frac{1}{2}}} \right)^{-1.15}$$

Formation Level	Distance from crown of Tai Koo Cavern	Charge Weight/delay (kg) ppv = 50mm/sec.		
		95% confidence curve derived from USBM expression and recommended by them	Arbitrary planning curve $V = 1753 \left(\frac{R}{W^{\frac{1}{2}}} \right)^{-1.86}$	84% confidence curve obtained from regression $V = 1253 \left(\frac{R}{W^{\frac{1}{2}}} \right)^{-1.5}$
50	44	8.6	39	30
42	36	5.8	26	20
36	30	4.0	18	14
30	24	2.6	12	9
24	18	1.4	7	5

Table 1

6. No definite conclusion could be reached as to whether cube root scaling or square root scaling was more appropriate. Where there was a consistent rock mass (rarely), cube root scaling appeared to be more appropriate.
7. It was observed that for surface monitoring of blasts (at Sai Wan Terrace) the square roof scaling gave better correlation. This confirms experiences recorded by USBM and in Sweden, related to buildings in hard rock.
8. In all the regression analyses the standard error was 0.2 or more, reflecting the variable geology and vagaries in implementing the blasts as designed. More rogue results, determined by the scale factor test, occurred east of the fault zone, possibly due to the subvertical major seams having different transmission characteristics.
9. It had been suggested that the subvertical major seams at the east end of the Station could cause "focussing" of the vibrations. This was not observed. Correlation was similar for this zone, with the regression line remaining flat for large scale distance blasting.
10. It has been suggested that for underground blasting, providing that the square root scaled distance is not less than 23 (in $kg.m^{-\frac{1}{2}}$ units) then buildings will not be damaged. Similarly, with surface blasting, for 'safe' explosions the square root scaled distance should be not less than 32. For the Kornhill site these suggested guides were found to be very conservative. The square root scaled distance 23 corresponds to a point on the 95% confidence line (derived from the standard USBM expression) for a peak particle velocity of 25mm/sec. For a ppv of 50mm/sec. the scaled distance would be 15. For Tai Koo Station within the Kornhill granite rock mass, no damage occurred when the blast designs were monitored against a square root scaled distance of 8 (for a ppv of 50mm/sec.). Because the standard error was never better than 0.2, the 84% confidence line was used for the Kornhill site, and this proved to be a safe limit line.

ACKNOWLEDGEMENT

The Authors wish to thank their colleagues who have assisted in the preparation of the paper and the Hong Kong Mass Transit Railway Corporation for permission to publish it. The authors would especially thank Mr. K.A. Gunaratne for his significant contribution to the vibration monitoring analyses.

The works were carried out by Aoki Corporation. Their co-operation and efforts ensured the success of the project.

The project was assisted in many ways by various Government departments. In particular the Mass Transit Office and the Mines Division made a positive contribution to solving the problems associated with project.

Client: Mass Transit Railway Corporation/ Headstar Ltd. Joint Venture

Employer: Mass Transit Railway Corporation

Consultant: Freeman Fox & Partners (Far East)

Project Management and Construction Supervision: Mass Transit Railway Corporation

Contractor: Aoki Corporation

REFERENCES

Blasters Handbook 16th Edition 1980: Du Pont Company

U.S. Bureau of Mines Bulletin No. 656, 1971: Blasting Vibrations and their Effects on Structures: H.R. Nicholls, C.F. Johnson and W.I. Duvall

Modern Technique of Rock Blasting, 2nd Edition 1973: U. Langefors and B. Kihlstrom: John Wiley & Sons

Damage from explosions - real or imaginary: J.D. McCaughey, In The Structural Engineer, May 1984

Review of Standards for blasting noise and vibration: K. Broadhurst, T. Wilton and

D. Higgins, In article Hong Kong Contractor, June 1984

A study of blast vibrations: I.D. Isaac, C. Bubb, In Tunnels and Tunnelling, July and September 1981; part of Engineering Aspects of Underground Cavern Excavation at Dinorwic

A Study of rock slope reinforcement at Westfield Open Pit and the effect of blasting on prestressed anchors: G.S. Littlejoln, P.J. Norton, M.J. Turner: Proceedings of Conference on Rock Engineering Newcastle, 1977

Dynamic Behaviour of Rock Masses: N.N. Ambraseys and A.J. Hendron Jr., In Rock Mechanics in Engineering Practice edited by Stagg & Zenkiewicz, 1968

U.S. Bureau of Mines Report of Investigations 8507, 1980: Structure response and damage produced by ground vibration from surface mine blasting: D.E. Siskind, M.S. Stagg, W.J. Kopp and C.H. Dowding

Airblast: a technical bulletin from AECI Ltd., June 1981

Code of Practice on Wind Effects Hong Kong - 1983

Rock slope study and stabilization in Pittsburgh, Pennsylvania: a case history

R.C. Speck
Department of Mining and Geological Engineering, University of Alaska-Fairbanks, Fairbanks, Alaska, U.S.A.
R.W. Bruhn,
GAI Consultants, Inc., Pittsburgh, Pennsylvania, U.S.A.

SYNOPSIS

The Duquesne Bluff, located near the center of the city of Pittsburgh, Pennsylvania, is 4240 feet long and 60 to 115 feet high. A 4-lane boulevard runs directly along the crest of the slope and a 6-lane highway parallels its base. In plan, only 50 to 95 feet separate the near curbs of the two roadways. A major rock-fall in April, 1978 ultimately resulted in the inclusion of the slope in the Pennsylvania Department of Transportation's Parkway Safety Update Program.

Failure modes occurring along the length of the slope included plane, wedge, toppling, and rock-fall types. General disintegration of the rock surface was also present. Two sets of tectonically-induced joints, as well as vertically-oriented stress-relief joints, and joints formed by the wet-dry, freeze-thaw weathering cycles were responsible for the various modes of failure.

From a cost effectiveness standpoint, no single type of corrective measure was found to be appropriate for the entire slope. Consequently, the slope was divided into a series of design zones based on a judgment of the potential for particular modes of rock failure, the possible consequences of rock failure, and the most promising methods for correction. Five categories of design zones were thus established. For comparative purposes, several candidate treatment programs were developed in the context of the design zones. The program ultimately adopted included construction of a reinforced concrete band to retard weathering of soft rock strata which historically had led to the development of large overhangs. The program also included rock trimming by mechanical methods (crane-mounted hydraulic hammers and a crane-mounted continuous miner), and the use of rock bolts and wire mesh and other techniques. The entire program is expected to cost 1.5 to 2.5 million US dollars.

INTRODUCTION

Site Description

The central portion of the City of Pittsburgh, Pennsylvania, the "Golden Triangle" lies between two major rivers - the Allegheny River to the north and the Monongahela River to the south. The rivers converge at the point of the Triangle to form the Ohio River and have produced prominent slopes within this important and heavily utilized portion of the city. The subject of this paper is the section of one of the slopes called the Duquesne Bluff, which parallels the north bank of the Monongahela River for a distance of approximately 4240 feet and varies from 60 to 115 feet in height (Figure 1).

Paralleling and immediately adjacent to the crest of the Duquesne Bluff is the four-lane Boulevard of the Allies (Pennsylvania Route 885). Along the base of the bluff is the six-lane Penn-Lincoln Parkway East (Interstate 376/U.S. Route 22-30). The horizontal distance between the toe and crest of the bluff, as measured between curbs of the Boulevard and Parkway East, varies from 50 to 95 feet. The present river channel lies approximately 300 feet to the south. Both the Parkway East and Boulevard are major thoroughfares, the Parkway carrying approximately 70,000 vehicles per day and the Boulevard approximately 35,500 vehicles per day at speeds of 55 and 45 miles per hour, respectively. The Parkway is the most heavily travelled highway in the Greater Pittsburgh area.

The Problem

The south-facing Duquense Bluff, formed by a sequence of nearly flat-lying sedimentary rock units, rises at approximately 70 degrees to the horizontal and represents a natural erosional surface created by the Monongahela River. Through the years, the rock face has been slowly modified by weathering, falls and slides. Occasionally, dislodged rock reaches the Parkway.

The largest rock fall in recent times occurred in April 1978 near the west end of the study section. At this time, several large overhanging blocks of sandstone and shale encompassing several hundred cubic yards of rock parted from the slope, slid downhill, and blocked the west-bound lanes of the Parkway (Figure 2). The Parkway pavement, rock-retention wall, guard rail, and a directional sign sustained damage. Fortunately, the fall took place in early morning before rush hour and injuries to motorists were avoided. Corrective measures to remove the remaining, most precarious rock overhangs in the general vicinity of the rock-fall were implemented in the latter half of 1978. These measures included drilling, controlled blasting and scaling, and construction of a catchment area at the base of the slope.

Minor rock falls have taken place periodically since that time. Authorities of the Pennsylvania Department of Transportation, recognizing that the slope posed a continuing threat both to the integrity of the Parkway East and the Boulevard, and to the safety and property of passersby, concluded that corrective measures were warranted. They therefore authorized GAI Consultants, Inc. to study the problems and design corrective measures as part of the Department's Parkway Safety Update Program.

SITE DESCRIPTION

Regional Setting

The Greater Pittsburgh area lies in the Appalachian Plateau physiographic province, which is characterized by a series of sedimentary rock units that include alternating sequences of shales, claystones, siltstones, sandstones and limestones, as well as occasional

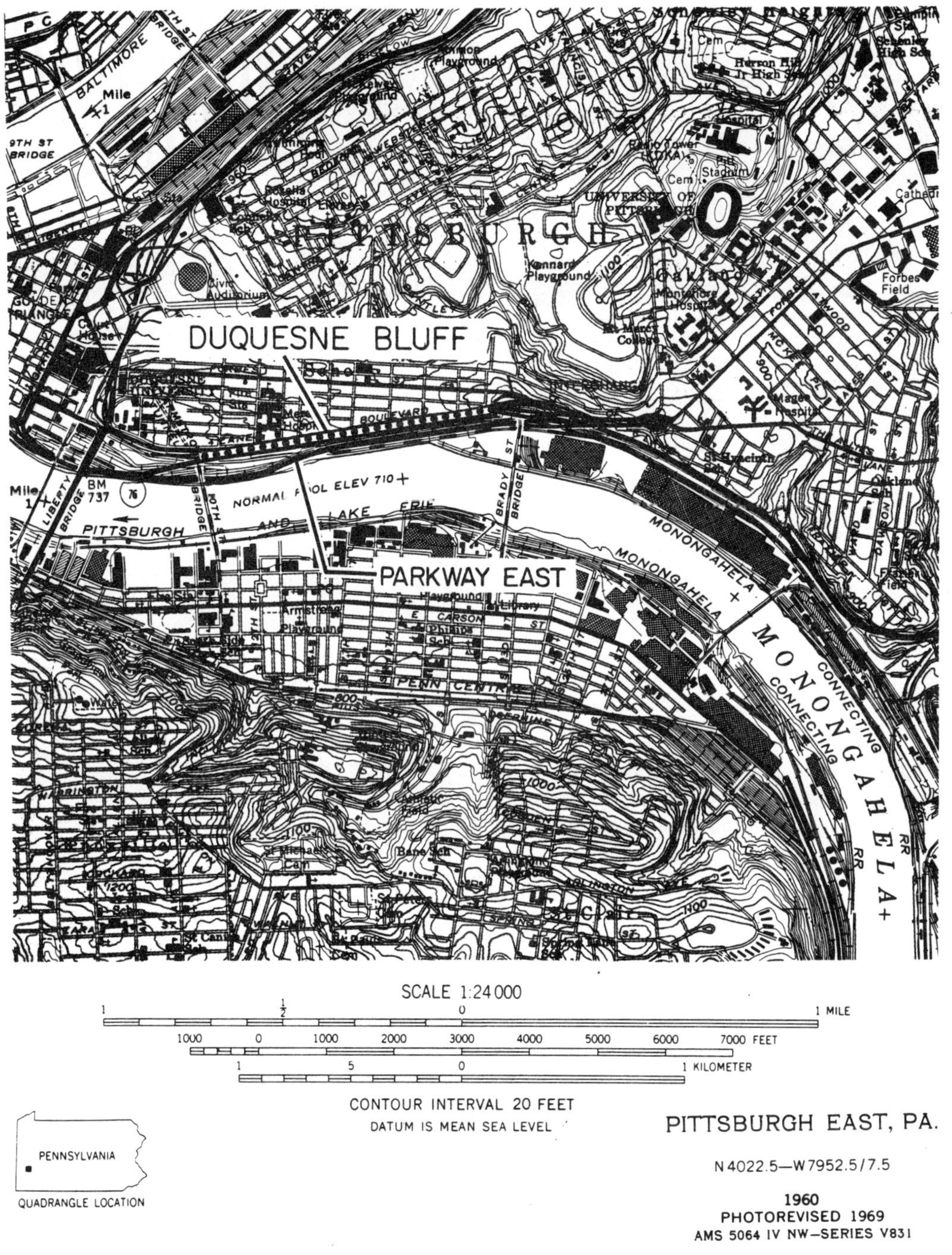

Fig. 1 Site Location of the Duquesne Bluff

coal seams. The slight southerly dip imparted to the strata by the Pittsburgh-Huntington basin is almost imperceptible owing to the great width and low structural relief of the basin.[1] Moderate to very steep hillsides and relatively flat hilltops result from deep erosion in the essentially flat-lying rock beds.

Stratigraphy

Rock strata comprising Duquesne Bluff are members of the Pennsylvanian System Conemaugh Group. The topmost (youngest) unit is the Morgantown sandstone from the middle interval of Casselman Formation. The lowermost (oldest) units are the Pittsburgh Redbeds of the upper

Fig. 2 Clean-up of April, 1978 Rockfall

internal of Glenshaw Formation (Figure 3). The Birmingham unit of the Casselman Formation forms the main section of the slope. The Birmingham unit is composed of six subunits that include siltstone at the top, a variable thickness (0 to 50 feet) of channel sandstones and shales at the base, and a thick section of silty, clayey, and carbonaceous shales in the middle.[2] Most potential rock failures occur within this middle interval. Overlying the Birmingham unit toward the west end of the slope is the poorly-indurated Wellersburg claystone, which is overlain by the Morgantown sandstone. The Wellersburg and Morgantown units are not present in the eastern half of the slope. The Duquesne claystone, the Ames limestone and the Pittsburgh redbeds lie below the Birmingham shale and are covered almost everywhere along the slope by a veneer of fine-grained talus that has accumulated from the shales above.

Rock Structure

Folds

The axis of an unnamed south-plunging anticline passes through the midpoint of the section of slope. The plunge of this fold is less than

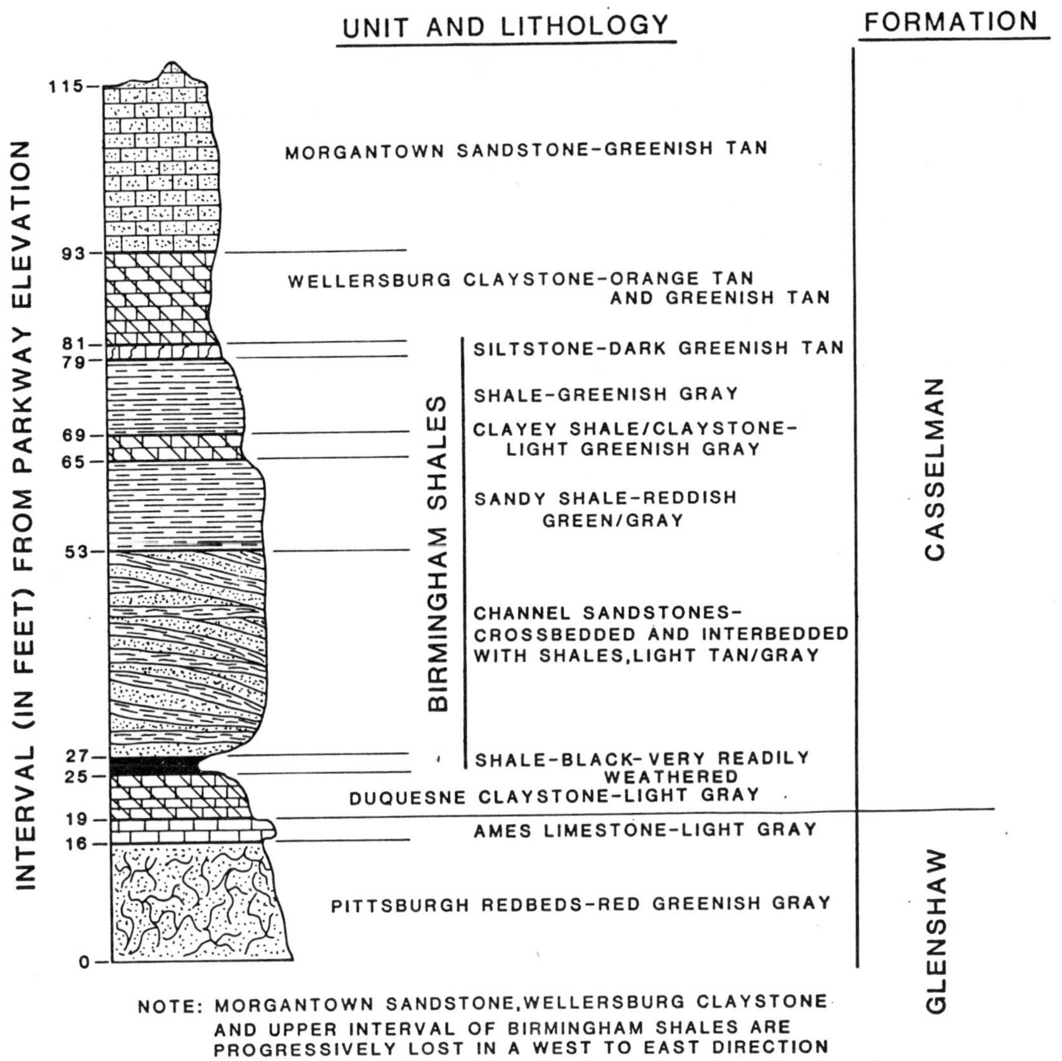

Fig. 3 Generalized Geologic Section of the Duquesne Bluff

one-half degree. Measurements from oblique aerial photographs show the vertical distance between the essentially level Parkway and the Ames limestone varies between 1 foot and 25 feet, confirming the generally mild structural relief of these flexures. Localized and measurable variations that occur in bedding attitudes are gentle dips of one to three degrees that reflect uneven erosion and sediment deposition through geologic time.

Faults

Faulting has not been observed anywhere along the slope although many rock fractures are vertically continuous through several rock units.

Joints

Three types of joints have been identified in the rock slope--(1) tectonically-induced joints, (2) stress relief joints, and (3) joints formed by the wet-dry and freeze-thaw weathering cycles.

Tectonically-induced joints are commonly attributed to lateral compressive deformation of the earth's crust. They are typically systematic planar, near vertical fractures that occur in sets roughly perpendicular to one another.[3]

More than one set is present in the slope. Most of the joints are found to intersect the slope face at angles of 30 to 60 degrees, the four most prominent strike trends being N58°W, N31°W, N24°E and N51°E. Nonsystematic tectonic joints were not measured and plotted inasmuch as they occur on a much smaller scale than the systematic joints and are not continuous or planar.

Stress relief (extension) joints are near vertical rock fractures that have developed parallel or subparallel to the face of the slope in response to downcutting and widening of the Monongahela River Valley by river erosion. The joints generally strike within ten degrees plus or minus of due east, roughly parallel to the slope face. Ferguson (4) indicates that the spacing between successive stress relief joints increases with distance behind the slope face, presumably to some limit beyond which such joints are absent. This limiting distance has not been identified.

Wet-Dry/Freeze-Thaw joints are the result of seasonal weather cycles and are associated with the continuing degradation of the rock face by weathering. These small scale joints are of limited continuity, are often irregular or curved and occur primarily in shales and claystones.

FAILURE MODES

Four primary failure modes are found along Duquesne Bluff--falling, toppling, sliding, and general degradation by weathering. These four failure modes characterize the manner in which the unstable rock first comes free of the rock adjoining it. Once in motion, the rock may continue downslope by rolling, bouncing, sliding, or free-falling, depending on the profile of the slope, the shapes and sizes of the blocks and other factors. Figures 4 and 5 show three locations on the slope where wedge and toppling particular modes of failure might occur. In examining the figures, it will be recognized that both stress relief joints and tectonic systematic joints create potential for failure by toppling or rock falls. Non-systematic joints do not, in themselves, offer potential for major slope failure, although they promote the progressive degradation of the slope and in so doing, the development of rock/debris falls.

Very important in the slope failure process at this site is the fact that differential rates of weathering of the various rock strata have resulted in substantial overhangs along much of the slope. The intersection of the joint surfaces in the rock slope, the continued general degradation of the exposed rock due to weathering, and the continued undercutting as a result of the differential weathering have created a variety of potentially unstable rock blocks and a variety of potential failure modes.

In relation to the overall dimensions of the slope, individual areas of potential failure are relatively small, with one of the larger areas measuring 50 by 30 feet and encompassing perhaps 800 to 900 cubic yards of rock, and one of the more typical areas measuring 15 by 40 feet, encompassing perhaps 200 to 300 cubic yards of rock. Where zones of instability are clustered together, as they are in some

A. STATION 9 + 80. NOTE EMPTY WEDGE MARKING SITE OF PREVIOUS FAILURE.

B. STATION 10 + 30.

Fig. 4 Potential Wedge Failures

intervals of the slope, failure in one location could potentially precipitate failure of adjacent rock, releasing a larger volume of rock than might have been anticipated. In no case is a single failure expected to reach the full height of the slope or extend hundreds of feet along the slope.

CORRECTIVE MEASURES

Choice of Method

In choosing the most appropriate methods of hazard mitigation, the following cost/performance trade-offs had to be borne in mind:

1. Prevention systems are generally the most positive means of reducing the hazard of local rock failures. Initial cost may be high but maintenance is generally minimal;
2. Control systems generally have a lesser initial cost, but require more maintenance, and do not prevent rock failures;
3. Warning systems are generally the least costly of the three options, but

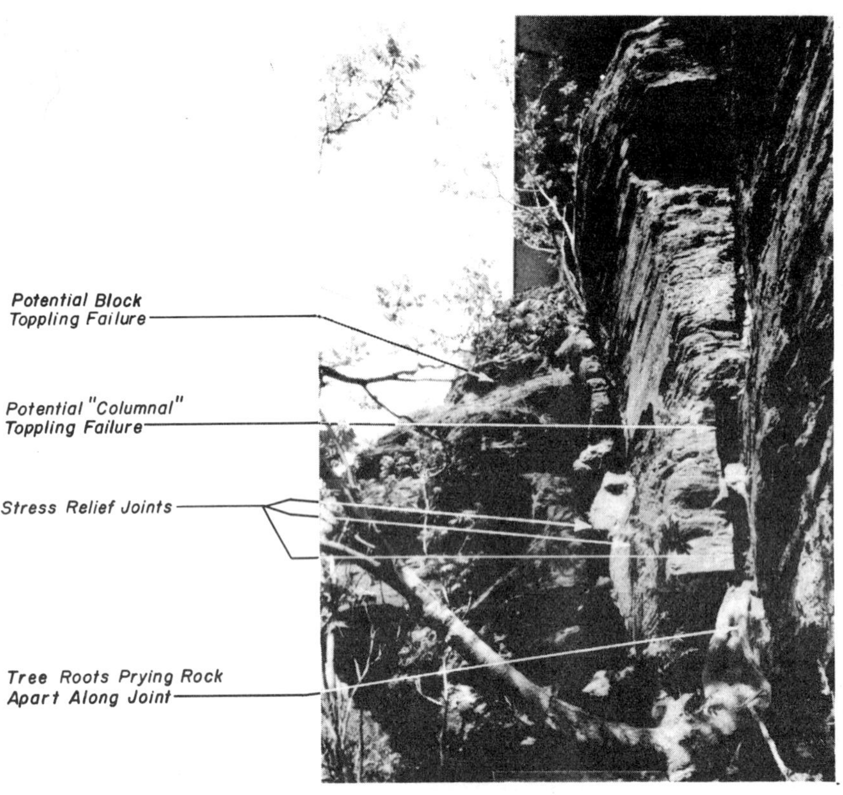

Fig. 5 Potential Toppling of the Reddish Green/Grey Shale of the Birmingham Unit

do not reduce the magnitude or frequency of falls, and do not reduce maintenance costs associated with clean-up. They are of limited usefulness where rock failures are localized and sudden. Electrically operated systems require periodic monitoring and data interpretation and are prone to false alarms and other malfunctions.

Slope characteristics also had to be considered in choosing the most appropriate methods of hazard reduction:

1. The slope is in a state of dynamic equilibrium. The stability of individual blocks of rock is continually changing, albeit slowly. Without corrective action, local rock failures can be expected periodically and at indefinite frequency throughout the life of the Parkway and the Boulevard. Achievement of a final, stable, hazard-free slope cannot be expected from natural processes;
2. Due to the broken and sometimes weathered nature of the rock, rock movements could take place at other than the expected locations.
3. In many locations jointing may allow rock failure by any of several modes,

and not in a manner amenable to rigorous stability analyses;

4. Most failures will probably be sudden with little or no forewarning;
5. Attention must be directed toward unstable blocks of all sizes. Whereas a large block could precipitate a clean-up operation more costly than a smaller block, a small rock bounding into the Parkway could create a safety hazard virtually as great as a large mass. The high traffic density of the Parkway increases the probability of a fallen rock creating a safety hazard;
6. The overall stability of a slope could probably be enhanced by cutting it back to a more stable configuration (reducing the slope inclination and creating benches to intercept loose rock). However, due to the location of the Boulevard directly at the crest of the slope and the need to maintain it in that location, there is no opportunity to modify the slope profile except by local trimming. It was necessary, therefore, to maintain the slope in essentially its existing configuration.

In view of the foregoing considerations, it was concluded from a cost-effectiveness standpoint that no single type of corrective measure would be appropriate for the entire slope. Consequently, the slope was subdivided into a series of intervals or design zones based on a judgment of:

1. The potential for particular modes of rock failures;
2. The possible consequences of rock failure; and
3. The most promising methods for correction.

Design zones

Five categories of design zones were established and were denoted by numbers 0 to 4--the higher numbers corresponding, in a broad manner, to a potentially greater consequence of failure and to the necessity for more elaborate methods of correction. The characteristics of these zones are summarized in Table 1, along with the station intervals of the slope assigned to each design zone. A total of 12 such intervals were defined -- one interval of Zone 0, four of Zone 1, two of Zone 2, two of Zone 3, and three of Zone 4:

- Design Zone 0 -- Degradation of the rock face by weathering is expected. Minor rock failure and deposition of some rock debris on Parkway East is possible with little or no potential for attendant damage to Parkway structures.
- Design Zone 1 -- Potential exists for all modes of rock failure. Some rock might reach the Parkway and cause minor damage to Parkway structures.
- Design Zone 2 -- Rock/debris fall and toppling represent the most probable mode of failure. The potential consequences of rock failure are expected to be the same as for Design Zone 1.
- Design Zone 3 -- Rock/debris fall represents the most probable mode of failure. Rock falls could potentially

Table 1. Prominant Failure Modes, Most Probable Consequences of Failure, and Station Intervals in Each Design Zone

	Prominent Failure Modes					Most Probable Consequences of Rock Failure			Design Zones	
Design Zone	General Degradation	Toppling Failure	Rock/ Debris Fall	Plane Failure	Wedge Failure	Rock Debris[a] Resulting From Slope Failure	Damage to[b] Parkway Structures	Recession[c] of Rock-Slope Surface	Station Intervals (feet)	Aggregate Length (feet)
0	X					1	1	1	36+50 to Brady Street Interchange	500
1	X	X	X	X	X	1	1,2	1	1+00 to 5+00 12+00 to 15+00 23+00 to 25+00 31+50 to 36+50	1,400
2	X	X	X			1	1,2	1	Armstrong Tunnel to 1+00 5+00 to 7+50	440
3	X		X			1,2	1,2,3	1,2	15+00 to 19+00 25+00 to 30+00	900
4	X	X	X	X	X	1,2	1,2,3	1,2,3	7+50 to 12+00 19+00 to 23+00 30+00 to 31+50	1,000

Note: Due to the nature of the rock slope, there is some possibility for rock failure in all design zones.

a. Potential for:
   1. Some debris on parkway;
   2. All westbound lanes of parkway closed.
b. Potential:
   1. None;
   2. Minor;
   3. Major.
c. Potential:
   1. Minor;
   2. Major;
   3. Possible Involvement of Boulevard of the Allies.

Note: Any failure that involves movement of rock debris onto the parkway could potentially result in injury to motorists.

block all westbound lanes of the Parkway and cause major damage to Parkway structures.

- Design Zone 4 -- Potential exists for all modes of rock failure. Failures could be major. The potential consequences of failure are the same as described for Design Zone 3, with possible involvement of the Boulevard of the Allies as well. Immediate slope stabilization measures might be required in the event of a failure.

Corrective measures as designed

As part of the initial phase of the project, potential treatments were identified for the various Design Zones (Table 2). These and other remedial measures (including tie-back retaining walls positioned close to the slope) were recommended to stabilize the slope and retard general rock degradation. The estimated cost for a program incorporating such stabilization measures ranged from 4 to 10 million US dollars. Due to the restricted construction funds available to the Pennsylvania Department of Transportation for this project, less extensive measures (discussed below) were adopted. These measures are expected to significantly contribute to the stability of the bluff and the safety of passersby. At the same time, the less extensive measures may not be as effective as the originally proposed measures nor will they likely to totally halt slope degradation or necessary eliminate the need for additional corrective action in the future. The stabilization program adopted is expected to cost 1.5 to 2.5 million US dollars.

Final measures incorporated into the stabilization program include:

1. trimming overhanging or potentially unstable blocks of rock;
2. placement of wire mesh over trimmed rock surfaces exhibiting closely spaced joints and other fractures;
3. construction of a reinforced concrete band to cover the soft, rapidly degrading rock strata where severe undercuts develop.
4. placement of dental concrete in small, local undercuts.
5. excavation of a drop zone along the base of the slope to collect small debris.
6. construction of gabion barriers behind drain inlets along the base of the slope; and
7. construction of a rock protection fence along the top of an existing concrete wall at the base of the slope.

Particulars concerning these design features are as follows:

Rock trimming

Areas of rock that were not firmly keyed into the slope or protrude from the slope were scaled with a blasting mat suspended from a crane or trimmed using a crane-mounted hydraulic ram. No blasting was employed. The undercut into which the concrete band was constructed was dressed with a crane mounted continuous miner (Figure 6). Laborers dressed localized areas using hand tools.

Isolated or individual blocks of sandstone and shale that required trimming were identified

Table 2. Potential Corrective Measures Recommended for eadh Design Zone

Design Zone	Scaling or Trimming of Loose/ Joint-Bounded Rock	Bolted Mesh Blanket or Shotcrete on Bolted Wire-Mesh	Rock Bolts	Rock Anchors	Buttress	Tie-Back Retaining Wall (approximate height in feet)			Drainage+	Removal of Trees and Bushes
						30 feet	50 feet	60 feet		
0									X	X
1	X								X	X
2	X	X	X	X					X	X
3	X			X*	X				X	X
4	X			X*		30+00 to 31+50	19+00 to 23+00	7+50 to 12+50	X	X

Note: In some intervals of design Zone 4, a gravity retaining wall is a possible alternative to a tie-back retaining wall.

*Used only in conjunction with the tie-back retaining wall or buttress.

+Ditches should be cleaned out and the drainage pattern should be reestablished.

△These measures are more extensive than those adopted.

at 73 locations on the slope, with the typical location measuring 10 to 25 feet square. Altogether, block trimming was expected to involve about 12,000 tons of rock excavation.

Fig. 6 Crane-mounted Milling Machine Used to Dress the Slope Surface (Reference Dick Corporation)

Extensive overhangs and large blocks that could not be removed in their entirety were to be trimmed back. Crane-mounted hydraulic rams were employed exclusively for this purpose. (Figure 7). Smooth wall blasting was permitted but not employed owing to obvious problems associated with blasting in an urban area. Five overhang large block areas were delineated, ranging in length from 150 to 530 feet and were in height from 10 to 35 feet. Together, these areas encompassed approximately 4000 square yards of slope face.

Wire mesh

Galvanized wire mesh fabricated from No. 11, Grade 60 steel wire and having 3-inch by 4-inch hexagonal openings was installed in overlapping vertical strips over highly jointed section of the rock face that were trimmed by smooth-wall blasting. The strips were joined to one another

Fig. 7 Crane-mounted Ram Removing Loose Blocks and Trimming Overhangs

by steel hog rings and by steel tie wire threaded through the overlaps. The strips were to be supported on a grid of horizontal and vertical aircraft cable spaced on 25-foot centers that were fastened to the slope by 5-foot long steel dowels on 5-foot centers. (Figure 8). Approximately 1200 square yards of wire mesh was required for the section of the slope where extensive rock jointing was observed.

Concrete band

To prevent continued severe weathering and undercutting, an 8-inch thick, 7- to 12-foot high band of concrete was constructed over a 1500-foot long exposure of the carbonaceous shale at the base of the Birmingham shale unit (Figure 9). Vertically, the band was extended from the Duquesne claystone directly below the carbonaceous shale to the Birmingham shale directly above and was constructed of PennDOT Class AA concrete or pumped mortar reinforced

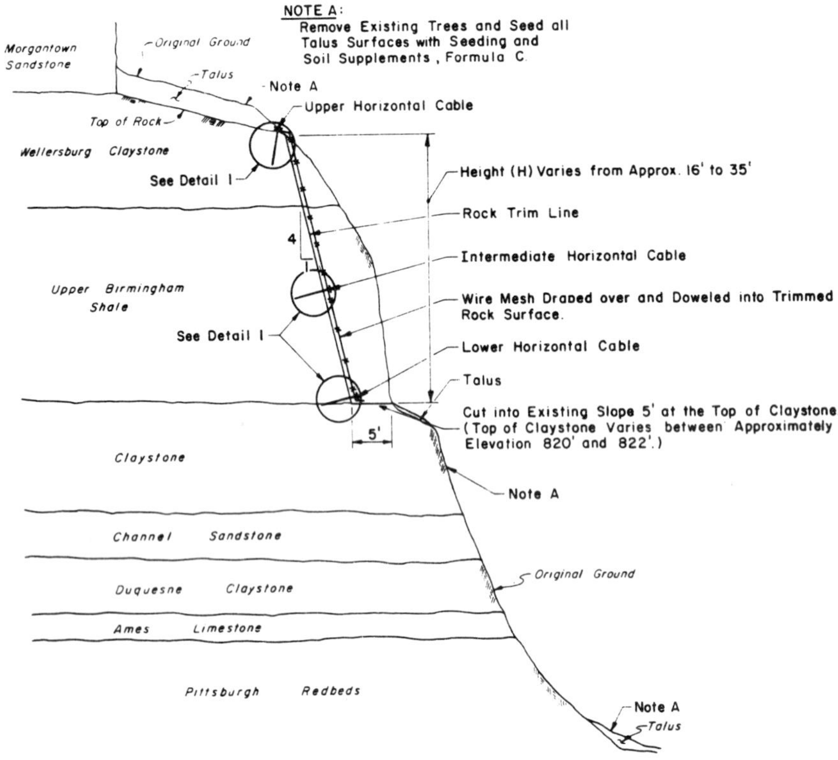

Fig. 8 Wire Mesh as Installed Over the Highly Jointed Trimmed Rock Face

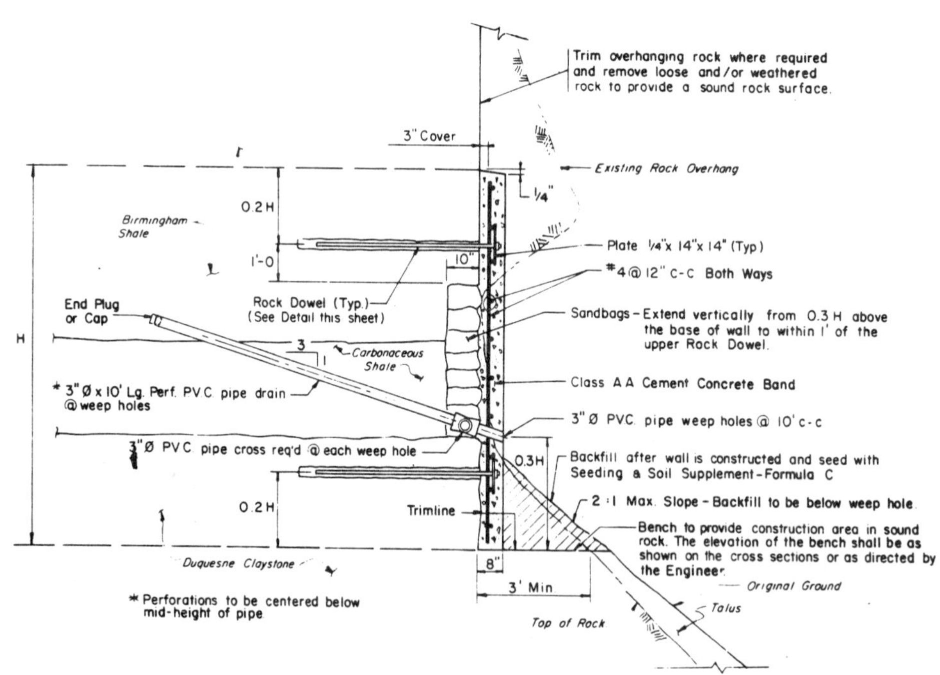

Fig. 9 Concrete Band as Installed Along the Carbonaceous Shale Horizon

with no. 4 steel bars on 12-inch centers each way. The band was secured to the slope by two rows of 5-foot long, 0.75-inch diameter Dywidag Grade 60 threadbars on 10-foot centers, which were anchored full depth by resin. Drainage of strata behind the concrete band was accomplished by a stack of sand bags and by 10-foot long, 3-inch diameter PVC pipe drains on 10-foot centers (Figure 10).

Dental concrete.

Small quantities of dental concrete were used where necessary to fill local undercuts, which, if left untreated, would produce dangerous overhangs. Dental concrete was envisioned for use in undercuts beneath exposed concrete piers of the Boulevard of the Allies, among other places. The dental concrete was PennDOT Class AA and was secured to the slope where necessary by steel rock dowels of the type used for the concrete band (Figure 11).

Drop zone.

A nominal 8-foot open area was created along the toe of the slope to collect debris shed by rock strata uphill. Construction of the 3300-foot long drop zone was accomplished primarily by mechanical methods (with light blasting) and involved approximately 2600 cubic yards of excavation in talus and soft rock.

Gabion barriers.

PennDOT Type B corrosion resistant gabions--rectangular, compartmental, wire baskets made of heavily galvanized, PVC-coated steel wire mesh and filled with Type C rock or with sandstone blasted from the slope--were used to prevent talus and other debris from covering drainage inlets.

Protection fence.

To prevent small rocks from bounding over the existing concrete retaining wall along the base

Fig. 10 Concrete Band Construction in Progress

Fig. 11 Excavating Indentations for Dental Concrete

of the eastmost quarter of the slope, a 3.5-foot high galvanized chain link fence was constructed atop the retaining wall. The installation involved approximately 1200 lineal feet of fence.

In addition to the foregoing hazard reduction measures, post-tensioned steel rock bolts were utilized on an as-needed basis. The bolts were 5-to 20-foot long No. 9 and lengths were determined in the field and installed at the discretion of the Engineer.

Construction Schedule

The rock slope stabilization work began April 1, 1985 and was completed October 31, 1985. During the period of construction, the three west-bound lanes of the Parkway were closed to traffic. The Boulevard of the Allies remained open.

CONCLUSION

Due to the large dimensions of the rock slope, its position between two existing closely spaced, high volume highways, the variety of rock-block sizes and failure modes, the differing consequences of failure from point to point, and economic constraints, no single design strategy was appropriate for complete stabilization of the slope. Division of the slope into distinct Design Zones based upon the consequences of failure and the need for varying degrées of design effort, allowed the formulation of design schemes appropriate for each discrete area.

The stabilization program implemented is expected to significantly enhance the stability of Duquesne Bluff.

REFERENCES

1. Wagner W. R. et. al. Geology of the Pittsburgh Area. General Geology Report G59. Pennsylvania Geological Survey, 1979, 145 pp.

2. Price M. L. The Stratigraphy and Environments of Deposition of the Birmingham Shale (Upper Pennsylvania) of the Pittsburgh Area. unpublished Masters thesis, 1970, 103pp.

3. Nicholson R. P. and Hough, V.D. Jointing in the Appalachien Plateau of Pennsylvania. Geological Survey of America Bulletin, Vol. 78, 1967, p. 609-630.

4. Ferguson H. F. Valley Stress Release in the Allegheny Plateau. Bulletin of the Association of Engineering Geologists, Vol. 4, No. 1, 1967, p. 63-71.

Interaction of underground mining and surface development in a central city environment

T.R. Stacey D.SC.(ENG.), D.I.C., M.I.M.M., M.S.A.I.M.M.
Steffen, Robertson and Kirsten, Consulting Engineers, Johannesburg, South Africa

SYNOPSIS

The City of Johannesburg is a large, modern, developing city that is probably unique in having extensive mining very close to its centre. Development of the shallowly undermined land is restricted by the authorities, but these restrictions may be relaxed if satisfactory stabilisation measures are implemented. Several case studies of alternative methods of stabilisation are described in this paper. Finally, a case study of subsidence associated with undermining is described to illustrate situations in which considerable caution is required if surface development is to proceed.

INTRODUCTION

The City of Johannesburg is probably unique in the world in that it is a large, modern, developing city with extensive undermined areas in its central district. These gold mining activities include both old defunct operations, as well as active small scale extraction and, in the future, may also involve re-establishment of deep level operations within close proximity of the city centre.

The gold deposits occur in thin conglomerate reefs within the bedded Witwatersrand quartzites. These strata strike in the east-west axis of the city, with dip at outcrop varying from near vertical to about 25 degrees. The quartzites vary from massive to heavily jointed, and are usually weathered to a depth of some 50 m. Though weathered the quartzites are generally very competent, even at surface. There are numerous diabase intrusions in the quartzites, often associated with faulting. The diabase is usually deeply weathered to a clay, often to depths exceeding 25 m.

With regard to mining there are three significant reefs in the Central Johannesburg area. The separation between them varies from zero to about 15 m. They have been mined as distinct entities or, as in many areas where two of the reefs are very close together, two of the three reefs have been mined together as a composite reef. This mining has left extensive narrow tabular open stopes, supported with occasional solid reef pillars, timber packs and waste rock packs. Typical stoping widths are 1.5 m. Mining began in the late nineteenth century and, in many areas, was in the form of outcrop mining. Owing to the geometry, however, the methods changed quickly and shafts were sunk for ore hoisting and underground access. The common geometry of mining layouts in the shallow zone displays strike drives established within the reef horizons, the first level usually being at a depth of 30 to 40 m, and subsequent levels at about 30 m depth spacings. Small solid reef pillars were usually left just below these levels to provide support.

In the early years of mining poor records were kept and available mine plans are often inaccurate. In addition, when areas were due to be closed, unrecorded pillar robbing or extraction often took place and hence plans may indicate non-existent support. This presents an additional complication when considering the stability of the overlying ground.

In more recent years reclamation mining has taken place. This consists of re-entering old accessible stopes and removing the sweepings (loose material on the footwall surface which is often high in gold content). Some solid areas of reef, now payable, are also removed. Reclaimed areas are usually supported with "stone walls" or waste rock cribs.

Mining in the Central Witwatersrand area was not halted because the gold reef was exhausted, but owing to flooding, consequent pumping costs rendering the operations uneconomic. It is likely that new deep level operations, protected from the water in the shallow workings by a barrier pillar, will be established in the future.

Restrictions on surface development

A set of guidelines, identifying the restrictions to be imposed on surface development, was drawn up by the governmental authorities in 1965. This followed several previous documents, much less comprehensive, dating back to the end of the 19th Century. In summary, the guidelines in force at present allow no development where undermining is shallower than 90 m, with progressive relaxation to a depth of 240 m, whereafter no restrictions are imposed. The guidelines are flexible, and the extent of mining, the number and separation of reefs mined, the stoping widths, the type and amount of artificial support and the dip of the reefs, are taken into account in determining the actual restrictions to be imposed.

There is no clear indication of the basis for the guidelines, but it is probable that they are simply based on records of subsidence. These indicated (Hill, 1981) that very few subsidences occurred where the mining depth was greater than 240 m.

Mechanisms of subsidence

Four types of subsidence have occurred as illustrated in Fig 1 :

i) sinkholes at the outcrops often associated with collapse of the immediate hangingwall. These normally only occur very close to the outcrop location.

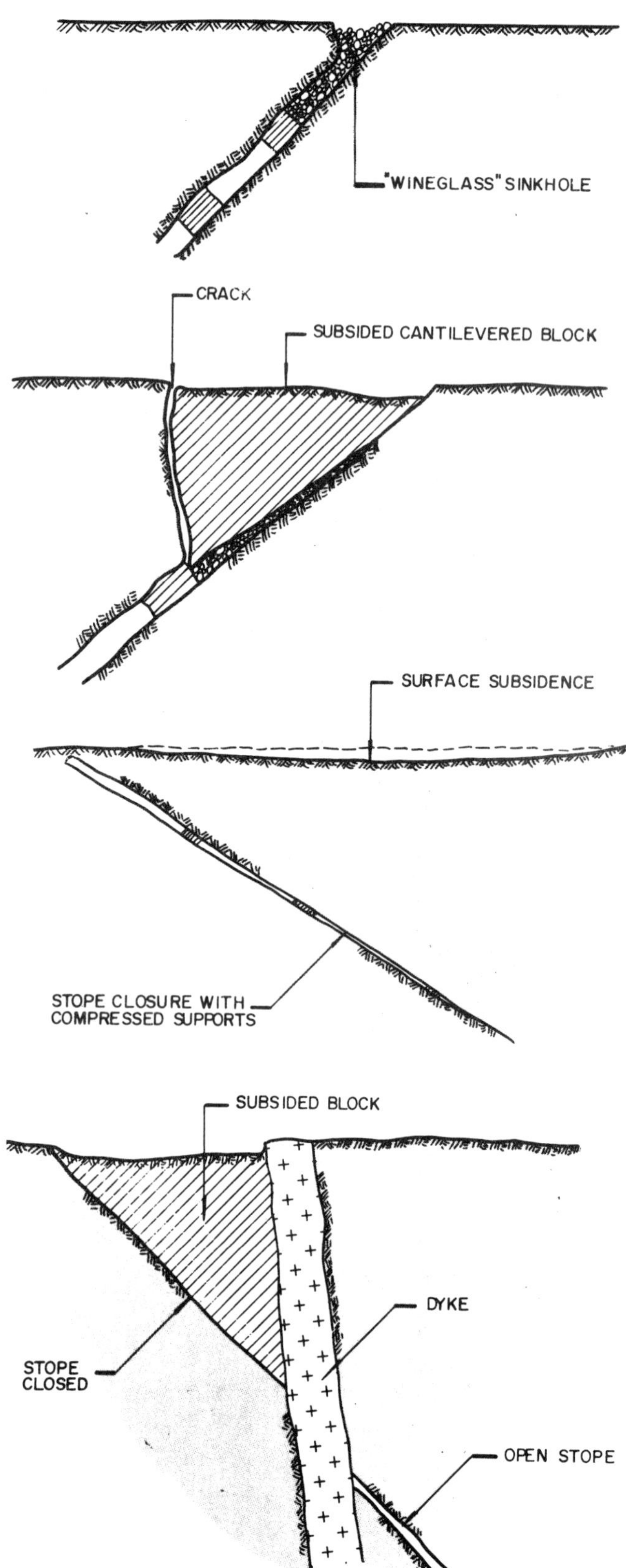

Fig 1 : Mechanics of Subsidence over Shallow Gold Mine Workings

ii) cantilevering of the hangingwall, resulting in cracks on surface. This may be associated with step subsidence across the cracks.

iii) a subsidence trough resulting from deformation of the hangingwall.

iv) block subsidence controlled by faults and dyke contacts.

The mechanisms identified in (i) and (ii) above can be prevented or remedied by suitable stabilisation measures. Case studies are described in the following sections illustrating both localised remedial works and very substantial measures to prevent types (i) and (ii) subsidence and permit development (Stacey, 1983). A further case study is then presented illustrating block subsidence on a faulted dyke contact.

The form of the surface subsidence profile of type (iii) was identified by means of finite and displacement discontinuity element stress analyses of tabular stopes dipping at 80°, 50° and 20°. These analyses simulated typical mining layouts, in which small solid pillars are left to provide support at each mining level. Analyses were both linear and non-linear, modelling partial collapse of the pillars and stope hangingwall, which might occur after a long period of time as a result of decay of support. The results of these analyses are shown in Fig 2. They illustrate the levels of horizontal strain and tilt which must be considered for structural design.

The "oscillations" which can be clearly seen in the results for the stope dipping at 20° are due to the presence of the pillars. They show a minor effect for this stope, but no effect is evident for the more steeply dipping stopes.

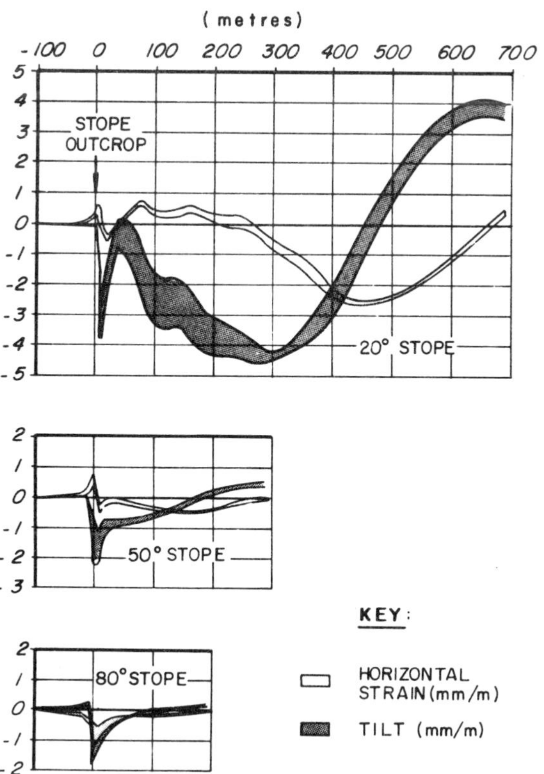

Fig 2 : Surface Deformation over Open Stopes

SHALLOW OUTCROP STABILISATION

Much of the area in Johannesburg where the outcrops occur is zoned for commercial or light industrial development. As such it is suitable for the erection of low rise, relatively low cost, warehouses and factories, with limited office accommodation. Such development cannot justify great expense on stabilisation measures, and the concept of the remedies is to provide local support which will prevent collapses. The design of flexible structures then will accommodate any subsequent minor surface deformation which may occur. The aim of the shallow stabilisation methods is to produce a competent surface zone to promote arching across the stopes. This can be achieved in several ways, two of which are described in the following sections.

Dynamic compaction

In this example, mining had taken place on three reef horizons as shown in Fig 3.

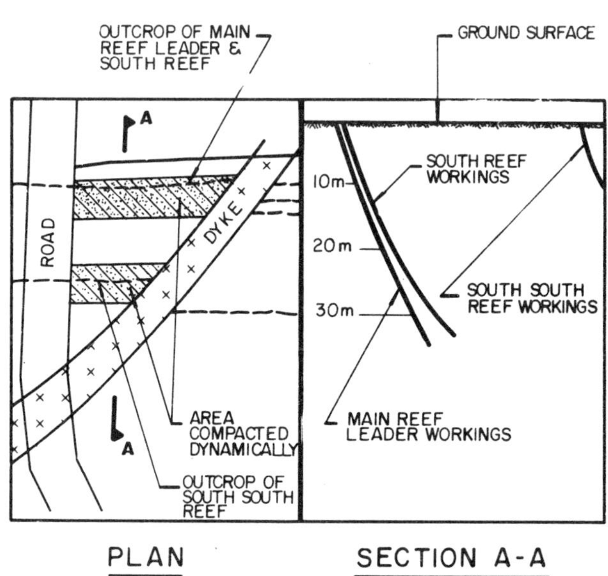

Fig 3 : Shallow Undermining Plan and Section

The condition of the ground over the reefs was inspected by excavating trenches with a backhoe across the strike of the outcrops. The ground consisted of very loose sandy gravel and imported ash fill. Residual quartzite occurred at depth of approximately 2.5 m. The mine workings themselves were found to have been filled with loose rockfill and sand, probably tipped into the open workings from surface. It was believed that dynamic compaction would improve the characteristics of the soil down to bedrock and would also close any cavities which existed near the ground surface as a result of the mine workings. The dynamic compaction would in addition form an in situ ground arch spanning over the reef outcrops.

The dynamic compaction was carried out by the landowner, dropping a 10 ton mass from a height of 18 m. The mass consisted of a concrete block encased in steel, a suitable projection being welded to the bottom to keep the mass vertical as it hit the ground. Backlash from the boom on release of the mass was rectified by stabilising the boom.

The settlement of the ground surface which resulted from the dynamic compaction was as much as 1 m near the reef outcrop, with an average value of nearly 0.5 m being achieved. The effect of the compaction was inspected by trenching through the compacted ground and was found to be satisfactory. Several large diameter auger holes were drilled to a maximum depth of 14 m to inspect the hangingwall strata as well as the effects of the compaction. No cavities were found, although no obvious effects of the dynamic compaction could be seen in the hangingwall quartzite below the immediate surface zone.

It was concluded that the dynamic compaction had induced settlement of the ground well beyond the settlement that would have resulted from the proposed buildings. These stabilisation works satisfied the authorities and permission was granted for the erection of structures spanning across the reef outcrops.

Outcrop plugging

A far more common method of providing the necessary local support is to plug the stopes, and then place compacted backfill above this plug to provide the stabilising ground arch. This approach has the added advantage that stopes are exposed for inspection during the construction phase. This ensures that construction can be varied with great flexibility to suit local variations in the stope conditions. The procedure is to excavate along the strike of the outcrop until the narrow throat of the stope is exposed with competent hangingwall and footwall surfaces. The plug, usually concrete, is then formed in this throat. The dimensions of the plug must be such as to span stably across the stope throat and to provide a suitable base for the backfill.

An example of three adjacent stopes, remedied in this way, is shown in Fig 4. In this case reinforced concrete beams, from a demolition site, were placed as "plums" in the concrete plug to reduce the quantity of concrete required.

Fig 4 : Shallow Outcrop Stabilisation by Concrete Plug

The use of a concrete plug also fulfills another very important function, namely to seal the stope against ingress of water. Many sinkholes and subsidences, which have occurred adjacent to stope outcrops, have been initiated by inflow of water from surface, causing gradual ravelling of backfill from the stopes.

DEEP OUTCROP STABILISATION

For major developments a different approach must be adopted, since the consequences of any subsidence require that the risk be minimised. Owing to the higher cost of the development, a greater expense on remedial works can easily be accommodated and can usually be offset against the lower cost of the "restricted" land. The shortage of land suitable for development close to the central business district of Johannesburg has resulted in these "restricted" areas becoming very attractive. The following case study deals with a development, completed in 1983, and involving an investment at the time of the order of 50 million dollars (Stacey et al, 1983).

The requirement for the Standard Bank Administration Centre was for a very extensive floor areas of some 10 000 m^2 on each level. An area of "restricted" land, which included the reef outcrops, was available to meet these requirements. For the restrictions on development to be lifted it was essential to design and implement stabilisation measures to the satisfaction of the statutory authorities.

Mining had taken place on two reefs dipping at approximately 80° and separated by 15 m. A thorough investigation was carried out, involving considerable drilling, as well as re-opening of the old workings to gain access to a depth of nearly 50 m below surface (Stacey et al, 1983). The purpose of this was :

- to examine the condition of the stopes and their state of filling.
- to assess the competence, degree of weathering and disturbance of the rocks.
- to investigate major gouge-filled joints identified by the drilling.
- to evaluate the possibilities of construction of in-stope stabilisation measures.

It was concluded from the investigation that satisfactory stabilisation measures could be designed to provide security against subsidence. As for the shallow outcrop stabilisation, the concept was to provide rigid support between footwall and hangingwall surfaces, such that in the event of collapse or instability at greater depth, the near-surface rigid zone would form an arch. In-stope support comprising formed pillars was contemplated, and several alternatives were considered :

- concrete dip pillars
- concrete strike pillars
- concrete distributed "blob" pillars
- distributed "blob" pillars formed by grouting of backfill

The last was rejected owing to doubt regarding the likely effectiveness of grouting and the problem of proving the solution. Concrete "blob" pillars were rejected owing to difficulty of access for construction. Of the other two alternatives, dip pillars were chosen owing to their greater ability to withstand shear deformations in a vertical plane normal to the reefs. Design calculations were carried out to check on the stability produced by the remedial method. With the stabilising pillars in place, two-dimensional stress analyses were used to simulate :

- collapse of mine workings below the base of the dip pillars
- the effect of potential failure surfaces formed in the footwall, on a major joint and on the dyke contact.

These two situations are illustrated diagrammatically in Fig 5. The results of the analyses showed that the pillars will adequately support the ground surface as a foundation for the

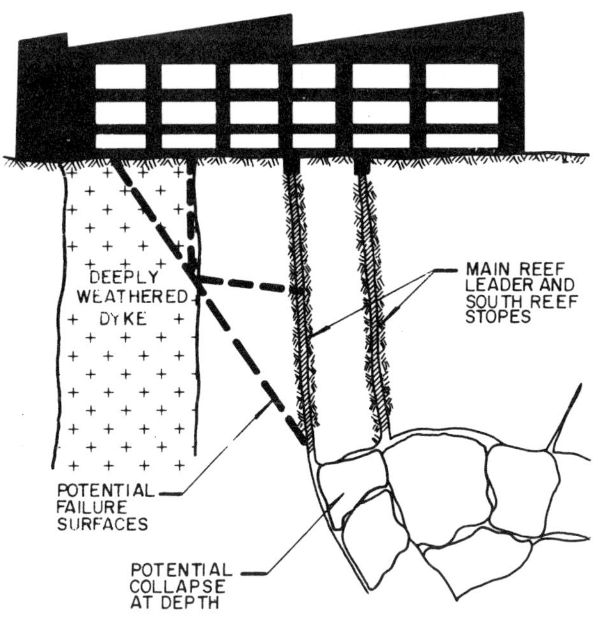

Fig 5 : Deep Outcrop Stabilisation

development, even under the most adverse conditions. The results were quantified as follows :

- a maximum horizontal surface movement as a result of the dyke/joint/ stope interaction of 5 mm.
- a maximum differential surface settlement resulting from underground collapse of 1 mm.

The design of the building was such that movement of this order could easily be accommodated by construction joints.

The second phase of the design calculations was the sizing of the concrete dip pillars. Three-dimensional mining simulation stress analyses (Diering, 1980) were used, and the stresses which could potentially develop in a series of equally spaced dip pillars satisfying practical spacing and construction requirements were calculated. It was found that, with this geometry, the lower sections of the pillars could be overstressed. The layout was therefore modified to include a strike pillar connecting the bases of the dip pillars, thus increasing the area of concrete and reducing the stresses at the base level. Further to this, it was possible to reduce the depth of alternate pillars without resulting in overstress of the concrete. The base strike pillar also had the advantage of ensuring the retention of fill in the stopes, which was required to prevent slabbing from hangingwall and footwall surfaces between the pillars.

The construction method involved the sinking of winzes in the stopes and backfilling them with mass concrete. The mine workings were conveniently open along strike at the second mining level (approximately 60 m depth), easing the construction of the strike pillar. A concrete crown pillar in each of the stopes was finally constructed before commencement of building operations across the stope outcrops.

SUBSIDENCE ON A FAULTED DYKE CONTACT

Numerous cases of subsidence associated with diabase dykes have occurred (Brink, 1979). The following case study briefly summarises such a subsidence situation (Stacey and Rauch, 1981) unusual owing to the depth of mining - significant subsidence has occurred where the depth exceeds the "no restriction" limit of 240 m.

The first indication of subsidence was a report by a locomotive driver of an irregularity on the rail tracks. A programme of levelling to monitor the subsidence was implemented immediately. The results of this programme, continued over a period of many months, are shown in Figs 6 and 7.

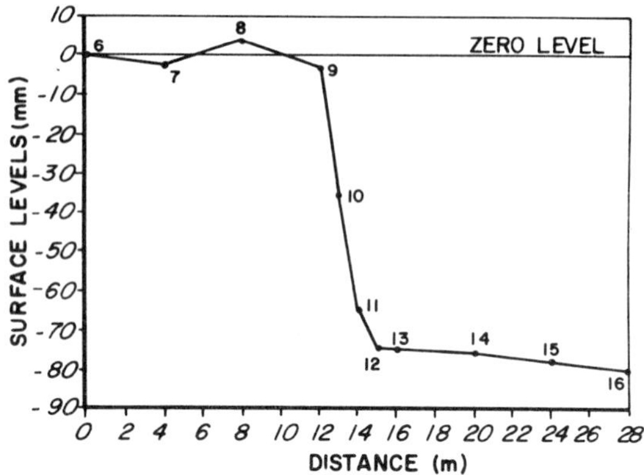

Fig 6 : Surface levelling Profile

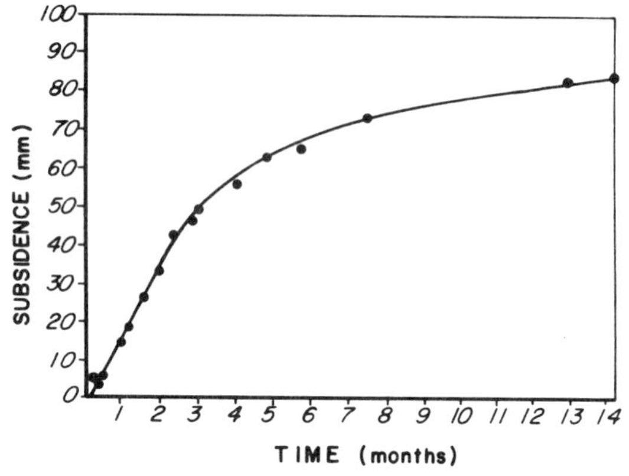

Fig 7 : Subsidence with Time

The location of the sharp step in the subsidence profiles corresponds with the location of the dyke contact which was exposed during a surface investigation of the area. The area affected includes a six lane freeway, a major secondary road, four main railway lines and a main water pipeline, as shown in Fig 8.

Three reefs had been mined, the shallowest being at a depth of 200 m below the railway. The reefs dipping at approximately 30° had been mined with stoping heights of 0.85 m, 0.6 m and 0.75 m, the stopes separated by middlings of 4.5 m and 3 m. Full scale mining

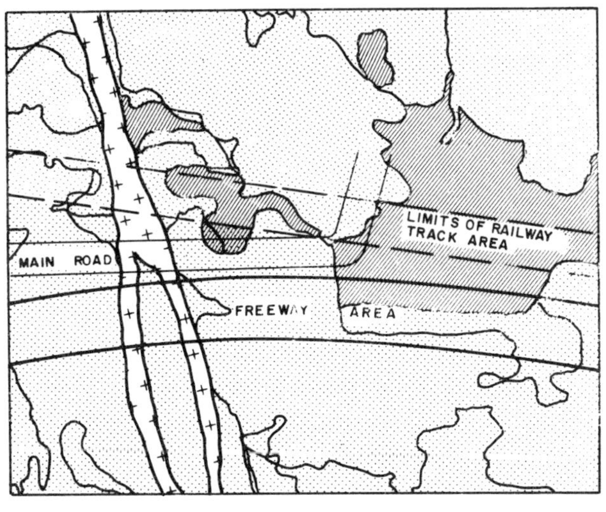

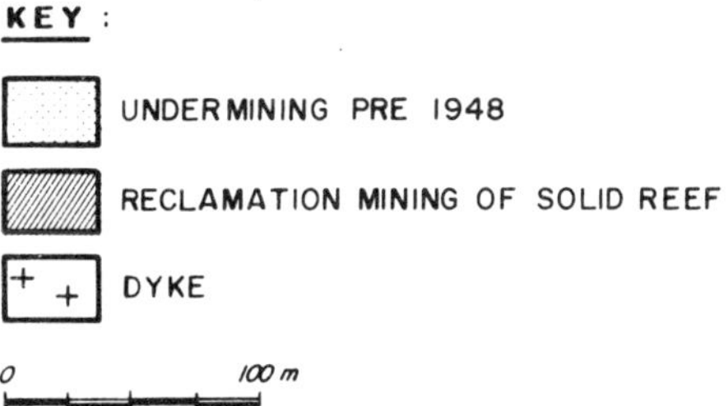

Fig 8 : Plan of Subsidence Area

operations ceased in 1948, but reclamation mining has recently taken place. An inspection of the dyke contact below the site was carried out in the two lower stopes. This showed a very clear contact with a small amount of gouge material on the surface. There was no evidence of significant movement and the artificial support, in the form of waste rock packs, was clearly not under any substantial load. No reclamation work was carried out adjacent to the dyke in the shallowest stope since a fairly extensive local collapse had rendered access too difficult.

As a result of the depth of the mining and the lack of visible evidence of movement in the lower two stopes, it was suggested that a possible explanation for the subsidence was that, as a result of collapse and movement in the shallowest stope, one or several cavities may have formed at a higher level within the strata. To investigate the hangingwall strata a diamond cored hole was drilled down dip over the full depth. Numerous fractured zones occurred, associated with drilling water losses. However, no significant cavities were evident above a depth of 210 m. Below this depth two large cavities, respectively 0.35 and 1 m in extent, were encountered. The latter cavity was probably the shallowest stope or a cavity in the immediate hangingwall resulting from collapse into the stope.

An extensometer was installed in the borehole to establish the form of hangingwall movements. Anchor locations were chosen such that adjacent points bridged zones of significant drilling water loss or poor rock quality. The results indicated that the hangingwall strata are not moving en masse, but that relative movements are occurring. This confirms the concept of a cavity effectively progressing upwards through the hangingwall. However, it is apparent that there is not a single cavity but several smaller zones where relative movements are taking place.

The case study shows that, even if the mining depth exceeds the limit at which statutory restrictions on development are imposed, it is necessary to appraise the site carefully to ensure that no development is placed unknowingly across an adverse feature such as a dyke contact.

CONCLUSIONS

Experience gained over many years has shown that development across the gold reef outcrops in Johannesburg can take place with safety. The scale of the stabilisation works required depends on the scale of the development and hence on the risk of subsidence. This risk is minimised by careful investigation, careful and appropriate design of stabilisation measures and structures, and careful implementation of the construction. Development across significant dyke contacts should be avoided.

REFERENCES

1. BRINK, A.B.A (1979) Engineering Geology of Southern Africa, Building Publications, Silverton, South Africa, 319 pp.

2. DIERING, J.A.C (1980) An improved method for the determination, by a MINSIM type of analysis, of stresses and displacements around tabular excavations. Jl. S.Afr. Inst. Min. Metall., Vol. 80, No. 12, pp 425-430.

3. HILL, F.G (1981) The stability of the strata overlying the mined-out areas of the central Witwatersrand, Jl. S. Afr. Inst. Min. Metall., Vol. 81, No. 6, pp 145-191.

4. STACEY, T.R (1983) Development over old gold mining operations, S. Afr. Inst. Civil Engrs, Continuing Education Lecture Course, Sub-Group G1 : Geotechnics in the Mining Environment.

5. STACEY, T.R., KIRSTEN, H.A.D and WIID, B.L (1983) Major development on an undermined site in a central city area, Proc. 5th Int. Cong. Int. Soc. Rock Mech., Melbourne, Section E, pp E101-E105.

6. STACEY, T.R and RAUCH, H.P (1981) A case history of subsidence resulting from mining at considerable depth, Trans. S. Afr. Inst. Civil Engrs., Vol. 23, No. 2, pp 55-58.

Rock excavation for a major underground cavern in Hong Kong

I.M. Thoms B.SC., C.ENG., M.I.C.E., M.H.K.I.E., M.I.H.T.
R. Coates A.C.S.M., C.ENG., M.I.M.M., M.H.K.I.E.
Mass Transit Railway Corporation, Hong Kong
V.D. Turner B.SC., C.ENG., M.I.C.E., M.H.K.I.E., M.CONS.E. (H.K.)
Charles Haswell & Partners (Far East), Hong Kong

SYNOPSIS

As part of the Island Line of the Hong Kong Mass Transit Railway, a large station cavern has been constructed in the granite of the Kornhill. The design development is summarised, with particular reference to exposed geological conditions and the unique unloading of the cavern by excavation of the overlying hillside. Excavation and primary support methods are described, with a review of the environmental restraints on the former. The paper concludes with a review of the methods used to monitor cavern stability, together with problems and results to date.

INTRODUCTION

Construction of the Mass Transit underground railway system (Figure 1) commenced in 1975 and comprises:-

Modified Initial System (fully opened February 1980)	(MIS)	15.6km
Tsuen Wan Extension (opened May 1982)	(TWE)	10.5km
Island Line (11.8km opened in May 1985)	(ISL)	12.5km

Construction has been under the supervision of the Mass Transit Railway Corporation, who also own and operate the Railway.

The Island Line includes Tai Koo underground station, which was designed as a large rock cavern, the first such excavation in Hong Kong. Concurrent with station construction, extensive site formation works were carried out above the cavern to provide for the Kornhill Development, a large residential and commercial complex.

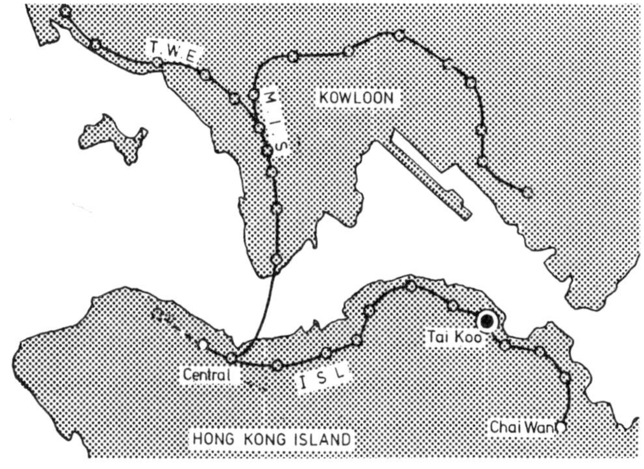

Fig. 1 H.K. Mass Transit Railway Route Plan

Overall dimensions, inside the cavern lining, are:-

Length	251.00 m
Width at track level	19.82 m
Width between arch springings	24.16 m
Height from invert to crown	13.95 m

The station has two levels. The upper, concourse level contains fare paying and entry facilities, plant rooms and accommodation rooms: the lower level has a 182.5m long island platform and plant rooms. At both levels, plant rooms are concentrated at the east and west ends of the station, leaving central sections clear for public use.

Three entrances to the station concourse have already been constructed, of which one, Entrance C, directly connects to an existing development at Tai Koo Shing, as well as the present King's Road. Two additional entrances are to be formed, to better serve the Kornhill

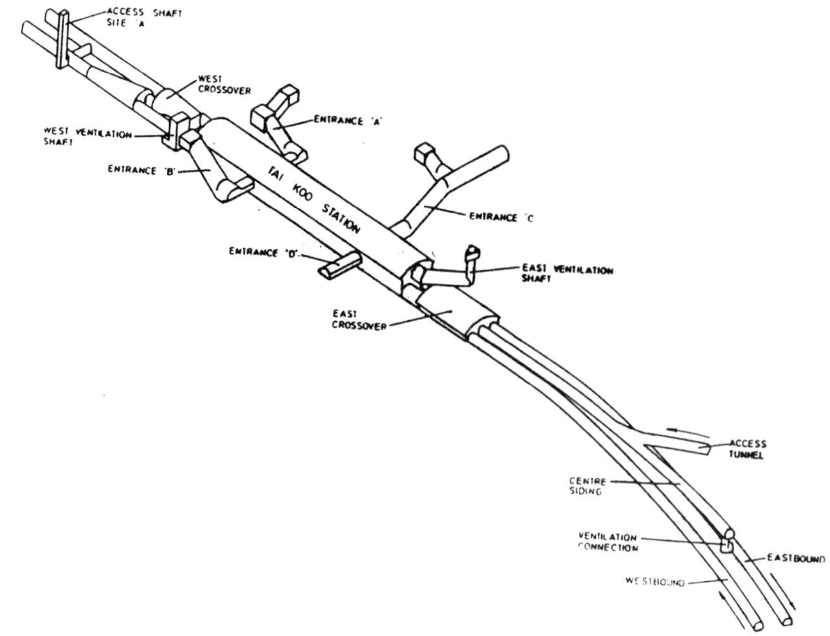

Fig. 2 Layout of Tai Koo Station and Tunnels

Development, which is currently under construction : one is Entrance D, presently a stub adit at concourse level, and one is a surface extension to the existing Entrance A.

The contract for the station also included construction of running tunnels, entrance and ventilation tunnels, which is not dealt with in this paper. The overall layout is shown in Figure 2.

CAVERN DESIGN

The location and layout of the station, which is orientated exactly east-west, was dictated by the geometric constraints of the railway, passenger utilization factors, overall planning considerations related to land use and adjacent developments and local geology.

The rock types present in the Kornhill area were known to consist of fine, medium and coarse grained Hong Kong granite with the presence of a significant fault crossing the station site. In general, rockhead level was substantially above the cavern and excavation was expected to be in fresh to slightly weathered rock with the possibility of localised zones of moderately decomposed granite due to penetrative weathering (see Figure 3). The position of the western end of the cavern was constrained by a rapidly falling rockhead at the Nam Fung valley, characterised as a major photogeological lineament on the Hong Kong Geological Map. The existing align. King's Road marks the edge of the exposed granite forming the Kornhill and constrained t. north-south position of the cavern.

The above mentioned predictions of competent rock permitted the large cavern excavation to be designed to the profile shown in Figure 4. The relatively flat arch was necessary due to the minimal rock cover of approximately 11 metres which will exist at the western end of Tai Koo Station when the existing King's Road has been realigned directly over the station.

The structural design of the cavern has progressed through four stages and continues in the final phase.

(a) Initial design using preliminary site investigation data. The final excavated profile remained largely unaltered from the initial design.

(b) Examination of the rock exposed during excavation and instrumentation results to assess excavation stability.

(c) Re-evaluation of ground conditions, instrumentation results and overhead development proposals to assess long-term stability.

(d) Monitoring of cavern behaviour and blasting vibrations during overhead excavation and construction of Kornhill Development.

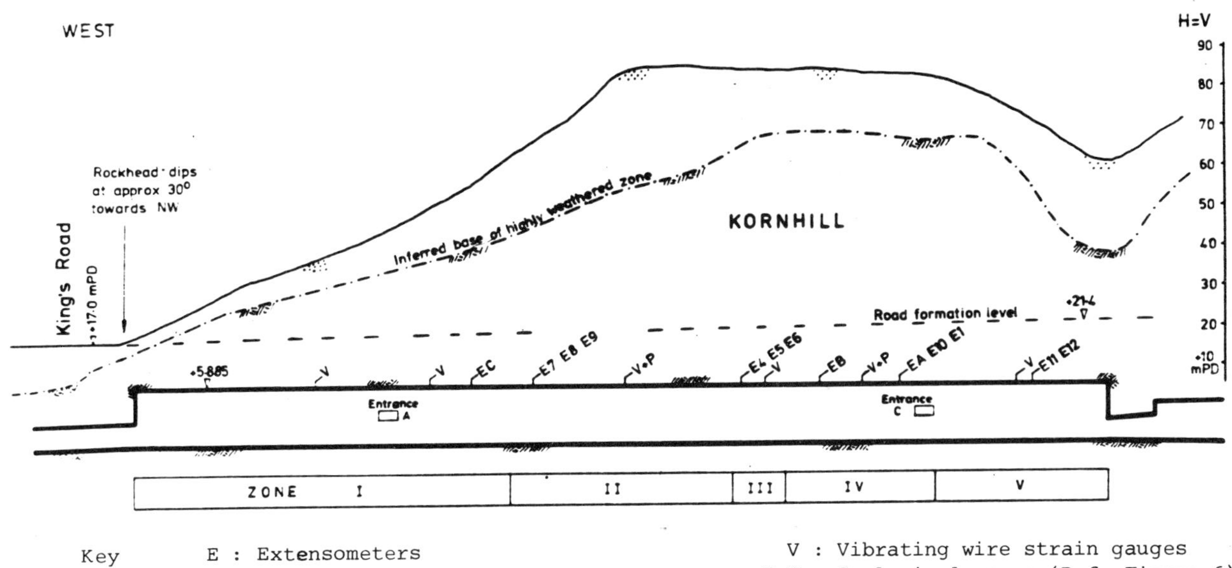

Key E : Extensometers
P : Pressure cells

V : Vibrating wire strain gauges
I-V : Geological zones (Ref. Figure 6)

Fig. 3 Longitudinal section of Tai Koo Station

(a) Initial Design

Rock conditions were deduced from boreholes carried out on the Kornhill, although only one of these penetrated the cavern itself. Information on jointing was supplemented by examination of surface rock exposures along King's Road and engineering parameters assigned from published and test data.

Primary support requirements were estimated on the basis of precedent practice (NGI, Q System)[1] and checked where possible using available structural data from the Kornhill mapping.

The following typical support arrangements were derived:

. Tensioned rockbolts : 3, 5 and 7m length on a 1.5 to 2.0m grid.

. Mesh reinforced shotcrete : 50 - 100mm thickness.

Boundary Element Analyses were undertaken of the completed excavation to determine typical displacements and stress changes.

The above support system would ordinarily have sufficed without the addition of a secondary structural lining. However, in view of the planned overhead excavation and the need for absolute security, together with aesthetic and operational reasons, a secondary concrete lining was decided upon. For this lining, Finite Element Analysis was applied using highway loading and full overburden load (loosened rock).

(b) Stability during Excavation

The rock conditions encountered are shown schematically in Figure 6. With the exception of sub-vertical, weathered joints at the east end of TAK the original assumptions concerning rock conditions were largely proved correct. These joints and their associated zones of decomposition influenced both the stability of the cavern excavation and the potential long term loadings onto the cavern lining.

After pilot headings had been completed, extensive mapping of the cavern arch and sidewalls was undertaken during excavation, with particular attention directed to joints. Extensometer information during this phase showed small movements (maximum 6.28mm at E5 - see Figure 8) in keeping with low levels of stress redistribution in an essentially elastic rock mass.

(c) Long-term Stability

Geological features, particularly rock jointing, determined long-term loadings to a far greater degree than the stresses induced in the rock, which were considered to be low compared to the rock strength.

In-situ stress measurements were not made in the cavern, therefore a range of horizontal to vertical stress ratios with values from 0.6 to 1.0 were examined as bounding conditions. Data from the Kornhill Development concerning rock mass moduli and shear strength of decomposed joint filling were used, together with joint conditions exposed in the cavern, to re-assess stability in the five zones shown in Figure 3, reflecting the presence and persistence of:-

Sub-horizontal sheet joints.

Near vertical joints, trending sub-parallel to the cavern long axis.

Near vertical joints, with locally thick decomposition products, trending sub-parallel to the cavern long axis.

Brecciated fault material, crossing the short axis of the cavern in approximately five metre wide band.

Also of importance was the condition of the haunch when excavated, where significant over-break existed in parts due to adverse geology and blast damage. In contrast, excavation and pre-splitting on the cavern sidewalls were generally successful. Negligible amounts of water entered the cavern, demonstrating the generally tight condition of the jointing.

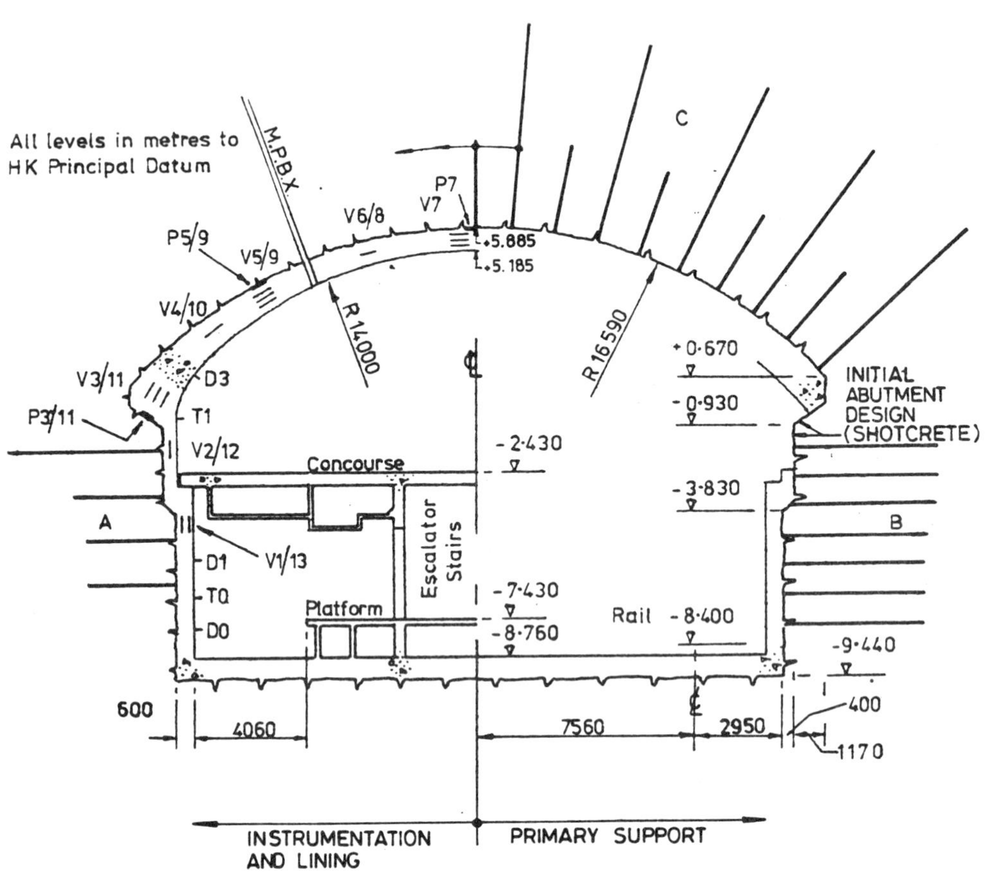

Key:-

A - Standard sidewall rockbolting pattern (top row 5m bolts, other rows 3m bolts at 1.5m centres)

B - Heavy rockbolting pattern, with anchor heads linked to wall reinforcement (5m bolts at 1.0m centres plus shotcrete)

C - Varying patterns of 3, 5, 7 or 9m rockbolts

D - Demec surface strain gauge

P - Pressure cell, north wall/south wall

T - Telemac Distomatic taping point

V - Vibrating wire strain gauge, north wall/south wall

Fig. 4 Tai Koo cross-section with instrumentation

The design re-appraisal concluded that, in zone V, which contained decomposed vertical joints, ground loadings in excess of 4000kN/lin. metre of abutment could occur. As the arch abutment at the decomposed areas would be inadequate to withstand this force, provision was made for the load to be directed through the sidewall concrete and rock. Based on the generally stiff, competent nature of the granite arch and the influence, in certain sections of the cavern, of sheeting joints, it was also considered doubtful that the concrete arch lining would act as a true arch. Consideration of the likely rock behaviour as a stiff beam and the sequencing of underground and surface excavations thus contributed to the revised lining arrangement as shown on Figure 4. For ease of construction, the amended arch shape was adopted throughout the cavern, thereby also negating problems with the over-excavated arch haunch.

(d) Monitoring of Cavern Behaviour during Overhead Excavation and Construction

Results are discussed in "INSTRUMENTATION".

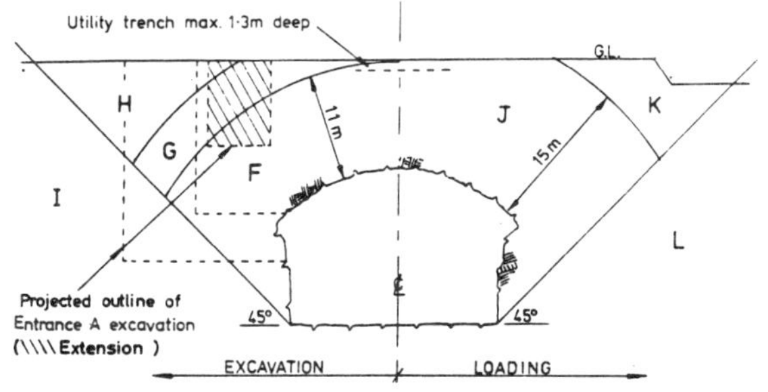

Fig. 5 Constraints on overlying development

Key:-

Exc. F - No excavation permitted

G - Excavations to be individually assessed and monitored*

H - Excavations to be monitored*

I - Deep excavations to be monitored*

* - Blasting vibrations restricted

Loads J - Vehicle loading only

K - Max. 650KPa (design load for entrance adits). Loads up to 2000KPa permissible, depending on depth of application (but not directly over entrance adits)

L - Unrestricted

The restrictions imposed on the surface development are summarised in Figure 5. As can be seen, a short length of the extension to Entrance A lies within the eleven metre envelope (zone F) and this was only accepted after consideration of local geology. It should be noted that in the relatively poorer ground conditions at the east of Tai Koo, entrance adit excavation is not parallel to the cavern (Entrance C) or will be further away (future Entrance D).

CAVERN EXCAVATION

Due to limited access at the location of the station cavern, all excavation and concreting had to be carried out from the Sai Wan Ho reclamation area and/or the Access Shaft at Site 'A'. A short Access Tunnel was, therefore, constructed from Sai Wan Ho Reclamation to the Centre Siding, prior to the commencement of tunnelling on the main Tai Koo Contract.

The main departures from the Engineer's envisaged construction sequence were the adoption of twin side headings with a central pillar, instead of a central heading, during cavern excavation and deferment of concrete arch construction until after all excavation was completed and invert and concourse slabs and side wall concrete had been cast.

Starting in mid July 1982, over a period of 14 weeks, the Access Adit was extended on its shared route with the permanent Centre Siding and the East Crossover was enlarged sufficiently to gain access to the taper tunnel sections and hence, to the east end of the cavern at track slab level. By late October 1982, excavation of the short section of twin running tunnels and associated trailing crossover to the west of Tai Koo Station had been completed, from the Site 'A' Shaft and a pilot access ramp (at 10% grade) advanced some 60m into the cavern, on the south side from the west end, as shown in Figure 7.

From the east end twin, 10% grade ramps, 60m in length, were driven in heading along the cavern sidewalls up to concourse level. Side headings were then driven, adopting the final arch profile, with the north side heading in advance of the south side heading.

Simultaneously, the east-bound pilot from Site 'A'

shaft was continued as a diagonal pilot to intercept the north side heading and break-through was achieved within 7 weeks of entering the cavern at the east end.

The enlargement of the cavern to full width was started in the central area of the station and progressed westwards. Bench excavation followed working from west to east, in two stages, to allow concreting to commence at the west end. This operation was serviced through the Site 'A' shaft. The removal of the remaining central pillar, between the two side headings, was completed in the eastern half of the station, contemporaneously with the concreting in the western half.

The access ramps were alternately backfilled in order to slash the roof to the correct profile. Of most concern was the stability of the full height pillar between the access ramps, due to the presence of sub-vertical jointing and decomposed material in some of the joints. The final thirty-four metres of slashing over the south side ramp to the east station headwall was, therefore, combined with central pillar removal. Adit and bench excavation followed, such that the overall excavation of approximately 80,000 cu.m. was completed in July 1983, only 40 weeks after the initial west end pilot tunnel entered Tai Koo Cavern. The scale of the completed cavern excavation may be gauged from Figure 9.

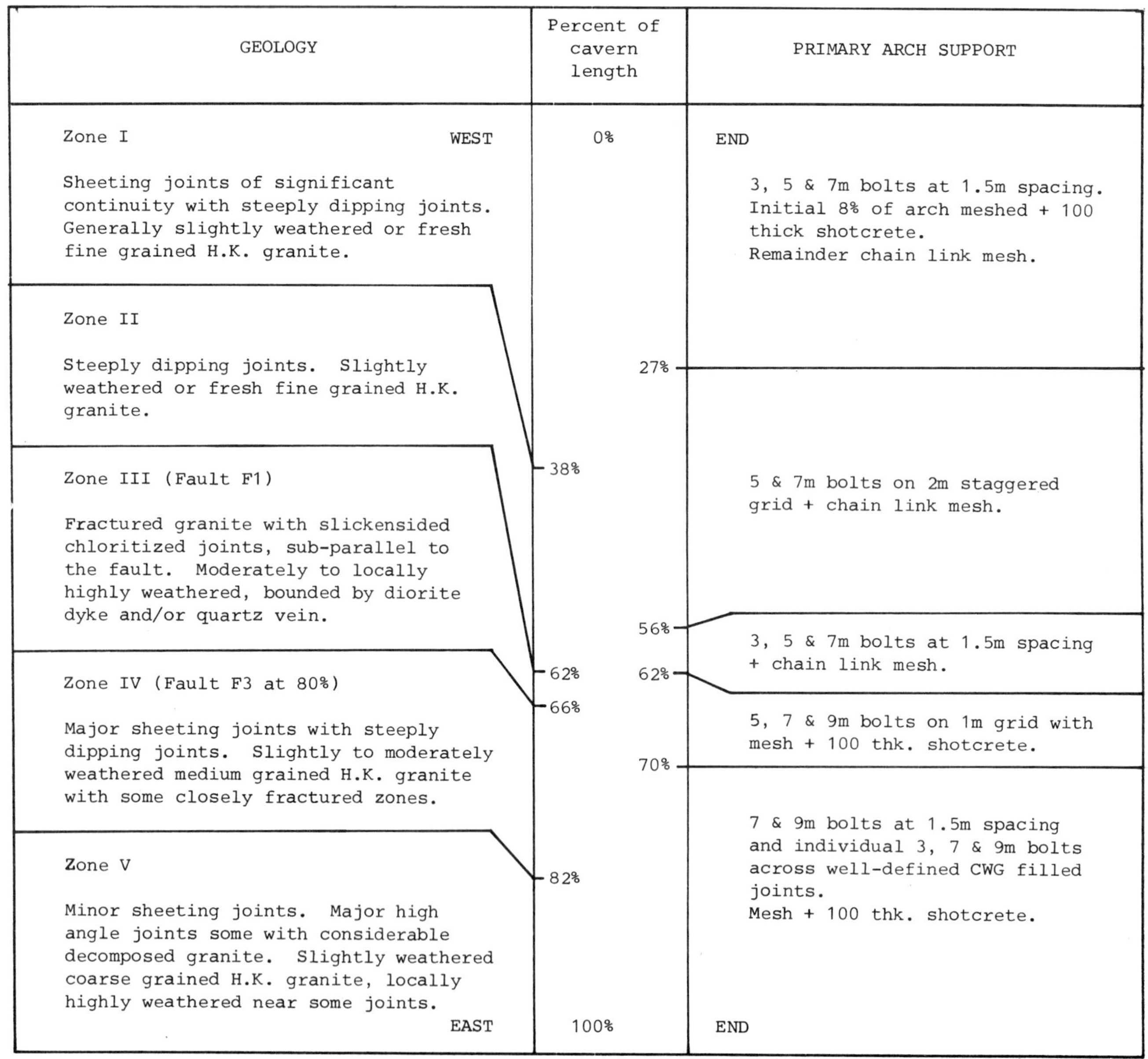

Fig. 6 Schematic diagram of arch geology and primary support

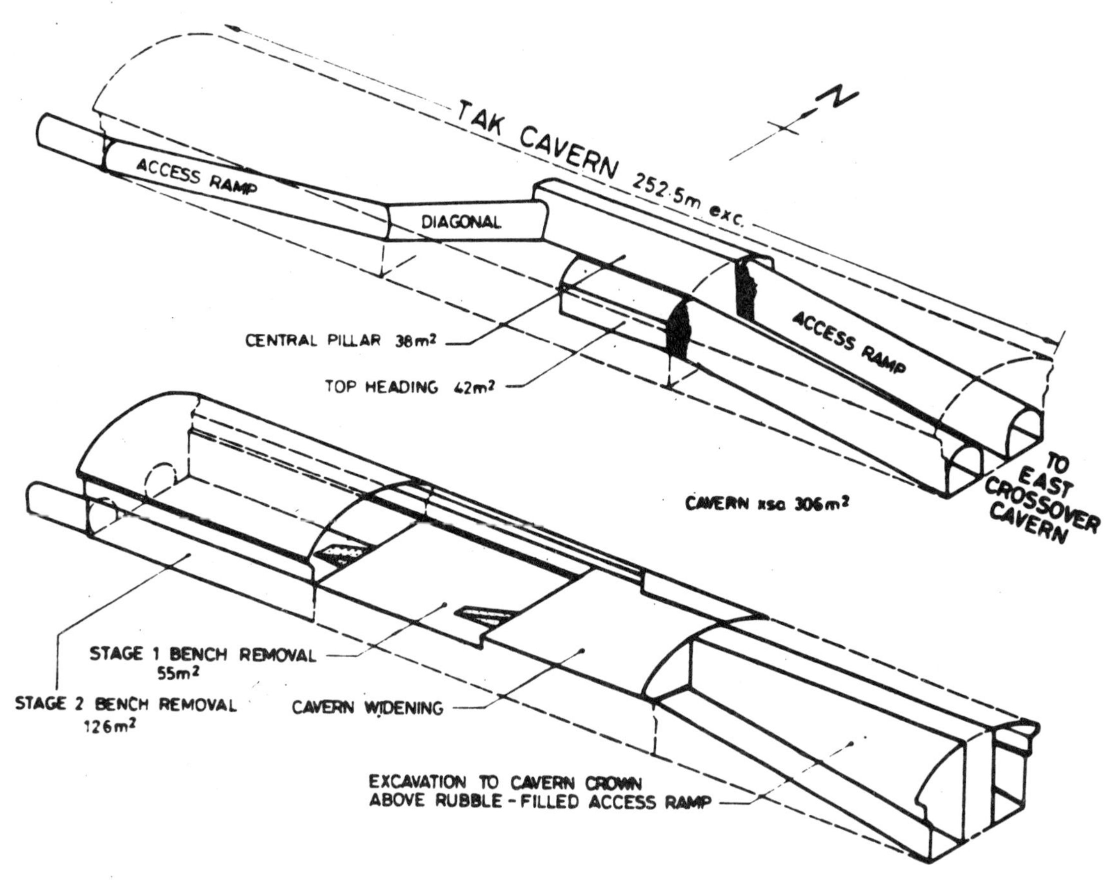

Fig. 7 TAK : Schematic Excavation Diagram

PRIMARY SUPPORT

Primary support to the arch excavation was provided in the form of fully grouted, tensioned 25mm dia. rockbolts with shotcrete and mesh reinforcement, according to the conditions encountered and as summarised in Figure 6.

In Zones I and II, where the best rock conditions were encountered, the lightest bolting densities were employed. Within Zone III, the area of Fault F1, 50mm of shotcrete was sprayed onto the rock surface immediately after blasting and prior to the bolt installation. A further 50mm of shotcrete with mesh reinforcement was applied before recommencing the excavation cycle. This procedure was also followed at the location of joints filled with weathered material, in Zones IV and V. Some problems with underbreak were encountered where this procedure was followed.

Patterns of 5m, 7m and 9m rockbolts were placed in Zones IV and V with additional 7m and 9m bolts being installed to "stitch" the weathered seams to avoid possible wedge failure.

Stability of the cavern sidewalls below arch springing level was assured by use of 3m and 5m long rockbolts with shotcrete as necessary. The dense pattern of sidewall bolts was used in the zone affected by the high angle, weathered joints. The sidewall bolting was not installed immediately following each blast, except where conditions dictated otherwise, this operation being relegated to non-blasting hours. (see Figure 4).

The primary support installed in the cavern is summarised below:-

		Cavern Arch	Sidewalls + Headwalls
Rockbolts	3m	667 no.	877 no.
	5m	913 no.	932 no.
	7m	1153 no.	188 no.
	9m	647 no.	-
Shotcrete	50mm thick	164m^2	274m^2
	75mm thick	-	911m^2
	100mm thick	3075m^2	85m^2
Mesh reinforcement		3075m^2	967m^2

(twinned D83 to BS 4483)

For comparison, the excavated surface areas of the arch and sidewalls are $6833m^2$ and $5860m^2$ respectively. The total length of rockbolts amounts to nearly 29 kilometres.

CONSTRUCTION TECHNIQUES

The excavation was carried out using normal drill and blast techniques, utilising 3 No. MONTABERT jumbos, with 1, 3 and 4 booms, and 1 No. ATLAS COPCO 3 boom jumbo for drilling of the headings, benches and rockbolt installation. A FURIKAWA crawler rig was used for drilling the pre-split holes for the sidewalls. Rock was removed from the cavern using 5 No. AVELING BARFORD SL 320 dump trucks, serviced by 2 No. CATERPILLAR 980 rubber tyred front-end loaders and 1 No. each CATERPILLAR 950 and 966 tracked excavators. During bench excavation a CATERPILLAR D6 dozer also assisted the mucking out operation, when not employed on stock pile control at Sai Wan Reclamation.

Blasting was carried out mainly by use of TOVEX 210 and 90 cartridged water gel explosive detonated by DuPONT blasting caps, the lower velocity gel being used to charge perimeter holes where these were on the final cavern profile (smooth blasting being a requirement of the Contract). In driving the various headings the "burn cut" method was generally adopted with 89mm dia. relief holes being drilled at the centre of the burn cut. Over 100 tonnes of explosives were used to excavate Tai Koo Cavern and the associated entrance and ventilation tunnels, with an average consumption of 1.11 kg/m^3. For comparison, consumption in the TAK-SWH running tunnels (29.2 sq.m face area) averaged 1.95 kg/m^3.

The Contractor elected to use LENOIR ET MERNIER shell anchors with threaded 25mm dia. high yield deformed reinforcement, in place of the specified resin anchor bolts. These were installed in 45mm dia. holes drilled by jumbo, immediately after completion of mucking, and tensioned, by a calibrated pneumatic torque wrench, to a working load of 100KN. Random lift-off tests were performed with a hollow jack to check that the tensioning was satisfactory and 2% of all bolts were tested to 1.5 times the working load. Rockbolts were re-tensioned where necessary and fully cement grouted after mortaring of the face plate. This operation was carried out remote from blasting operations using a grout mix of 135kg OPC : 0.681 kg "CONBEX 100" : 67 l. water. At TAK east, drill flushings were monitored to ensure positioning of the anchorage shells in sound rock.

Dry mix shotcrete was applied through 2 No. electrically driven MEYNADIER shotcrete machines. "SIGUNITE" accelerator, in powder form, was used at a weight dosage of 2%, dispensed onto the dry shotcrete conveyor. Although shotcrete for primary support was specified to have $25N/mm^2$ minimum concrete cylinder strength at 28 days, production shotcrete generally achieved strengths between $15N/mm^2$ and $20N/mm^2$, based on coring results. Strengths achieved varied considerably amongst operators.

Arch concreting operations commenced from the west concourse slab using one, later two, full-width shutters. Twenty-nine, nine-metre long pours were completed in the twelve weeks ending late January 1984. The concrete mix contained 23% OPC replacement by local PFA, with a maximum internal temperature of 68^oC recorded (43^oC above ambient).

A PFA/OPC mix was used in the cavity grouting of the arch. A total volume of $226m^3$ of grout was injected at a maximum pressure of 3 bars measured at the grout hole. This represents an average void depth of 60mm over the 252m length and the central 15m width covered by the grout injection points. All holes were re-drilled, water tested and re-injected where necessary. Grout cube strengths after 60 and 90 days all exceeded the required $30N/mm^2$.

ENVIRONMENTAL RESTRAINTS

In the Contract emphasis was placed on avoiding damage to surrounding properties and on minimising nuisance to the public.

(a) Settlement Monitoring

A number of standpipes were installed in surficial deposits in the vicinity of the temporary access shaft, to monitor the water table.

An extensive network of levelling points on

roads and buildings was installed and monitored at a frequency depending on any movements detected.

No significant movements in either system were detected, nor significant water ingress into TAK.

(b) Noise

Noise levels were restricted to less than 65 decibels (dBA) between the hours of 2300 and 0600 hours measured at the nearest inhabited structure. This precluded blasting between those hours, as "air blast" from the Access Tunnel portal registered up to 80 decibels (dBA) during blasting in the station cavern, some 400 metres away : blasting noise was also easily audible by transmission through the rock.

All rock handling at the surface dump was prohibited during these hours, leading the contractor to use temporary storage within the East Crossover cavern.

For the same reasons, surface air compressors and portal fans were also silenced, in the former case by acoustic screening.

(c) Traffic Control

Since the rock dump had to be reached by crossing a road heavily utilized by ferry passengers each morning, a system of barriers and lights was instituted, to control both vehicles and pedestrians.

(d) Spoil Size and Disposal

Although specifically noted as being required for use as reclamation fill elsewhere, the contract did not define a maximum size of rock spoil. This led to oversize rocks being a matter for dispute on a number of occasions, although amicably resolved. Spoil removal from the rock dump was undertaken by the contractor for the Kornhill formation works and was subject to barge availability.

A larger dumping area would have eased problems here - the original area was 514 sq.m., later enlarged to 700 sq.m., which was also insufficient on occasions.

(e) Vibrations

The peak particle velocity (PPV) , from blasting, permitted in the vicinity of occupied property was 25mm/sec. This was never exceeded during cavern blasting.

A lower PPV of 12.5mm/sec was placed on waterworks structures : for pipes, this could not be monitored directly without excavation and so surface transducers were used, where entrance excavation was carried out.

Concern was expressed by owners of a computer installation near the temporary access shaft. Vibrations and performance were monitored, but no problems were experienced (max. PPV recorded 2mm/sec.).

(f) Fly-rock/flooding

In open-cut Entrance blasting, substantial protective "birdcages" were provided to eliminate dangers from fly-rock : these also incorporated 1.2m high, concrete flood-protection walls, totally encompassing the excavation, to prevent surface flood water entering the underground excavations. A flood gate was also provided at the Access Tunnel portal for operation during heavy rains or typhoons.

INSTRUMENTATION

The objective of installing comprehensive instrumentation in Tai Koo cavern was to provide data for the following purposes:-

(i) To confirm the validity of the design.

(ii) To check the overall behaviour of the excavation and to monitor the response of the excavation to the mining of the adjacent development.

(iii) To provide long term assessment of the structural behaviour of the station.

To this end preliminary requirements for Tai Koo instrumentation included multiple point borehole extensometers (MPBX) with a maximum array length of 30m and up to 5 readout points, convergence measurement devices and hollow inclusion stress cells. Provision was also made in the Contract for investigatory boreholes (logged) and rock testing. Initial trials were carried out during the East Crossover cavern excavation with MPBX and convergence stud installation, to assess requirements and familiarize personnel with procedures, expected magnitudes and reporting methods.

Amendments to the initially planned requirements were made, following further review of foundation loads from the Kornhill Development and assessment of the rock conditions exposed in the cavern. After thorough discussions between all parties involved in design and installation the final list of instrumentation comprised:-

- Multiple rod extensometers (up to five readout points)
- Convergence taping
- Pressure cells
- Vibrating wire strain gauges
- Surface strain gauges (Demec)

No attempt has been made to deduce actual stresses induced in the structure or rock. The raw data, comprising dimensions, pressures or strain units, are examined for continuity. Any abrupt changes, or indications of changing trends, are examined in the light of the surrounding circumstances. In particular, the cavern geology was totally mapped during excavation and, besides being used to determine primary support requirements, instrument locations and behaviour were and continue to be related to these geological conditions.

The installed instrumentation is listed in Table 1.

Table 1 - Instrumentation installed in Tai Koo Cavern

ITEM	QTY.	REMARKS
Rod Extensometers (Rock Instruments)	2 No.	Used in East Crossover (one horizontal, one vertical) : mechanical (dial gauge) readout. Three anchors installed at depths up to 15m (horizontal) and four anchors to 15m (vertical) in percussive holes.
Rod Extensometers (Sinco-Slope Indicator Co.)	10 No.	Five point anchors installed at depths up to 15m (one to 30m) in cored holes. Electrical readout, with battery charger and site fabricated calibration gauge.
Rod Extensometers (Sinco)	3 No.	As above but three point. Anchors installed at depths up to 10.6m in percussion holes.
Convergence tape with studs, calibration frame and spare 30m band (Rock Instruments)	2 No.	For use during excavation. Lack of repeatability and extreme measures required to obtain readings led to abandonment and reliance upon extensometers.
Acoustic pressure cells (Soil Instruments Type 3P18, non-standard 175mm active diameter)	35 No.	Mercury filled, with compensating tube for concrete shrinkage. No failures recorded, although two cells in Entrances required repressurizing.
Telemac "Distomatic" (Soil Experts)	1 No.	Battery powered, dynamometer tensioned Invar wire with digital display. Taping between station sidewalls, concourse and platform levels at nine cross-sections. Separate wires required for concourse and platform.
Demec surface strain gauge (600mm long gauge bar) (Soil Instruments)	1 No.	1.8m long, six arrays, each with four studs, at cavern sidewalls and arch shoulder at nine cross-sections.
Vibrating Wire Strain Gauges (Gauge Techniques type TES/5.5/WP, 200mm long with integral temperature gauges)	189 No.	Settings : To measure 2500 microstrain compression, 500 microstrain tension or 1500 microstrain compression, 1500 microstrain tension. Electrical read-out of strain and temperature. Usage : 147 in cavern : 42 in entrances. (Five no. Cedec TEST/200/WP gauges also used).

Precise levelling was carried out on MPBX heads during early phases of the excavation and showed no discernible displacements. Levelling could not continue once bench excavation had been completed in the vicinity and was found to be impracticable even when construction was completed. Because the longest anchor has been the reference in each crown MPBX and has been affected by overhead excavation, no absolute rock displacements can be computed.

Installation and Operation

The supply and installation of all instruments was the responsibility of the Main Contractor, who employed Soil Mechanics Ltd. as specialist sub-contractor. An MTRC technician was assigned full-time to record results.

With the exception of the first array of vibrating wire strain gauges, where four gauges failed to give readings immediately after concreting and three subsequent failures occurred, no problems were experienced during installation. The failures were attributed to problems at joints in the wiring carried within conduit to the readout position at Concourse level and subsequent routings were simplified. Elsewhere, only five gauges have failed in use, one in an entrance, out of 189 installed.

Initially, ten multiple rod extensometers were installed during cavern excavation - three horizontally in the sidewalls and seven monitoring arch deformations. Once excavation had been carried out to a maximum of five metres beyond the proposed extensometer position, continuous working was carried out to core the hole, determine rod anchorage locations and install the extensometer. No blasting was allowed in the vicinity until 24 hours after grouting the rods in position.

During excavation and sidewall concreting the MPBX's functioned normally. However, during arch concreting, readings from two of the instruments became unstable, taking the form of a continuously varying figure at the readout. Possible water ingress into the measurement transducers at the instrument head, which were encased in the concrete lining, appeared to be the possible cause and extraordinary measures were taken to seal the covering box against water ingress on two remaining extensometers. Whilst one remained unaffected, the second exhibited similar reading instability. Readouts from all seven crown instruments became unstable following cavity grouting.

At this time, March 1984, Kornhill Development excavation remained incomplete and rock movements still had to be monitored. Three replacement extensometers with similar readouts were therefore installed, radial to the arch, to give as balanced a picture as possible, whilst concentrating on areas where most movement was expected. These were functional in early August 1984 and no further problems with readout have been experienced.

Pressure cells have functioned normally and distance measurements have been routine but the completion of the station and installation of permanent equipment have led to the loss of some reading positions.

Instrumentation Results

The locations of all instruments are indicated in Figures 3 and 4 and the results obtained up to July 1985 are summarised as follows:-

(A) Extensometers (Table 2)

As can be seen the more competent rock in Zones II and V has shown the least movement generally, although crown MPBX's in Zone IV also show relatively small movements. (E1 in Zone IV is commented upon later).

The poorer rock conditions in III & IV are illustrated by the dilation plots for crown MPBX's E5 and EB in Figure 8. These highlight a number of points:-

i. Movement is generally initiated by central pillar excavation (E5, week 52/82).

ii. Bench excavation also affects E5 but to a very slight degree.

iii. Overhead mining began to affect the cavern at the position of EB in week 48/84 with an accelerating influence from week 5/85 when the surface had been excavated to a level approximately 28 metres above the cavern. The sudden convergence in week 13/85 is interpreted as instrument response rather than true convergence.

iv. It is not always possible to correlate readings with surrounding activities as exemplified by the disturbance in E5 commencing at week 14/83. This may be due to

Table 2 Summary of MPBX results

Zone Location	Number	Max Movement on (mm)	Anchor Length m	Remarks
I Replacement Crown, N. of centreline	EC	2.59	11	2.30mm dilation between all anchors and head due to overhead excavation.
II S. sidewall	E7	1.60	10	Not affected by bench excavation.
II Crown, N. of centreline	E8	1.42 at failure	15	Dilation of 0.6mm between 15m and 9m anchors.
II Crown Diagonal crossover	E9	1.61 at failure	15	Nearly all movement within four weeks of initiation.
III S. sidewall	E4	2.10	15	Nearly all movement between 1m and 3m anchors, half within 4 wks. of installation.
III Crown, S. of centreline Fault F1	E5	6.28 at failure	15	Movement initiated by central pillar excavation. See fig. 8
III Crown Centreline	E6	2.99 at failure	16	
IV Replacement Crown N. of centreline	EB	See fig. 8	11	Dilation 2.93mm between 2m and 11m anchors, followed overhead mining.
IV S. sidewall	E1	6.03	12	Close to weathered seam. See text.
IV Crown N. of centreline	E10	0.95 at failure	15	
IV Replacement Crown N. of centreline	EA	0.90	11	
V Crown N. of centreline	E11	1.06 at failure	30	*Steady rise from 0.08mm in 40 weeks prior to failure.
V Crown N. of centreline	E12	1.00 at failure	4/11	*Steady rise from 0.55mm in 33 weeks prior to failure.
-	E2, E3			Not installed.

*The steady rise in readings is typical although more marked with E11 and E12 due to installation through a weathered joint.

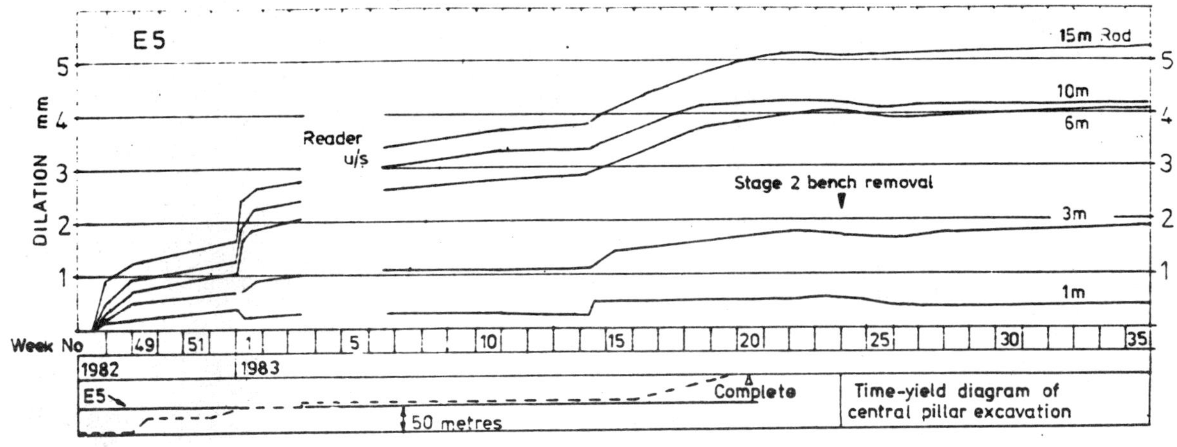

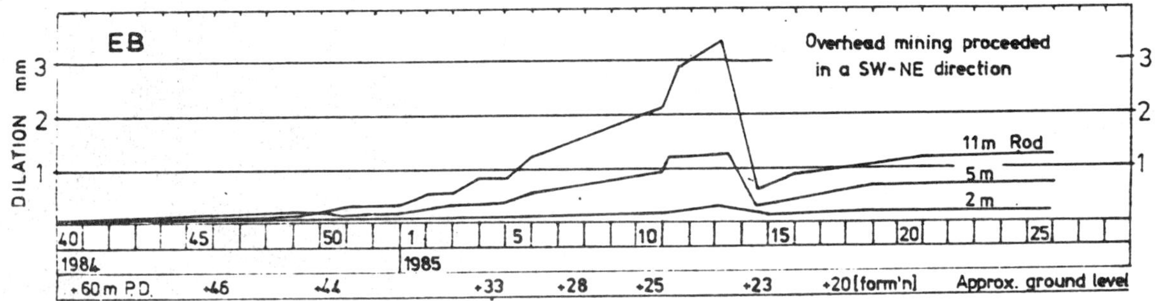

Fig. 8 Displacement plots for extensometers E5 and EB

a combination of:

Blasting in Entrance C (50m eastwards)

Bench blasting (110m westwards)

Surface blasting (70m northwards, along the strike of Fault F1)

Fault plane lubrication by excessive rainfall

Other unrecognised influences.

In regard to sidewall MPBX E1, this passed through high angle joints, with chloritized and limonitic surfaces, sub-parallel to the cavern. The influence of nearby blasting and bench excavation was very marked : the reading on the 12m anchor showed an accelerating increase from 0.6mm to 4.8mm within the seven week period attending these activities and the remaining increase evolved over the next thirty-four weeks.

(B) Vibrating Wire Strain Gauges (VWSG)

Generally, no bending in the concrete lining has been inferred and, except for one gauge across a crack, induced strains are generally slight (less than 50 microstrain). No difference in performance of the four briquetted gauges was noted.

VWSG have been placed in six sections throughout the cavern, covering all geological zones (see fig. 4 for configuration). Stable readings were obtained, at most, six weeks after emplacement and, in the majority of cases, slight compressive trends are noted, except for VWSG at positions V2/12 at Concourse slab support. Here the trend is slightly tensile, except at the east end, where steady readings are obtained. Since overhead mining is now complete, the different zonal response to the reduction of overburden is interesting. Any response at all has been very small.

(C) Pressure Cells

No abrupt loading of the arch has been recorded at the two arrays where cells are installed in Zones II & IV (see fig. 3).

A steady increase to 4.5 bar has been recorded at the cavern centreline in Zone IV since overhead mining began. No similar increase has been recorded in Zone II.

The two, near-surface arrays in each of Entrances A and B (i.e. four in all), show significant reductions in contact pressure, ranging

Fig. 9 Completed cavern excavation

from zero to -8 bars, with an isolated cell recording an increase of +5 bars. The former entrance shows greater reductions, reflecting the greater amount of surface excavation in the vicinity. This is reflected also in VWSG readings in these Entrances, although strain changes are less significant than pressure cell readings (all within 75 microstrain).

A single instrumented array in Entrance C, which has been subjected to the greatest degree of overburden removal, shows a maximum change in contact pressure of 2 bar and a maximum VWSG change of +150 microstrain at one shoulder : elsewhere, microstrain changes are less than 30.

(D) Telemac Distomatic

Although measurement was interrupted occasionally by the necessity to re-locate taping heads, the cavern sidewalls have exhibited a general closure trend, particularly at concourse level and platform east end, approximating to 0.4mm over the cavern width.

CONCLUSIONS

In the densely populated, ever-expanding urban areas of Hong Kong and Kowloon, it is highly conceivable that rock caverns could present an economical solution for underground storage, car-parking and other facilities, without encroaching upon prime land areas. It is considered that the design and construction of Tai Koo Station has provided invaluable experience in the field of large-scale underground excavation under Hong Kong conditions and it is hoped that this brief paper may prove a useful reference for future work of similar nature.

The fundamental lesson learned from Tai Koo Station must surely be that design and construction of such major rock caverns are inextricably interrelated and both aspects must be thoroughly reviewed on a regular basis in relation to the conditions revealed during excavation.

Continual attention is needed with respect to quality control in rockbolting, shotcreting, concreting and cavity grouting. Similarly, with regard to excavation, it is essential that blasting operations are closely monitored and that pre-splitting and blasting techniques are amended in a timely manner to suit actual conditions. In the case of Tai Koo Station works, these aims were accomplished by close contact and cooperation between the Contractor and the Engineer's staff and necessitated the submission and discussion of numerous method statements for carrying out the Works.

The history of the selection and installation of instrumentation within Tai Koo Cavern illustrates the need for extremely careful consideration of instrumentation requirements at all phases of construction. Whilst very few problems developed overall with the instruments installed, the failure of the majority of the extensometers, following cavity grouting of the arch lining, was exceedingly unfortunate and highlights the need for special attention in cases of future, similar application. It should also be stressed that one of the main reasons for installing instrumentation is the long-term monitoring of the performance and stability of the cavern and the recording of instrumentation results will therefore be ongoing. This demands particular care in choice of location and instrument type to ensure reasonable access to take readings.

Also of considerable importance during the construction period is the inevitable conflict between progress of the works and stoppages to permit installation of instrumentation. This may prove a particular problem with MPBXs, where the extensometers need to be installed close to the face and made fully operational prior to further advance. This, of course, can only be achieved by arresting the particular drive. In such instances, delays may be overcome or minimised by judicious siting and speedy installation or careful phasing of the excavation programme.

The successful execution of such a major underground excavation obviously requires the development of a strong spirit of cooperation on site and demands flexibility by all parties involved in the design and construction. It is satisfying to be able to record that such was the case in the completion of Hong Kong's first major underground rock cavern at Tai Koo.

References

1. Underground Excavations in Rock. E. Hoek & E.T. Brown. The Institution of Mining and Metallurgy, London 1980.

Acknowledgements

The authors wish to thank the Mass Transit Railway Corporation for permission to publish this paper. Acknowledgement must also be given to the many individuals who contributed to the success of the Contract, in particular to the Contractor's Project Manager, Mr. S.E. Fery and his staff and Mr. C.J. Mundy, geologist, who was responsible for geological mapping of the cavern.

Employer

Mass Transit Railway Corporation, Hong Kong.

Principal Consultant:

Charles Haswell and Partners (Far East).

Geotechnical Review Consultant:

Geo-Engineering, Jersey (Channel Islands).

Contractor:

A joint venture of Dragages et Travaux Publics with Coignet Enterprise, both of France.

Project Manager and Supervisor:

Mass Transit Railway Corporation, Hong Kong.

Blast vibration monitoring on anchored retaining walls and within boreholes

T.J. Wilton B.A., M.SC., M.I.EXP.E.
Rock Environmental, Ltd., Derby, England
R.L. Hills A.R.S.M., B.SC., M.I.C.E., M.H.K.I.E., M.I.Q.
Charles Haswell and Partners, London, England

SYNOPSIS

The method and results of the monitoring of blast vibrations at an anchored retaining wall located at Tai Koo Shing, Hong Kong, are described. The results of this monitoring together with results of other recorded blast monitoring within the same location were analysed to predict vibration levels. These predictions were used, during the construction of a tunnel which passes through the anchored wall, to relate vibration measured at the anchor heads to the vibration level in the anchorage zone. The methods of prediction are discussed together with the results of monitoring during construction.

New instrumentation for the measurement of blast vibration from within boreholes is described together with its practical application. The authors describe recent experience with down-the-hole measurements at a site in the United Kingdom and discuss the results.

The accuracy and methods of determining blast vibrations at specific locations are discussed and comment made on the confidence limits of prediction methods, with particular reference to blasts at relatively close distances to monitoring locations.

DESCRIPTION AND LOCATION OF THE TAI KOO SITE

As part of Tai Koo Station on the Island Line of the Hong Kong Mass Transit Railway (HKMTR) an entrance to the City Plaza shopping complex was constructed.

The entrance was tunnelled beneath Kings Road to enter City Plaza at second floor level through an anchored retaining wall. The City Plaza complex is part of the redevelopment of the old Tai Koo Dockyard. This redevelopment of the site included a retaining wall along Kings Road. In the section adjacent to City Plaza the wall was anchored back with a series of 15 metre long multistrand 100 tonne rock anchors.

The construction of the Station entrance necessitated the replacement of some of these anchors and the monitoring of other adjacent anchors. The monitoring consisted of load and lift off tests together with measurement of blast vibration levels.

Object of Vibration Monitoring

The object of the vibration monitoring was to satisfy the Geotechnical Control Branch (GCB) of the Building Ordnance Office (BOO) of the Hong Kong Government that the permitted vibration levels at the anchorage of the rock anchors was not exceeded.

Because of the difficulties involved in measuring the vibration levels at the anchorage zone, it was accepted that the vibration levels at the anchor heads would be monitored.

The monitoring was done at the anchor heads primarily to reduce cost and also at that time because of the difficulty of obtaining suitable instrumentation for down the hole monitoring. It was also felt that the permitted levels of vibration were so conservative that with precise measurements of vibration at the anchor heads combined with standard prediction methods this was sufficient.

For the maximum permitted peak particle velocity, ppv, at the anchorage zone of the closest anchor to the blast the maximum charge weight was predicted. This charge weight together with the distances from the anchor heads to the blast was then used to predict the ppv at the adjacent anchor heads. As long as the measured ppv at these anchor heads did not exceed these predicted levels then it was assumed that the permitted level at the anchorage would not be exceeded either.

Method of Prediction of Vibration Levels

The initial assessment of charge weights per delay and prediction of ppv's at the anchor heads was based on the following relation ship between the peak particle velocity ppv of ground motion, the weight of the charge w and the distance d from the blast to the measuring point of ppv.

$ppv = kw^{x}D^{-y}$ equation 1

where k, x and y are empirically determined constants which are influenced primarily by geological characteristics between the blast and receiving sites and the type of confinement of the blast.

The more common form of equation 1 is

$ppv = k(D/\sqrt{w})^{-y}$ equation 2.[1]

Where $D/\sqrt{w}$ is known as the scaled distance (SD).

Numerous field measurements from quarry and surface mining sites have shown that the propogation equation 2 can be typically expressed as:-

$ppv = 1143\ (D/\sqrt{w})^{-1.6}$ equation 3

Where ppv is the peak particle velocity $mm.s^{-1}$

D is the distance between the blast and the recording site in m

w is the maximum instantaneous charge weight in kg.

It is important to realise that equation 3 should only be used to provide typical ppv values in the absence of on site measurements. The predicted charge weights and ppv's at the anchor heads for selected distances between the blast and the anchorage zone are given in Table 1. The maximum permitted ppv in the anchorage was taken as 25 $mm.s^{-1}$ at the closest anchor point.

Site Measurements

To establish the seismic constants, k and y, from equation 2 particular to the site, on site measurement is necessary. It was recommended in this case that the site seismic constants be determined by a regression survey using a minimum of two but preferably three vibrographs spaced over a range of distances from the blast similar to those from which predictions were to be made.

For the monitoring at the anchor heads it was recommended that VME Seistectors be used. These would continuously monitor vibrations at the anchor head, and would be installed prior to the construction of the entrance tunnel when access to the anchors was made for the lift off testing of the anchors and the installation of load cells.

Results of Site Measurements and Regression Analysis

Due to delays in implementation of the site measurement monitoring a regression survey was not carried out. Instead the results of monitoring from the associated Kornhill Development together with some tunnel blasting on the Tai Koo tunnels were analysed to try to establish the site seismic constants.

These results were assembled from various sources and a large variety of types of blasts which included :-

quarry type production blasts from 6m high benches using ANFO explosives

'popping' of large insitu covestones

'pre-splitting' shots for formation of permanent slope profiles

tunnel blasting using packaged slurry explosives

open trench excavation blasts for station entrances

Table 1

Predicted Vibration Levels at anchor head which should not exceed Permitted Level at closest anchor to blast

Distance (m) to anchor zone	Maximum Charge Weight (kg)	Maximum ppv mm/s measured at anchor heads									
		Y1	Y2	Y3	U7	U8	U9	N1	N2	N3	N4
5	0.37	3.33	3.41	3.24	3.25	3.31	3.25	2.94	2.79	2.94	2.80
10	1.12	5.91	6.00	5.79	5.87	5.95	5.89	5.37	5.17	5.36	5.42
15	2.02	7.30	7.39	7.19	7.34	7.39	7.32	6.77	6.56	6.74	6.53
20	3.36	8.81	8.91	8.67	8.87	8.96	8.88	8.26	8.06	8.22	8.02
25	5.42	10.78	10.89	10.67	10.85	10.98	10.97	10.03	10.01	10.19	9.96
30	7.87	12.00	12.07	11.93	12.19	12.41	12.38	11.48	11.25	11.41	11.16

Assumes maximum peak particle velocity at closest anchor of 25 $mm.s^{-1}$.

The majority of the results were from at least double incident monitoring i.e. at least two monitoring machines for each blast at different distances. The large variation in the location of the monitors meant that sets of data related to a particular location could not be identified as a source of sampling the data.

Least squares analysis was employed to obtain the values of the seismic constants k and y, see Table 2, file T. An extremely low Coefficient of Determination (CD) of 0.385 was obtained. The data ranged in terms of scaled distances from 5 to 50.

It was considered that the wide scatter of results could be due to the type and range of blasts being monitored.

Because of this further analysis was undertaken. This consisted of firstly selecting results where the recorded ppv was several orders of magnitude different to the predicted ppv using equation 3. These results were mainly recordings from 'pre-split' blasts and blasts measured within Tai Koo Station cavern. These results were then omitted and the remaining data re-analysed. Finally the remaining pre-split blast data was also discounted. The results of these three stages of analysis are tabulated in Table 2.

From Table 2 it can be seen that at each stage the coefficient of determination, CD, increased indicating a greater degree of correlation. However the values of CD were still considered to be too low for the accurate determination of k and y.

A second batch of data consisting of vibration records collected by the Seistectors mounted adjacent to the anchor heads on the City Plaza Wall was then assessed. This batch of results were produced by blasting in the vicinity of the City Plaza Entrance. In the batch the scaled distances ranged from 20 to 40 compared to the initial range of

Table 2

Assessment of Original Data

File Name	Type of Data	No. of Points	CD	Ground k	Constant y
T	Total Original Data	107	0.385	53	-0.74
T1	Original Data less Certain results	66	0.461	135	-0.99
T2	T1 data less further pre-split blast results	46	0.627	192	-1.05

from 5 to 50. Again the results were analysed in terms of ppv versus scaled distance using least squares regression analysis. The results for each location were analysed individually. These sets of results are tabulated in Table 3.

Comparing the results there is a good correlation of the seismic constant y at a value of -2.24 and a good CD for the results at anchor 1. However, for the prediction of vibration the results were based upon observations all within the scaled distance range of 20 to 40. The seismic constants needed to be validated over the full range of scaled distances to be used of 5 to 50. The critical scaled distances in this particular case being the range of 10 to 20.

The majority of information available on the prediction of vibration levels is related to quarrying and opencast mining, there is relatively little published data on vibration from tunnelling or underground construction projects and particularly in hard crystalline rock.

Major differences between surface and tunnel blasting are due to

1) The generally higher confinement of blasts in tunnels

and

2) Normally a need to monitor at much closer distances to the blast source.

Both of these factors were particularly relevant in the case at City Plaza. Another complication was the need for the monitoring point to be further from the blast source than the prediction point.

Table 3

Assessment of Additional Data

File Name	Type of Data	No. of points	CD	Ground k	Constant y
A1	Data from Anchor U6	11	0.83	4896	-2.24
A2	Data from Anchor U7	26	0.506	2828	-2.24

The combination of both sets of data gives SD values over the desired range and results in k and y values similar to those used for the initial predictions. The CD is relatively low reflecting a large scatter and variation in the results.

Thus an assessment of the complete sets of data was made and is given in Table 4. The plot of the vibration levels versus scaled distance is given in Figure 1.

One method of assessing data, but not fully investigated, was the selection for analysis of results from blasts with similar charge weights. The other method not pursued was the use of the relation of the cube root of the charge weight. The USBM[2] concluded that the cube root of scaling was more pertinent for tunnel blasting. However, there appears to be as yet insufficient data to fully substantiate this.

Table 4

Assessment of Data (Total)

File Name	Type of Data	No. of points	CD	Ground k	Constant y
TCPW	T2 data plus A1 & A2	83	0.509	1132	-1.77

Conclusions

The analysis of the data demonstrates the variation of the constants between ranges of scaled distance values and the type of blast being monitored.

Safe Levels of Vibration

The recommended value of ppv at the City Plaza wall of 25 $mm.s^{-1}$ was considered very conservative as the anchorage zone would be subject mainly to higher frequency body waves, compared

FIGURE 1

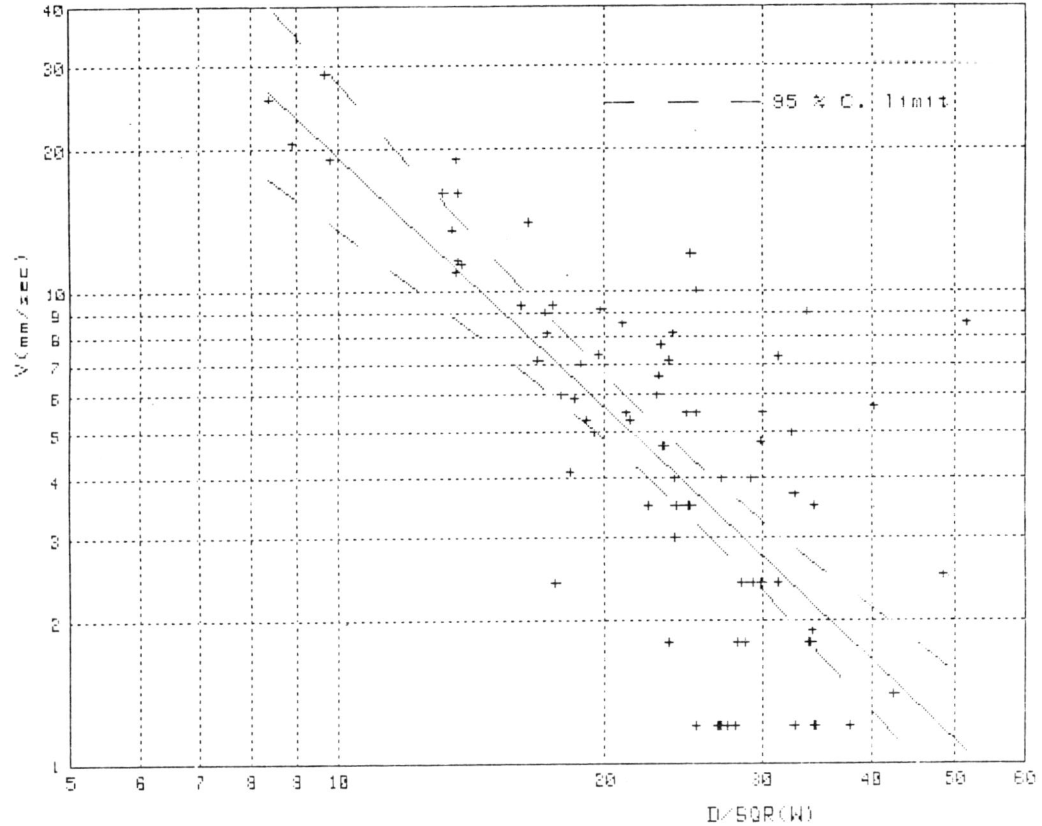

DOF = 81
Confidence limit= 95 %
t = 1.990

k = 1132
y = -1.77

Charles Haswell & Partners [Ba 12/83] [FILE= TCPW]

to the combination of body and surface waves being monitored at the anchor heads.

The actual levels were of the order of 12 mm.s^{-1}, which is probably a safety factor of at least 10 and probably nearer 12 to 15. With a proper monitoring programme and suitable supervision of blasting it is considered that a factor of safety of 5 is more than adequate. Thus safe levels of the order of 75 mm.s^{-1} should be used for rock anchors, especially in hard crystalline rock where the frequency of vibration is relatively high.

VIBRATION LIMITS AND COST

Whilst the necessity for blast induced vibration limits on services and structures is clearly very great in many civil engineering projects, the cost of these limits to such projects should not be overlooked.

Figures given by the Swedish Council for Building Research[3] show how the relative costs increase as vibration limits decrease for an example of excavation at 10m distance. In this case a limit of 70 mm.s^{-1} doubles the cost relative to no vibration restriction. This is further doubled at a 35 mm.s^{-1} limit and for 25 mm.s^{-1} is 7 times the unrestrictive costs.

Clearly such figures will vary in detail depending upon the exact local circumstances, however, it is equally clear that great value should be put on establishing realistic damage criteria based upon a sound knowledge of vibration theory. Once these limits have been set then the blasting should be well supervised and well monitored in order to maximise blast results which, in many cases, will mean approaching close to these vibration limits.

Down Hole Measurements

One, relatively recent technique that can help considerably in projects such as that described at Tai Koo Station is that of within borehole measurement of vibration. Such measurements can be used to overcome uncertainties often introduced by extrapolation of data, whether this is between sites, inter-site or between different blast types.

The in-hole measurement of vibration from blasting was an essential requirement of a project within the UK where it was necessary to establish the effects of blast induced vibration on an area of land slip. The response of the land slip zone to explosives detonated beneath it in competant rock was required in order to be able to establish safe blasting criteria for stabalising works. Measurements of this vibration at the ground surface above could not be expected to give adequate information for the rigorous analysis which was to follow. Thus, a series of in-hole recording packages were developed for this project to enable three-component geophone sets to be located at known depths within a series of boreholes. Permanent records were made at the surface and initially onto high speed UV paper recorders then subsequently using micro-processor controlled thermal paper/magnetic tape recorders.

Instrumentation Details

The in-hole packages comprised three component geophone packs assembled in plaster of paris within a UPVC pipe. Also assembled within the pipe was a rubber bladder capable of being injected with air from the ground surface via a pneumatic nylon tube. On expansion through a slit cut in the pipe the bladder would firmly hold the package in place at the required depth and allow for easy relocation. Permanent location would have been possible by grouting once the package was in the desired location.

Two sizes of package were designed with external diameters of 67 mm and 79 mm. This was to enable two such packages to be located within the one borehole whose internal diameter in turn needed to be between 80 mm and 140 mm. Their lengths were 800 mm and they weighed approximately 2 kg. Depths of up to 40 m within boreholes were attainable.

The overall frequency response of the system was designed as 5 Hz to 350 Hz with vibration in terms of peak particle velocity of up to 150 mm.s^{-1} being possible.

During initial trial blasting at this site the required total of 20 three-component recordings were successfully undertaken with no channel failures which clearly illustrated the robust nature of the equipment under what were fairly harsh field recording conditions. Vibration values from around 1 $mm.s^{-1}$ to 40 $mm.s^{-1}$ and frequencies 50 Hz to 300 Hz were recorded during these trials.

Subsequent blasting associated with the actual stabilising works requires long term recordings over successive 24 hour periods with rapid data reduction in terms of both peak particle velocity, amplitude and acceleration. Thus the same packages have now been linked into microprocessor controlled recorders capable of such analysis. Permanent recordings are made via thermal printout and digitised data can also be recorded onto magnetic tape for easy storage, replay through the thermal printer or spectral analyser if desired.

These recorders are also self-triggering and internally powered for up to several days of recording.

System Advantages

The great advantage of being able to use such a system is that recordings are now possible at, or very close to, the area in which they are required.

The Tai Koo project showed the necessity to measure the right type of blast over the same scaled distance ranges for which predictions are required and illustrated the practical problems associated with this need. As blasting approaches a monitoring location it is invariably found that vibration levels increase at an ever increasing rate and therefore this must be accounted for in such operations. To have an in-situ recording device to chart the actual rise in vibration as blasting approaches would enable the operators to maximise each blast. Data scatter is bound to be minimised which again would give rise to a greater degree of confidence in blasting operations and thus enable a realistic vibration limit to be set and for blasting to be accurately designed to approach but not exceed such a limit.

References

1. Nicholls H.R., Johnson C.F., and Duvall W.I. Blasting Vibrations and their effects on structures. United States Bureau of Mines, Bulletin 656, 1971.

2. Olson J.J., et al, Ground Vibrations from tunnel blasting in granite : Cheyenne Mountain (NORAD), Colo. United States Bureau of Mines, RI7653, 1972.

3. Holmberg R., et al, Vibrations generated by traffic and building construction activities. Swedish Council for Building Research, 1984.

City Development and Supply of Building Materials – Delhi, India

G.K. Sarin B.SC.(HONS), A.I.S.M., F.C.C.(COAL & METAL RESTRICTED), F.I.E.(INDIA), F.I.I.M.(INDIA), M.A.I.M.E.(USA), F.M.G.M.(INDIA)

Ex-General Manager Bailadila/Kiruburu Iron Ore Projects, National Mineral Development Corporation Ltd.
Dy General Manager (Mines & Quarries) Bokaro Steel Plant (Steel Authority of India Ltd.)

1. INTRODUCTION

Delhi is a sprawling city in the north-western juxtaposition of Indian map and is the capital of India. It had many ups and downs and its history is a graphic account of rise and fall of many empires. The great visionary and first Prime Minister of India, Pandit Jawaharlal Nehru described Delhi in the following words :

"Even the stones of Delhi whisper in our ears of the ages of long ago, and the air we breathe is full of the dust and fragrance of the past as also of the fresh and piercing winds of the present. It is not the narrow lanes and houses of old Delhi, nor wide spaces rather pretentious buildings of New Delhi that count, but the spirit of the ancient city of Delhi has been the epitome of India's history with its succession of glory and disaster and with its great capacity of absorb many cultures and yet remain itself. It is a gem with many facets, some bright and some darkened by age, presenting the course of India's life and thought during the ages. Here the tradition of millenia of our history surrounds us at every step, and the procession of innumerable generations passes by before our eyes".

When we talk of Delhi, we do not talk only of its legacy, traditions, cultural varieties etc. but also of its people and their problems, the needs of 5.45 million people, the challenges the future poses for about 14.42 million inhabitants by the end of this century. The graph below indicates the population growth of Delhi, (Fig.A).

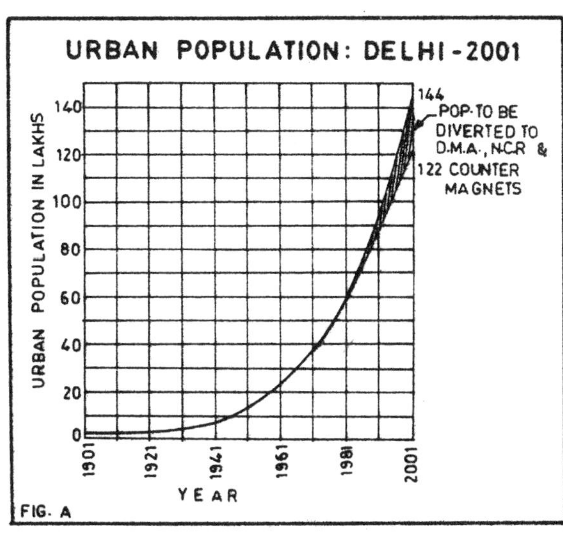

FIG. A

The urban population of Delhi has been rising at a phenomenal rate of 4.69% annual growth rate during 71-81 and the rural population of union territory has grown at 0.77" of annual growth rate. The slow growth of rural population is attributable to gradual shifting of the rural area and its merger with urban area. If the same rate of population growth continues, the urban population by the year 2001 would be about 14.426 million and rural 52.7 million. Based on assumptions of a more balanced development, the population for the union teritory of Delhi has been projected as under :

Population within Delhi Urbanisable limits - 2001	:	12.17 million
Population outside the urbanisable limits - 2001	:	0.64 million
Total	:	12.81million

The Government of India is fully conscious of the problems of Delhi and has enacted the "National Capital Region Planning Board Act-1985" with a view to ensure balanced development of the core city as well as creating infrastructures for sattelite towns so as to relieve pressure of high population density from the core city. Though the average gross residential density prescribed during 61-81 was 187 persons per hectare with overall city level density of about 100 pph : the same has been updated and with a careful planning it is possible to achieve gross residential density of 350-400 pph, and at city level overall density of 180-200 pph by 2001 AD. The boundary of Delhi Metropolitan Area has been redefined and a NCR (National Capital Region) created (Fig.B).

The civic amenities and dwelling needs are largely governed by the type of people and importance of the place. the classification of anticipated work-force in Delhi by 2001 A.D. is as given in Table 1.

The total area of the union territory of Delhi is 148639 hectares, out of this, 44,777 hectares had been earlier included in the urbanisable limits prescribed in plan. To accommodate about 12.2 million urban population by the year 2001 the NCR Board had identified a two pronged strategy : (i) To increase the population holding capacity of the area within urbanisable limits declared till 1981 and (ii) to extend the present urbanisable limits to the extent necessary.

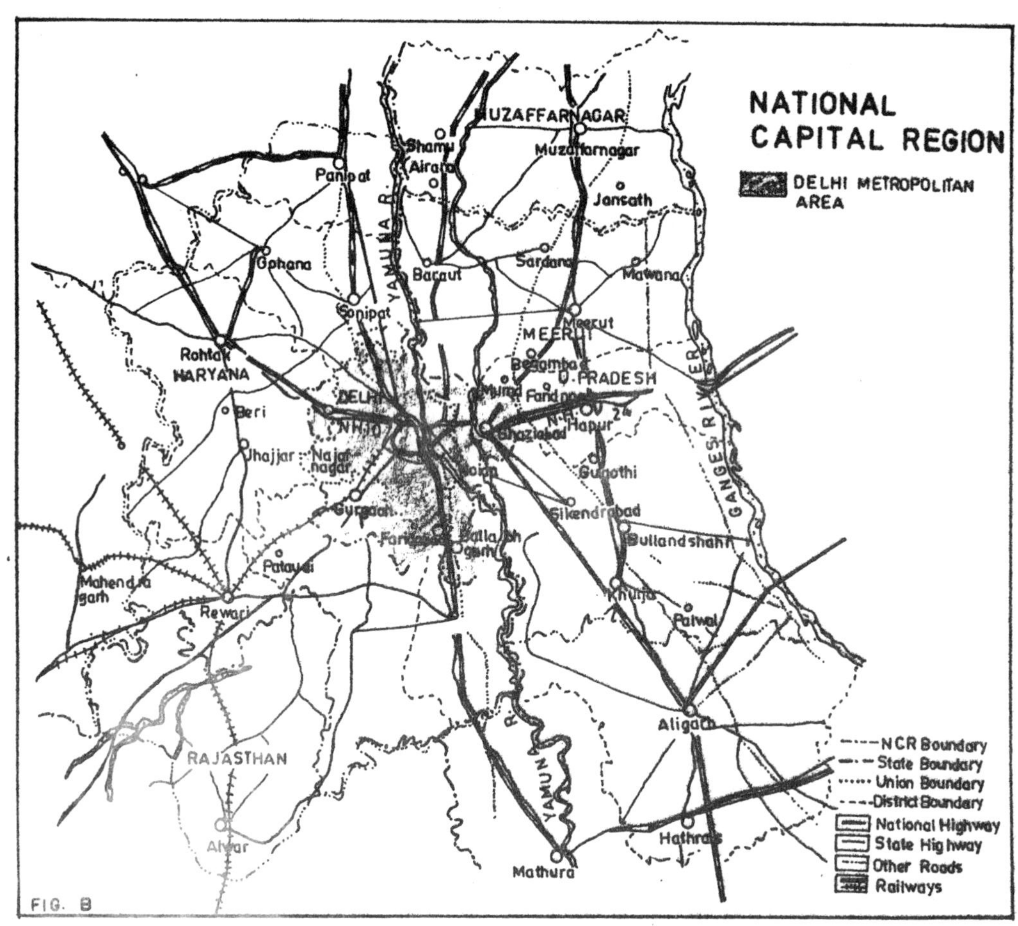

Table - 1

Sector	Within urban limits (in '000)	Outside urban limits ('000)	Total (in '000)
Agriculture	13.00	59.00	72.00
Manufacturing			
(i) Establishment	1071.00	39.00	1110.00
(ii) Non-establishment	214.00	6.00	220.00
Construction :	227.00	4.00	231.00
Trade & Commerce	964.00	12.00	976.00
Transport	488.00	16.00	504.00
Other services	1306.00	61.00	1367.00
Total :	4283.00	197.00	4480.00
Floating work force :	428.00	-	428.00
Total :			4908.00

Holding capacity of an area depends mainly on:

(i) Residential development types and their potential for higher absorption.

(ii) Availability/possibility of infrastructure physical and social.

(iii) Employment areas/centres, capacity and potential.

(iv) Transportation network capacity.

It has been ascertained that DUA-81 urbanisable limits would be able to accomodate about 8.2 million population in next 15 years by judicious infill and selected modification of densities. The remaining 4.0 million population is proposed to be accommodated in the urban extensions. To accommodate 4.0 million population, the DUA which could systematically hold 8.2 million population approximately needs to be extended by about 24,000 hectares over the next two decades to effectively respond to the growth of the capital. Land required for various developments in the extended time frame by the year 2001 may be acquired from time to time, with due regard to the balanced development of the city.

The land in the urban extension would approximately be distributed in the different land uses in the following manner :

Land use	% of Land
Residential	45-55
Commercial	3-4
Industrial	6-7
Recreational	15-20
Public & Semipublic facilities	8-10
Circulation	10-12

Town and country planning organisation has depicted proposed land use as shown below (Fig.C).

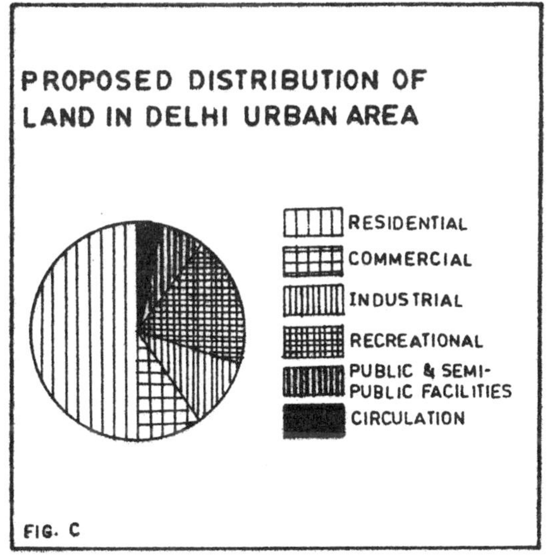

FIG. C

Urban Delhi at present accommodates about 1.15 million house-hold in different housing developments, resettlement, squatter, plotted, multi-family, unauthorised villages, traditional and others. Next two decades would add another approximately 1.3 million house-holds. Emphasis must be given both on the development of new housing areas as well as on conservation, improvement and revitalization of the existing housing areas. Housing shortage, at present, has been estimated to be about 0.3 million which includes (i) squatters and shelterless (ii) families sharing houses in the congested built-up areas and (iii) houses requiring immediate replacement. Thus about 1.62 million new dwelling units would be required during next two decades (1981-2001) divided in 5 yearly intervals as is given in Table 2.

A graph showing dwelling units requirements at 5 years interval is shown in Fig.D.

Table - 2

	New housing required ('000)	Average per year ('000)
1981-1986	323	65
1986-1991	379	76
1991-1996	434	87
1996-2001	483	97
	1619	

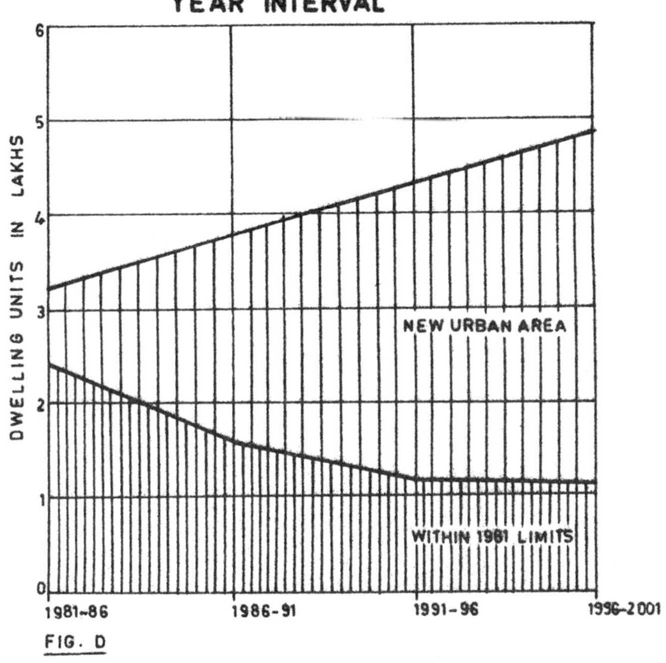

FIG. D

2. GEOLOGY

To meet above housing needs and civic amenities requirements, it is necessary to evaluate adequancy or otherwise of the available land, geological conditions and ground water conditions.

Delhi may not be an eminently suitable place for location of a capital city due to high seismic intensity. The area is techtonically disturbed and earthquakes of minor intensity are very frequent. However, no major damages have been encountered so far due to any appreciable ground movement. The city overlies a competent rock bed and drainage conditions are by and large satisfactory.

The Geological studies conducted in and around Delhi has established that the rocks belonging to this area are of Alwar and Ajabgarh series of Delhi system and structurally undergone much deformation and subjected to intrusions by acidic and basic magma.

2.1 The union territory of Delhi lies between 28° 29' and 28° 45' latitude and 76° - 5' and 77° 20' longitude covering an area of 1484 sq.kms. North and East sides are covered by Indogangetic plains, West by the Thar Desert and South by the Malwa plateau.

The area has a mature topography with hillocks of quartzite and Schists having longitudinal extension in N-S direction. These hillocks are intruded by pegmatitic solutions which were later weathered into clay. The synclinal depression near Marudpur due to second phase of deformation indicates the structural control over outcrop pattern. Drainage system is also controlled by structure through nallas flowing in SSW-NNE direction.

The climate is semi-arid. During summer the temperature exceeds 41°C and dust-storms are common. During winter, the temperature falls upto 4°C. The annual precipitation is about 40-60 cm.

A promontory of Aravalli Hills enters Delhi from South and expands into a rocky table land running North-Easternly across Delhi. The ridge has got a strike direction of NNE-SSW and terminates on the right bank of Yamuna on the North. The strike direction of the rocks vary from N-S to NE-SW. Dips are high and on the northern part towards East and South-East and on the South, towards West and North-West.

The greater part of Delhi lies on alluvium, but the small hills and ridges in and around New Delhi consist of Alwar quartzite of Delhi system. These quartzites are highly bedded, recrystallised and non-porous having joints of rectangular pattern.

2.2 Stratigraphy

The rock type in Delhi Area belongs to the Delhi system which constitute the North-Eastern part of the Aravalli belt running in the general direction of NNE-SSW. The Delhi system is divided into Alwar and Ajabgarh series. The former, mainly arenaceous constituting lower part and latter argillaceous and calcareous forming upper part. The startigraphic sequence of Delhi system is as given in Table 3.

Among the rocks of the Delhi system, Kushalgarh limestone and Hornstone breccia intervene between Alwar and Ajabgarh series. Heron considers that Hornstone breccia is of autoclastic origin and has been formed by shattering of their quartizite veins and srikanthan (1956-66) suggested the modification that the Hornstone breccias do not have any definite stratigraphic levels.

2.3 Geological Setting and Lithology

The geology represented in the present area is the continuation of Alwar Series of Delhi System. The most predominent rock types are :

(i) Quartzite
(ii) Pegmatite
(iii) Schist

The general trend of the rocks is N-S and it coincides with the trend of Alwar Series of Delhi System. The rocks of the area are profusely intruded by pegmatitic veins, which were weathered into clay. Schists are the product of Contemporaneous volcanic rocks. The quartzite and pegmatites belonging to the Alwar Series trend in the N-S direction and are found exposed in isolated exposures in the Northern part of the area near Munirka. However, further South near Marudpur, alluvium conceals the quartzites and such isolated exposures are rare.

Table - 3

Delhi System (Purana)	Intrusive pegmatites and quartz-veins, granite & Amphibolite	Ajabgarh Series	Slates, phyllites,quartzites, sandstone and quartzite. Impure limestone (several hundred mts. thick). Hornstone Breccia.
			Kushalgarh Limestone (500 mts.)
		Alwar Series	Quartzite, Arkose, grits, Conglomerates, limestone, mica-schists, contemporaneous volcanic rocks. (4000 mts.)
			Railo limestone and quartzite (600 mts.)

------------------------------------- UNCONFORMITY ---------------------------------------

2.4 Qyartzite

Quartzite occurs in the form of anticlines and synclines with axial trend as N-S, SE-SW. Quartizites are grey to bluish grey on fresh surface. On weathering, the colour varies from yellowish pinkish to reddish brown depending upon the extent of leaching of iron oxide. Weathering is usually prominent arnong joint planes giving rise to sub-rounded blocks. Schists bands are found inter beded with quartzite. Garnet and pyrite specks are found in the quartzite giving rise to amygdaloidal appearance.

Due to differential solution action of inhomogenities of the rocks, quartizites bands near Munrika show pitted appearance.

2.5 Pegmatite

Pegmatite bodies lie between Munrika in the North and Marudpur in the South. Out of the three pegmatite bodies located so far, two lie on SW of Munrika and run parallel to each other. The third band lie SE of Marudpur. The pegmatite bodies are lensoid in shape, measuring 1.5 to 0.25 km in length and width varying from 200 mts. The pegmatite bodies occupy structurally weak planes such as faults and joints, and are sheared.

The pegmatites show strong structural control. They are localised along the axial zones of NNE-SSW trending folds.

Thick and coarse book of mica are rare and mostly muscovite of 2" to 3" is observed. Along with muscovite, vermiculite are also seen. The mineral constituents seen are quartz, relict felspar, clay, mica and tourmaline. Ethedral crystals of beryl is reported to have been recovered.

2.6 Clay

The area lying between Munrika in the north and Mahipalpur-Marudpur in the south having an area of 7 sq.m. is extensively mined for clay. Clay is formed due to the residual deposits derived from the alteration of felspars present in the pegmatite.

27. Schist

Schist occurs to the SW of Munrika inter-layered with quartzite in thin layers varying in thickness from 60 cm to 1 m. They possess well-defined Schistosity due to alignment of muscovite flakes. Schistosity is often puckered indicating the inverment of beds. The rock is light greyish to green in colour and friable.

2.8 Ground Water Conditions

The mean annual rainfall for the last 75 years is 64.1 cm. The maximum recorded rainfall is 153.2 cm. and minimum is 26.3 cm. The rainfall is erratic, and the precipitation is fairly high.

The slope of the ground is at gradients of 2.84 mts/km. from west to centre and from north and south at 1.33 mts/km. The eastern portion has a gradient of 1.38 mts/km. All these gradients are towards centre.

The ground water in the area is derived from rain-fall. Water table below the land and surface vary greatly. The ground water is restricted to alluvial area.

Out of the 62 wells where water tables were observed, 48% lie between 6' and 10', 34 percent between 11 and 20 feet and 12 percent lie between 21' and 40'. The remaining 6% of wells show water table between 1' and 5'.

The abundance of building materials in close proximity satisfies the growth pattern planned for Delhi in coming decades.

3. EXCAVATION PLAN

3.1 Building Standards - The modern world is witnessing a concrete culture with high rising sky-scrapers in urban areas. The concrete jungle may appear imposing and give satisfaction to egoistical expression of status and authority but it is associated with many social, aesthetic and technological problems especially with regard to fire hazards and pollution problems.

Keeping above factors into considerations, sky-scrapers in Delhi are allowed on selective basis and in selected areas. Most of the residential buildings are planned on 3-storeyed and 5-storeyed basis.

3.2 Building Materials Requirements

The annual rock mass required for meeting planned growth pattern of Delhi is assessed to be of the following order :

i) Boulders, gravels, chips etc. - 5 mil.M^3

ii) Sand - 6.6 mil.M^3

iii) Earth for brick making - 15.0 mil.M^3

Over and above, large scale excavation is contemplated within the urbanisable limits for following purpose :

i) Site levelling including filling
ii) Sewarage development
iii) Laying underground cables and pipelines
iv) Foundation for buildings
v) Road and rail facilities
vi) Cellars, ponds and underground storage buildings.

Excavations for the above jobs may amount to roughly 3-3.5 million m^3 per year.

Coupled with above is the need for dislodging dilapitated structures which have outlived their lives. This would amount to planned destruction of about 25,000 dwellings and allied structures by end of this century and would call for disposal of about 4 to 6 mill.M^3 of rubble and debris.

Taken together, the total rock mass handling and excavation in and around the urbanisation area may amount to about 35 mil.M^3 per year.

The mining plan to meet the above excavation requirement is as shown in Fig.E.

Large scale mechanised mining in the urban environ is ruled out. Small scale mining using 100 mm drills, portable crushers and 15-25 tonne haulers have been planned. Conventional N.G. explosives are used for dislodging rocks from the quarry and consumption of explosives may go up to about 1500 tonnes by the turn of next century. Directional blasts with low V.O.D. explosives may be developed for excavation in the urban surroundings.

4. INFRASTRUCTURE PLANNING

Other than roads, rails, train and air traffic, the infra-structure requirement include water supply, sewarage, power, solid waste manage-

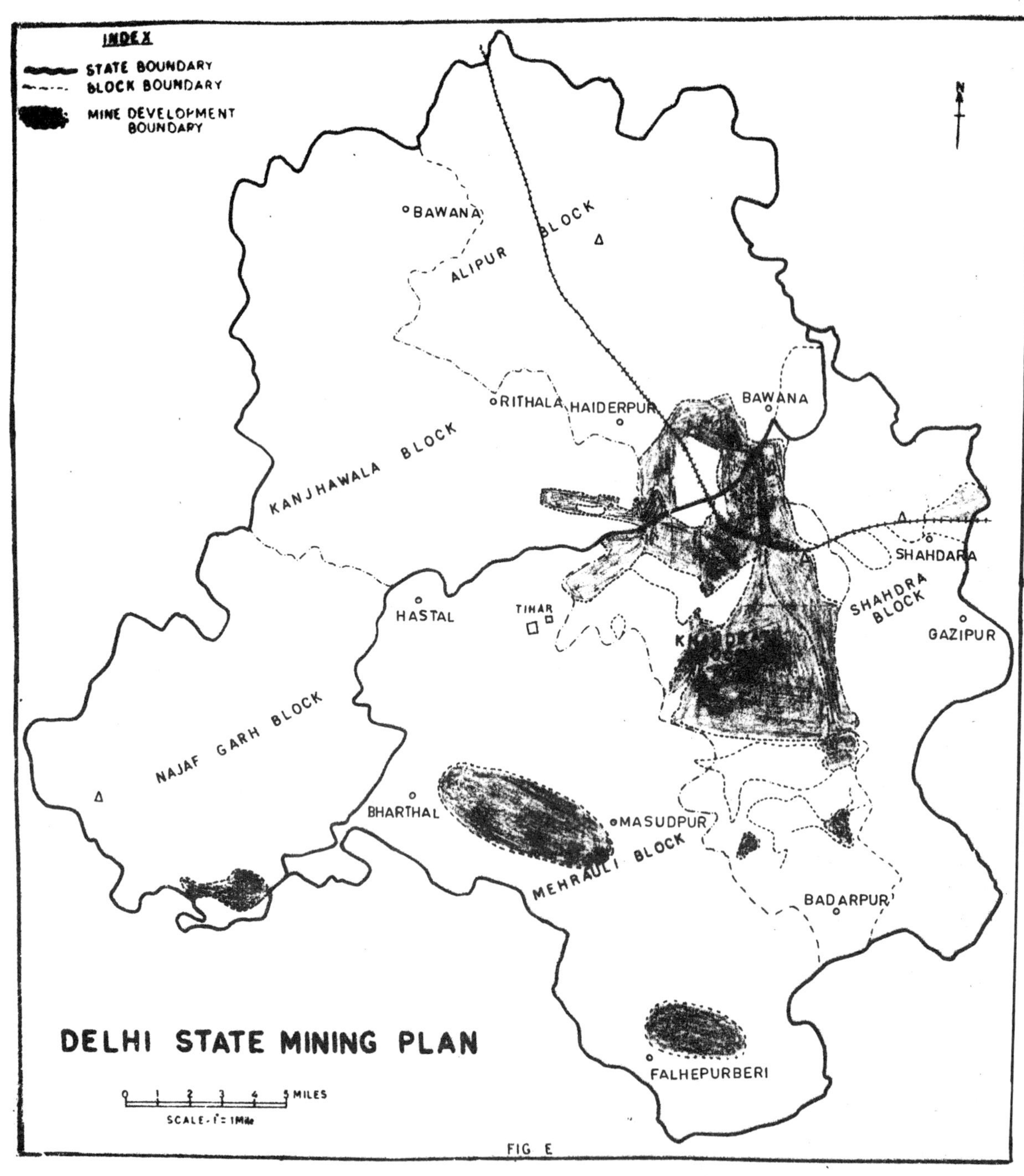

FIG E

ment and drainage and channelisation.

The quality of life in settlement very much depends on the level of availability, accessibility and quality of infrastructure that we provide. The rapidly growing population would necessitate the augmentalisation of water, sewerage, drainage and solid waste management. With the existing availability and projected need, water supply, sewerage, power and solid wastes is as indicated in Table 4.

4.1 Water - Delhi has to depend on river Yamuna for raw water though partial supply of water in trans-Yamuna area is being provided from River Ganga. Tehri Dam in UP and Kishau, Lakhwar and Giri Dams in Himachal Pradesh when complete could provide a major share of Delhi water requirements upto 2001. Balance could be met through exchange of waste water from Delhi with Haryana.

To prove additional water supply of 671 mgd, the existing water treatment plants would require augmentation and also construction of a new ater treatment plant in North-West by the year 2001 is as shown in Table 5.

Table -4

	Solid Wastes Ton/day	Water in Mil./gal/day	Sewerage Mil./gal./day	Power MW
Earlier target for 1981	2300	250	200	558
Present requirement	2568	496	397	650
Present availability	2058	303	118	-
Projection for 2001	6735	1127	902	2500

Table - 5

Water Treatment Plant	Existing capacity in mgd 1981	Recomended capacity in mgd 1985	Recomendded capacity in mgd 2001
Chandrawal I & II	90	90	150
Wazirabad	80	80	150
Haiderpur I & II	50	100	150
Shahdara	-	100	200
New Plant in North-West Delhi	-	-	300
Okhala	6	-	-
Renney Well	20	61	67
Local Tube Well	7	7	7
Total	253	438	1024

4.2 Sewerage

Sewerage treatment is essential to check environmental decay, as well as, to maintain the healthy living conditions. It is noted that the existing capacity of sewerage system in Delhi is grossly inadequate, as about 70% of present population does not access to regular municipal sewerage. The increasing pollution in the river Yamuna is also a major indicator for lack of sewerage treatment facilities. By augmenting the capacity of existing treatment plants, as well as, through two new sewerage treatment plants, one in North and other in West Delhi of 125 mgd capacity each, the liquid waste in Delhi in 2001 could be taken care of is as given in Table 6.

4.3 Power

Delhi's requirement of power in the year 2001 is estimated to be 2500 Mega Watts (MW). Delhi may be able to add only a limited capacity to its existing power generation because of increasing air pollution, scarcity of water and problematic coal transportation. It would have to bank upon sources of supply away from Delhi. Upto 1991, requirement of power shall be met is as given in Table 7.

To meet the targetted demand of 2500 MW by the year 2001, the power distribution network would be required to be taken over to 400 KV grid from existing 220 KV grid. To cater to this, a power net work has been worked out providing three major 400 KV electric sub-station in the North of Wazirabad barrage in (i) trans-Yamuna area; (ii) near Bawana in West Delhi and (iii) near Barthal in South-West Delhi. This would be fed from the Northern Grid and further power distribution system in Delhi would be from this grid and existing 220 KV grid.

Table - 6

Sewerage treatment Plant	Existing Capacity in mgd 1981	Recommended capacity in mgd 1985	Recommended capacity in mgd 2001
Okhala	66	128	150
Keshopur	32	70	170
Coronation	20	20	20
Rithala	-	75	150
Shahdara	-	60	160
New Plant in North Delhi	-	-	125
New Plant in West Delhi	-	-	125
Total	118	353	900

The areas where immediate regular sewerage is not available, low cost sanitation system for individual family should be adopted as a short range provision. The area should be planned in such a way that in the long range regular sewerage could be provided.

Table - 7

Source	Firm Capacity	
	1982-83 (MW)	1989-90 (MW)
DESU Local Generation	176	236
Badarpur thermal power station	500	500
Singrauli Super Thermal Power Station (MP)	15	150
Baira Suil (UP)	17	45
New Super Thermal Power Station to be provided in Murad Nagar (UP)	-	500
Total	708	1431

4.4 Solid Waste Management

Considering the nature of solid waste and the economic aspects of disposal, major part of solid waste has been proposed to be disposed off in sanitary landfils. The sites proposed are;

Site Description	Area in hactre
Site near Hasthal village in WD West Delhi	26
Site on Ring Road near village Sarai Kale Khan	20
Site in the North West Delhi	58.5
Site near Gazipur Dairy Farm, Trans-Yamuna Area	52
Site near Timarpur existing Landfil	40
Site near Gopalpur village in North Delhi	20
Site near Jahangirpuri	12

The sanitary landfilling on Ring Road is being done satisfactorily, however, it could be further improved by providing water prevention lever at the bottom to avoid water contamination.

At present, there are two compost plants, one each run by M.C.D. and the N.D.M.C. located near Okhala Sewerage Treatment Plant.

Waste from vegetable and fruit markets having higher organic contents should be used in these compost plants. No further sites have been identified for compost plant. The experience of the compost plants should be reviewed in 1992 and if necessary, policy changes could be done.

Special care is required in the disposal of waste from hospitals, slaughter houses, fruit and vegetable markets, diary farms and congested areas of old Delhi. Hospital waste which contains harmful micro organism should be handled separately and be incinerated. To avoid bird menace special care in the form of covered dust bins and quick removal of waste should be taken in the areas within five kilometers of airport.

To workout the requirement of dust bins, dhallaos, the following norms of solid waste may be adopted :

N.D.M.C. : 0.67 Kg.c.d.*

M.C.D. Area : 0.60 Kg.c.d.

*Kilogram per Capita Per Day.

4.5 Drainage and Channelization

Drainage : Drainage has two aspects, flood protection and storm water discharge, which are inter-related. The storm water and flood protection in Delhi are not local but have regional bearing including areas of Haryana and Rajasthan. Najafgarh drain and the Barapula Kushak drains which take storm water discharge in the urban areas, run to their full capacity during peak discharge periods. The required extensions of the present urbanisable limits would cause change in the surface run-off in the areas significantly, and thus the discharge would increase and there would be need of remodelling of existing drains and provision of additional drains. Possibility of a new major drain in the South through Harayana or Delhi to take discharge from Sahibi basin needs to be examined on priority.

Channelization of River Yamuna : Rivers in the major metropolitan cities of the world like the Thames in London and Seine in Paris, have been channelized providing unlimited opportunities to develop the river fronts. The possibilities in respect of river Yamuna have been studied in depth and indications are that it could be channelized within about 550 metres width and an area of about 3,000 hectares could become available for river front development. The Central Water and Power Research Station, Pune are conducting model tests for the same. After these studies are finalised, development of river front should be taken up by the D.D.A. as a project of special significance for the city.

5. POLLUTION CONTROL AND ECOLOGY

The N.C.R. is fully conscious in the problems of enormous pollution likely to threaten the lives and safety of this great city of India. Stretch of river Yamuna in Delhi has high level of water pollution. It has been planned to divert the discharge of waste water from Najafgarh, Barapula, Tuglakabad, Trans-Yamuna MCD, Sen Nurshing Home, Maharanibagh and Kalkaji drains through appropriate sewerage system followed by adequate waste water treatment so that the drain effluent conforms to the effluent standards prescribed by the Central Water Pollution Control and Prevention Board. Attempts may also have to be made to treat the waste water at the drain out fall through deep-shaft seration process. In the meantime, chlorination of above dams are being started. Sewerage systems are being extended in areas not served by sewerage. There are about 82 water polluting industries which together generate over 25 KL/day or more effluent. These industries will compulsorily make arrangements for treatment of pollutants collectively or individually.

There are about 55,000 industrial units and 0.64 million vehicles of various types and thermal power station which are polluting the atmosphere. The suspended particle matters ranges from 750 - 1500 mg/M^3 whereas permissible limit is 150 mg/M^3 only. Toxic-traces of elements like lead, mercury, manganese, calcium, arsenic etc. are dispersed in deep bio-spheres in addition to sulphur oxides and efforts are on way to minimise air-pollution before prolonged exposures cause various health hazards. The carbon monoxide and hydro-carbons emitted from heavy vehicles are proposed to be controlled to within less than 3% and 100 PPm respectively.

Noise pollution is rising rapidly with increasing traffic and industrial activities. The other cause of noise is the unabated use of loud-speakers.The warning limit of 85 DBA is suggested. Prolonged exposures to noise levels of 100 db can cause physiological changes and damages.

Smog during winter after sun-set in the early morning hours reduces visibility considerably. The semi-arid climate plays a significant role in impairing visibility. Large-scale use of sodium vapour lamp is planned to minimise visibility pollution.

The large-scale urbanization has its own advantages and disadvantages. Every effort is being made to regulate excavation, construction and pollution for capital of India for improving quality of life.

ACKNOWLEDGEMENT

I am thankful to Shri S Tiwary, but for whose assistance this paper would not have been completed.

I am thankful to management of SAIL, Bokaro Steel Plant for according kind permission fro presentation of this paper.

Session 1: Site investigation and geophysics

Chairman: W.G. Yuill
Rapporteur: W.N. Ridley Thomas

Discussion

W.N. Ridley Thomas,* in introducing the three papers† in session 1, said that they were concerned with promoting the design of more balanced investigation, during which consideration was given to the several disciplines of terrain evaluation, engineering geology, geotechnical investigation and geophysics.

Rapid changes were taking place in the application of geophysical methods to rock excavation site investigation. The appearance of cheap and powerful computing facilities, new electronics that enabled the development of shallow reflection equipment and the use of seismic measurements down boreholes to predict the dynamic properties of the intervening material provided the engineering gephysicist with fresh weapons to assist the rock excavator.

In Hong Kong all the best sites had been occupied long ago and most developments therefore took place either further and further up the harsh topography or on reclaimed land. The urban environment was, of course, geophysically more difficult, particularly in Hong Kong, where residents knew about living in the middle of a construction site and where almost all of the city was covered with concrete. It could be argued that conventional site investigation was not hampered by the urban environment — indeed, ease of access probably made aspects of it more efficient. Geophysical methods on both land and at sea were, however, invariably hampered by the urban environment — the former by noise, access and concrete and the latter by the fact that city/harbour waters were increasingly and adversely affected by pollution and a disturbed sea-bed.

Darwish and Hanif had described the investigation, design and some constructional aspects of a series of road, pedestrian and utility tunnels in Mecca, Saudi Arabia. The tunnels were being excavated in generally competent, very strong, granites and gneisses with slight to moderate weathering mainly at the surface; the rocks were intruded by thin dolerite dykes. A series of rock test data was presented for the strata and had been used to design lining thickness. Temporary support was designed on the basis of the NGI rock classification system.

The papers by McDowell and by McFeat-Smith and co-workers discussed geophysical applications to the problems of site investigations. The main theme of both papers was the absolute limitation of land geophysical surveys in the urban environment. The use of geophysics under conditions where acceptable records were obtainable was reported: clearly, there were conditions of noise in urban environments, access and other restrictions in which no seismic applications stood any chance of succeeding with the technology that was available to the engineering geophysicist.

The third paper presented a model or package site investigation for driven rock tunnels and illustrated very well how the use of an indirect 'seismic' method had served to point the drilling investigation towards problem areas, as well as stressing the importance of carrying out the investigation in the right order. The need for a different site investigation approach to the main types of terrain for such tunnelling was discussed and the principal site investigation techniques that were available and their application to Hong Kong conditions were outlined.

*Electronic and Geophysical Services, Ltd., Hong Kong.

†Darwish M.A. and Hanif M. Characteristics of rocks and their excavation in Makkah tunnels; McDowell P.W. Seismic refraction surveying in urban areas; and McFeat-Smith I. Nieuwenhuijs G.K. and Lai W.C. Application of seismic surveying, orientated drilling and rock classification for site investigation of rock tunnels.

G.W. Lovegrove,* in commenting on the paper by McFeat-Smith and co-workers, said that they were to be complimented on a very informative paper, but he believed that their statements regarding acoustic tomography (or cross-hole geophysics) should be qualified. In his experience acoustic tomography was not as simple or as successful for shallow investigations (up to 70-m depth) as the authors had suggested. He had used the technique during a recent investigation in Hong Kong in an attempt to identify a major fault and the extent of its associated brecciated zone. The rock type was fine- to medium-grained granite with 4–18 m of decomposed granite near the surface. The fault, as penetrated by drill-holes, comprised about 3 m of stiff to very stiff dark grey clayey silt with closely spaced slickensided shear planes.

Acoustic tomography was attempted with an array of four drill-holes, 60–70 m apart in line, spanning the fault normal to the strike. Drill-holes had been cased with PVC and filled with water. Two types of seismic sound source had been used — bolt PAR air gun model 5500 activated by 5–5.9 MPa air pressure and disintegrating electrolytic capacitors. The receiver was a 12-channel signal enhancement refraction seismograph (ABEM 'Trio' with SAS enhancement) together with associated hydrophones. The air gun had frequently ruptured the PVC casing.

The multi-element hydrophone cable (hydrophones at 5-m intervals) had been lowered to the bottom of one hole and the sound source was fired from a neighbouring hole starting at the bottom. Sound travel times between source and receivers had been measured and the process was repeated at 5-m intervals until the sound source was less than 5 m from ground surface.

Data from the pair of holes that spanned the known position of the fault were very difficult to interpret. The range of seismic velocity in the immediate vicinity of the fault was 3.9 to 4.5 m/s, compared with an expected range of 4.0 to 5.0 m/s for fresh and slightly decomposed granite. No distinct velocity profile had emerged to identify the fault location. Remote from the fault it was noticeable that the velocity pattern was reasonably regular, showing a clear profile of increasing velocity with depth, as might be expected. Close to the fault the pattern was erratic, suggesting either the existence of extraordinary ground conditions or, alternatively, confused ray paths that could not be interpreted sensibly. A drill-hole had indicated that rock conditions were not extraordinary, except that the material was fault-brecciated. It had been concluded that the erratic pattern of seismic velocity represented the extent of fault brecciation.

In summary, acoustic tomography had failed to identify the location of a major fault and the extent of brecciation only could be inferred.

*Mott, Hay and Anderson Hong Kong, Ltd., Hong Kong.

He would like to know if others had been able to identify the location and extent of a geological structure of major engineering significance solely by acoustic tomography.

Dr. R.L. Langford* said that the Geological Survey was currently mapping in the northwestern New Territories of Hong Kong. The tuffs of the Repulse Bay Volcanic Group were metamorphosed in that area, varying from augen schist to weakly foliated, hydrothermally altered tuff. The schist outcropped in a belt about 1 km wide, trending northeast – southwest. To the northwest were metamorphosed sediments of the Lok Ma Chau Formation, and the contact between the schist and sediments was a major shear belt.

To the southeast, away from the schist, the metamorphism became localized in hydrothermally altered shears, ranging from 5 to 500 m in thickness. Within those shears the tuff was altered to sericite and calcite, with accessory pyrite. Only the quartz pyroclasts remained to indicate the original tuff rock type. Rock exposures of the altered tuff were characteristically hard, angular and only slightly moderately weathered. The surrounding unaltered tuff was deeply weathered, with large, cuboidal corestones of fresh rock on hill tops. That difference in weathering and material properties could be used to explain some of the changing ground conditions in rock tunnels. As the pattern of altered belts was quite complex, accurate mapping of the outcrop would be useful in predicting changes in proposed tunnels.

Authors' replies

P.W. McDowell I should like to take the opportunity to comment on questions that were raised during the discussion but which have not been submitted for publication.

In regard to recent developments to improve the signal to noise ratio for seismic surveys in urban area, most modern seismographs have the facility for signal enhancement and/or signal averaging by use of repetitive sources that increase the signal to noise ratio, and also filters to remove unwanted frequencies from a recorded waveform. The number of traces that can be added together for signal enhancement or averaging depends on the memory of the computer in the seismograph. Another practical consideration is that signal definition and timing accuracy may be reduced if the energy source coupling changes and the signal is not reproduced faithfully at each impact. The largest drop-weight system possible should, therefore, be used to reduce the number of repeat blows and to obtain optimum signal to noise ratio quickly.

Another approach is to use a continuous vibration source with variable frequency (e.g. Vibroseis) or a rapidly repeated impact from a road tamping device (e.g. Mini-Soisi). These are commonly used for high-resolution seismic reflection surveys and have been employed for large-scale seismic refraction investigations in urban areas. The client must, of course, be prepared to pay the additional cost of recording and processing the more complicated data from such sources.

Seismic refraction surveying has several advantages over drilling. (*a*) Access is usually far easier for a seismic crew with a portable seismic system than for a drilling rig, especially in difficult terrain. (*b*) Seismic refraction results can be used to assess the geological variation at a site much more quickly and at far lower cost than a close pattern of vertical boreholes. The use of a multi-channel seismograph and sufficient impact points for time–depth analysis at each geophone station is assumed in making this statement. (*c*) Seismic velocity values from refraction surveys are closely related to vertical fracturing and seismic spreads can be orientated to map the pattern of vertical fracture systems. Vertical boreholes, of course, only provide information regarding non-vertical fractures and give little indication of the continuity and openness of fracture systems. (*d*) Seismic velocity values can be used to determine the elastic properties and the rippability of the rock mass.

It should be stressed that seismic surveys are used to complement rather than to replace the drilling programme, as both methods have their advantages and limitations.

The most significant development in the seismic evaluation of rock mass quality has been in the measurement of shear wave velocities, as well as compressional wave velocities, both in the laboratory and in the field. Shear wave velocities are not affected by the presence of groundwater, at seismic frequencies, and provide a more accurate assessment of fracture state and *in-situ* modulus for the rock mass below the water-table. Recent research has indicated that shear wave measurements can also be used to assess the shear strength of engineering soils and that the frequency of shear wave events can, for some rock types, provide a scaling factor between the elastic modulus of the fractured rock mass and that of unfractured rock samples. Borehole measurements of compressional and shear wave velocities are commonly used to assess the mechanical properties of oil-bearing formations and could be advantageous in many geotechnical investigations.

Cross-hole shooting is a well-established technique for the investigation of ground conditions at depths and locations unsuitable for surface seismic refraction surveying. More interest has been shown in the procedure in recent years as computer software has become available for acoustic tomography, i.e. the mapping of velocity zones between adjacent boreholes. Unfortunately, the practical difficulties in carrying out accurate measurements between boreholes negate the advantages of the sophisticated interpretation procedures. One problem is that there is usually a limited number of boreholes available and these are not always at suitable spacings or locations for the cross-hole investigations. Often, only two adjacent boreholes are used and coupling errors may be included in the transit times that exceed the time variability produced by ground conditions. Accuracy and definition can be improved by use of several boreholes in line in similar fashion to surface seismic measurements and borehole to surface measurements as well as cross-hole shooting. An array of boreholes can be used to map the extent and orientation of sub-vertical fault and fracture zones.

Dr. I. McFeat-Smith, G.K. Nieuwenhuijs and W.C. Lai As a general comment, the principal geological features of interest for rock tunnelling locally are the following.

(1) Thickness and stratification of surficial deposits, e.g. fill, marine deposits, alluvium and colluvium (ill-sorted hill-wash and landslip deposits).

(2) Anomalies in these deposits, e.g. boulders.

(3) Variations in weathering of the local igneous rocks.

(4) Rockhead.

*Geotechnical Control Office, Kong Kong.

(5) Rock material conditions, e.g. granites and volcanic rocks are generally strong to very strong, whereas the welded tuffs can be extremely strong. Layers of weak sedimentary rocks occur within the massive volcanic sequences.

(6) Rock mass condition, e.g. joint spacing averages about 0.3 m, but increases in the vicinity of contact zones and minor faulting that occur commonly.

(7) Water inflows, particularly for rock tunnels under reclamation areas that can dewater overlying soils.

(8) Major discontinuities.

With reference to items (3) and (4), sub-tropical weathering in the local igneous rocks together with well-developed jointing has resulted in a notoriously erratic rockhead profile. Large core boulders of fresh unfractured rock (say, up to 12 m in diameter) are found adjacent to zones of highly to completely weathered rock, and thick horizontally layered bands of weathered rock occur below a well-defined rockhead. The effect, as far as tunnelling is concerned, is to give rise to hazardous conditions when tunnelling close to rockhead. This has particular implications for excavation in developed areas.

In addition to standard field and laboratory testing, the principal techniques that are available for investigation of these features are noted below.

Geological mapping/terrain evaluation techniques Data search, geological mapping and aerial photography are a good start to any site investigation. The return from geological mapping is highly dependent on the amount of rock outcrops and deep cuttings for roads in Hong Kong, which can be considerable. In massive igneous rock terrain the lack of stratification gives rise to difficulties in determining the structural geology.

Drilling methods Vertical drilling is normally best to determine the stratification and locate rockhead in flat-lying areas and for sub-aqueous crossings. Inclined drilling is better, particularly for locating the many sub-vertical discontinuities, and gives some lateral coverage along the line of the tunnel. At portal areas horizontal drilling gives the best return as the total length drilled is relevant to the tunnel, and water outflows can be measured easily.

Seismic surveying Conventional refraction surveying is commonly used to locate rockhead and to measure rockhead velocity. Variations in the latter can provide data on rock mass conditions and the location of discontinuities. Anomalies can hinder the interpretation. Borehole correlation improves the interpretation greatly. Acoustic tomography is a new technique that can be used to investigate suspected major discontinuities for a more detailed understanding. Seismic reflection surveying is generally applied successfully to determine the stratification and rockhead levels for sub-aqueous crossings.

Rock classification Of the classification methods that are available, that presented has several advantages for use in Hong Kong. It does not rely on the use of index testing and has been specifically designed for local conditions (i.e. massive igneous rocks). It is used to give a quantitative summary of the geological conditions expected from site investigations for costing and programming purposes and, hence, to compare the feasibility of different alignments and carry out sensitivity analysis. It can also be used for payment purposes during tunnelling for risk-sharing contracts.

In reply to G.W. Lovegrove, it is not our intention to present acoustic tomography as a simple site investigation tool. It can, in our opinion, however, be an extremely effective method.

The technique can be used to complement a seismic refraction survey. The conventional seismic refraction survey has proved to be a cheap site investigation method well suited to reconnaissance work for excavation purposes. It is, however, only effective when the subsurface closely resembles the layered earth model used for the interpretation, as was discussed in the paper.

Like refraction surveying, tomography uses acoustic measurements to investigate the geological properties of the subsurface. Unlike refraction surveying, it uses no pre-determined model for the interpretation. Thus, tomography could detect irregular geological features. It is suggested that the technique be used in the final stages of a site investigation to obtain detailed information of a particular anomaly.

It is true that extreme care has to be taken in the preparation, execution and interpretation of a tomographic survey. The preparation and the lining of the boreholes are critical for successful execution; the accurate measurement of travel times is critical for the successful interpretation. In addition, it is important to have a clear understanding of likely and possible errors. This was borne out by the results of the survey mentioned by Lovegrove in which some of the authors were also involved. The survey in this case was not totally successful in determining the location of a fault. With hindsight there are several reasons for that: the tomographic investigation was in an experimental phase, the boreholes were not very suitable for acoustic measurements (the lining was too thin and shattered at an early stage, limiting the extent of the intended programme) and the grid spacing of the tomographic interpretation (5 m) was wider than the fault width (3 m).

Lovegrove states that 'the extent of the brecciation only could be inferred': it must be mentioned that it was fortuitous that one of the boreholes was drilled through the fault. Had this not been the case the tomography could possibly have been the only indication available to the engineer of adverse conditions.

Numerous successful tomographic experiments have been carried out in the U.S.A., Sweden and Finland.* The present authors hope to publish a paper shortly of a similar tomographic experiment carried out in Hong Kong for underground works in more favourable conditions.

*Peterson J.E. Paulsson B.N.P. and McEvilly T.V. Applications of algebraic reconstruction techniques to crosshole seismic data. *Geophysics*, **50**, 1985, 1566–80.
Cosma C. Determination of rockmass quality by the crosshole seismic method. *Bull. Int. Ass. Engng Geol.*, no. 26–27, 1983, 219–50.

Session 2: Deep excavations and foundations

Chairman: W.S.Y. Chan
Rapporteur: Dr. B. Littlechild

Discussion

Dr. B. Littlechild* summarized the contents of the three papers† that had been presented for discussion, outlining their principal conclusions and posing questions for further consideration.

K. Morton‡ said that he would like to comment on the assumptions made for water pressure distribution on joints in excavations of the type under discussion. In doing so he wished to use information obtained by Earth Sciences (Asia), Ltd., at the site of a 40-m deep vertical rock face excavation for the MTRC station and concourse at North Point, Hong Kong.

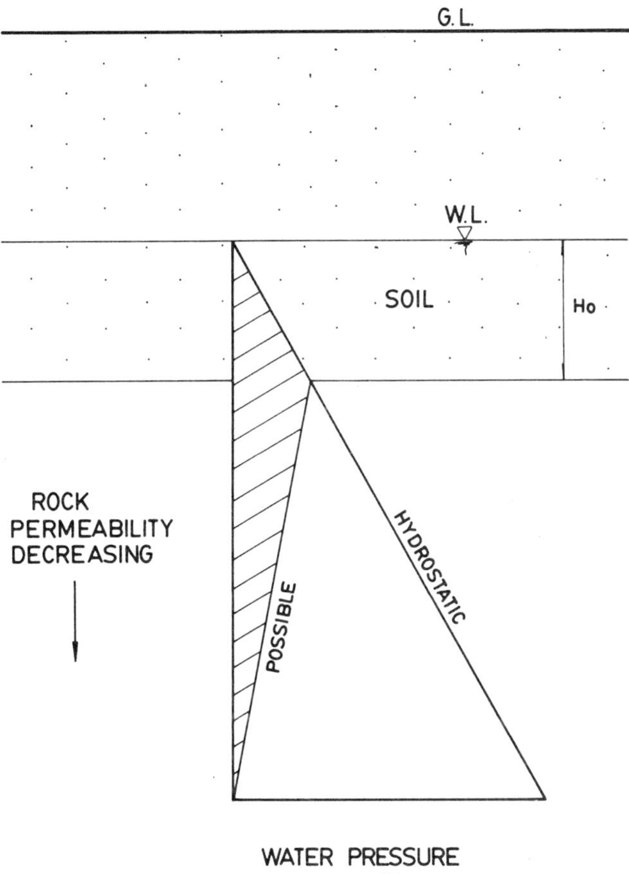

Fig. 1 Comparison of possible water pressure versus hydrostatic

The first aspect concerned the *in-situ* water pressure in the rock prior to excavation. It was traditional to assume that water pressure was hydrostatic, but where rock permeability decreased with depth it was possible that surface and ionic phenomena invalidated that and the water pressure reduced to zero with depth (Fig. 1). Examples of movements of water pressures in rock were rare, but an instance of such a phenomenon had been given [1] for a borehole at a mine in Zambia (Fig. 2). In that case there appeared to have been under-drainage at depth and then an increase to hydrostatic, probably in fractured rock. Clearly, the actual water pressure was a function of the local hydrogeology and not always hydrostatic.

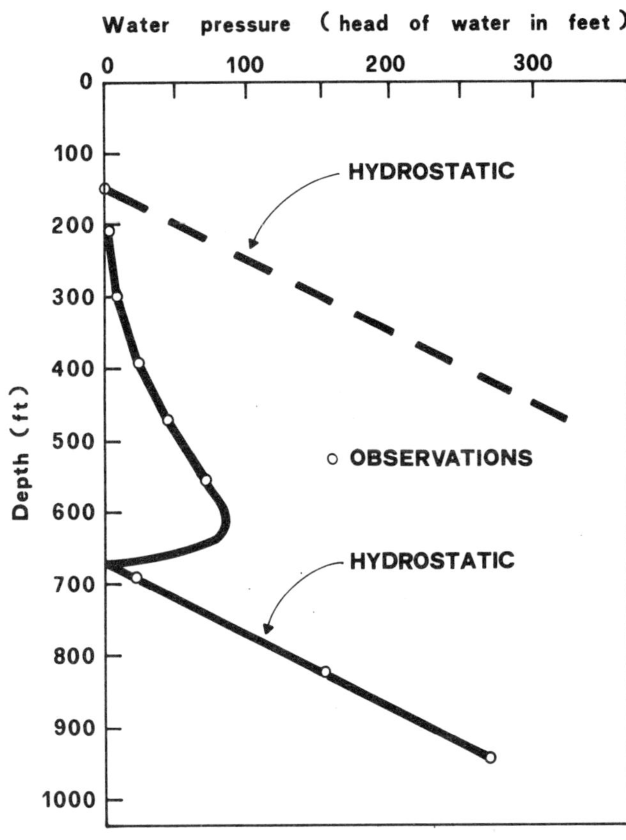

Fig. 2 Water pressures in borehole at Nchanga mine

When one was considering excavation into rock a further factor – release of water from joints at the face – had to be considered. Hoek and Bray [2] had proposed that a build-up of hydrostatic to mid-depth of excavation followed by a decrease to zero at the excavation toe was valid. Other groundwater models were possible and had been considered by Matson and co-workers.[3] Those were reproduced in Fig. 3: given known persistent joint patterns, the potential variation of forces due to water pressure was very large.

In considering that problem at North Point they had, in common with the other contributors, little data at the time of design. Their assessment of likely water pressure was therefore based on observed fracture frequency in boreholes (Fig. 4). They had then made a prediction of likely water pressures during excavation based on a method described previously.[4] The effect of using that assumption of water pressure, compared with the Hoek and Bray approach, could be seen in Figs. 5 and 6. For that case, when the excavation was relatively shallow Hoek and Bray was a good approximation, but at great depths it was a large overestimate.

On the basis of that evidence they had decided to use their predicted model, but to check it by *in-situ* observations during excavation. To do that they had installed

*Ove Arup and Partners, Hong Kong.

†Costello J. Mak L.M. and Chan M.T.S. Geotechnical designs for the construction of Fortress Hill Station, Island Line, Hong Kong; Greenway D.R. Powell G.E. and Bell G.S. Rock-socketed caissons for retention of an urban road; and Matson C.R. Choy H.H. and Gibson A.M. Rock support prediction in the deep basement of Cuaseway Bay East Concourse for the Mass Transit Railway, Hong Kong.

‡Earth Sciences (Asia), Ltd., Hong Kong.

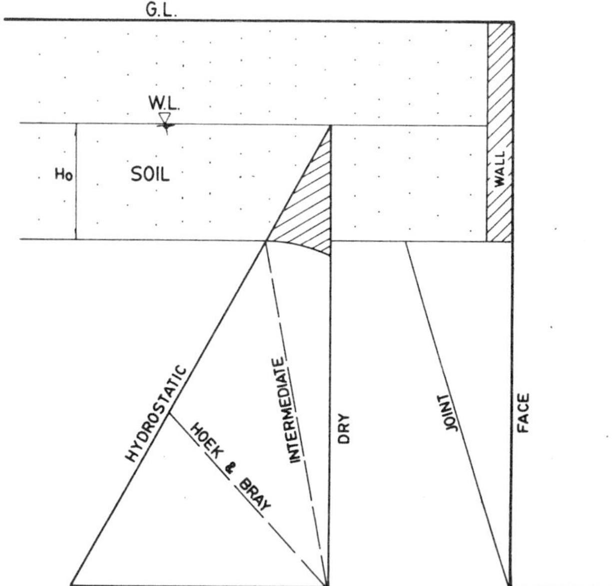

Fig. 3 Possible groundwater pressure models for excavated rock face

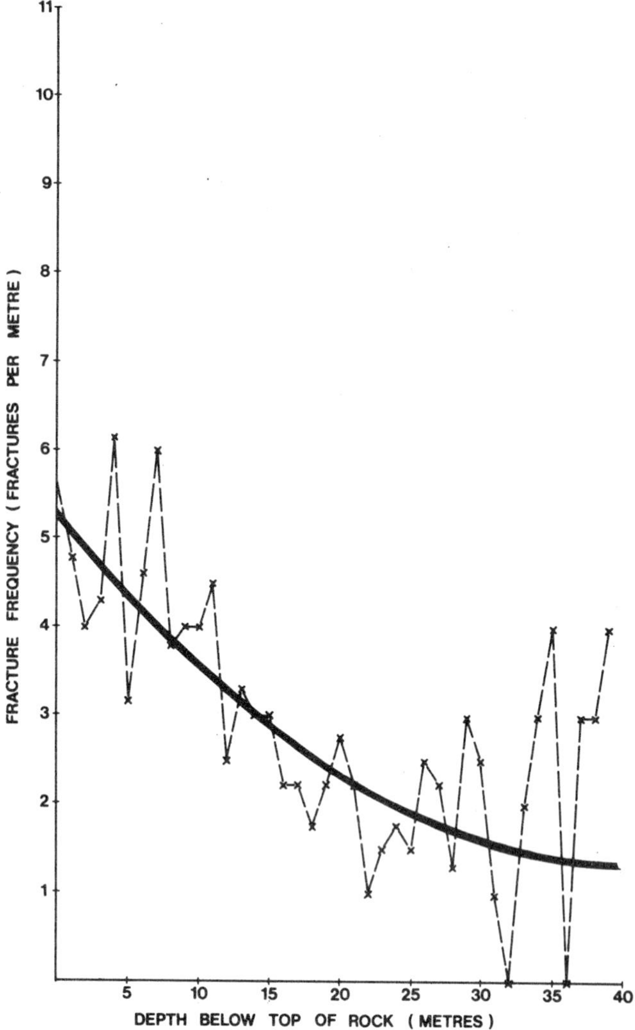

Fig. 4 Fracture frequency versus depth

inclined pneumatic piezometers with a wide sand observation zone placed close to potential joints. The results of measurements were shown in Fig. 7. Two main conclusions could be drawn from the results: (1) that, within a reasonable scatter, the assumption of water pressure was confirmed

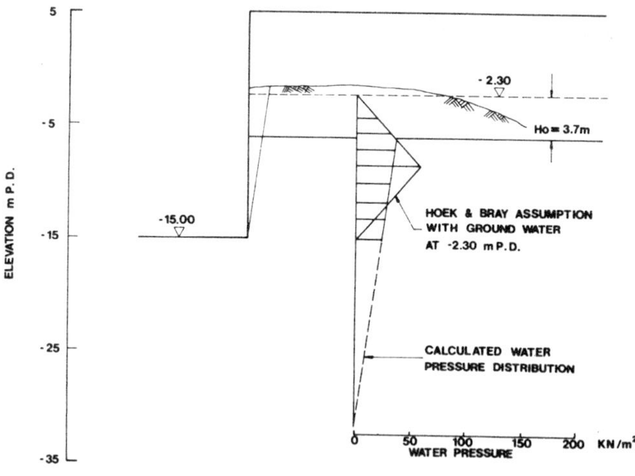

Fig. 5 Water pressures predicted during excavation

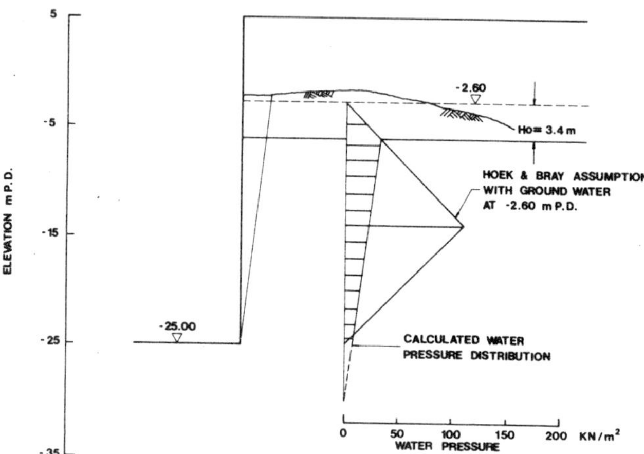

Fig. 6 Water pressures predicted during excavation

and (2) the pressure in deeper piezometers reduced as the excavation proceeded and allowed relief from the face.

In conclusion, it was noted that a variety of assumptions for water pressure had been made by contributors to the conference and in other documented cases. Very few measurements of water pressure in rock either before or during excavation had, however, been made. It was clear from their data that the local hydrogeology and the excavation procedure both influenced the water pressures present on joints. As those were likely to significantly influence the stability of the excavated faces it was recommended that more data of that kind should be collected and reported in future to advance the state of the art.

References

1. Steffen O.K.H. and Klingman H.L. Slope stability at the open pit of Nchanga Consolidated Copper Mines Limited. *J.S. Afr. Inst. Min. Metall.*, **67**, 1966–67, 140–71.
2. Hoek E. and Bray J.W. *Rock slope engineering, 3rd edn* (London: IMM, 1981), 358 p.
3. Matson C.R. Choy H.H. and Gibson A.M. Rock support prediction in the deep basement of Causeway Bay East Concourse for the Mass Transit Railway, Hong Kong. This volume, 285–97.
4. Morton K. *et al.* Temporary support for a deep excavation at North Point, Hong Kong. In *Design and performance of underground excavations: ISRM symposium, Cambridge, 1984* Brown E.T. and Hudson J.A. eds (London: British Geotechnical Society, 1984), 347–52.

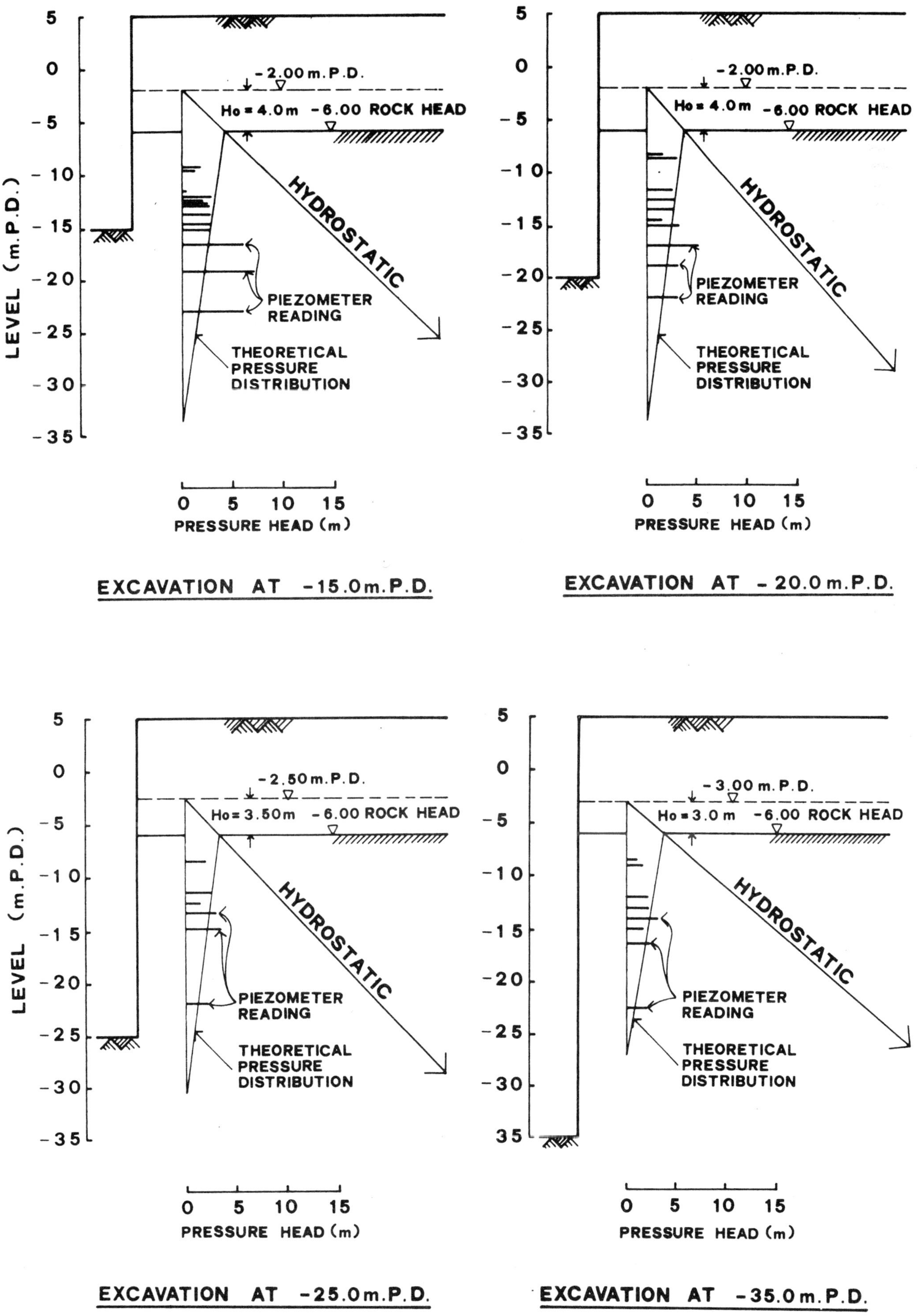

5
0
-5
-10
-15
-20
-25
-30
-35
LEVEL (m.P.D.)
-2.00 m.P.D.
Ho = 4.0m -6.00 ROCK HEAD
HYDROSTATIC
PIEZOMETER READING
THEORETICAL PRESSURE DISTRIBUTION
0 5 10 15
PRESSURE HEAD (m)

EXCAVATION AT -15.0m.P.D.

5
0
-5
-10
-15
-20
-25
-30
-35
-2.00 m.P.D.
Ho = 4.0m -6.00 ROCK HEAD
HYDROSTATIC
PIEZOMETER READING
THEORETICAL PRESSURE DISTRIBUTION
0 5 10 15
PRESSURE HEAD (m)

EXCAVATION AT -20.0m.P.D.

5
0
-5
-10
-15
-20
-25
-30
-35
LEVEL (m.P.D.)
-2.50 m.P.D.
Ho = 3.50m -6.00 ROCK HEAD
HYDROSTATIC
PIEZOMETER READING
THEORETICAL PRESSURE DISTRIBUTION
0 5 10 15
PRESSURE HEAD (m)

EXCAVATION AT -25.0m.P.D.

5
0
-5
-10
-15
-20
-25
-30
35
-3.00 m.P.D.
Ho = 3.0 m -6.00 ROCK HEAD
HYDROSTATIC
PIEZOMETER READING
THEORETICAL PRESSURE DISTRIBUTION
0 5 10 15
PRESSURE HEAD (m)

EXCAVATION AT -35.0m.P.D.

Fig. 7 Pneumatic piezometer results compared with theoretical pressure distribution

Written contribution

K.L. Lai and K.H. Li* Greenway and co-workers are to be congratulated on a very interesting paper, which has particular relevance to Hong Kong conditions, where space is at a premium and conventional methods of improving the stability of existing slopes by cutting back and constructing standard retaining walls are not always feasible. The paper is especially welcome as there is little published data on the type of rock-socketed caisson wall described by the authors. In addition, the monitoring carried out as part of the project is of particular interest.

In his presentation (unpublished) Powell gave the results of some of the strain gauge monitoring, and compared the measured bending moments with those allowed in the design. Because of the rather inconsistent results, he sought ideas that might help to explain them.† In this contribution an attempt to examine the results is made, and some ideas are put forward. Certain assumptions have had to be made and these are listed in the Appendix and illustrated in Fig. 8.

Deflection Assuming three different loading conditions on the wall ($Ka = 0.228$, $Ko = 0.371$, $Ko^* = 0.5$), the corresponding structural deflections of the reinforced caisson have been calculated (Table 1).

Table 1

Loading condition	Computed structural deflection for duration of loading, mm			Remarks
	0 day	100 days	∞	
Ka	1.2	3.0	4.5	(1) Only creep effect taken into account
Ko	2.0	5.0	7.5	
*Ko**	2.7	6.5	10.0	(2) Assume full fixity at rock-head

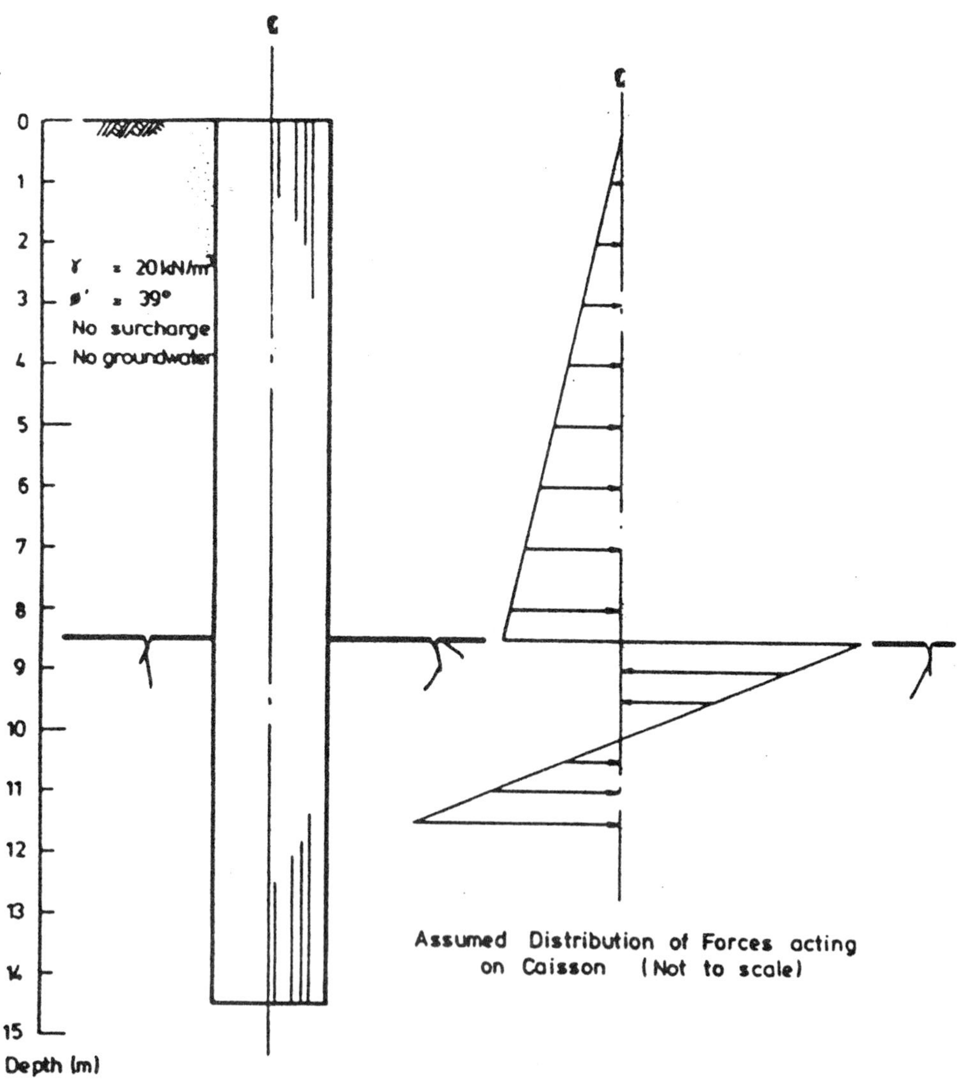

Fig. 8 Assumed force distribution

*Scott Wilson Kirkpatrick and Partners, Hong Kong.

†G.E. Powell proposes to present the monitoring results in a further paper and regrets that he is unable to comment on the Lai–Li contribution here. – Editor.

According to Powell, the actual deflection was very difficult to measure, but he suggested a figure of 2–4 mm. This measured deflection, therefore, compares with the computed structural deflection due to loading under active stress conditions. Nevertheless, conventional theory suggests that

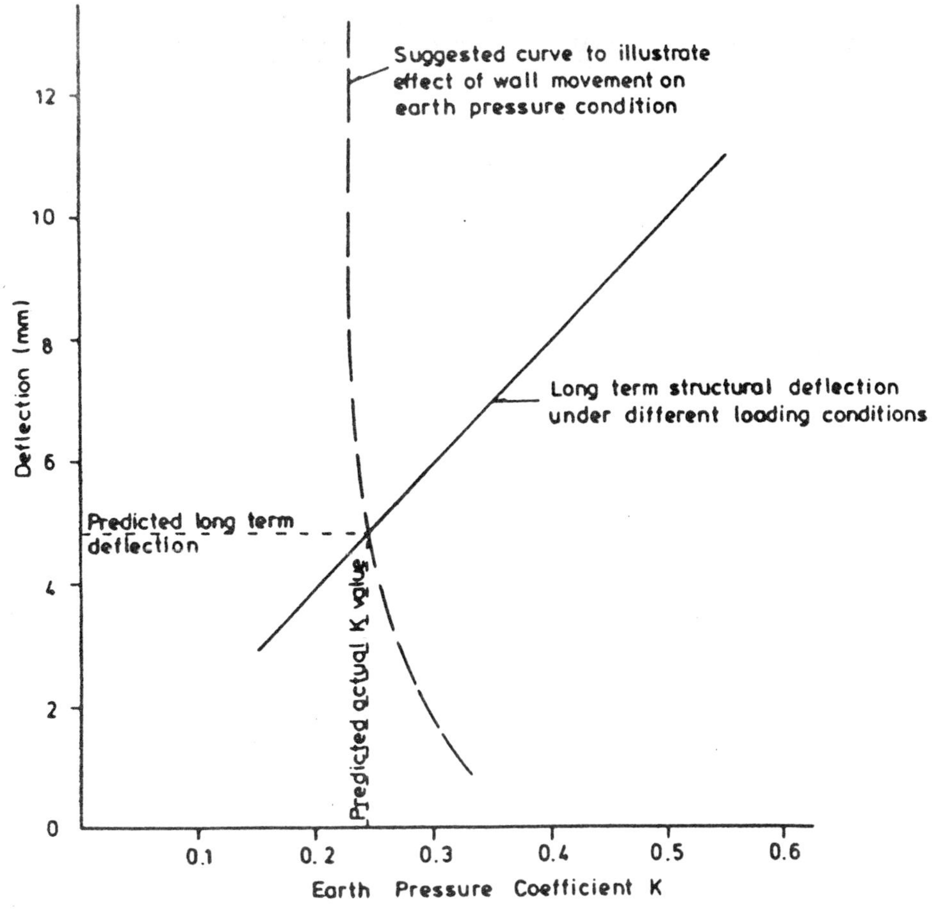

Fig. 9 Deflection—loading condition relationship

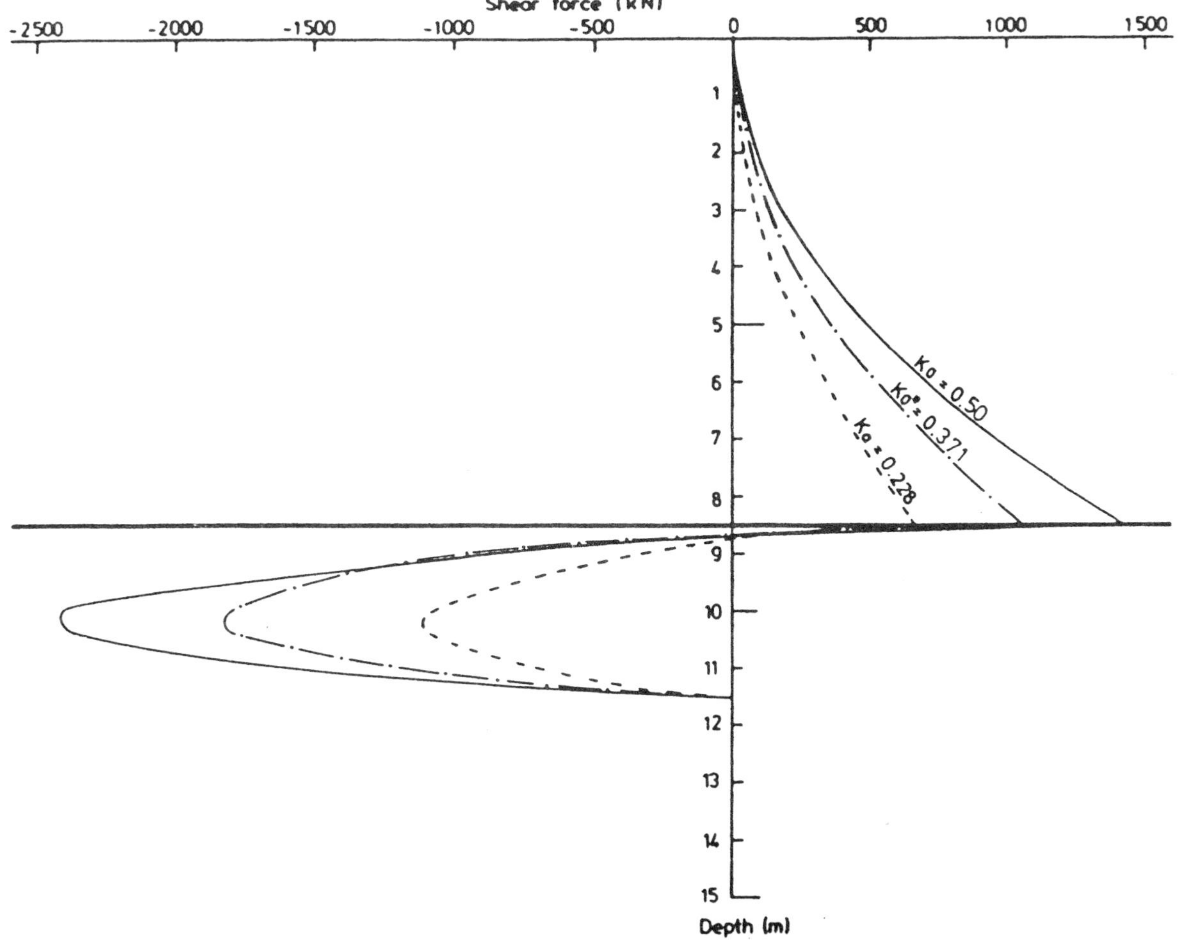

Fig. 10 Shear force diagram for different loadings (assuming effective socket depth of 3 m)

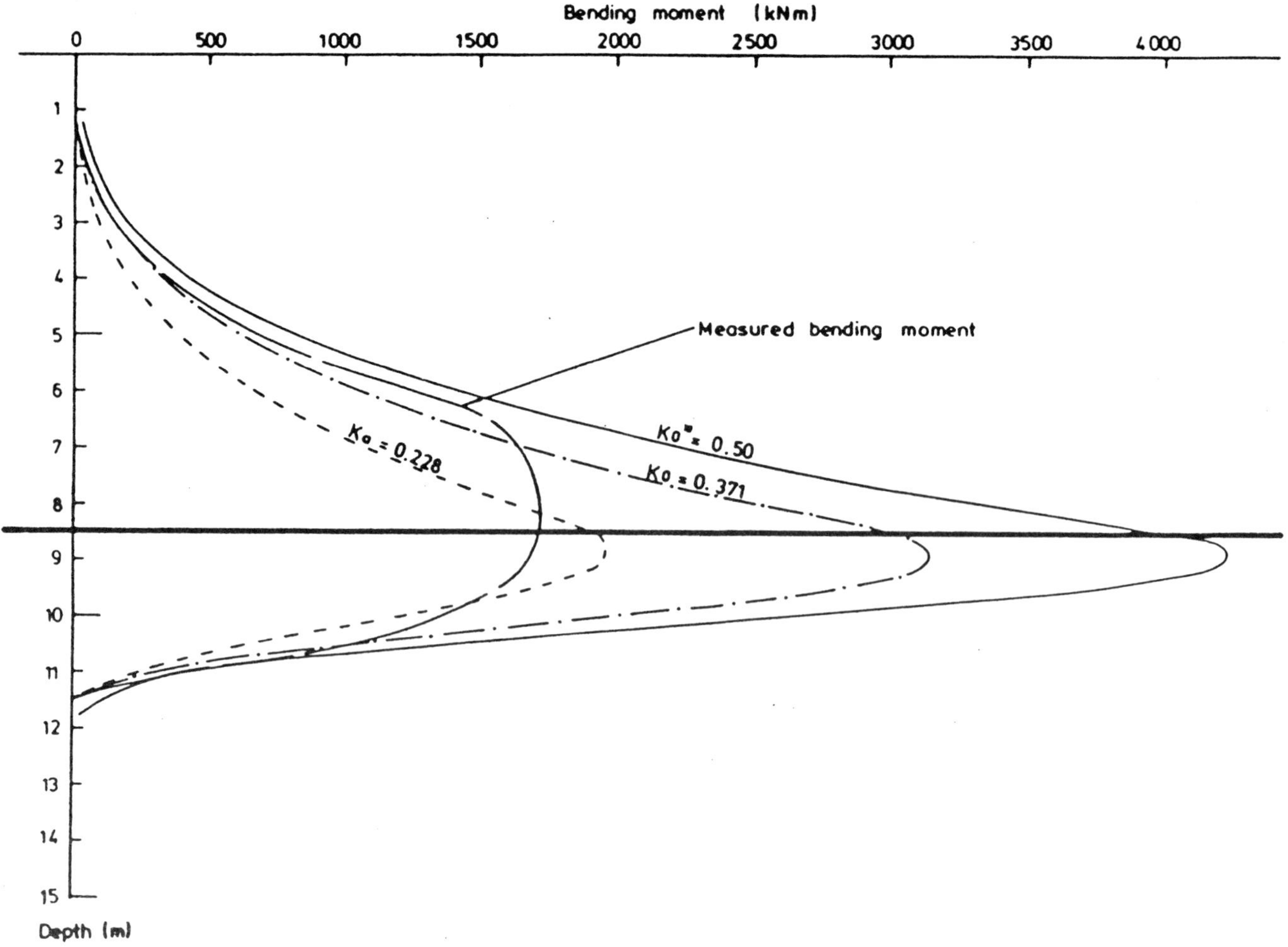

Fig. 11 Comparison of bending moment diagram for different loadings with measured values

the necessary rotational displacement for a rigid wall about its base to develop an active pressure condition is about 0.001*H* (where *H* = height of wall), though this could be less for a medium dense cohesionless material. In this case this expression computes as 8.5 mm. Fig. 9 illustrates the form of a suggested relationship between deflection and loading condition.

Thus, from a consideration of deflection alone, it does not appear that the earth pressure acting on the wall is necessarily represented by the at-rest loading condition.

Bending moment diagram Based on the force distribution defined in Fig. 8, a series of bending moment and shear force diagrams has been plotted (Figs. 10 and 11). As was illustrated by Powell, the calculated bending moment represented by a *Ko* factor of 0.5 shows remarkable similarity to the 'measured' bending moment derived from the strain gauge readings for the upper part of the caisson. Nevertheless, the maximum calculated bending moment for a *Ko* of 0.5 is considerably in excess of the 'measured' value. In preparing these diagrams the effective rock socket was taken as 3 m, as opposed to the actual depth of 6 m, based on the results of the strain gauge monitoring. It is, however, interesting to compare the bending moment and shear force diagrams for different rock sockets (Figs. 12 and 13).

From Fig. 12 it can be seen that the smaller the effective rock socket, the greater the shear force and the nearer its line of action to the rockhead. The strain gauge results might, however, lead one to think that the lower part of the socket was not contributing to the effectiveness of the wall. Obviously, there must be a minimum socket length, and parallels can be drawn to the 'bond length' in reinforced concrete design. Fig. 14 shows a suggested mechanism.

Returning to Figs. 10 and 11, an explanation was sought by Powell to reconcile the differences between the 'measured' bending moments and those calculated from the assumed loading on the wall. The shape of the 'measured' bending moment diagram tends to suggest that some force, in addition to the cantilevering effect of the rock socket, is restraining movement. Although there is some soil above the rockhead, the passive resistance of this is negligible. What other explanations are there?

The following points are put forward for further discussion:

(*a*) In any experiment of this type the first queries must be addressed to the monitoring itself: how was the correct orientation of the strain gauges ensured during installation and does the strain gauge on the expected neutral axis give any tensile strain reading?

(*b*) How were the bending moments derived from the strain gauge readings?

(*c*) Is it possible that some arching effect between caissons is occurring and what is the effective rockhead level in the caissons on either side of the instrumented caisson?

(*d*) Could the arrangement of the reinforcement in the caisson affect the measured value of the bending

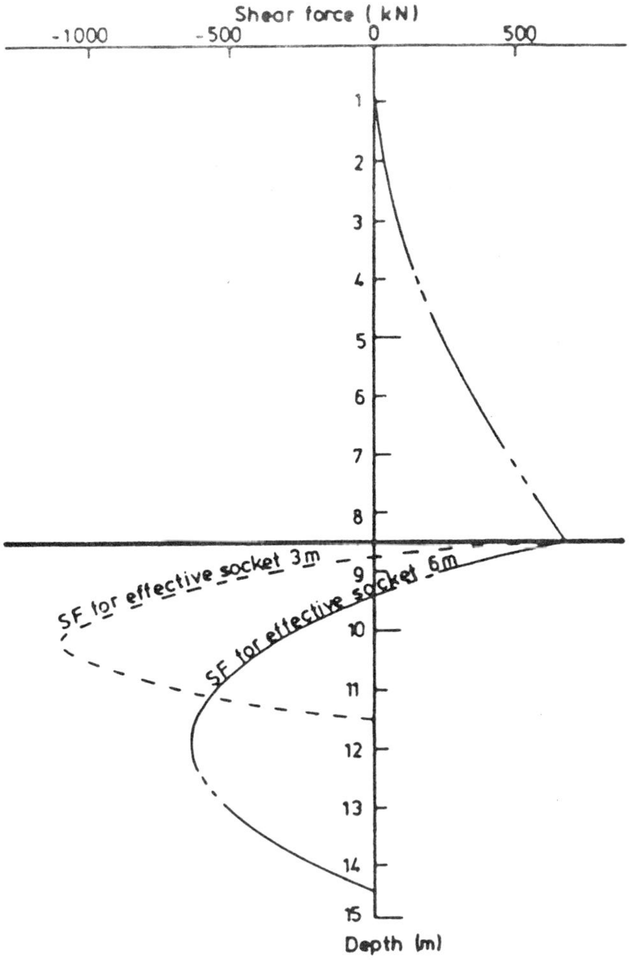

Fig. 12 Shear force diagram for active loading with different socket depths

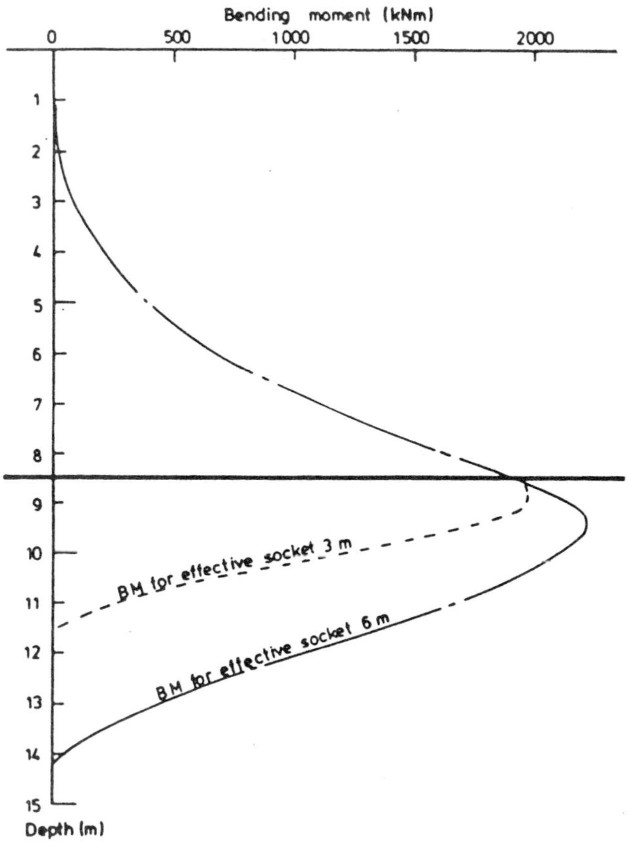

Fig. 13 Bending moment diagram for active loading with different socket depths

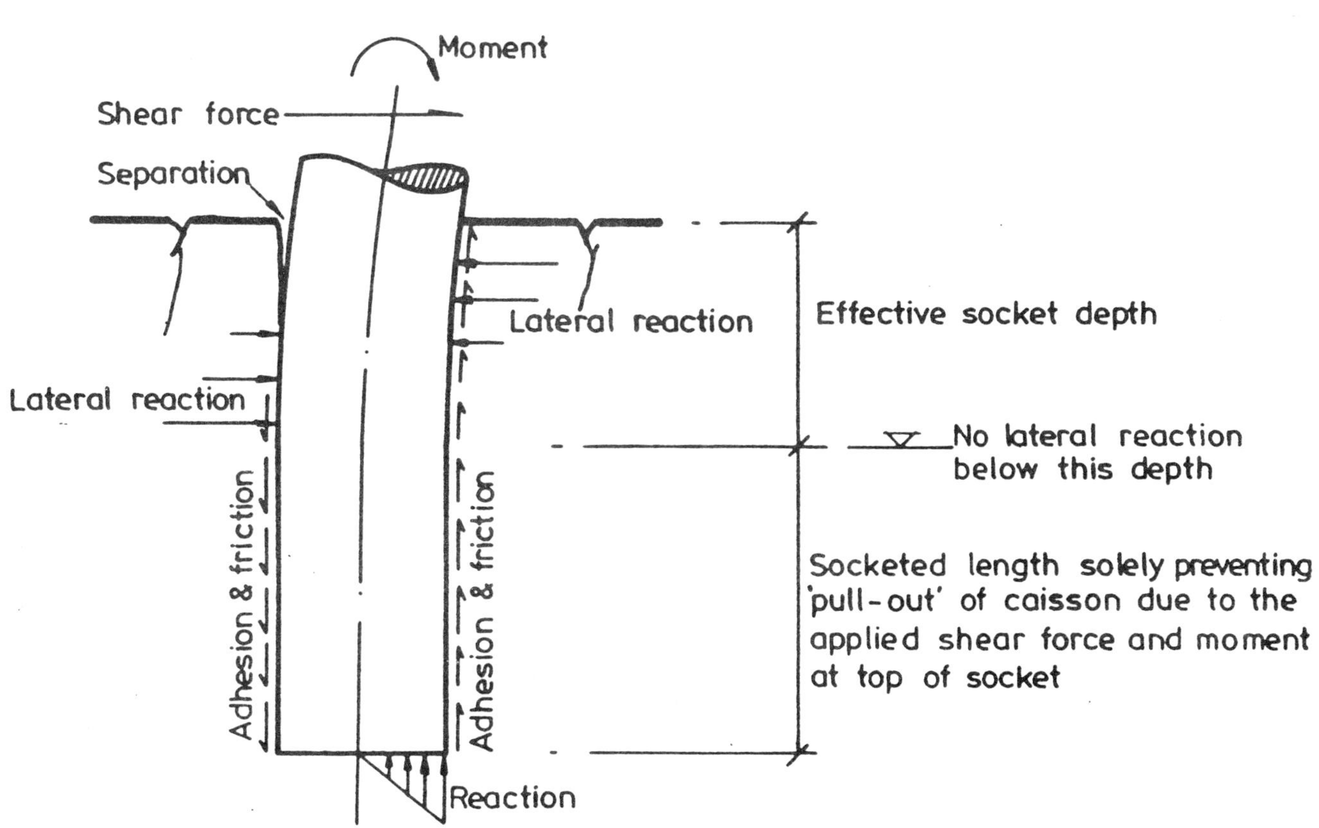

Fig. 14 Forces and reactions acting on caisson socket

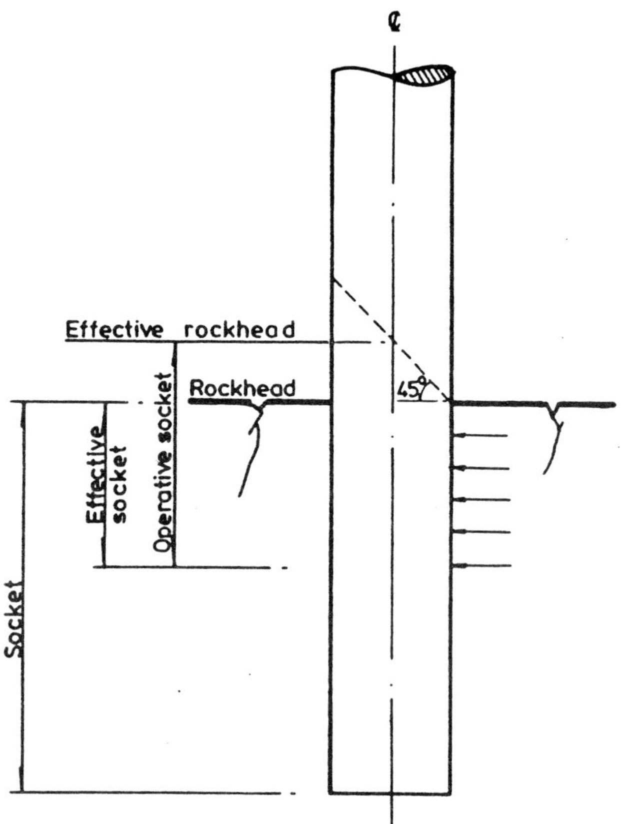

Fig. 15 'Effective' and 'operative' sockets

moment (e.g. lapping or curtailment of reinforcement near the strain gauge)?

(*e*) Is there some sort of load distribution effect at the rockhead (see Fig. 15)?

Although unanswered questions arise from the monitoring, the data do provide an interesting insight into the design problems with this sort of structure. As the authors remark in their paper, a 1-m increase in depth from 18 to 19 m gives rise to a 63% increase in main reinforcement. Although a *Ko* value of 0.5 may have overestimated the maximum bending moment, the use of a *Ka* value of 0.228 would have underestimated the measured value for a significant depth of the caisson.

Appendix The following assumptions were made for the purpose of this note:

(*a*) No groundwater behind or in front of the caisson wall (i.e. groundwater table below founding level).

(*b*) No vertical surcharge on retained soil.

(*c*) Shear strength parameters of retained soil:

$c' = 0$

$\phi = 39^\circ$

$\gamma = 20 \text{ kN/m}^3$

(*d*) Active state of stress

$$Ka = \frac{1 - \sin\phi'}{1 + \sin\phi'} = 0.228$$

At-rest state of stress

$$Ko = 1 - \sin\phi' = 0.371$$

$$Ko^* = 0.5 \text{ (value assumed in design)}$$

(*e*) No friction between caisson and soil or caisson and rock, i.e. $\gamma = 0$.

(*f*) Elastic modulus of SDG, $E_v = 40 \times 10^3$ MPa.

(*g*) Elastic modulus of grade 30 concrete, $E_c = 28 \times 10^3$ MPa.

(*h*) Stress distribution within rock socket to be triangular.

(*i*) Coefficient of horizontal subgrade reaction of SDG to be constant with depth.

(*j*) Each caisson retained a 4-m width of soil (i.e. 2 m on either side of central line of caisson).

H.T.M. Insley* I should like Greenway and co-workers to explain what water pressures were included in the design. Bearing in mind the restricted flow path resulting from the constitution of the caissons, what rise in water-table was predicted and could that be checked by monitoring?

Perhaps the authors would describe the nature of the drainage layer between and behind the caissons.

Dr. T.R. Stacey† and A. Haines† In surface and shallow underground excavations in rock it is usually the jointing in the rock mass that controls the stability of the excavation. Consequently, considerable attention is usually paid to the collection of adequate joint data. The joint characteristics that should be recorded as part of a joint mapping programme are dip and dip direction, spacing, continuity, roughness and joint surface strength and infill.

A sufficient number of observations should be made to define the statistical variations of the characteristics. This is important as it is essential to take the variations into account in the design process. Interpretation of jointing in a variety of rock types has shown that the distributions of geometrical factors are commonly as illustrated in Fig. 16.[1,2,3]

The actual extent and characteristics of joints are *known* only on the exposed rock surface at the locations of the joint measurements. At the level of an underground opening these data will therefore only be known for certain once the opening has been excavated. The design engineer needs, however, to have a realistic idea of the jointing at the underground opening in advance of excavation. A common approach is to assume that the joint sets are ubiquitous and that joints are continuous, and hence define potentially unstable wedges. This approach is, however, extremely conservative. A much more realistic approach is to use all the measured joint characteristics, and their statistical variations, to extrapolate the jointing geometry to the underground locations. Although this should be a three-dimensional operation, it can be simplified by extrapolating in three orthogonal directions to

*Scott Wilson Kirkpatrick and Partners, Hong Kong.

†Steffen, Robertson and Kirsten, Johannesburg, South Africa.

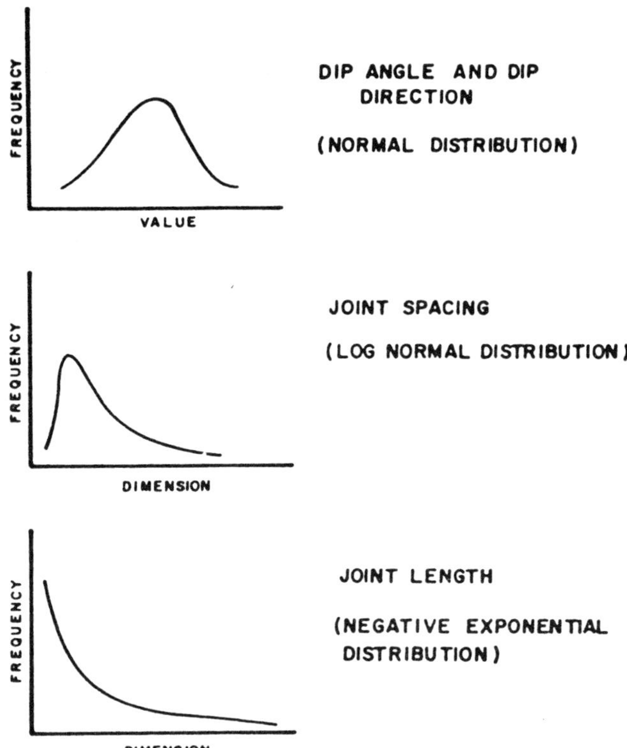

Fig. 16 Common distributions of joint characteristics

generate distributions of joint traces in three orthogonal planes. Care must be taken to ensure that this extrapolation only takes place within structural regions and not across their boundaries. The resulting 'likely' computer-generated joint distribution maps provide the engineer with an extremely useful design tool[5] – he can picture the rock mass, the likely extent of joint intersections and the potential for unstable block or wedge formation. A typical joint distribution map is shown in Fig. 17 with an excavation, at the same scale, superimposed. This allows potentially

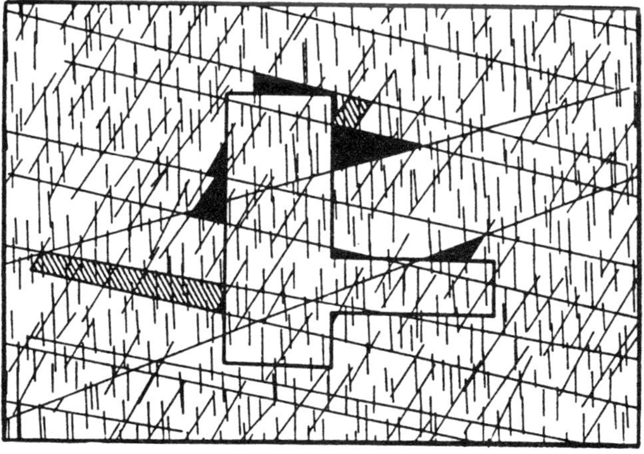

Fig. 17 Joint trace map, excavation outline and potentially unstable wedges

unstable wedges to be identified visually, as shown in Fig. 17, which is again a practical engineering approach. Equivalent views for the other planes can be generated as shown, for example, in Fig. 18, and the approach is equally applicable to any type of excavation, surface or underground. Note that in the three different distribution planes the blocks identified are completely independent, i.e. they are not physically related to one another. By superimposing the excavation geometry in many different positions on the joint trace map it is possible to determine the likely range in size of blocks, both in surface area and depth behind the exposed face, and the likely spacing between adjacent blocks. This will immediately give an indication of the required length and spacing of support elements, such as rockbolts or cables.

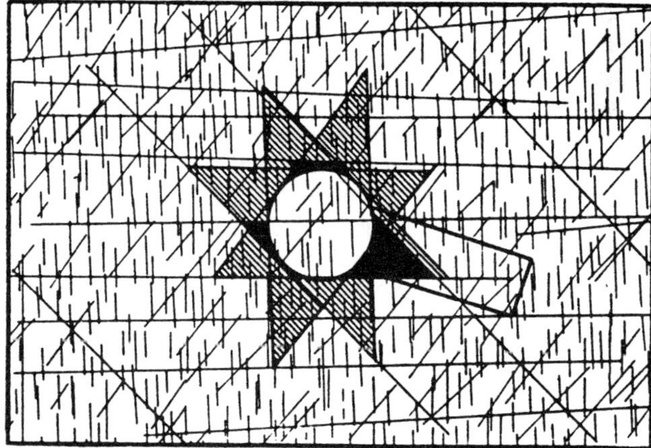

Fig. 18 Joint trace map showing unrealistic wedges from ubiquitous joint assumption

In Fig. 18 the unrealistic wedges that would result from an ubiquitous joint assumption are shown with hatched shading. The more likely unstable blocks and wedges, identified from the generated joint distribution map, are shown with solid shading.

The above approach may be described as a manual approach to define the probability of *occurrence* of potentially unstable blocks. It is a much simpler approach than to attempt to determine the probability of *failure* of blocks, particularly when it is not known in advance of excavation whether such blocks actually occur.

References

1. Steffen O.K.H. Kerrich J.E. and Jennings J.E. Recent developments in the interpretation of data from joint surveys in rock masses. In *Soil mechanics and foundation engineering: Proceedings of the sixth regional conference for Africa, Durban, 1975* (Rotterdam: Balkema, 1975), 17–25
2. Priest S.D. and Hudson J.A. Discontinuity spacings in rock. *Int. J. Rock Mech. Min. Sci.,* **13,** 1976, 135–48
3. Hudson J.A. and Priest S.D. Discontinuity frequency in rock masses. *Int. J. Rock Mech. Min. Sci.,* **20,** 1983, 73–89.
4. Haines A. The application of generated rock mass discontinuity patterns. In *Soil mechanics and foundation engineering: Proceedings of the eighth regional conference for Africa, Harare, 1984* (Rotterdam: Balkema, 1985), 13–21.
5. Stacey T.R. and Haines A. Design of large underground openings – an integrated approach. In *Design and construction of large underground openings: SANCOT seminar 1984* Giles E.L. and Gay N. eds (Johannesburg: South African National Committee on Tunnelling, 1984), 17–25.

Authors' reply

D.R. Greenway, G.E. Powell and G.S. Bell In reply to H.T.M. Insley, based on the final configuration (Fig. 3, p. 174), groundwater levels at the ground line in front of the wall and at a height of $H/3$ behind the wall were included in the design. This was considered conservative in light of the groundwater levels monitored before construction and the expected rise due to restricted flow paths. One piezometer that is still being monitored also indicates this to be the case.

After excavation between caissons was completed (see Fig. 11, p. 179) granular drainage material was placed between caissons in a 'bottom up' sequence in conjunction with the concrete infill panels described in the paper. Temporary bracing of the soil cut face was employed until the drainage layer was completed. The granular drainage material was selected to be in proper filter relationship with the *in-situ* soil, and weep holes were placed at 1-m centres in the concrete infill panels for drainage. The successful construction of this drainage layer was due in part to the selection of a spaced caisson design rather than a contiguous caisson design.

Session 3: Interaction between surface and underground workings

Chairman: W.S.Y. Chan
Rapporteur: Dr. F.G. Bell

Discussion

Dr. F.G. Bell* said that of the three papers† in session 3, one considered the influence of surface blasting on an existing tunnel and two were concerned with subsidence. Both the latter dealt with the old workings, for gold and for limestone, respectively, and how their presence had influenced development at the surface.

The paper by Clover dealt with an investigation that had been carried out for a proposed scheme to remove 3 000 000 t of fill for reclamation from a site that lay above a tunnel that supplied water to an important industrial area of Hong Kong. Accordingly, if the scheme was to go ahead, blasting operations would not have to affect the tunnel. Indeed, restrictions on blasting near tunnels were contained in the Provisional Standing Order, no. 1038, issued by the Hong Kong Water Supplies Department. Those regulations specified, first, that the maximum particle velocity and amplitude of ground movement in such instances should not exceed 13 mm/s and 102 μm, respectively. Secondly, no blasting could be carried out within 55 m of a tunnel. Thirdly, blasting design was determined by the Hong Kong Commissioner of Mines after site trials (which in that case could not be carried out!).

The investigation by necessity involved an extensive desk study, but five deep cored drill-holes sunk alongside the tunnel and the tunnel itself had been inspected and a video record taken of its condition. The evidence obtained from the drill-hole cores indicated that the granite into which they went, and in which the tunnel was located, was of fair to good quality. Furthermore, the tunnel, which was lined for about 10% of its length beneath the proposed borrow area, was in good condition, showing no signs of distress.

The scheme proposed that the minimum cover that would remain above the crown of the tunnel would be 93 m. Subsequently, the investigation had shown that at that distance from blasting dynamic mechanisms would not have any adverse influence on the tunnel.

Because trial blasting could not be carried out at the site, a survey of blasting records from Hong Kong (from areas with similar geology and topography) and abroad had been undertaken. The same approach had been adopted in that study as was used by the Hong Kong Division of Mines, which was the square root scaled distance relationship to predict peak particle velocities. The analysis of those records indicated that the maximum peak particle velocity of 13 mm/s, as contained in PSO no. 1038, was probably extremely conservative. In addition, it showed that with that value the maximum permissible charges per delay, at 50 and 90% confidence levels, would be 31 kg and 6.8 kg, respectively. It was admitted that the latter figure would give rise to a slow and, no doubt, uneconomic rate of working. Nonetheless, the charge could be reduced with depth as the tunnel was approached. Moreover, the effect of vibrations on a tunnel could be reduced if the direction of blasting was orientated more or less at right angles to the direction of the tunnel.

Turning to the paper by Stacey, gold had been mined from quartzite reefs in the Johannesburg area since the nineteenth century. The workings took the form of extensive narrow tabular stopes with occasional pillars left to support the roof. Those pillars had frequently been robbed when a mine neared the end of its working life. In common with all old areas of mining activity, the records of early workings were poor and inaccurate, if, indeed, they existed. In more recent years some reclamation mining had occurred.

The presence of old workings had led to restrictions being placed on development at the surface. In other words, no development had been allowed where mine workings existed at less than 90 m below the surface. On the other hand, there were no restrictions on development in those areas where the workings occurred at a greater depth than 240 m. The surface area underlain by workings located at depths between those two limits was controlled in terms of its development. Perhaps control was too strong a word as the guidelines laid down were not rigid and depended on a number of factors. Nonetheless, in that zone low-rise, relatively low-cost structures, which usually incorporated some degree of flexibility, were the common forms of construction. The basis of that zonation was mainly historical in that subsidence had been associated more frequently with workings at shallow depth: however, the upper limit that defined such a safe zone should not be taken as absolute. With time, subsidence affecting the surface could occur due to movements in workings at greater depths. In fact, surface subsidence had occurred in the Johannesburg area where workings were located at depths in excess of 240 m.

Four mechanisms had been identified as leading to surface subsidence. Three of those types of subsidence were associated with the hanging-wall — with its collapse, with cantilevering of the hanging-wall or with its deformation. The remaining type of subsidence was attributed to block movements at fault or dyke contacts and gave rise to a step at the surface.

A number of stabilization measures had been taken in an attempt to minimize the effects of subsidence. Shallow stabilization had taken the form of either dynamic compaction or outcrop plugging, its aim being to produce a competent surface zone by promoting arching across the stopes. Dynamic compaction had been used to improve the soil conditions above bedrock and to close any cavities near the ground surface that had developed as a result of mining activities. Stoped plugging was more common and consisted of exposing the throat of a stope by excavation, sealing the throat with concrete and then placing and compacting backfill above the concrete plug.

Because of the competition for land, large expensive structures would need to be built in the restricted zone. That, in turn, would necessitate more expensive remedial works to mitigate the subsidence problem. Such remedial works included the provision of rigid support between hanging- and footwalls, which took the form of pillars. The function of such pillars was to provide a near-surface rigid zone that would arch across stopes if collapse or instability occurred at depth within the workings. Analysis had shown that dip pillars, the bases of which were connected by strike pillars, appeared to be the most effective form of support. Concrete crown pillars had been constructed across stopes that outcropped at the surface.

*Department of Civil and Structural Engineering and Building, Teesside Polytechnic, Middlesbrough, England.

†Clover A.W. Appraisal of potential effects of surface blasting on a rock tunnel, with particular reference to the Tai Lam Chung water tunnel, Hong Kong; Stacey T.R. Interaction of underground mining and surface development in a central city environment; and Braithwaite P.A. and Cole K.W. Subsurface investigations of abandoned limestone workings in the West Midlands of England by use of remote sensors.

Mining of limestone of Silurian age in the West Midlands dated back to the mid-eighteenth century. Nonetheless, the areas affected by abandoned gallery and pillar and stall workings in limestone (over 1.6 km^2 of mines), investigated by Braithwaite and Cole, formed only a small part of the total urban area and the risk of subsidence damage to property in the areas affected was low. For example, prior to 1978 about 100 surface disturbances had been noted, affecting around 50 000 m^2. Such a figure represented approximately 5% of the area underlain by limestone mines. The situation had been complicated in some areas by subsidences due to mining in coal measures above the limestone, it being difficult, if not impossible, to distinguish between subsidence due to workings in one or the other formation. The collapse of Cow Pasture mine in 1978, however, had led to the DoE, together with the West Midlands County Council and the Borough Councils concerned, commissioning consulting engineers to assess the subsidence risk and to recommend remedial measures.

Before 1978 the existence of old mines at shallow depth had influenced development at the surface. Development generally had been avoided where the overburden above workings was less than 30 m. Between 30 and 60 m some precautionary measures had usually been taken to minimize the effects of any future subsidence on structures, but it had been assumed that movements in workings deeper than 60 m would not affect surface structures. Only three cases in which mines were at depths greater than 100 m, including that of Cow Pasture mine, had given rise to surface subsidence.

The investigation introduced the concept of the 'consideration zone', defined as one in which consideration should be given to the need to investigate the ground for subsidence that could be caused by the deterioration of old limestone workings. The boundary of such a zone was defined in terms of the distance from an individual mine where horizontal ground strains produced by subsidence were 0.2%. That was based on the method of subsidence prediction given in the *Subsidence engineer's handbook.** The angle of draw at a rib face in longwall extraction of a coal seam did vary, however, depending on the proportion of hard and soft rocks above the same concerned, from less than 20° to more than 45°. Thus, if the boundary of a consideration zone was drawn by using an angle of draw of 35° (the usual angle quoted for Coal Measures strata in Britain), the boundary could be in error. The workings in the limestone were partial not total extraction and so subsidence profiles were likely to be complicated by the existence of intact pillars.

The investigation of those abandoned mine workings was undertaken by both direct and indirect methods. Indeed, access was available to some mines. Rotary drilling rigs had produced core of 92-mm diameter for examination and testing. Logging had included the incidence of rock fracturing, the rate of penetration and water flush return. The condition of the rock had also been assessed in terms of *in-situ* permeability tests. The drill-holes had been subjected to geophysical logging, which had helped to identify fractured rock. Usually, rock became more disturbed as a drill-hole neared the mine workings. Furthermore, the dip and direction of strata in a drill-hole had been recorded, as had the depth of workings. The size, shape and location of pillars and stalls were surveyed with the aid of closed-circuit television (of limited value) and ultrasonic equipment. The latter proved to be an effective tool to map water-filled workings.

Subsidence engineer's handbook, 2nd rev. edn (London: National Coal Board, 1975), 111 p.

The investigation had identified two types of subsidence. First, crown holes had appeared at the surface above shallow workings and had resulted from roof collapse and void migration. Secondly, general trough-like subsidence had occurred as a consequence of pillar failure. Usually, however the limestone pillars were strong enough to carry the overburden load. It was the upward extension of the pillars, as a result of roof collapse, into the overlying shales that gave rise to the problem in that it was the shale 'pillars' that collapsed. Failure of one pillar threw a greater load on to those about it, which, in turn, might cause their failure. The collapse in Cow Pasture mine had affected a surface area of 200 m by 300 m and subsidence at the centre of the trough had amounted to some 1.2 m.

A number of methods had been suggested for treating those abandoned mine workings, ranging from the use of artificial supports, through infilling to induced collapse, and were outlined in the DoE pamphlet.* That also gave a brief summary of treatment of existing structures to minimize the effects of subsidence.

All three papers under discussion gave outlines of detailed investigations, one, for example, highlighting a painstaking desk study, another sophisticated site operations. One of the most interesting points was the use of an ultrasonic survey technique to map abandoned mine workings, which would appear to be very effective. It was also shown that the establishment and use of subsidence hazard zones based on historical events needed to be an evolving process in that boundaries had to be revised in the light of subsidence events. Lastly, mitigation of subsidence effects on surface structures could take the form of treating the ground and/or incorporating special design features into the structures themselves, both representing additional cost.

Written contributions

R. Coates† Coates‡ has used the results from blasting trials, carried out to assess the impact on tunnels of 'suface' nuclear explosions, to give an indication of the damage that would be inflicted on an unlined tunnel near conventional bench blasting. In his example, if a tunnel was 117 ft away from 5103 lb of explosive fired at one time, it would be at the extreme boundary of the zone where it would suffer discontinuous rock failure, probably arising from the shaking down of previously loosened material.

It would appear, therefore, that the Tai Lam Chung tunnel would suffer no damage whatsoever from similar blasting carried out at the final formation level of the borrow operations (93 m (300 ft) from the tunnel).

It is interesting to speculate on the potential instability in the Tai Lam Chung tunnel walls caused by the hydrostatic head (48 m?) remaining in the rock after the, presumably rapid, dewatering of the tunnel for the inspection.

C. Laughton§ Having recently blasted directly against an existing tunnel in which was installed a range of highly

*See reference 1 on page 39.

†Mass Transit Railway Corporation, Hong Kong.

‡Coates D.F. Rock mechanics principles. *Can. Mines Branch Monogr.* 874 (revised 1970), 1970, 8-16.

§CERN, Geneva, Switzerland.

sensitive laboratory equipment, we found ourselves in a similar 'Catch 22' situation to that faced by Clover. Trial blasting, with the use of minimal charges, with vibration monitoring in proximity to the laboratory equipment, was carried out to establish site parameters for use in a tested formula* and production blasts were prepared accordingly.

This procedure reassured the laboratory staff prior to the debut of production blasting and provided a valid set of criteria for the blast design.

Author's reply

Dr. T.R. Stacey Mining in the central Johannesburg area has effectively occurred from surface, as outcrop mining, to a depth in excess of 1000 m. Regional subsidence of type (iii) in Fig. 1 (p. 398) is known to occur, but rarely results in any problems, and is hence not often recorded. Local collapse at the outcrops is common in open and undeveloped areas, both in the form of sinkholes or crown holes and subsided cantilevered blocks within 20–30 m of the outcrops. The majority of recorded subsidences, however, are associated with movement on dyke contacts.

Percentage extraction is usually very close to 100. Dyke intrusions, which are quite common, are usually unmined unless they are very narrow. Reclamation mining takes place in previously mined areas that are accessible without significant 'making safe' operations. Hence, unsupported movements can take place in the inaccessible areas. Reclamation mining is likely to result in reactivation of subsidence, but this may not always be the case.

Stone walls can provide an effective support against local collapse, but will require to be compressed by some 30% before generating significant overburden support.

Table 1

Depth of mining, m	Building permitted, storeys
0–90	No building
90–120	One with one basement
120–150	Two with one basement
150–180	Three with one basement
180–210	Four with one basement
210–240	Five with one basement
240+	No restrictions unless mining situation is exceptional

The guidelines presently applied to proposed developments over shallow gold mine workings are noted in Table 1. Damage to buildings and roads as a result of subsidence has been recorded, but there is no known instance of loss of life as a result of subsidence in this environment. The depth limit of 240 m, beyond which restrictions on surface development are not usually imposed, was set historically since records indicated that there were very few cases in which subsidence was observed when the mining depth was greater than 240 m. A government commission is presently reviewing the guidelines described in Table 1.

With reference to the old gold mine workings, 'sinkhole' is colloquially used to describe a hole that appears at the previously mined stope outcrop location. It usually results when loosely tipped backfill ravels away over a period of time, or, more commonly, is washed down the stope by rain water, and is a minor form of instability.

Type (i) subsidence is very localized, affecting the immediate outcrop area, with dimensions of several metres. The extent of the affected area in type (ii) subsidence is usually limited to 20–30 m from the outcrop on the hanging-wall side. The area of surface affected by type (iii) subsidence can be very extensive, but it usually passes unnoticed since horizontal strains and tilts are such that damage does not result. It should also be noted that mining in the shallow areas usually took place before substantial buildings were erected in these areas, i.e. probably after much of the potential subsidence had taken place.

The area influenced by type (iv) subsidence is very variable and may be localized or very extensive. The area of sudden subsidence associated with the dyke in the case study described in the paper was estimated to be 30 000 m^2. Movement on the dyke contact was visible in the form of a surface crack over a length of approximately 200 m.

There is not much evidence of pillar deterioration in shallow workings. Very small pillars have been observed to have remained competent for the past 70 years. Void migration does not appear to occur except in the loose backfill material. Many old stopes have been observed to be standing open over large spans without support, and apparently perfectly stably. With behaviour such as this, bulking is not a significant factor. Should some hanging-wall collapse occur, a small amount of block or slab rotation would result in uncompacted bulking that would provide sufficient support to prevent any further collapse. This effect has not yet been quantified.

There is a blanket restriction zone to which the guidelines described above are applied. Investigation of individual sites and their history of undermining is required to motivate relaxation of the restrictions.

The dynamic compaction stabilization was applied in one particular case in a small area 'compartmentalized' by the outcrops, the road and a dyke, as is shown in Fig. 3 (p. 399). Waste rock packing had taken place many years previously beneath the road, and the stopes had been loosely filled from surface. The dyke intersection served as a barrier to prevent possible ravelling of any fill or collapse rock to greater depths. Hence, the dynamic compaction was only required to deal with the near-surface zone. The superficial cover of soil and fill extended to a depth up to 2.5 m, being underlain by weathered quartzite. The maximum depth of excavation by mechanical backhoe was of the order of 4–5 m. Dynamic compaction effects were observable to a depth of approximately 4 m.

Stopes are not completely filled by plugs. Their function in both shallow and deep outcrop stabilization is twofold: to provide rigid support between hanging-wall and footwall to promote the formation of an arch and to provide a barrier against subsequent ravelling or washing away of backfill in the stope. In shallow operations the plug takes the form of a concrete strike pillar approximately 2 m in dip dimension, commonly at a depth below surface of 4–10 m. In deep operations dip pillars are formed as the main support elements, with strike connections to contain fill.

Dykes that have not been stoped through appear to be unaffected by subsidence and have been observed to have been left standing proud with subsidence on either side. On substantial dykes considerable relaxation of the guidelines described above has been permitted for developments to within 5 m of the contacts.

*Chapot P. Etude des vibrations provoquées par les explosifs dans les massifs rocheux. *Rapport de Recherche du LCPC* no. 105, 1981, 56 p.

The results obtained from the extensometer indicated that relative movements were occurring in the hanging-wall strata, i.e. a section of strata would tend to separate with the deeper portion moving downwards, and then the deeper portion would stabilize and the upper portion move down to 'catch up' with the lower portion.

Session 4: Rock slopes – investigation

Chairman: Xiao Huan Jie
Rapporteur: Dr. L.J. Endicott

Discussion

Dr. L.J. Endicott* said that his presentation on the papers† in session 4 was made in the context of the state of the art of site investigation for rock slopes as practised in Hong Kong.

As other sessions were concerned with other aspects of rock slopes, such as design, construction and groundwater, it was necessary to provide a definition of site investigation to avoid an overlap. Site investigation, once started, could be considered, from one point of view, never to be over. Whatever the project, there were always some areas that were unknown or some uncertainty as to the nature of the ground, even when the project might be considered to have been completed. *BS 5930,* the code of practice for site investigation, provided a similar, far-reaching definition for site investigation extending from conception of the project, through its realization, into long-term maintenance. For the purposes of the present session, and to avoid transgressing into subjects to be included in other sessions, 'investigation' was to be limited to the site investigation stage of rock slopes.

Review

He would, first, like to review some of the recent developments of site investigation of rock slopes in Hong Kong, commenting on the extent to which rock slopes were being used, and the techniques and the approaches that were being adopted for site investigation. That was not a comprehensive review of the current state of the art but a subjective highlighting of significant features of local practice in order to provide an introduction to the session.

Scale of development in Hong Kong In the last decade Hong Kong had experienced some remarkable developments. A demand for land and a reasonably healthy economy had resulted in very extensive site formation works and construction on sites that would have been considered previously to have been too unattractive or too dangerous to develop. Years ago, for single building developments, the extent of site formation and forming slopes had been limited. The consequences of failure had been similarly limited. In recent years extensive estates had been built, both in the public and private sectors, that had necessitated extensive site formation works, on sites of many hectares in area, and many of them on hillside sites. Such was the demand for building land that even the major quarries were being considered for residential development use in the future and the steep rock slopes that were being formed as part of the quarrying operations might affect residential buildings in the future. An example of a major site formation project, including a large rock slope, was given in the paper by Beggs and McNicholl.‡

Terrain evaluation and mapping A recent development in the practice in Hong Kong had been the increasing proper use of air photos for systematic terrain evaluation, coupled with systematic geological or engineering-geological mapping. Several of the papers described a little of the geological mapping of the projects covered. He could find no reference in the papers to the use of air photos. In large-scale projects such studies were very necessary before any field work was designed.

Geophysical and remote sensing Geophysical methods were the subject of the first session of the conference. Their use for the investigation stage of rock slopes was still at a stage of being developed. Work with Landsat imagery and local infrared photography were also at the stage of development. Useful results might be expected to be forthcoming.

Designing the site investigation Considerably more thought and preparation were being put into deciding the nature and extent of the site investigation. He hoped that the days of commencing a site investigation by merely ordering drill-holes spaced at equal centres across the site, without any other preparation, and then deciding what to do with the results, were over.

If the cost of one drill-hole was comparable with one month's salary, how much time should be spent on investigation of the site before designing the first borehole? In some cases the results of the drill-hole should be predicted before it was carried out. Its execution should be used to test an hypothesis – the working model of the geological structure.

Drill-holes were being carried out for specific purposes and at specified orientations. Papers by Beggs and McNicholl, Buttling and McFeat-Smith and co-authors all cited the use of inclined drill-holes.*

It might be noted, on the one hand, from a statistical point of view, that drill-holes intercepted very few features that were sub-parallel to the axis of the drill-hole and many that were normal to the axis. On the other hand, for purposes of obtaining samples, or of investigating the lateral extent of a feature, drill-holes orientated along the plane of the feature might be more useful.

It should be noted that reliance on drill-holes could lead to rogue features being overlooked. Clover described such a feature and commented on its appearance in the original drill-hole samples. The paper by Richards and Cowland† included another feature, described as a sheeting joint, which had required a specific type of investigation. The recognition of such features had required the adoption of improved drilling and sampling techniques. As was mentioned by Powell and Irfan, and again by Richards and Cowland, the use of foam drilling had permitted the taking of good-quality samples that included both soil and rock material within the same sample.

Risk evaluation Finally, in his review he would like to include comment on the increasing use of the evaluation of risk in relation not only to the safety standards to be used in design but also to the approach to be adopted for investigation. Clover and Powell and Irfan mentioned the consid-

*Brickell, Moss and Partners, Hong Kong.

†Clover A.W. Slope stability on a site in the volcanic rocks of Hong Kong; Poisel R. and Eppensteiner W. Control of a large natural slope in a suburb of Vienna; and Powell G.E. and Irfan T.Y. Slope remedial works in weathered rocks for differing risks.

‡Beggs C.J. and McNicholl D.P. Formation of a high rock slope at Ap Lei Chau, Hong Kong. This volume, 1–14.

*Beggs C.J. and McNicholl D.P. *Op.cit.* Buttling S. Remedial works to a major rock slope in Hong Kong. This volume, 41–8. McFeat-Smith I. Nieuwenhuijs G.K. and Lai W.C. Application of seismic surveying, orientated drilling and rock classification for site investigation of rock tunnels. This volume, 249–61.

†Richards L.R. and Cowland J.W. Stability evaluation of some urban rock slopes in a transient groundwater regime. This volume, 357–63.

eration of the risk involved in relation to the projects described.

Session papers

The paper by Clover was set in Hong Kong and dealt with an extensive failure during construction. Had the construction been completed, the failure would have caused extensive damage. Some 3800 m^3 of material had fallen on to the construction site. The site investigation had been directed towards identification of the cause of the landslide. That had led to a redesign of the site formation work. The investigation had taken the following stages: survey of the landslip area, geological mapping, boreholes, a groundwater study, trial pits and block samples of the failure surface and *in-situ* shear box testing.

The investigation had identified a weak zone, some 0.7 m thick, of silt and clay located below the zone of fractured rock. By re-examination of the original site investigation there had been some evidence of its existence, but core recovery had been too poor to be certain.

The remedial works had included the use of permanent support from steel bar rock anchors. A total of 139 had been installed and 12 had been monitored with load cells.

During construction another extensive discontinuity had been found that had necessitated a further 12 anchors to be installed. In his conclusions Clover cited the variable nature of volcanic rocks, which was made more difficult by the effects of tropical weathering. He emphasized the need, under such conditions, to extend the investigation to embrace the construction design review stage.

In that case the investigation had been somewhat pathological – after the event. The groundwater studies should be noted, however (see session 8). The paper provided much information on the performance of the early stage of monitoring the ground anchors.

The paper by Poisel and Eppensteiner concerned a study that had been carried out on steep rock slopes that comprised the sides of a river gorge. The expansion of the suburbs of the city of Vienna had resulted in the gradual development of the valley floor for residential purposes. After several rockfalls the stability of the slopes had had to be investigated.

The rock was dolomite and the slopes rose steeply to a height of some 80 m. The length of the gorge was about 1 km.

The investigation had been based on geological mapping of the two sides of the valley. Geophysical investigations included multiple spacing electronic logging, refraction seismic and infrared geothermy, but the results of those aspects of the study were not described. The investigation included an assessment of the stability of the slope considered as sub-areas.

No evidence had been found to suggest that a major landslip was likely to occur. Several block and block structures had been identified to be in danger of falling and recommendations were made to study the area further.

Powell and Irfan's paper was set in Hong Kong and included not only the investigation but also the design and assessment of risks for slopes in weathered rocks. The paper provided a review of the problem and it cited three case histories at the end.

It introduced the geology of Hong Kong and the effects of weathering. The paper included soil, soil and rock, and rock slopes, the stability being governed by the strength of the mass or by the properties of discontinuities present as the case might be.

For the investigation state the authors stressed the prerequisite to carry out a proper identification mapping, including the rock types, degree of weathering, the jointing and the properties of the joints, including those joints which occurred in the residual weathered materials. They referenced the need to use foam drilling to obtain samples of joints and joint infill materials, especially for joints in materials that were located above the rock surface, introduced the considerations of relating the investigation and design processes to the degree of risk or loss in the event of failure, and quoted three case histories of slopes that had been investigated.

Other conference papers

It was noticeable that papers submitted to the conference included a number of well-described case histories and several rock slopes. It was not surprising that several included a description of the investigation stage.
Of the other papers that dealt with rock slopes, he would like to draw attention to the following.

That by Beggs and McNicholl (*op. cit.*) described a very large slope and a detailed study and investigation, including reference to the significance of sampling of orientation of drill-holes and surface mapping.

Buttling's paper (*op. cit.*) indicated the benefit of an extensive history of an existing slope. The author made interesting comment on clay-filled joints and the need to recognize that reliance on statistical evaluation of plots could obscure the importance of widely spaced joints. The investigation had included drill-holes at different orientations.

The paper by Dubin and co-workers* addressed the question of what to investigate, and how, in the case of stabilization of many existing slopes. That by Matheson† noted the difference between blast-induced and blast-affected joints that should be recognized when exposures were being investigated.

Muir and co-workers‡ described an extensive and comprehensive site investigation. In particular, he noted their reference to the use of rapid probing to determine bedrock profile. Richards and Cowland (*op. cit.*) described a detailed investigation of some existing slopes where sheeting joints with soft inclusions were present. A photograph of a cored sample that included such a joint and its soft infilling materials was given (Fig. 5, p. 359).

Written contributions

D.P. McNicholl§ Directional bias is a major source of errors in joint surveys because some joint sets may be so aligned that unidirectional boreholes or scanlines miss or under-represent them. Terzaghi [6] and Hudson and Priest [2] have already mentioned the importance of directional bias in rock joint surveys.

This biasing effect is best illustrated in Fig.1. The smaller the intercept angle between the scanline/borehole direction and the mean discontinuity plane of a set of dis-

*Dubin B.I. Watkins A.T. and Chang D.C.H. Stabilization of existing rock faces in urban areas of Hong Kong. This volume, 155–71.

†Matheson G.D. Design and excavation of stable slopes in hard rock, with particular reference to presplit blasting. This volume, 271–83.

‡Muir T.R.C. Smethurst B.K. and Finn R.P. Design and construction of high rock cuts for the Kornhill Development, Hong Kong. This volume, 309–29.

§Hong Kong Housing Authority, Hong Kong.

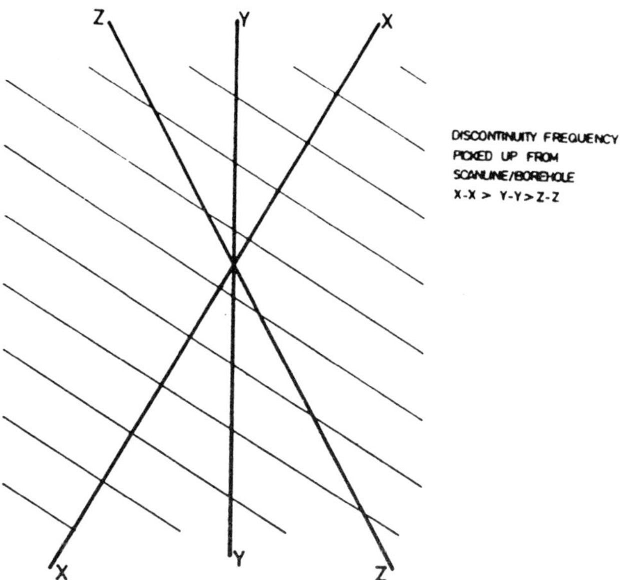

Fig. 1 Vertical section through scanlines/boreholes in discontinuous rock

continuities, the less representative the obtained results become. It follows that vertical boreholes will underestimate the significance of the sub-vertical joint sets that are common in Hong Kong.

The purpose of this contribution is to study the effect of survey orientation on discontinuity frequency. The physical orientation of discontinuities is ignored.

Directional bias of discontinuity spacing survey

Discontinuity spacing is one of the major and measurable indices of a rock mass. It is an indication of the state of fracturing of the rock mass and, hence, the volume involved in a potential failure; it also influences the probable failure mode. In large rock excavations the index may be used to assess face treatment or finishing works and may also be used in the selection of excavation techniques.

Tnere are several ways of expressing the characteristic spacing of discontinuities. Many engineers and geologists use Rock Quality Designation, RQD,[1] as an index of rock mass and rock quality. In Hong Kong mass RQD is commonly estimated from a series of identically orientated boreholes, normally vertical. Many engineers are not aware that RQD obtained from boreholes is to some extent an arbitrary index and is very sensitive to drilling quality and, particularly, drilling malpractices. Moreover, a survey of several hundred geotechnical reports has indicated that directional bias is often overlooked and that it is uncommon to correct raw data as recommended by Robertson[5] and Terzaghi.[6]

In some instances it is useful to express the typical spacing of discontinuities as Discontinuity Frequency, a measure of the number of discontinuities per unit length of line in a particular direction.[7] Priest and Hudson [4] discussed the theoretical distributions of discontinuity frequency and, based on a negative exponential spacing distribution, derived a theoretical RQD (designated as RQD*) in terms of spacing data

$$RQD^* = 100e^{-\lambda t}(\lambda t + 1)$$

where λ is number of discontinuities per unit length and t is threshold value, taken as 0.1 m for conventional RQD.

When λ falls within 6 to 16 the relationship can be simplified to

$$RQD^* = -3.68\,\lambda + 110.4$$

Priest and Hudson also pointed out that, for a reasonable estimate, the following requirements should be satisfied: (1) the scanline length should be at least 50 times the mean discontinuity spacing and (2) at least 200 measurement values are required.

DISCONTINUITY FREQUENCY

SLOPE REF NO.
Sheet 1 of 3

Collected by: PF Wallis, DP McNicholl
Method: Scanline Survey
Location: Aberdeen HOS.

Date: 31.10.83
Plunge: Horizontal
Trend: E - W
Length:

Sketch:

←N—

n = 271
$\bar{x}$ = 39.90
λ = 25.01.

SPACING COUNT

	00	05	Total/%		00	05	Total/%		00	05	Total/%
00		9	9/3.3	100	3		3/1.1	200			
10	13	31	44/16.2	110	2		2/0.7	210			
20	15	38	53/19.6	120	1		1/0.4	220			
30	26	26	52/19.2	130			0/0	230			
40	23	18	41/15.1	140			0/0	240			
50	18	5	23/8.5	150	2		2/0.7	250	2		2/0.7
60	5	9	14/5.2	160			0/0	260			
70	10	5	15/5.5	170			0/0	270			
80	4	1	5/1.8	180	1		1/0.4	280			
90	3	1	4/1.5	190			0/0	290			

LARGE SPACINGS - LIST

	00	10	20	30	40	50	60	70	80	90		00	10	20	30	40	50	60	70	80	90
300											800										
400											900										
500											1000										
600																					
700																					

PROJECT:
SLOPE NO.:

HONG KONG HOUSING AUTHORITY

Fig. 2

RQD* may be used as a mass characteristic and data from a differently orientated scanline can be aggregated. The measurement of joints intercepting the scanline on an incremental basis is found to be quick and simple. A 'standard' booking form for logging is reproduced in Fig. 2.

Case histories

The Aberdeen HOS case history studied by the author in 1984 illustrated the effects of directional bias. Fig. 3 shows the layout of the study area.

The measured discontinuity frequency was considered in two categories – the predicted and the as-encountered. The prediction could be said to be equivalent to a normal site investigation prior to construction. It consisted of core logging and impression packer tests taken from the vertical borehole T1 and limited face mapping on the existing exposed rock face. The as-encountered readings were logged by traversing two vertical and one horizontal scanlines on a new rock cut plus a horizontal scanline inside the north portal of the tunnels. The results are given in Table 1.

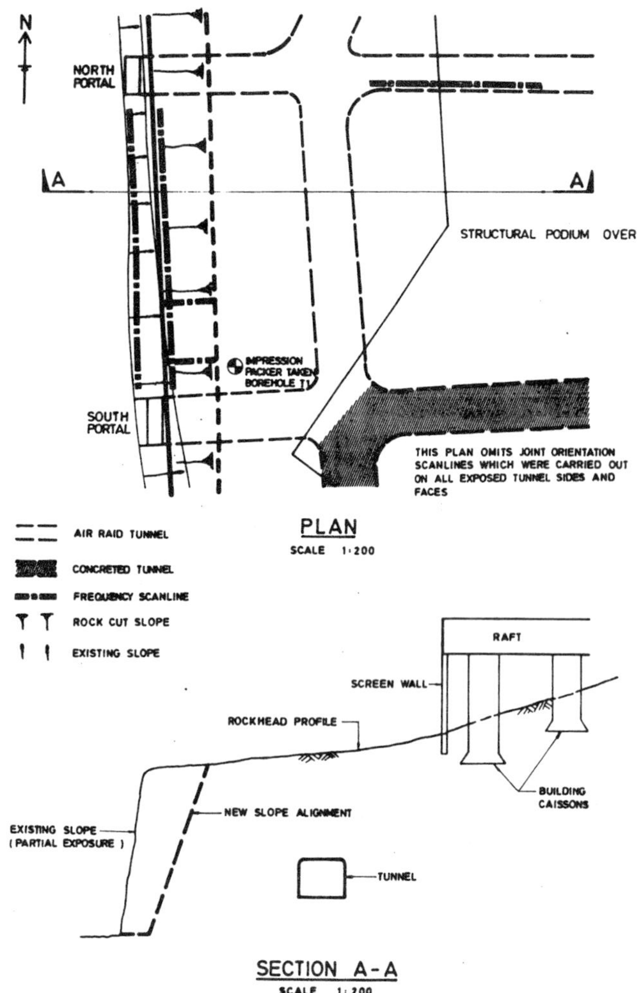

Fig. 3

From Table 1 the following observations can be drawn: (1) the as-encountered frequency picture was significantly different from that predicted; (2) the figures from the vertical borehole were overoptimistic; and (3) there was no relation at all between the vertical borehole (or scanline) readings and horizontal scanline readings. Moreover, vast difference existed between east–west and north–south running horizontal scanlines.

Observations from this case history confirm the published literature in that vertical boreholes offer erroneous or biased information on joint spacing and rock mass quality. This biasing effect can be minimized if joint surveys are carried out in at least three orthogonal scanlines/boreholes and the data can be pooled to obtain a representative mass index. This approach is made possible with the use of RQD*. The RQD from boreholes can be converted into λ values and combine with raw spacing data from scanlines to obtain a composite λ value and, hence, a less biased mass RQD*.

The writer has explored the practicality of this approach by means of a case study[3] at Lam Tin North. Pre-construction measurements were obtained from two differently orientated horizontal scanlines on the perimeter of the study area together with two vertical boreholes located near a future rock face. The as-encountered readings were measured on areas on the rock slope, which were progressively revealed by construction. The mean spacing was calculated from the readings and the RQD* deduced. The results are summarized in Table 2. Pooled readings for pre-construction and as-encountered conditions are plotted on semi-log graphs shown in Fig. 4 in the style of Priest and Hudson.

There is a reasonable agreement between the pre-construction and as-encountered figures as shown in Table 2. In the semi-log graphs the pooled discontinuity frequency readings are scattered about the theoretical distribution line. The as-encountered RQD* index correlates fairly well with the predicted figures.

It may be concluded from the above case history that, with sufficient redings taken at different orientations, it is practicable to use RQD* to predict the actual spacing conditions in the rock mass.

Conclusions

Directional bias is a major source of errors in discontinuity frequency surveys. This biasing effect can be minimized with the use of RQD* with frequency data obtained from at least three orthogonal directions.

Table 1

	Location	Number of readings	RQD* or RQD	Mean spacing, mm	λ
As-encountered figures	Northern tunnel				
	Horizontal scan	271	29*	40	25
	Cut face				
	(1) Horizontal scan	197	44*	53	19
	(2) Vertical scan	45	93*	239	42
	Pooled	513	52*	63	16
Predicted figures	From borehole T1				
	(1) At tunnel level		81		
	(2) For whole borehole		64		
	Pooled data from face		88*	160	6.25
	Mapping and borehole				

Table 2

	Scanline	RQD* or RQD	Mean spacing, mm	λ
Discontinuity spacing from pre-construction data	10	84*	138	7.3
	11	61*	75	13.4
	10 and 11	72*	106	10.4
	Borehole 25	88		
	Borehole 26			
	Domain	**RQD*, mm**	**Mean spacing, mm**	**λ**
Discontinuity spacing from as-encountered data	I	64*	80	12.5
	II	85*	148	6.8
	III	84*	141	7.1
	Pooled	79*	117	8.6

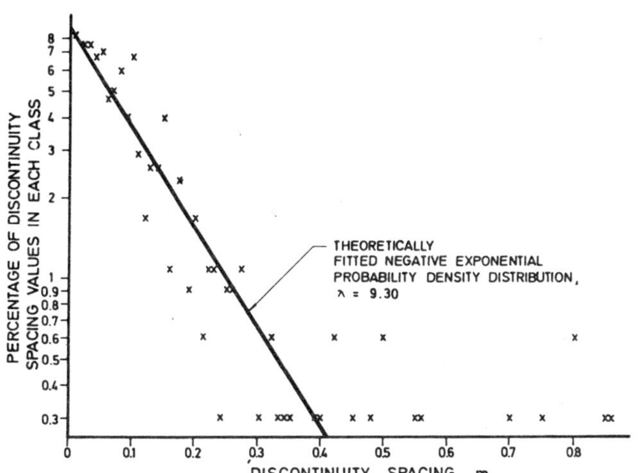

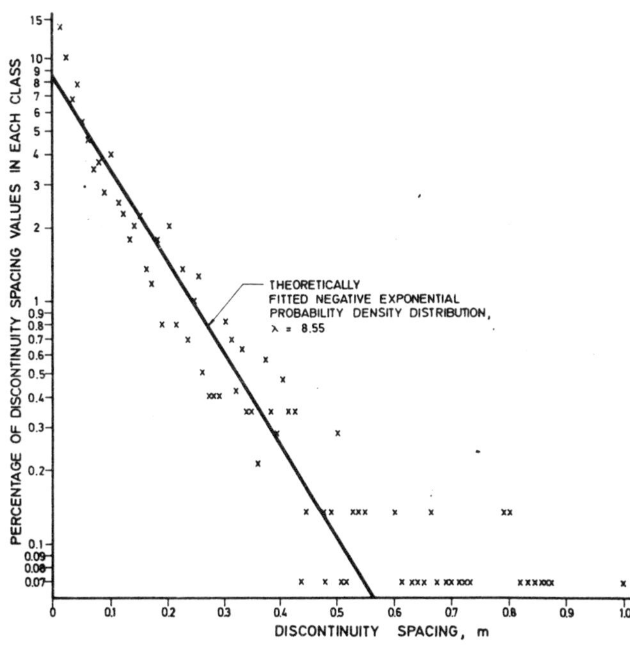

Fig. 4 Discontinuity frequency from pre-construction survey (*top*) and encountered in construction (*bottom*)

References

1. Deere D.U. Geological considerations. In *Rock mechanics in engineering practice* Stagg K.G. and Zienkiewicz O.C. eds (London, New York, Sydney: Wiley, 1968), 1–20.
2. Hudson J.A. and Priest S.D. Discontinuity frequency in rock masses. *Int. J. Rock Mech. Min. Sci.,* **20,** 1983, 73–89.
3. McNicholl D.P. The assessment of rock mass characteristics prior to the excavation of large cut slopes in Hong Kong. M.Sc. dissertation, University of Leeds, 1984.
4. Priest S.D. and Hudson J.A. Discontinuity spacings in rock. *Int. J. Rock Mech. Min. Sci.,* **13,** 1976, 135–48.
5. Robertson A. MacG. The interpretation of geological factors for use in slope theory. In *Planning open pit mines: proceedings of the symposium, Johannesburg, 1970* Van Rensburg P.W.J. ed (Cape Town, Amsterdam: Balkema, 1971), 55–71.
6. Terzaghi R.D. Sources of error in joint surveys. *Geotechnique,* **15,** 1965, 287–304.
7. Wallis P.F. and King M.S. Discontinuity spacings in a crystalline rock. *Int. J. Rock Mech. Min. Sci.,* **17,** 1980, 63–6.

S.W.C. Au* In Hong Kong dolerite dyke intrusions into the country rocks (mainly granitic and volcanic rocks) are not uncommon. When weathered *in situ,* they turn into a material very different in engineering properties from the surrounding weathered country rock – in particular, the permeability is very much lower owing to their high clay content and plasticity. Their thickness generally varies from tens of millimetres up to 1 m. Fig. 5 shows a decomposed dolerite exposure in a soil slope failure in Hong Kong.

Three major soil slope failures in Hong Kong (in grade *W* IV and *W* V decomposed granite) have been reported to be associated with the presence of decomposed dolerite dyke(s);† It is of interest to note that the failures all

*Geotechnical Control Office, Hong Kong.

†Au S.W.C. Decomposed dolerite dykes as a cause of slope failure. *Hong Kong Engineer,* **14,** no. 2 1986, 33–6.

Fig. 5

occurred in granitic areas and during or after extremely heavy rainfall. In Hong Kong precipitation in the form of rainfall is extreme and plays a predominant role in slope failure. If a decomposed dolerite dyke exists in a slope, rainfall infiltration into as well as groundwater flow through the slope will be impeded as it constitutes a flow barrier. Perched water-tables or water-table back-up will result during or following heavy rainfall if it is persistent over a sufficient area. If the pore water pressure is high enough, a slope failure may occur. This type of pore water pressure induced failure should probably be included under materials of types *W* IV and *W* V in Table 3 (p. 350).

Like relict joints, these thin seams, especially when they dip very steeply, are difficult to locate by means of ordinary site investigation techniques. It is therefore proposed that in the site formation contracts in areas where the presence of decomposed dolerite dykes is expected provisions should be included for inspections of the slope by experienced geotechnical personnel during excavation and for possible redesign during the contract.

Authors' replies

Dr. R. Poisel and Dr. W. Eppensteiner In response to general discussion (unpublished) on our paper, the geological mapping was directed to the parting plane structure (attitude as well as outcrop on the slope surface of bedding joints, fault zones, master and minor joints) as rock slope stability is influenced, in particular, by parting planes. That necessitated intensive work with a geological compass, mostly roping down because the significant areas were difficult to reach. In addition to the master parting planes, the depth of the rock floor under the loose sediments in the valley bottom (are there any undercut rock slope areas?) and eventually existing – but not perceptible – fault zones or zones of loosening in the upper slope surface behind the crest were to be investigated. Thus, refraction seismic and multiple spacing electronic logging techniques were chosen. The additional application of infrared geothermy was suggested in an attempt to locate outcrops of joints parallel to the slope. The local administrative board, the client for these investigations, welcomed the application of this method.

Refraction seismic measurements explored the solid rock at shallow depths. Fault zones at the slope crest located by geophysical methods could not be traced, despite intensive work. Further investigations have been ordered because of the great importance of the stability of the rock slope. Unfortunately, these investigations have still to be completed.

For the potentially dangerous areas, which were identified, further investigations must show the trends in the movements of the rock masses and whether they are progressive. All are conscious of the present danger so evacuation of the valley is avoidable, the extent of the long-existing hazard being revealed by measurements.

Financial considerations affected the duration and intervals of the investigations, but not their type and order.

G.E. Powell and Dr. T.Y. Irfan In reply to S.W.C. Au, in Hong Kong it is unusual to find a slope that does not have some geological or weathering anomaly that affects the water regime under both steady and transient conditions. The estimation of the pore water pressures in such slopes is the major uncertainty in the design, and knowledge of the location and orientation of such continuous features as dykes would certainly help. To our knowledge, however, no design in Hong Kong has taken into account the effect of such a feature on the water regime.

Dykes and sills are generally sub-vertical or sub-horizontal in Hong Kong and are rarely orientated so as to provide the sliding plane in a failure. Nevertheless, they are one of the features that should be mapped once the highly weathered zone is exposed in any cutting.

Session 5: Rock slopes – design

Chairman: Xiao Huan Jie
Rapporteur: Dr. G.D. Matheson

Discussion

Dr. G.D. Matheson* said that session 5 contained two papers from widely separated geographic localities and contrasting geologic environments. Both, however, showed a remarkably similar logic in tackling complex geotechnical problems arising from natural and man-made excavations, respectively, and relied heavily on the expert use of engineering geology for their success.

The paper by Speck and Bruhn described geotechnical conditions and rock failures along a natural slope formed by river erosion and the remedial measures that had been carried out to improve the stability. The rock slope was located on the south-facing slope of Duquesne Bluff, within the city of Pittsburgh in the east of the U.S.A., and had been formed by the Monongahela River. A rock face some 1300 m (4250 ft) in length, varying in height from 18 to 35 m (60—115 ft) and with an overall dip of 70°, had thus been formed. Immediately above lay a four-lane highway and immediately below the face a six-lane highway; the stability of the slope was therefore critical to both structures. There had been a history of rockfalls on to the Parkway East (the lower of the two highways), culminating in a major fall in 1978. Minor remedial work had been carried out since, but major remedial work had not been started until 1985. The last mentioned work was the subject of the present paper.

The rock face was formed from alternations of sandstones, claystones and shales, a thin limestone band outcropping near the base. The dip of the bedding, very nearly horizontal, and differential weathering, mainly on the shales, had resulted in overhangs. That suceptibility to weathering and the presence, orientation and importance of discontinuities, identified as belonging to three distinct genetic types, controlled the stability. A stability assessment, made prior to designing the remedial work, indicated that failure according to plane, wedge and toppling modes had taken place and could be expected in the future.

Redesign of the slope angle had clearly been impractical because of the proximity of the upper highway; the slope had had to be retained at its existing angle. Remedial action to strengthen the face and reduce the level of instability had therefore been necessary. The paper described the corrective measures used, the cost-effectiveness of the procedures adopted and the logic employed to decide the most appropriate. Because of the diversity of the stability problems no single design strategy had been possible. The solution adopted had been to define five design zones in terms of risk to Parkway users, and to identify regions of the face falling into each zone. The remedial treatment required to strengthen the rock face and reduce the instability in each zone had then been itemized and the work contracted out. The extremely varied techniques reflected not only the complex nature of the work but also the care taken to ensure cost-effectiveness. Those included scaling and trimming, local blasting and the use of mesh, shotcrete, concrete, bolts, anchors, drains, removal of trees and bushes, and the erection of gabion barriers and retaining fences.

The authors believed that the stabilization programme had minimized the risk to highway users and significantly enhanced the stability of the rock face. That had been accomplished even though expenditure was limited to only \$1 500 000—\$2 000 000 of an original estimate of between \$4 000 000 and \$10 000 000.

The paper by Muir and co-workers described design methods and the techniques adopted to excavate and render stable large rock slopes associated with the site formation of a housing and commercial development. The locality lay on a 22-hectare site in the north of Hong Kong Island and would contain forty-four 30-storey tower blocks housing some 30 000 people and two commercial centres. It, in turn, was part of a larger complex that involved roads, schools, an underground railway station and a service reservoir. Maximum utilization of space had been the main factor in deciding to construct the 60-m high rock face at as steep an angle as practical. The stability of the final face was critical as the tower blocks lay within 3—15 m of the toe of the face.

The local geology consisted of medium- and coarse-grained granites, which made up most of the host rock, dolerite dykes and a fault zone. Weathering was locally very intense and a residual soil, typically 10 m thick, overlay most of the site. Site investigation techniques had included aerial photographic interpretation, boreholes, trial pits, seismic refraction and detailed surface mapping. The last-mentioned technique had been used primarily to obtain discontinuity data for stability analysis. Friction angles, obtained by shear box testing, had then been combined with the data and used to suggest a preliminary design for the rock slope. The basic design for 95% of the slopes consisted of a slope angle of 65—70°, 12-m lifts and 2-m benches. Adequate monitoring equipment had been installed behind the final design face to measure pore water pressures.

Presplit blasting had been used to form the design slope to optimize stability and to minimize remedial requirements. The method that had been adopted employed Tovex 210 explosive in 76-mm diameter drill-holes at 0.6-m spacings. Each hole was charged with 25 mm x 400 mm explosive cartridges taped at 1-m centres to detonating cord, with two cartridges at the base. Approximately 1.5 m of granular material had been used as stemming at the top of each hole; that length was, of course, uncharged. Detonation was by individual electric detonators.

Drilling accuracy was identified as being the most important factor to success in presplitting. The accuracy claimed during drilling was 10 mm per metre of depth (0.57°), using hydraulic drifter rigs. That was remarkable considering that the most popular method of rig alignment was by traditional methods, setting dip by use of a template and spirit level and azimuth by eye, using a plumb line on a bamboo tripod. In one area, where blast vibration had been an extremely critical factor, non-explosive agents had been used in presplitting. The use of a chemical rock-breaking (CRB) slurry had increased the cost of presplitting some 20 times, but was still regarded as being 'advantageous'.

In spite of the obvious success obtained by presplitting, considerable remedial work had been required in specific regions and a design team had been set up to assess and treat face instability. The technique employed had consisted of detailed field mapping, plotting and evaluation as work had progressed; treatment of instability had been regarded as an ongoing process. Remedial techniques had included hand and mechanical scaling, and the use of mesh, shotcrete, no-

*Transport and Road Research Laboratory, Livingston, West Lothian, Scotland.

†Speck R.C. and Bruhn R.W. Rock slope study and stabilization in Pittsburgh, Pennsylvania: a case history; and Muir T.R.C. Smethurst B.K. and Finn R.P. Design and construction of high rock cuts for the Kornhill Development, Hong Kong.

fines concrete, local buttresses, drains, dowels, bolts and catchment fences. Local geotechnical conditions had been used to indicate the most cost-effective technique.

The authors regarded the site investigation to have been adequate and the design and construction work to have been carried out in an optimum fashion. They suggested that the procedures followed might be applicable, with some changes, to excavation projects on all scales.

One of the most difficult aspects in the analysis of stability was to predict time to failure: he would like Speck and Bruhn to elaborate on the rate of development of instability in the various failure modes described and give some idea as to both the design and the effective life of the different forms of remedial work chosen.

A number of distinct genetic groups of discontinuities had been identified: he asked the authors to indicate the characteristics of each group and whether they had diagnostic features that allowed them to be recognized individually in the field.

Turning to the Kornhill paper, the aspects that were raised in the first paragraph on the Pittsburgh paper applied equally there and he would welcome the authors' comment on the effective design life of their remedial solutions.

A trial of different presplitting techniques had been carried out: he would like the authors to describe their views on how success in presplitting could be objectively quantified and to indicate the techniques that had been used to judge the relative success of the different trial panels. In their response he would like them to include CRB presplitting.

Emphasis had been placed on accurate drilling of presplit holes. Perhaps the authors would describe how accuracy had been specified in the contract, monitored during construction and evaluated.

Both papers illustrated the importance of engineering geology to the design, excavation and treatment of instability in rock slopes. They also demonstrated convincingly that geotechnical evaluations should not be confined to the site investigation phases of a project but should be regarded as part of an ongoing process of assessment and evaluation. Data collection and evaluation during construction were vital to efficient design and cost-effective remedial work.

In natural rock slopes, such as that described in Pittsburgh, instability in hard rock was largely a geometric problem relating the orientation of important discontinuities to that of the ground surface. Success in carrying out remedial work was therefore mostly a problem of recognizing the geometric field parameters responsible for the instability and dealing with them. Weathering was also a very important aspect in the development of instability and the protection of critical areas of a face was often necessary. Collection and evaluation of field data were therefore of fundamental importance and had been a key issue in the success obtained. The paper demonstrated that such evaluations, although often of necessity subjective, were still vital in assisting the engineering geologist to appraise the failure potential.

A large rock face could contain a number of different stability domains. The success of the remedial work at Pittsburgh was largely attributable to the recognition of that variability and the fact that no single design strategy had been possible. The introduction of the concept of design zones, linked to individual stability domains, each requiring a particular type of remedial treatment, had therefore been instrumental in defining the remedial action necessary to strengthen the rock face and overcome the natural instability.

In man-made rock excavations both natural instability and instability induced into the final face by the method of excavation were possible. Design of a rock face must therefore take both into account. The excavations at Kornhill illustrated those principles admirably and demonstrated techniques that could be used to optimize the design slope in terms not only of the geotechnical conditions present but also in terms of the space constraints imposed when working in an urban environment. The importance of using blasting methods, in that case presplitting, designed to limit disturbance and thus minimize induced instability in the final face, was clearly emphasized.

The success of the Kornhill Development excavations must be attributed largely to a methodical engineering approach to the initial design, to the technique of continuous assessment during and after excavation and to the selection of cost-effective remedial measures where instability occurred.

Both the Pittsburgh and the Kornhill projects demonstrated the importance of the process of optimization as applied to the design, to the method of excavation and to the selection of remedial techniques, and were excellent examples of the practical application of engineering geology.

Authors' replies

R.C. Speck and R.W. Bruhn Prediction of the 'time to failure' at this or any other rock exposure is virtually impossible, given the complexities of the weathering process in a jointed rock mass. Duquesne Bluff is largely a natural erosional rock face formed over tens of thousands of years by downcutting of the Monongahela River, some local shaping of the bluff having accompanied construction of the Boulevard of the Allies in the 1920s and the Parkway in the 1950s. Some falls have involved rocks that became unstable only a few weeks after the roadways were placed into service, whereas others have involved rocks that became unstable only after several decades had passed. Rock weathering and attendant destabilization is, of course, a continuous process – a fact confirmed in part by the distinctly audible 'raining' of fine rock particles from the slope as we worked beneath the rock overhangs during the field mapping operations as well as by the perceptible degradation of the claystone units. Erosion of the claystone units in a hillward direction is the principal reason why overhangs have developed in the more resistant overlying rock units. At one location in the Pittsburgh area this claystone erosion process has been found to progress at rates of 2–7 in per year.*

In recognition of the foregoing factors the main thrust of our hazard-reduction programme was to slow, if not altogether prevent, the weathering process. The initial concept was to construct a series of retaining walls across a significant portion of the lower slope that would serve to seal the rock as well as provide positive structural support. This approach proved to be too costly. Consequently, our efforts were redirected towards a more modest and, in our opinion, optimal solution commensurate with the funds available. This approach involved trimming off the most significant rock overhangs and sealing with a concrete band only the highly weatherable claystone beds, which, if left unprotected, could lead once again to the development of significant overhangs. Construction of the concrete band

*Philbrick S.S. Engineering geology of the Pittsburgh area. In *Guidebook for field trips, Pittsburgh Meeting, 1959* (New York: Geological Society of America, 1959), 189–203. (*Field Trip* no. 6)

and the implementation of the other control measures, such as wire mesh, are expected to effectively reduce the rockfall hazard for years to come. With normal maintenance we expect the control measures to be effective for several decades.

The three genetic groups of discontinuities identified in the study – stress relief joints, tectonic joints and weathering (wet/dry/freeze/thaw) cracks – were identified primarily on the basis of their orientation in relation to the slope face, the stress relief joints being nearly vertical and generally paralleling the rock face, the tectonic joints being nearly vertical and occurring in nearly perpendicular families that intersect the rock face at oblique angles, and the weathering cracks being randomly orientated. Whereas the stress relief and tectonic joints subdivide the rock mass into blocks several feet on a side, the weathering cracks subdivide the rock mass (primarily the weak rocks) into particles that are of gravel size or smaller.

T.R.C. Muir, B.K. Smethurst and R.P. Finn Rock slope failure modes are generally relatively 'fast' and occur when unfavourable geotechnical conditions coincide. One such failure occurred in an old (90 years) slope on the northern edge of the site. Just when such a combination will occur is not easy to predict and all the designer can do is make conservative assumptions for slope treatment with due regard for any improvements that the excavation activity may have – for example, lowering groundwater table, reduction of large overburden pressures – the philosophy being that failure could occur at any time in the life of the slope (in Kornhill's case the same as the Development, about 25–30 years), although it is a generally held belief that new slopes in hard rock tend to 'settle down' after two or three years. The slopes and slope treatment are subject to routine maintenance and inspection, although the slope treatment has been designed and installed in accordance with current Hong Kong standards and there is no reason to believe at this time that it will not be effective through the life of the slope.

Cooling discontinuity sets tended to comprise one flat-lying and two vertical orthogonal sets in both granite phases present on the site. Tectonic sets were sub-vertical and clearly developed for a short distance on either side of the dolerite zones, although they may have been present elsewhere on site. There was no characteristic surface texture, mineralization or weathering to enable different sets to be distinguished, but some sets were clearly related to the dolerites, crosscut other sets or were controlled by the contact of the granite phases.

The contract required that all rock slopes were to be presplit and production blasting was prohibited within 20 m of any presplit face before presplitting was carried out. A series of presplit trials was undertaken early in the contract, as was described in the paper, and the pattern that was adopted was chosen as the best balance between satisfactory slope surface quality between the half barrels, in terms of tolerance and subsequent slope treatment works, and drilling progress, which had to be kept in advance of bulk production and programme requirements.

The specified tolerances for presplit drilling were that holes should be no more than 50 mm from their true location at the top of the holes and a maximum of 1° divergence from the true orientation in the line of the hole. Tolerances with regard to slope profiles were specified as being ±150 mm in general and ±300 mm in highly fractured rock. Deviation was a source of concern in some areas where the bottom 1–2 m of the holes tended to 'kick out'. This resulted in some subsequent hand excavation with jackhammers to position and construct the berm drainage channels. No problems were encountered in drilling the shorter holes or the vertical holes for building foundation pockets.

Generally satisfactory results were obtained in presplitting. In some areas where it was considered that the ppvs produced by presplit blasting would exceed the imposed limits or cause damage to adjacent slopes or structures, chemical rock-breaking compounds were used to produce the presplit. Satisfactory results were generally obtained.

Finally, it may be of interest to note that in general discussion it was suggested that 'kick-out' at the slope toe could arise purely as a result of the blast itself rather than hole alignment. The comment was made that in the United Kingdom it is now common practice to omit a bottom charge in presplit holes and that approach was recommended. The question was also raised as to whether bottom charging should be specifically excluded if trial splitting did not indicate any damage to the toe of the slope.

On a different note it was said that shotcrete had a limited life in freezing conditions and that was a difficult problem to resolve. One solution might be to exclude groundwater by grouting well behind the face, or substantially strengthening the shotcrete, which might only be justifiable for permanent works. That problem does not, of course, arise in Hong Kong.

Session 6: Rock slopes – construction and remedial works

Chairman: Dr. E.W. Brand
Rapporteur: P.G.D. Whiteside

Discussion

P.G.D. Whiteside* commented that the session title was something of a misnomer in that rock slopes were not really constructed but exposed and then, if necessary, stabilized, i.e. the natural state of the rock material was revealed and assessed. The four papers† in the session dealt with the problem of assessing the likelihood of failure of an already exposed rock face in a high-risk urban situation.

In each of the four papers observations, measurements and calculations showed that there was some possibility that failures might occur. All the possible failures that were discussed were discontinuity-controlled and, therefore, the starting point in all the papers had been the gathering of rock joint data. Those data were, of course, variable, and when calculations were carried out some indicated failure and some did not. Therein lay the problem. As Evert Hoek had pointed out in his keynote address,‡ the small computers that were available currently meant that individual calculations were straightforward. The real problem was how to deal with the variability of the input date.

As an example of that variability they might consider the parameter rock joint orientation. In a recent paper§ by Hencher more than 400 readings were presented of the orientation of a single joint 100 m^2 in area. Those were shown on an equal-area plot in Fig. 1.

The scatter of points was comparable with concentration of poles found in the papers under discussion and was typical of natural variations found in joint sets elsewhere. Fig. 2, however, showed a frequency distribution of those same data and emphasized the quite wide range of orientations in such clusters of poles – a fact that was all too often glossed over when average values from contoured density plots were used for stability calculations.

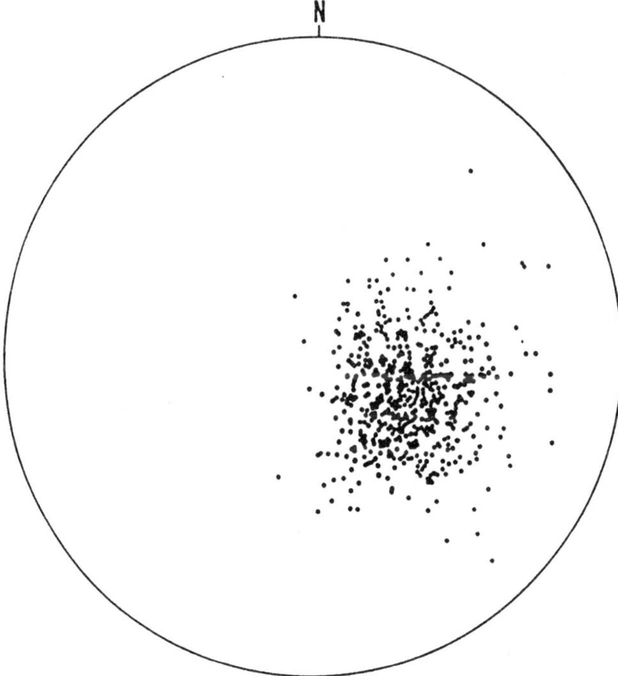

Fig. 1 Poles measured at 0.5-m centres on single joint 100 m^2 in area. From Hencher, *op. cit.*, Fig. 4

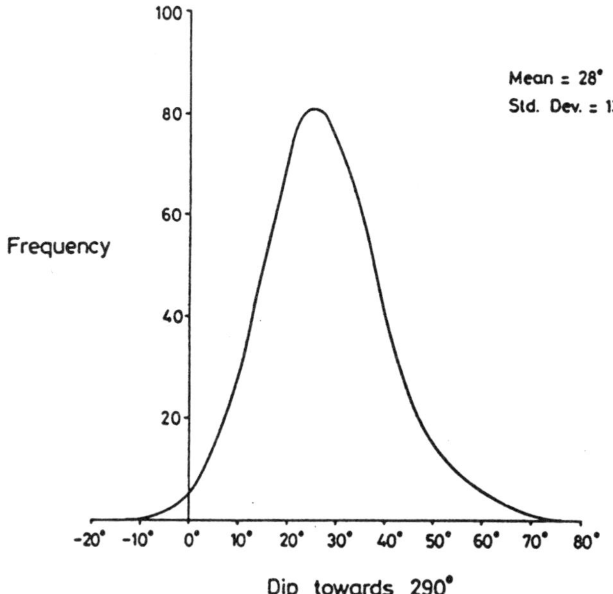

Fig. 2 Frequency distribution of Fig. 1 data (dipping towards 290°)

To illustrate what that range of orientation might mean in terms of a calculated factor of safety Fig. 3 had been drawn. Clearly, the approach to the problem of stability and risk assessment was not straightforward. So how did they assess the likelihood of failure?

The current state of the art as it was most commonly practised was still exemplified by Hoek and Bray's *Rock slope engineering,* in which, in reference to that problem, it was stated that 'This is one of the most controversial questions in rock engineering and many eminent engineers have argued that, because of the uncertainty associated with the input data for a factor of safety calculation, the value obtained is too unreliable to have relevance in engineering design.

Some authors have suggested that a *probabilistic* approach is more meaningful . . . Although this approach has many attractive features, it has two drawbacks . . . the difficulty of obtaining adequate input data for a meaningful statistical analysis . . . The second drawback is associated with . . . statistical concepts . . . How does one relate the adequacy of a design to a probability of failure of 1 in 100,000? Indeed, to some clients, the admission, by a consulting engineer, that there is a possibility of failure, however small, is totally unacceptable'.*

*Scott Wilson Kirkpatrick and Partners, Hong Kong.

†Beggs C.J. and McNicholl D.P. Formation of a high rock slope at Ap Lei Chau, Hong Kong; Buttling S. Remedial works to a major rock slope in Hong Kong; Dubin B.I. Watkins A.T. and Chang D.C.H. Stabilization of existing rock faces in urban areas of Hong Kong; and McCracken A. and Jones G.A. Use of probabilistic stability analysis and cautious blast design for an urban excavation.

‡Hoek E. Keynote address: Practical rock mechanics – developments over the past 25 years. This volume, ix–xvi.

§Hencher S.R. Limitations of stereographic projections for rock slope stability analysis. *Hong Kong Engr,* 13, no. 7 1985, 37–41.

*Hoek E. and Bray J.W. *Rock slope engineering* (London: IMM, 1974), 27–8.

All the papers in the present session had had to tackle that problem and each had resolved it in a different way.

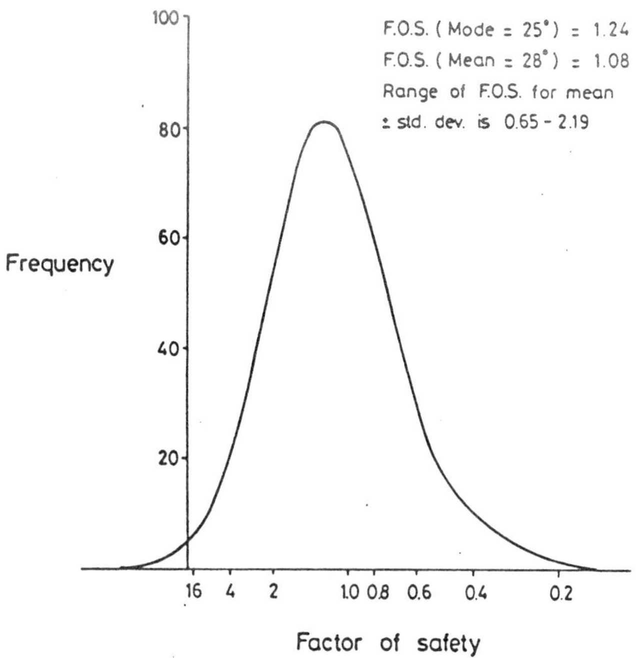

Fig. 3 Factor of safety against planar failure (towards 290°, dry, ϕ = 30°). Data from Fig. 1

Summary of papers

Beggs and McNicholl described the design and construction of a 70 m high slope cut in very strong, well-jointed tuff to provide a building platform for high-rise housing. Central to the paper was the design review that took place during construction once the actual conditions had been exposed. The authors' original design had recognized that there was a natural variability of the joint orientations, persistence, etc., and they had selected a 70° presplit face as likely to result in a manageable number of potential failures needing to be stabilized during construction. Inherent in the design for that high risk to life area was the provision to stabilize all potential wedge failures. In the event, exposure of the actual rock conditions had revealed that minor joint sets were better developed than had been expected and, as a result, a considerably greater number of potential wedge failures would need to be treated. The slope had therefore been redesigned to 60° to reduce that number to a level that was acceptable both technically and financially. In other words, the variability of the rock had been assessed and the slope had been designed so that the likely number of potential failures in need of treatment had been kept to a practical level. It was not clear from the paper just how the authors had quantified that 'probability of failure', although it certainly had been done.

The paper by Buttling covered remedial works to a large rock slope where progressive wedge failures were occurring in a partially weathered rock mass with clay-lined joints. It had become apparent that after heavy rain joint water build-up was causing slipping. To tackle the problem of variability of input data sensitivity analyses had been carried out with the use of different joint orientation, tension crack depths, water levels and clay shear strengths. It was worth noting that, in the absence of acceptable statistical methods, Hoek and Bray (*op. cit.*) cited a sensitivity analysis as the most satisfactory solution to the problem of assessing variability.

As a result of those analyses a grillage of groundbeams and rock anchors had then been designed. The calculations gave the theoretical range of stability levels that resulted from the natural variability of the rock mass and the uncertainty as to what water pressures might develop. The scale of support that had finally been designed was then based on the maintenance of a theoretical limiting equilibrium under the worst possible conditions and the achievement of a factor of safety of 1.2 under those most likely. (There was, in fact, a safety margin in the worst case since the ultimate strength of the anchors could reliably be taken as twice their design load.)

It appeared from the paper that the requirement for a factor of safety over unity for the worst circumstances had been the more onerous condition for the overall stability.

To summarize, the author had recognized and measured the range of parameters affecting stability and designed remedial works that would cater for the worst conditions envisaged, while also achieving, under conditions reasonably expected to occur, a factor of safety of 1.2, which for such remedial works was the guideline laid down by the Hong Kong Government.

Dublin, Watkins and Chang described the approach that the Hong Kong Government adopted for the stabilization of existing urban rock slopes, many of which had been formed some time ago by simple bulk blasting methods.

As was briefly referred to in the paper, the first stage of dealing with those slopes was an assessment based on a ranking system. The rank assigned depended on slope geometry, history of stability, location, etc., and had proved to be a reliable method of providing an initial measure not only of innate stability but also of the need to carry out remedial works. Once a slope was selected for remedial works a design/build approach then minimized the nuisance value of those works. The preponderance of small localized instabilities lent itself to that pragmatic approach, most on-site decisions being made on an *ad-hoc* basis by an experienced engineer. So the problem of assessing the risk of failures and the need or otherwise for remedial work were essentially approached in two stages. First, a slope was reviewed, both in terms of its theoretical stability and its own and other similar slopes' past performance. Secondly, subjective on-site decisions were made to identify not only those instabilities which were expected but also those which slope clearance had revealed.

McCracken and Jones discussed the design of a large city hospital to be located in an old quartzite quarry. The design included the retention of a high existing face as a permanent feature with vehicle and pedestrian access at its toe. A large amount of field information on the existing face had been gathered, particularly concerning the range of joint strengths and orientations. It was clear from that information that plane failures in that blocky rock mass were theoretically possible and had to be assessed.

A standard deterministic analysis with the use of values for the rock mass had given a satisfactory factor of safety even within the range of ± the Standard Deviation. The authors had recognized, however, that there was still a possibility of the occurrence of failures if the values of rock parameters were more extreme than the Standard Deviation. They had fitted standard normal distributions to their field data and thence obtained a probability for a failure. The authors had overcome the difficulty quoted from Hoek and Bray of deciding the meaning of the probability value by referring to the acceptability of different levels of probability as published by various authors. In comparison with those published values the calculated probability had been thought acceptable.

Slope reinforcement

An important aspect of the papers that time did not permit to be discussed at the conference concerned the design of stabilizing works. Some of stability problems that had been referred to in the papers had been resolved by adding a restraining force — either actively prestressed, such as anchors, or in passive mode, such as dowels. It was worth noting that calculations of increased factors of safety as a result of such slope reinforcement were to some extent arbitrary in the sense that additional resisting forces could either be added to existing resisting forces or subtracted from existing driving forces. The resulting factor of safety was different in each case, being greater if, as Hoek and Bray suggested, part of the additional force was taken as reducing the driving force — an approach that was difficult to justify since a gravitational driving force could not be reduced but only counteracted. Hoek and Bray (*op. cit.*) quoted a personal communication from P. Londe, who suggested the avoidance of that problem by the use of individual safety factors on each parameter. That approach had appeal, but it required a difficult choice to be made of what value of safety factor to adopt for each parameter.

In contrast to the above, the probabilistic approach entirely bypassed that problem, the reduction in probability being precisely the same no matter whether the additional force was wholly or partly allocated to the resisting or driving forces. Thus, by use of Fig. 9 (p.236) it made no difference whether the capacity curve moved right or the demand curve moved left or a combination of both since the relative increase in separation (and thence reduction in probability) were the same. That was a further attraction of the probabilistic approach.

Conclusions

The four papers illustrated the considerable difficulty that was involved in assessing the significance of the variability within a rock mass and, in particular, a rock mass that was well exposed. The individual approaches to the problem exemplified the current state of the art. The sentiments quoted earlier from Hoek and Bray of a definite appeal in the probabilistic approach seemed to be echoed, but, despite the elegant demonstration of probabilistic methods by McCracken and Jones, it also seemed that the civil engineering world was not yet ready to adopt the full statistical approach. They might, nevertheless, be encouraged by the high degree of computational ability that was being developed for stability analyses in the mining world, as had been described by Evert Hoek in his opening address, and they might foresee a time when similar techniques were used in civil rock engineering. Perhaps a step in the right direction would be to stop referring to a small probability of failure and start talking of a small possibility of failure.

Written contribution

Dr. S.M.M. Rice* In general discussion it was queried whether the observed failure mechanism could have been predicted from an apparently similar joint-controlled failure that occurred at the southwest of the slope referred to by Buttling during 1982 and 1983.

The slope discussed has a complex history of investigation and design, but the two failures can be conveniently compared by examining stability analyses carried out in 1983 for a design report submitted to the Government Control Office. Figs. 4 and 5 are taken from that report, but simplified. The analyses were carried out by use of best estimates of strength parameters and groundwater conditions available at the time.

The 1982–83 failure could have been predicted without recourse to unusually weak zones or perched water-tables, i.e. although maybe joint-controlled, it was not 'joint-caused'.

By contrast, the 1984 failure is inadmissible without preferentially orientated weak planes.

This interesting case history and the ensuing discussion highlighted the difficulties of applying data obtained from one part of the site elsewhere.

* Scott Wilson Kirkpatrick and Partners, Hong Kong.

Authors' replies

C.J. Beggs and D.P. McNicholl The risk of failure was assessed as indicated below.

(*a*) By mapping and joint surveys carried out on the upper faces as they were formed. These showed the same four dominant joint sets that were identified in the original investigation. Of the four minor sets that had originally been identified, only one was observed in the upper part of the cut slope, but an additional minor set (*H*) was seen. This could combine with the other minor set (*J*) to form wedges that could fail.

(*b*) Detailed examination of the minor joints exposed showed that joints were typically planar to slightly undulating, joint surfaces were typically smooth and iron-stained and commonly coated with kaolin up to 5 mm thick and the joint spacing was 1–2 m.

(*c*) A back check on previous joint surveys showed that joint sets *H* and *J* were likely to be present throughout the rock mass.

(*d*) An assessment of rock quality based on exposure and drill core showed that the overall joint spacing was close (60–200 mm), giving rise to spalling. Blasted rock was small-size and rectangular. Only minor secondary breaking was needed. Drill core showed poor-quality rock at depth.

(*e*) Wedge analysis was carried out with $\phi = 40^\circ$ and $c = 0$. The volume of rock in a wedge would depend on the persistence of the unfavourable joints forming the wedge. With the observed persistence of 4 m the volume of a wedge would be about 2.5 m^3. If the persistence were great enough to involve an entire inter-berm face, the volume of the wedge would be up to 100 m^3.

By laying back the face to 60° the volume of a probable failure would be reduced to about 1 m^3. To eliminate failure altogether the face would have to be cut at 45°.

B.I. Dubin, A.T. Watkins and D.C.H. Chang In addressing points regarding the approach adopted by the Geotechnical Control Office (GCO) to remedial works to existing rock faces in Hong Kong it may be helpful to reiterate some of the observations made in our paper.

First, existing slopes differ from newly formed rock slopes, or even *old* slopes associated with new infrastructure projects. One of the main differences involves the constraints that are placed on remedial works by physical, environmental, public safety and economic factors. Fig. 6 shows a slope that illustrates a number of these constraints. The rock

m PD
80
60
40
20
SLIP SURFACE | FACTOR OF SAFETY
S 1 | 0.99 (back analysis)
S 2 | 1.05
S 3 | 1.22
1982/83 Failure area
Loose completely decomposed granite
ø' = 35°
c' = 6kPa
γ = 19kN/m³
Original profile
After failure profile
Design ground water level
S1
S2
S3
Medium dense to very dense completely decomposed granite
ø' = 40°
c' = 10kPa
γ = 20kN/m³
Bedrock
Road TY1

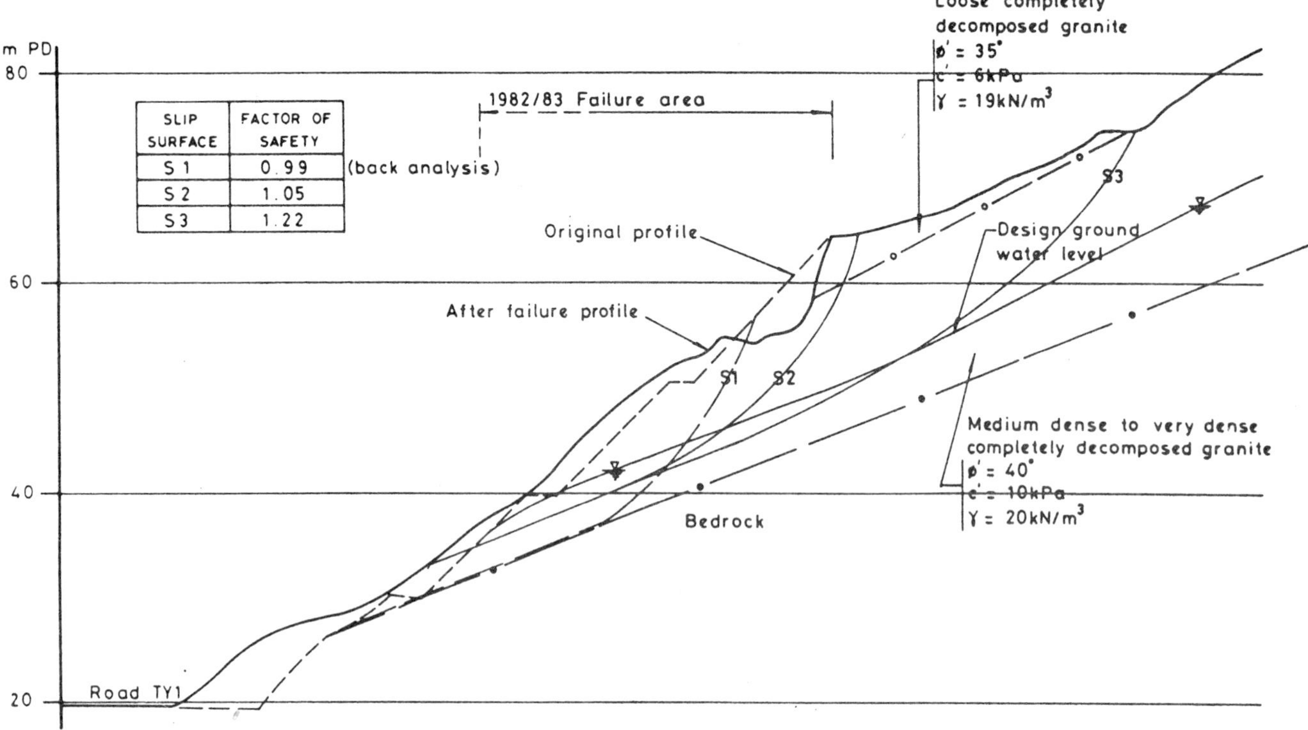

Fig. 4

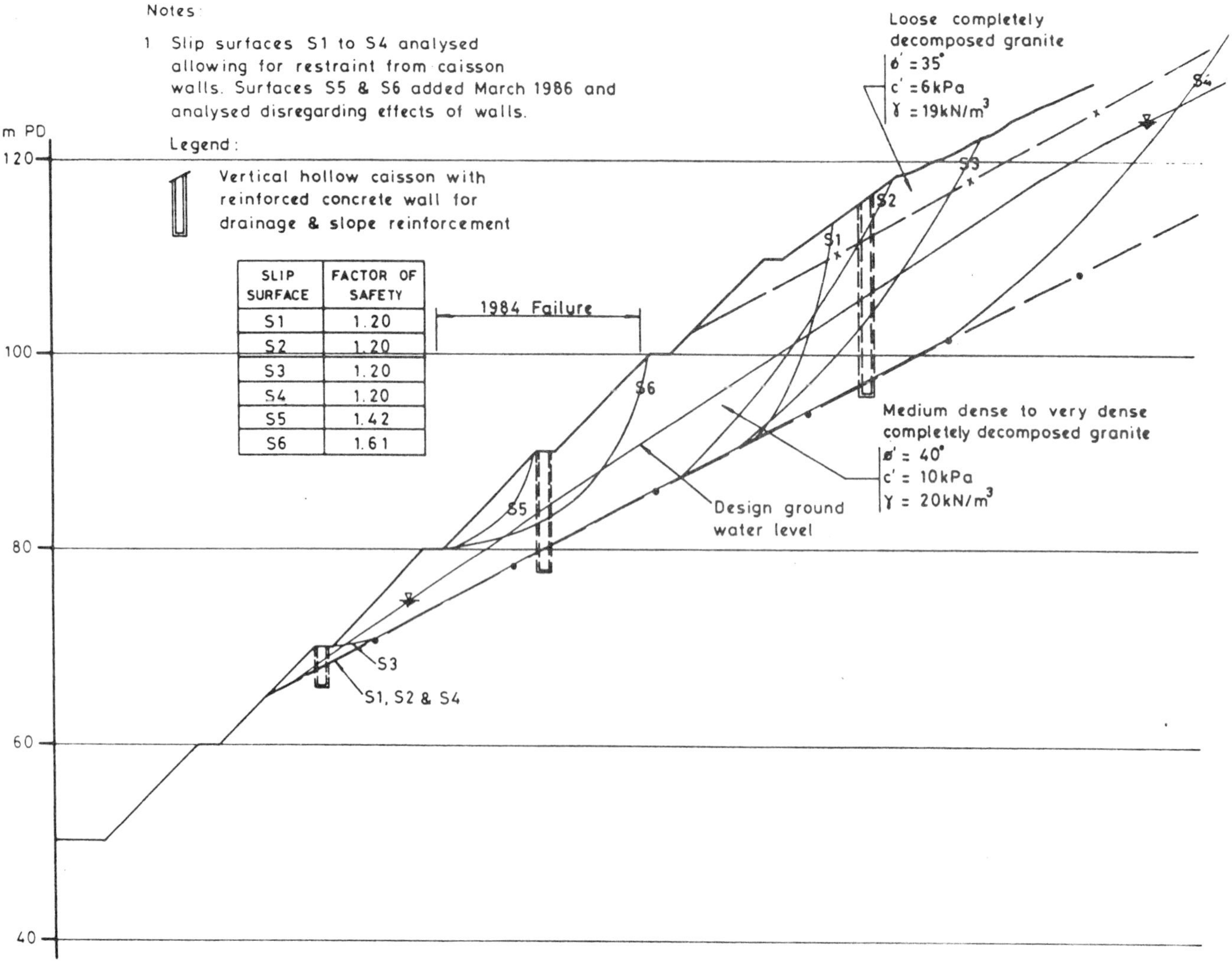

Notes:
1 Slip surfaces S1 to S4 analysed allowing for restraint from caisson walls. Surfaces S5 & S6 added March 1986 and analysed disregarding effects of walls.
Legend:
Vertical hollow caisson with reinforced concrete wall for drainage & slope reinforcement
SLIP SURFACE | FACTOR OF SAFETY
S1 | 1.20
S2 | 1.20
S3 | 1.20
S4 | 1.20
S5 | 1.42
S6 | 1.61
m PD
120
100
80
60
40
1984 Failure
Loose completely decomposed granite
ø' = 35°
c' = 6kPa
γ = 19kN/m³
Medium dense to very dense completely decomposed granite
ø' = 40°
c' = 10kPa
γ = 20kN/m³
Design ground water level
S1
S2
S3
S4
S5
S6
S3
S1, S2 & S4

Fig. 5

Fig. 6 Slope prior to remedical works

Fig. 7 Slope following remedial works

Fig. 8 Rock slope threatening temporary structures

slope is a fairly heavily trafficked road, which was required to remain open during the commission of the remedial works, and across the road is a housing estate, which necessitated that noise and dust during construction be limited and absolutely precluded the use of explosives. Fig. 7 shows the works in an advanced state. As the soil slope required flattening, and was constrained by the presence of the road barracks above, it was not feasible to alter significantly the geometry of the rock slope.

Remedial works sometimes have one freedom that new works do not. In some cases the final form that remedial works take is not constrained in any way by other civil engineering requirements. For example, Fig. 8 shows a rock slope threatening a temporary structure. Bearing in mind that GCO's Landslip Preventive Measures are aimed at elimiating threat to life situations, in this case the first stage of the remedial works took the simplest form — the structure was evacuated and demolished and the area below the slope was made secure to prevent public access.

One of the points queried was how it is determined whether or not stabilizing measures are required. In what fashion are slopes selected for GCO's Landslip Preventive Measures programme? Slopes are picked for LPM works based on the Government's ranking system. Briefly, virtually all cut slopes in Hong Kong have been ranked in terms of consequences of failure and potential for instability, the instability rating being a function of many factors — in particular, the geometry of the slope, including slope height and slope angle. Slopes are considered for stabilizing measures under the LPM programme based on these rankings, in conjunction with such other factors as history of past failure, present condition and evidence of present instability, which is based on a fairly straightforward visual assessment.

At this stage in the selection process no attempt would be made to carry out numerical factor of safety analysis as any such answers would probably be highly unreliable owing to the lack of hard data. Further, this would be a daunting prospect, as a typical LPM selection exercise can involve hundreds of slopes. Comparisons are made, however, with empirical studies such as the 'Chase' plots.[1] Fig. 9 shows one of the result plots from this study, which attempted to provide semi-empirical guidelines for the design of soil cut slopes. Although, strictly speaking, it is not applicable to a rock slope, it can be used for guidance in the early stages of the selection process. The plot shows height versus slope angle for 117 granite slopes, the dotted lines being failure paths. The Chase study indicated that slopes within the lower-bound envelope were stable in the long term. Virtually all LPM cut slopes, including soil cut slopes and the

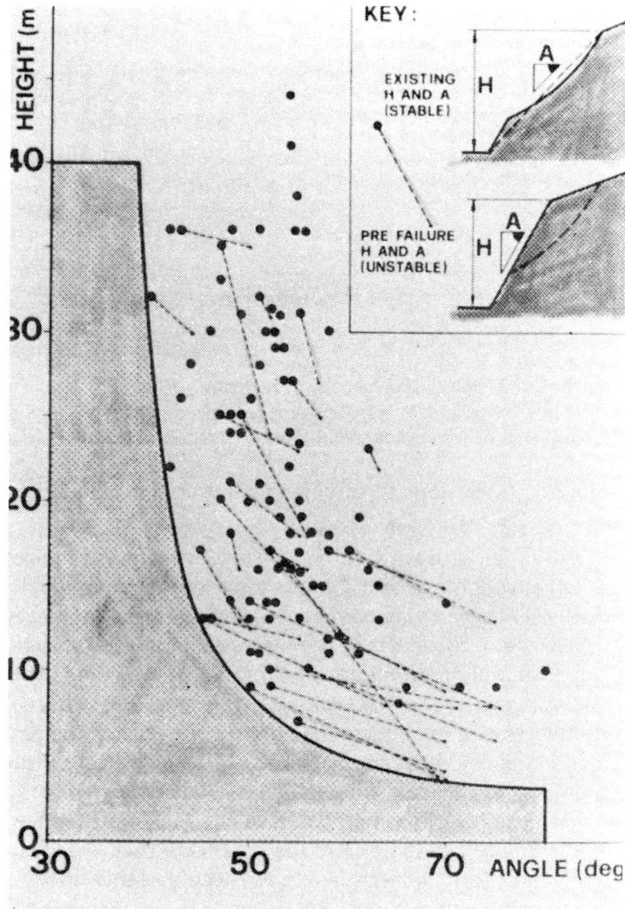

Fig. 9 'Chase' plots

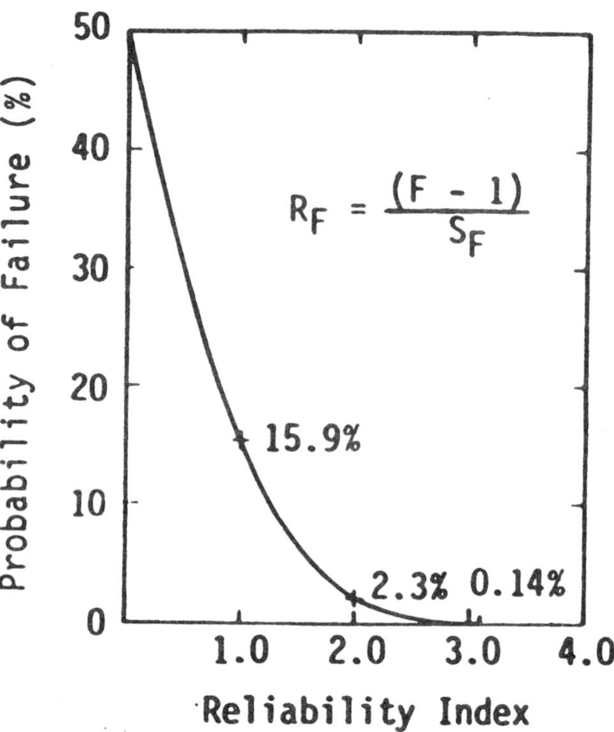

Fig. 10 Reliability index

more typical soil overlying rock slopes, plot outside this envelope, about 50% being within the region bounded by 15 and 25 m on the height axis and 60° and 70° on the slope angle axis.

Although the ranking system is the basis for the current selection process, GCO will continue to try to find means of better quantifying various aspects of the selection process, but it seems unlikely that a factor of safety type of approach can be adopted. Of course, during the course of tackling a particular problem on a particular slope, any number of numerical 'factor of safety' types of analysis would be used if it was felt that they were required, or if the other branches of GCO that review LPM works designs so requested. In most cases sophisticated rock slope analysis is undertaken during construction, or even after, rather than during the selection process. In any case these analyses must be used with an understanding of the reliability of the answers because of the uncertainties associated with the design parameters.

In principle, and as was pointed out by McCracken and Jones,[2] in practice, with the use of modern computer techniques it is possible to estimate the probability of failure of a design by regarding all parameters as variables.

A simpler method is to estimate the mean value of the safety factor (F) and its standard deviation. From these, a 'reliability index'[3] can be determined, being the difference between the mean safety factor and a safety factor of unity, divided by the standard deviation. Fig. 10, taken from the *Manual,*[3] shows the significance of this parameter in terms of the probability of failure. As the reliability index increases, the probability of failure decreases.

In his keynote address Hoek[4] stressed the importance of probabilistic studies and sensitivity analysis. It is hoped that these techniques can be applied to LPM works in future. It is, however, unlikely that these techniques can eliminate completely the need for the design–build approach, particularly when detailing works on site.

As is shown in Fig. 11, the statistical centre of a joint set may not reflect all of the data, and even though the set pole may be stable, some of the joints that make up the set may not be.

One excellent technique for coping with this natural variation in data is to number all joints in the field so that the critical joints that fall within the unstable areas can be located. Another method, which is complementary, is for an experienced engineer, who knows what modes of failure he is looking for from the stereoplots, to examine the entire face.

It has been noted that a considerable number of steel bolts and dowels are used in LPM work and that steel can be subject to corrosion. As was mentioned previously, one of the constraints on LPM works is physical. Often, a slope cannot be cut back to a flatter angle because there are structures upslope. Often, to remove a potentially unstable block would create a risk to structures below during the operation, or would undermine other blocks. Concrete buttresses are very effective but, often, placement of concrete support is impractical because of geometry and, in any case, most buttresses require steel dowels into the rock. Therefore steel is used because there is no practical choice. To avoid the problem of corrosion oversize bars are used, not wire strands, with an allowance for sacrificial diameter. Generally, dowels or bolts are used only in dry rock. In the simplest case of untensioned dowels the bars are protected by a grout annulus with centralizers on the bar (Fig. 12). A very high-strength, low-shrinkage group specification is used to reduce the possibility of cracking and resultant corrosion problems. A conservative design approach is adopted. Bolts, being tensioned, are designed with two- and even three-stage corrosion protection.

Very little evidence seems to be available regarding the lifetime of these devices, particularly in dry rock at or near surface. This question was addressed to Dr. Evert Hoek and

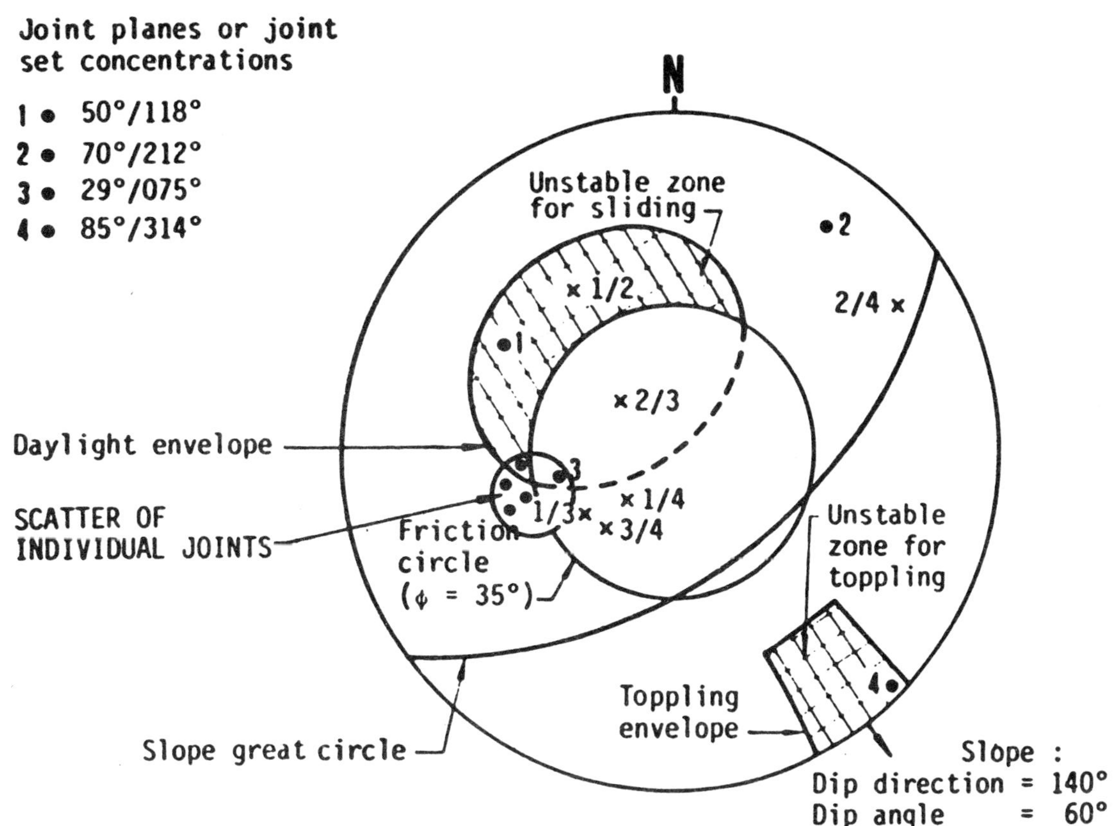

Fig. 11 Variation of joint set data

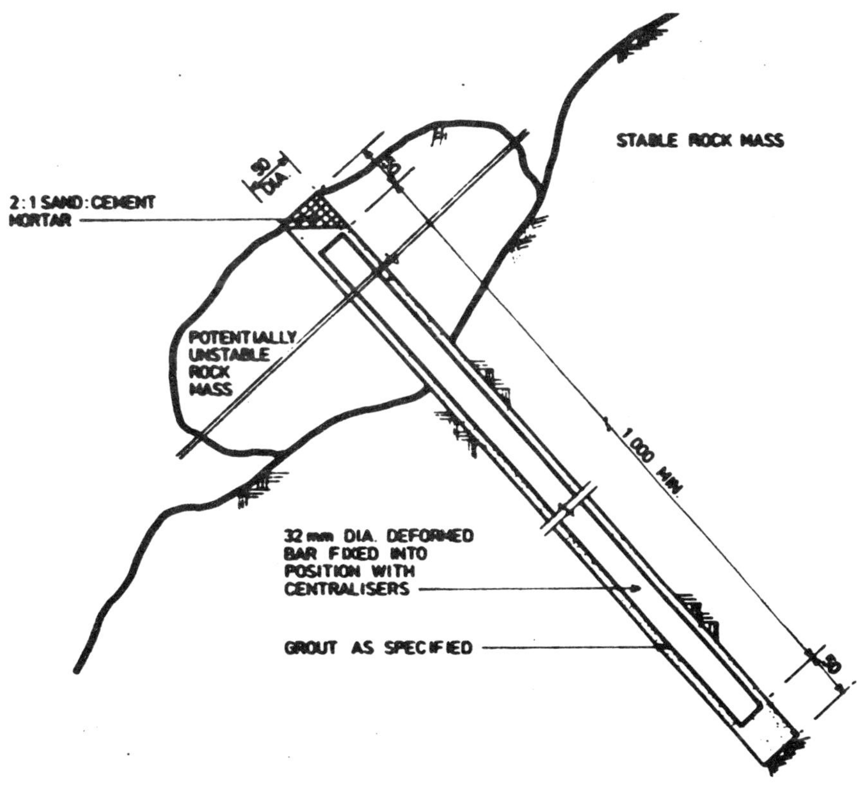

Fig. 12 Typical details of rock dowel

in reply (personal communication, 23 February, 1986) he stated that although in some corrosive underground environments the lifetime of a bolt can be very short, he felt that under conditions such as those described above, with very tight control of the grout, their working lifetime may be as much as 40 years. At a recent informal discussion of the Ground Engineering Group of the Institution of Civil Engineers Littlejohn[5] stated that although millions of highly stressed tendons had been installed around the world, there had been very few incidents of failure in service. Dowels are defined as a passive reinforcement. The fact that there are few recorded failures can be considered as passive evidence in their favour.

If anyone has any evidence with regard to the lifespan of dowels or bolts in Hong Kong, GCO would welcome a chance to share it.

References

1. Brand E.W. and Hudson R.R. CHASE – an empirical approach to the design of cut slopes in Hong Kong soils. *Proceedings seventh Southeast Asian geotechnical conference, Hong Kong, November 1982,* vol. 1, 1–16; Discussion, vol. 2, 61–72; 77–9.
2. McCracken A. and Jones G.A. Use of probabilistic stability analysis and cautious blast design for an urban excavation. This volume, 231–40.
3. Geotechnical Control Office. *Geotechnical manual for slopes* (Hong Kong: Hong Kong Government Printer, 1979), 252 p. (Reprinted, with corrections, 1981)
4. Hoek E. Keynote address: Practical rock mechanics – developments over the past 25 years. This volume, ix–xvi.
5. Littlejohn G.S. Ground anchorages – case histories – informal discussion of the Ground Engineering Group. *Proc. Inst. civ. Engrs* pt I, **78,** 1985, 1500–3.

A. McCracken and G.A. Jones The probability density function for multi-variable equations may be derived by (1) taking the statistical parameters of the distributions of the relevant individual factors and combining them to create the composite distribution, i.e. lower bound with lower bound, mean with mean, upper bound with upper bound (this approach tends to be conservative in respect of the lower tail) or (2) if the parameters are independent variables, by sampling randomly the distributions of each of the participating parameters and combining them in the relevant equations to form a resulting combined distribution.

Our approach is to use the Monte Carlo sampling technique to sample the various distributions and combine them as in (2). In the paper the capacity–demand model for the quarry is simple, and this approach was not followed. Since the dip angle and peak friction angle were determined as couples at the same measurement point, the 'true' distribution of safety margin could be obtained.

There is likely to be a specific relationship between probability of failure and factor of safety for a particular structure. A general relationship, however, cannot be defined, as it will depend on how well the distributions of capacity and demand are defined and, hence, what the form is of the distribution of probability of failure against factor of safety. The latter distributions are shown in Fig. 13. The steeper the curve, the better defined is the capacity–demand model.

The support assessment was, in fact, partly probabilistic in that the NGI system was applied to the results of the probabilistic stability analysis. The NGI system has an element of probability within the Excavation Support Ratio (ESR) value, which is a number chosen on the basis of the function, size and permanence of the excavation. N. Barton explained the number as being related to the inverse of the factor of safety, but it could perhaps be related better to an acceptable probability of failure.

In response to general points, all the joint data are made use of and sampled to form the input for stability analysis. This is a much more representative approach than to choose the more critical joint parameters for input, which would result in a more conservative design.

A probabilistic approach is much more realistic than a unique value approach. The probability of failure may be interpolated linearly.

A time-dependence of slope stability in itself is intriguing. The stability of a slope is reduced owing to the reduction of joint strengths, among other factors. This must be recognized when one is designing a support system.

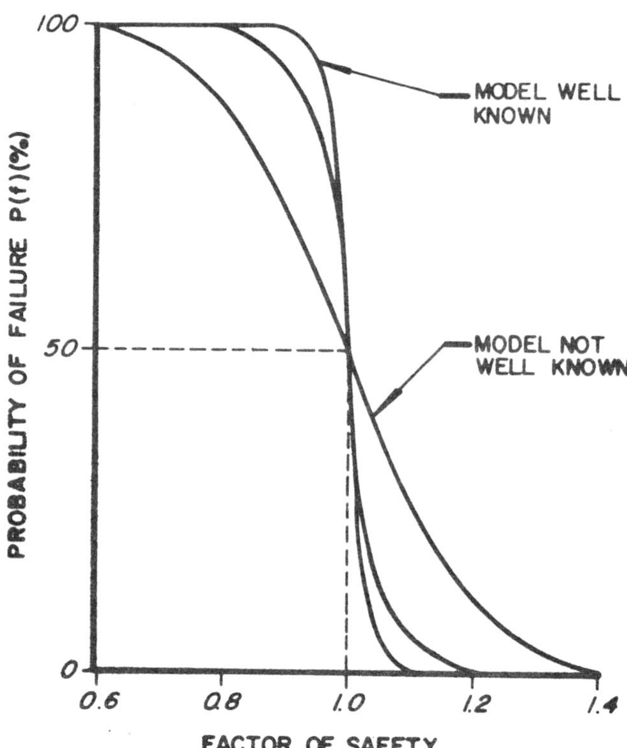

Fig. 13

The corrosion of the steel in a support system can be a problem – generally, we do not consider a support system to be permanent unless the bolts or cables are fully grouted. The function, public access, etc., of the excavation must be evaluated and the support system designed accordingly. Corrosion protection (in addition to grouting) may have to be recommended in critical long-life situations.

Table 2 (p. 236) goes some way towards quantifying the risk component in an empirical sense. The question of desirability is more vexing. There is no doubt that a probabilistic approach is preferable in a technical sense, but this is not necessarily so in a practical sense, as recognition of a probability of failure may be unacceptable to an owner, whereas a factor of safety in excess of 1.0 may be acceptable.

Session 7: Rockfall protection methods

Chairman: Dr. E.W. Brand
Rapporteur: Dr. L.R. Richards

Discussion

K. Morton,* in presenting the rapporteur's comments in the unavoidable absence of Dr. L.R. Richards,† said that the problem of rockfalls from natural and excavated slopes, with which the two papers‡ under discussion were concerned, was one that was common to a variety of situations from open-pit mine slopes to highway cuttings and building platform excavations on steep slopes. The most commonly quoted reference on that subject was the work done in 1963 by Ritchie [1] for the Washington State Highway Commission. He had noted that there was a need for a better understanding of the criteria for estimating rock mass strength and predicting the stability of the material on the surface of a rock cut, commenting that, so far, those factors remained elusive and many engineers approached the problem with apathy, as though walking up to a stone wall and halfheartedly demanding that the wall give up its secrets and come under their slide rule. Ritchie had carried out a series of experiments on falling rocks and produced guideline relationships between width of fallout and height and angle of slope. Those guidelines had tended to become enshrined in the literature and accepted without reservation by the succeeding generation of engineers. It was therefore of great interest to learn of the work that was being done on that subject in Hong Kong. Ritchie would be delighted to see his stone wall being demolished by sledge hammers instead of coerced with slide rules (or their current equivalents).

Since the date of Ritchie's paper there had been a number of excellent reviews of methods of controlling rockfall problems. Notably, those included the work by Piteau and Peckover,[2] Fookes and Sweeney [3] and Peckover and Kerr,[4] which provided specific guidance on methods of preventing rockfall problems, but all relied implicitly on Ritchie's early empirical studies. Peckover and Kerr's paper [4] provided a clear summary of the logic required to select the appropriate type of slope treatment. The various possibilities comprised stablization methods (e.g. rockbolting), protection methods (e.g. catch) and warning methods (e.g. electric fences).

Stabilization methods were generally expensive, but required minimal maintenance and were a long-term solution. Protection methods generally cost less, but required continuing maintenance. Typical types of protection included wire mesh, shaped ditches, catchment areas, catch nets or fences, catch walls and rock sheds or tunnels. Warning methods might be appropriate in certain circumstances, but they did not reduce maintenance costs and had no effect on the source of dangers. References 2 and 4 should be consulted for specific practical details of rockfall protection methods.

With the replacement of Ritchie's slide rule by the computer it was possible to simulate rockfall trajectories down slopes of varying geometry. Those programs required assumptions as to the coefficient of restitution for rocks in the bouncing mode and frictional values in the sliding and rolling modes. The simplest programs modelled rock blocks as point masses, the more complex being able to model individual block geometries. Because of generally inadequate data for input parameters (particularly coefficient of restitution) and lack of calibration against real field examples, those models were not yet design tools. The ability to model hundreds of rockfalls on a given slope in a very short time, however, made them ideal for sensitivity analyses and they might therefore provide useful guidelines as to possible rockfall behaviour and effectiveness of protection methods.

Mak and Blomfield described field experiments that involved dropping 13 000 blocks over presplit slopes. That was possibly the first comprehensive test of that type since Ritchie's original experiments. The authors presented a number of graphs to summarize the experimental data. In practical terms, however, the conclusions could simply be summarized thus: 1- and 1.5-m high barriers at 1.5 m from the slope toe would trap 95 and 100%, respectively, of the boulders falling from a 12-m slope (for the experimental range of slope angles studied).

The range of slopes studied was pertinent to many of the excavated slopes in Hong Kong and the results were certain to be of great benefit in allowing more confident design of rock traps and fences in the territory.

The paper by Chan and co-workers described the design of a fence to retain boulders falling from colluvial slopes. In that case the problem of boulder movement was far more complex than that studied by Ritchie or Mak and Blomfield. In view of the lack of field data the authors had assessed their design loads on the basis of mathematical simulation and small-scale models. The authors provided useful data on the design of yielding rock fences.

Both papers were a welcome addition to the literature on rockfall problems. The organizations concerned deserved acknowledgement for having undertaken those trials and the authors congratulation for their presentation.

References

1. Ritchie A.M. Evaluation of rockfall and its control. *Highway Res. Record* no. 17, 1963, 13–28.
2. Piteau D.R. and Peckover F.L. Rock slope engineering. In *Landslides – analysis and control* (Washington, D.C.: Transportation Research Board, 1977), 192–228.
3. Fookes P.G. and Sweeney M. Stabilization and control of local rock falls and degrading rock slopes. *Q. Jl Engng Geol.*, **9**, 1976, 37–55.
4. Peckover F.L. and Kerr J.W.G. Treatment and maintenance of rock slopes on transportation routes. *Can. geotech. J.*, **14**, 1977, 487–507.

C.M. Guilford* said that he had found both papers very interesting, but did feel that their authors should have stressed the desirability of having an effective energy-absorbent medium in the fall zone at the bottom of a slope.

Mak and Blomfield referred (p. 264) to the ground at the toe of each slope being covered with a layer of compacted rockfill 'to provide similar energy absorption characteristics at every location'. Surely, uncompacted rock fill would have met the above requirement yet been more effective and, incidently, less expensive! The effectiveness of loose gravel as an absorbent material was well illustrated on escape roads positioned at suitable locations on steep hills. The

*Earth Sciences (Asia), Ltd., Hong Kong.

†Golder Associates, Maidenhead, Berkshire, England.

‡Chan Y.C. Chan C.F. and Au S.W.C. Design of a boulder fence in Hong Kong; and Mak N. and Blomfield D. Rock trap design for presplit rock slopes.

*Scott Wilson Kirkpatrick and Partners, Hong Kong.

authors had also brought up another important aspect of rock trap design — the percentage retention to be adopted for barrier design. In most cases in an urban environment authorities were likely to require a theoretical 100% retention, perhaps with an additional safety margin, to prevent a boulder clearing the barrier and thus reduce to an absolute minimum the possibility of an accident.

Again from the paper by Chan and co-workers it would be noted that no energy-absorbent material covered the surface of the large exposed concrete foundation. From a practical point of view there would appear to have been little difficulty in incorporating into the design a suitable loose gravel blanket to cover the unslope section, possibly by means of a layer of gabions.

Fig. 1 Granular-lined artificial boulder ditch in Hong Kong

On one Hong Kong site an artificial boulder ditch 3 m wide had been formed by means of a 0.75-m high tapered reinforced-concrete wall dowelled into the underlying rock, and extended by a 2.6-m high steel supported chain-link fence, to afford protection to a public footpath and roadway adjacent to a 20-m high rock face. To dissipate energy from falling boulders loose material had been left at the toe of the slope and granular backfilling positioned at the back of the RC wall (Fig. 1).

Written contributions

P.G.D. Whiteside* The design of rock traps, perhaps more than most other areas of rock engineering, is amenable to field trials. These, however, are rarely carried out or at least are rarely reported in the literature and so the paper by Mak and Blomfield is a particularly welcome addition to the limited data presently available.

It should be noted, however, that the method of experiment adopted by Mak and Blomfield is different from that originally used by Ritchie [1] and this appears to have resulted in some confusion in the authors' comparison with Ritchie's original work. A further confusion arises from the method by which Fookes and Sweeney [2] drew up their design chart based on Ritchie's data. Fookes and Sweeney were conservative in their summary of Ritchie's work. For example, Ritchie quoted for a 0.5:1 slope a trap width of 10 ft for a slope height of 15–30 ft, 15 ft for 30–60 ft, 20 ft for 60–100 ft and 25 ft for heights of more than 100 ft. Fookes and Sweeney interpreted this as 10 ft wide for 15 ft high, 15 ft wide for 30 ft, 20 ft for 60 ft and 25 ft for 100 ft high, whereas Ritchie implied that the 10-ft width was adequate for up to and including 30 ft high, 15-ft for up to 60 ft, etc. The effect of this is that each of Fookes and Sweeney's W lines is 5 ft too high in value. Similarly, there is an overestimate in D values in that Ritchie's original ideas were of a fallout zone W wide in which all rocks impacted and of a barrier, height D, sloping at 1.25:1, which prevented fallen boulders rolling out of the fallout zone. The base of his sloping barrier was, it appears, positioned at the edge of the fallout zone. In this case we may use a vertical barrier and reduce W by $1.25D$.

Redrawing Ritchie's data on the above basis results in the chart given in Fig. 2. This redrawing was prompted by the present writer's experience — in common with many other rock engineers — that Fookes and Sweeney's chart usually gives a reasonable D value but an unnecessarily large W value.[3]

Returning to the results of Mak and Blomfield, it is clear that they are actually in fairly close agreement with the results of Ritchie, as shown in Fig. 2. To illustrate this we can take two examples.

Fig. 3(*b*) (p.265) shows that all falling rocks on an 8-m high 55° slope are retained by a 1-m high barrier placed 1.5 m from the toe. In Ritchie's terms it means that if the ground was at the level of the top of the 1.0-m high barrier, a fallout zone of width (1.5 + 1.0/tan 55) would initially catch all falling rocks from the (8.0–1.0) m high 55° slope (Fig. 3). Thus, for slope height 7 m and angle 55° the fallout zone is 2.1 m wide. Fig. 2 gives a fallout zone width of 1.7 m for this slope. Fookes and Sweeney, in contrast, would give W = 3.9 m.

Curve C in Fig. 6(*b*) (p.267) similarly suggests that for a (12.0–1.8) m high 55° slope a fallout zone (1.5 + 1.8/tan 55) m wide would contain all initial impacts, i.e. a trap 2.8 m wide for a 10.2-m high slope. Fig. 2 gives W = 2.2, whereas Fookes and Sweeney would give W = 4.8 m.

Other comparisons between Mak and Blomfield's results and Fig. 2 also show a reasonably good correlation.

It is probably more useful in future experiments to keep to Ritchie's original concept of a rock trap — that is, a fallout zone being that width in which all rocks initially land and a barrier being a method of stopping fallen rocks from escaping from the fallout zone. There are alternatives to barriers — a soft earth, landscaped area, for example, can help to prevent escape after initial impact.

It should be noted that it is unlikely that Mak and Blomfield's intention to allow 5% of boulders to pass over the rock trap would be acceptable to engineers designing for an urban environment if there was a risk to life.

Design charts should always be used with caution and particular thought is needed for rock traps when slopes become high because very minor variations high up in the trajectory of rocks can have a major effect on their final landing place. Nevertheless, Fig. 2 appears to give relatively reliable guidelines for rock trap design for low to intermediate height slopes. Comments on Fig. 2 by other designers of rock traps would be very useful, especially since it is not completely clear from Ritchie's paper (*a*) if 100% retention was achieved and (*b*) whether his sloping barrier was internal or external to his fallout zone.

*Scott Wilson Kirkpatrick and Partners, Hong Kong.

References

1. Ritchie A.M. Evaluation of rockfall and its control. *Highway Res. Record.* no. 17, 1963, 13–28.

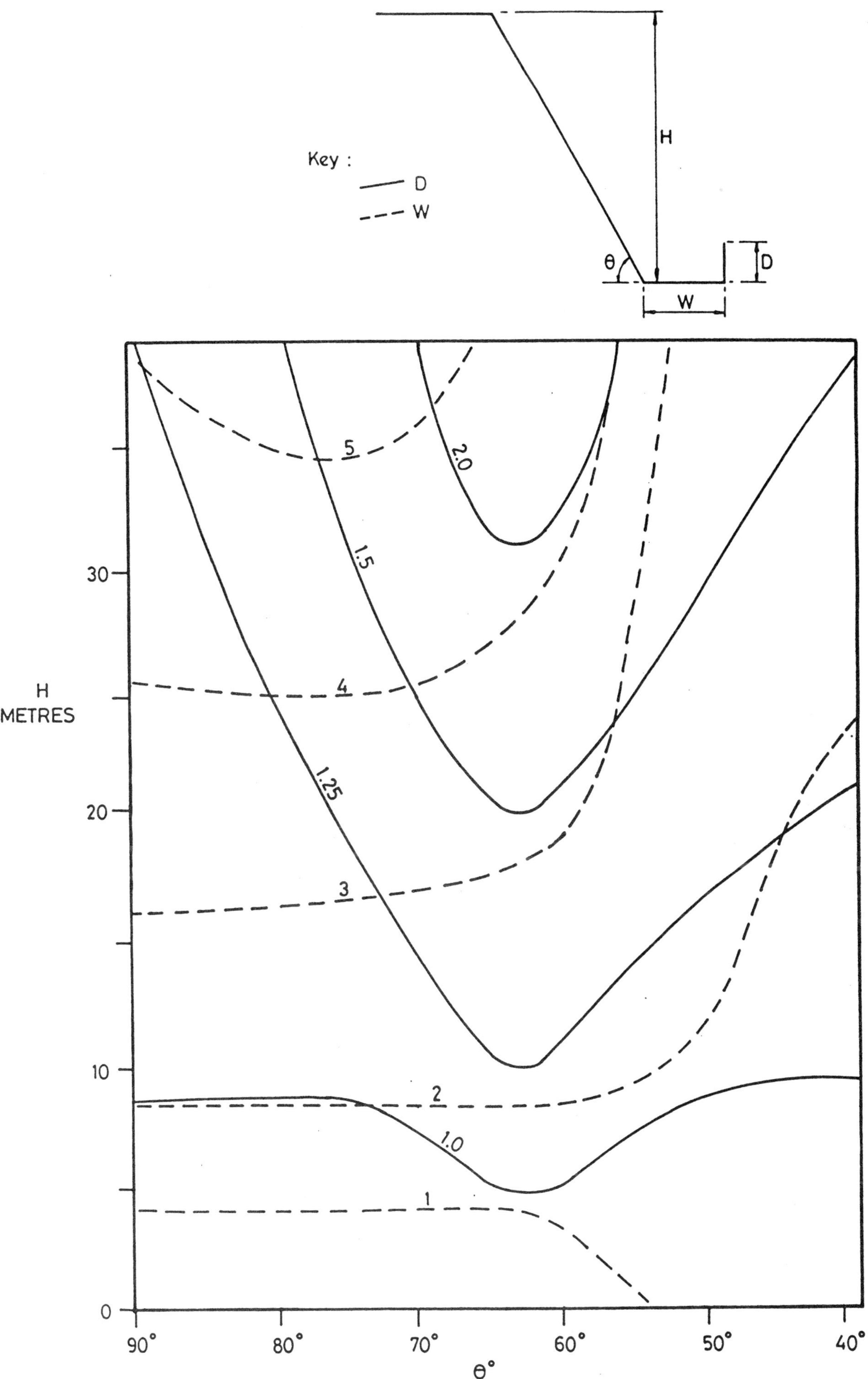

Fig. 2 Rock trap guidelines (revision of Fookes and Sweeney [2] based on Ritchie [1])

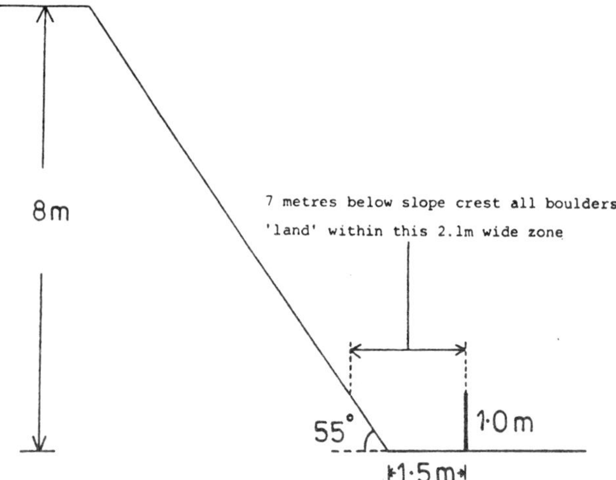

Fig. 3 Geometry for 100% boulder retention in Fig. 3(*b*) of Mak and Blomfield

2. Fookes P.G. and Sweeney M. Stabilization and control of local rock falls and degrading rock slopes. *Q. Jl Engng Geol.*, **9**, 1976, 37–55.
3. Whiteside P. Rockfall and its control. *Hong Kong Engr.*, **12**, no. 6 1984, 12.

Dr. H.A.D. Kirsten,* Dr. O.K.H. Steffen* and Dr. T.R. Stacey* In South Africa there have been several cases of damage being caused by boulders rolling or bouncing down hillsides. Two of these occurrences are shown in Figs. 4–6 and Figs. 7–9, respectively. The first case is a granite boulder in the Cape Town area and the second a lava boulder in the Johannesburg area. The form of these occurrences was similar: the source of the boulder was a relatively low cliff, initiation of movement occurred after a period of heavy rainfall and the boulders bounced down the slope below, gouging out 'footprints' in the surface.

In the latter case the surface of the slope was surveyed to define the footprint locations, and the trajectory of the boulder was back-analysed by use of a slightly modified version of the rockfall computer program developed by Piteau and Clayton.[1] This program is based on a simple model: spherical boulder, initial horizontal and vertical boulder velocities and boulder obeys conventional laws of reflection, modified using the coefficient of restitution approach to energy loss for each bounce (this coefficient can be different in the tangential and normal directions).

Despite this very simple model, it can be seen from Fig. 10 that the agreement between the calculated and surveyed trajectories is remarkable. The conclusion to be drawn from this is that, if correctly used, this simple model provides a very satisfactory design tool. The following case study is an example of a practical application of the analysis to a road cutting and the subsequent final solution adopted.

The problem involved the investigation of the requirements for a catch barrier to protect a road cutting plagued by substantial boulder falls. The profile of the slope and its division into 'cells' for analysis purposes is shown in Fig. 11. As it is impossible to simulate all the parameters correctly a probabilistic approach was adopted to take into account the variability of the parameters. The result was that more than 1700 possible boulder trajectories were generated to predict the probable distribution of boulder paths at the

*Steffen, Robertson and Kirsten, Johannesburg, South Africa.

Fig. 4 Source of granite boulder and footprint

Fig. 5 Boulder

Fig. 6 Damage caused

Fig. 8 Boulder path

Fig. 7 Source of lava boulder

Fig. 9 Damage and boulder

base of the slope. The following parameters were varied: initial horizontal and vertical boulder velocities, boulder start location, slope angle and roughness (although boulders were assumed to be spherical, the introduction of local variation in the slope angle had the effect, in a statistical sense, of taking into account the angularity of boulders) and geometry of the base zone, i.e. location of catch barrier and inclusion of a sand layer between barrier and base of slope.

From the series of analyses it was possible to define the effectiveness of a barrier, in probabilistic terms, as follows: a barrier 7.5 m high, 3 m from the base of the slope with sand fill between barrier and slope, provides a 93% probability of retaining all the boulders, and the barrier heights that are required to provide probabilities of 98 and 100% of retaining all the boulders are, respectively, 12.5 and 18.0 m.

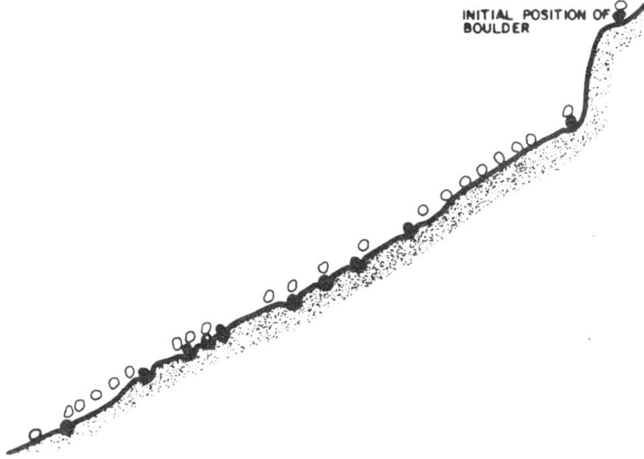

Fig. 10 Actual (solid) and calculated (open) 'footprints'

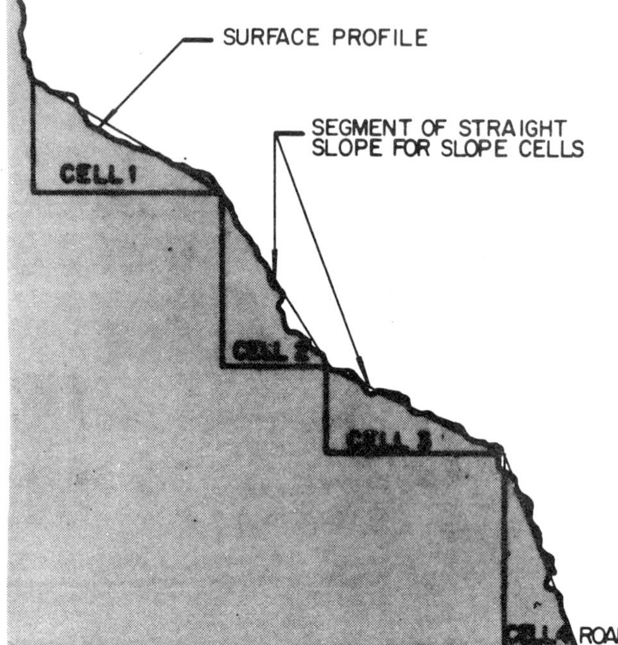

Fig. 11 Rock slope profile

Fig. 13 Rockfall shelter

The catch barrier therefore would not provide a complete solution to the problem.

Subsequent to the barrier investigation, boulder instability occurred along a greater length of the cliff face. Some of this area had previously been secured with rock anchors, and an extension of this approach was considered. However, as anchoring also did not provide a complete solution, a rockfall shelter was chosen as an alternative, being only slightly more expensive. The design of this shelter has been described in detail by Kirsten.[2] A typical section through the structure is shown in Fig. 12, which illustrates the structural components and the soil backfill above to dissipate the energy resulting from the impact of a boulder. The pleasing aesthetic appearance of the structure is illustrated in Fig. 13. The shelter has been in operation for nearly six years and has achieved completely the maintenance-free requirement that was specified.

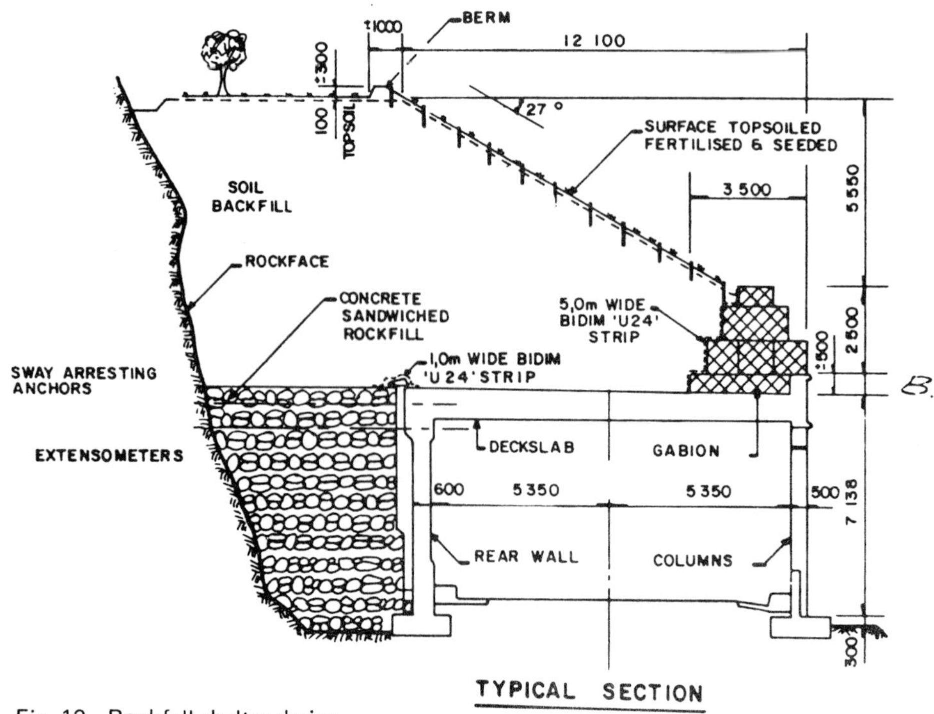

Fig. 12 Rockfall shelter design

References

1. Piteau D.R. and Clayton R. Description of the slope model computer rockfall program for determining rockfall distributions. D.R. Piteau and Associates, November 1976.
2. Kirsten H. Design and construction of the Kowyn's Pass rockfall shelter. *Trans S.Afr. Instn civ. Engrs.*, **24**, 1982, 477–92.

Authors' replies

Y.C. Chan, C.F. Chan and S.W.C. Au In response to the general points that were put to us by the rapporteur, as a part of our further research a full-scale boulder experiment was carried out following the completion of our paper. Two remote sites were selected for this purpose, one being a 30° sloping rock outcrop of moderately decomposed rock with thin vegetation cover in places and the other a 30° sparsely vegetated slope of highly to completely decomposed rock with some boulder deposits. About 70 boulders of sizes ranging from 30 kg to more than a tonne, and of differing angularity and shapes, were rolled down the slopes. Their motions were monitored by two video cameras, each with a built-in timer. With the aid of pre-fixed colour marks at 5-m intervals on the slopes, boulder velocities were measured from the television screen when the tapes were played back.

Among the many observations in the experiment, the following were of significance: (1) the boulder movement pattern was generally comparable with that predicted by the mathematical model; (2) in many runs the boulder velocities varied down the slope in an erratic manner; and (3) the boulder velocities were, on the whole, lower than those predicted by the mathematical model (Figs. 14 and 15).

We believe that points (2) and (3) are attributable to the effect of the uneven slope surface, the vegetation cover and existing boulder deposits on the slope. These have not been taken into consideration in our mathematical model.

By use of the mathematical model we have derived a set of graphs to show the relationship between velocity and distance travelled of the moving boulders based on characteristics relevant to the boulder field concerned (i.e. slope angles, boulder shapes and sizes, coefficient of restitution and coefficient of friction): Fig. 15 is one of these graphs. It was intended that they should serve only as a reference to facilitate estimation of the velocity of a boulder reaching the fence.

All boulders on the site were inspected and their characteristics identified; then, with reference to the graphs, the corresponding boulder velocities were established. If the velocity estimate yielded an energy greater than 100 kN m, the boulder would be stabilized *in situ*. In areas where surface irregularities of the slope, vegetation cover and boulder deposits appeared to be such that they could have a substantial effect on boulder velocities, however, the velocities obtained from the graphs were adjusted to suit the particular situation. This aspect of the design was based on engineering judgement. Such adjustments were later found to be in close agreement with the results of the full-scale boulder experiment (see above).

A number of methods were employed for boulder *in-situ* stabilization. Owing to the difficult access, they were mainly labour-intensive methods, which required only light mechanical plant and the minimum amount of construction materials. They included the combination of one or more of the following: dowels, concrete or masonry buttresses, reinforced concrete tie/strut beams, wire rope lashing, rock splitting (complete or partial), surface protection, surface drainage and sealing of rock joints. Figs. 16–19 show some of these treatments.

Owing to the varying site characteristics across the boulder field, in some areas particular methods were preferred.

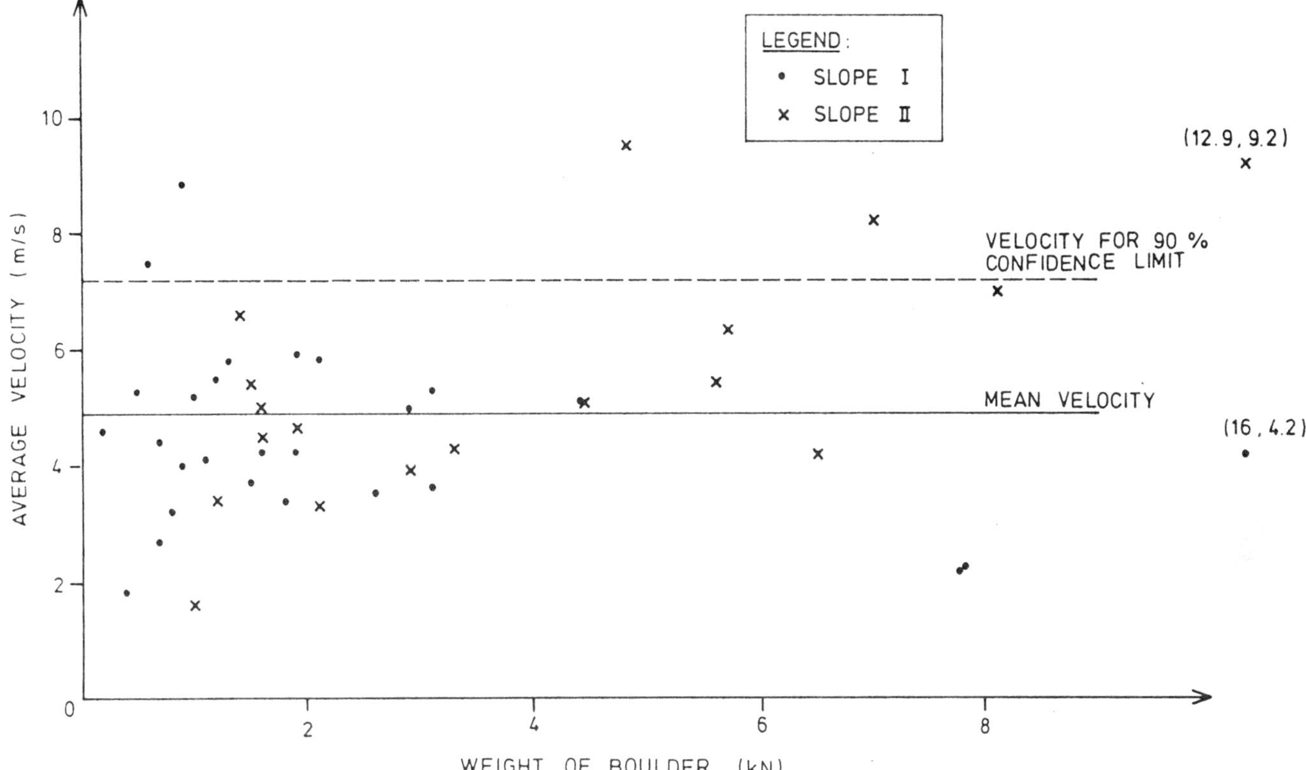

Fig. 14 Full-scale boulder experiment: velocity versus boulder weight

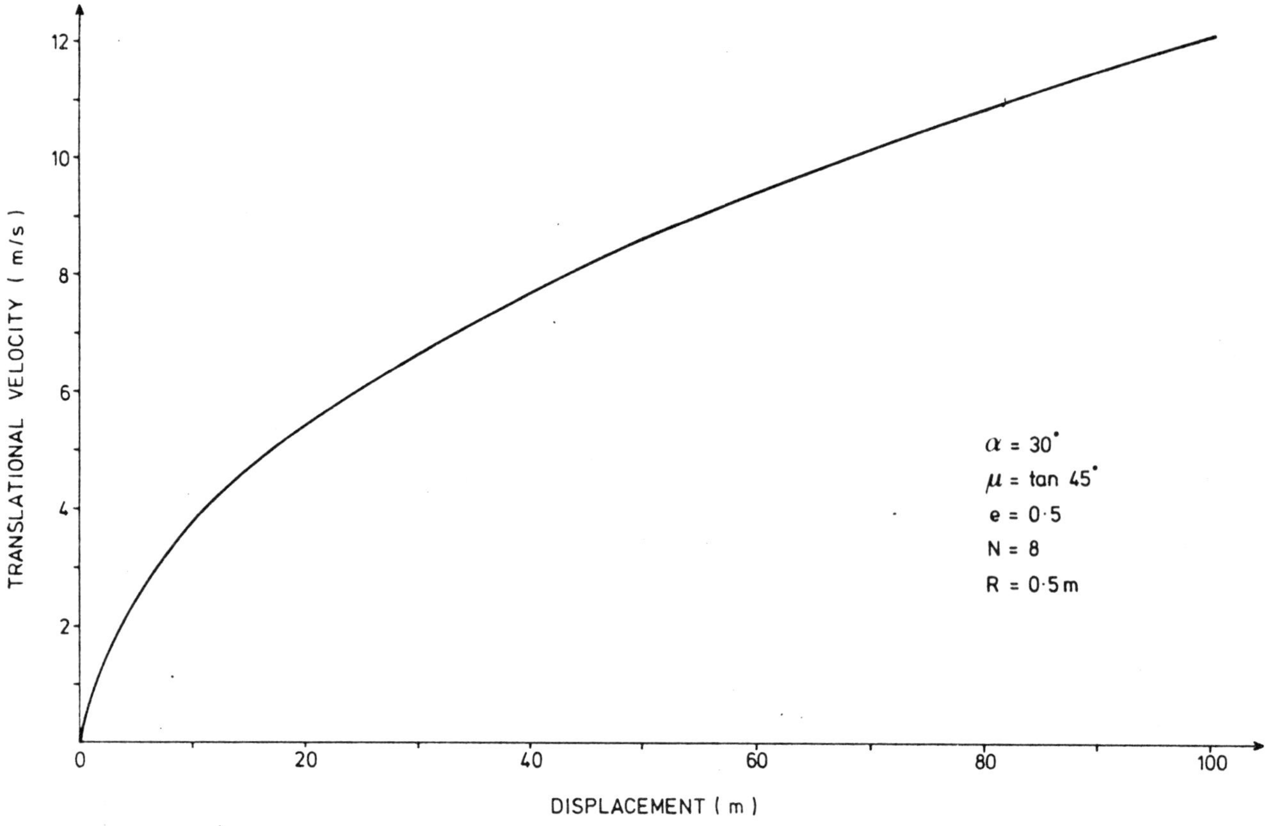

Fig. 15 Velocity–displacement relationship of polygonal prism derived from mathematical model

Fig. 16 Masonry buttress supporting overhanging boulder

Fig. 18 Rock splitting plus masonry buttressing by use of split fragment

Fig. 17 Concrete buttress supporting group of boulders

Fig. 19 Reinforced concrete tie beams

For example, rock splitting plus masonry buttressing with use of the split fragments was widely employed at higher elevations where transportation was extremely difficult (see Fig. 18).

The present mathematical model simulates movement of polygonal prismatic boulders down a planar slope. This model forms the basis for an understanding of boulder behaviour and its validity has been verified by the laboratory experiment. To cater for the surface irregularities of the slope, the vegetation cover or the presence of boulder deposits, however, boulder velocities derived from the model had to be modified. This was done on the basis of engineering judgement. In this regard the results of the full-scale boulder experiments are promising and are a useful supplement to the mathematical model. To simulate an irregular slope surface the model can be refined to handle a slope that comprises a series of planes of different gradients and minute irregularities can be introduced as small steps along these planes (see Fig. 20).

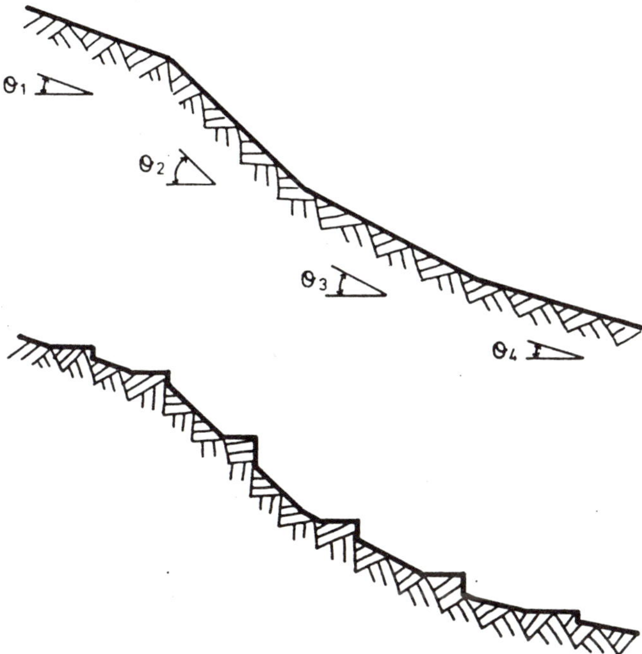

Fig. 20 Slope profiles to be simulated by mathematical model

We must, however, reiterate that in boulder field design the mathematical models can only be regarded as a guide to aid the engineer in the formulation of his design. Because of the inherent difficulties in simulating the various factors that will affect the motion of a boulder down a natural slope, sound engineering judgement always plays an important role.

The construction of the boulder fence has just been completed. No monitoring records are yet available.

In reply to C.M. Guilford, the provision of a gravel bed on the concrete surface was considered in the project. It was not adopted for the following two main reasons: (1) because of the gentle inclination of the slope (35° approximately) the main component of kinetic energy of falling boulders is horizontal (a gravel bed is only effective in absorbing energy in a vertical direction); and (2) the surface of the concrete foundation (approximately 4 m wide) is divided by the two rows of fences into three strips. The middle 1.5-m wide strip is used as an inspection access. Many steps had to be provided as, in most places, the longitudinal profile is steep. Therefore it would not have been easy to cover this strip with a gravel bed. The only strip that could have been covered with ease is the inner strip, but this is limited in width to approximately 1.5 m, which is too narrow for a surface gravel bed to be cost-effective.

N. Mak and D. Blomfield In general comment Dr. L.R. Richards queried why we considered Ritchie's work would be conservative when consideration of coefficient of restitution between natural slopes and our formed slopes would make the converse true. It was not from consideration of this aspect that we were led to the view that perhaps Ritchie's findings would be conservative for presplit slopes. Ritchie did his work on natural slopes, where slope faces were often rough and irregular, inclined surfaces acting as launching pads for falling boulders, which is clearly not so for the smooth presplit face with which we experimented. Furthermore, the boulders in Ritchie's work were often considerably larger than would normally be left unstabilized on a presplit face. It was on this basis that we considered that Ritchie's data may be conservative and perhaps even inappropriate for the type of boulder and slope under consideration.

The shape of the boulders that were used in the experiments was angular and up to 300 mm in the largest dimension. Within this range site observations have shown that the shape made no noticeable difference in the falling characteristics observed.

In regard to the significance of the difference in the result between Ritchie's recommendation and our own — in particular, the relevance of a 2.0-m reduction in ditch width — in an urban area such as Hong Kong, where usable space is limited and very expensive, this reduction in space can amount to a considerable saving in either land cost or amount of excavated material.

As to concern that the angular momentum of the boulders on hitting a rock trap surface may carry the boulder beyond the trap, with small boulders and a vertical barrier we believe that the effect is minimal. It was our intention, however, that once the barrier was constructed, selective panels would be tested and modification of the barrier by the incorporation of a small inward return would be made, if necessary.

The reason for recommending a lower percentage of retention of boulders for those boulders which stopped naturally rather than were stopped by a barrier is because boulders of the size that we examined when stopped naturally are likely to be travelling at a lower velocity and height than those which are stopped by a barrier, and therefore do not present a serious risk to life.

It has been queried whether rolling 1000 boulders is essential for each experiment, especially in view of the time taken to complete the experiment. We were striving to obtain minimum barrier height so as to provide an aesthetically acceptable and safe solution; we believe that 1000 boulders was a realistic number to generate the degree of confidence that we were looking for. The time taken was only five or six days for each experiment, and the material for the experiment was readily available. For the two-year site formation work where this investigation was carried out the time required was insignificant.

We agree with C.M. Guilford's comment that absorbent material is a more effective means of dampening the travelling boulders in the drop zone, and would welcome any refinement to specific geometry of drop zone by use of varying material.

Session 8: Groundwater

Chairman: Professor P. Lumb
Rapporteur: G.W. Lovegrove

Discussion

G.W. Lovegrove* said that the subject 'groundwater' was extremely broad and, even when confined to the context of 'rock engineering and excavation in an urban environment', it was not possible to do it full justice in a single one-hour session. Two papers† had been submitted for discussion, but several others‡ submitted to other sessions of the conference also discussed groundwater.

Nelson described the excavation of a large cavern that formed part of the University of Minnesota's Civil and Mineral Engineering (CME) Building. The torus-shaped opening measured about 53 m by 48 m, height 8 m, with a central pillar 27 m long and 14 m wide. The cavern was mined in soft sandstone and a 1-m thick impermeable shale layer. The shale was located directly below a capping layer of 9 m of limestone. Two shafts, 8 m and 13 m in diameter, penetrated the limestone; they had been constructed as work shafts during mining and provided access from the main part of the building constructed by cut and cover in a 12-m thick soil stratum above the limestone. The groundwater regime comprised a water-table in the soil and limestone perched on the impermeable shale layer, and a base groundwater table in the sandstone just below floor level of the mined space. The author described methods to control groundwater flow from the perched water-table into the cavern by way of (*a*) predominantly horizontal joints exposed in the limestone walls of the access shaft and (*b*) steeply inclined joints exposed in the roof when the impervious shale stratum had been removed.

Horizontal joints daylighting in the shafts had been sealed. Sub-vertical joints daylighting in the cavern roof had been allowed to drain into metal troughs because all attempts to seal them had failed.

Richards and Cowland described studies that had been carried out to quantify the transient nature of groundwater in one of a suite of rock slopes in urban Hong Kong. The slope that they had investigated was about 50 m high overall and comprised a 65° cut face in rock below a 30° fill slope. Monitoring installations included both Casagrande standpipe and pneumatic piezometers located in sheeting joints that dominated the stability of the slope. Both types of piezometer were monitored automatically, but the pneumatic piezometers had malfunctioned after about one year. Data presented in the paper were from the Casagrande standpipe piezometers, which were monitored at 40-min intervals during heavy rainfall and less frequently at other times. It was interesting to note that the Casagrande standpipes had been installed as a back-up system to the pneumatic piezometers.

The fluctuation of the base groundwater level between wet and dry seasons was of the order of 1 m. Transient rises during and after heavy rainfall were typically 5 m, with a maximum recorded rise of nearly 10 m. Durations of transient rises were from less than 40 min to 30 h (of the order of 5 h, typically).

The authors were unable to formulate a mathematical model to relate rainfall and groundwater response. They noted the apparent randomness of piezometer response and the fact that on no occasion did all piezometers respond to the same rainfall event.

Both papers reflected the difficulty of assessing and predicting groundwater conditions. At the University of Minnesota CME building only some of the vertical fissures in the limestone were water-bearing (although it was unlikely that the designer would have known that for certain at the design stage) and, when it had proved impossible to seal them, the author had adopted a pragmatic approach to groundwater control. There were several interesting aspects to that type of approach: what had led the designer to believe that that approach would be successful in the long term and would not lead to large-scale drawdown and associated settlements and how did the designer formulate the hydrogeological model and boundary conditions for his groundwater predictions and how did performance compare with those predictions?

As flow was through a small number of discrete fissures, one must ask if the drawdown zone of influence was very irregular due to rapid drainage through those fissures and if that affected groundwater levels to a greater degree than would otherwise have been expected. For example, a tunnel driven in Hong Kong Granite several years ago had resulted in modest drawdown except where it encountered a suite of highly permeable joints. Pore pressures close to the joints, far outside the normal zone of influence of the tunnel, had dropped dramatically. The consequences of that in rock overlain by compressible strata in an urban environment were clearly unacceptable.

Apart from groundwater response in the limestone it would be interesting to know the long-term performance of the drainage system. When drainage measures were incorporated in permanent works one must consider if chemical or other deposits were likely to form where seepage occurred and if those would adversely affect performance of the drains. In that respect it had been noticed in Hong Kong that chemical deposits built up over a period of years in some unlined tunnels in the urban area, whereas they were largely absent in tunnels and mine workings in similar materials in rural areas. Analysis of those deposits showed that they were mainly carbonates, possibly resulting from leaching of concrete foundations or other structures in the region of the tunnels.

In contrast to the pragmatic approach to groundwater control for the CME building, Richards and Cowland had gone to great lengths to assess groundwater response for their stability study. There were ample high-quality site investigation and monitoring data, but, in the end, predictions had had to be based on a considered judgement of the available information because it was impossible to derive reliable correlations. The analytical techniques were available, but in that case it had not been possible to formulate the geological and hydrological model and boundary conditions. Considering first, the geological model, the authors had 10 drill-holes within a relatively small site and the considerable advantage of an exposed rock face to assist mapping. Dominant sheeting joints within the study area varied in dip, width, infill type and smoothness. When drill-holes indicated such significant variations it was probable that in reality the situation was much more complex.

The authors had been unable to correlate groundwater data with rainfall. For apparently similar rainstorms, piezo-

*Mott, Hay and Anderson Hong Kong, Ltd., Hong Kong.

†Nelson C.R. Problems and control of groundwater in fissures: a case history; and Richards L.R. and Cowland J.W. Stability evaluation of some urban rock slopes in a transient groundwater regime;

‡See, for example, those by Buttling, Caiden and co-workers, Clover, Costello and co-workers and Matson and co-workers in this volume.

meters responded in totally different ways, leading to the belief that groundwater flows and piezometric levels were being affected by the uneven pattern of rainfall within the area surrounding the site, constrictions within the joints and possible variations of those with time and variations of groundwater flow paths and flow type with time. With those unknowns it was not surprising that the authors had been unable to develop a reliable set of boundary conditions.

In that situation he wondered if it were worthwhile to attempt to derive an analytical solution for a relatively small site. Certainly, investigations should be carried out and monitoring data collected so that there was ample information on which to base a judgement, but accurate analysis of the situation must be impossible.

On the other hand, analytical solutions could be appropriate to studies of large areas where local variations in the hydrogeological model were of less significance. In the Hong Kong context Leach and Herbert [1] had described a numerical model that had been used to study groundwater flow in a hillside, albeit for the slightly simpler case of flow in soils, concluding that for large-scale studies such models could be used to simulate the probable responses of a hydrogeological system to extreme rainfall conditions.

Predictive models had been used successfully for two decades in the mining industry and, more recently, particularly in Scandinavia, to predict flow between underground oil storage caverns, where it had been possible to develop reasonably accurate hydrogeological models from mapping carried out during mining. One such example was described by Braester *et al.*[2] for caverns at Gavle in Sweden.

Two other papers [3,4] were worthy of note in the context of groundwater prediction. Matson and co-workers[3] considered predictions based on drained, undrained, dry and full hydrostatic conditions and adopted the undrained model for their preliminary assessment. Costello and co-workers [4] chose full hydrostatic conditions initially, but modified that to drained conditions when a drainage system was included at base slab level. Both of those predictions contrasted with the approach for a similar situation at North Point Station,[5] where it was confirmed by monitoring that groundwater pressures in the rock during and after excavation were less severe at North Point than those which would have been predicted by the methods initially adopted by either Costello and co-workers or by Matson and co-workers.

References

1. Leach B. and Herbert R. The genesis of a numerical model for the study of the hydrogeology of a steep hillside in Hong Kong. *Q. Jl Engng Geol.*, **15**, 1982, 243–59.
2. Braester C. *et al.* Depression of groundwater level around rock storages. In *Rockstore 77: storage in excavated rock caverns, proceedings of the first international symposium, Stockholm, 1977* Bergman M. ed. (Oxford, etc.: Pergamon, 1977), vol. 3, 691–6.
3. Matson C.R. Choy H.H. and Gibson A.M. Rock support prediction in the deep basement of Causeway Bay East Concourse for the Mass Transit Railway, Hong Kong. This volume, 285–97.
4. Costello J. Mak L.M. and Chan M.T.S. Geotechnical designs for the construction of Fortress Hill Station, Island Line, Hong Kong. This volume, 135–44.
5. Morton K. *et al.* Temporary suppor for a deep excavation at North Point, Hong Kong. In *Design and performance of underground excavations, ISRM symposium, Cambridge, 1984* Brown E.T. and Hudson J.A. eds (London: British Geotechnical Society, 1984), 347–52.

M.D. Howat* said that Nelson had suggested that if a slope could be shown to stand at a given angle, it might be assumed that its factor of safety was at least unity. If the slope were then cut back to a shallower angle, he postulated that a realistic minimum factor of safety could be calculated for the final geometry. That approach might be considered analogous to the New Austrian Tunnelling Method of temporary lining design, where tunnel deformation was measured and controlled empirically with an increasing thickness of shotcrete support. He was doubtful whether that approach could safely be used in slope design. At the Po Shan Road disaster of 1972, for example, the first clear sign of creep had been the observation that cracks in Po Shan Road could no longer be kept sealed by the Highways Office, and the final disaster had occurred only 60 h after that.[1] Such a short time did not allow for effective remedial measures to be completed.

Secondly, the statistical link between rainfall and slope instability was clear,[2–5] but Richards and Cowland showed that the response of groundwater pressures to individual rainstorms was not reproducible. A slope might be stable during one rainstorm event, but failure could occur during another event with equal intensity.

Finally, stability could be decreased by internal erosion of planes of weakness of groundwater flow and gradual strain-softening from traffic vibrations, adjacent civil engineering activity and seismic events, so the factor of safety of a given slope decreased with time.[6]

References

1. *Rainstorm disasters 1972: Final report of the Commission of Enquiry* (Hong Kong: Government Printer, 1972), 91 p.
2. Crozier M.J. Earthflow occurrence during high intensity rainfall in Eastern Otago. *Engng Geol.*, **3**, 1969, 325–34.
3. Lumb P. Slope failures in Hong Kong. *Q. Jl Engng Geol.*, **8**, 1975, 31–65.
4. Nilsen T.H. Taylor F.A. and Brabb E.E. Recent landslides in Alameda County, California (1940–71): an estimate of economic losses and correlations with slope, rainfall and ancient landslide deposits. *Bull. U.S. geol. Surv.* 1398, 1976, 21 p.
5. Fukuoka M. Landslides associated with rainfall. *Geotech. Engng*, **11**, 1980, 1–29.
6. Howat M.D. A regressive and time dependent landslide in Hong Kong. In *Proceedings of the first international conference on geomechanics in tropical lateritic and saprolitic soils, Brasilia, 1985*, vol. 3, 395–9.

Dr. L.J. Endicott† said that he would like to congratulate Richards and Cowland on their paper. The important lesson was that groundwater behaviour, even on one major joint, was not a simple matter and was certainly not unique for a given rainstorm condition. The study of, in effect, one

*Mass Transit Railway Corporation, Hong Kong.

†Palmer and Turner, Ltd., Hong Kong.

plane had produced results that were indicative of general hillslope hydrology. A recent study, of which he hoped to be able to publish the results soon, had shown similar anastomotic behaviour, with varying response such as was developed by the opening and closing of water paths at different times and at different locations in three dimensions. Obviously, in the end, an engineer had to decide on a 'design value'. Such studies were very useful in assisting the evaluation of hydrological data.

Written contributions

P.G.D. Whiteside* The problem of groundwater flow through anisotropic jointed rock masses is extremely complex and it is vital to the development of our understanding of the problem to have detailed records of actual case histories. In this respect Richards and Cowland are to be thanked for providing such detailed records of their study at North Point. There are, however, two aspects of their study that prompt comment.

First, the authors make reference to vertical joints that may link the sheeting joints and help to provide part of the network through which groundwater flows. Given that such joints are present, and if their occurrence and permeability are of the same order as the sheeting joints, it is possible that the groundwater flow, and hence water pressure regime on a sheeting joint, is best related to a notional phreatic surface. In three dimensions this phreatic surface would then develop, during rainstorms, in a form that reflected the routes along which rainwater had infiltrated in each particular storm. Uplift forces inferred from a phreatic surface could be greater than are obtained from a simple triangular pressure distribution such as that used by Hoek and Bray† and could be less than those which the authors calculated by use of the method of Stimpson (see Fig. 9, p. 362). Rather than employ either of these idealized water pressure distributions would it not be better to calculate the factor of safety by use of a three-dimensional phreatic surface derived from actual piezometer data?

The second comment concerns the piezometer readings. It is unfortunate that the piezometers that ceased to function were the pneumatic types, because the open standpipes that provided the bulk of the authors' data can require what is, in terms of rock joint volume, a considerable quantity of water to register. The actual volume would, of course, depend on the design of the piezometer response zone, etc., but it would be interesting to learn if it is possible or likely that any of the occasions on which individual piezometers did not register a transient increase was due to there being an insufficient quantity of water available.

C. Laughton‡ When working in similar rock, marl and sandstone strata, CERN normally insists on the application of a primary shotcrete layer on the final rock surface within 24 h of excavation. Was a similar constraint placed on the contractor on this project and, if not, did any marked deterioration of the exposed rock surface occur?

*Scott Wilson Kirkpatrick and Partners, Hong Kong.

†Hoek E. and Bray J.W. *Rock slope engineering, 3rd edn* (London: IMM, 1981) 358 p.

‡CERN, Geneva, Switzerland.

Authors' replies

Dr. C.R. Nelson In reply to M.D. Howat, the proposal to utilize a steeper slope for a 'test' of the strength of the deeper rock and joint system was intended to be just that – a test. The data obtained from this large-scale test would be used in the design of the final slope geometry with the usual rock slope design methods. Of course, the data obtained from the test would be for the conditions at the time of the test and would have to be used with the necessary professional judgement in arriving at the final design.

I believe that this type of field testing could, if used as part of a rational slope design procedure, improve slope stability reliability. It may also reduce the need for some drilling and other field sampling that is used to estimate the strength behaviour of the material deep within the slope.

In reply to C. Laughton, the Contractor was required to apply shotcrete over the shale, shaly sandstone and shaly limestone within 30 days of exposure. This has proved adequate for the materials and job conditions to date.

Dr. L.R. Richards and J.W. Cowland In reply to the questions that were posed by G.W. Lovegrove, before discussing the variations it should be noted that the sheeting joints in the slopes in the study area also showed consistent features. They were all parallel, or sub-parallel, to the original ground surface (resulting in a dip of between 20° and 40°), and their spacing increased with depth from that surface. They were frequently persistent over long distances, although some showed a small translation on either side of some vertical joints. The majority of the sheeting joints were clean, although some had varying thickness of soft infill. The basic friction angle of the sheeting joints was found to be remarkably consistent, regardless of the weathering grade of the rock material.

Perhaps the greatest variations were in waviness, some joints showing considerable waviness and others being more planar, and in number, one slope containing no sheeting joints at all. Indeed, the complete absence of sheeting joints in that one very closely jointed slope, and the presence of soft infill in some sheeting joints, led to an interesting geological debate as to the origin of these 'sheeting joints'.

As far as the formulation of a hydrogeological model was concerned, however, the most significant variations were the width of the joints and the distribution of soft infill. The majority of the sheeting joints were clean and tight, but some were clean and open, and some were filled with up to 300 mm of soft infill.

The problems encountered with the pneumatic piezometers were associated with both the piezometers and their automatic recording system.

It was believed initially that the phreatic surface would always be above the piezometer tips located in the sheeting joints, since these had been observed seeping water at the end of the previous dry season. Unfortunately, the first two years of the study were significantly drier than normal, and the joints did not seep continuously even in the wet seasons. Some of the readings indicated that certain tips probably became de-aired during the first dry season. Other pneumatic tips, however, remained well below the groundwater table. Their failure to record significant transient rises is thought to be associated with sticking of the mechanical linkages and the chart recorder in the automatic recording device as a result of the very high local humidity.

The initial aim of the groundwater study was to establish a relationship between rainfall and groundwater response. This could then be extrapolated to the worst predicted rainfall and used in subsequent slope stability calculations. Examination of the observed peaks, however, showed that such a relationship is complex and not amenable to simple mathematical treatment. The situation is further complicated by the extreme areal variation of rainfall in Hong Kong.

In 1978 the Geotechnical Control Office established twelve automatic rain gauges throughout Hong Kong, all linked to one teleprinter. These rain gauges contain a simple tilting bucket, connected to a telephone line, which provides information on how much rain falls on the rain gauge every 15 min. Following the experience of wide local variation in rainfall, the system has been expanded to 46 gauges, most of which are located in the urban area. Perusal of the rainfall data shows that rainfall at one gauge can be as much as ten times higher than the rainfall at an adjacent gauge only 1 km away during a particular storm. In the next storm, however, the situation can be reversed. This is a particularly interesting finding, as to the casual human observer this fact is not immediately obvious. If one travels from the location of one gauge to another during a storm, the presence of rain at both gauges (or at least a wet ground surface) can lead to a false assumption that both locations have received approximately the same amount of rain. The reason for this extreme areal variation of rainfall seems to be associated with local variations in wind direction, temperature, topography and rain-bearing clouds.

During the North Point Rock Slope Study a rain gauge was installed near the top of the most critical slope. However, it has unfortunately been one of the few automatic rain gauges to have suffered electronic problems. The nearest automatic rain gauge to have functioned correctly throughout the study was 2 km away, and situated in a quite different topography (a flat valley at sea-level instead of a ridge of a hill). Although there may be some doubt about the applicability of the available data for specific events, it is considered that the groundwater monitoring results overall are of significance because the latter two years of the study were much wetter than average.

The groundwater measurements showed that the sheeting joints experienced groundwater surges in times of heavy recharge. The most notable aspect of the surges was their complete randomness when compared with the piezometer locations. Despite a large amount of analysis, no correlation could be found between either the size, form or duration of the surges and the piezometer locations, joint thickness or joint infill.

First, from consideration of the extreme areal variation of rainfall, it is apparent that different parts of the catchment upslope may have experienced quite different amounts of rainfall during individual storms. Thus, each rainstorm may have introduced quite different amounts of water into each groundwater flow path, which, in turn, caused apparently random piezometer responses to rainfall events.

Secondly, it is conceivable that constrictions in the joints could produce different flow patterns, for the same upslope infiltration pattern (though different infiltration rates), because the effect that each particular constriction has on the flow of water probably changes with an increase of flow. It may also be possible that the constrictions could vary with time, either through opening up or closing with changes of water pressure in the joint, or through possible movement of soft infill through the joint.

Looked at from the point of view of the groundwater travelling down the flow path, varying constraints may make infiltration from one rainstorm pass along a flow path that has a piezometer in it, whereas infiltration from the next storm may pass along a flow path that does not contain a piezometer.

After analysis of the groundwater measurements, a considered judgement was made of the maximum water pressure that might be experienced by the total surface area of the most critical sheeting joint. It is clear from the measurements that the total surface area of the large sheeting joints was never subjected to large simultaneous transient groundwater rises. Indeed, from the many 'zero' readings it is evident that at any one time during a period of transient rises large parts of the sheeting joints were experiencing no change of water pressure above their seasonal base level.

When the results of the monitoring started to become available considerable concern was generated by the magnitude of the transient groundwater peaks. Use of bilinear phreatic surfaces (as shown in Fig. 9, p. 362), with the position of the surfaces determined by the maximum groundwater level recorded, produced low factors of safety for the initial sensitivity studies. However, it was also very noticeable that two piezometers in the same, and most significant, sheeting joint rarely both responded to the same storm, and even then they never both recorded high transient groundwater pressures at the same time. The final stability studies were, therefore, carried out with the water pressure distributions as measured over the whole of the potential failure surface.

The effect of using the measured pressures is illustrated in the following diagrams — first in two dimensions and then in three. Fig. 1(*a*) shows the water pressure acting on the failure surface for piezometer condition 3 in Fig. 9 of the paper. This water pressure leads to a calculated uplift pressure on the potential failure surface of $U = 3400$ kN per metre width of the slope.

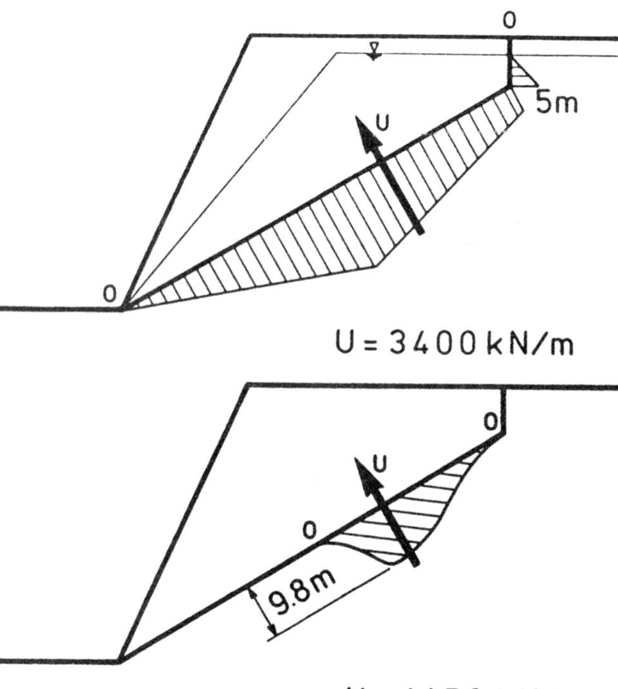

Fig. 1 Water pressure distribution based on bilinear phreatic condition 3 in Fig. 9 (p. 362) ((*a*), *top*) and possible transient water pressure ((*b*), *bottom*)

In reality, there were many occasions when a piezometer recorded a zero increase of water pressure immediately downdip of another piezometer that recorded a large increase of water pressure. Consideration of these readings leads to a possible transient water pressure distribution of the shape shown in Fig. 1(*b*). It is thought that this distribution of water pressure could be produced by a constriction in the joint. Purely for the purposes of a simple illustrative comparison, the maximum recorded transient rise of 9.8 m of water has been shown in Fig. 1(*b*). Measuring the volume under the water pressure surface produces an approximate uplift pressure on the potential surface of U = 1150 kN per metre width of slope.

experience an increase in water pressure of a form indicated by the pressure contours. In reality, this could represent the situation where water travelling on discrete flow paths, along a variety of joints, meets the critical sheeting joint. Simply summing the pressure represented by the contours produces an uplift pressure on the critical sheeting joint of about U = 65 MN. It is interesting to compare this value with that produced by the bilinear phreatic surface estimate of U = 238 MN, as illustrated in Fig. 3.

In response to the comments by P.G.D. Whiteside, the vertical joints encountered in the study were generally tight, whereas the sheeting joints varied from also being tight to being open or infilled. The width of some sheeting joints

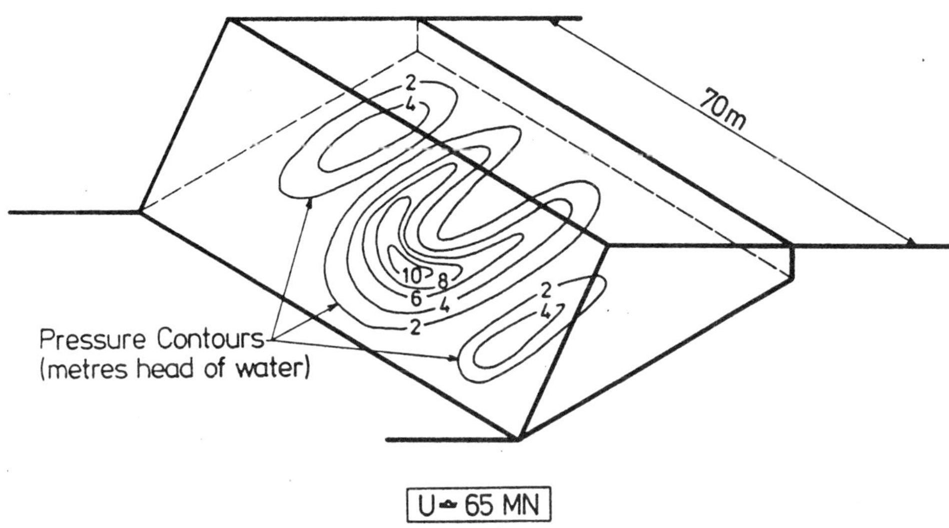

Fig. 2 Example of transient water pressure on total surface of critical sheeting joint

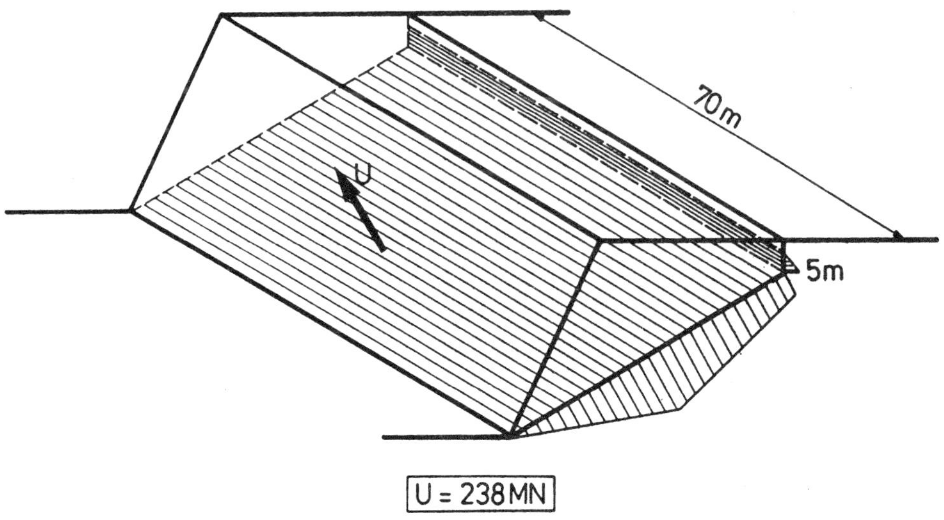

Fig. 3 Water pressure on total surface of critical sheeting joint for bilinear phreatic condition 3 in Fig. 9 (p. 362)

A three-dimensional approach was adopted for the final stability analysis. The concept that was adopted is illustrated in Fig. 2. The numbers shown are for simple illustration only, as the real situation is complicated by the addition of transient water pressures to the base groundwater levels. Considering the maximum transient groundwater rise of 10 m, the more typical smaller rises and the many 'zero' rises, it is thought that the critical sheeting joint could

(i.e. from joint wall to joint wall) was as much as 300 mm. The permeability of the vertical joints was, therefore, less than the permeability of the sheeting joints.

Given the frequent occurrence of infill in the sheeting joints in the critical slope, and the very peaky transient flows that were recorded, it is thought that the groundwater flow occurs along a multiplicity of specific conduits through the slope. These conduits probably twist and turn through various lengths of vertical joints and sheeting joints.

The need for a large flow of water into a standpipe piezometer in order to record a change in water pressure around the tip is fully appreciated and was the reason for the choice of pneumatic piezometers as the primary instrumentation system. However, after consideration of the large transient groundwater rises recorded in some standpipes adjacent to others where there was no recorded increase, and the constant reversal of which standpipes responded and which did not, it is thought that the lack of response by alternating standpipes to the sequence of storms was unlikely to be due solely to the insensitivity of the standpipes.

Session 9: Quarrying and large excavations

Chairman: Professor P. Lumb
Rapporteur: G. E. Pearse

Discussion

G.E. Pearse,* in introducing the three session papers,† expressed keen disappointment that neither Midea nor Sarin was able to be present. One of the objectives of the conference had been to seek contributions from countries other than Hong Kong; the Institution particularly encouraged engineers from developing countries to participate in conferences and discussions.

Midea's work described the establishment of acceptable norms for controlling blasting vibrations from quarries in the São Paulo area. Those studies were intended to assist in formulating regulations expected to be promulgated in 1986. The author's team at IPT had investigated flyrock occurrences and blast vibrations at some 50 quarries. Measurements had been made of the effect of blasting on cracks in the walls of houses near the quarries by use of a mechanical 'Tenso-test' instrument that had been calibrated against known vibrations in the laboratory. Temperature changes had been found to have a greater effect on the cracks than blasting.

A particular problem in São Paulo was the large number of sub-standard houses, which were affected more by blasting vibrations than were well-constructed buildings. That aspect arose again in the paper by Ciccu *et al.*

New regulations concerning blasting in Brazil were reported by Midea to be nearing agreement between the government authorities and the mining companies. In written communications Midea had stated that engineering seismographs would be used for routine measurements of particle velocities and teams of technicians would be trained to read and interpret them. Directives would be issued regarding training blasting technicians assisted by a team of one engineer and three technicians from IPT. The government commission accepted that fragments must not be thrown outside the quarry property areas; 15 mm/s vibration velocity and 130 dB should be maximum values.

The paper by Ciccu *et al.* described a case study of blasting work at underground water storage tunnels in Sardinia. The project was unusual because the geology of the area was complicated, being a zone of contact between fractured granite and metamorphic cover rocks with surface talus and alluvium. A large fault cut across the zone with a throw of several hundred metres.

Two of the three tunnels had been driven from east to west and intersected the fault, the other being driven northwards on the same side of the fault as the nearby village of Villacidro, the outskirts of which were only 300 m from the works. An inhabited house was 100 m from the operations and a hotel was about 150 m away. Preparatory open-cut blasting had utilized distributed charges and a pattern to cause sideways throw to minimize vibrations and flyrock. Air-blast effects had been reduced by the use of electric caps instead of detonating fuse, adequate stemming and burden being ensured. Selected atmospheric conditions were when the prevailing northwest wind was not affecting the site and temperature inversion periods in the afternoons had been avoided.

*Mining Journal, Ltd., London, England.

†Ciccu R. *et al.* Blasting studies in the excavation of an underground water reservoir in granitic rock; Midea N.F. Urban quarries in São Paulo city: operation, development, control and regulations; and Sarin G.K. City development and supply of building materials — Delhi, India.

Underground tunnelling was by a top semi-circular heading and a 4-m bench. For blasting vibration calculations a frequency for the granite had been selected as 50 Hz, which had proved to be reasonably accurate. Using the standard Langefors and Kihlström formula with a constant K for granite of 400, the calculated peak particle velocity was 7.6 mm/s at 300 m. Measurements at the outskirts of the village 300 m away had shown that the first three of seven wave trains peaked at 55 Hz, reducing to 23 Hz, but the maximum velocities were only 90 μm/s, or about 80 x less than calculated, with maximum displacements of 0.36 μm. That would indicate that the choice of 400 for K had been overly cautious.

Nevertheless, some damage complaints had been received from houses on the outskirts of the village. Those were three- or four-storey buildings on poor foundations and all were on the same side of the fault as the northerly tunnel that was being driven when the complaints were made. Some wave channelling and magnification might have occurred along the fault plane.

He would welcome, in particular, the authors' comments on the following points. (1) The cost penalty incurred by using cautious blasting. (2) The use of half-second delay detonators to decrease vibration. (3) K was assumed at 400 in the equation, whereas it appeared by measurements to be only about 4.7 for buildings on unconsolidated rubble. Would geotechnical measurements made before the operations commenced have assisted in making the assessments more accurately? (4) Would the authors recommend, in hindsight, a thorough pre-job inspection of all buildings that might be affected, especially regarding existing cracks, poor foundations, etc? (5) Could the fault zone have been avoided in the selection of the site?

G. Mortimer* said that it was unfortunate that the authors of the two papers relating to quarrying had been unable to attend the conference. The worldwide output of commercially usable quarried rock might be 'guesstimated' in the order of five billion tonnes per annum, a significant part of which was produced in urban environments, not least in Hong Kong itself, and, as such, was very relevant to the conference.

Papers had dealt in some detail with blasting and its environmental hazards, such as vibration, noise and flyrock, all of which were pertinent to commercial quarrying and had, for instance, almost eliminated blasting as a means of secondary breaking in modern European practice. Other environmental constraints had been touched on only briefly, and there was a degree of tolerance from the local community in recognition of the expected benefits from the water supply scheme concerned, as well as, no doubt, from an appreciation of their temporary nature in that vicinity. Commercial quarrying could not expect such consideration and, quite rightly, had to adapt its practices to urban locations, and to accept an increasing degree of regulation in the public interest. It was perhaps not widely appreciated how far that process had advanced in Western Europe, or fully understood that, once established in quarrying, similar controls would gradually but inexorably be applied to other

*Amey Roadstone Corporation, Ltd., Chipping Sodbury, Bristol, England.

engineering works involving rock excavation. That was already evident in Hong Kong, but was likely to become the future pattern even in less highly populated regions.

Matters to which attention might therefore need to be given at some future conference included the following.

Groundwater Problems associated with dewatering were well understood in the mining industry, and were equally relevant to surface operations. The handling and disposal of pumped or runoff water were increasingly problematic as required standards of discharge water quality were constantly rising, but the availability of space for treatment, particularly silt settlement, was often a constraint in urban operations.

Dust Dust nuisance was one of the most emotive issues in many quarrying neighbourhoods: despite the best modern practices in process control, dust inevitably emanated from large windswept rock exposures, roads, stockpiles, etc., and was often aggravated by vehicle movements. Regulations that concentrated on workforce exposure and respirable dust were a potential new area of difficulty, notwithstanding the almost total lack of evidence that that had been a health hazard in surface quarrying (as opposed to known problems in the closed mining environment).

Noise Subjective limits on noise from a site were increasingly being supplemented or replaced by specified boundary fence perceived noise measured levels, often of an arbitrary local nature established by officials with limited experience of that particular matter. Time limits during which even the imposed noise levels were permissible were being added in some cases and constrained working hours and operational economics as a result. Machinery was a greater problem than blasting, and although manufacturers had so far given relatively little attention to that aspect of design, both the equipment itself and operating practices would have to adapt to that new challenge. Within the workings, regulations and codes of practice were quite properly tightening provisions regarding employee exposure to noise, as, in contrast to dust-related illness, industrial deafness was a very real hazard arising in quarrying and many other industries.

Traffic Heavy traffic delivering from quarries could cause similar problems to the operations themselves, i.e. vibration noise, dust, mud on roads, etc., but extended them over a wide area, often coming much closer to individual homes. The subject was more emotive in consequence, and had led to imposed or agreed lorry routeing limitations on particular operations, as well as more recently to a generalized heavy vehicle night-time ban in central London. Early morning departures from quarries, necessary for long-distance deliveries, might well become a future target for controls, and could also affect rail-borne distribution, which tended to be a round-the-clock operation involving early or late loading and discharging.

Planning controls Local government planning controls were giving increasing attention to matters mentioned above, and to others such as visual impact, landscaping, restoration of workings, etc. Those were legitimate matters of public interest, and the stanards were generally those that leading quarrying companies would in any case wish to apply. Concern arose from a tendency to overlap with other regulatory bodies established to monitor and control environmental aspects, the longer and longer time periods required to determine planning permissions and a reluctance in some cases to grant sufficiently long-term reserves to allow the most efficient operating and, indeed, environmental arrangements to be made. It was also a matter of regret that planning control was not always as even-handed or farsighted as might be wished in relation to quarrying, and examples of housing developments being permitted up to the very boundaries of long-established working quarries were legion.

The conclusion of the foregoing in relation to the theme of the conference was that commercial rock excavation in urban environments now and in the future was giving rise to requirements for a wider range of professional and technical expertise and specialized disciplines than had previously been the case, and those would have to be regarded in the future as an integral part of the mining activities, requiring detailed consideration by professional bodies such as the IMM. The professional quarry manager would need to understand the basics personally, as well as knowing how to deploy the skills of experts as necessary. Above all, he would need to be skilled in the art of public relations, as anticipating problems and dealing effectively with complaints on the spot would always be the best and easiest means of ensuring satisfactory neighbourhood relationships and the minimum of conflict with regulatory officials.

Authors' reply

Professor R. Ciccu, Professor P.P. Manca, Professor G. Massacci and Dr. C. Pedemonte As regards the site of the underground reservoir, the choice was dictated by the elevation of the tunnel with respect to the water supply network as well as the need to avoid spoiling the landscape. Consequently, no alternatives were available.

Fig. 1

Fig.1 shows the outskirts of the town surrounded by a pinewood and the reservoir itself under construction.

The location of the tunnels in the immediate vicinity of the town posed a series of problems at the design and excavation stages in view of the need to minimize environmental disturbances. The main difficulties lay in the geological features of the zone, which proved very difficult to assess. In fact, the granite where the tunnels had to be driven is characterized by heavily jointed zones not clearly evidenced in the outcrops. Moreover, the region is crossed by a fault with a marked throw, bringing the otherwise overlying metamorphic rocks into tectonic contact with the granite; both formations are covered by talus and alluvial terraces.

The possible effects of blasting on the structures at relatively short distances from the blast were, therefore, not easy to predict.

An idea of the structural characteristics of the granite is given in Fig.2, which was taken during the bench excavation of the overburden.

Fig. 2

Before proceeding with trial blasting, the effects of vibrations and air blast were roughly calculated on the basis of the well-known relationships found in the literature. Because of the lack of direct information on the rock-related parameters conservative figures were adopted. In particular, the factor K in the Langefors formula was taken as 400 – higher than the value normally assigned to a partly weathered rock. Under this assumption the permissible charge per blast was lowered on account of the higher level of vibration intensity predicted. Moreover, the widely recognized fact that K decreases with distance was not taken into account. Experience confirms that measured peak particle velocities at increasing distances from the blasting site fit the model for lower values of that factor. Thus, calculated blasting vibrations were again overestimated and, consequently, the permissible charge per blast was further reduced. More realistic assumptions should have led to K values around 50 for distances of the order of 100 m from the nearest building and even lower values for those further away.

With the above precautions and assuming a limit of 50 mm/s for peak particle velocity, the maximum charge to be simultaneously detonated in order to protect the structures standing outside a radius of 100 m was estimated to be around 15 kg of explosive.

A further preventive measure concerned the frequency of particle vibration, directly involved in the calculation of wavelength together with propagation velocity of the seismic front. Wavelength was taken into consideration since the entity of any damage ensuing from differential displacements of the points of the structure traversed by the wave trains depends thereon. Waves of shorter length than the size of the structure are more likely to produce damage resulting from displacement stresses. Therefore the conservative approach followed in the calculations suggested a vibration frequency value of the upper limits for granite.

Moreover, the fact that vibration frequency decreases with distance was not taken into consideration. This should have meant a reduction by half of the original frequency value at the source for distances of the order of 100 m. On account of the above precautions, which led to 100 Hz as the assumed characteristic frequency of the rock, a wavelength around 40 m was calculated, which seemed somewhat reassuring at least for residential structures occupying a limited foundation area.

However, some additional safety measures were introduced. First, it was decided to adopt half-second delays in the firing pattern for fractioning the total charge of the blast to preclude any possible risk of the superimposition of vibrations, which geophysical measurements subsequently confirmed. The use of detonators with delay intervals longer than those required also allowed advantage to be taken of the favourable effect of charge interaction related to the statistical dispersion of effective priming times of the same numbers. In fact, detonators of the same delay number will explode with fractionally small random intervals: therefore the resulting wave motions have a phase difference that eventually results in decreasing vibration.

According to the manufacturers, the standard deviation of actual initiation times can be as much as 100 ms for half-second delays such as those used for the tunnels. Consequently, a relatively low reduction coefficient can be applied when calculating the permissible charge per detonator number in order to control the intensity of ground vibrations.

It is believed that the degree of coordination within the delay depends on frequency. This is evident since overlapping of the waves originated by each blasthole is more likely to occur at lower frequencies. When half-second delays are used, a reduction coefficient of 1/6 is usually adopted for consistent rocks such as solid granite with a typical frequency of 100 Hz, whereas for weathered rocks with a frequency of 50 Hz a conservative figure of 1/3 is generally used. As far as blasting vibration is concerned, this means that the same effect would be produced by a charge three times higher when distributed into a number of holes with the same delay compared to a charge loaded into a single hole.

Besides the very conservative figures adopted in the blasting plan regarding the choice of the rock-related parameters, the geometry of the blast was designed to minimize disturbances other than vibration.

The main problem in opencast excavation for overburden removal was considered to be the control of both air blast and flyrock, since in this case a larger proportion of blast energy is scattered in the atmosphere, whereas a minor portion is wasted in the form of ground vibration.

Regarding the energy vented as sound waves, calculations according to the commonly accepted criterion of cube-root scaled distances gave a peak overpressure of 900 Pa on the vertical walls of the structure some 150 m from the site. This is quite a safe level since only very large windows are likely to suffer breakage, which was not the case here. Of course, all suggested precautions were observed regarding charge confinement, stemming, hole burden to spacing ratio and ignition sequence. In addition, blasting was postponed in adverse weather conditions, especially when prevailing south-bound winds might focus air blast towards the town. Normally, blasting was planned in the early afternoon, corresponding to the generally most favourable time.

During tunnel excavation particular attention was devoted to ground vibration, since in this case a larger proportion of blast energy is transmitted to the rock.

Fig. 3 shows the tunnel face where small-diameter holes for a better distribution of the explosive are being drilled using light equipment.

On account of the very cautious criteria adopted, blasting procedures were deemed quite safe. However, towards the end of the excavation work, damage complaints were lodged by some residents, especially during the excavation of the third north-bound tunnel. The buildings concerned were a few storeys high and located about 300 m from the blast site. Some were constructed with supporting walls and in a poor state of repair; others in concrete.

A survey of damage revealed the presence of a series of cracks typical of settlement in structures with badly de-

Fig. 3

signed foundations; however, these cracks could spread owing to the effects of vibrations. What could potentially support the complaints was the fact that all the houses were located on the same side of the tunnel with respect to the fault line, which runs close to the alignment. This posed the question of the possibility of wave trains channelling along the fault. This kind of phenomenon, known as the Hattezburg effect, has been reported in a few instances, although only meagre data have been published.

To dispel any doubts and suppress the cause of discontent among residents it was decided to carry out a geophysical campaign to measure values of vibration variables and parameters and verify them in the context of safety factors. Transducers were applied to one of the concrete structures involved, its foundations being partly solid rock and partly on compacted detritus. Neglecting the slight dampening through the discontinuity planes, it can be assumed that measurements actually reflected the features of the rock mass underlying the foundation area.

Results were negative as far as particle velocity and associated acceleration and displacement are concerned. In fact, peak velocity of blasting vibration produced during excavation of the third tunnel was as low as 90 μm/s, which is by far below the permissible limit of 50 mm/s. Therefore any direct effect of blasting on the safety of the structures could be excluded without doubt. Moreover, the presence of the fault turned out to be of negligible importance: no channelling of the wave trains was evidenced.

The measured peak particle velocity fits the Langefors formula for values of K equal to 4.73 or 3.34 under the assumption of a charge coordination factor of 1/6 or 1/3, respectively; consequently, calculated K was roughly two orders of magnitude lower than the conservative value assumed in the blasting design.

This is not surprising if one considers K to decrease with distance, as reported in many cases described in the literature. For instance, Fig. 5 (p. 345) in the paper by Pöllä and his colleagues* shows a very sharp decline of K at increasing distances. A value around 5 at 300 m seems to fit the extrapolated trend of the curve.

*Pöllä J. Kähönen Y. and Raudasmaa P. Geotechnical maps, risk management and vibration limits in tunnelling in Helsinki. This volume, 341–6.

A series of systematic tests in Italy confirmed this phenomenon quite clearly. It was found that the agreement between predicted and measured velocities was good for K equal to 400 within a radius of 40 m from the blast, whereas at a distance of 100 m a value of 200 gave a better fit. Of course, the relationship between factor K and distance greatly depends on the geological features of the rock, which, in turn, controls vibration dampening. It seems reasonable to say that lower negative gradients of K as a function of distance should be observed in solid rocks and much higher gradients in weathered, jointed or scarcely consistent terrains. On the basis of available data, a formula can be proposed for the calculation of the appropriate value of K: it consists of three additive terms (a constant, a term inversely proportional to distance and the third proportional to the propagation velocity of the elastic waves). Constant and coefficients depend on the geological situation and may vary from site to site.

Besides the overestimation produced by the inadequacy of the Langefors formula at increasing distances, two aspects related to the topography of the zone could have contributed to the discrepancy between predicted and measured peak particle velocities. The first is the higher elevation of the blast site with respect to the buildings: it is well known that in similar situations vibration can be considerably reduced. The second aspect consists in the presence of a small valley between the tunnel and the town. This feature could have influenced the mode of propagation of elastic waves, causing a more pronounced dampening of vibrations. However, neither of these aspects was included in the calculations since they were very difficult to assess.

Besides the direct effect of seismic waves transmitted to the structures, the possibility of settlement involving the foundation ground was also considered. In fact, damaged houses were built on the downward slope (30°) of a hill and are partially set on dump ground. Therefore either differential subsidence or downhill sliding of this wedge-shaped mass of detritus could be caused by blasting vibrations, resulting in cracks in walls and ceilings. Again, however, geophysical measurements allowed these possibilities to be refuted. In fact, peak particle acceleration was as low as 2 cm/s^2, which is far below the threshold for subsidence phenomena in loose gravel. Moreover, the acceleration coefficient calculated for the bedrock was one order of magnitude lower than the stability threshold for the overlying ground, thus leading any conjecture of sliding to be ignored.

On the basis of our experience we positively recommend a pre-job inspection of all buildings likely to be affected. although this may be costly and time-consuming. This could also help to smooth the way with the population, making them better disposed to the inconvenience caused by excavations.

Finally, cautious blasting of trenches and tunnels naturally increases excavation cost compared with normal blasting where no limitations are imposed. The following factors contribute to making blasting more expensive: increase in specific drilling, because of the need to drill holes of smaller diameter, which, in turn, implies a greater overall length of holes for blasting a given volume of rock; increase in charging costs, since bulk explosives cannot be used (these are less expensive and can be more easily loaded into the hole); more complicated ignition pattern owing to the need to limit the amount of charge being simultaneously blasted; cover and other protective measures to prevent flyrock; inspections and ground vibration measurements; and insurance costs.

It is difficult to provide figures of general validity regarding the higher costs incurred in cautious blasting, since the prevailing conditions vary widely for each situation. For small-diameter blastholes and limited bench height only qualitative diagrams are available in the technical literature: they show the relative blasting cost at different levels of permissible ground vibration as a function of distance; the shorter the distance, the sharper is the increase in blasting cost.

Session 10: Drilling and blasting

Chairman: R.G. Roberts
Rapporteur: R.H. Lloyd

Discussion

R.H. Lloyd* said that the first of the four papers† under discussion – that by Berry and Dantini – related to the construction of a 5.5-km tunnel, 47 m^2 in section, which in itself would have presented no more than normal excavation problems: in that case, however, the construction problems had been compounded by the incidence of a 33 year old penstock tunnel that crossed the line of the proposed new tunnel and, because of the economics of the situation, the new tunnel could not be excavated by mechanical means so drilling and blasting had had to be used.

The two major constraints on blasting related to the unknown quantities or the quality of the concrete forming the penstock lining and the state of the rock that surrounded it, bearing in mind the possibly less-controlled blasting methods used more than 30 years ago. The paper described very clearly the methods of preliminary trial and assessment and the systems of monitoring and adjusting successive blasts as the penstock was approached, and as it receded from the danger zone, and also made some very pertinent comments on the standard ppv formula used in most situations. The comprehensive tables and diagrams with actual test results were related to the decisions made in controlling blasting to effectively and safely carry out the tunnelling. The conclusions of the paper emphasized the importance of pre-contract investigations.

The paper by Hagan concerned the actual design of underground blasts in an urban environment. The author underlined the importance of experience in the implementation of any blast design, and considered, essentially, the actual drilling, the condition of the rock faces, drilling patterns and the allocation of delay detonators. Although Hagan's main thrust was related to blasting in tunnels, he also briefly covered the questions of shaft-sinking and excavation of chambers/caverns by horizontal and downhole benching. The paper also covered what were termed (p. 187) the 'undersirable side effects of blasting' and, he was glad to note, emphasized the necessity for accurate drilling. Some of the many interesting features of Hagan's paper related to the usage of different types of cut in tunnelling to control vibrations, and also to comments on scatter in the timing of delay detonators. That latter point was mentioned a number of times in the paper and obviously constituted a very significant influence in designing blasts with use of the available range of delay detonators. Perhaps Berry and Dantini and Hagan would enlarge on that. Hagan's paper contributed an effective practical guide to the methods that were available to contractors to carry out their work efficiently, without becoming too involved in theory – the emphasis was on experience in putting theory into good practice.

Matheson's paper should be of particular interest to those who were concerned with the formation of slopes in Hong Kong – an ever present problem since, in particular, the disasters of 1972. Matheson explained quite clearly the distinction between natural and induced instability in rock slopes, and suggested an approach of forming slopes that involved assessment of data available both before and during excavation, and the use of an excavation technique that would not have an adverse effect on the final face. Emphasis was placed on optimizing the excavation so that effects on natural (i.e. geometric) and induced instability were minimized. The author stated quite strongly that the mere collection of voluminous data would not in itself produce the right answers: the accuracy, relevance and representativeness of data were of prime importance.

The paper outlined the effective methods of evaluating data, and covered design methods for rock slopes in an urban environment, again emphasizing that a knowledge of the type and extent of possible failures in slopes was very important to the design process. Matheson concluded with some very real comments and suggestions on the design of presplit blasting, including the conflicting theories on the actual formation of fractures that constituted the presplit, and a description of the various practical methods used and available to contractors.

In the speaker's experience of tender documents in Hong Kong, whenever slope formation was mentioned in relation to site development, the documents invariably specified presplit blasting: not only that, they also specified the actual spacing of holes, method of charging, and so on, rather like a direct extract from a textbook. Cushion blasting was sometimes mentioned, as almost a poor relation, and he would welcome an evaluation of the different methods of controlling slope formation. He noted the author's somewhat pungent comment to the effect that the knowledge that any instability in the final face could be rectified by remedial work afterwards appeared to give confidence to the design engineer.

The paper by McAllister and co-workers related to the construction of a rapid transit line in Buffalo, N.Y. State. The particular features of that project were the comparatively shallow depth of construction and the unusually difficult rock conditions, highlighted by excessive rock movement and settlement. The paper described in some detail the reasons for using tunnel-boring machines in the running tunnels, and the reasons why they could not be used in forming the stations, and he had found the very detailed geological description of the ground conditions particularly enlightening – not to say frightening! – and he had to record his admiration for the diver who had conducted a tactile survey under water. He must have been, in the Americans term, something else!

The paper described in great detail the constraints placed on the contractor in his blasting operations, and the penalties that he would incur through inaccurate or inefficient blasting. It also gave considerable detail on the methods of providing rock support, the design of blasts and vibration levels recorded. He noted with particular interest the authors' comments on the non-electric delay detonators, which were apparently detonating almost simultaneously, and the change to electric detonators, because it appeared that in spite of that change, and other measures that related to the rockbolts, it had still not been possible to achieve the limits of vibration that were allowed in the contract. They appeared to have reverted to the comments of Hagan and Berry and Dantini on the inaccuracy and limited range of millisecond delay detonators that were available from manufacturers.

*Blasting Consultants (Asia), Ltd., Hong Kong.

†Berry P. and Dantini E.M. Role of blasting control in excavation works near a pre-existing tunnel; Hagan T.N. Design considerations for underground blasts in an urban environment; Matheson G.D. Design and excavation of stable slopes in hard rock, with particular reference to presplit blasting; and McAllister D.S. Nelson C.R. and Strassburg R.J. Drill and blast tunnel excavations for the Buffalo Rapid Transit System, U.S.A.

Dr. T.R. Stacey* said that the use of presplit blasting required the simultaneous detonation of all presplit charges, which could result in large vibration levels. Although that produced a good final surface, it was believed that that might result in greater damage to the rock mass than would be caused by cautious blasting, for example. The authors' comments on their experience in practical situations would be welcomed.

Written contribution

R. Coates† Both Hagan and Matheson stressed the influence of site workmanship, particularly in hole direction. Fig.1 illustrates how difficult it is to correctly collar *inclined* drill-holes (and drill them to the correct depth) in a non-planar face. As is apparent, the collaring position of hole *A* is unaffected by face irregularities, whereas the collar of hole *B* must be adjusted. If this is not done, the burden becomes excessive and vibrations are likely to be higher. The figure is drawn for a tunnel, but the same difficulties exist where a bench surface is very irregular.

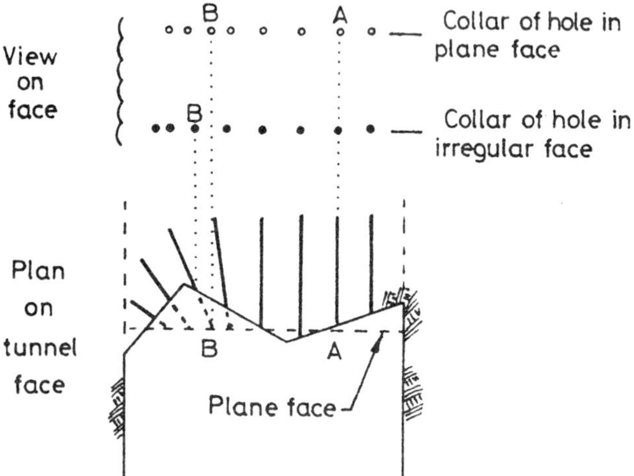

Fig.1

Table 4 (p. 280) gives variations in bulk strength between explosives. Coates,‡ in referring to Atchison and Roth,§ stated that 'a five-fold difference in effects can be produced by varying the explosive type . . . for a first approximation this is usually ignored'.

In Matheson's experience has the variation in bulk strength (or any of the other 'measures' of explosive 'power') produced any attributable variation in vibration levels? It would appear that most reports concerning vibration levels do not comment on the explosive strength. This may require recognition in carrying out desk-top studies before actual, site-specific, trials.

Authors' replies

Professor P. Berry and Professor E.M. Dantini In regard to the ppv criterion, it covers a wide range of situations and can be used both to secure the safety of structures and to ensure that the materials are sound. In the first case

*Steffen, Robertson and Kirsten, Johannesburg, South Africa.

†Mass Transit Railway Corporation, Hong Kong.

‡Coates D.F. Rock mechanics principles. *Mines Branch Monograph* 874 (revised 1970), 8–23.

§Atchison T.C. and Roth J. Comparative studies of explosives in marble. *Rep. Invest. U.S. Bur. Mines* 5797, 1961.

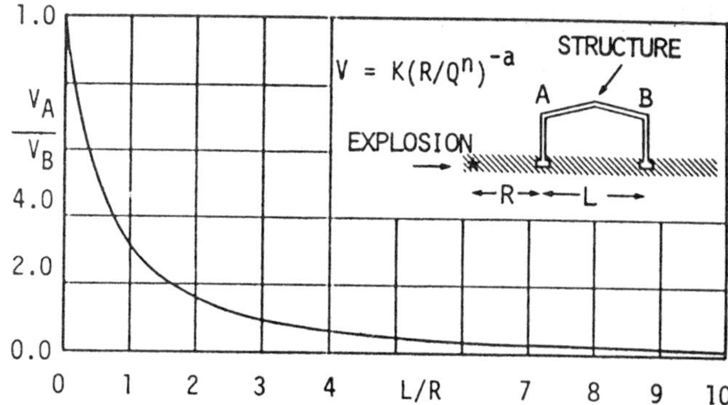

Fig. 2 Ratio between particle velocities at *A* and *B* as a function of ratio between size of structure (*L*) and its distance from blasthole (*R*)

the criterion works, essentially, in empirical and statistical terms in that it does not analyse the stresses induced in the structure by the transient. In the second case the induced stresses are directly proportional (via the characteristic impedance of the materials) to particle velocity. The problem that still remains to be solved is when to use one and when the other.

In fact, the statistical approach is unsatisfactory for explosions close to the structures because, owing to attenuation, the seismic transient induces high particle velocities only in a portion of the structures involved (Fig. 2). A possible solution to this problem could be that of indicating the amount of explosive and the distance between the explosion and the structure.

As regards the issue of the range of delay of the detonators, we think that practical solutions at present are still not feasible because the needs of blasting engineers clash with the interests of manufacturers in that, in practice, each blasting operation is different. Moreover, it is extremely difficult to manufacture detonators that actually match the nominal delays. Bearing this in mind, together with the fact that the overlapping of waves is a function of scaled distance when the object to be protected is not very far away, it is necessary for the spreading to be very high to avoid an overlapping of effects. Perhaps what could be done, and this would be quite an achievement, is to have the manufacturers focus on the production of detonators that come closer to the nominal value of delay that they are expected to have. This would cut down the probability of mistakes in designing the rounds.

We should like to add some comments on the measurements of the compressional waves induced by the air blast described in our paper. The phenomena involved are rather complex: in fact, in the case dealt with in our paper, we would obtain higher pressures values that those actually measured if these formulae are used:

$$\sigma_{t,c} = \rho c_1 v \qquad (1)$$

$$\tau = \rho c_2 v \qquad (2)$$

where σ and τ are the state of stress acting locally on the penstock, ρ is the density of the concrete, c_1 and c_2 are the velocities of the *P* and *S* waves in the lining of the penstock and v is the maximum particle velocity. Despite the fact that account was taken of three factors (the overlapping of seismic and compressional waves, the fact that in our paper we refer to pseudo-vectors and reflection at the air–rock interface), we obtained higher velocity values than those expected on the basis of the above relationships. It is quite likely that the higher values are to be ascribed to the fact that the measurements were carried out within the plastic zone of the rock mass fractured by the explosions and disturbed by the borings required to set the geophones. Moreover, because of the permeability of the rock mass, quite probably the pressure of the gases spread out to the innermost portions of the rock and was not merely restricted to its surface. The ensuing particle movement may have been greater than the $P-V$ relationship suggests. But, in any case, we think that this topic requires further investigation, especially for those materials which have low characteristic impedance and high permeability.

Turning to the comparison of blasting techniques used in the 1950s and their probable effect on the mechanical characteristics of rock with current methods, the system of fractures induced by blasting in the rock mass around a tunnel and their dimensions are conditioned by the characteristic parameters of the explosive and of the rock mass, by the geometry of the explosive and by decoupling. It was only after the 1950s that it was felt that more research was needed to restrict fracturing and the seismic effects induced by blasting. In fact, the earliest results on smooth blasting date back to 1953 in Sweden.[1] The extensive use of presplitting dates back to the late 1950s, when it was used in the Niagara Project.[1]

Thus, up to the 1950s, the explosive would be used without decoupling. For example, a 1-m hole with a diameter of 3.2 cm would be filled by rod tamping with about 0.8 kg of dynamite. Under these conditions the detonation of an explosive charge in a homogeneous and isotropic rock mass would induce three main systems of fractures in the rock. The extent of these three main zones (crushed, radial fracturing and damage zone) can be assessed on the basis of the experimental results and of simplified physical models. In fact, the maximum extension of the crushed zone and of the radial fracturing zone that can be obtained in homogeneous and isotropic rock masses can be calculated by the relationships [2]

$$\text{Crushed } R_1 = \tfrac{1}{2}A_1 Q^{0.5} \quad (3)$$

$$\text{Radial } R_2 = 1.25A_1 Q^{0.5} (A_2 B_c^n + 1) \quad (4)$$

where A_1 can be estimated on the basis of experimental results and depends on the type of explosive and the type of rock, A_2 depends on the shape of the wave and the attenuation, and varies from 0.22 to 0.5. B_c is the 'blastability' coefficient and n is a function of the attenuation of the stress wave and varies from 0.67 to 1.25. Q is the amount of explosive per metre of hole. If we choose average values for these parameters, we obtain

$$R_1 = 0.05\, Q^{0.5} \quad (5)$$

$$R_2 = 0.62\, Q^{0.5} \quad (6)$$

The extent of the damage zone can be evaluated by means of empirical relationships.[3] The extension of the fractures is therefore related to the particle velocity by means of relationships (1) and (2), which are valid for plane waves and for spherical and cylindrical wave fronts. Therefore, it is possible to establish critical vibration velocity ranges for the damage in different rock mass types.[3]

$v = 2 \div 1$ m/s (hard rock, strong joints)

$v = 0.1 \div 0.7$ m/s (medium-hard rock, no weak joints)

$v < 0.7$ m/s (soft rock, weak joints)

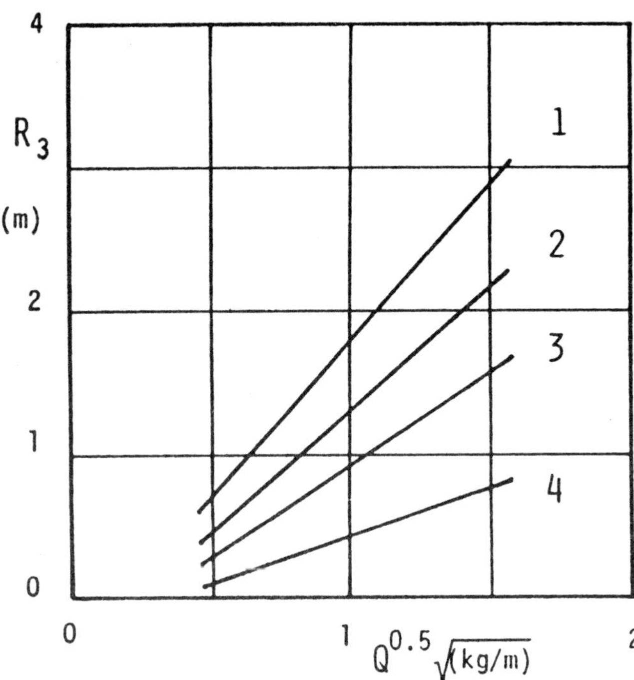

Fig. 3 Tunnel blasting: maximum extension of damage zone as a function of amount of explosive per metre of blasthole. 1, v = 0.5 m/s; 2, v = 7 m/s; 3, v = 1.0 m/s; 4, v = 2.0 m/s (where v is peak vibration velocity)

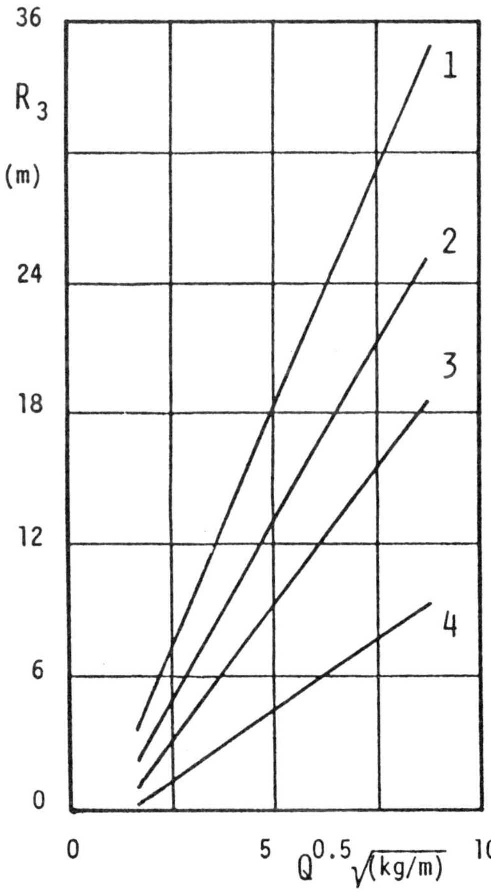

Fig. 4 Bench blasting: maximum extension of damage zone (curves refer to different characteristics of rock, see Fig. 3)

If we reprocess the relationships of Holmberg and Persson [3] we obtain Figs. 3 and 4, which give the extension of the damage zone as a function of the square root of the charge.

The presence of natural discontinuities that cross the rock mass and, more generally, the inhomogeneity of the elastic characteristics of the medium influence the transmission of the stress wave induced in the rock mass by the explosion. In particular, the stress wave may be reflected by the joint and therefore we obtain a further extension of the radial fractures.

An evaluation of the influence of the joints on the maximum extension of the radial fractures can be obtained through the following relationship, which has been worked out from the Langefors law [1] used for surface blasting:

$$R^{j}_{2} = 2.7\,Q^{0.5} \qquad (7)$$

In fact, this value is underestimated. To evaluate the further extension of the damage zone (R^{j}_{3}) in the presence of discontinuities a relationship that is derived from the following relationship can be used:

$$\sigma_{cd} = \sigma_{fd}\,(R_1/R^{j}_{3}) \qquad (8)$$

where σ_{cd} is compressive stress, σ_{fd} is strength, R_1 is the maximum extension of the crushed zone and R^{j}_{3} is the maximum extension of the damage zone in the presence of discontinuities. At the joint the compressive stresses become tensile stresses and so

$$R^{j}_{3} = 6\,Q^{0.5} \qquad (9)$$

Fig. 5 gives a summary of the extension of the fracturing zones that can be obtained in the case of an explosion without decoupling. To restrict the fracturing explosion techniques have been developed that require the use of limited amounts of explosive per metre of hole – principally, smooth blasting, presplitting and line drilling. All these

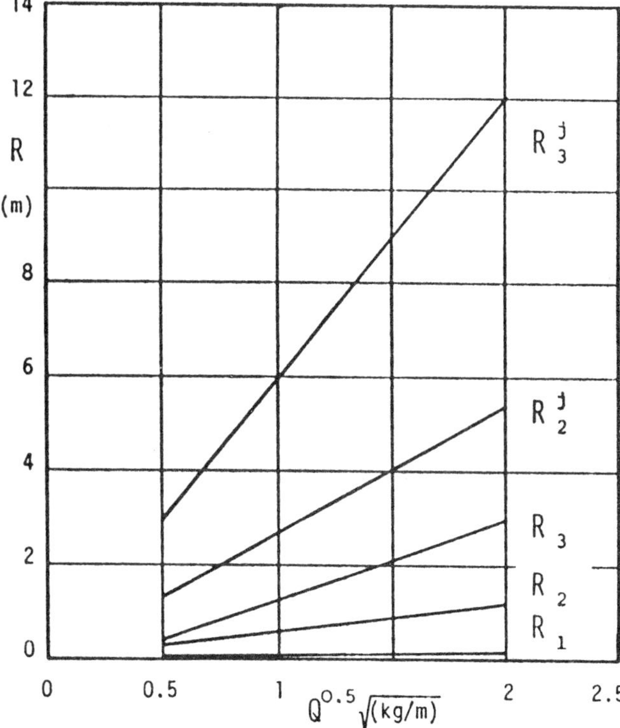

Fig. 5 Maximum extension of crushed zone (R_1), radial zone (R_2 and R^{j}_{2}); damage zone (R_3 and R^{j}_{3}) for tunnel blasting and for rock with mean mechanical characteristics (v = 0.7 m/s)

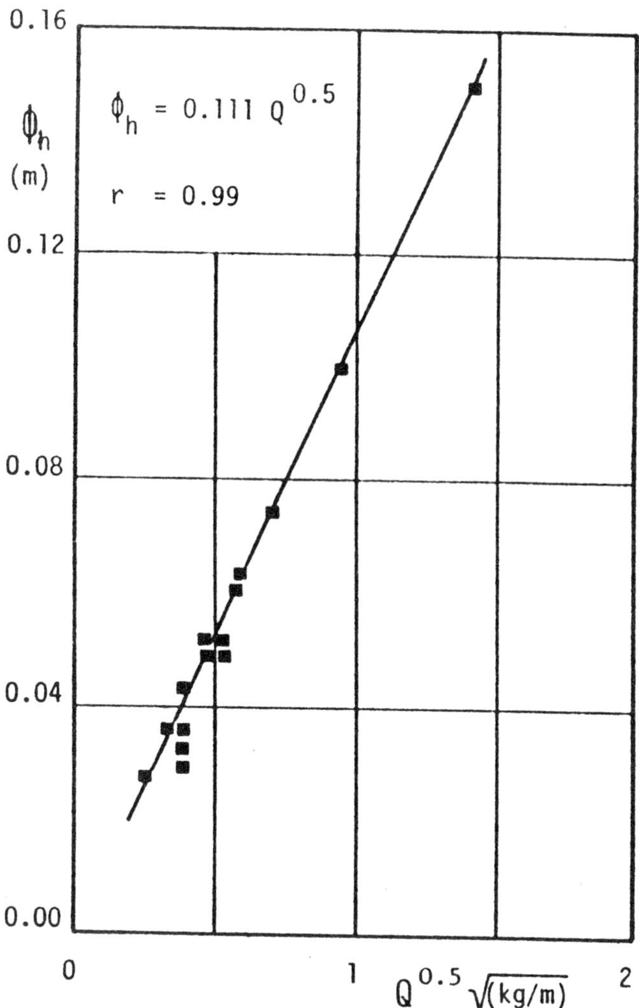

Fig. 6 Suggested diameters for blastholes in control blasting techniques as a function of amount of explosive charge

techniques require a high decoupling and so there is no crushed zone. In fact, the model suggested here shows that to eliminate the crushed zone under equal amounts of charge it is sufficient for the diameter of the hole to be equal to the crushed zone. Moreover, if we look at the diameters suggested on empirical bases by other authors,[1,4] there appears to be a good correspondence (Fig. 6) with relationship (5).

The extension of the radial fractures ($R_{2,d}$) can be calculated from relationship (6). The elimination of the crushed zone by use of decoupling entails a saving of energy that serves to lengthen the fractures by about 20%:

$$R^{j}_{2,d} = 0.74\,Q^{0.5} \qquad (10)$$

and where joints are present the extension of the radial fracturing zone is

$$R^{j}_{2},d = 3.24\,Q^{0.5} \qquad (11)$$

But it must be pointed out that the presence of a burden (s) may lead to a drop in pressure inside the hole and so

$$R^{j}_{2,d} \simeq 1.3\,s \qquad (12)$$

These calculations, however, do not include the line drilling technique, in which case the extension of the radial fractures coincides with the distance between the explosion

Table 1 Example of maximum extension of fracturing zones around blastholes (*A*) and (*B*) of Fig. 7

Zones		Extension, m *A*	*B*
Crushed	R_1	0.035	0.04
Radial	R_2	0.490	0.52
Damage	R_3	0.760	1.05
Radial (joint)	R_2^j	2.010 (0.8)	2.10 (0.8)
Damage (joint)	R_3^j	4.020	5.00

point and the line drilling, whereas the damage zone can be calculated by use of relationship (9). It must be pointed out that the damage zone is underestimated because in presplitting some of the charges must detonate simultaneously and so there is an overlap of seismicity.

Table 1, which shows the extent of induced fracturing around holes (*A*) and (*B*) in the round illustrated in Fig. 7, gives an example of how the relationships proposed can be applied. The burden of blastholes (*A*) and (*B*) is equal to 0.6 m and so from relationship (12) maximum extension of the radial fractures can be obtained (given in brackets in Table 1) and which is equal to

$$R_2^j = 0.8 \text{ m} \tag{13}$$

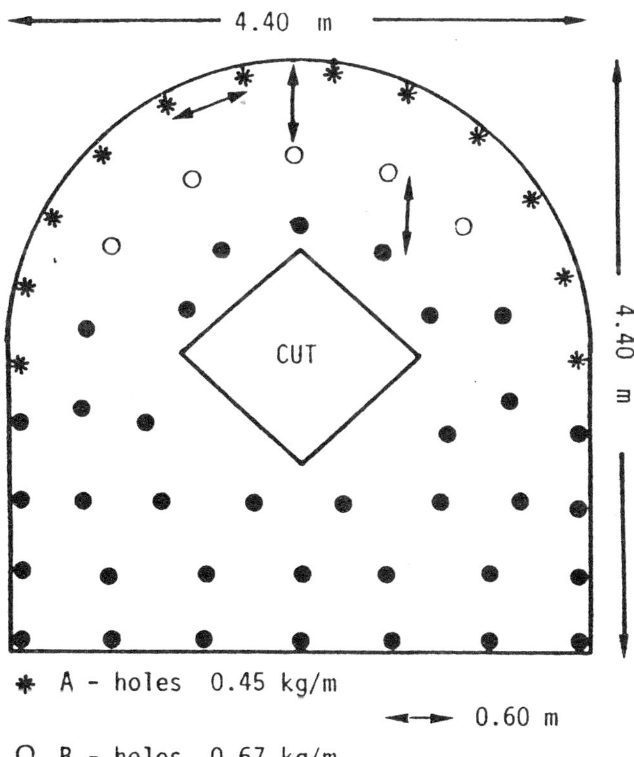

Fig. 7 Round used to calculate maximum extent of fracturing zones (Tables 1 and 2)

To assess the entity of the fracturing induced around the rock mass surrounding the roof of the tunnel it must be noted that the holes (*B*) are 0.6 m away from the profile of the tunnel. The values in Table 2 therefore give the maximum width of the fracturing zone calculated from the profile of the tunnel.

Table 2 Maximum extension of fracturing zone inside rock mass on tunnel roof (Fig. 7)

Zones		Extension, m *A*	*B*
Crushed	R_1	0.035	–
Radial	R_2	0.490	–
Damage	R_3	0.760	0.45
Radial (joint)	R_2^j	0.800	0.20
Damage (joint)	R_3^j	4.000	4.40

The example shows that the greater extent of the fracturing zone produced by the explosions without controlled blasting (1950s) is determined, as far as the damage zone and the crushed zone (with or without joints) are concerned, by the $Q^{0.5}$ ratios with and without controlled blasting; as to the radial zone, the extent depends on the value of the burden.

References

1. Langefors U. and Kihlström B. *The modern technique of rock blasting 2nd edn* (New York, etc.: Wiley, 1967), 405 p.
2. Berry P. Cataldi C. and Dantini E.M. Geothermal stimulation with chemical explosives. Paper presented to Energy technology conference, Houston, Texas, 1978.
3. Holmberg R. and Persson P.A. The Swedish approach to contour blasting. In *Proceedings 4th conference on explosives and blasting techniques, New Orleans, February 1978* (Montville, Ohio: Society of Explosives Engineers, 1978), 113–27.
4. Gustafsson R. *Swedish blasting techniques* (Nora: Nora Boktryckeri A.B., 1973), 163–80.

Dr. T.N. Hagan In reply to R. Coates, the advantages of modern drill jumbos can be exploited to the full only in rounds designed around parallel-hole cuts. The difficulty associated with drilling angled cuts in irregular tunnel faces is another good reason for selecting a parallel-hole cut. Unfortunately, it is not as easy to control ground vibrations when firing rounds based on parallel-hole cuts.

In my experience the effect of explosive type on ground vibrations is much less than that stated by Coates and Atchison and Roth, to whose work the contributor referred. Because maximum charge weight per delay is a principal parameter in vibration control, explosives should be compared on a weight strength rather than bulk strength basis. Knowing the ranges of weight strengths and reaction rates of commonly used explosives, I estimate that changes in explosive type have up to a twofold effect on ground vibrations. The ground vibration intensity can be increased by a factor as high as five or even six where the degree of confinement of a charge increases from its normal level to its maximum possible level.

Dr. G.D. Matheson Before I reply to the rapporteur I consider it necessary to define the characteristics of the types of blasting mentioned. This is of particular importance as there appears to be some variation in local and international terminology. That adopted here is believed by both rapporteur and author to be accepted usage. All three are methods of controlled blasting used to form the final face in rock excavations. They consequently differ in objective from main, or bulk, blasting, which is designed to fragment the rock and facilitate removal. The main features of each method are summarized in Fig. 8.

simultaneously – before those of the main fragmentation blast.

In addition to the timing of detonation relative to the main blast, there are significant differences in the burdens required. Presplitting depends on the realignment of principal stress parallel to the presplit plane. This favours fracture propagation along the presplit plane and requires adequate burden to allow such stresses to develop. Both line drilling and cushion blasting are, however, 'post-split' techniques designed to improve on the face achieved by bulk blasting by removing disturbed rock, and require minimum burdens.

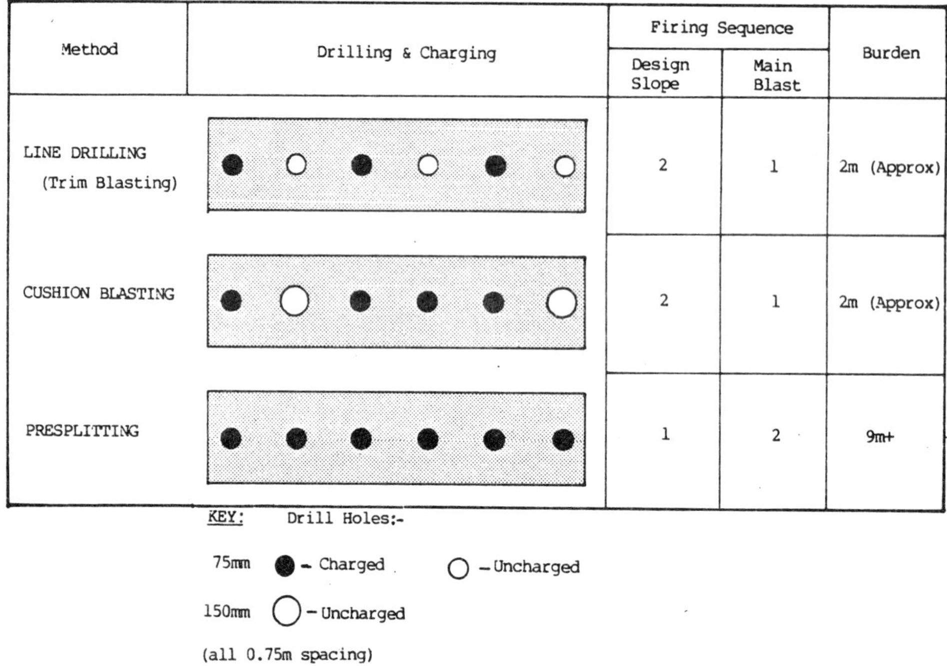

Method	Drilling & Charging	Firing Sequence		Burden
		Design Slope	Main Blast	
LINE DRILLING (Trim Blasting)		2	1	2m (Approx)
CUSHION BLASTING		2	1	2m (Approx)
PRESPLITTING		1	2	9m+

Fig. 8 Main characteristics of three methods of controlled blasting

Line drilling (or trim blasting as it is sometimes referred to) involves the drilling of relatively small-diameter (typically 75 mm), closely spaced (typically 0.75 m) holes near and parallel to the proposed design slope. Only alternative holes are fully charged, the intervening holes being either uncharged or only very lightly charged. All charges in a single panel are detonated simultaneously – after those in the main fragmentation blast.

Cushion blasting involves the drilling of mainly small-diameter (typically 75 mm), closely spaced (typically 0.75 m) holes near and parallel to the proposed design slope. Larger-diameter holes (typically 150 mm), however, replace the smaller holes at regular intervals. A common sequence is three of small diameter followed by one of larger diameter. Only the small-diameter holes are fully charged. Detonation of a panel, containing a number of such sequences, is simultaneous and after those in the main fragmentation blast.

Presplitting involves the drilling of small-diameter (typically 75 mm), relatively closely spaced (typically 0.75 m) holes along the proposed design slope. Every hole is fully charged and a panel of adjacent holes is detonated

Table 3 Relative comparisons of cost and of disturbance (+ increase; – decrease)

Method	Cost Drilling	Charging	Disturbance
Line drilling		–	+
Cushion blasting	+		
Presplitting		+	–

The relative value of each method can be judged by the cost-effectivness in forming a final face with minimum induced instability. A simple appreciation of this can be obtained from Table 3, which compares the relative costs involved in drilling and charging and the relative extent of the induced disturbance produced. Cushion blasting is normally the most expensive to drill as it involves two sizes of drill-holes, one notably larger than the other. Presplitting can be the most expensive in terms of explosives as every hole is charged; line drilling is usually the least expensive as alternate holes are not charged. This simple relationship is, however, complicated if additional charge weights are needed to compensate for increased charge spacings.

The effectiveness of different methods of blasting in reducing the disturbance induced into the final face has been summarized elsewhere* in connection with highway excavations; presplitting in that context is clearly the most effective. A direct comparison in terms of cushion blasting and line drilling predicts, in theory, that disturbance would be relatively high with the former and highest with the latter. The methods, however, are not directly comparable with presplitting as they are intended to remove rock disturbed by main blasting rather than limit blast disturbance. Line drilling is therefore not considered as an alternative

*Matheson G.D. The stability of excavated slopes exposing rock. In *Failures in earthworks: proceedings of Institution of Civil Engineers symposium, London, March 1985* (London: Thomas Telford, 1985), 295–306.

to presplitting but a separate method. For line drilling to work in a presplit mode charges would have to be detonated prior to, not after, the main blast and the charges in individual holes increased, or the spacing significantly reduced, to allow fractures to extend between holes. In such circumstances the intervening, uncharged holes could be omitted. Where the burden was adequate the end result would be identical to presplitting.

It is relevant to note that smooth blasting, a technique akin to line drilling but having every hole charged, is similar in drilling and charging to presplitting but is a post-split technique carried out with low burdens. So-called 'presplitting' performed with inadequate burden is, in effect, a type of smooth blasting.

The objective of presplitting is to form a stable face with minimal induced instability. This is achieved by presplitting at the optimum design angle and forming fractures between presplit drill-holes, which will limit the disturbance caused by the main blasting. Pre-existing natural discontinuities can therefore understandably have an important bearing on success both by influencing stability on the final face and by interfering with fracture propagation between holes. The strength of the rock, including the degree of fracturing, and the orientation of the important planes of weakness are therefore significant aspects that have to be considered when a presplit face is being designed.

A relatively lower explosive charge weight is required in the presplitting of highly fractured, weathered or weak rock. Field experience has shown that optimum linear charge weights (calculated in terms of 100% blasting gelatine) vary from 200 g/m in schistose and other highly fractured, weak rocks to 400 g/m in relatively poorly fractured psammites and granites. The aim is to use the minimum charge required to produce an effective presplit plane. A clear indication of excessive charging is an extensive surface crater formed along the presplit line.

The orientation of natural discontinuities is of major importance to success in presplitting. The influence of

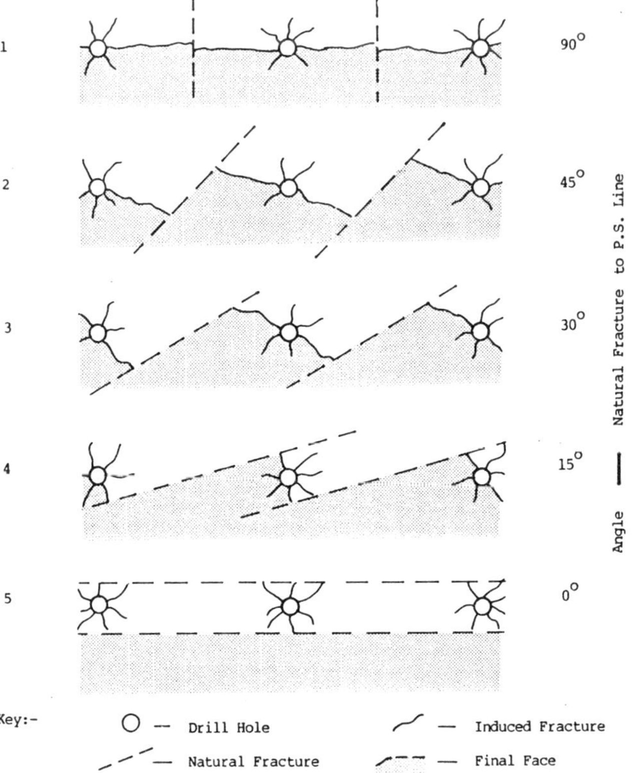

Fig. 9 Effect of discontinuities of formation of a presplit face

such discontinuities in fracture propagation during presplitting is summarized in Fig. 9. This shows diagrammatically the fractures formed in the presence of planes at angles of 90, 45, 30, 15 and 0° to the presplit. The final presplit surface clearly results from fractures propagating along the paths of least resistance between adjacent holes. Progressively stepped or serrated faces result as the angle decreases from 90 to 30°; at 15° the final face is dominated by the natural fracture planes until at 0° there is no indication that presplitting was attempted.

In practice natural fractures only become important when their orientation falls within approximately 15° of that of the proposed presplit plane. At such low angles important natural fractures can be expected to take over and form the final face. It should be noted that the objective limiting the disturbance produced by the main blast will still have been achieved to a certain degree, although there will be local overbreak behind the intended design plane. There is also an important corollary — if important fractures are known to lie within approximately 15° of the proposed design slope, they should be used as natural presplit fractures, the design slope being altered to coincide with them; presplitting as such can therefore be dispensed with. If this cannot be done, significant remedial work can be expected.

Although it is perfectly possible to presplit even very weak, highly fractured or weathered rock, a resultant high-angle presplit face may not be stable at this angle. The effect of weathering on the rock exposed on the final face should therefore always be considered. Presplitting at steep angles may not always be the most cost-effective method in the long term.

Adjustments to presplit charge calculations are therefore of relatively minor importance and involve slight modifications to the linear charge weights when relatively strong or relatively weak rock is encountered. Usually, a single linear charge weight capable of successfully presplitting the strongest rock is chosen (this is normally defined following a presplit trial at the start of the contract). The existence of important discontinuities on or near the proposed design slope is much more significant; if these fall within approximately 15° of the proposed slope, either a redesign or extensive remedial work may be required.

In reply to Dr. T.R. Stacey, in blasting both the confinement of the explosive charge and the confinement of the rock mass are important in determining vibration levels. Maximum levels can be expected where fully coupled charges are detonated in a confined rock mass.

Results of trials specifically designed to measure vibration levels produced by different charges in various degrees of coupling are given in Fig. 8 (p. 281). These show that decoupling can reduce the vibration levels by 3 to 6 times. This is entirely in accord with theoretical expectations. The point is made that if water is present and takes over from air as the decoupling agent, this will increase the degree of coupling and give rise to relatively higher vibration levels.

Bulk blasting is generally carried out to a free face with the use of coupled charges. The rock mass is poorly confined and energy is consumed in moving the rock burden. Vibration levels lower than that expected in a confined rock mass can therefore be expected. On a scaled distance basis, peak particle velocities from presplit blasting with coupled charges can therefore be expected to be higher than those in bulk blasting.

In the type of presplitting advocated the charges are highly decoupled and charged holes are grouped into panels of adjacent holes for simultaneous detonation.

Vibration levels are therefore lower than those with confined charges; the presplit face can also be sectioned to reduce total charge weights per detonation. Experience with highway excavation in Scotland indicates that the resultant maximum peak particle velocities from presplitting can be consistently lower than for associated bulk blasting. An example, from an excavation in schistose gneiss, is given below:

	Presplitting	Bulk blasting
Mean resultant ppv	5.64 mm/s	13.99 mm/s

This suggests that vibration levels on this site have been affected more by a reduction in the degree of confinement of the presplit charges than by a reduction in the degree of confinement of the rock mass during bulk blasting. A similar relationship has been found on other highway sites. The constrasting experiences in opencast mining described by Hagan suggest fundamental differences either in charge coupling, or in confinement of the rock mass, or in the total charge weights detonated per delay.

Studies into the effects of blasting on highway projects in Scotland indicate that the damage to adjacent rock is deminantly fracture dilatation. New fractures are mostly confined to the vicinity of the drill-holes and seldom propagate far into the rock mass. Such damage has been termed 'blast disturbance' and is described in the paper. Blast disturbance is associated with induced instability and both are minimal with successful presplit blasting. This is not the situation with other types of blasting and significant disturbance is usually associated with even the cautious, controlled methods of bulk blasting. Field studies that employed surface observations, seismic refraction techniques and closed-circuit television in holes drilled into the rock mass indicate that the type of presplitting advocated results in less damage to the adjacent rock mass. Presplitting is, consequently, to be recommended as a method of minimizing induced instability in final faces.

In reply to R. Coates, variations in explosive strength can be expected to affect the levels of vibration following detonation, and the formation and propagation of fractures during presplitting. Successful presplitting relies on optimizing explosive type, charging, coupling and detonation in terms of the geotechnical conditions. When different explosive types are being considered or their effects evaluated strengths must be reduced to a common base.

The characteristics of explosives that are normally used in presplitting in the United Kingdom are given in Table 4 (p. 280). Two explosives types have been used in their manufacture, each having a different VOD, density, weight strength and bulk strength. To enable comparisons to be made strengths have been recalculated to a common base of 100% Blasting Gelatine and the values used in the calculation of linear strength for presplit charging. A similar exercise is recommended whenever the effects of different explosive types are being evaluated.

Session 11: Blast vibrations

Chairman: R.G. Roberts
Rapporteur: Dr. T.N. Hagan

Discussion

Dr. T. N. Hagan* said that in the first of the three papers† under discussion Čermák described the work that was being done by his Institute with two other technical bodies in Czechoslovakia to determine the shortcomings of the Czechoslovak Standard on blast-generated ground vibrations and to improve that Standard. He rightly stated that when blasts were to be fired on a new urban site ground vibrations could be predicted (*a*) by using a general empirical equation or (*b*) by developing a site-specific equation by measuring the vibrations generated by one or more small experimental blasts.

At that point he would like to give two notes of caution. First, if they used a general empirical equation to predict vibrations, that should be as compatible as possible with the geology. If, as was likely, they had doubts about the suitability of that equation, they should err on the consevative rather than on the adventurous side. His second note of caution related to experimental blasts. Any experimental blast should be as realistic as possible. Assume, for example, that bench blasts were to be fired on a site. If the experimental blast consisted of vertical blastholes that did not have a sub-vertical free face to which to shoot, ground vibrations would be unrealistically high. If possible, any experimental blast should shoot to a free face. That made the experimental blast more expensive. But they gained what they paid for.

Čermák stated that the time interval between successive detonations in a blast was important. If that interval was less than 250 ms, he doubled the charge weight per delay in the equation

$$\text{Vibration amplitude} = \frac{\text{Constant} \sqrt{\text{charge weight per delay}}}{\text{Distance}}$$

That critical time interval (250 ms) contrasted sharply with the U.S. Bureau of Mines' use of 8 ms as an effective delay. Although there was evidence to demonstrate that the USBM's 8 ms was by no means a universal constant (effective delays for large-diameter blastholes, especially in weak strata, were appreciably more than 8 ms), no one in the western world used critical time intervals anywhere close to 250 ms.

Currently, all delay detonators exhibited scatter, i.e. delay time variability. In Czechoslovakia it would appear that two charges on consecutive delay numbers could detonate simultaneously. With delay detonators manufactured in many western countries the probability of simultaneous detonation increased with the delay number, especially in non-expanding series of detonators. (For example, where the nominal delay between consecutive numbers was 25 ms, the probability of simultaneous detonation was greater for numbers 17 and 18 than for numbers 1 and 2.)

Quite rightly, Čermák stated that in order to avoid damage the peak particle velocity (ppv) should be lower when the frequency of vibration was less than 10 Hz. That agreed with the latest results and views presented by the USBM. For frequencies more than 50 Hz Čermák stated that higher ppvs could be used without causing damage. To prevent damage Čermák advised contractors to avoid the use of charges that had insufficient or too much power. *Prima facie,* reducing the power of the charge caused a decrease in ground vibrations. But logical analysis of the situation proved that that was not so. If a charge had insufficient power, its gases had difficulty in breaking and displacing the burden rock. Therefore, the explosion gases within each blasthole were bottled up for an excessive period of time. Because the gases were retained at high pressures for longer periods and because less of the explosion energy was converted into kinetic energy of rock movement, each kilogramme of explosive produced abnormally high ground vibrations. Where charges were too powerful, their gases burst through the burden rock with ease. When they escaped into the atmosphere the considerable amounts of energy that remained in the explosion gases were manifested as noise (the audible portion of the air vibration spectrum) and air blast (the sub-audible portion of that spectrum). As one would expect, those escaping gases were still at appreciable pressures and, therefore, had the ability to propel flyrock in front of them.

The paper also dealt with spherical expansion of the strain wave generated by a charge. When one was doing calculations of spherical expansions it was important to remember that the ground vibrations that affected buildings consisted very largely of Rayleigh waves. Rayleigh waves were surface waves and, therefore, could diverge and dissipate only in two dimensions. On the other hand, body waves expanded spherically and, therefore, underwent spherical (or three-dimensional) divergence and dissipation. That explained why ground vibrations at a certain distance from a surface blast were greater when measured on the surface than when measured, say, in a tunnel beneath.

The paper by Wilton and Hills described the method of monitoring blast-generated ground vibrations at an anchored retaining wall at Tai Koo Shing, Hong Kong. A tunnel had been blasted through the retaining wall. At the time it had not been possible to measure vibrations in the anchorage zone, so plans had been made to measure vibrations at the anchor heads, and then vibrations at the points of interest (i.e. the anchorage zones) had been predicted. The results of the monitoring together with other recorded vibrations from the same site had been analysed to predict vibration levels.

The peak particle velocity of vibration was given by

$$\text{ppv} = k \left\{ \frac{D}{\sqrt{W}} \right\}^{-y}$$

where D was the distance from the blast to the recording location, m, W the maximum charge weight per delay, kg, and k and y were site-specific constants. Quite rightly, the authors stated that the values of k and y were influenced by the degree of confinement of the blast.

For hard-rock quarrying operations the USBM had developed the general empirical equation

$$\text{ppv} = 1143 \left\{ \frac{D}{W} \right\}^{-1.6}$$

The authors had considered the use of that equation but, quite rightly, had realized that that should be used to

*Golder Associates Pty, Ltd., Melbourne, Australia.

†Čermák F. Seismic effects near blasting operations; Wilton T.J. and Hills R.L. Blast vibration monitoring on anchored retaining walls and within boreholes; and Smith M.C.F. and Morton D.G. Construction and blasting methods at Kornhill site formation works.

provide typical ppv values only in the absence of on-site measurements.

The intention had been to take on-site measurements of vibrations by use of two or three vibrographs, but operational delays had prevented that. When those delays occurred the results of vibration monitoring from the associated Kornhill Development, together with some tunnel blasting on the Tai Koo tunnels, had been analysed in an effort to determine site-specific constants.

The results of that analysis emanated from quarry-type production blasts, presplit blasts, tunnel blasts, trench-type blasts, etc. For those data, collected from very different types of blasts, the scatter of results had been found to be very considerable. The magnitude of that scatter was quite understandable. If they wanted to obtain meaningful values of k and y in the above equation, they must use data from blasts that were relevant. (For a given maximum charge weight per delay and at a certain distance the vibrations produced by tunnel blasts and especially presplit blasts were appreciably greater than those generated by quarry-type blasts that fired to a free face.)

The limiting value of ppv at the anchor heads had been at 25 mm/s. Actual ppv values had been about 12 mm/s or less. The authors considered that limit of 25 mm/s to be very conservative and he supported their view. The feeling that the vibration limit was very conservative had led the authors to the very important point of the effects of vibration limits on overall costs: 'great value should be put on establishing realistic damage criteria based upon a sound knowledge of vibration theory' (p. 426).

In regard to vibration limits that were too conservative he would like to digress and relate what had tended to occur in certain western countries over the past 20 years or so. Consider a country with Federal, State/Provincial and Local Governments: suppose that a Federal body, after having spent considerable time and money in reviewing the world's accumulated knowledge and data, decided on a ppv limit of v mm/s. The State/Provincial body, probably without knowing the wealth of literature reviewed by the Federal body, decided to 'cover' itself and declare a ppv limit of v/2 mm/s. A Local Government body within that State/Province knew of the State/Provincial limit of v/2 mm/s and (without knowing of the background to or perhaps even the existence of the Federal Government's limit of v mm/s), therefore, decided to 'cover' itself by declaring a ppv limit of v/4 mm/s. He used to call that process of dilution 'progressive conservatism', but it was certainly not progressive for anyone. He preferred to call it 'successive conservatism'. It was his firm view that such successive conservatism was needlessly costing contractors and quarrying companies, in the western world at least, millions of dollars annually. He encouraged explosives users to oppose the establishment of such conservatism in their regions of activity. In all of that they, as engineers, must remember that they lived in a world not of mechanical efficiency but of cost-effectiveness. And, furthermore, they should not blindly assume that highest cost-effectiveness was always associated with zero damage. End of digression!

The authors had wanted to measure vibrations at the anchorage zones, but because access to those zones had not been possible, vibrations had been measured at the anchor heads. But, in the United Kingdom, in-the-hole measurements of vibrations had been made. The in-hole packages consisted of three-component geophone packs assembled in plaster of Paris within PVC pipes. The package could be grounted at the selected position in a borehole. The diameter of the package was typically 80 mm, its length about 800 mm and its weight about 2 kg. Such packages had been located as deep as 40 m in boreholes. They could measure ppvs up to 150 mm/s at frequencies in the 5–350 Hz range and could be self-triggering.

The third paper, by Smith and Morton, described the Kornhill Project, in Hong Kong, excavation of which had been completed recently. The project had been very challenging in that 5 000 000 m^3 of material, 75% of which was rock, had had to be removed at a constant rate in just 2½ years to create platforms for housing for about 35 000 residents. The site was surrounded on three sides by buildings and by a road that carried very high-density traffic. To add to those major constraints, an underground station, the Tai Koo Station of the Hong Kong Mass Transit Railway, was, at the time, being excavated and constructed in a rock cavern directly beneath the site. The hillside had to be removed before the underground railway station opened.

So, they had, essentially, a high-production granite quarry in downtown Hong Kong, with either buildings or roads around it and with an important cavern/railway station beneath. That meant (*a*) no flyrock and (*b*) controlled ground vibrations, noise, air blast and blast-generated dust.

The session was concerned with blasting vibrations. But, of the undesirable side effects of blasting (i.e. flyrock, ground vibrations, air vibrations and dust), flyrock was the most important when operating in an urban environment. Flyrock had been and remained the major cause of injuries and fatalities due to blasting. How many people had been killed by vibrations?

On Kornhill flyrock had been controlled (*a*) by putting lots of effort into blast designs and, subsequently, into the implementation of those designs, and (*b*) by using roof-over screens on some parts of the site.

Because of the large quantity of explosives required (about 1500 t) and the wetness of many blastholes, the use of a bulk pumpable slurry (i.e. water gel) explosive had been envisaged. But, for bulk slurry explosives, the Hong Kong Mines Department required that the entire site be covered by roof-over screens to control flyrock. Because total roof-over screening had been impracticable, the contractor had chosen to use (*a*) a cartridged water gel explosive for wet sections of blastholes and (*b*) site-mixed ANFO for dry sections of blastholes.

Initial bench heights and blast designs had been based on ground vibrations predicted by the USBM equation:

$$\text{ppv} = 1143 \left\{ \frac{D}{\sqrt{W}} \right\}^{-1.6}$$

The ppv values given by that equation indicated that the bench height should be 6 m or less. Early blasts had been monitored and had shown that that USBM equation was too conservative. The site-specific equation developed was

$$\text{ppv} = 790 \left\{ \frac{D}{\sqrt{W}} \right\}^{-1.5}$$

Because that site-specific equation was less restrictive than the USBM equation, the bench height had later been increased to as much as 12.5 m (more commonly, 10 m). The increase in bench height had enabled production rates to be increased.

Two flyrock incidents had occurred during the project. After the first the stemming length had been increased. That change had reduced the probability of flyrock, but also resulted in poorer fragmentation and, therefore, an increase in the need for secondary breakage and a decrease in production rates. After the second flyrock incident mobile screens, each 8 m wide and 15 m high, had been placed within 10 m of the bench face. Those screens consisted of strong wire mesh attached to tubular steel frames. The relocation of the screens had been very time-consuming.

Final faces had been presplit and it had been recognized that the charge weight per delay had to be reduced (cf. charge weight per delay for free-face blasts) to remain within the ground vibration limits that had been set. The ppv limit for surrounding residences was 25 mm/s.

Almost all of the bulk excavation had been completed by the opening date for the railway. Some blasts had been fired after that date and, as a precautionary measure, trains had been stopped at either end of the station. Passengers had been informed of each impending blast but had not been cleared from the station. The ground vibrations experienced by the passengers lay essentially in the 10–25 mm/s range. Complaints had been expected but had not been received.

The authors suggested that because blasts occurred frequently and because people had received advance warning and had been advised of what was going on, there had been a higher degree of tolerance. He totally supported their view. Many complaints about blasting were based on a fear of the unknown.

A major sub-vertical seam at the east end of the station did not cause focussing of ground vibrations. He was rather pleased that that observation was made, as it added weight to the evidence to explode the myth that vibrations could propagate along seams and faults with abnormally small amounts of attenuation.

M. Y. M. Yau* said that scatter and a larger variation in the results of blast vibration had been reported by Wilton and Hills. Such vibration was largely due to the types and sources of blasts. A higher level of vibration was expected with blast under a higher degree of confinement such as underground blasting. In view of the little published data on vibration from underground construction projects, it was worth presenting the results of vibration measurement from caisson blasting in mass crystalline rock at a site in Hong Kong. The relationship between peak particle velocity and the scaled distance ($D/W^{1/2}$) that had been derived was somewhat different from that obtained by Wilton and Hills and Smith and Morton.

Fig.1 showed a section through part of the site. The bedrock was of moderately to slightly decomposed granite. Blasting had been carried out to loosen the rock in order to form the bell-outs at the toes of the caissons. There was concern in regard to adverse effects due to blasting on the adjacent building (Park Villa) and the dry masonry walls.

Generally, blasting had been performed in 2.2- and 2.5-m diameter caissons. The explosive that had been used was Gelatine Dynamite '75', charges ranging from 0.25 to 0.40 kg per delay. Blasting had occasionally been carried out in 1.2-m diameter caissons with a charge weight of 0.2 kg per delay. Vibration measurements had been taken at critical locations, including the top and toe of the masonry walls, with the portable VS1600 Sprengnether seismograph.

*Ove Arup and Partners, Hong Kong.

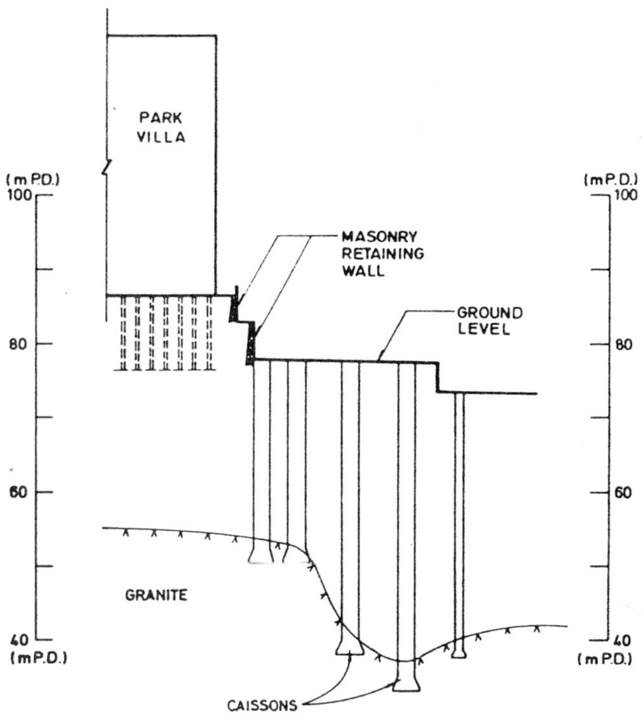

Fig. 1 Section through site

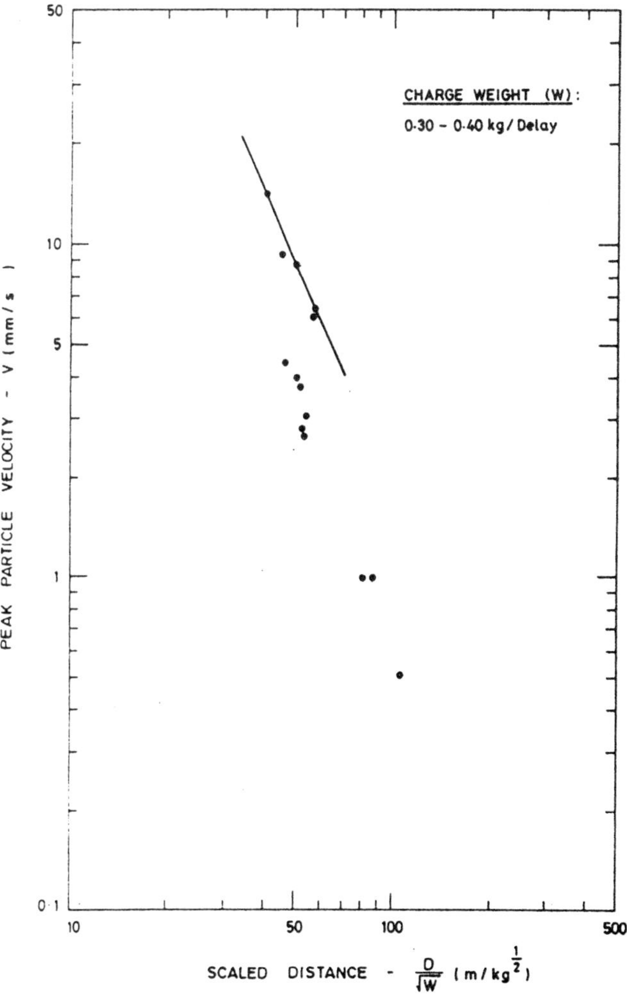

Fig. 2 Results of trials

A trial blasting had been performed prior to the production blasting to ascertain the level of vibrations on the surrounding ground (Fig. 2). The ppv shown was the measured maximum directional particle velocity.

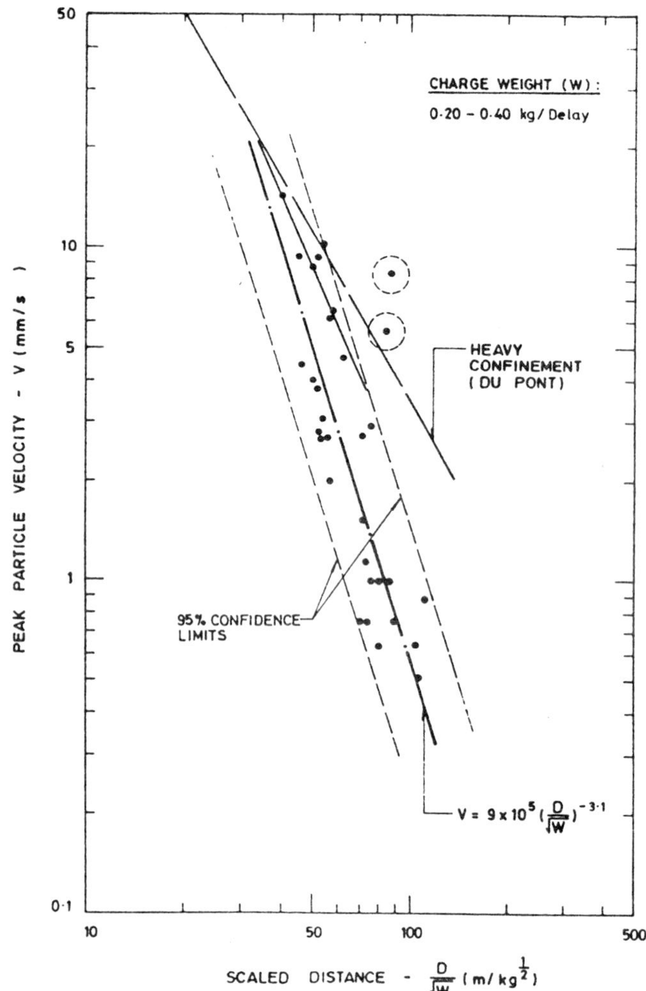

Fig. 3 Results obtained during works

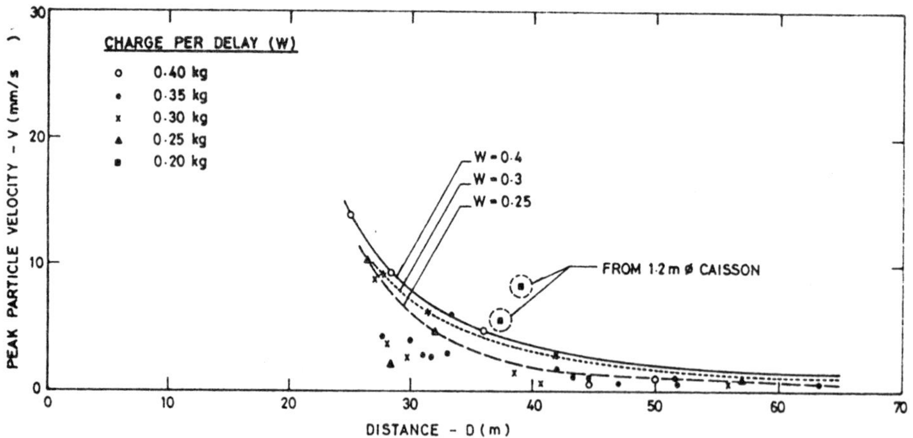

Fig. 4 Upper bound of peak particle velocity for each charge weight

With the trial blast results the charge weights had been limited in relation to the distance between the structure and point of blasting. A conventional plot of the monitoring results was shown in Fig. 3. A regression line for the data had been determined as followed by use of the common power law:

$$V = 9 \times 10^{-5} (D/W^{1/2})^{-3.1}$$

where V was peak particle velocity, mm/s, $D/W^{1/2}$ was scaled distance, D was distance from point of blasting, m, and W charge per delay, kg.

The data ranged between 40 and 100 in terms of scaled distance. The coefficient of determination for the data was 0.9, which was a high degree of correlation. In the interpretation it also showed that there was no distinct better fit of the data by adopting the relationship between peak particle velocity and the cube-root scaling ($D/W^{1/3}$). Fig. 3 also showed the typical values for the heavy confinement case provided by the du Pont *Blasters' handbook,** which could be compared with the measured data.

Fig. 4 showed a plot of the same data in the natural scale of peak particle velocity against distance. The upper bound curves for three different charge weights had been given. The results from the 1.2-m diameter caissons stood out clearly in that plot.

Owing to the limited data from the 1.2-m diameter caissons it was not possible to conclude whether the charge weight per cubic metre of rock or the confinement of small caisson diameter caused that increase in peak particle velocity. It would be of interest to compare other available data in similar blasting conditions.

As concluded by Wilton and Hills and Smith and Morton, the results did not support the cube root of scaling as a more appropriate relationship for underground blasting. Together with the monitoring results from inclinometers installed in the retaining wall and settlement markers on structures and the ground, it was found that there was no significant movement of the building and the masonry walls with those levels of vibration.

Written contribution

R. Coates† As Resident Engineer on Tai Koo Station, I should like to provide brief details of the excavation work carried out for the City Plaza entrance. The tunnel is approximately 27 m long with a face area of 20 m^2. The face was advanced in 0.8- to 1.0-m lengths, with three or four blasts for each advance. A burn cut was used, the relief holes being 200 mm in diameter, five in number, overlapping in a vertical line. These holes were drilled in 4- and, later, 7-m lengths. Explosive charges were initially 0.4 kg/delay, reducing to 0.20 kg/delay towards the City Plaza end of the tunnel.

No problems were encountered when tunnelling past the rock anchors: the inferred ppv in the anchorages did not

**Blasters' handbook* (Wilmington, Delaware: E.I. du Pont de Nemours & Co., 1980).

†Mass Transit Railway Corporation, Hong Kong.

exceed 25 mm/s. Blasting was, however, terminated prematurely near the City Plaza end of the tunnel, as this vibration limit was exceeded in a water main over the tunnel. Approximately 40 m^3 of rock was then removed by hand and chemical breakers.

I should also like to endorse a comment in the paper by Berry and Dantini.* Vibration levels generated by hydraulic hammer have been measured in excess of allowable limits from the use of explosives. Indeed, on a number of occasions, this kind of work (and passing traffic) swamped vibrations recorded from explosions, which were the prime cause of concern in particular locations.

As hydraulic hammers are almost invariably used without consideration of vibration limits, this illustrates the rather irrational fears attendant on the use of explosives.

Authors' reply

M.C.F. Smith and D.G. Morton In response to discussion comments (unpublished), there is little guidance on the effects of blast-induced ground vibrations on fresh concrete. It is generally accepted that in a concrete section subjected to a vibration wave the maximum radial strain due to shear can be reliably estimated from the radial peak particle velocity of oscillation and the seismic propagation velocity of the transmitting medium.

Commercial blasts are generally at a frequency, at distances requiring checks for probable damage, of 50 Hz. A vibration wave of 50 Hz and velocity of 200 mm/s has an amplitude of nearly 1 mm. Morris[1] suggested that this motion will cause strains, leading to cracks in concrete. He further suggested that an amplitude of 0.76 mm should not be exceeded to ensure that concrete sections are not cracked. This limiting amplitude can be related to various peak particle velocity values, depending on whether the wave motion is at the low frequency (less than 20 Hz) or high frequency.

Nitro Consult AB[2] tested fresh concrete by subjecting test cylinders to varying vibrations. They found that 6 h old concrete suffered a decrease in strength when subjected to a vibration measured as a ppv of 150 mm/s. Isaac and Bubb[3] found that 40 N/mm^2 concrete subjected to vibrations as high as 187 mm/s (at 76 Hz) was not damaged. It can be inferred from the results published in both the above papers that 10 N/mm^2 concrete can safely withstand vibrations measured as a ppv of 25 mm/s.

A suggested guide for fresh concrete is given in Table 1.

Table 1

Age/strength	Vibration restriction
0–2 h	Unrestricted
2–48 h	Vibrations less than 5 mm/s ppv (2–12 h, preferably no vibration)
After 2 days (or 10 N/mm^2 strength)	Vibrations 25 mm/s ppv
30 N/mm^2	Normal limits of 50–70 mm/s ppv

*Berry P. and Dantini E.M. Role of blasting control in excavation works near a pre-existing tunnel. This volume, 15–25.

One part of the theme of the conference was excavation in an urban environment: the Kornhill paper concerns itself with this theme and is a description of a very large site excavation, adjacent to a very densely populated area and with a very tight programme for completion. The major consideration for working in an urban environment is the environmental impact. Environmental impact can be considered under two topics – pollution and damage. To control pollution requires restraints on noise, dust and interference. To control damage the safety of personnel, public, buildings and utilities has to be considered.

Conventional above ground gyratory or jaw-crushers and lorry haulage would have caused major pollution problems and led to an unacceptably long time to carry out the removal of the excavated material. The Kornhill works had to be completed in 1985, four years from inception. The more normal programme for Hong Kong Housing Authority sites involves 1 000 000–2 000 000 m^2 in a three-year period. As a matter of interest, to remove 15 000 m^3 of material in a week required 30 lorries for 12-h days. So, to ensure the programme and minimize pollution a material crushing and underground conveying system was set up. This system took 18 months to become operational by the end of 1982. The shafts and tunnel were completed in one year and the equipment was installed, overlapping, in nine months.

The system was designed for a nominal throughput of 2000 t/h of crushed material. At peak periods each crusher operated in excess of its designed capacity at 1500 t/h: this assisted in retrieving the delays to the works. The overall cost of this equipment and related civil works equated to a cost of approximately \$70/m^3 (bank) of material excavated and placed into the reclamation.

The major potential source of damage in a quarry is from blasting – by flyrock to persons, by vibrations to buildings. The measures to limit flyrock are explained in the paper. That there were only two flyrock incidents in a three-year blasting period, with as much as 20 000 kg of explosives being used weekly, is a remarkable record that indicates the care taken in blasting and precautions.

Limits on ground vibrations at the adjacent buildings had been specified. As a result of delays, up to 20 weeks, these limits were reviewed so that production could be increased by bigger blasted/excavated volumes. This review indicated that the previously specified vibration limits could be safely increased in conjunction with a vibration monitoring programme. Therefore, the blast design criteria were relaxed in stages sufficient to ensure that the optimum production of 50 000 m^3 (bank) could be achieved weekly.

Cautious blasting is a prerequisite of rock excavation in an urban area. However, this cautious blasting should be tempered with engineering assessments of the risk of damage and with confidence in the predictions of vibrations. Engineering assessments to determine structure response from vibrations can be estimated reasonably well by use of the mass, stiffness and dampening of relative displacement characteristics of the buildings. A confident means of prediction can only be attained from a proper monitoring programme. And even this requires very careful interpretation of the results because the geology and distance both affect the transmission and charge/distance factors (note that fixing of the monitoring instruments also requires care).

Even if these two stages are carried out, there is still a third factor to consider – the blast operator's skills. All regression analyses of monitoring programmes, or the use of the conservative USBM formula, for prediction require modification by use of confidence limits to allow for the vagaries inherent in blasting.

A proper assessment and monitoring programme can give production benefits. For example, free-face bench blasts on the surface structure 20 m distant and underground; the charge weights/delay can be increased from 3 to 5 to 8 kg for increases in allowable ppv of 13 mm/s to 25 mm/s to 50 mm/s, respectively. This results in increased height of benches, less drill downtime, fewer primers/detonators and bigger blasted volumes. It is estimated that relaxing the vibration limits from a ppv of 25 to 50 mm/s within Tai Koo Station saved approximately \$2/m^3 for a considerable quantity of the material above the station. It also ensured completion on time.

Consideration of the environmental impact in early planning can aid the production of major site formation excavations in an urban area.

References

1. Morris G. Vibrations due to blasting and their effects on building structures. *Engineer,* **190,** 1950, 394–5; 414–8.
2. Nitro Consult AB. Report cited in reference 3.
3. Isaac I.D. and Bubb C. A study of blast vibrations. *Tunnels Tunnell.,* **13,** 1981; July, 35–41; Sept., 61–5.
4. Oy Finnrock AB. Report cited in reference 3.

Session 12: Tunnelling in urban areas

Chairman: Dr. R.K.F. Nemitz
Rapporteur: Dr. J.C. Sharp

Discussion

Dr. J.C. Sharp* said that session 12 was dedicated to the discussion of problems of underground construction in an urban environment. The papers submitted† included a valuable report on the construction of the Monte d'Oro tunnel, south of Trieste, Italy, an interesting and detailed account of a new rockbolting concept from China and a report on the Baishan project, also from China.

The account of the Monte d'Oro tunnel indicated the practical problems associated with blasting in an urban environment and the limitations of non-explosive excavation methods for tunnelling in varying rock conditions. The report gave valuable geological data and reviewed the practical difficulties of assessing permissible vibration limits based on a variety of differing European recommendations. The importance of vibration frequency in assessing permissible peak particle velocities was also illustrated. A good example of how site-specific data should be used to calibrate the effects of blasting was given and trends from site measurements provided a useful addition to the data base on vibration measurements.

The Monte d'Oro paper also contained an interesting account of pre-reinforcement usage for crossing of a road tunnel in close proximity to the railway tunnel alignment. The structural reinforcement of the rock mass gave a good example of how the mass properties of weak rocks could be improved.

A comprehensive survey of rockbolting principles was provided by Luo Bangzhao with fundamental considerations of bolt/rock mass interaction mechanisms and bolting arrangements for various underground excavation geometries. The paper concluded with a description of the wedge-pipe (WP) bolt system, developed and widely used in China. The bolt had a wedging action both at the anchorage location and close to the hole collar, the latter leading to a reduction in stress below the conventional face plate. Though somewhat complicated, many advantages were claimed.

Unfortunately, nobody was able to discuss the major Baishan hydroelectric project with its many different underground and foundation arrangements all constructed from a relatively narrow gorge. The tight project dimensions required novel blasting from small pilot headings to develop the foundation excavations.

The essential conclusions from the session were that by use of controlled excavation techniques, including blasting, and with timely and adequate rock reinforcement techniques, underground excavation in an urban environment could be safely executed with the minimum of disruption to the above-ground area.

*Consultant in rock engineering, Jersey, United Kingdom.

†Carastro M. Ribacchi R. and Soccol G. Blasting and support problems in the Monte d'Oro tunnel, Trieste, Italy; Guo Zongyan and Zhou Yukun. Segment-shaped blasting in Baishan hydroelectric station excavations, China; and Luo Bangzhao. Rockbolting for excavations in sensitive urban areas.

Session 13: Tunnelling in urban areas

Chairman: Dr. R.K.F. Nemitz
Rapporteur: R.J. Whalley

Discussion

R.J. Whalley* said that the first of the three papers† that were to be presented described how the Helsinki authorities planned and constructed hard rock tunnels and caverns in the city. The starting point for Helsinki's design process was the centralized geotechnical data bank, where geotechnical maps and borehole data were compiled and stored by the Geotechnical Department of the City. Maps of underground space were also kept by the City Survey Department. Those maps provided invaluable information both to tunnel designers and to contractors and, according to the authors, reduced significantly the element of risk that contractors would otherwise include in their tender price.

That concept of a centralized comprehensive data bank, accessible to all, was a recent trend in international terms and he supposed that Helsinki's size and relatively homogeneous tunnelling medium made it an ideal city for that approach. In Hong Kong the Geotechnical Control Office had gone some way to that end by setting up its own data bank for Government projects. It would be interesting to hear the authors' views and those of other contributors on the contractual implications of geotechnical data availability or disclosure prior to tender.

The authors also described further methods of risk-sharing during construction based on rock classification. When the site investigations were considered adequate no specific rock classification *per se* was adopted. Essentially, two price categories were included for rock reinforcement (rockbolts, shotcrete). One applied to rock reinforcement interrupting the excavation or blasting operation and the second applied to reinforcement installed some time after the excavation cycle and without interrupting it.

When site investigations were deemed to be inadequate or insufficient a rock classification system was adopted. Generally, two classes of rock only were considered and appropriate billed items were provided that enabled the Contractor to price accordingly. Those items were subject to remeasurement in the normal way. That relatively simple approach deserved some elaboration.

Several conference papers discussed methods of monitoring and limiting vibration due to blasting. In the present paper the authors described the method that was commonly used in Helsinki, where assessments of vibration values were made in the usual way. During construction, charge weights were adjusted based on the results of monitoring, and an interesting feature of the Helsinki authority's contractual arrangement was that separate billed rates were provided for differing levels of charge weights, again reducing the Contractor's risk element.

The paper concluded by describing an interesting case history of careful construction of a sewage tunnel, which was required to be blasted within 2 m of an existing water pipe tunnel.

The paper had as its main theme the concept of risk management: that was a common thread that was worth elaborating on for all three of the session's papers.

*Mott, Hay and Anderson Far East, Hong Kong.

†Pölla J. Kähönen Y. and Raudasmaa P. Geotechnical maps, risk management and vibration limits in tunnelling in Helsinki; Thoms I.M. Coates R. and Turner V.D. Rock excavation for a major underground cavern in Hong Kong; and Caiden D. Haines P. and Turner V.D. Construction and design of 'Island Line' rock tunnels for the Hong Kong Mass Transit Railway.

From Helsinki, the home of cavern construction, he would like to move to the paper by Thoms and co-workers, which described a specific case history of the excavation for a large rock cavern for Hong Kong's Mass Transit Railway Island Line. The cavern construction was particularly interesting in that a large quantity of rock overburden was due to be removed, principally after cavern formation. Blasting works for the Kornhill site formation had been very carefully controlled and closely monitored to limit blast vibrations in the cavern below.*

The design for the cavern was described in four stages: (1) initial design based on preliminary site data; (2) assessment of stability from rock exposures and from instrumentation during excavation; (3) re-evalaution of instrumentation and overhead development proposals to assess long-term stability; and (4) monitoring of cavern behaviour during overhead excavation for the Kornhill Development and subsequently to confirm long-term stability.

In the initial design primary support requirements had been estimated by use of the NGI system. Boundary-element analyses had been carried out to determine typical displacements and stress regimes for the completed excavation sections. For the secondary (structural) lining finite-element analysis had been used, both the overburden rock load and highway loadings being taken into consideration as appropriate. During excavation, examination of the rock exposures had generally confirmed the original assumptions regarding excavation stability. Pilot headings, which had permitted extensive geological mapping, and extensometer readings had also helped to complete the picture at that stage.

Although no *in-situ* stress measurements had been attempted, by careful re-examination of the exposed rock conditions and instrumentation results an assessment of the long-term stability could be made.

A very comprehensive programme of instrumentation had been set up for the Tai Koo cavern and the paper described details of the extensometer arrays, convergence installations, strain gauges and other instruments that had been installed to assess the cavern behaviour during mining and construction for the Kornhill Development above. The monitoring had also been required to confirm the validity of the original design and to provide a long-term assessment of the structural behaviour of the station. The authors described in some detail the installation and operation of the various instruments in the cavern and reported on the results that had been obtained. In general, movements had been small, as might have been expected for competent rock.

The paper also gave an interesting description of the excavation sequence for the cavern, which had been excavated in a relatively short period of time (approximately 80 000 m^3 had been excavated in 40 weeks). That achievement typified the effective manner in which all the MTR contracts had been completed successfully within a very tight programme.

The authors described the different primary support systems that had been used for the different parts of the cavern development, noting the extensive use of rockbolts and shotcrete. For such a complicated development it appeared that the primary support had been installed largely on the basis of examination of the conditions encountered:

*Smith M.C.F. and Morton D.G. Construction and blasting methods at Kornhill site formation works. This volume, 365–80.

again, it would be interesting to have the authors enlarge on the contractual arrangement for installation of primary support during the cavern development. Of particular interest would be the interaction between the Engineer's design concept and that finally undertaken by the Contractor.

The paper concluded that continual reassessment and monitoring during construction and careful coordination and cooperation between all parties involved were essential for the successful excavation of such a large cavern.

Dr. C.R Nelson, in the discussion of his joint paper,* had suggested that, in Hong Kong, the production of usable underground space by the construction of further caverns such as those planned in Minnesota should be considered: that would be an interesting topic for discussion.

diameter driven tunnels being selected for most of the MTR stations, those being connected to off-line cut and cover concourses by driven adits.

The varying geology and generally short driven lengths had ruled out the use of tunnel-boring machines. Conventional drill and blast techniques had been used for the rock lengths and, where contracts had also included soft ground sections, provision had been made for compressed-air working. The authors briefly discussed what they termed primary support, which, more strictly, should perhaps have been termed initial support. Combinations of rockbolts, shotcrete and, in some cases, steel arch ribs, both lagged and unlagged, had been used. It would be helpful, once again, for the authors to elaborate on the contractual

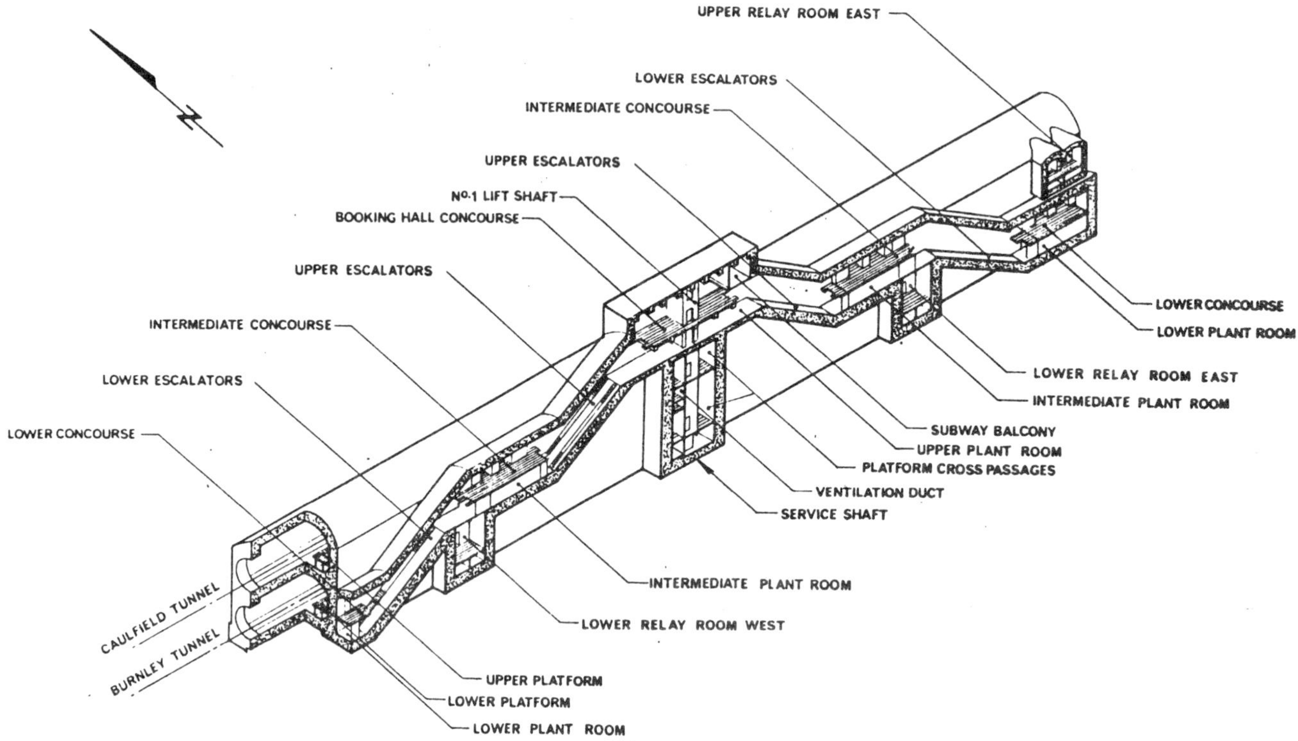

Fig. 1 Flagstaff Station, Melbourne underground rail loop

He would like to draw attention to a similar-size cavern (90 000 m^3) constructed in Melbourne for the Melbourne Underground Railway Loop in the mid-1970s (Figs. 1 and 2). As no competent stratum had been available for founding the cavern arch, the Contractor had opted to construct four temporary drifts. He had then raise-bored from the lower to upper drift to enable concrete columns to be cast to support the arch off the lower drift, founded in competent Silurian mudstone.

That was a rather novel approach to cavern construction in less than favourable ground conditions.

Caiden and co-workers described the design and construction of rock tunnels for the Hong Kong Mass Transit Railway's new Island Line, focussing in more detail on a specific contract that included running tunnels and a crossover.

The authors explained the background to the chosen alignment for the route, which had largely been dictated by the requirement to relieve an already congested transport corridor, minimal environmental disruption being caused during construction. That philosophy had led to large-

*McAllister D.S. Nelson C.R. and Strassburg R.J. Drill and blast tunnel excavations for the Buffalo Rapid Transit System, U.S.A. This volume, 211–30.

arrangement for the selection of type and timing of the initial support and where the responsibility lay for that support.

A brief description was given of design methods adopted for the *in-situ* concrete linings, usually unreinforced. Conventional Terzaghi rock pressure theory had formed the basis for the design. Certain assumptions had been made regarding water pressure on the linings both for shallow and for deep tunnels, and the authors described several methods of dealing with that problem.

The specific contract described in the paper (Contract 404) was for twin tunnels in variable ground between Admiralty, close to Hong Kong's Central business district, and Causeway Bay, a busy hotel and shopping area 1.6 km to the east. The particular part of the contract that the authors described included a large crossover and approximately 900 m of twin running tunnel. All tunnelling had been carried out from a small works area through a large rectangular shaft. The authors described the particular problems associated with urban tunnelling from a restricted site with its attendant environmental constraints. Of a total of approximately 55 000 m^3 of excavation on that part of the contract, almost half was in rock, approximately 10 000 m^3 being excavated in the crossover. The authors showed a typical running tunnel blasting pattern and gave

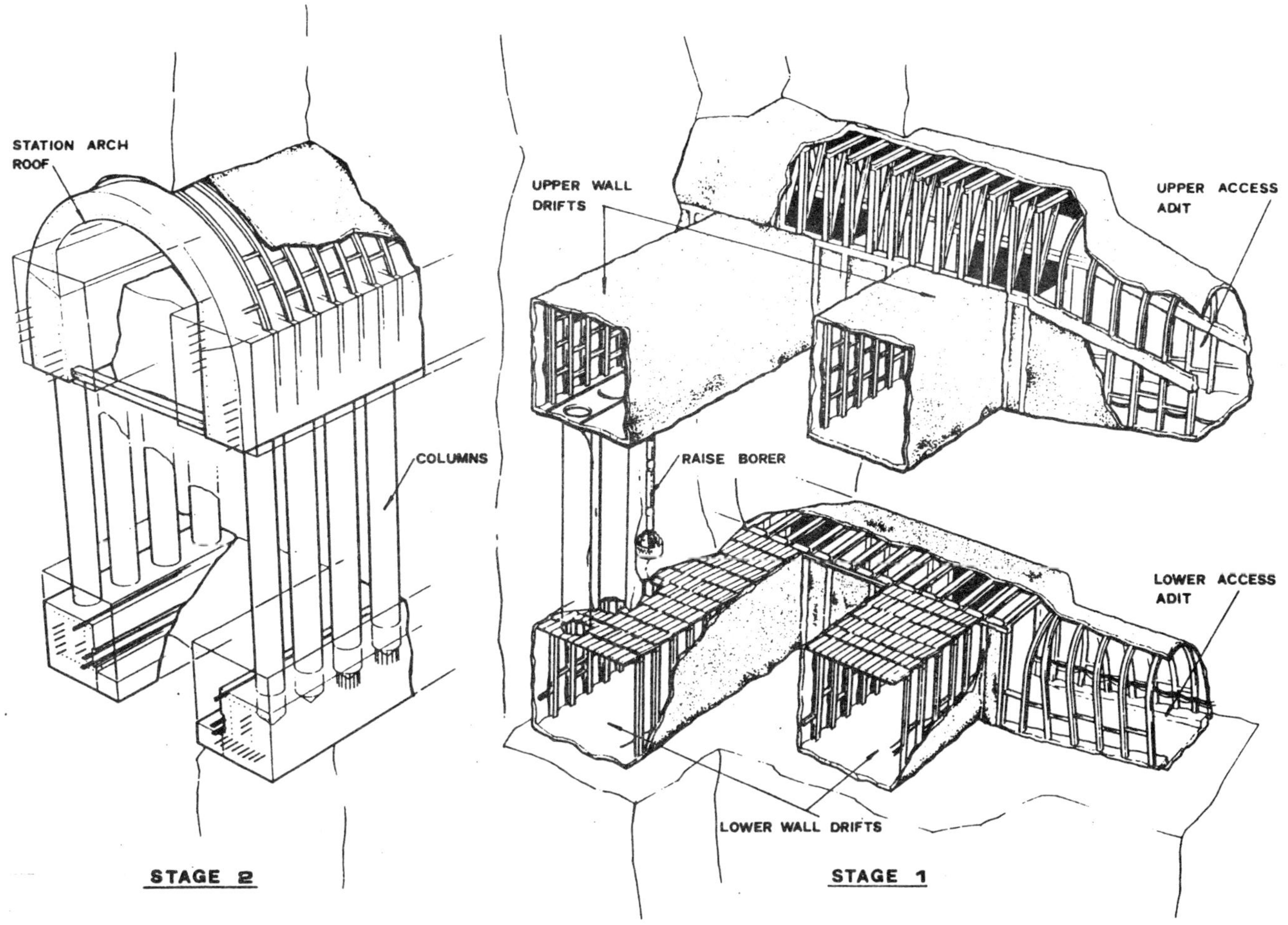

Fig. 2 Construction method

details of typical explosives, noting that perimeter blasting had been used to limit overbreak to about 10%.

The MTR specification for the blast vibrations at the adjacent structures had never been exceeded, according to the authors, despite the close proximity of many buildings. It would be interesting to know what typical reduced charge weights had been used when tunnelling near the buildings that were closest to construction and whether that reduction, in fact, had hindered the Contractor's progress.

A description was given of the spoil-handling plant, which had a total capacity that approached 200 m^3/h with cable-winched hoists in the main shaft. The authors also described how interfaces between rock and soft material had been dealt with in the shield drives and particularly mentioned the use of advance pilot drives to enable accurate inverts to be placed to guide the shields.

The authors referred to the presence of several old wells along the route and although generally during compressed-air working air loss through the wells had not proved to be a serious problem, some wells had been sealed with either temporary or permanent caps. The authors noted that water ingress had been minimal, in general, apart from the upper sections of the shaft and where tunnelling had been close to the soft ground interface. Settlement due to water drawdown did not appear to be a problem, although some wells had been reported to have reduced flows.

In summary, the authors had described the general approach to MTR rock tunnelling on the Island Line and by their description of a specific contract had given an insight into how a very complex project had been executed to a very tight programme.

He would welcome the authors' responses to the following points.

Perhaps Pöllä and co-workers would elaborate on the City Geotechnical Department's geotechnical data bank, noting how that minimized the risk element in tunnelling projects. Some schools of thought considered full disclosure of geotechnical information inadvisable in tunnelling contracts where an unforeseen ground or 'latent conditions' clause was included in the Contract. Comment on the Helsinki rock classification system as a basis for payment and on why it was preferred to the more detailed alternatives that were used elsewhere would be helpful.

Thoms and co-workers were invited to discuss in more detail the cavern design concept, comparing the approach adopted with European (NATM) current practice. He asked if a more traditional horseshoe profile had been considered for the cavern rather than a vertical sidewall and if that would have led to economies in rockbolting and secondary lining.

An explanation of the contractual arrangement for primary support and comment on departures from the Engineer's envisaged construction sequence would be useful. He asked if that change involved the Engineer in additional primary support design or if that had been designed by the Contractor.

The very extensive monitoring had produced a large amount of data, indicating, generally, rather small movements of the rock. Perhaps that could be put into perspective with comment on what the results implied about the original design. Was the monitoring more of a research exercise in that particular case or were there very real economies or benefits to be gained directly from confirmation of rock behaviour, other than an increased confidence in the adequacy of the design? In retrospect, should more or less monitoring have been called for?

Turning to the paper by Caiden and co-workers, although the initial philosophy for the Island Line had been to construct the works with the minimum of surface disruption, in the event extensive ground treatment and some cut and cover work had been carried out, in some cases resulting in traffic diversions. Comment on the degree of departure from the original concept and how that, in fact, had affected traffic patterns would be useful, as would that on the cost implications for the project as a whole.

Details of the extent of the site investigation programme, noting the cost as a percentage of the civil works cost, would be valuable.

He would like the authors to discuss the contractual arrangement for choice and timing of initial support and whether the responsibility varied at different locations and to explain how on-site decisions had been made regarding initial support and which rock classification scheme had been used in that determination.

He wondered if the assumption that hydrostatic head reduced from full head at the crown to 25% head at invert required the lining to accept water. Information on the forms of back drainage that had been provided to ensure that that condition obtained in the long term and on the measures that had been taken to ensure that drainage was not destroyed by cavity grouting would be appreciated. Details of the methods that had been chosen to assess the water pressure acting on tunnels in water-bearing fissured rock and on the methods of designing for such pressures should be given. He asked if those pressures had been confirmed by monitoring during and after construction.

One of the duties of the rapporteur, one might say in his 'Conditions of Contract', was to put the papers in perspective in terms of the 'state of the art'.

Rock tunnelling might use tried and tested procedures, such as drilling and blasting, shotcrete and rockbolts, but the three papers under discussion had indicated particular features that showed that the approach to tunnelling was making steps forward.

The paper by Pöllä and co-workers was a good example of contractual risks being reduced significantly by the establishment of a comprehensive data bank of geotechnical and geological information. The availability of that information both to contractors and planning authorities must help to minimize the risk element in tunnelling projects.

Thoms and co-workers had given an example of how a large cavern could be constructed by careful planning, monitoring and re-evaluation during the construction process. Dr. C.R. Nelson had suggested that in some circumstances large caverns, creating useful underground space, could be more attractive than surface developments. Planning authorities in Hong Kong and elsewhere should consider that option carefully.

Caiden and co-workers had demonstrated that a very complex urban tunnelling project could be completed in a very tight time-frame by determination and careful planning.

Written contribution

Dr. I. McFeat-Smith* B.I. Dubin (Geotechnical Control Office, Hong Kong) has queried why the site investigation for such a major underground work as that described by Thoms and co-workers had consisted of only one borehole. Having been responsible together with MTRC geologists for the investigations, I should like to take this opportunity to explain the situation more fully.

The cavern was to be constructed below Kornhill, a dumpling-shaped hill, with an overburden cover of about 50–90 m. It was surrounded on three sides (west, north and east) by the old route for King's Road for which near-continuous rock cuttings had been made. In addition, a conveyor tunnel, for removal of spoil from the overhead development, was being driven in rock on the south side of the cavern sub-parallel to the alignment.

Preliminary investigations for the cavern included aerial photography studies and detailed geological mapping of the cut slopes and in the tunnel. This work, together with data from a series of short vertical boreholes for the development, showed the presence of consistently fresh to slightly weathered, close to moderately jointed granite at construction level for the cavern. Several minor faults were located. The most significant of these (termed *F1*) was mapped in both the conveyor tunnel and road cuttings, and later proven in the cavern. The zone for *F1* consists of a fractured and altered zone 6 m wide, including a minor dolerite dyke.

During the design stage we proposed that two long horizontal boreholes be drilled from King's Road at either end of the cavern parallel to the alignment. This work would, however, have impeded traffic on King's Road and, in the event, proved to be impractical owing to fast tracking of the project. As an alternative two long, 45° inclined boreholes were drilled from the top of Kornhill. One was initiated from directly above the excavation area and drilled parallel to it. The other had to be located to one side of the cavern owing to access problems. Both holes showed that conditions within the massive granitic hill were similar to those which had been expected from the geological mapping.

Conditions monitored during excavation proved to be generally in line with those expected. The only variations of any note were two vertical joints that contained up to 1 m of highly to completely weathered granite. These were intercepted in part of the roof and in the walls trending sub-parallel to the alignment and required minor modification of the lining design. It appears unlikely to me that the drilling of a substantial number of additional boreholes would have located the presence and orientation of these relatively minor features.

*Charles Haswell and Partners (Far East), Hong Kong.

Authors' replies

J. Pöllä, Y. Kähönen and P. Raudasmaa The Geotechnical Department's data bank consists of different types of maps, including 46 sheets on the scale of 1:10 000, soil and bedrock maps on the scale of 1:10 000 (in colour), 150 sheets on the scale of 1:2000, 500 sheets on the scale of 1:500, numerous maps and sections on different scales made for city planning or construction purposes, engineering geological maps based on information about rock outcrops and underground caverns and tunnels, and maps of foundations of buildings in downtown areas with groundwater problems. In these areas geotechnical planning and supervision is a complex technical and administrative task.

There are also data on all investigation holes, on levels from the groundwater table and settlements as well as full information on underground projects consisting of, for example, reinforcement measures and leakage water observations.

All this information is available when a new tunnelling project is started. This is, of course, of great value, because there is always a lack of money. This system helps the engineers to direct their investigations to critical points along the tunnel line. As the very difficult areas are known beforehand, there is no risk of encountering them by accident. In this way a high standard of investigations and good technical results can be achieved with only a small investment.

Table 1 shows the element of risk that contractors include in their tender prices. When the quality of the investigations is normal, there is a 5% risk reserve.

Table 1 Risk reserve coefficient

Quality of investigations	Coefficient
Very good	0.95
Good	0.97
Normal	1.00
Poor	1.04
Very poor	1.15

Ground investigation information is available to the public and it is offered for sale at a price covering only the selling expenses. The normal price is about $U.S. 3.5 per copy. The City of Helsinki is not responsible for the information sold to the public. Most of the material is provided with the text 'Not accurate enough for detailed planning'. Persons using this information have to decide whether to make more investigations or not. If no further investigations are carried out and the project fails due to unexpected ground conditions, the municipality city is in no way responsible.

In our opinion the contractors should obtain all available information before they make their tenders. As there are almost always places that have not been investigated, and therefore areas of unknown rock quality, it is vital to settle beforehand the conditions of payment. We have solved this by dividing the rock types and reinforcement works into different classes. It must be remembered that we do not classify the supporting work but apply payment classification when poor rock is met. This kind of classification must be very simple and correspond to the difficulties in excavation. The normal geological or other classifications (RQD, NGI, etc.) do not meet this demand. Also, the classification must be described by terms that cannot be misunderstood.

This system demands that the engineer making the estimates is familiar with the difficulties caused by different rock types. It is also important that the estimates be made by the planning engineer, because he can spend more time on the task and he knows the rock conditions much better than the contractors, who normally have only four weeks to present their tender.

I.M. Thoms, R. Coates and V.D. Turner We should, first, draw attention to the following corrections to the printed text of our paper:
Fig.3, p.407: *For* extensometer EC in Zone I *read* EA; *for* extensometer EA in Zone IV *read* EC
Table 2, p.416: EA and EC together with movements and remarks should be transposed as indicated below (Table 2)
p.418, line 21: *For* 0.4 mm *read* 4.0 mm

Further information on the Kornhill Development formation works above Tai Koo cavern may be obtained from Muir T.R.C. Smethurst B.K. and Finn R.P. Design and construction of high rock cuts for the Kornhill Development, Hong Kong; and Smith M.C.F. and Morton D.G. Construction and blasting methods at Kornhill site formation works (this volume, pp. 309–29 and pp. 365–80, respectively).

The subject of exactly defining what constitutes New Austrian Tunnelling Method has been a hotbed for discussion for some time, several of the world's experts holding different views on certain features of NATM. It is, therefore, probably sufficient to record that the Tai Koo Cavern has been constructed and monitored on the basis of rock–concrete lining interaction and has many of the accepted elements of NATM.

On the matter of utilizing a more traditional horseshoe profile for the cavern shape, it is doubtful that any economies would have been derived. Bearing in mind the essentially 'square' shape of MTRC trains, the vertical sidewall configuration is preferable in stations. Curved sidewalls would have necessitated more excavation, resulting in unutilized space, and probably would have required the same number of rockbolts and similar applications of shotcrete to secure the sidewalls owing to the persistence of adverse steeply dipping joints throughout the cavern. Curved sidewalls would also have required more costly curved formwork. In short, the profile adopted for Tai Koo was one that was basically tailored to suit operational railway constraints as well as the prevailing geological conditions and, furthermore, would present a cavern that was economical in construction terms.

Design of the primary support in the cavern on the final excavated profile was the total responsibility of the Engineer, under the Contract. This differed from the normal MTRC contractual arrangement for support in running tunnels and similar excavations, where the Contractor is responsible and has to propose types and lengths of rockbolt (or ribs or other types of support) to suit the range of expected ground conditions. These measures, together with a method statement that fully describes installation, are required to be submitted for consent by the Engineer prior to excavation being commenced. Generally, payment for ribs, laggings, shotcrete and rockbolts is made under

Table 2

Zone Location	Number	Max Movement on (mm)	Anchor Length m	Remarks
I Replacement Crown, N. of centreline	EA	0.90	11	
IV Replacement Crown N. of centreline	EC	2.59	11	2.30mm dilation between all anchors and head due to overhead excavation

Provisional Bill Items, which are included in the Contract for use at the 'direction and discretion of the Engineer'. In practice the Contractor would propose the support that he feels is required, depending on his assessment of conditions at the face, and this would be agreed on site by the Resident Engineer's site staff on an ongoing basis. Disagreements inevitably occur, but are usually resolved at site level provided that a reasonable approach is adopted by both parties.

In Tai Koo cavern bolts and shotcrete were specified in detail and instructed in writing by the Engineer's staff and then measured and paid for. The exception to this was temporary support in the headings driven by the Contractor during the early stages of development of the the cavern excavation (i.e. support to rock surfaces that were not on the final profile). Such support was the Contractor's responsibility and liability.

There are two main departures from the Engineer's envisaged construction sequence. The first was the Contractor's proposal to drive two sidings as opposed to a central pilot at the crown. This did not result in a need for any additional support and was permitted on condition that one side heading was driven in advance of the other. The early exposure of the arch in the top headings permitted a 'zonal' approach to be made in specifying the arch support (see Fig. 6, p. 410) and also identified potential problem areas in the sidewalls. In this respect, therefore, replacing the Engineer's single, central, top heading by two side headings was of some assistance. Sidewall primary support was instructed as rock was exposed, during which original perceptions were confirmed or amended according to rock conditions. This placed a particular onus on the Engineer's staff to issue suitable instructions, without delaying the progress on site: sidewall support in Zones III, IV and V was particularly subject to this daily assessment procedure. The other basic departure was the deferment of concreting the arch lining until after all cavern excavation was completed. Again, no additional support was required. Safety mesh was installed in non-shotcreted areas by the Contractor, but this measure had been included within his offer.

Both departures benefited the Contractor's programme considerably, the former giving more flexibility to the upper excavation with two headings for access and the latter obviating interruption to the critical bench removal by arch concreting operations.

One improvement that possibly could have been made with respect to primary support would have been the use of larger-diameter rockbolts. This would have reduced the number of bolts required to provide the restraint necessary for the rock to remain 'self-supporting'. For most of the Tai Koo cavern this could have been done without the risk of failure by blocks dropping out from between the rockbolts.

Although it is agreed that the comprehensive arrays of instrumentation produced a large amount of data, generally indicating small rock movements, it must be stressed that the monitoring was definitely not a research exercise. Basically, the instrumentation was required to confirm the design and check the behaviour of this large cavern during and after the Kornhill formation works and in the long term. What has to be borne in mind is that Tai Koo Station is a major underground excavation, heavily utilized by the public, with a crown pillar reduced to the order of ½ to ¾ the arch span at completion of the Kornhill excavation works. As far as is known, this is without precedent. In view of these circumstances there is an obvious need for a very careful approach — hence the adoption of conservative lining design and implementation of an adequate monitoring scheme. Another contributory factor in deciding on the scope of the instrumentation was the relative timing of Tai Koo Station works and the Kornhill Development, the railway being opened just as final formation levels were reached, directly above.

The monitoring results summarized in the paper covered the period from installation to July, 1985. Subsequent readings to the end of 1985 have continued to confirm a basically elastic response to the Kornhill overburden removal by the integrated RC lining/rock arch. Reasonably good correlation between expected movements and recorded values is apparent, especially in the western half of the cavern, where the relatively better-quality rock was expected to behave in an elastic manner. In the eastern half of Tai Koo, and particularly in the fault-affected zone III, transfer of some load to the arch lining is discernible from the strain gauge arrays. These indications, together with the slighly higher dilatations recorded on the MPBX installations near the fault, reflect a tendency to depart from an elastic response where the granite is highly fractured and altered.

On the question of whether more or less monitoring should have been employed, it is considered that the scope of the instrumentation employed in Tai Koo was of the right order, given all the parameters associated with the cavern works. Notwithstanding this statement, it is felt, in retrospect, that the number of extensometers could probably have been reduced. In practical terms this did happen, owing to the loss of several MPBXs during cavity grouting of the lining.

The shape of the concrete works to the cavern was modified as a result of design development. It was recognized that the concrete lining and adjacent rock surround would act compositely to produce a relatively stiff constrained cavern lining. Owing to the relatively poor sidewall and haunch conditions a slender upper sidewall connection was introduced to connect the main arch and sidewalls.

The design was considered adequate to cater for the most severe long-term loadings, taking into account time-dependent stress redistribution. Composite lining/rock behaviour over the cavern arch was ensured by cavity/low-pressure consolidation of the rock behind the lining.

D. Caiden, P. Haines and V.D. Turner It is true that some deviation from the 'minimal disruption' philosophy occurred during construction. This was due to the adoption of Contractors' alternative construction methods. Perhaps the question has been raised because of the small cut and cover section of work carried out on Contract 404. This was for a shield chamber for construction of the eastbound Wan Chai Station tunnel. The Engineer's design was to mine the chamber from an adjacent shaft. The cut and cover alternative was carried out beneath decking with traffic running above. This particular alternative was probably allowed because trams were locally absent from the route.

In other areas road space was occupied for ground treatment work. Although the disruption could be considered greater than envisaged by some of us at design stage, it was, however, in no way comparable with that which would have occurred with full-length cut and cover station boxes. We therefore think that the bored tunnel approach was still a distinct advantage over that used on previous lines and probably the only acceptable method of construction.

We believe that the general MTRC philosophy was to consider Contractors' alternatives only if cost savings or programme advantages were offered.

There was not a single 'site investigation programme' as the question suggests. The design of the line had, over the years, passed through several stages and alternatives.

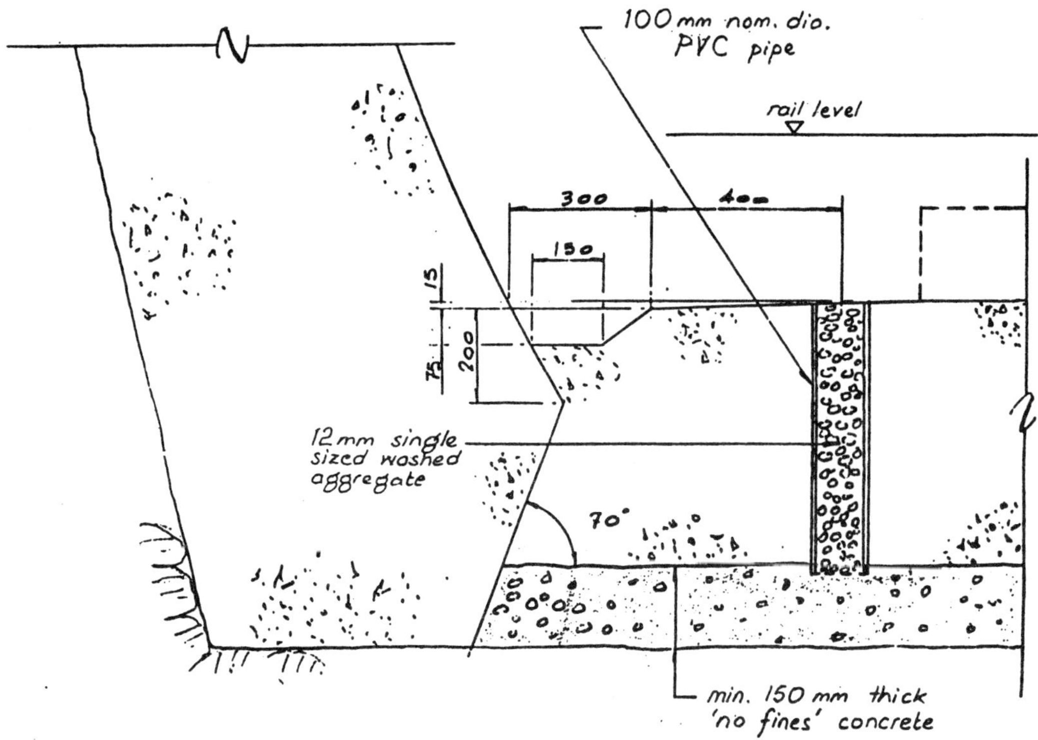

Fig. 3 Contract 404 crossover: detail at pressure relief hole (scale, 1:10)

One of these was for a light rail system that only passed underground at congested areas. There were therefore several site investigation programmes and associated feasibility studies with a number of route variations. This means that some boreholes were not relevant to the final alignment. In total, however, there were some 1500 boreholes undertaken. We do not have access to costs for all these contracts, but estimate this to represent, with laboratory testing, something like 2% of total costs. Considering the complexity of the project and the rapid changes in tunnelling conditions and ground types, this is not considered excessive.

The general contractual arrangements for initial support have been discussed by Thoms and co-workers in regard to the special conditions pertinent to Contract 408.

Decisions were made on the basis of experience, and Contractors and the MTRC employed staff with the necessary abilities. Of course, opinions of Contractors' staff and Engineer's supervisory staff may not always have been the same; compromises were sometimes necessary. This ability to communicate and make reasoned and combined decisions is the essence of a successful contract.

No rock classification system was used in standard Island Line Contracts (the particular Contract mentioned above was a special exception). This is not regretted: classification systems may be said to solve a number of problems, but sometimes replace those with more of their own. (D. Caiden stresses that this is a personal opinion that is not necessarily shared by his co-authors.) It is believed that there is no substitute for experience coupled with a human affinity with the ground. This combination sometimes manifests itself as survival instinct, when miners rush from a face moments before there is any apparent sign of impending collapse.

The '25% head' assumption relates to tunnels in good rock with adequate cover. Running tunnels were designed to be dry. In all other rock, linings were designed for full water pressure to ground level. Running tunnel inverts were modified as described in the paper to resist this pressure.

For larger tunnels, in either of the above cases, flat invert thicknesses would be too great (or heavily reinforced) to resist the pressure. In such cases weep holes were provided in the invert. Fig. 3 shows the standard weep hole detail. The plastic pipe was used as a former in the base slab concrete. Subsequently, a drill was placed in the hole to drill just into the no-fines concrete. The hole was then filled with the gravel.

Name index

Subject index